Nanostructure Science and Technology

Series Editor

David J. Lockwood, FRSC, National Research Council of Canada, Ottawa, ON, Canada

Nanostructure science and technology now forms a common thread that runs through all physical and materials sciences and is emerging in industrial applications as nanotechnology. The breadth of the subject material is demonstrated by the fact that it covers and intertwines many of the traditional areas of physics, chemistry, biology, and medicine. Within each main topic in this field there can be many subfields. For example, the electrical properties of nanostructured materials is a topic that can cover electron transport in semiconductor quantum dots, self-assembled molecular nanostructures, carbon nanotubes, chemically tailored hybrid magnetic-semiconductor nanostructures, colloidal quantum dots, nanostructured superconductors, nanocrystalline electronic junctions, etc. Obviously, no one book can cope with such a diversity of subject matter. The nanostructured material system is, however, of increasing significance in our technology-dominated economy and this suggests the need for a series of books to cover recent developments.

The scope of the series is designed to cover as much of the subject matter as possible – from physics and chemistry to biology and medicine, and from basic science to applications. At present, the most significant subject areas are concentrated in basic science and mainly within physics and chemistry, but as time goes by more importance will inevitably be given to subjects in applied science and will also include biology and medicine. The series will naturally accommodate this flow of developments in the sciences and technology of nanostructures and maintain its topicality by virtue of its broad emphasis. It is important that emerging areas in the biological and medical sciences, for example, not be ignored as, despite their diversity, developments in this field are often interlinked. The series will maintain the required cohesiveness from a judicious mix of edited volumes and monographs that while covering subfields in depth will also contain more general and interdisciplinary texts.

Thus the series is planned to cover in a coherent fashion the developments in basic research from the distinct viewpoints of physics, chemistry, biology, and materials science and also the engineering technologies emerging from this research. Each volume will also reflect this flow from science to technology. As time goes by, the earlier series volumes will then serve as reference texts to subsequent volumes.

More information about this series at http://www.springer.com/series/6331

Thimmaiah Govindaraju • Katsuhiko Ariga
Editors

Molecular Architectonics and Nanoarchitectonics

Springer

Editors
Thimmaiah Govindaraju
Bioorganic Chemistry Laboratory, New Chemistry Unit and School of Advanced Materials (SAMat)
Jawaharlal Nehru Centre for Advanced Scientific Research
Bengaluru, Karnataka, India

Katsuhiko Ariga
International Center for Materials Nanoarchitectonics (WPI-MANA), National Institute for Materials Science (NIMS)
Ibaraki, Tsukuba, Japan

Department of Advanced Materials Science
Graduate School of Frontier Sciences, The University of Tokyo
Chiba, Japan

ISSN 1571-5744 ISSN 2197-7976 (electronic)
Nanostructure Science and Technology
ISBN 978-981-16-4188-6 ISBN 978-981-16-4189-3 (eBook)
https://doi.org/10.1007/978-981-16-4189-3

This Springer imprint is published by the registered company Springer Nature Singapore Pte Ltd.
The registered company address is: 152 Beach Road, #21-01/04 Gateway East, Singapore 189721, Singapore

Preface

Introduction: Molecular Architectonics to Nanoarchitectonics

Many current problems, including environmental, energy, and biomedical problems, must be solved with rational developments of science and technology. The performances depend not only on the quality of materials but also on the rational organization of the internal structures of materials. Therefore, structural studies of objects at nanoscopic, atomic, and molecular levels are a major concern of recent sciences and technologies. Fortunately, the research developments of past decades enable us to observe, analyze, and manipulate nanoscopic objects precisely, which is of course known as a great success of nanotechnology. However, functional materials with a rational organization cannot be created directly by top-down nanotechnology-related approaches alone. Therefore, a new research paradigm must be developed alongside nanotechnology to create functional materials from molecular and nanoscale units. This task can be assigned to an emerging concept, molecular architectonics and nanoarchitectonics.

Bengaluru, Karnataka, India — Thimmaiah Govindaraju
Ibaraki, Tsukuba, Japan — Katsuhiko Ariga
Kashiwa, Chiba, Japan

Contents

Part I
Molecular Architectonics and Nanoarchitectonics

Chapter 1
Molecular Architectonics

Mouli Konar and Thimmaiah Govindaraju

1.1 Introduction

The generation of advanced molecular and material architectures with novel functional properties and applications by custom design and engineering of molecular assemblies of functional building blocks is referred to as molecular architectonics [1, 2]. The scheme of molecular architectonics enables the construction of 0D, 1D, 2D, and 3D architectures with a wide range of shapes and sizes from molecular, nano-, micro-, to macroscale by the judicious manipulation of subtle noncovalent forces at molecular level. The functional molecular building blocks custom designed by conjugating the tailorable functional core with appropriate assembly-directing auxiliary units are integrated into functionally organized state under meticulously planned experimental conditions [1, 2]. Nature has been evolved with numerous elegant and sophisticated molecular and material architectures of DNA, proteins, carbohydrates, and lipids, which serve as inspiration and guiding principles in the field of molecular architectonics [1–4]. The biomolecular assemblies help in understanding the intricate molecular recognition and host-guest interactions and facilitate the creation of complex and delicate architectures with emergent properties and applications [1, 2, 4, 5]. In the scheme of molecular architectonics, the molecular organizations are executed through solution-processable self-assembly, co-assembly, hetero-assembly, templated assembly, and hierarchical assembly processes by controlled manipulation of weak noncovalent forces such as hydrogen bonding, hydrophobic, van der Waals, aromatic, π-π stacking, dipole-dipole, metal coordination, anion-π, cation-π and, electrostatic interactions [1, 2, 6, 7]. The

M. Konar · T. Govindaraju (✉)
Bioorganic Chemistry Laboratory, New Chemistry Unit and School of Advanced Materials (SAMat), Jawaharlal Nehru Centre for Advanced Scientific Research, Bengaluru, Karnataka, India
e-mail: tgraju@jncasr.ac.in

T. Govindaraju, K. Ariga (eds.), *Molecular Architectonics and Nanoarchitectonics*, Nanostructure Science and Technology,
https://doi.org/10.1007/978-981-16-4189-3_1

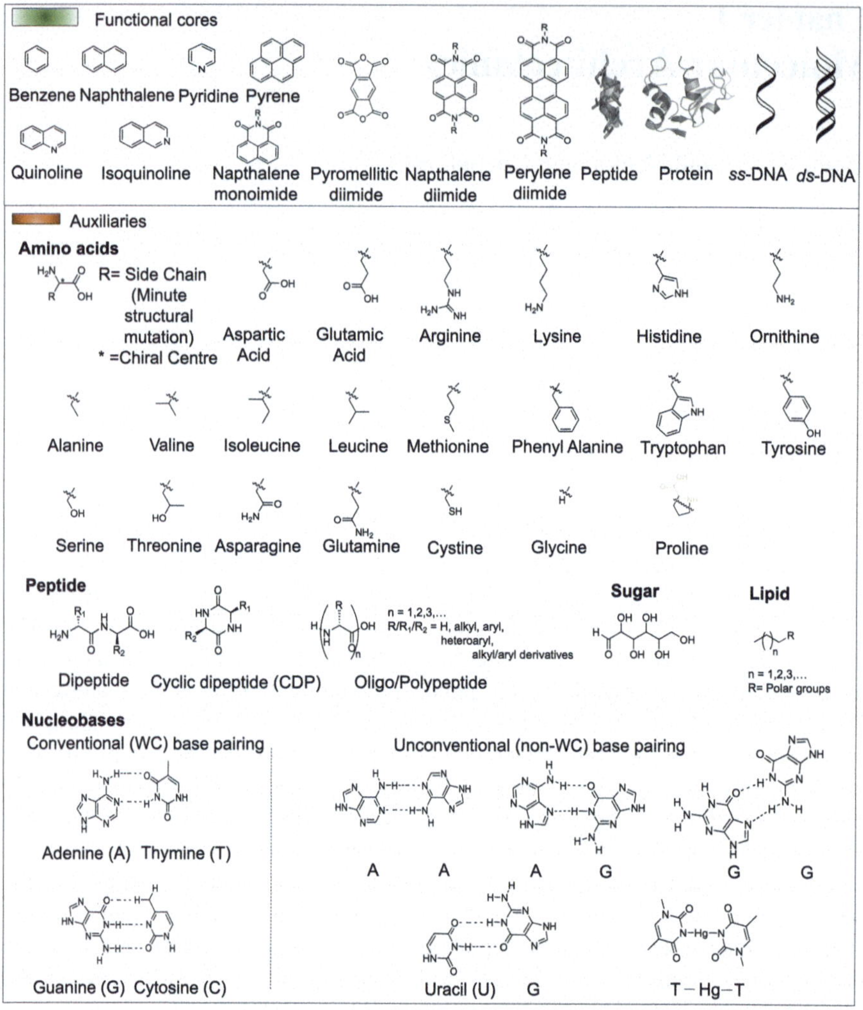

Fig. 1.1 Chemical structures of representative functional cores and biomolecular auxiliaries such as amino acids, peptides (dipeptide, cyclic dipeptide, oligo-/polypeptide), nucleobases, sugar, and lipids and their synthetic/artificial derivatives or mimetics, including metal/organic nanoparticles and quantum dots. The Watson-Crick (WC) or conventional and representative non-WC or unconventional base pairing of nucleobases is shown

controlled assembly of functional molecules is orchestrated by incorporating biomolecular auxiliaries with built-in information for molecular recognition. The biomolecules like amino acids, peptides, nucleobases, sugars, and lipids and their derivatives and mimics serve as the assembly-directing auxiliaries through their unique molecular recognition, minute structural mutations, and chirality (Fig. 1.1) [2–4, 8–13]. The biomolecule-armed functional core assembly generates the molecular and material architectures with functional properties and applications in biological and non-biological systems across the inter-related domains of health, energy, and environment [1, 2, 5, 14]. Furthermore, molecular architectonics has the potential to contribute immensely to the field of nanoarchitectonics [6, 15–17]. In general, the simplicity, unique design principles, robustness, cost-effectiveness, biological relevance, and structure-function tunability of the reductionistic and bioinspired molecular architectonics strategies resulted in the creation of diverse and fascinating architectures in 0D, 1D, 2D, and 3D with functional properties and multitude of applications (Fig. 1.2).

In this chapter, we discuss various molecular architectonics strategies with the help of representative examples reported from our laboratory to advance the engineering of molecular assemblies by precisely controlling the interplay of noncovalent interactions. Numerous molecular systems such as benzene, pyrene, naphthalene, quinoline, naphthalene monoimides (NMIs), pyromellitic diimides (PMIs), naphthalenediimides (NDIs), and perylenediimides (PDIs) are employed as functional cores because of their inherent opto-electronic, redox, and/or assembly characteristics [2, 8]. The special structural characteristics of these functional cores, viz., molecular planarity, HOMO-LUMO tunability, optical, redox, π-acidity, π-basicity, and semiconductive properties, can be tailored by biomolecular auxiliaries [1, 2, 8]. The diverse bioinspired strategies that are adapted to engineer the assembly of designer functional building blocks suitably armed with auxiliary units through skilled manipulation of experimental conditions are presented. The well-ordered assembly architectures cater to various applications including self-cleaning, sensing, opto-electronics, catalysis, bioimaging, drug delivery, and tissue engineering, among others. By systematically highlighting a number of strategies, we underline the potential of molecular architectonics to address problems related to the interconnected domains of health, energy, and environment. Finally, the chapter is concluded by providing future directions with emphasis on harvesting the scheme of molecular architectonics through innovative ideas to produce advanced and multifunctional material architectures.

1.2 Self-Cleaning Materials

Self-cleaning materials exhibit special property and ability to keep their surface clean without any human intervention. The special property of self-cleaning material depends on the surface morphology or roughness, the water contact angle, and the contact angle hysteresis [18]. In nature, the self-cleaning surfaces work in two different ways: (i) water wets the exposed surfaces completely and the contaminants are washed off with water and (ii) the exposed surfaces are not wetted by the water, rather the water beads up into droplets and rolled off the surface along with the contaminants or dust [19]. These two self-cleaning procedures with contrasting surface properties are commonly known as superhydrophilic (extremely wettable) and superhydrophobic (non-wettable) surfaces, respectively. Both superhydrophilic

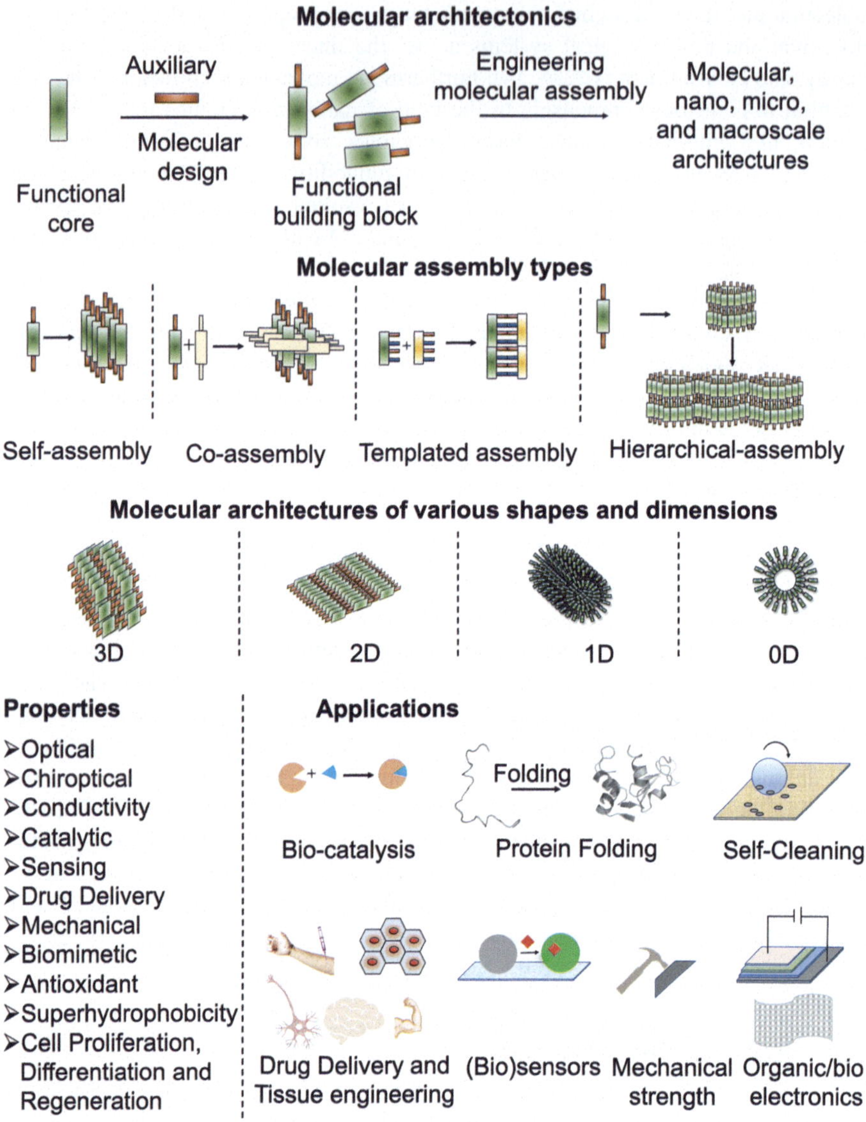

Fig. 1.2 Custom design of functional building blocks by the conjugation of functional core with appropriate auxiliary units. The molecular assembly-engineering (self-assembly, co-assembly, templated assembly, and hierarchical assembly) of functional building blocks in the scheme of molecular architectonics to produce molecular and material architectures of various shapes, sizes, and dimensions (0D, 1D, 2D, and 3D) with diverse functional properties and biological and non-biological applications

and superhydrophobic property-based self-cleaning phenomena are found in the nature [20]. For instance, the self-cleaning effects of fish scale and pitcher plant are due to their superhydrophilic property, and that of rose petal, lotus leaf, and lizard skin are attributed to their superhydrophobic property. The self-cleaning phenomenon of lotus leaf is termed as "lotus effect" and has been the subject of current research interest due to its interesting hierarchical surface architectures and technological implications [21]. An extensive research on various types of natural self-cleaning surfaces has revealed that the basic criterion of a superhydrophilic or superhydrophobic surface to exhibit self-cleaning behaviors is their hierarchical surface roughness with the water contact angle (WCA) values $< 10°$ and $>150°$, respectively. A number of modern theories have been proposed including Wenzel theory and Cassie-Boxter theory to establish a plausible relationship between the surface roughness and WCA of lotus leaf self-cleaning surface (Fig. 1.3a) [25]. Although both the superhydrophilic and superhydrophobic surfaces can be mimicked artificially, the latter is industrially preferred as non-wettable self-cleaning surface. The numerous advantages of superhydrophobic surfaces include low-cost fabrication procedure, biocompatibility, and diverse application possibilities in the preparation of anti-corrosion, antibacterial, and anti-fogging coatings, among others [26].

We have mimicked the self-cleaning property of lotus leaf by adapting a solution-processable molecular architectonics-guided bottom-up strategy to fabricate superhydrophobic and self-cleaning surface. In this method, the well-known breath figure technique (BFT) was explored to fabricate large area honeycomb-type superhydrophobic surface by using the solution of NDI-based small functional molecule in dichloromethane (DCM) (Fig. 1.3b) [22]. The functional building block (**1**) was designed by the symmetrical functionalization of NDI core with two N-*t*Boc (*tert*-butoxycarbonyl)-protected guanidinocarbonyl pyrrole (GCP) auxiliaries. The BFT-driven self-assembly of **1** in DCM was engineered into honeycomb-like rough hydrophobic surface on a substrate. The microscopy and optical profiler investigations of the honeycomb array surface revealed that the hydrophobic hexagonal microwells (~10–15 μm) were comprised of stacked nanobelt architectures deposited around the condensed water droplets (formed by the evaporative cooling of volatile solvent, DCM). The surface roughness, hydrophobicity, and overall wettability of honeycomb array surface were made comparable to that of lotus leaf surface upon successive gold sputtering followed by functionalization with 2H-perfluorodecanethiol (FSH). Interestingly, the FSH-functionalized surface exhibited superhydrophobic property and the self-cleaning effect with WCA of 156° and contact angle hysteresis value of only 1° (Fig. 1.3c–g). This unique study is a rare example of fabricating superhydrophobic self-cleaning surfaces employing small functional molecule in BFT. The molecular design strategy was further extended to fabricate the colored honeycomb array surface of **1** by uniformly incorporating the fluorescent dyes rhodamine B and isoleucine methyl ester-appended perylenediimide (Fig. 1.3h and i), which may find use in bioelectronics, cell sorting, and miniaturized bioanalytical assays. Based on the molecular self-assembly approach, Lee et al. have reported a biomimetic surface

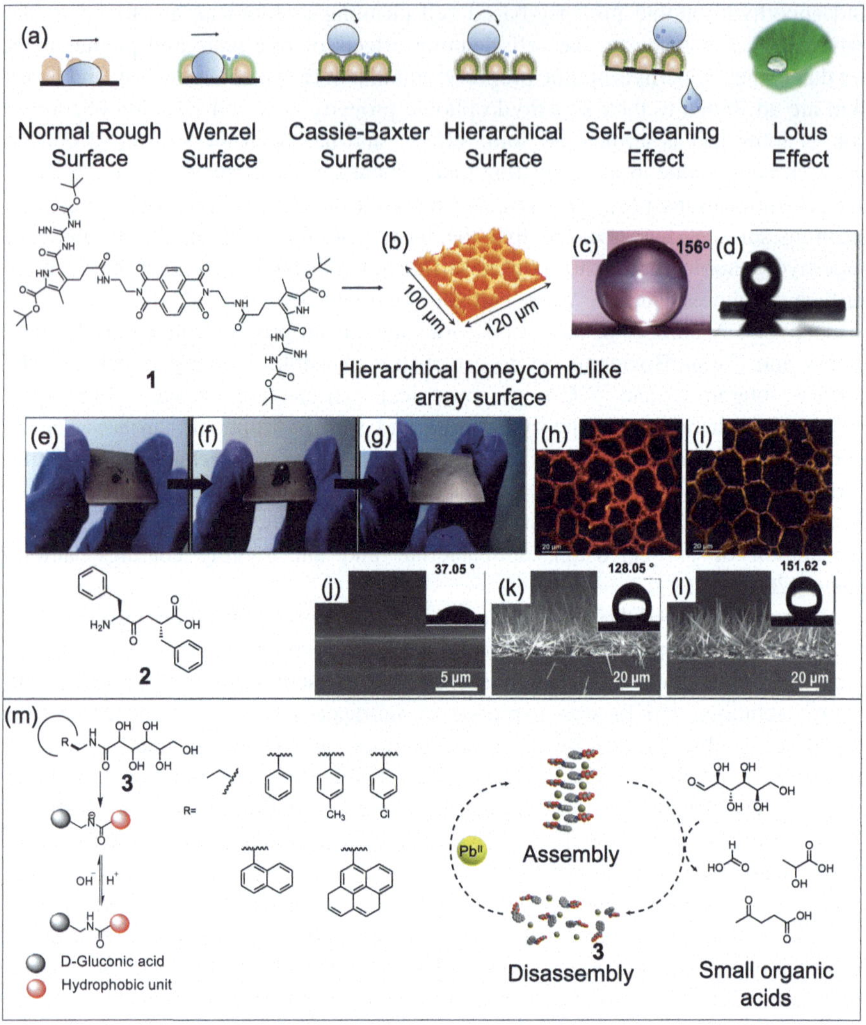

Fig. 1.3 (**a**) Dynamics of water droplets on various rough surfaces (normal, Wenzel, Cassie-Baxter, and hierarchical rough surfaces), self-cleaning effect on the hierarchical rough surface, and natural lotus effect. (**b**) Optical profiler image of the honeycomb-like hierarchical rough surface generated by the BFT-induced self-assembly of **1**. (**c and d**) WCA (156°) on the FSH-functionalized honeycomb-like surface and contact angle hysteresis (1°). (**e–g**) Self-cleaning effect on the FSH-functionalized honeycomb-like array surface. (**h** and **i**) Confocal micrographs of honeycomb-like arrays of **1** incorporated with rhodamine and isoleucine methyl ester-appended PDI. Panels (**b–i**) are adapted with permission. Copyright 2012, Wiley-VCH [22]. (**j–l**) Surface morphologies of the amorphous film and aniline- and pentafluoroaniline-treated films of diphenylalanine (**2**), inset showing the respective WCAs. Panels (**j–l**) are adapted with permission. Copyright 2009, Royal Society of Chemistry [23]. (**m**) Chemical structures of D-gluconamide (**3**) with variable substituents, and the schematic shows the dynamic assembly architectures of 3 offering coordination center (deprotonated amide NH) for Pb^{2+} to catalyze the glucose hydrolysis. Panel (**m**) is adapted with permission. Copyright 2020, American Chemical Society [24]

strategy to fabricate self-cleaning superhydrophobic material. The self-assembly of diphenylalanine (Phe-Phe, **2**) from anhydrous hexafluoroisopropanol (HFIP) formed hierarchical surface with enhanced hydrophobicity upon treatment with aniline and pentafluoroaniline solvent vapors [23]. The hierarchical surface roughness reduced the wettability of the pentafluoroaniline-treated film to impart superhydrophobic property (WCA ~151°) (Fig. 1.3j–l).

These studies highlight the potential of molecular architectonics to guide the fabrication of hierarchical assembly architectures and surfaces over multiple scale lengths with desired surface hydrophobicity (or hydrophilicity), superhydrophobicity (superhydrophilicity), surface roughness, and functional features among other characteristic properties for advanced applications in self-cleaning, surface coatings, optoelectronics, (bio)electronics, cell sorting, and miniaturized bioanalytical device fabrications.

1.3 Biomimetic Catalysis

Mimicking the function of enzyme is considered a holy grail in the field of catalysis. The meticulously designed molecular assembly architectures encompassing all the catalytic components are anticipated to mimic enzyme active center with improved catalytic efficiency under milder conditions. We have designed enzyme mimics by adapting the scheme of molecular architectonics by orchestrating the assembly of metal-coordinated amphiphiles (D-gluconamide-Pb^{II} complex) and effectively improved the catalytic activity toward glucose hydrolysis in aqueous alkaline solution [24]. Notably, Pb^{II} alone acts as Lewis acid catalyst in glucose hydrolysis to form value-added formic acid (FA), lactic acid (LA), and levulinic acid, among others, while the reactions are performed under extremely harsh conditions including high temperature [27]. To improve the catalytic efficiency of Pb^{II} toward glucose hydrolysis at ambient temperature, we designed a number of D-gluconamide amphiphilic ligands (**3**, R = propyl, phenyl, naphthalene, and pyrene units). These amphiphilic ligands were found to be capable of coordinating with the lead ions (Pb^{II}) through amide-NH deprotonation under the alkaline conditions. The amphiphiles were self-assembled to create second coordination spheres around the Pb^{II} center. This dynamic assembly architecture exhibited superior catalytic efficiency toward the glucose hydrolysis at ambient temperature compared to dispersed state of amphiphile or Pb^{II} alone (Fig. 1.3m). Such biomimetic catalyst systems designed based on the dynamic metal-coordination assembly architectures of functional amphiphilic molecules can serve as perfect model systems for developing advanced catalyst to convert the chemical or biowaste into value-added chemicals.

The second coordination spheres are considered as essential criteria for the functioning of metalloproteins or metalloenzymes. The second coordination sphere influences the microenvironment of the metal ion and therefore has the profound ability to modulate the reactivity and selectivity of metal complex (metalloprotein). The second coordination sphere in confined space around the active catalytic center (usually metal ion) is therefore considered as the basis of designing the synthetic metal complexes as advanced catalytic systems. In this context, Borovik and coworkers have significantly contributed toward mimicking the second coordination sphere of natural metalloenzymes for synthetic systems [28]. They have shown that the binding and activation of dioxygen by transition metal (Fe^{II} and Mn^{II}) complexes to form highly reactive metal-oxo complexes is controlled by the second coordination sphere of the metal complexes, as exemplified with the respiratory proteins, oxygenases, and synthetic complexes [29]. The dynamic process of formation of metal-oxo systems by binding of dioxygen to metal complexes is governed by the noncovalent interactions such as intramolecular hydrogen bonds involving dioxygen-derived ligands. Such noncovalent interaction-assisted metal complexes and their assemblies can be architectonically evolved and applied for a variety of reaction types in large-scale industrial purposes.

1.4 Organic Electronics

Engineering the molecular assemblies of π-conjugated small organic molecules leading to electrically conductive materials is the basis of advanced organic electronics. Currently, organic electronic materials are potentially used in the fabrication of photovoltaic devices, organic field-effect transistors (OFETs), flexible displays, sensors, and bioelectronics [30–32]. The major advantages of organic electronic materials over the traditional inorganic conductive or semiconductive materials are their low production cost and tunability with respect to conductive, mechanical, and thermal properties. Further, the fabrication of bioelectronic materials and devices using biomolecule or biomimetic functional molecules through bottom-up molecular architectonics strategies can effectively improve their biocompatibility and biomedical applications. In the scheme of molecular architectonics, we have meticulously tailored the molecular assembly properties of the organic semiconductor NDI by conjugating with D- and L-phenylalanine methyl esters as auxiliaries (**4** and **5**). By the minute structural alteration in the α-substituents of amino acids, we were able to engineer the self-assemblies of **4** and **5** into free-floating crystalline 2D nanosheets with micrometer lateral dimensions (Fig. 1.4a and b) [33]. The striking features of these NDI nanosheets was their high metallic conductivity (1.6 Scm^{-1}) comparable to heavily doped conducting polymers and small molecules, which was the highest reported value among the undoped organic semiconductors (Fig. 1.4c). Interestingly, the molecular ordering of **5** was further engineered into container architectures such as nanocup, mesocup, and bowl (macrocup) architectures by employing chlorinated co-solvents (Fig. 1.4d). The container architectures with attoliter volume capacity can be used as miniaturized containers for biological assays apart from their utilization as bioelectronic materials. It has been seen that the lateral organization of **4** or **5** into nanosheet was attributed to the balanced π-π stacking interactions between the NDI chromophores initiated and reinforced by the hydrophobic methyl ester moiety of phenylalanine, whereas strong halogen bonding interactions between NDI and

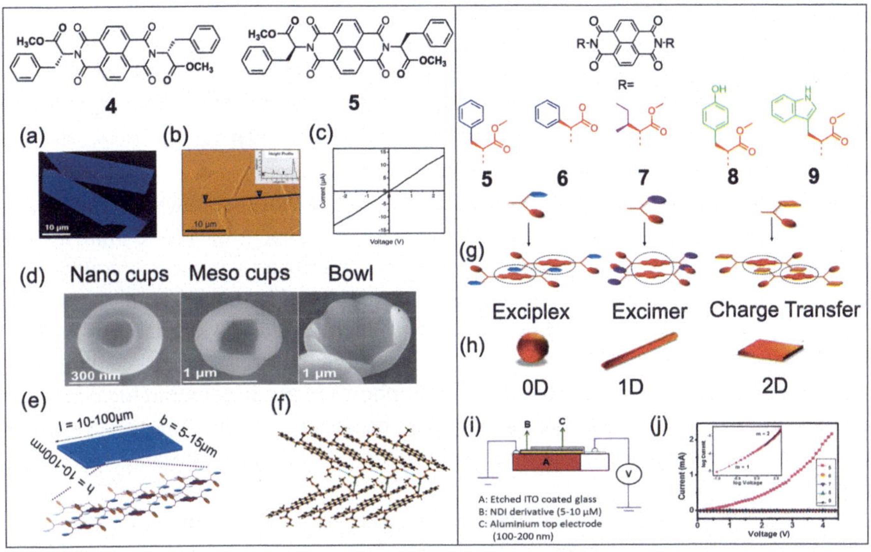

Fig. 1.4 (**a–c**) Microscopy (fluorescence and AFM) images of the large 2D nanosheets formed by the self-assembly of NDI derivative **5** (or **4**) and its current-voltage (*I*-*V*) plot showing the metallic conductivity of 1.6 Scm^{-1}, as determined by conductive AFM measurements. (**d**) Morphological alteration of assembly of **5** into nanocup, mesocup, and bowl architectures in the presence of halogen co-solvents. (**e** and **f**) Proposed 2D molecular organization of **5** and the proposed halogen bonding patterns in crystalline state, which drive the formation of container architectures (nanocups, mesocups, and bowl). Panels (**a** and **f**) are adapted with permission. Copyright 2011, Wiley-VCH [33]. (**g** and **h**) Formation of exciplex (**5** and **6**, 2D), excimer (**7**, 0D), and CT complexation (**8**, 2D, and **9**, 1D) upon respective self-assemblies of different amino acid-conjugated NDI derivatives **5–9** to access the influence of minute structural mutations in the α-functionalities of amino acids. (**i** and **j**) Schematic diagram of SCLC device and the corresponding *I*-*V* plots of the self-assembled architectures of **5–9**. Panels (**g–j**) are adapted with permission. Copyright 2016, American Chemical Society [34]

chlorinated co-solvents were responsible for the formation of container architectures (Fig. 1.4e and f). The metallic conductivity of the nanosheets of **5** motivated us to undertake the structure-function relationship study by evaluating various structurally related NDIs having conductive property [34]. Comparative spectroscopy studies of **5** and other NDIs appended with L-phenylglycine methyl ester (**6**), L-isoleucine methyl ester (**7**), L-tyrosine methyl ester (**8**), and L-tryptophan methyl ester (**9**) revealed their assembly interactions in 90% aqueous acetonitrile media, viz., exciplex (**5** and **6**), excimer (**7**), and charge transfer (CT) (**8** and **9**) complexation modes (Fig. 1.4g). These different modes of molecular interactions of the individual NDI cores orchestrated their molecular assemblies into morphologically and dimensionally distinctive architectures such as 0D spheres (**7**), 1D fibers (**9**), 2D flakes (**8**), and nanosheets (**5** and **6**) (Fig. 1.4h). A space-charge-limited current (SPLC) studies showed the electron mobility of all NDIs within the range of 10^{-8} to 1 $cm^2 V^{-1} s^{-1}$, while the highest electron mobility (1 $cm^2 V^{-1} s^{-1}$) was observed for **5** (Fig. 1.4i and j). This bulk electronic mobility further supported the metallic conductivity (1.6 Scm^{-1}) of L-phenylalanine methyl ester-functionalized NDI derivative (**5**) determined by the conductive atomic force microscopy (cAFM) study.

In another study, asymmetrically functionalized NDIs with ethyl and D- or L-alanine methyl ester moieties as imide substitutions (**10** or **11**, respectively) were synthesized to investigate their self-assembly in 80% aqueous dimethylsulfoxide (DMSO). Molecular engineering of self-assembled NDI **10** produced hierarchical crystalline green fluorescent nanosheets of 10–40 mm lateral dimensions. The green fluorescent NDI nanosheets showed a nonlinear conductive behavior with conductance value of 2.22 nS as measured by the scanning tunneling microscopy (STM) technique (Fig. 1.5a). These hierarchical 2D sheets were formed due to chiral self-assemblies assisted by the extensive intermolecular π-stacking and π-extending interaction modes of NDI chromophores to produce hierarchical layered nanosheets (Fig. 1.5b) [35]. The molecular architectonics-guided unique crystalline molecular packing with semiconductive property is useful in organic electronic applications.

Further, we developed amino acid- and peptide-based semiconducting materials with NDI core in an attempt to master the art of custom design and engineering the assembly of advanced and complex molecular systems. These amino acid- or peptide-containing functional molecular systems were anticipated to impart extended intermolecular interactions in their assembly process. Manipulating the electronic properties of NDI and pyrene aromatic cores, we accomplished the molecular co-assembly (two-component system) of alanine (L or D) methyl ester-functionalized NDI (**12**) as acceptor (A) and di-pyrene tweezer (**13**) as donor (D) through CT interactions. The chiral mixed-stack CT interaction-driven co-assembly resulted in the formation of crystalline nanofibrous hydrogel through the hydrogen bonding-supported tweezer-inclusion-sandwich (TIS) process (Fig. 1.5c and d) [36]. The unique hydrogen bonding and CT interactions allowed spontaneous unidirectional macroscopic polarization in the nanofiber architecture under ambient conditions. For the first time, the extended TIS-CT molecular co-organization of **12** (A) and **13** (**D**) allowed to fabricate thin-film ferroelectric capacitor with a Ps ~ 4 at room temperature with potential applications in multistate memory devices and smart multicomponent organic electronics (Fig. 1.5e).

In a crystallography-guided study, peptide-conjugated NDIs were crystallized to understand the influence of minute structural mutations on the molecular ordering and conductive properties of their assembly architectures [37]. The NDI derivatives functionalized with non-proteinogenic achiral peptide (Aib-Aib **14**, Aib – α-aminoisobutyric acid) and proteinogenic chiral peptide (L-Ala-L-Ala or D-Ala-D-Ala, **15** or **16**, respectively) auxiliaries were crystallized with 1D and 2D molecular ordering. The 1D and 2D molecular orderings were found to be the result of face-to-face (**14**) and edge-to-edge (**15/16**) NDI-NDI core interactions. The solution-processable achiral face-to-face and chiral edge-to-edge NDI-NDI molecular orderings of **14** and **15/16** (H-type and J-type, respectively) were found to be stabilized by CH----CO and NH----O (for **14**) and β-bridge-like CO---NH hydrogen bonding interactions along the respective crystallographic axes (Fig. 1.5f). The

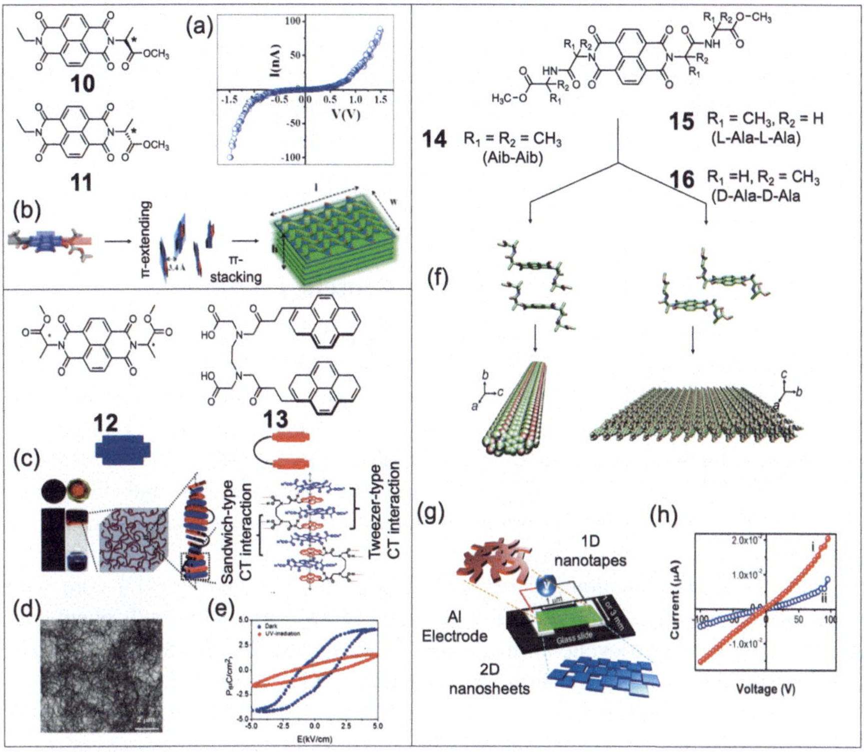

Fig. 1.5 Chemical structure of the asymmetrically functionalized NDI derivatives with ethyl and D- or L-alanine methyl ester moieties as imide substitutions **10** and **11**. (**a**) The characteristic I-V plot obtained for nanosheets of **10/11** using STM technique. (**b**) Schematic representation of molecular organization of **10/11** to form hierarchical green fluorescent nanosheets. Panels (a and b) are adapted with permission. Copyright 2013, Royal Society of Chemistry [35]. (**c**) Schematic representation of co-assembly of guest **12** and host **13** (tweezer) via tweezer-inclusion-sandwich (TIS)-charge transfer (CT) (TIS-CT) complexation to form nanofibrous hydrogel supported by the unidirectional hydrogen bonding and CT interactions. (**d** and **e**) Field emission scanning electron microscopy (FESEM) image of the xerogel film and the photo-responsive ferroelectric switching measurements of the xerogel thin film. Panels (**c–e**) are adapted with permission. Copyright 2016, American Chemical Society [36]. (**f**) Space-filling 1D and 2D molecular ordering representation of the peptide (with non-proteinogenic amino acids)-conjugated NDI **14** (Aib-Aib) and (with proteinogenic amino acids) **15/16** (Ala-Ala) in crystalline state. (**g** and **h**) Schematic representation of *I-V* device fabricated by the nanotapes of **14** and nanosheets of **15** and the corresponding *I-V* plot. Panels (**f–h**) are adapted with permission. Copyright 2015, Royal Society of Chemistry [37]

detailed studies have suggested chiral transcription from the peptide backbones to the molecular organization of the NDI core both in solution and solid states. The investigation of conductive properties by two-probe method showed two times more conductivity (3.5×10^{-6} Sm^{-1}) of 1D fiber architectures of **14** compared to that of 2D sheets of **15** (1.6×10^{-6} Sm^{-1}) (Fig. 1.5g and h). This suggests that the face-to-face molecular arrangement maximizes the electronic coupling among the NDI molecules (**14**) in comparison to the edge-to-edge NDI organization (**15**).

These amino acid and peptide auxiliary-mediated engineering of NDI assemblies and co-assemblies exemplifies the potential of molecular architectonics to produce complex architectures and fine-tuning of structural and functional outcomes with implications in optoelectronics and biomedical applications.

1.5 Chirality, Homochirality, and Protein Folding

Nature has a mysterious bias for molecular homochirality (L-amino acids and D-sugars) to guide the life processes by translating the complex assembly information from the basic building blocks to the molecular and material architectures. Inscription of chirality on achiral functional molecule by employing a suitable chiral auxiliary to induce handedness in the assembly architectures and to extract chiroptical properties is fast becoming an actively pursued research activity in the advanced materials engineering. To understand the chiral transcription from chiral auxiliary to achiral functional core, we considered incorporating NDI core with chiral auxiliaries, viz., aliphatic (D-/L-alanine, **17/18**), aromatic (D-/L-phenylalanine, **19/20**), and heterocyclic aromatic (D-/L-tryptophan, **21/22**) amino acid auxiliaries, while an achiral ethylamine-functionalized NDI (**ene**, **23**) was used as a control probe [38]. Engineering of their molecular self-assemblies led to all the four modes of aromatic interactions of NDI core, viz., cofacial, brickwork, herringbone, and slipped stacks (Fig. 1.6a). Detailed photophysical studies have indicated that chiral substitution (amino acid) induced the preferential helical bias to NDI resulting in the four distinct crystal packing modes of NDI core in their solid or solution states. Morphological analyses revealed the self-assembly of NDI-amino acid conjugates into 0D nanospheres (**17/18** and **21/22**) and 1D nanofibers (**19/20**) in their gel state, whereas the achiral NDI (**23**) produced laterally extended 2D nano- and mesosheet architectures from its solution (cyclohexene/chloroform, 80/20, *v*/*v*). On account of these unique properties, the material architectures derived from NDI-amino acid conjugates cater to a wide range of applications from materials to biomedicine.

The characteristic chirality transcription and helical assemblies of the NDI-amino acid conjugates prompted us to investigate the assembly properties of homochiral, heterochiral, and achiral NDI-peptide conjugates. These functional building blocks were designed by symmetrically functionalizing the NDI core with homochiral (LL, **24**, and DD, **25**), heterochiral (LD, **26**, and DL, **27**), phenylalanine dipeptide (Phe-Phe), and achiral (AA, **28**) glycine dipeptide (Gly-Gly) auxiliaries [39]. Photophysical investigation unveiled the chirality transcription, amplification, and retentive helical memory within the respective noncovalent supramolecular assembly architectures. Remarkably, the helical ordering of homochiral (**24** and **25**) and heterochiral (**26** and **27**) NDIs is organized into hierarchically piled 1D nanobelt and 0D mesosphere architectures, respectively, with the retention of

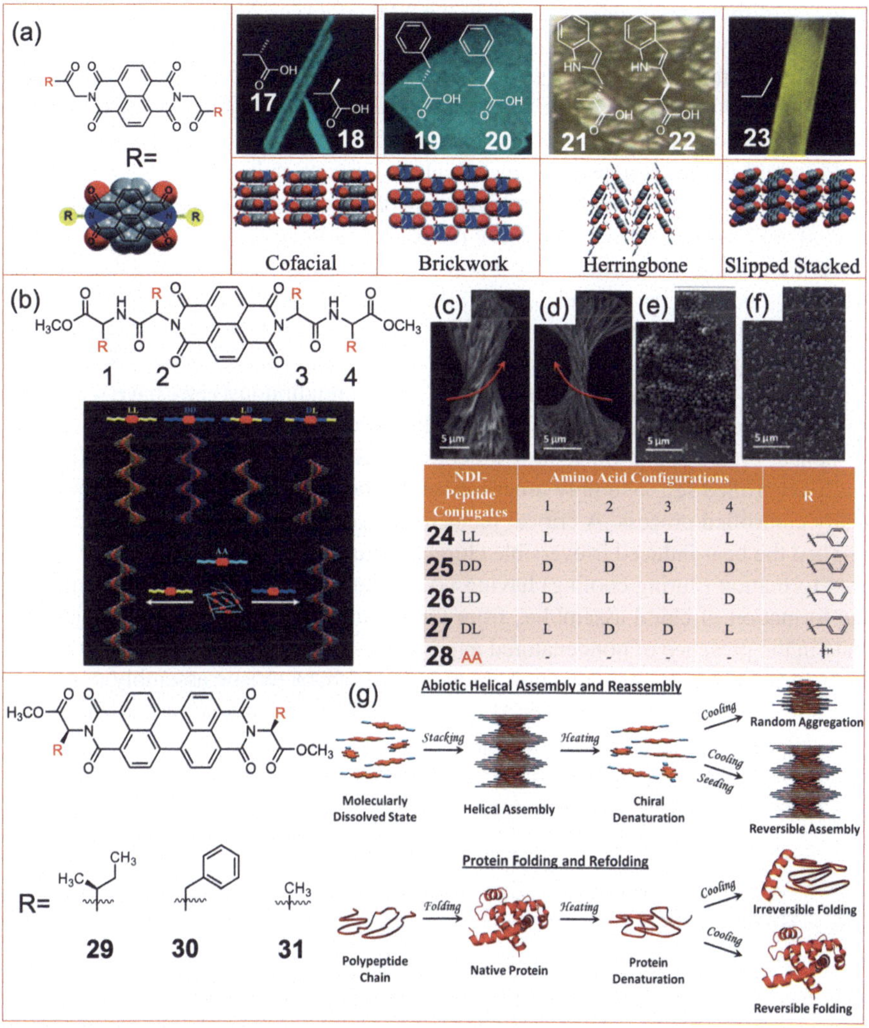

NDI-Peptide Conjugates	Amino Acid Configurations				R
	1	2	3	4	
24 LL	L	L	L	L	
25 DD	D	D	D	D	
26 LD	D	L	L	D	
27 DL	L	D	D	L	
28 AA	-	-	-	-	

Fig. 1.6 (**a**) NDI derivatives with different auxiliaries **17–23** and their molecular organization leading to the formation of all possible crystal packing modes of NDI core such as cofacial, brickwork, herringbone, and slipped stacked and the respective photographs of the solid-state emissions. Panel (**a**) is adapted with permission. Copyright 2014, Royal Society of Chemistry [38]. (**b**) Proposed model of M helicity and P helicity (with the schematic illustration of sergeants-and-soldiers effect) of the NDI-peptide conjugates. As shown in table, homochiral (LL and DD, **24** and **26**), heterochiral (LD and DL, **25** and **27**), and achiral (AA, **28**) dipeptide-conjugated NDIs are studied. (**c-f**) FESEM images of the self-assembled architectures of **24**, **25**, **26**, and **27**, respectively. Panels (**b–f**) are adapted with permission. Copyright 2012, Wiley-VCH [39]. (**g**) PDI derivatives with different nonpolar auxiliaries (**29–31**) and the schematic representation demonstrating the abiotic helical assembly-reassembly phenomenon of **29** and its correlation with the protein folding pathway (biotic). Panel (**g**) is adapted with permission. Copyright 2013, American Chemical Society [40]

chirality of first stereocenter adjacent to the NDI core (i.e., M helicity for **24** and **27** and P helicity for **25** and **27**) (Fig. 1.6b–f). On the other hand, chirality was introduced within the assembly of **28** with achiral auxiliary upon mixing with either of the homochiral NDIs. These unprecedented observations support the biological homochirality mechanism of spontaneous deracemization and amplification.

Protein folding is a process by which a nascent polypeptide attains its 3D functional state consisting of secondary structural features of α-helix, β-sheet, and random coil, among others. The entire process of protein folding takes place within merely a narrow time scale limit involving several complex intermediate steps, which remains to be a challenging task to understand. For the comprehensive understanding of the biotic protein folding mechanism, we demonstrated an abiotic design strategy with PDI core functionalized with nonpolar isoleucine (**29**), phenylalanine (**30**), and alanine (**31**) methyl esters [40]. Serendipitously, we have witnessed an extremely slow assembly dynamics of **29**, needed over a period of 1 day toward its helical organization. Such slow kinetics is quite unusual for any thermodynamically controlled self-assembly processes of synthetic molecules and suggests kinetically controlled process. A critical focus on the temperature-dependent CD studies implied the heat-induced irreversible chiral denaturation followed by random aggregate formation (during cooling) having no bias for specific handedness. Eventually the formation of chiral assemblies from the thermally denatured random aggregate state in the presence of nondenatured aggregated seeds of **29** revealed its correlation to the protein folding phenomenon (biotic). Such characteristic assembly dynamics was not observed in either of **30** or **31** (Fig. 1.6g). This protein folding-like intriguing slow helical assembly and temperature-induced irreversible denaturation of **29** (abiotic) suggests its plausible implications (reductionistic approach) in understanding the protein folding and disease-associated misfolding phenomena.

1.6 Biosensors

Engineering biomolecule-directed programmable molecular self-assemblies, co-assemblies, and templated assemblies is one of the simple and cost-effective routes to fabricate advanced biosensors with ultrasensitivity and selectivity. The scheme of molecular architectonics aids in the design of tailorable biomimetic architectures used in the development of ultrasensitive biosensors for the detection of environmental, biomedical, food, and defense-related analytes. Mercury (Hg) is a globally widespread toxic heavy metal and one of the top six hazardous environmental pollutants. Therefore, developing a reliable and ultrasensitive method for the detection and monitoring of organic and inorganic forms of Hg accumulation in drinking water is of prime interest. We developed a dual-responsive (chiroptical and conductivity) and templated DNA nanoarchitectures for the detection of Hg by exploiting its strong and specific interaction with thymine (T-Hg-T) [41]. Adenine (A)-conjugated NDI (BNA) molecule and the complementary deoxyribo-oligothymidine (dT_n, $n = 6, 10, 20$) were subjected to mutually templated co-assembly in buffer medium (pH 7). The templated co-assembly between BNA and dT_n resulted in the formation of hybrid DNA ensemble of the type [BNA_n-dT_n] which undergo hierarchical organization to form 2D nanosheets. [BNA_n-dT_n] exhibit interesting chiroptical and conductive property, and the lateral dimensions of the 2D nanosheets were found to be directly dependent on the length of the dT_n sequence. Remarkably, addition of Hg^{2+} displaces BNA units from [BNA_n-dT_n] ensemble due to the formation of metallo-DNA duplex of the type $[dT\text{-}Hg\text{-}dT]_n$, which result in distinct change in the chiroptical and conductivity properties (Fig. 1.7a). This observation has enabled us to fabricate a field-effect transistor (FET) device using [BNA_{10}-dT_{10}] material for the selective detection of inorganic/organic Hg in water with high selectivity and sub-nanomolar sensitivity ($\geq$0.1 nM, 0.02 ppb). The chiroptical method of detection by circular dichroism spectroscopy measurements also revealed the ultrasensitivity of [BNA_n-dT_n] for the detection of Hg. The observed detection limit is 100 times lower than the US Environmental Protection Agency (USEPA) tolerance limit (10 nM, 2 ppb) of the Hg^{2+} in drinking water (Fig. 1.7b and c). We employed FET device for the detection of Hg in real samples from various water bodies, and the method can be used for the detection of Hg in biological samples.

In another interesting design, a histidine (H)-functionalized perylenediimide (PDI) bolaamphiphile (HPH) was developed as the fluorescence- and surface-enhanced Raman spectroscopy (SERS)-based dual-responsive optical marker for the ultrasensitive detection of Hg [42]. The self-assembly of HPH led to J-type fluorescent aggregate that transformed into H-type non-fluorescent metallo-supramolecular aggregates upon binding to Hg^{2+}. This fluorometric change allowed selective detection of Hg^{2+} in water with a detection limit of 5×10^{-9} M (0.1 ppb). Remarkably, a detection limit of as low as 60×10^{-18} M or 60 attomolar (0.01 parts per quadrillion, ppq) for Hg^{2+} was achieved by SERS measurements, which is the highest ever reported sensitivity for the detection of Hg. A significant change in the SERS signal was observed upon interaction of Hg^{2+} with the marker (HPH) immobilized on SERS-active gold (Au) surface and facilitated the host-guest complexation-driven assembly between the adjacent HPH molecules (Fig. 1.7d). The SERS-based ultrasensitive method allows the detection of Hg^{2+} levels 10 order lower than the USEPA tolerance limit in drinking water. These two simple and advanced molecular architectonics-based molecular engineering strategies are expected to inspire the development of numerous unique biosensor platforms and opto-electronic devices for the ultrasensitive detection and investigation of various analytes.

The changes in the optical properties of self-assembly or co-assembly architectures of functional chromophore by the external agents (analyte) were harvested to detect benzene, a well-known aromatic carcinogen. In this optoelectronic architectonics strategy, we investigated the photophysical properties of molecular assembly architectures of **10** (discussed in "Organic Electronics" section) in different aromatic solvents, namely, benzene, toluene, and xylene. The fluorometric studies of **10** in polar (DMSO and water) and nonpolar aromatic (benzene, toluene, and xylene) solvents suggested distinct self-assembly-induced fluorescence response attributed

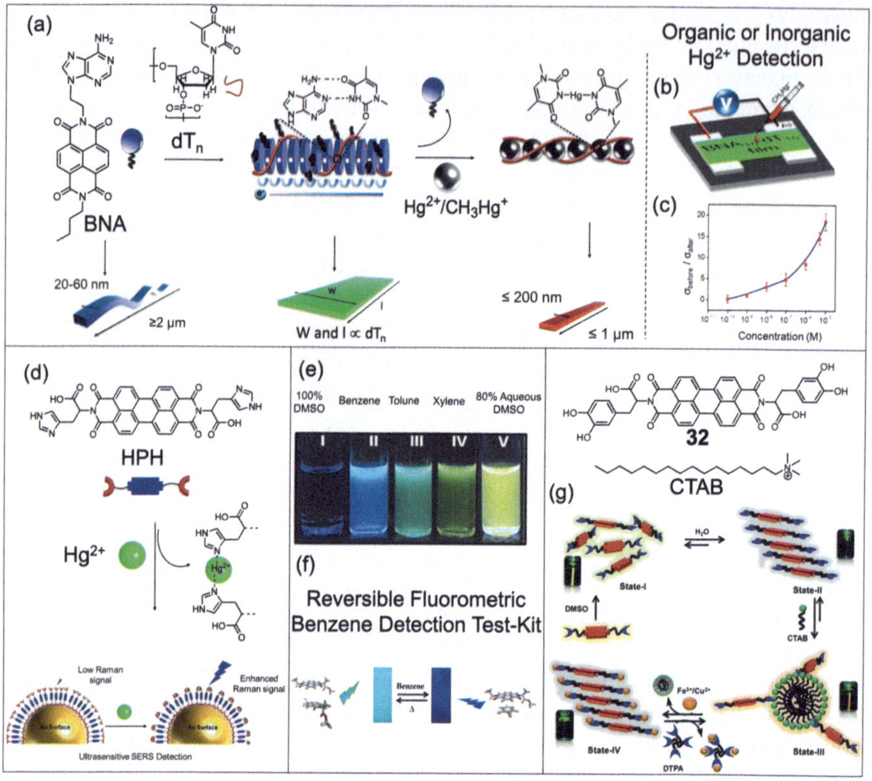

Fig. 1.7 (**a**) Molecular architectonics-guided co-assembly of NDI derivative BNA and dT_n to form hybrid DNA ensemble [BNA_n-dT_n] and their hierarchical organization to 2D nanosheets. Hg displaces BNA from [BNA_n-dT_n] to form metallo-DNA architectures [dT_n-Hg_n-dT_n] through strong Hg-T interactions. (**b** and **c**) Schematic of field-effect transistor (FET) device fabricated using [BNA_n-dT_n] for the ultrasensitive detection of organic and inorganic Hg^{2+} and the concentration-dependent conductometric sensing of CH_3Hg^+. Panels (**a–c**) are adapted with permission. Copyright 2016, American Chemical Society [41]. (**d**) Attomolar detection of Hg^{2+} by SERS upon interaction with the HPH immobilized on the SERS active gold surface. Panel (d) is adapted with permission. Copyright 2018, Elsevier [42]. (**e** and **f**) Sensing of aromatic solvents by the different fluorescence responses (colors) of self-assembled **10** in polar (DMSO and water) and nonpolar aromatic (benzene, toluene, and xylene) solvents. Lower panel showing specific detection of carcinogenic benzene by reversible fluorometric response using filter paper-based test kit. Panels (**e** and **f**) are adapted with permission. Copyright 2013, Royal Society of Chemistry [35]. (**g**) Schematic representation of biosensing of CTAB and Fe^{2+}/Cu^{2+} by the assembly-disassembly modulation of PDI derivative **32** with various aggregation and disaggregation states (State I, State II, State III, and State IV). Panel (**g**) is adapted with permission. Copyright 2014, American Chemical Society [43]

to excimer and exciplex formation [35]. The detailed studies revealed that **10** is capable of selectively sensing benzene, toluene, and xylene through distinct and visible fluorescence color and response within blue to yellow-green region. This selective spectroscopy response encouraged us to develop a filter paper-based reversible fluorometric test kit for the detection of carcinogenic benzene (Fig. 1.7e and f).

We further established an assembly-disassembly platform for the optical detection of cationic surfactants and metal ions like Fe^{3+}/Cu^{2+} in water media [43]. The fluorescence property of self-assembly architectures of an amphiphilic PDI derivative functionalized with L-3,4-dihydroxyphenylalanine (L-DOPA) auxiliary (**32**) was reversibly modulated to realize switching off-on states. The monomeric **32** showed bright yellow fluorescence in DMSO (State I), which transformed into weakly fluorescent aggregated species upon self-assembly in water (State II). Interestingly, the presence of a cationic surfactant cetyltrimethylammonium bromide (CTAB) disrupted the assembly pattern of **32** in water as a result of host-guest interactions among the PDI and CTAB molecules, which modified the assembly of **32** into fluorescent State III (Fig. 1.7g). In other ward, CTAB-mediated modulation of assembly property of **32** restored the bright yellow fluorescence color of monomeric **32** under the micellar environment. The disaggregated fluorescent state in micellar medium was switched into the metal-bound non-fluorescent off state (State IV) in the presence of Fe^{3+}/Cu^{2+} due to metal chelation by L-DOPA moieties of **32**, which resulted in the formation of metallo-supramolecular aggregates (Fig. 1.7g). The metal-driven fluorescence off state was reversed into the fluorescence on state (State III) by addition of external metal ion chelator, diethylenetriaminepentaacetic acid (DTPA). This study explicitly pointed out the significant role of external agents in controlling the architectural outcomes and functional properties through reversible assembly-disassembly modulation.

The art of creation of co-assembly architectures was exploited to selectively recognize and detect G-quadruplexes (GQs). GQ is a non-canonical guanine (G)-rich secondary structure of DNA found in the 3′ end of chromosomes (telomere) and promoter regions of oncogene. GQs are known to play important biological functions including telomere maintenance, transcription, and replication, while considered important as diagnostic and therapeutic targets for cancers [44]. The pyrene-conjugated polyethylenimine (PEI-Py, **33**) showed GQ-specific fluorescence emission attributed to pyrene excimer formation (Fig. 1.8a). The selective excimer emission of **33** for GQ containing oncogene sequences can be developed into diagnostic and therapeutic tools in cancer therapy.

The co-assembly architectures of two pyrene-dipeptides (**34** and **35**) of opposite polarity were developed for the efficient detection of nitroaromatics (NACs) such as nitrophenol, trinitrobenzene, and trinitrotoluene [45]. Development of reliable chemical sensor tools for the selective and sensitive detection of toxic NAC explosives is essential to secure the public health and environment. The designer pyrene-dipeptide conjugates **34** and **35** co-assembled through antiparallel β-sheet interactions, which resulted in the formation of 3D fluorescent hydrogel with fibrillar network. This hydrogel system with fibrillar network of pyrene chromophore served as a host for the selective intercalation of NAC guests. The electron-rich pyrene chromophore formed sandwich-like ground-state (pyrene-NAC-pyrene) assembly upon exposure to electron-deficient NACs causing long-range exciton migration along the extended

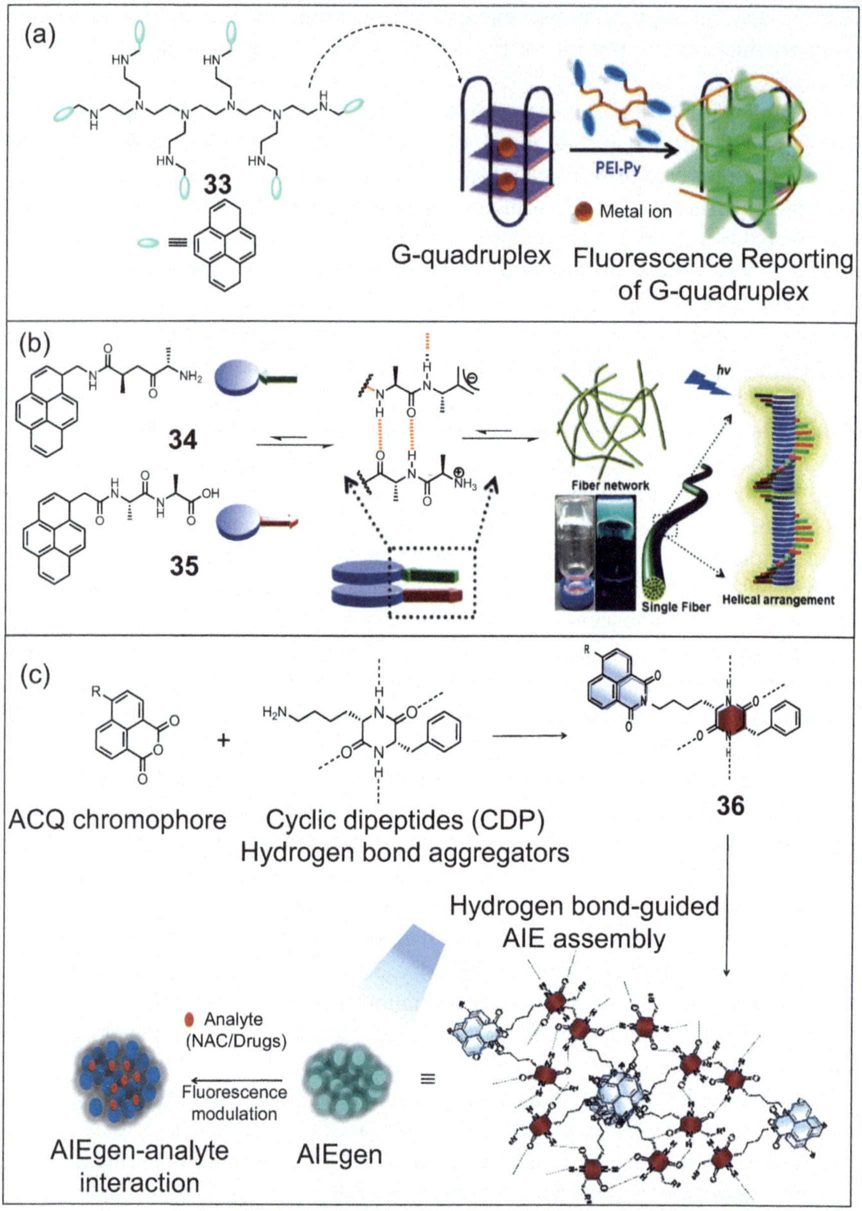

Fig. 1.8 (**a**) Schematic representation for the detection of G-quadruplex by PEI-Py conjugate **33** upon the interaction with G-rich DNA sequence. Panel (**a**) is adapted with permission. Copyright 2015, Elsevier [44], Copyright 2018. (**b**) Molecular co-assembly of pyrene-conjugated dipeptide amphiphiles **34** and **35** with opposite polarities to form 1D nanofiber architecture supported by antiparallel β-sheet formation. Photographs of hydrogels (fluorescent) as shown by the inverted gel vials and the schematic of the helical bicomponent assembly of dipeptides (**34** and **35**) in nanofibers. Panel (**b**) is adapted with permission. Copyright 2018, Royal Society of Chemistry [45]. (**c**) Schematic illustration showing the design strategy of CDP-NMI conjugate **36** and CDP hydrogen bonding-guided AIEgenic assembly behavior and the corresponding applications in the detection of

and π-conjugated nanofibers within the hydrogel (Fig. 1.8b). This aids in the tunneling of excited-state electrons of pyrene into the low-lying LUMO of NAC in a non-radiative decay process, which results in the fluorescence off state and thus suggests selective response toward NACs. To amplify the NAC sensitivity of the hydrogel system, we prepared a thin film of the hydrogel, which enabled visual detection of NACs. This bioinspired co-assembly architecture design strategy provides a newer insight into the fabrication of novel sensors for the detection of explosives and other analytes.

Extending our peptide-chromophore assembly architecture-based strategies for explosive sensor, we rationally designed a series of cyclic dipeptide-tethered naphthalimides (CDP-NIs) (for detailed discussion on CDP, see "Drug Delivery and Tissue Engineering" section) [46]. The CDP units are tethered to NI core to modulate the aggregation-induced emission (AIE) property of the latter to explore its applications. The intriguing intermolecular hydrogen bonding ability of CDP auxiliary was anticipated to modulate the self-assembly of NI chromophore in transforming its aggregation-caused quenching (ACQ) state to aggregation-induced emission (AIE) state in aqueous media. Interestingly, the fluorescence intensity of the AIEgenic assembly of CDP-NI **36** was found to exhibit maximum AIE quenching in the presence of picric acid (NAC) with a limit of detection (LOD) of ~0.4 ppm (1.7 μM). Further exploration revealed selective modulation of AIEgenic property of **36** in the presence of phenolic drugs, doxorubicin and rifampicin. The sensitivity of AIEgenic **36** toward doxorubicin and rifampicin exhibited LOD in nanomolar range, viz., ~18 nM (9.7 ppb) and ~202 nM (164 ppb), respectively (Fig. 1.8c). CDP modulated AIEgenic property of chromophores is expected to inspire the design of smart sensor systems for various bioactive analytes.

1.7 Drug Delivery and Tissue Engineering

Development of novel molecular and material architectures as biomaterials for drug delivery and tissue engineering applications is of prime importance to improve human health [3]. A considerable progress has been achieved in designing advanced biomaterials in the past two decades. In this section, biomolecule-derived biomaterial fabrication strategies for drug delivery and specific tissue engineering applications involving electrically active cells and tissues are discussed.

CDPs are six-membered heterocyclic lactams and are the smallest cyclic peptides [11]. CDP derivatives are found in many living organisms as secondary metabolite natural products. CDPs form intermolecular hydrogen bonding in the form of molecular chains and layers, which guide the formation of diverse molecular

Fig. 1.8 (continued) NAC (picric acid) and phenolic drugs (doxorubicin and rifampicin) in water media. Panel (**c**) is adapted with permission. Copyright 2020, American Chemical Society [46]

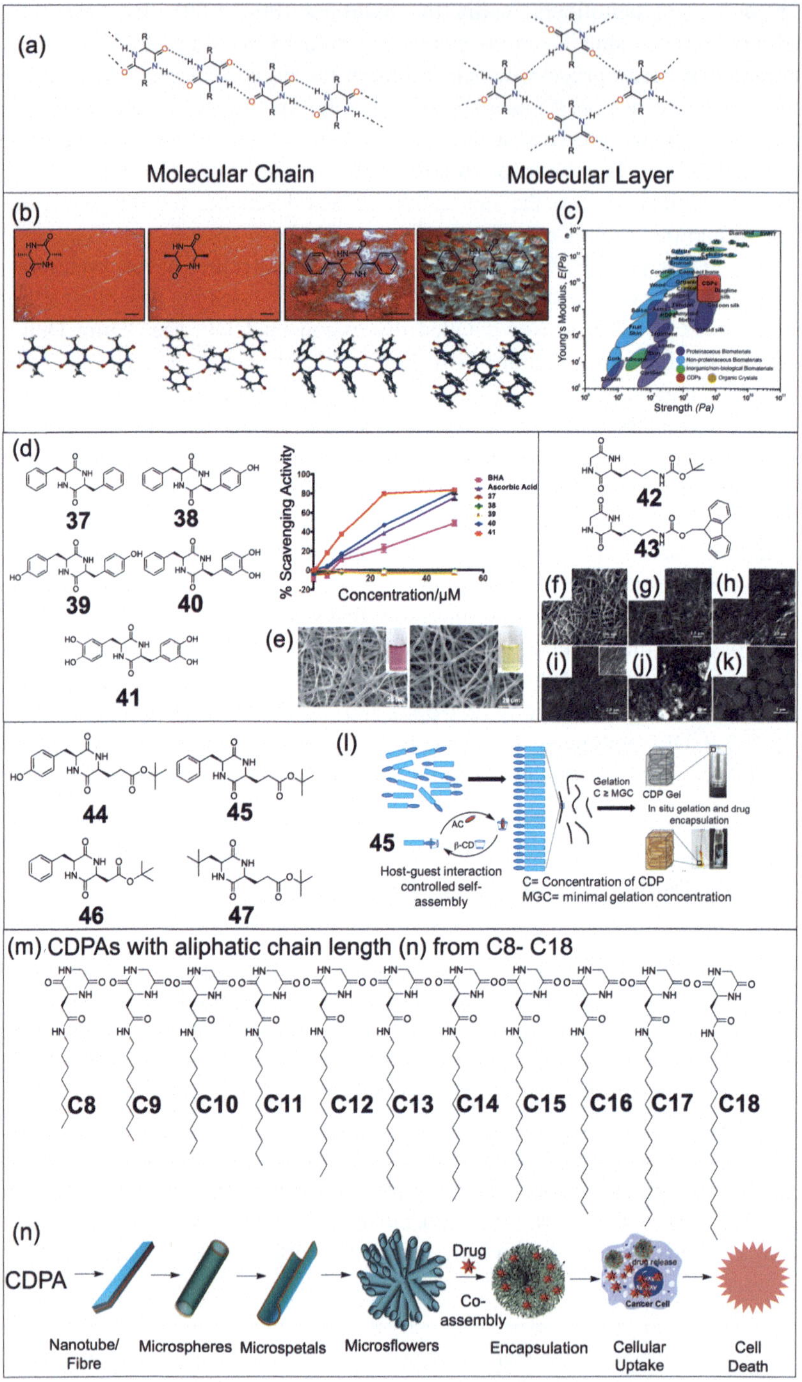

Fig. 1.9 (**a**) The unique molecular self-assembly of CDP to form molecular chains and molecular layers through multiple hydrogen bonding sites. (**b** and **c**) Single-crystalline self-assembled architectures of different CDPs (L,D-Phe, L,L-Ala, cyclo(L-Ala-D-Ala), and L,L-Phe) and the plot of

architectures (Fig. 1.9a). The unique self-assembly propensity and inherent structural rigidity made CDP a versatile building unit to produce material architectures that are mechanically robust and resistant toward proteolytic cleavage. In a bioinspired reductionistic strategy, we exploited the synergistic interactions of multiple hydrogen bonding networks and aromatic interactions by studying a series of CDP derivatives cyclo(L-Phg-D-Phg) (L,D-Phe), cyclo(L-Ala-L-Ala) (L,L-Ala), cyclo(L-Ala-D-Ala), and cyclo(L-Phg-L-Phg) (L,L-Phe) for their mechanical strength (Fig. 1.9b) [47]. The detailed investigation on the mechanical properties of CDP architectures using nanoindentation technique revealed superior mechanical stiffness and strength of cyclo(L-Phg-D-Phg) similar to natural silk (Fig. 1.9c). This study demonstrated the possibility to produce large-scale low-density organic materials with excellent mechanical strength following CDP-based reductionistic approach, which are sought-after biomimetic materials in the biomaterial research. We have also employed the CDP moiety as the functional core or auxiliary candidate in the scheme of molecular architectonics for drug delivery and tissue engineering applications. We have designed a number of proteinogenic and non-proteinogenic aromatic amino acids containing CDPs (**37–41**) to evaluate their antioxidant property (Fig. 1.9d) [48]. The thorough structure-activity-based investigation revealed the superior antioxidative property of CDPs containing DOPA units (CDPs **40** and **41**) compared to ascorbic acid and food additive butylated hydroxytoluene (BHT). Cellular studies have indicated that these CDPs are non-cytotoxic in murine adrenal PC12 cells. Further effort to prepare a biocompatible electrospun SF scaffold incorporated with antioxidant CDP **40** (CDP/SF composite) resulted excellent radical scavenging properties and therefore can potentially be utilized as therapeutic and tissue engineering scaffolds (Fig. 1.9e). We developed CDP-based low molecular

Fig. 1.9 (continued) elastic modulus (E) versus strength showing the high mechanical strength of CDP architectures in comparison to various materials including silk. Panels (**b** and **c**) are adapted with permission. Copyright 2015, Nature Publishing Group [47]. (**d**) Chemical structures of antioxidant CDPs (**37–41**) studied to assess their respective radical scavenging properties and a comparative plot of antioxidant property of all CDPs. (**e**) Antioxidant property of CDPs upon incorporating into electrospun silk fibroin (SF). SEM micrographs of native SF and CDP/SF electrospun scaffolds (left to right). The inset showing the respective radical scavenging abilities of the SF and CDP-SF by the change in DPPH coloration. Panels (**d** and **e**) are adapted with permission. Copyright 2017, Wiley-VCH [48]. Chemical structure of CDP gelators **42** and **43** and the FESEM images of xerogels formed by **42** in (f) $CHCl_3$, (**g**) CCL_4, (**h**) hexane, and (**i**) toluene and that of **43** in (**j**) toluene and (**k**) $CHCl_3$ showing ordered microporous hierarchical arrays. Panels (**f–k**) are adapted with permission. Copyright 2012, Royal Society of Chemistry [49]. Chemical structure of another set of CDP gelators **44–47** and (**l**) schematic representation of self-assembly of **45** showing the gel formation in the presence of guest molecules β-cyclodextrin at their minimal gelation concentrations. Panel (**l**) is adapted with permission. Copyright 2017, American Chemical Society [50]. (**m**) Representative chemical structures of CDPAs (**C8–C18**) and their morphological evolution into different nano- and micro-architectures and use of CDPA architecture as drug delivery vehicle to deliver drug and induce cancer cell death. Panel (**m**) is adapted with permission. Copyright 2020, American Chemical Society [51]

weight gelators (LMWGs) due to their mechanical robustness and utility in biomaterial applications [49]. Asymmetric cyclo(Gly-L-Lys) ε-amino derivatives (**42** and **43**) with tert-butyloxycarbonyl (tBoc, **42**) and fluorenylmethoxycarbonyl (Fmoc, **43**) Lys ε-amino protection were designed. CDPs **42** and **43** were subjected to self-assembly in both aqueous and organic solvents by sequential heating-cooling methods to evaluate their propensity to form hydrogels and organogels. The results showed that **42** form organogels (in hexane, carbon tetrachloride, toluene, and chloroform) and **43** formed both organogel (toluene) and hydrogel (Fig. 1.9f–k). Interestingly, the Fmoc group of **43** imparted high mechanical strength to the organogel formed in chloroform with ordered microporous hierarchical morphology through extensive intermolecular hydrogen bonding and π-π aromatic interactions. We demonstrated the encapsulation of dye and drug molecules such as rhodamine B and curcumin within the gel network without affecting the gel architecture, which infer possible utility as drug delivery vehicle. To understand the structure-function relationship, we studied the gelation property of four CDPs, namely, cyclo(L-Tyr-L-Glu(OtBu)) (**44**), cyclo(L-Phe-L-Glu(OtBu)) (**45**), cyclo(L-Phe-L-Asp(OtBu)) (**46**), and cyclo(L-Leu-L-Glu(OtBu)) (**47**) [50]. CDP **45** showed the superior tendency to form cytocompatible and thermostable ambidextrous gels with a minimal gelation concentration of 0.3 wt%, qualified as supergelator (Fig. 1.9l). The hydrogel forming self-assembled 2D fibrous networks was stabilized by intermolecular N–H···O hydrogen bonding, while t-Boc and phenyl group introduced hydrophobic and π-π stacking interactions. Notably, **45** formed an injectable hydrogel entrapped with curcumin under the ambient conditions, which highlights the possible use as injectable hydrogel formulations in drug delivery-related applications.

CDP-based molecular architectonics strategy was adapted to develop a drug delivery system by utilizing robust and cytocompatible CDP amphiphiles (CDPAs). A number of cyclo(Gly-L-Asp) functionalized with alkyl chains of varying length (n = C8–C18) were designed (**C8–C18**) (Fig. 1.9m) and judiciously engineered their molecular self-assembly into diverse nanoarchitectures such as nanotubes, nanospheres, nano-/microsheets, and flowers in organic or aqueous solvents [51]. The formation of various molecular assembly architectures was found to be influenced by the hydrophobic interactions of alkyl chains and hydrogen bonding interactions of CDP. The facile co-assembly tendency of CDPAs with nucleosides and nucleotides through specific noncovalent interactions was studied to assess their ability to act as potential drug delivery vehicles. The *in cellulo* studies of CDPA co-assembly with an anticancer drug 5-bromo-2′-deoxyuridine (BrdU) resulted in the BrdU-triggered cellular cytotoxicity to HeLa cells (Fig. 1.9n). The cellular studies together with the computational modelling of CDPAs and their co-assembly with the drug molecule confirmed the molecular level interactions of CDPAs and the drug molecules, which confirmed the utility of CDPA architectures as robust platforms for drug delivery applications.

In another interesting study, we designed a CDP-containing peptidomimetic system with improved cell-penetrating and excellent cytocompatibility properties for DNA delivery applications. The customized design strategy was conceived specifically to impart properties that are required for effective DNA delivery into cells and transfection by overcoming the trivial drawbacks of cytotoxic polycationic peptides. CDP peptidomimetics were synthesized by incorporating lysine and aspartic acid-based unnatural CDP (**kd**) amino acid units at the alternative positions in lysine (K) decapeptide and glutamic acid (E) decapeptide to obtain $\mathbf{K_5kd_5}$ and $\mathbf{E_5kd_5}$, respectively. The optimized and minimum cationic charge of five K units and rigidity attributed to **kd** units rendered $\mathbf{K_5kd_5}$ with high serum stability, bioavailability, and cytocompatibility compared to known cationic cell-penetrating peptides (CPPs) such as stearyl R8. In fact, $\mathbf{K_5kd_5}$ effectively combine the characteristics of CPPs by reducing their cytotoxicity and the rigid rod-like structure of polyproline-based CPPs to promote specific noncovalent interaction-driven DNA delivery and transfection. The CDP peptidomimetic $\mathbf{E_5kd_5}$ was ineffective for DNA delivery application due to its anionic nature. The NMR and SAXS studies have revealed efficient interaction of $\mathbf{K_5kd_5}$ with DNA and membrane lipid (DOPC) through specific hydrogen bonding of backbone CDP units (**kd**) in addition to electrostatic interactions imparted by the K units, which promoted specific and selective delivery of DNA into cells in comparison to the nonspecific interaction-driven delivery by cationic CPPs and polymers. As a result, $\mathbf{K_5kd_5}$ showed effective delivery of plasmid DNAs in HeLa and HEK293 cells, and the transfection was confirmed by the expression of green fluorescent protein (GFP) and β-galactosidase (Fig. 1.10a) [52].

In this section, we discuss innovative molecular architectonics design strategies based on small molecules targeting DNA. Threading intercalators are small molecules that bind to DNA through dual mode of interaction, viz., intercalation and groove binding, which are useful molecular tools in the development of advanced DNA-based architectures with biomedical applications. In the scheme of molecular architectonics, threading intercalators recognizing long sequences of DNA compared to traditional intercalators or groove binders are capable of modulating the DNA condensation-decondensation process with tremendous implications in DNA- or gene-targeted diagnostics and therapeutics. We have designed a set of imidazole-functionalized NDI derivatives to investigate their threading intercalation ability [53]. In this design strategy, NDI chromophore with planer geometry was chosen as an intercalating scaffold and functionalized with imidazolyl auxiliary to facilitate the groove-binding ability. Among all the compounds studied, NIm6 showed the best threading intercalation ability with a long half-life of 51 min. The threading process was achieved by the intercalation of NDI chromophore between the DNA base pairs and simultaneous binding of imidazolyl moieties in the adjacent major and minor grooves. The imidazolium moieties aided the DNA groove binding through electrostatic and hydrogen bonding interactions, which in turn led to favorable enthalpy change for the threading intercalation process (Fig. 1.10b). Furthermore, the cytotoxic effect of NIm6 in HeLa cells and inhibition of topoisomerase I activity upon binding to the supercoiled DNA inferred the biological implication of threading intercalating agent. In an interesting study aimed at understanding the various factors influencing the DNA condensation and decondensation process, we exploited the threading intercalation property in the design of DNA condensing agent. A set of threading intercalators (BIN, BIHQ, and BIAQ) with ethyl bridge

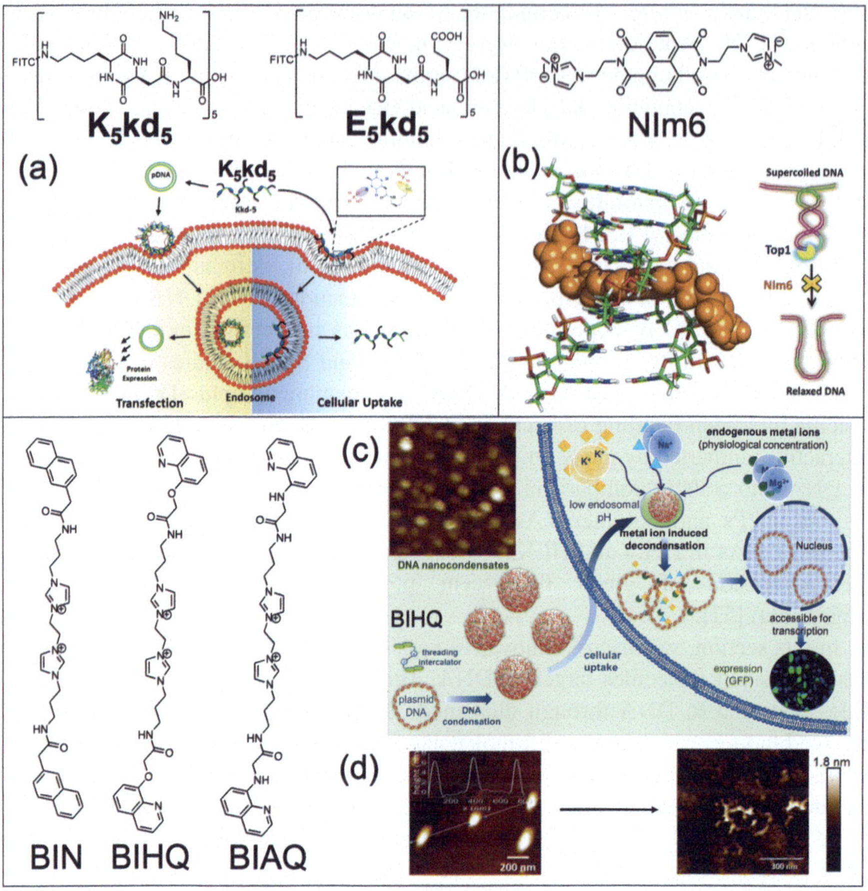

Fig. 1.10 Chemical structures of CDP peptidomimetics $\mathbf{K_5kd_5}$ and $\mathbf{E_5kd_5}$. (a) Schematic representation showing cellular uptake, DNA delivery, and transfection efficiency of $\mathbf{K_5kd_5}$. Panel (**a**) is adapted with permission. Copyright 2017, Royal Society of Chemistry [52]. (**b**) Geometry-optimized structure of DNA complex with NIm6 which shows the threading intercalation mode of interaction of NIm6. Panel (**b**) is adapted with permission. Copyright 2016, Wiley-VCH [53]. Chemical structures of BIN, BIHQ, and BIAQ. (**c**) Schematic showing threading intercalation of BIHQ-induced DNA nanocondensate formation and the endogenous metal ion-induced decondensation to release the DNA and its effective transfection. Panel (**c**) is adapted with permission. Copyright 2020, American Chemical Society [54]

connecting bisimidazolium system to the terminal naphthyl or quinolinyl aromatic moieties via amide or ether linkages were designed [54]. The design strategy of introducing naphthyl or quinolinyl moieties at the termini of bisimidazolium linker was conceptualized to promote DNA intercalation, while imidazolium moiety is expected to provide specific hydrogen bonding interaction, in addition to minimum electrostatic interactions. The specific intercalation and hydrogen bonding-driven binding interactions reduced the complete dependence of molecules on nonspecific electrostatic interactions for DNA recognition. The AFM imaging illustrated the thread-like morphology of the free DNA that transformed to distinct nanoarchitectures with an average size of 100 nm upon treatment with BIHQ. The distinguishable architectural form of free DNA and BIHQ-bound DNA indicated threading intercalation-driven formation of DNA nanocondensates. Further, the acidic pH-dependent turn-on fluorescence response of quinoline moiety of the ligand BIHQ allowed the tracking of cellular uptake and delivery of DNA nanocondensates into cell through the endocytic pathway. Remarkably, further study revealed decondensation of nanocondensate by the physiological concentrations of endogenous metal ions such as Mg^{2+} and K^{+}, followed by effective transfection (Fig. 1.10c). This strategy highlights the design and potential of small molecule-based threading intercalators for the development of tailorable nonviral vectors for the DNA or gene delivery applications. The threading intercalation-based condensation of DNA into nanocondensates and their decondensation by endogenous metal ions have potential implications in understanding the chromatin condensation-decondensation dynamics.

Recently, we have formulated an injectable silk fibroin (SF) hydrogel (iSFH) for controlled and sustained delivery of insulin in the treatment of diabetic conditions [55]. The sustained insulin delivery is highly desirable and sought-after to maintain the physiological glucose homeostasis, as it avoids the complications related to multiple subcutaneous insulin injections leading to poor patient compliance and insulin amyloidosis at the site of injection. To address this long-standing issue, we utilized biocompatible SF protein to produce injectable mesoporous hydrogel formulation within 50 min in the presence of biocompatible viscous additives (ethylene glycol and triethylene glycol). It is worth noting that the high mechanical stiffness of natural SF is extremely useful for hydrogelation, electrospinning, and preparation of composite scaffolds with excellent biocompatibility to cater to biomedical applications including drug delivery, cell transfection, tissue engineering, vaccine thermostabilization, and biocidal coating materials, among others [56]. Interestingly, the mesoporous hydrogel prepared by SF showed excellent insulin encapsulation ability. The subcutaneous injection of the insulin-encapsulated SF hydrogel in diabetic T1DM Wistar rats and follow-up studies have demonstrated potential for slow and controlled release of the encapsulated insulin over a period of 4 days with effective maintenance of normal physiological glucose levels and homeostasis (Fig. 1.11a and b). The encapsulation, storage, and controlled release of active insulin indicate the possible use of silk hydrogel to develop oral delivery formulations of insulin in the near future for the treatment of diabetic conditions. This study demonstrates the advanced engineering of macromolecular organization supported by small molecule additives for technologically advanced drug delivery formulations to treat chronic and metabolic disorders such as diabetic conditions.

We rationally designed a protein mimetic di-block copolymer of tryptophan-functionalized methacrylic acid (PTMA) polyampholyte [57]. The zwitterionic PTMA was found to serve as an artificial chaperone and protect protein from aggregation-inducing conditions with possible therapeutic implications in

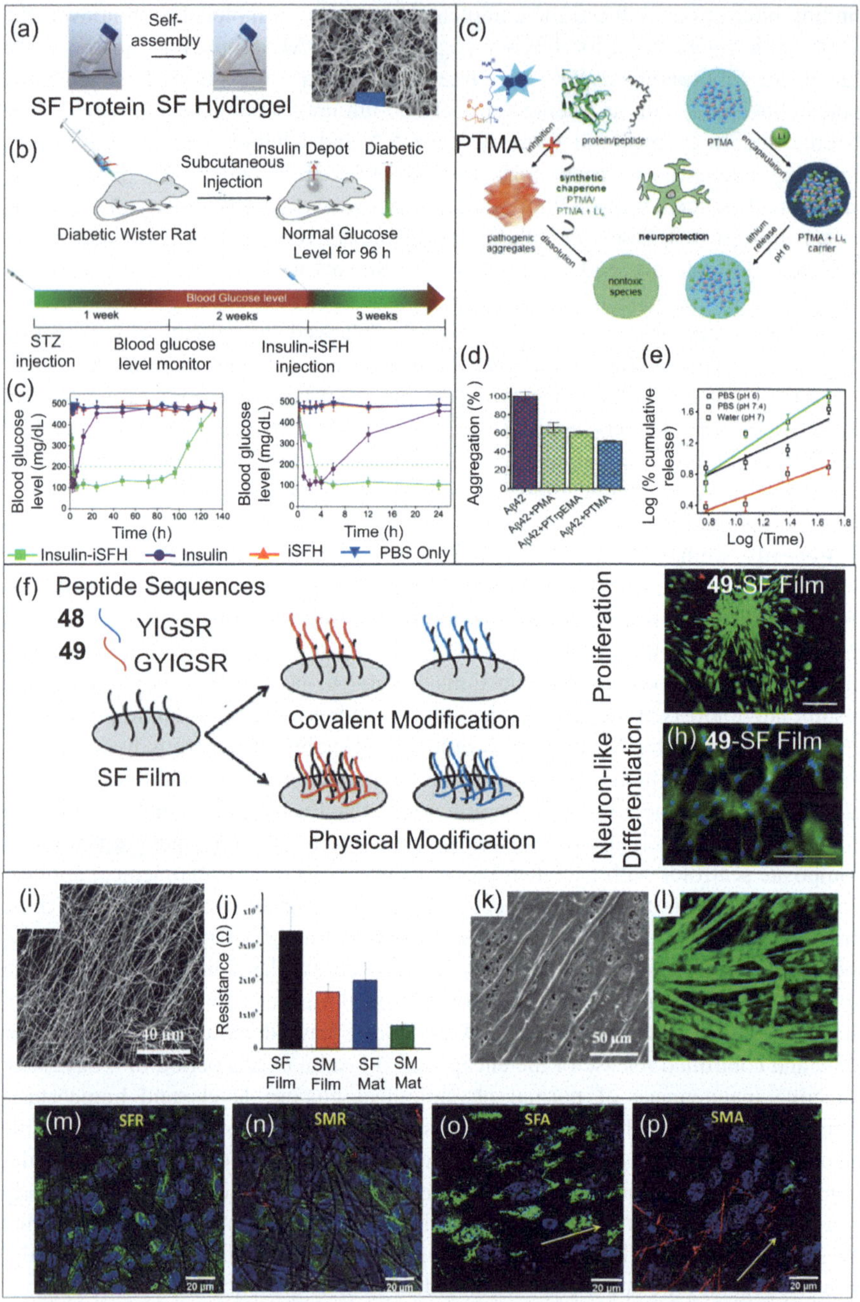

Fig. 1.11 (**a**) Photograph of the formation of injectable SF hydrogel (iSFH) and FESEM image of the corresponding xerogel. (**b**) Schematic illustration of in vivo delivery and therapeutic efficacy of insulin-loaded iSFH evaluated in streptozotocin (STZ)-induced diabetic Wistar rat through the subcutaneous injection of iSFH-loaded insulin and monitoring the slow release of the entrapped insulin to maintain normal glucose level over a period 4 days. The controlled release data show the blood glucose levels monitored after subcutaneous injection of iSFH in diabetic and control groups.

Alzheimer's disease (AD). The experimental studies have shown that PTMA protect heat-induced aggregation of lysozyme and modulate amyloid-beta (Aβ) aggregation, an important pathological event in AD. These observations and results confirmed the credentials of PTMA to serve as synthetic chaperone to protect the structure of essential proteins and prevent the aggregation and accumulation of toxic peptide or proteins (Fig. 1.11c and d). In addition, PTMA efficiently sequester lithium in neutral pH (~ pH 7) conditions and release under acidic conditions (~pH 6) (Fig. 1.11e). Lithium is used as a neuroprotective agent to treat various depression-related conditions including AD. The absence of effective delivery options resulted in the use of very high concentrations of lithium causing numerous side effects. The PTMA-guided lithium encapsulation, storage, and pH-responsive (therapeutic window) release of lithium ions are highly desirable for therapeutic value. This synergistic and dual-responsive biomimetic polymer-based protein aggregation modulation and controlled lithium delivery strategy has the potential in the development of therapeutic options for AD.

We explored the engineering of SF scaffolds for tissue engineering applications involving cells and tissues that require electrical impulses for their growth and differentiation. The skeletal muscle tissue engineering (SMTE) is a process of repairing and regeneration of impaired muscle tissue [59]. The electroactive scaffolds are used to support the assembly and differentiation of myoblasts into high aspect ratio functional myotubes [59]. In order to avoid the risk of oxidative stress-mediated immunogenic response and poor biocompatibility and biodegradability issues of synthetic polymers as conducting scaffolds for SMTE, we focused on the design and fabrication of novel extracellular matrix (ECM) mimicking and natural biopolymer-based customized scaffolds. In this context, natural SF protein with

Fig. 1.11 (continued) Panels (**a** and **b**) are adapted with permission. Copyright 2020, American Chemical Society [55]. (**c–e**) Chemical structure of polyampholyte PTMA and its protein aggregation (Aβ) modulation and encapsulation, storage, and pH stimuli-responsive release of lithium ions (monitored by the quenching and recovery of fluorescence, respectively). Panels (**c–e**) are adapted with permission. Copyright 2020, American Chemical Society [57]. (**f**) Schematic showing the surface functionalization of SF films through physical adsorption and covalent grafting of peptides **48** (YIGSR) and **49** (GYIGSR). (g and h) Stem cell proliferation and differentiation into neuronal-like cells on **49** SF film (fluorescence images). Panels (**f–h**) are adapted with permission. Copyright 2016, American Chemical Society [58]. (i) FESEM image of the electrospun fiber mats of silk-melanin (SM) composite which show aligned nanofibers. (**j**) Electrical conducting properties (as measured by the reduction in resistance) of electrospun scaffolds (SF film, SM film, SF mat, and SM mat). (**k** and **l**) TEM and fluorescence images of myotubes on the SM scaffold mat confirm the myogenic cell differentiation of C2C12 mouse myoblast cells. Panels (**i–l**) are adapted with permission. Copyright 2016, Wiley-VCH [59]. (**m–p**) Neuronal cell differentiation (SH-SY5Y cells) visualized by fluorescence imaging upon β-tubulin staining on the surface of the random and aligned nanofibers of SF and SM scaffolds. Neuronal cell differentiation monitored on silk fibroin random nanofibers (SFR), silk fibroin/melanin random nanofibers (SMR), silk fibroin aligned nanofibers (SFA), and silk fibroin/melanin aligned nanofibers (SMA) after 3 days of cell culture study. Panels (**m–p**) are adapted with permission. Copyright 2019, Elsevier [60]

excellent mechanical property and biocompatibility can serve as the base scaffold material [59, 61]. Taking advantage of the superior scaffold characteristics of SF, we have demonstrated silk composite biomaterial scaffold-based fabrication strategies and their applications in stem cell differentiation and SMTE. The free-standing SF films surface functionalized with integrin-binding laminin peptides YIGSR (**48**) and GYIGSR (**49**) by physical adsorption and chemical covalent attachment were fabricated [58]. The functionalized SF films exhibited good cytocompatibility to bone marrow-derived human mesenchymal stem cells (hMSCs) and promoted their proliferation and cell maintenance in an undifferentiated pluripotent stage. The SF film showed differentiation of hMSCs into neuronal cells in the presence of a chemical cue, retinoic acid (Fig. 1.11f–h). The detailed study revealed that SF film covalently tethered with **49** showed superior proliferation and neuronal cell differentiation as compared to SF films with physically adsorbed (**48** or **49**) or covalently tethered peptide epitope **48**. The sustained undifferentiated stem cell maintenance and on-demand differentiation capabilities of suitably functionalized SF films make them excellent tissue engineering scaffolds. In a unique study, we prepared SF composite scaffold by the physical incorporation of a naturally occurring polymeric skin pigment melanin to impart conducting and antioxidative properties [59, 62]. The electrospun fiber mats of silk-melanin (SM) composite were found to exhibit antioxidant and conducting property (Fig. 1.11i and j). The SF mats were evaluated for their effectiveness in promoting in vitro myogenesis. The combined influence of surface topography (aligned nanofibers in the mat) and antioxidant and conductivity property of SM scaffold films has successfully promoted the formation of aligned myotubes in vitro through effective cell attachment, proliferation, and differentiation processes (Fig. 1.11k and l). This study validated the potential of SM scaffolds in SMTE with promising applications in wound healing and wear and tear-related skeletal muscle tissue repair. We have also demonstrated the utility of radical scavenging property of SM composite in the protection and differentiation of neuronal cells. The human neuroblastoma cells (SH-SY5Y) were grown on random and aligned electrospun nanofibrous SF and SM scaffold films (such as SF random nanofibers (SFR), SF/melanin random nanofibers (SMR), SF aligned nanofibers (SFA), and SF/melanin aligned nanofibers (SMA)) [60]. The cellular viability, proliferation, adhesion, and differentiation assays have shown that SM scaffolds with aligned nanofibrous morphology promoted improved neuronal growth and differentiation characteristics, which could be a highly promising strategy for effective neuronal differentiation- and regeneration-related applications (Fig. 1.11m–p).

1.8 Conclusion and Future Prospects

We have summarized the principles of molecular architectonics in producing advanced molecular and material architectures with novel functional properties and practical applications. The scheme of molecular architectonics advocates the custom design and engineering of assembly and co-assembly of functional building blocks through the controlled and meticulous manipulation of noncovalent interactions. The biomolecules and their derivatives or mimetics conjugated to functional core serve as assembly-directing auxiliaries in the bottom-up fabrication of desired 0D, 1D, 2D, and 3D molecular and material architectures with well-defined nano- to macroscale sizes and shapes. In molecular architectonics, judicious and skilled experimental conditions including external stimuli play a key role in controlling the molecular organization into smart architectures. The sophisticated molecular assembly-engineering strategies provided access to advanced, mechanically robust, flexible, and cytocompatible molecular and material architectures with exciting optical, electrical, chiroptical, catalytic, sensing, and surface properties and applications. The detailed structure-property-function relationship studies resulted in diverse applications across self-cleaning, biomimetic catalysis, organic electronics and bioelectronics, homochirality and protein folding, sensors, bioimaging, drug delivery and tissue engineering, diagnostics, and therapeutics, among others. As a whole, the principles of molecular architectonics are anticipated to aid in the art of mastering the programmed molecular assemblies and co-assemblies with tailorable structure, property, and function.

The biomimetic and bioinspired molecular architectonics strategies envision new avenues for solution-processable bottom-up fabrication of advanced and emergent materials. The practice of the molecular architectonics is futuristic in nature and provides the opportunity to gain deeper insights into the subtle and complex interplay of poorly understood noncovalent interactions at the molecular level, which in turn allow the innovative and preprogramed molecular assembly engineering to create advanced molecular and material architectures with emergent properties and functional applications to solve long-standing problems related to health, energy, and environment. Apart from the design and construction of functional organic materials and biomaterials, the principles of molecular architectonics guide the generation of a wide range of well-defined inorganic, organic-inorganic hybrid, nanoparticle, quantum dot, and macromolecular architectures.

References

1. Avinash MB, Govindaraju T (2018) Architectonics: Design of Molecular Architecture for functional applications. Acc Chem Res 51:414–426
2. Roy B, Govindaraju T (2019) Amino acids and peptides as functional components in Arylenediimide-based molecular architectonics. Bull Chem Soc Jpn 92:1883–1901
3. Datta LP, Manchineella S, Govindaraju T (2020) Biomolecules-derived biomaterials. Biomaterials 230:119633
4. Ghosh D, Datta LP, Govindaraju T (2020) Molecular architectonics of DNA for functional Nanoarchitectures. Beilstein J Nanotechnol 11:124–140
5. Yu G, Jie K, Huang F (2015) Supramolecular Amphiphiles based on host–guest molecular recognition motifs. Chem Rev 115:7240–7303
6. Avinash MB, Govindaraju T (2014) Nanoarchitectonics of biomolecular assemblies for functional applications. Nanoscale 6:13348–13369
7. Whitesides GM, Grzybowski B (2002) Self-assembly at all scales. Science 295:2418–2421

8. Avinash MB, Govindaraju T (2012) Amino acid Derivatized Arylenediimides: a versatile modular approach for functional molecular materials. Adv Mater 24:3905–3922
9. Moorthy H, Datta LP, Govindaraju T (2021) Molecular architectonics-guided Design of Biomaterials. In Chem Asian J 16:423–442
10. Govindaraju T (2019) Templated DNA nanotechnology: functional DNA Nanoarchitectonics, 1st edn. Pan Stanford Publishing, New York City, NY, U.S.A.
11. Manchineella S, Govindaraju T (2017) Molecular self-assembly of cyclic dipeptide derivatives and their applications. ChemPlusChem 82:88–106
12. Roy B, Ghosh D, Govindaraju T (2019) Functional molecule–templated DNA Nanoarchitectures. In: Govindaraju T (ed) Templated DNA nanotechnology: functional DNA Nanoarchitectonics, 1st edn. Pan Stanford Publishing, New York City, NY, U.S.A., pp 69–106
13. Roy B, Ramesh M, Govindaraju T (2019) DNA-based Nanoswitches and devices. In: Govindaraju T (ed) Templated DNA nanotechnology: functional DNA Nanoarchitectonics, 1st edn. Pan Stanford Publishing, New York City, NY, U.S.A, pp 365–408
14. Avinash MB, Samanta PK, Sandeepa KV, Pati SK, Govindaraju T (2013) Molecular architectonics of Stereochemically constrained Π-complementary functional modules. Eur J Org Chem 2013:5838–5847
15. Govindaraju T, Avinash MB (2012) Two-dimensional Nanoarchitectonics: organic and hybrid materials. Nanoscale 4:6102–6117
16. Aono M, Bando Y, Ariga K (2012) Nanoarchitectonics: pioneering a new paradigm for nanotechnology in materials development. Adv Mater 24:150–151
17. Ariga K, Ji QM, Hill JP, Bando Y, Aono M (2012) Forming nanomaterials as layered functional structures toward materials Nanoarchitectonics. Npg Asia Mater 4
18. Konar M, Roy B, Govindaraju T (2020) Molecular architectonics-guided fabrication of Superhydrophobic and self-cleaning materials. Adv Mater Interfaces 7:2000246
19. Bhushan B, Jung YC (2011) Natural and biomimetic artificial surfaces for Superhydrophobicity, self-cleaning, low adhesion, and drag reduction. Prog Mater Sci 56:1–108
20. Bhushan B (2009) Biomimetics: lessons from nature - an overview. Philos T R Soc A 367:1445–1486
21. Ensikat HJ, Ditsche-Kuru P, Neinhuis C, Barthlott W (2011) Superhydrophobicity in perfection: the outstanding properties of the lotus leaf. Beilstein J Nanotechnol 2:152–161
22. Avinash MB, Verheggen E, Schmuck C, Govindaraju T (2012) Self-cleaning functional molecular materials. Angew Chem Int Ed 51:10324–10328
23. Lee JS, Ryu J, Park CB (2009) Bio-inspired fabrication of Superhydrophobic surfaces through peptide self-assembly. Soft Matter 5:2717–2720
24. Roy B, Pal S, Govindaraju T (2020) Intrinsic role of molecular architectonics in enhancing the catalytic activity of Lead in glucose hydrolysis. ACS Appl Mater Interfaces 12:14057–14063
25. Koch K, Barthlott W (2009) Superhydrophobic and Superhydrophilic plant surfaces: an inspiration for biomimetic materials. Philos Trans R Soc A Math Phys Eng Sci 367:1487–1509
26. Khanmohammadi Chenab K, Sohrabi B, Rahmanzadeh A (2019) Superhydrophobicity: advanced biological and biomedical applications. Biomater Sci 7:3110–3137
27. Wang Y et al (2013) Chemical synthesis of lactic acid from cellulose catalysed by Lead(ii) ions in water. Nat Commun 4:2141
28. Borovik AS (2005) Bioinspired hydrogen bond motifs in ligand design: the role of noncovalent interactions in metal ion mediated activation of dioxygen. Acc Chem Res 38:54–61
29. Shook RL, Borovik AS (2010) Role of the secondary coordination sphere in metal-mediated dioxygen activation. Inorg Chem 49:3646–3660
30. Katz HE, Johnson J, Lovinger AJ, Li W (2000) Naphthalenetetracarboxylic Diimide-based N-channel transistor semiconductors: structural variation and thiol-enhanced gold contacts. J Am Chem Soc 122:7787–7792

31. Sakai N, Bhosale R, Emery D, Mareda J, Matile S (2010) Supramolecular N/P-heterojunction photosystems with antiparallel redox gradients in electron- and hole-transporting pathways. J Am Chem Soc 132:6923–6925
32. Anthony JE, Facchetti A, Heeney M, Marder SR, Zhan X (2010) N-type organic semiconductors in organic electronics. Adv Mater 22:3876–3892
33. Avinash MB, Govindaraju T (2011) Engineering molecular Organization of Naphthalenediimides: large Nanosheets with metallic conductivity and Attoliter containers. Adv Funct Mater 21:3875–3882
34. Avinash MB, Swathi K, Narayan KS, Govindaraju T (2016) Molecular architectonics of Naphthalenediimides for efficient structure–property correlation. ACS Appl Mater Interfaces 8:8678–8685
35. Pandeeswar M, Govindaraju T (2013) Green-fluorescent naphthalene Diimide: conducting layered hierarchical 2D Nanosheets and reversible probe for detection of aromatic solvents. RSC Adv 3:11459–11462
36. Pandeeswar M, Senanayak SP, Narayan KS, Govindaraju T (2016) Multi-stimuli-responsive charge-transfer hydrogel for room-temperature organic ferroelectric thin-film devices. J Am Chem Soc 138:8259–8268
37. Pandeeswar M, Khare H, Ramakumar S, Govindaraju T (2015) Crystallographic insight-guided Nanoarchitectonics and conductivity modulation of an N-type organic semiconductor through peptide conjugation. Chem Commun 51:8315–8318
38. Pandeeswar M, Khare H, Ramakumar S, Govindaraju T (2014) Biomimetic molecular Organization of Naphthalene Diimide in the solid state: tunable (chiro-) optical, viscoelastic and nanoscale properties. RSC Adv 4:20154–20163
39. Pandeeswar M, Avinash MB, Govindaraju T (2012) Chiral transcription and retentive helical memory: probing peptide auxiliaries appended with Naphthalenediimides for their one-dimensional molecular organization. Chem-Eur J 18:4818–4822
40. Avinash MB, Govindaraju T (2013) Extremely slow dynamics of an abiotic helical assembly: unusual relevance to the secondary structure of proteins. J Phys Chem Letters 4:583–588
41. Pandeeswar M, Senanayak SP, Govindaraju T (2016) Nanoarchitectonics of small molecule and DNA for ultrasensitive detection of mercury. ACS Appl Mater Interfaces 8:30362–30371
42. Makam P, Shilpa R, Kandjani AE, Periasamy SR, Sabri YM, Madhu C, Bhargava SK, Govindaraju T (2018) SERS and fluorescence-based ultrasensitive detection of mercury in water. Biosens Bioelectron 100:556–564
43. Dwivedi AK, Pandeeswar M, Govindaraju T (2014) Assembly modulation of PDI derivative as a supramolecular fluorescence switching probe for detection of cationic surfactant and metal ions in aqueous media. ACS Appl Mater Interfaces 6:21369–21379
44. Narayanaswamy N, Unnikrishnan M, Gupta M, Govindaraju T (2015) Fluorescence reporting of G-Quadruplex structures and modulating their Dnazyme activity using Polyethylenimine-pyrene conjugate. Bioorg Med Chem Lett 25:2395–2400
45. Madhu C, Roy B, Makam P, Govindaraju T (2018) Bicomponent β-sheet assembly of dipeptide fluorophores of opposite polarity and sensitive detection of nitro-explosives. Chem Commun 54:2280–2283
46. Balachandra C, Govindaraju T (2020) Cyclic dipeptide-guided aggregation-induced emission of Naphthalimide and its application for the detection of phenolic drugs. J Org Chem 85:1525–1536
47. Avinash MB, Raut D, Mishra MK, Ramamurty U, Govindaraju T (2015) Bioinspired Reductionistic peptide engineering for exceptional mechanical properties. Sci Rep 5:16070
48. Manchineella S, Voshavar C, Govindaraju T (2017) Radical-scavenging antioxidant cyclic dipeptides and silk fibroin biomaterials. Eur J Org Chem 2017:4363–4369
49. Manchineella S, Govindaraju T (2012) Hydrogen bond directed self-assembly of cyclic dipeptide derivatives: gelation and ordered hierarchical architectures. RSC Adv 2:5539–5542

50. Manchineella S, Murugan NA, Govindaraju T (2017) Cyclic dipeptide-based ambidextrous Supergelators: minimalistic rational design, structure-gelation studies, and in situ Hydrogelation. Biomacromolecules 18:3581–3590
51. Pandurangan K, Roy B, Rajasekhar K, Suseela YV, Nagendra P, Chaturvedi A, Satwik UR, Murugan NA, Ramamurty U, Govindaraju T (2020) Molecular architectonics of cyclic dipeptide Amphiphiles and their application in drug delivery. ACS Appl Bio Mater 3:3413–3422
52. Madhu C, Voshavar C, Rajasekhar K, Govindaraju T (2017) Cyclic dipeptide based cell-penetrating Peptidomimetics for effective DNA delivery. Org Biomol Chem 15:3170–3174
53. Suseela YV, Das S, Pati SK, Govindaraju T (2016) Imidazolyl-Naphthalenediimide-based threading Intercalators of DNA. Chembiochem 17:2162–2171
54. Pratihar S, Suseela YV, Govindaraju T (2020) Threading Intercalator-induced Nanocondensates and role of endogenous metal ions in Decondensation for DNA delivery. ACS Appl Bio Mater 3:6979–6991
55. Maity B, Samanta S, Sarkar S, Alam S, Govindaraju T (2020) Injectable silk fibroin-based hydrogel for sustained insulin delivery in diabetic rats. ACS Appl Bio Mater 3:3544–3552
56. Tansil NC, Koh LD, Han MY (2012) Functional silk: colored and luminescent. Adv Mater 24:1388–1397
57. Datta LP, Samanta S, Govindaraju T (2020) Polyampholyte-based synthetic chaperone modulate amyloid aggregation and lithium delivery. ACS Chem Neurosci 11:2812–2826
58. Manchineella S, Thrivikraman G, Basu B, Govindaraju T (2016) Surface-functionalized silk fibroin films as a platform to guide neuron-like differentiation of human mesenchymal stem cells. ACS Appl Mater Interfaces 8:22849–22859
59. Manchineella S, Thrivikraman G, Khanum KK, Ramamurthy PC, Basu B, Govindaraju T (2016) Pigmented silk Nanofibrous composite for skeletal muscle tissue engineering. Adv Healthc Mater 5:1222–1232
60. Nune M, Manchineella S, Govindaraju T, Narayan KS (2019) Melanin incorporated electroactive and antioxidant silk fibroin Nanofibrous scaffolds for nerve tissue engineering. Mater Sci Eng C 94:17–25
61. Das S, Bora U, Borthakur BB (2014) 2 - applications of silk biomaterials in tissue engineering and regenerative medicine. In: Kundu SC (ed) Silk biomaterials for tissue engineering and regenerative medicine. Woodhead Publishing, Amsterdam, pp 41–77
62. McGinness J, Corry P, Proctor P (1974) Amorphous semiconductor switching in Melanins. Science 183:853–855

Chapter 2
Nanoarchitectonics

Katsuhiko Ariga

2.1 History of Nanoarchitectonics

The most important terminology in current science and technology is probably nano. Especially in application-oriented fields, nanotechnology has become an extremely influential magic concept. People may think that any dream applications become possible with nanotechnology. This key term, nanotechnology, is said to have originated thanks to the foresight of Richard Feynman, There's Plenty of Room at the Bottom, in 1959 [1–3]. He foresaw a multitude of scientific and technological possibilities in nanoscale spaces. Technological and engineering possibilities of nanotechnology were socially reevaluated in a declaration in 2001 under US President Bill Clinton to promote nanotechnology, called the National Nanotechnology Initiative. This historical flow is also supported by innovations regarding various observation and evaluation tools for nanoscale objects, such as the invention of scanning tunneling microscopy (STM) and nanofabrication techniques. The inventions of STM and related scanning probe microscopy (SPM) in atomic force microscopy (AFM) allow us to observe and manipulate objects at the level of atoms and molecules. In fact, the manipulation of artificial arrangements of Xe atoms to draw "IBM" has been demonstrated. In parallel, various micro- and nano-fabrication techniques have been developed.

Another conceptual proposal was brought by Kim Eric Drexler, in his book entitled *Engines of Creation: The Coming Era of Nanotechnology*, in 1986 [4]. As

K. Ariga (✉)
International Center for Materials Nanoarchitectonics (WPI-MANA), National Institute for Materials Science (NIMS), Ibaraki, Tsukuba, Japan

Department of Advanced Materials Science, Graduate School of Frontier Sciences, The University of Tokyo, Chiba, Japan
e-mail: ARIGA.Katsuhiko@nims.go.jp

T. Govindaraju, K. Ariga (eds.), *Molecular Architectonics and Nanoarchitectonics*, Nanostructure Science and Technology,
https://doi.org/10.1007/978-981-16-4189-3_2

imaginary machine construction, molecular and biomolecular nanotechnology and the design of assemblers to produce functional systems were described. Not limited to the concept of machine-type assembly, scientific developments in biology and chemistry, such as the promotion of the understanding and technological development of biocentric dogma from DNA to proteins, the initiation and demonstration of the concept of molecular machines, and the exploration of self-assembly and self-organization in supramolecular chemistry may support this idea as a core methodology. These trends are considered to be bottom-up types of nanotechnology, particularly in the fields of chemistry and biochemistry.

These concepts and strategies can be more generalized as fabrication of functional material systems from nanoscale building blocks. The term architectonics related to small objects was used in the paper "Architectonic Quantum Dots of Solids" written by Heath and colleagues in 1999 at the University of California, Los Angeles (UCLA) [5]. The use of a new conceptual term, nanoarchitectonics, was initiated in 2000 by Masakazu Aono, who organized an international symposium, the first International Symposium on *Nanoarchitectonics* Using Suprainteractions in Tsukuba, Japan. This is the first initiation of a new concept – nanoarchitectonics. Aono intends to combine the nanotechnology initiated by Feynman with other scientific disciplines to create a common methodology for creating functional materials with nanoscale units [6–8].

The first appearance of the term nanoarchitectonics in the titles of articles in scientific journals was made by Stefan Hecht of the Free University of Berlin, Germany, in his paper *Welding, Organizing, and Planting Organic Molecules on Substrate Surfaces–Promising Approaches towards Nanoarchitectonics from the Bottom up* [9]. Several research organizations and projects with the term nanoarchitectonics were initiated at the same time. In 2003, the "Interfacial Nanoarchitectonics" research center was established at the National Institute of Advanced Industrial Science and Technology in Tsukuba, led by Toshimi Shimizu. In 2003, the Functional Engineered Nano Architectonics research center was established at the University of California, Los Angeles. The founder of nanoarchitectonics, Masakazu Aono, established the International Center for Materials Nanoarchitectonics (WPI-MANA) at the National Institute for Materials Science in Tsukuba in 2007. Nanoarchitectonics was recognized as one of the key research areas in a report in 2017 by Institutes of Science and Development and The National Science Library, Chinese Academy of Sciences, with the aid of Clarivate Analytics.

These historical flows are summarized in Fig. 2.1. As clearly recognized, the concept of nanoarchitectonics emerged and evolved along with the beginning of the twenty-first century.

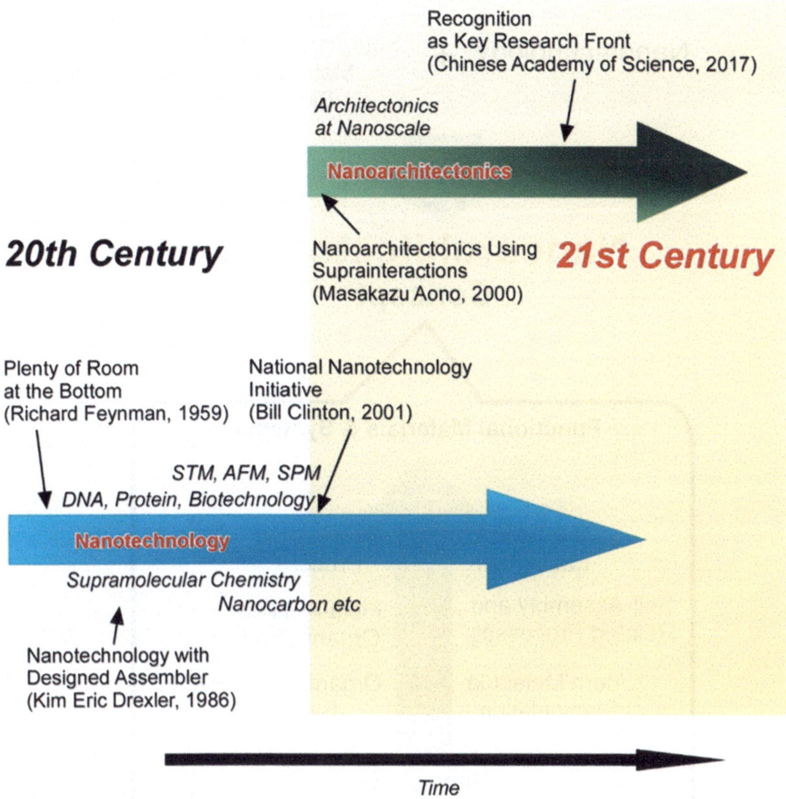

Fig. 2.1 Historical flows of nanotechnology and nanoarchitectonics

2.2 Essence of Nanoarchitectonics

A new concept called nanoarchitectonics is created by combining nanotechnology with other research fields such as organic chemistry, supramolecular chemistry, materials science, and biotechnology [10–12]. Nanoarchitectural approaches should produce functional materials using nanoarchitectonics units through combined and/or selected efforts of atomic/molecular manipulation, organic synthesis, self-assembly/self-organization, micro-fabrication, and biological process (Fig. 2.2) [13, 14]. These processes and effects are working in harmony with nanoarchitectonics uncertainties such as thermal fluctuations. The contribution of multiple principles to construction is beneficial for hierarchical architectural structures. The basic concept of nanoarchitectonics is general enough to be applied to various fields of research such as sensors [15–18], catalysts [19–22], devices [23–26], energy-related applications [27–30], environmental issues [31–34], and biomedical problems [35–38]. In addition to these practical fields, the concept of

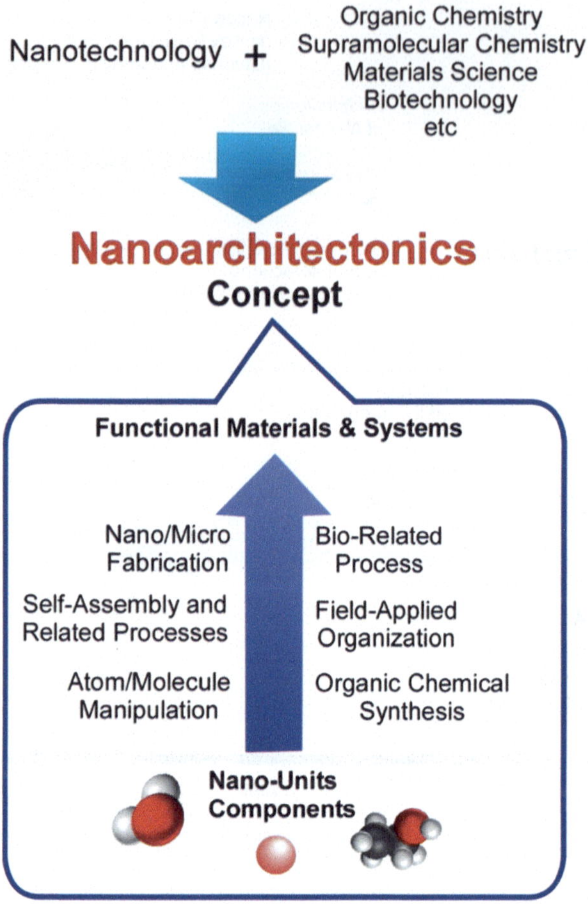

Fig. 2.2 Nanoarchitectonics concept created through combining nanotechnology with other research fields such as organic chemistry, supramolecular chemistry, materials science, and biotechnology

nanoarchitectonics is effective in the basic sciences including materials synthesis [39–42], structure controls [43–46], and the biological sciences [47–50].

Although self-assembly is also a central concept in organically based supramolecular chemistry, it can be widely applied in the assembly processes of various nanoblocks, including quantum objects, inorganic nanomaterials, and biological materials. Actually, self-assembly plays indispensable roles in nanoarchitectonics [51]. The basic idea of self-assembly processes is based on equilibrium, without energy input or external stimuli. Therefore, self-assembly processes tend to provide a symmetric and simple structure. In contrast, the notion of self-organization is used for the architecture of structures with energy flows from outside the system, and self-organization processes are far from equilibrium. This nature of self-organization benefits architects of asymmetric, vector, and hierarchical structures, which is universally used in biological structural organization. Nanoarchitectonics has a strong characteristic similar to self-organization. Nanoarchitectonics procedures can involve multiple energy-intensive processes in a staggered order and/or fashion harmonization. The integration of some artificial energy-intensive processes and multiple fabrication processes into the spontaneous self-assembly of nano units results in asymmetric and hierarchical architecture in the nanoarchitectonics process [52]. Nanoarchitectonics approaches are rather closer in nature to the processes of formation of biological structures than conventional self-assembly.

Nanoarchitectonics processes are not completely same as architectonics processes in macroscopic scales [53]. In contrast to macroscopic, architectonics processes of materials and structural systems on the nanometer or mesoscale (nanoarchitectonics processes) do not always proceed decisively according to their design and programs. Uncontrolled and unexpected disturbances and fluctuations have major impacts in nanoscale regions. The physical phenomena and behavior of materials in nanoscale regions cannot escape from the effects of thermal/static fluctuations and the mutual interactions that inevitably occur between atoms, molecules, and constituent materials. The results, such as material structures and properties, are not simply determined by known input data. At the nanoscale, an input signal to a particular target can also interfere with surrounding entities or components and cause additional mutual interactions. Therefore, the synthesis and fabrication of materials or the use of components at the nanoscale are quite different from the authoritative and predictable fabrication processes at the microscale. The fabrication of nanoscale materials based on nanoarchitectonics requires concerted coordination of the various interactions and effects, as well as many techniques and tricks to control the organization of the materials and stimulate spontaneous processes such as self-assembly.

2.3 Example of Nanoarchitectonics

Several nanoarchitectonics processes in various scales are exemplified below. As molecular-level nanoarchitectonics processes, nanotechnology-based organic chemistry has been recently demonstrated. Unlike the traditional approach to organic chemistry, the combination of nanotechnology and organic synthesis with the help of advanced probe microscopy techniques allows us to manipulate organic modification in molecules in site-specific manner [54]. Scanning probe microscopy has seen major advances in imaging technology, where probes can be used to remove specific atoms from a molecule and change its structure. It would be a starting point for the fabrication of functional materials from molecular units that can be considered an organic nanoarchitecture.

For example, Kawai and co-workers demonstrate site-specific substitution of Br atoms on a three-dimensional graphene nanotape by a fullerene molecule using a tip of scanning probe microscope (Fig. 2.3) [55]. The Br atoms were removed by

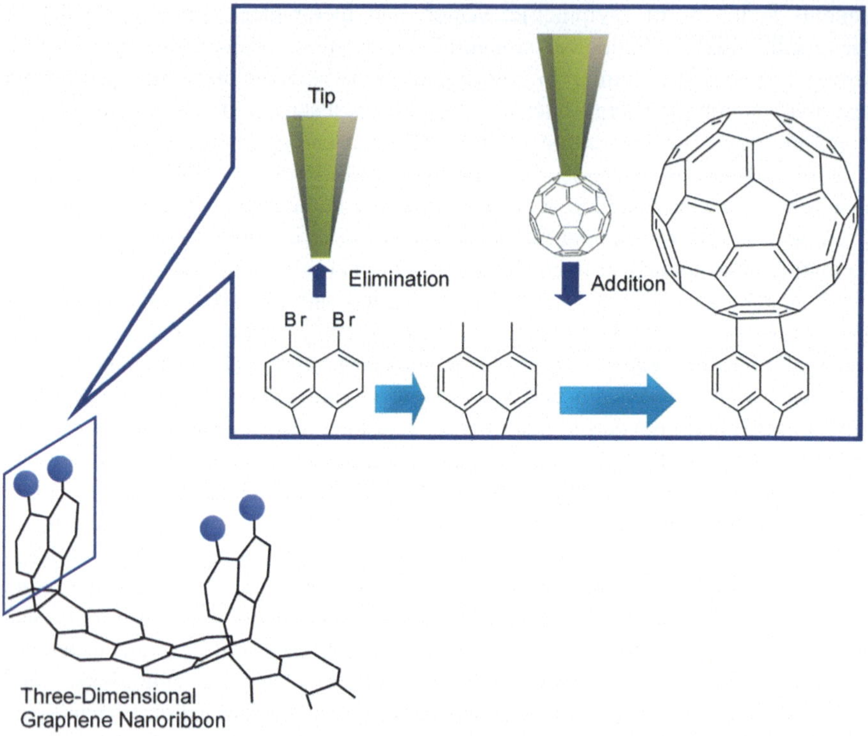

Fig. 2.3 Site-specific substitution of Br atoms on a three-dimensional graphene nanotape by a fullerene molecule using a tip of scanning probe microscope

moving the probe tip to a vertical position several hundred picometers away from the Br atoms and then applying voltage of about 2.5 V to the target site. The detachment of Br atoms leaves highly active unpaired electrons on carbon. The reactive and unstable electrons are retained under ultra-high vacuum at extremely low temperature. The three-dimensional nature of the graphene nanotape can also prevent unwanted reactions between the unpaired electrons and the surface. A fullerene molecule can then be attached to the active site with unpaired electrons. The fullerene molecule is previously adsorbed on the surface and is picked up by the probe tip. By pushing the fullerene molecule onto the active diradical site, the fullerene molecule is covalently attached to the desired positions of the graphene nanotape. In this organic nanoarchitectural approach, chemical reactions can be performed with a site-specific mode under selected conditions. It is a nice demonstration of the premise of the nanoarchitectural approach to prepare functional materials from the molecular bottom.

Figure 2.4 demonstrates a nanoarchitectonics example in micrometer scale. These objects can be architected from fullerene molecules. A new fullerene object, fullerene microhorns, was produced by in situ structural modification of fullerene microtubes [56]. Fullerene precursor microtubes were spontaneously formed by

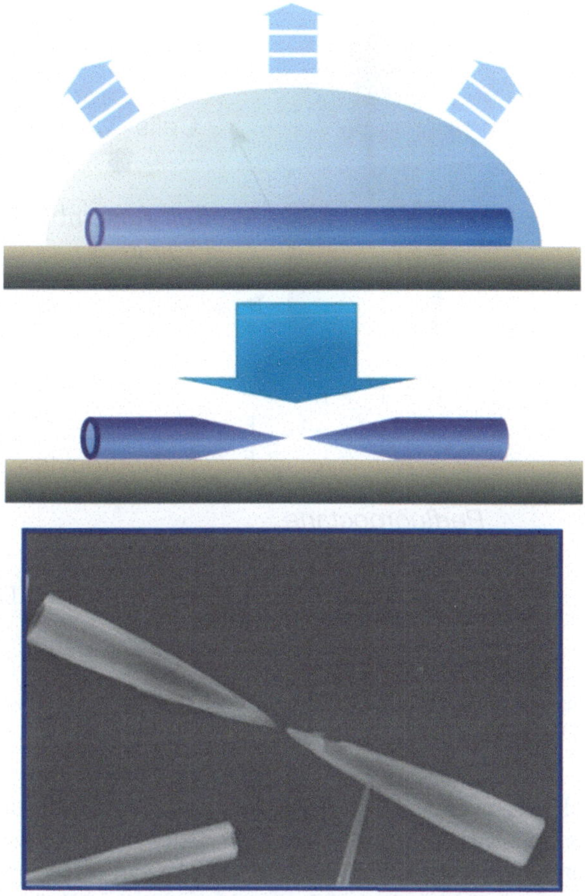

Fig. 2.4 Fullerene microhorns produced by in situ structural modification of fullerene microtube

adding *tert*-butyl alcohol to a mesitylene solution of a mixture of C_{60} and C_{70} fullerenes. The structural transformation of the formed fullerene microtubes into fullerene microhorns was carried out by adding a few drops of a mixture of *tert*-butyl alcohol and mesitylene to fullerene microtubes deposited on the surface of a silicon wafer and then slowly evaporating the solvent. After careful observation, it was found that one fullerene microtube disintegrates into two fullerene microhorns within seconds. Solvent dissolution occurred selectively in the central region of the fullerene microtube, forming two microhorns of length equal to half the length of the original fullerene microtube. Such a manipulable microscale hollow object, the fullerene microhorn, was then subjected to micron-sized object recognition. Among the microparticles studied, silica particles are preferentially trapped and attached to fullerene microspheres compared to other particles of similar size, such as C_{70} fullerene particles, polystyrene latex particles, hydroxylate particles, and carboxylate particles. The zeta potential measurement showed that the zeta potential of the

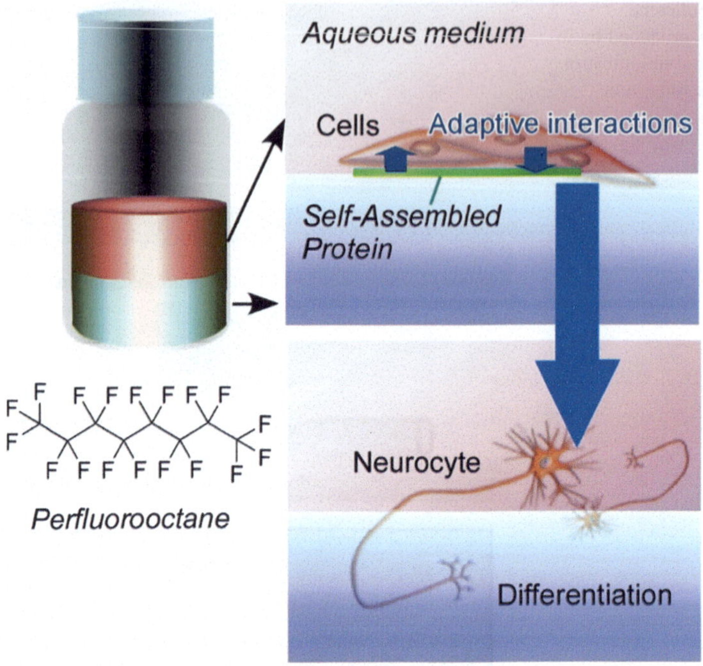

Fig. 2.5 Nanoarchitectonics of living objects at the liquid-liquid two-dimensional interface for differentiation of stem cells into neurocyte without the use of expensive reagents

fullerene microhorns was negative (−51.1 mV), while the silica nanoparticles had a positive value (+1.2 mV) for zeta potential under the same conditions. Electrostatic interaction has a key role in this selective particle recognition.

Nanoarchitectonics of living objects is also possible. Jia et al. applied this cell culture strategy at the liquid-liquid two-dimensional interface to human mesenchymal stem cells and demonstrated the differentiation of stem cells into neurocytes without the use of expensive reagents (Fig. 2.5) [57]. At the liquid interface between aqueous cell culture medium and fluorocarbon layer, protein two-dimensional nanolayers spontaneously formed from proteins contained in the cell culture medium together with externally added fibronectin. According to the transformation of the protein nanosheet into hierarchical fibrous structures, the mesenchymal stem cells cultured on the surface of the nanolayer were confirmed to differentiate into neurons without the addition of differentiation-inducing factors. The formation of prolonged focal adhesion with increased activation of focal adhesion kinase is promoted by adaptation to elongated fibronectin fibers, leading to neuronal differentiation of mesenchymal stem cells. Spatial rearrangement of extracellular matrix proteins in the presence of altered cell traction forces induces changes in cell fates. Instead of conventional solid surfaces, such as plastic films, the use of a liquid-liquid dynamic two-dimensional interface is essential to induce the formation of adaptive protein nanosheets that are able to deform and remodel in response to cellular mechanical forces associated with cell differentiation. The proposed approach does not require the use of expensive differentiation-inducing factors such as cytokines. This technique offers opportunities for cost-effective regulation of stem cell differentiation, which may also be valuable in regenerative medicine and other related fields.

2.4 Short Perspective

This short chapter briefly explained historical backgrounds and essences of the nanoarchitectonics concept as well as several examples of materials nanoarchitectonics. The nanoarchitectonics approaches can produce functional material systems in various scale ranges. However, total architectonics, from nano units (atomic or molecular) to macroscopic objects with complicated hierarchical structures such as living beings, still remains a difficult target. Strategies for highly sophisticated functional assemblies in biological systems, developed as the result of evolutionary processes lasting billions of years, are far superior to our science and technology created in just a few centuries. Filling this gap is the most important task of nanoarchitectonics [58].

References

1. Feynman RP (1960) Eng Sci 23:22
2. Roukes M (2001) Sci Am 285:48
3. Ariga K (2015) J Inorg Organomet Polym 25:177
4. Eric Drexler K (1992) Nanosystems: molecular machinery, manufacturing, and computation. Wiley, New York
5. Markovich G, Collier CP, Henrichs SE, Remacle F, Levine RD, Heath JR (1999) Acc Chem Res 32:415
6. Aono M (2011) Sci Technol Adv Mater 12:040301
7. Aono M, Bando Y, Ariga K (2012) Adv Mater 24:150
8. Ariga K, Ji Q, Hill JP, Bando Y, Aono M (2012) NPG Asia Mater 4:e17
9. Hecht S (2003) Angew Chem Int Ed 42:24
10. Ariga K, Ji Q, Nakanishi W, Hill JP, Aono M (2015) Mater Horiz 2:406
11. Ariga K, Minami K, Ebara M, Nakanishi J (2016) Polym J 48:371
12. Ariga K (2021) Small Sci 1:2000032
13. Ariga K, Li M, Richards GJ, Hill JP (2011) J Nanosci Nanotechnol 11:1
14. Ariga K, Li J, Fei J, Ji Q, Hill JP (2016) Adv Mater 28:1251
15. Ishihara S, Labuta J, Van Rossom W, Ishikawa D, Minami K, Hill JP, Ariga K (2014) Phys Chem Chem Phys 16:9713
16. Jackman JA, Cho N-J, Nishikawa M, Yoshikawa G, Mori T, Shrestha LK, Ariga K (2018) Chem Asian J 13:3366
17. Ariga K, Makita T, Ito M, Mori T, Watanabe S, Takeya J (2019) Beilstein J Nanotechnol 10:2014
18. Liu J, Zhou H, Yang W, Ariga K (2020) Acc Chem Res 53:644
19. Abe H, Liu J, Ariga K (2016) Mater Today 19:12
20. Ariga K, Ishihara S, Abe H (2016) Cryst Eng Comm 18:6770

21. Guo Y, Park T, Yi JW, Henzie J, Kim J, Wang Z, Jiang B, Bando Y, Sugahara Y, Tang J, Yamauchi Y (2019) Adv Mater 31:1807134
22. Deka RC, Deka A, Deka P, Saikia S, Baruah J, Sarma PJ (2020) J Nanosci Nanotechnol 20:5153
23. Wang KL, Galatsis K, Ostroumov R, Khitun A, Zhao Z, Han S (2008) Proc IEEE 96:212
24. Ariga K, Watanabe S, Mori T, Takeya J (2018) NPG Asia Mater. 10:90
25. Nayak A, Unayama S, Tai S, Tsuruoka T, Waser R, Aono M, Valov L, Hasegawa T (2018) Adv Mater 30:1703261
26. Ariga K, Ito M, Mori T, Watanabe S, Takeya J (2019) Nano Today 28:100762
27. Chen R, Zhao T, Zhang X, Li L, Wu F (2016) Nanoscale Horiz 1:423
28. Kim J, Kim JH, Ariga K (2017) Joule 1:739
29. Giussi JM, Cortez ML, Marmisollé WA, Azzaroni O (2019) Chem Soc Rev 48:814
30. Huang HJ, Yan MM, Yang CZ, He HY, Jiang QG, Yang L, Lu ZY, Sun ZQ, Xu XT, Bando Y, Yamauchi Y (2019) Adv Mater 31:1903415
31. Ariga K, Ishihara S, Abe H, Li M, Hill JP (2012) J Mater Chem 22:2369
32. Pandeeswar M, Senanayak SP, Govindaraju T (2016) ACS Appl Mater Interfaces 8:30362
33. Paul L, Banerjee B, Bhaumik A, Ali M (2020) J Nanosci Nanotechnol 20:2858
34. Pham T-A, Qamar A, Dinh T, Masud MK, Rais-Zadeh M, Senesky DG, Yamauchi Y, Nguyen N-T, Phan H-P (2020) Adv Sci 7:2001294
35. Ariga K, Naito M, Ji Q, Payra D (2016) Cryst Eng Comm 18:4890
36. Ariga K, Leong DT, Mori T (2018) Adv Funct Mater 28:1702905
37. Dutta S, Kim J, Hsieh P-H, Hsu Y-S, Kaneti YV, Shieh F-K, Yamauchi Y, Wu KC-W (2019) Small Methods 3:1900213
38. Banerjee S, Pillai J (2019) Expert Opin Drug Metab Toxicol 15:499
39. Ramanathan M, Shrestha LK, Mori T, Ji Q, Hill JP, Ariga K (2013) Phys Chem Chem Phys 15:10580
40. Cordier S, Grasset F, Molard Y, Amela-Cortes M, Boukherroub R, Ravaine S, Mortier M, Ohashi N, Saito N, Haneda H (2015) J Inorg Organomet Polym 25:189
41. Komiyama M, Mori T, Ariga K (2018) Bull Chem Soc Jpn 91:1075
42. Ariga K, Jia X, Shrestha LK (2019) Mol Syst Des Eng 4:49; Lee YJ, Park Y (2020) J Nanosci Nanotechnol 20:2781
43. Ariga K, Lee MV, Mori T, Yu X-Y, Hill JP (2010) Adv Colloid Interf Sci 154:20
44. Zhang L, Wang T, Shen Z, Liu M (2016) Adv Mater 28:1044
45. Ariga K, Matsumoto M, Mori T, Shrestha LK (2019) Beilstein J. Nanotechnol. 10:1559
46. Ariga K, Mori T, Kitao T, Uemura T (2020) Adv Mater 32:1905657
47. Nakanishi W, Minami K, Shrestha LK, Ji Q, Hill JP, Ariga K (2014) Nano Today 9:378
48. Zou Q, Liu K, Abbas M, Yan X (2016) Adv Mater 28:1031
49. Ariga K (2016) Chem Nano Mat 2:333
50. Stulz E (2017) Acc Chem Res 50:823
51. Ariga K, Nishikawa M, Mori T, Takeya J, Shrestha LK, Hill JP (2019) Sci Technol Adv Mater 20:51
52. Ariga K, Jia X, Song J, Hill JP, Leong DT, Jia Y, Li J (2020) Angew Chem Int Ed 59:15269
53. Aono M, Ariga K (2016) Adv Mater 28:989
54. Ariga K (2020) Trends Chem 2:779
55. Kawai S, Krejčí O, Nishiuchi T, Sahara K, Kodama T, Pawlak R, Meyer E, Kubo T, Foster AS (2020) Sci Adv 6:eaay8913
56. Tang Q, Maji S, Jiang B, Sun J, Zhao W, Hill JP, Ariga K, Fuchs H, Ji Q, Shrestha LK (2019) ACS Nano 13:14005
57. Jia X, Minami K, Uto K, Chang AC, Hill JP, Nakanishi J, Ariga K (2020) Adv Mater 32:1905942
58. Ariga K, Yamauchi Y (2020) Chem Asian J 15:718

Part II
Architectonics of Functional Molecules

Chapter 3
Topological Supramolecular Polymer

Yasuki Kato, Sougata Datta, and Shiki Yagai

3.1 Sixty Years of History of Catenanes

Although it is difficult to introduce mechanical flexibility in one-dimensional "hard" materials directly, chain-like structures formed by interlocking of rings enable fabrication of flexible one-dimensional structures from hard materials. Moreover, chain-like structures endow materials with advantage of not only mechanical flexibility but also ease of repairing of the damaged chains just by replacing only the damaged ring. Therefore, "catenanes," which are mechanically interlocked chain-like structures of small cyclic molecules, have attracted great attention of chemists in many fields because realization of such structures gives rise to materials exhibiting the molecular scale chain-like motion and properties [1].

The first synthesis of molecular catenane was reported by Prof. E. Wasserman in 1960 (Fig. 3.1) [2]. Prior to the development of supramolecular chemistry,

Y. Kato
Division of Advanced Science and Engineering, Graduate School of Science and Engineering, Chiba University,, Chiba, Japan
e-mail: y.kato@chiba-u.jp

S. Datta
Department of Applied Chemistry and Biotechnology, Graduate School of Engineering, Chiba University,, Chiba, Japan
e-mail: sougata.datta@chiba-u.jp

S. Yagai (✉)
Division of Advanced Science and Engineering, Graduate School of Science and Engineering, Chiba University,, Chiba, Japan

Department of Applied Chemistry and Biotechnology, Graduate School of Engineering, Chiba University,, Chiba, Japan

Institute for Global Prominent Research (IGPR), Chiba University, Chiba, Japan
e-mail: yagai@faculty.chiba-u.jp

T. Govindaraju, K. Ariga (eds.), *Molecular Architectonics and Nanoarchitectonics*, Nanostructure Science and Technology,
https://doi.org/10.1007/978-981-16-4189-3_3

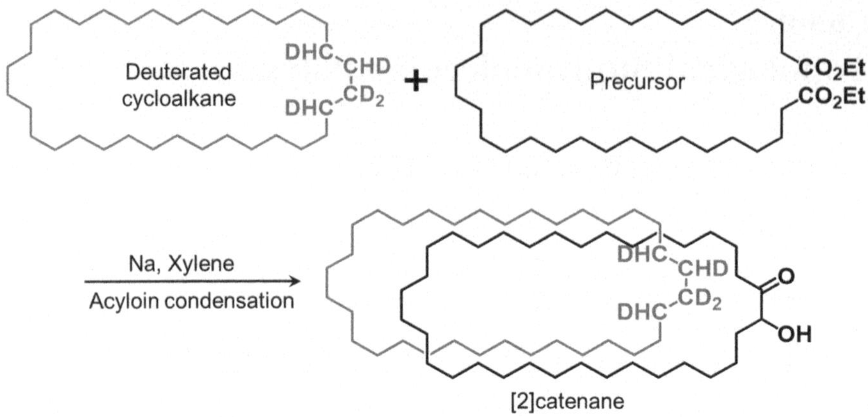

Fig. 3.1 Scheme of Wasserman's catenane synthesis

Wasserman attempted cyclization reaction in a solution of cyclic molecules hoping that they would statistically interlock. In order to increase the probability of threading of a linear molecule, which is the precursor of the cyclic molecule, through the cyclic molecule, the reaction must take place at high concentrations. However, intramolecular cyclization reactions are more likely to occur at lower concentrations. Due to these conflicting factors, the yield of two ring-linked [2]catenanes was less than 0.0001% [3, 4].

In 1983, Sauvage and co-workers discovered an efficient synthesis of [2]catenane through a supramolecular approach using metal-ligand complexation as shown in Fig. 3.2a [5]. This approach, so-called template directed synthesis, has been followed by various supramolecular strategies to synthesize a wide variety of catenanes. For example, Beer's group synthesized [2]catenanes by template-directed synthesis using halogen bonds. In this example, catenanes were obtained by preassembly of two cyclic precursors through Br¯Br¯Br interaction (Fig. 3.2b), which was followed by ring-closing metathesis (RCM) [6]. Soon after, in their another approach, the interaction between I- and pyridine was used to form pseudorotaxanes which could be converted to [2]catenanes by RCM (Fig. 3.2c) [7].

After about 10 years of Sauvage's epoch-making report, Stoddart and coworkers realized the synthesis of [5]catenane through properly orienting the precursor components using charge–transfer interactions, and they named the mechanically interlocked five molecules "olympiadane" (Fig. 3.3a) [8]. In the same year, Fujita and co-workers reported the use of hydrophobic effect, which is among the least specific interactions, to create a thermodynamic bias toward interlocked macrocycles at equilibrium. They reported that interlocking of macrocycles formed by palladium (II)–pyridine coordination driven self-assembly reached 99% in yield at high concentrations [9]. In contrast, the equilibrium shifted toward single macrocycles in dilute conditions (Fig. 3.3b). Interestingly, employing third row Pt(II) cation instead of the second row Pd(II) cation led the system to a kinetically locked state wherein the ratios of macrocycle and catenane became independent of concentration. Unlike

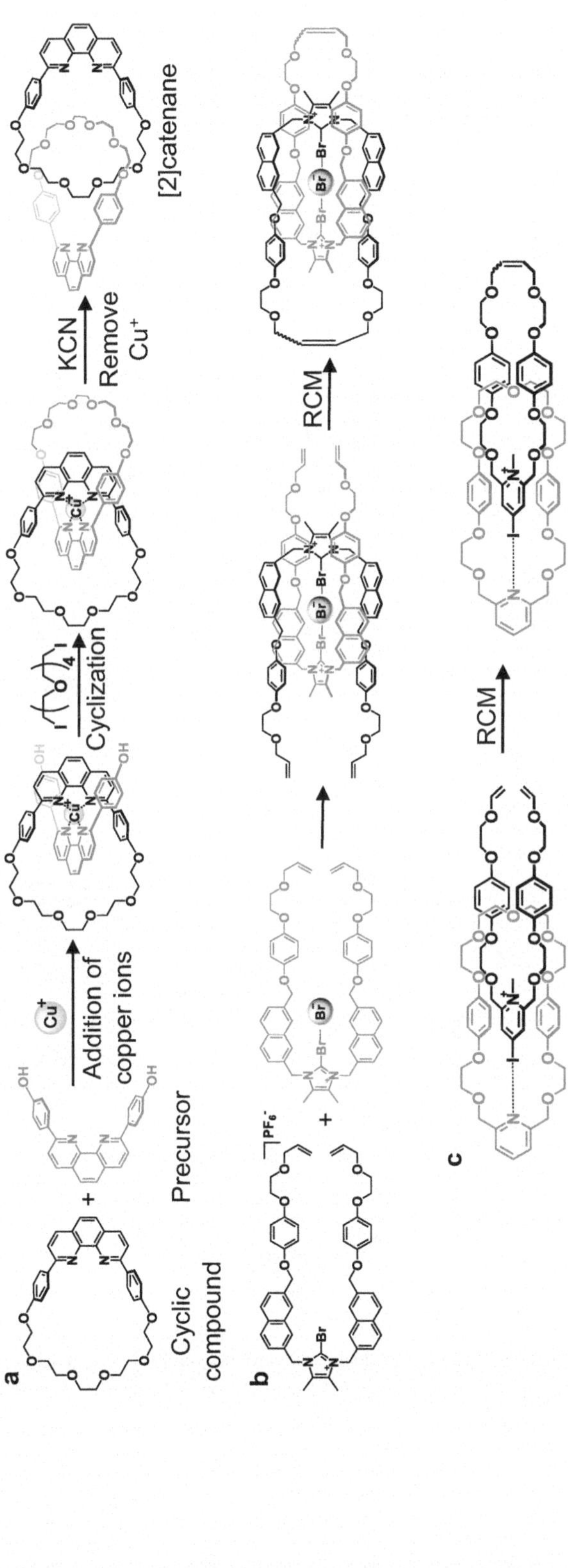

Fig. 3.2 (**a**) Template-directed synthesis of [2]catenane reported by Sauvage et al. Synthesis of [2]catenane using (**b**) halogen–halogen bonding or (**c**) iodide–pyridine interaction as template

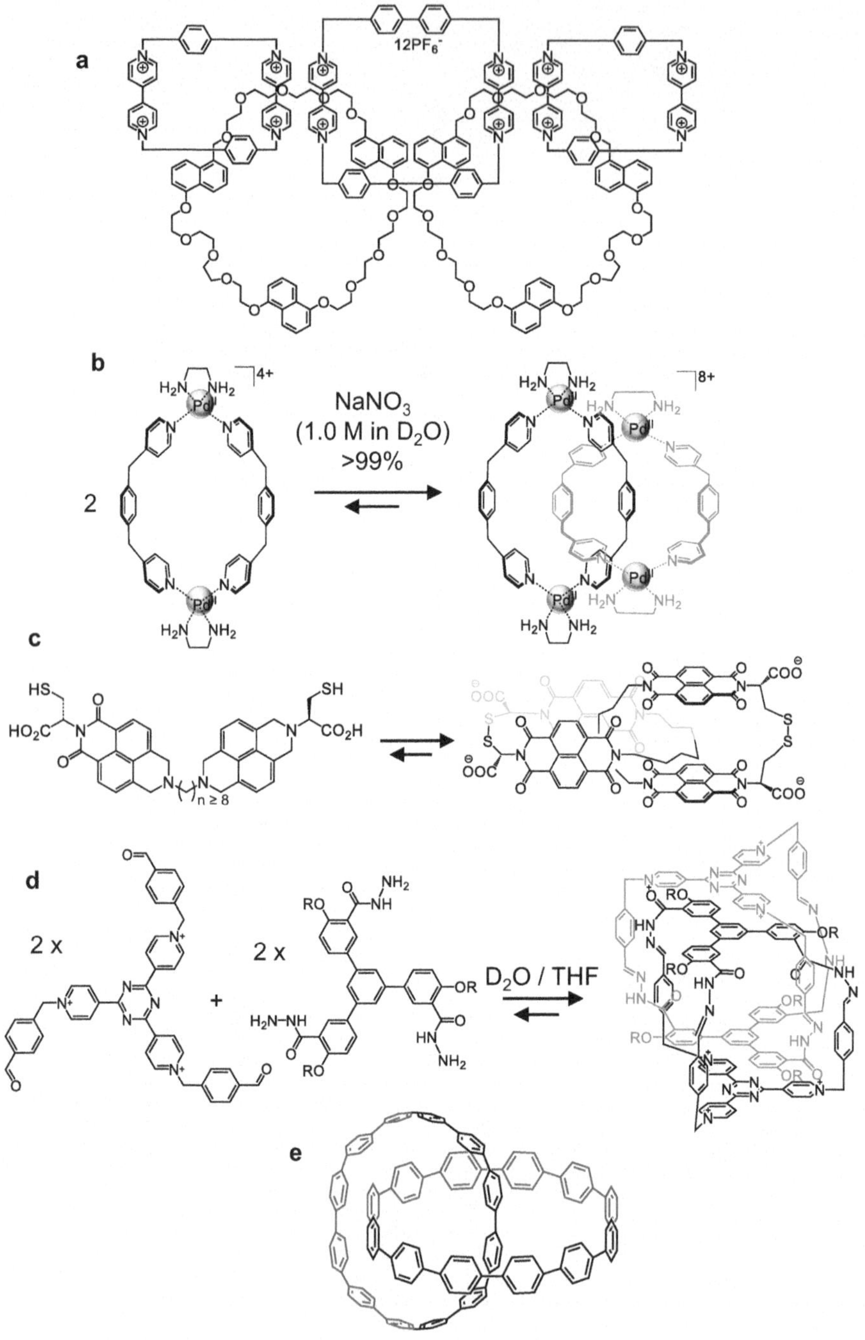

Fig. 3.3 (**a**) Olympiadane reported by Stoddart et al. (**b**) "Magic ring" [2]catenane synthesis. (**c**) NDI-based [2]catenane. (**d**) Three-dimensional [2]catenane. (**e**) All-benzene [2]catenane

Fujita's metallocycle system, Sanders' and co-workers investigated the formation of non-trivial topologies involving dynamic combinatorial chemistry of a series of gemini-type cysteine-naphthalene diimide (NDI) conjugate with alkyl chain spacer of variable length ($n = 2$–9) [10]. The gemini compounds with longer spacer ($n \geq 8$) afforded [2]catenanes composed of two mechanically interlocked monomeric rings as the major product (Fig. 3.3c), while macrocyclic adducts were predominantly obtained for shorter-spacer compounds ($n < 8$). The formation of the [2]catenanes was driven by hydrophobic interaction between NDIs in water, while the hydrophilic cysteine side chains were exposed to the solvent. Sessler, Li, and co-workers also exploited the combined benefits of dynamic combinatorial chemistry and hydrophobic effect to obtain a remarkable three-dimensional [2]catenane via one-pot synthesis (Fig. 3.3d) [11]. The three-dimensional interlocked structure was successfully characterized by single-crystal X-ray diffraction analysis, and HPLC-MS analysis confirmed almost quantitative yield in a mild acidic aqueous medium. Recently, all-benzene (CPP) catenanes, which have been one of the extremely challenging compounds, have been eventually realized by Segawa, Itami, and co-workers (Fig. 3.3e) [12, 13].

More than 10 years after the realization of olympiadane, Rowan and co-workers have finally succeeded in synthesizing poly[n]catenanes including linearly connected poly[26]catenanes. The precursors and cyclic molecules are alternately preassembled by coordination with Zn^{2+} ions to form metallosupramolecular polymers. The olefin metathesis reaction of the precursor units in the metallosupramolecular polymers followed by the removal of Zn^{2+} using tetrabutylammonium hydroxide finally afforded poly[n]catenanes (Fig. 3.4) [14]. It has also been also shown that the addition of Zn^{2+} ions causes the conformation of the linear poly[n]catenanes to become rigid, and thereby the glass transition temperature (T_g) increases from 97 to more than 160 °C. Moreover, the poly[n]catenanes behave differently from conventional linear covalent polymers. For example, while these poly[n]catenanes appear as powders, the linear covalent polymers of comparable molecular weight can be transformed to films easily. This is ascribable to the differences in intermolecular interactions and interchain entanglement. The interlocked structure of poly[n]catenanes inhibits $\pi - \pi$ stacking interactions between the chains, thereby reducing cohesive energy density. At the same time, they are unable to entangle like linear covalent polymers due to high flexibility relative to their size. This has been also supported by molecular dynamic simulation studies of poly[n]catenanes ($n = < 25$) in the melt state [15]. Using ring-opening metathesis polymerization of a zinc-templated [2]catenane, Yan and co-workers have recently reported preparation of woven polymer networks with high flexibility and ductile features, as well as high modulus and thermomechanical stability [16]. The reversibility of metallation and demetallation process further enabled modulation of mechanical and dynamic properties of the woven polymer networks without disrupting their structural integrity. Thus, polycatenanes have the potential to exhibit interesting physical properties that are different from those of ordinary polymers and are accordingly expected to be the next generation of polymer materials. In 2016, Prof. Sauvage, Prof. Stoddart, and Prof. Feringa, who contributed significantly to the

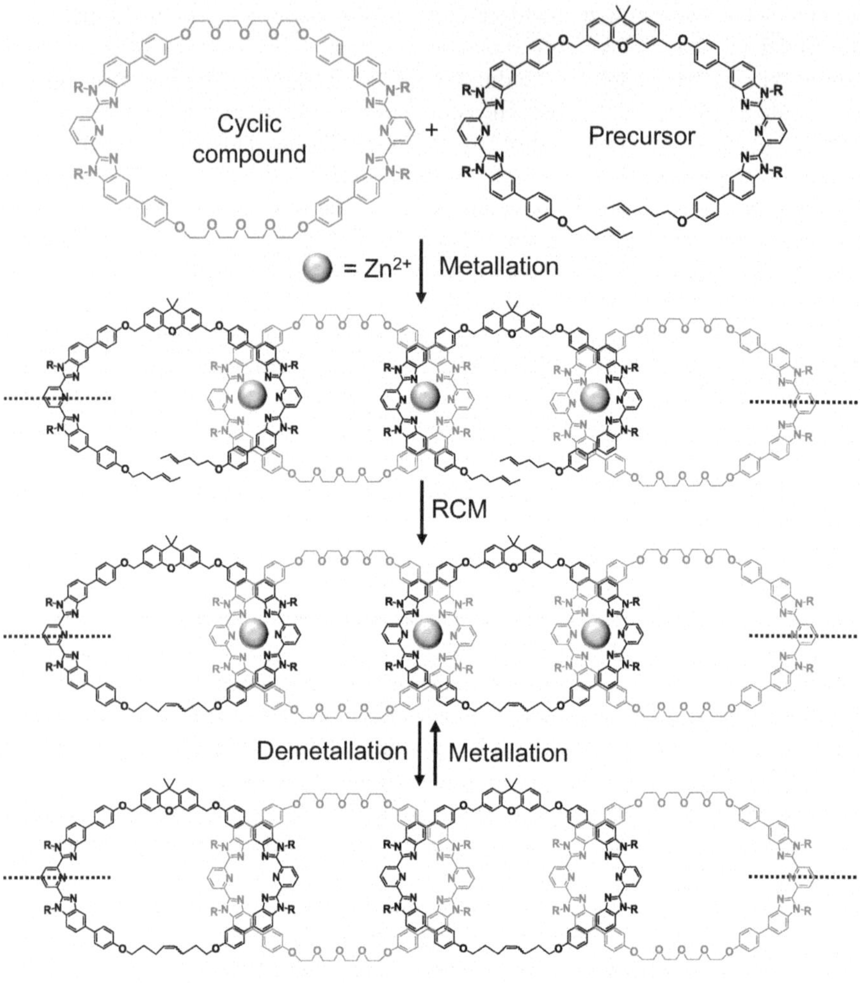

Fig. 3.4 Strategy for the synthesis of polycatenanes reported by Rowan et al.

development of catenanes, rotaxane, and other parts necessary to realize molecular machines, were awarded the Nobel Prize in Chemistry.

Although a number of catenanes composed of synthetic covalent macrocycles have been studied in the past, their characterization was performed by crystallization. However, crystallization of such interlocked compounds is difficult if the constituent circular molecules are large and flexible and sometimes polydispersed like polymers. Moreover, mixtures of poly[n]catenanes are intrinsically impossible to be crystallized. The development of microscopy techniques such as transmission electron microscopy (TEM) and atomic force microscopy (AFM) has increased the ease of characterization of not only synthetic covalent catenanes but also biomolecular catenanes by enabling unambiguous visualization of their structures. This attempt

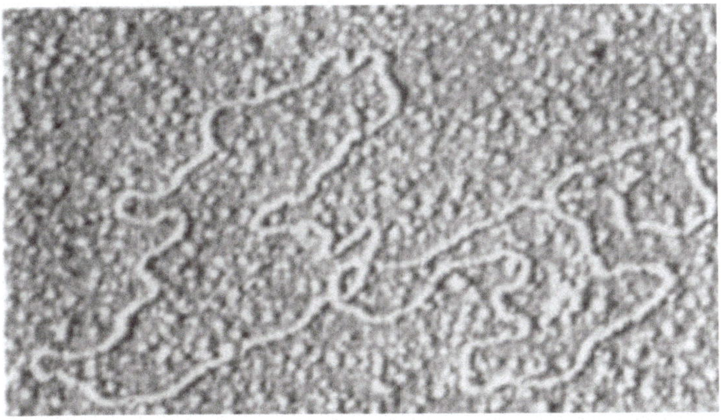

Fig. 3.5 Electron micrograph of circular DNA revealing a catenane topology. Reproduced from Ref. [17] with permission

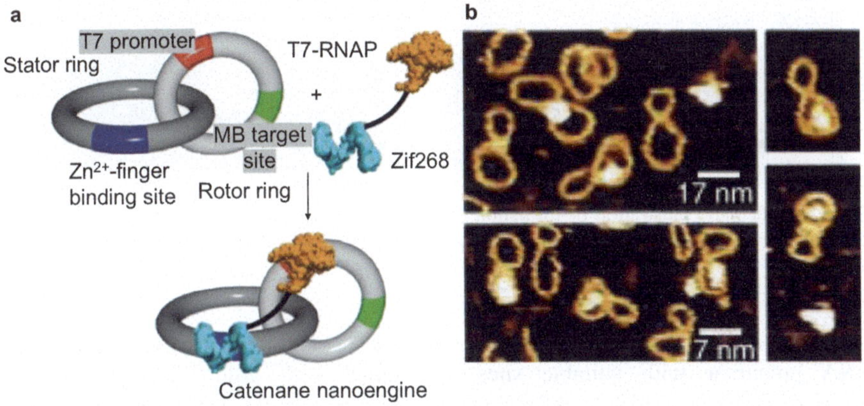

Fig. 3.6 (**a**) Synthesis of DNA-catenane nano-engine. (**b**) AFM image of catenane nano-engine. Reproduced from Ref. [19] with permission

has been already done in 1967 by Hudson and Vinograd, who first reported microscopic characterization of naturally occurring DNA catenane isolated from the mitochondria of HeLa cells (Fig. 3.5) [17, 18].

Recently, M. Famulok and co-worker reported a [2]catenane composed of the T7RNAP-ZIF protein and clearly visualized their structures by AFM [19]. They also reported nucleotide triphosphate-driven locomotion in a DNA-[2]catenane nanoengine comprising one dsDNA nanoring rotor and T7 RNA polymerase (T7RNAP-ZIF) as a power generator attached to the other interlocked dsDNA nanoring (Fig. 3.6).

Motivated by transition metal-templated approach, N. Liu et al. have used gold nanoparticles as a template to form not only [2]catenane but also oligocatenanes composed of DNA origami [20]. They have obtained DNA-[2]catenanes, DNA-[3]

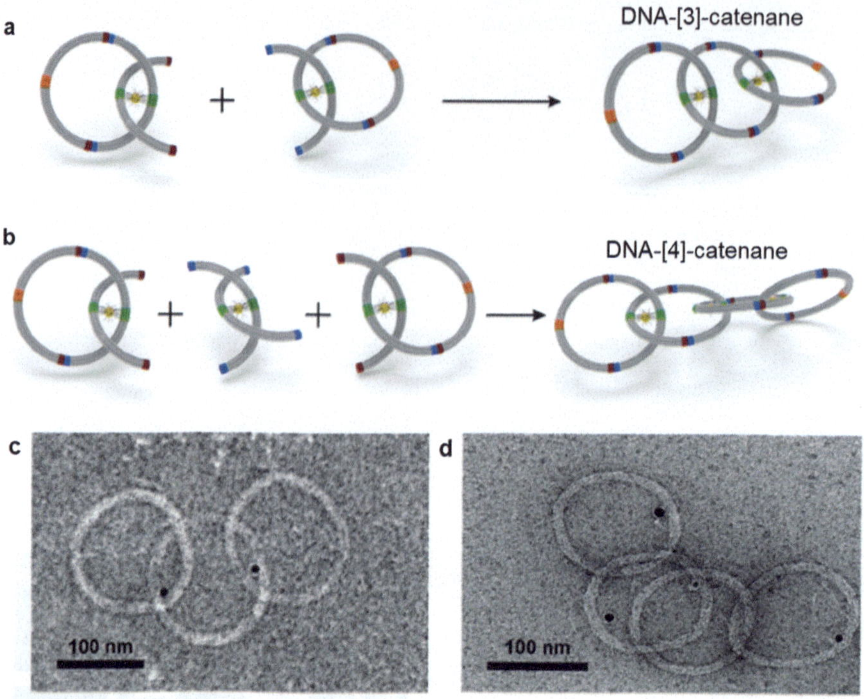

Fig. 3.7 Schematic illustration of formation of (**a**) DNA-[3]catenane and (**b**) DNA-[4]catenane. TEM images of (**c**) DNA-[3]catenane and (**d**) DNA-[4]catenane. Modified from Ref. [20] with permission

catenanes, and DNA-[4]catenanes by employing a cyclic DNA and a semi-annular DNA precursor with suitable sites to interact with gold nanoparticles for preorganization and subsequently cyclizing the semi-annular DNA precursor. The DNA catenanes have been successfully characterized by TEM (Fig. 3.7). Furthermore, the rings of the DNA catenanes have been unlocked to set them free by removal of the gold nanoparticles.

Thus, catenanes are currently expected to produce not only attractive structures but also unprecedented chemical and physical properties due to their unique structures. If such complex topologies can be constructed from molecular assemblies in a bottom-up manner, additional properties can be generated, which should contribute to the future development of nanotechnology. While various synthetic approaches for catenanes and catenane building blocks are being studied by many researchers around the world, we have also published a paper on [2]catenanes in 2011 [21]. What makes this system unique is that the [2]catenane was not made up of two cyclic molecules, rather an assembly of about 1000 small molecules with molecular weights of about 1000 to form a single ring-shaped, more specifically, "toroidal" supramolecular polymers (**SP**s). The size of the ring is about 20 nm, and the structure of this "nano-[2]catenane" could be clearly identified by AFM

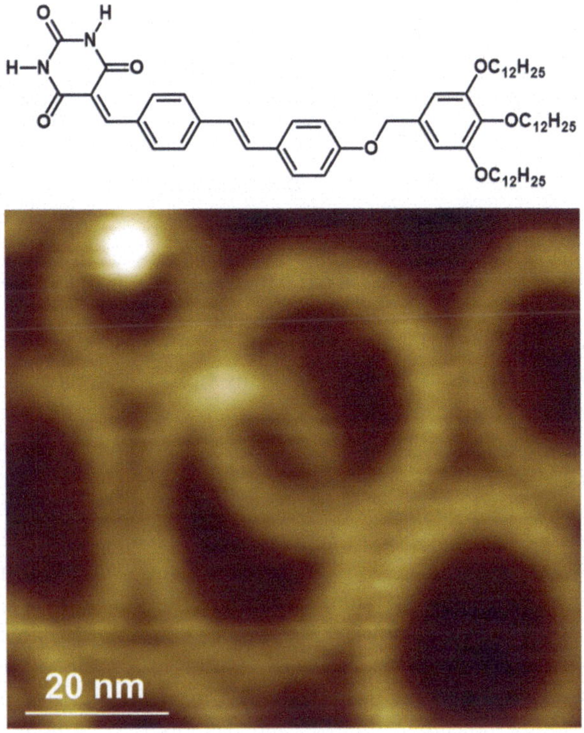

Fig. 3.8 AFM image of nano-[2]catenane, which was first reported by Yagai group

(Fig. 3.8). However, at this stage, yields of the nano-[2]catenane was not considerable and was worthy of being simply "discovered" with the benefit of AFM.

3.2 Supramolecular Polymer with Intrinsic Curvature

Supramolecular polymers, **SP**s, are recognized, in a broad sense, as one-dimensionally elongated molecular assemblies formed by polymerization through weak interactions (supramolecular polymerization, Fig. 3.9) [22]. **SP**s have attracted attention from industry to academia not only because of their ability to polymerize a variety of functional molecules without relying on chemical reactions but also their dynamic nature in their chain formation that underpins biological counterparts. **SP**s are the next generation of polymeric materials not only because they possess the functions and physical properties of conventional synthetic polymers but also their ability to respond to external stimuli (heat, light, etc.) and self-heal upon chain breakage. Typically, **SP**s are formed in solution through reversible association of two binding sites of a monomer linked via a suitable spacer. However, recently reported functional **SP**s include monomeric units with an electron-rich π-surface that endows the main chain with various fruitful properties related to

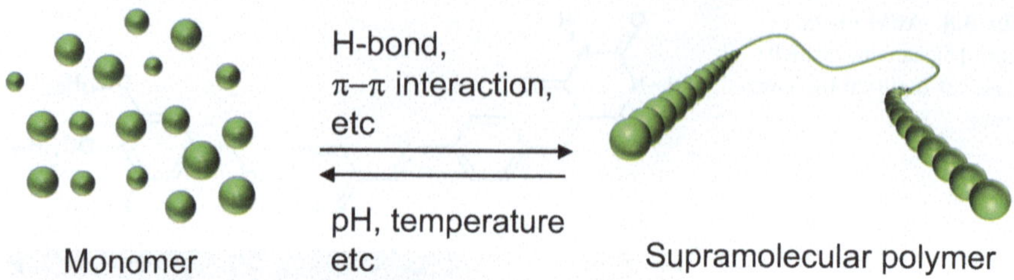

Fig. 3.9 Schematic representation of a supramolecular polymerization

optics and electronics [22]. However, **SPs** based on large π-conjugated molecules are mostly rigid and one-dimensionally elongated fibrous structures due to the rigidity of main chain. For **SPs** as well, the control of their geometric structures, i.e., topologies, is essential because their functional properties and practical applicability depend strongly on their higher-order structures.

The supramolecular monomer design which we have extensively studied in recent years consists of a barbituric acid moiety for hydrogen bonding, an aromatic motif as a π–π stacking unit, and a long-chain alkyl moiety as a van der Waals interaction unit for solubilization in nonpolar solvent. These monomers are not soluble in nonpolar solvents at room temperature, while they are solubilized upon heating to obtain molecularly dissolved state. Upon cooling such hot monomeric solutions, the monomer molecules initially form hydrogen-bonded aggregates of various molecular numbers and patterns [23]. Among them, the cyclic hexamer, which we call "rosette" [24], has a uniform structure, and readily stacks through π–π interactions and van der Waals force, generating "nuclei" composed of several stacked rosettes (Fig. 3.10a). Although the formation of rosettes by barbiturates

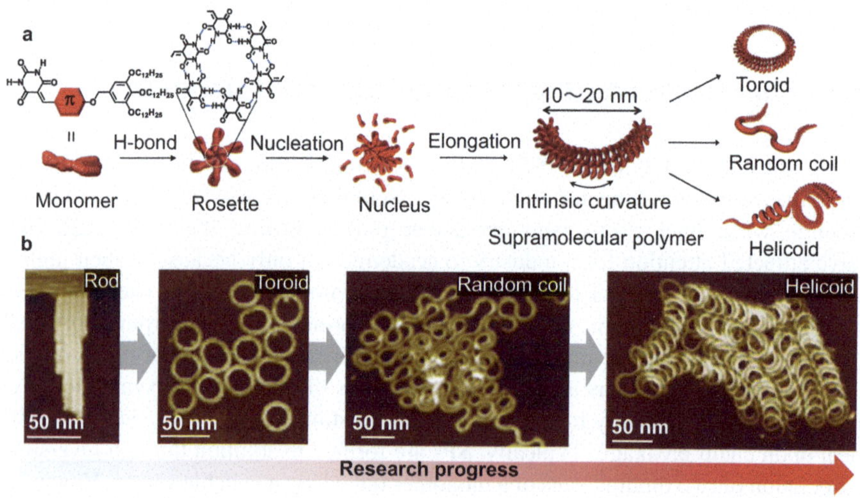

Fig. 3.10 (**a**) General molecular design of π-conjugated barbiturated monomers and a schematic representation of the unique stacking of hydrogen-bonded cyclic hexamers (rosettes) to generate intrinsic curvature. (**b**) Various nanostructures formed by π-conjugated barbiturated molecules

had not been elucidated by X-ray crystallography, we have successfully corroborated its formation using scanning tunneling microscopy (STM) of an oligothiophene-appended barbiturate at the solid–liquid interface [25–27]. The formation of nuclei, i.e., "nucleation," is an important concept in the supramolecular polymerization as well as in crystallization and is the formation process of "templates" for the final assembly. Although molecularly dissolved monomers in a poor solvent at high temperature or in a good solvent dynamically change their conformations, they are fixed in a certain conformation and stabilized when they assemble to some extent. This process is the nucleation and is difficult to occur due to high energy barrier. Once a stable nucleus forms, other monomers use the nucleus as a template to bind at the termini one after another, and thereby the subsequent elongation process occurs more smoothly. This type of supramolecular polymerization is called nucleation–elongation or cooperative supramolecular polymerization and is in clear contrast to isodesmic supramolecular polymerization wherein monomer-binding events occur identically at all steps of the polymerization process [22]. During the growth of the nuclei, the rosettes often stack with translational and rotational displacements depending on the molecular structure [28]. As a result, an intrinsic curvature is generated in the resulting fiber structure [29]. This intrinsic curvature has been used to form various supramolecular polymer topologies (Fig. 3.10a).

We briefly look back on how we have developed supramolecular polymers with various topologies (Fig. 3.10b). In 2012, we found that a barbiturated molecule bearing 1,4-naphthalene as the π-conjugated unit formed cylindrical supramolecular polymers, and a regioisomeric 2,6-naphthalene derivative gave toroidal supramolecular polymers [30]. After elucidating that both supramolecular polymers are formed via the formation of rosettes [31], we extended the π-conjugated unit of the toroid-forming molecular structure to increase the aggregation tendency. This structural extension led to the formation of not only toroidal structure as the final product but also randomly curved and helicoidal structures depending on supramolecular polymerization conditions [32].

The difference between randomly coiled and helicoidal structures is the presence or absence of structural defects. Under more thermodynamic conditions such as slow cooling and addition of polar solvents, the yield of helicoids increases through repairing of the defects generating upon supramolecular polymerization. For example, an azobenzene-functionalized monomer **Azo** afforded randomly coiled structures in MCH through supramolecular polymerization by natural cooling (~15 °C/min) [32]. On the other hand, when polymerized at a very slow cooling rate (0.1 °C/min), spirally folded **SP**s were obtained. Supramolecular polymerization under more thermodynamic condition was achieved by addition of a good solvent that can render the supramolecular polymerization reversible. Thus, in 15 v/v% chloroform–MCH mixture, **Azo** afforded helicoidal **SP**s at a cooling rate of 0.1 °C/min. UV light irradiation of the solution containing helicoidal **SP**s induced *trans*-to-*cis* photoisomerization of azobenzene chromophores in the **SP**s, which generate structural defects. Thus, upon photogeneration of 20% *cis*-azobenzene, helicoid **SP**s

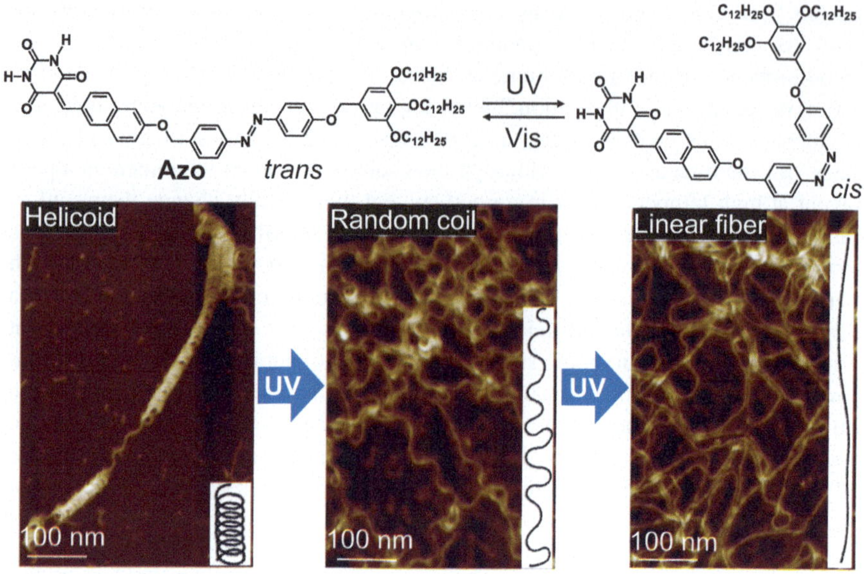

Fig. 3.11 UV-irradiated structural changes of **SP**s formed by **Azo**

unfolded into randomly coiled structures, and, with 30% *cis*-azobenzene, straight **SP**s were obtained due to the complete loss of intrinsic curvature (Fig. 3.11).

Although the above UV-irradiated **SP**s did not reversibly fold up into the original helicoidal structure even after complete *cis*-to-*trans* thermal conversion of azobenzene units probably due to conformational flexibility of the monomer molecule, another molecule **DPN** featuring diphenylnaphthalene π-conjugated unit exhibited self-folding behavior by spontaneously repairing defects (Fig. 3.12). When **DPN** was subjected to undergo supramolecular polymerization at a cooling rate of 1 °C/min from the molecularly dissolved state in MCH, **SP**s composed of misfolded and helicoidal domains in a main chain were obtained. Upon aging this solution at room temperature, the misfolded domains spontaneously folded to helicoidal structures using the preformed helicoidal domains as templates, and after a week, fully folded **SPs** were obtained [33]. It is noteworthy that such a higher-order supramolecular polymer topology is comparable to the tertiary structure of proteins. This self-folding process was also supported by the absorption spectroscopic measurements. The aforementioned mechanism of the self-folding process also raised a question: if **SP**s are composed of misfolded domains throughout the chain, would they self-fold? The fully misfolded **SP**s were thus prepared by cooling a hot solution of **DPN** at a cooling rate of 10 °C/min. Although UV-Vis spectra suggested identical strength of π–π stacking interactions in the fully misfolded and fully folded **SP**s, fully misfolded **SP**s never self-folded upon aging owing to lack of the helical domain as template in **SP** chain (Fig. 3.12). Mixing **SP**s composed of misfolded and helicoidal domains with the fully misfolded **SP**s, followed by aging

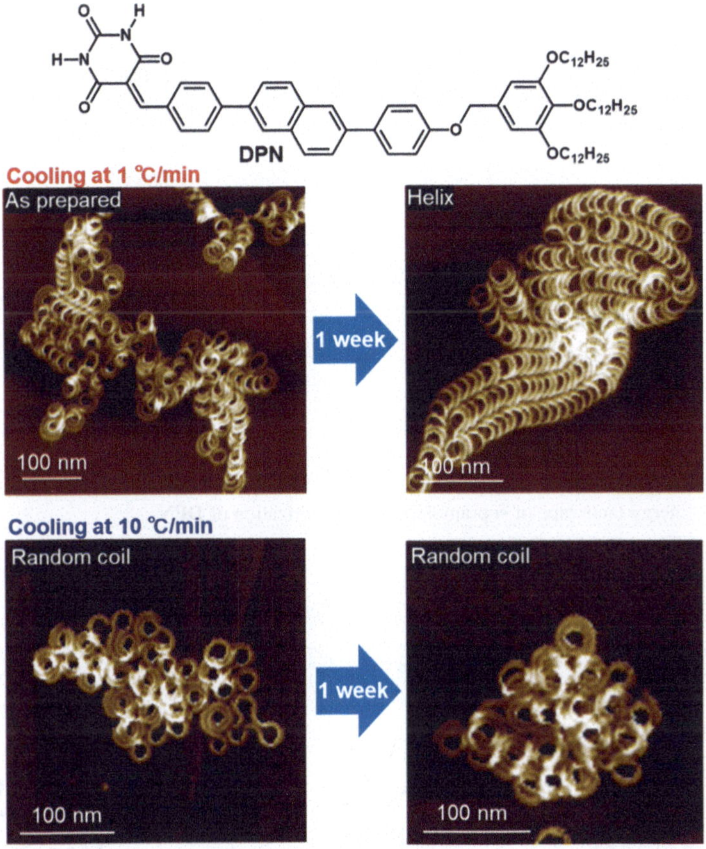

Fig. 3.12 Spontaneous self-folding of supramolecular polymers upon aging for 7 days

the mixture solution for 7 days, could not induce self-folding in the chains of the fully misfolded **SP**s, suggesting that self-folding occurs only in the presence of both misfolded and helical domains in the same **SP** chain. This result also excludes the possibility of a dissociation–reconstruction via the monomer (monomer exchange) during the self-folding of **SP** chain (Fig. 3.13).

We further developed an unprecedented type of supramolecular polymerization of two different monomer molecules. We found that anthracene monomer **DPA**, despite its structural similarity with naphthalene monomer **DPN**, self-assembles into linearly extended ribbon-like **SP**s lacking any curvature. Having **SP**s with distinct secondary structures (helicoids and ribbons) from structurally similar monomers, we have attempted to copolymerize them supramolecularly by simply cooling their monomeric mixtures. Owing to the kinetic mixing of **DPN** and **DPA** by forming heteromeric rosettes, we could easily obtain dumbbell-like chimeric supramolecular block copolymers with coexisting helical and ribbon domains in the main chain (Fig. 3.14) [34].

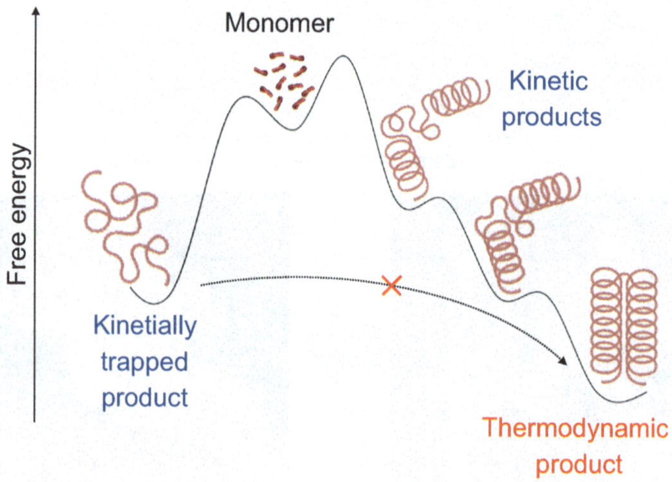

Fig. 3.13 Energy landscape of supramolecular polymerization of **DPN**

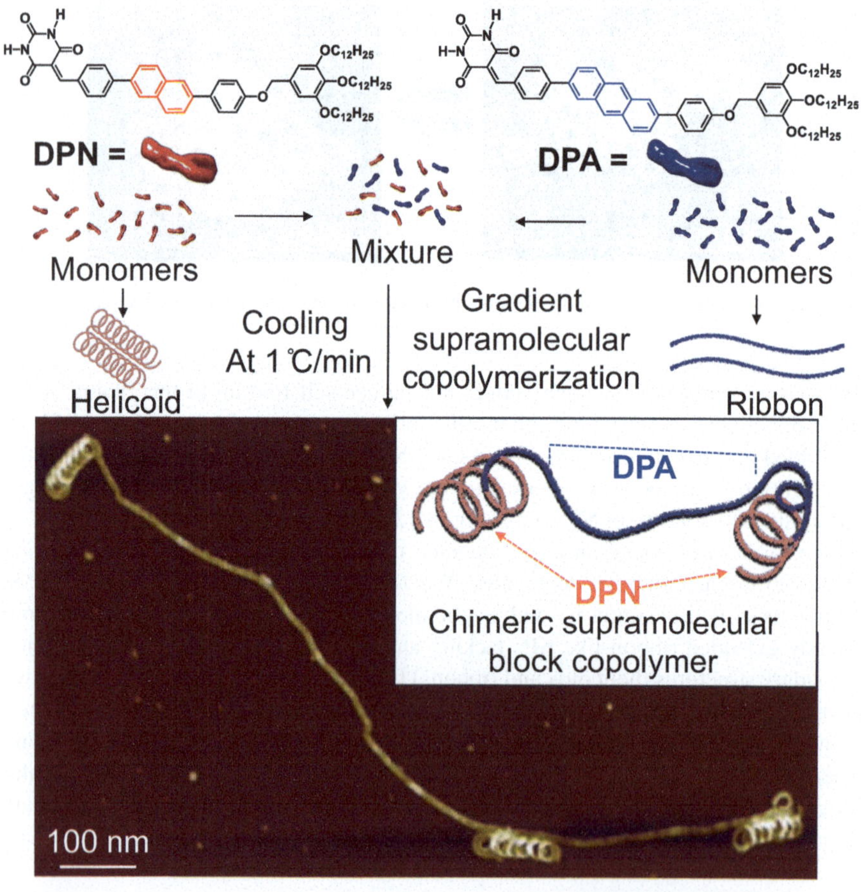

Fig. 3.14 Chimeric supramolecular block copolymer composed of **DPN** and **DPA**

In the barbiturate rosette-based **SPs** discussed so far, monomer exchange did not occur because the rosette–rosette interactions are sufficiently strong to impart significant stability in the **SP** chain. But, replacing the ether linkage by an ester group between the π-conjugated core and the tri(dodecyloxy)benzyl ("wedge") unit could destabilize the rosette–rosette interactions probably due to dipole–dipole repulsion between wedge units substituted with the ester group, leading to tapelike supramolecular chains through a rearrangement of hydrogen-bond pattern [35, 36]. In another example, upon mixing two barbiturate-naphthalenes with complementary electron densities, supramolecular alternating copolymerization occurred from randomly hydrogen-bonded amorphous agglomerates through monomer exchange, leading to helically elongated supramolecular polymers that undergo a specific thermal phase transition to the amorphous agglomerates [37]. Thus, we succeeded in controlling the topology of **SPs** with intrinsic curvature on a nanoscale for the first time by cleverly investigating supramolecular polymerization conditions and mixing of different molecules (Fig. 3.10b).

3.3 Nanolympiadane

As mentioned in this chapter's sect. 2, we explored a variety of nanosized architectures composed of supramolecular fibers with a width of 5–10 nm (corresponds to diameters of rosettes) since the discovery of **SPs** with intrinsic curvature. These "open-ended" fibers are expected to show specific physical properties arising from their topologies in solution and also in the bulk state. On the other hand, interesting results were also obtained from the closed topology of "toroids." For instance, supramolecular ring-opening polymerization occurred when a solution of toroids composed of an azobenzene monomer was irradiated with UV light [38]. Additionally, we found that heating a solution containing mixture of toroids and random coils resulted in selective disassembly of the latter open-ended structures, while the structural integrity of the toroids was retained due to their high thermal stability [39, 40]. These results motivated us to further study physical properties of toroidal **SPs** composed of monomers with expanded π-conjugated systems.

As briefly mentioned in this chapter's sect. 2, supramolecular polymerization of monomer **DPN** (Fig. 3.12) from the molecularly dispersed state at a slow cooling rate of 1 °C/min led to a supramolecular polymers in which random coils and helices coexist in the main chain. In order to selectively obtain toroidal **SPs** by **DPN**, we applied a more kinetic supramolecular polymerization protocol, that is, a fast cooling of the hot monomer solution in ice water bath (quenching). The yield of toroidal **SPs** on the basis of monomer was however found to be only about 20%. This is because, even after quenching, a certain temperature gradient can occur during the decrease of solution temperature, and the supramolecular polymerization becomes more or less under thermodynamic control. For more kinetic supramolecular polymerization to occur, it was necessary to assemble the molecules more kinetically.

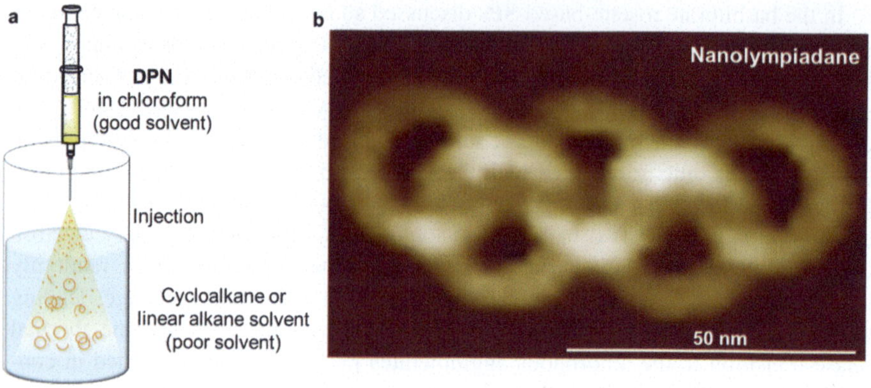

Fig. 3.15 (**a**) Depiction of the injection protocol. (**b**) AFM image of a nanolympiadane

To attain such condition, we attempted a solvent mixing protocol (Fig. 3.15a). In this protocol, the compound was dissolved in a good solvent and the solution was injected into a poor solvent at once. It was found that, by this protocol, the supramolecular polymerization underwent more kinetically in comparison to the quenching protocol and led 44% of the monomers to self-assemble into toroidal **SPs** (the rest was random coil structures). To our surprise, AFM observation showed that a large number of toroids existed as mechanically interlocked state to form nano-[2] catenanes. By detailed AFM investigation, it was found that approximately 150 out of 5000 toroids existed as nano-[2]catenanes. The yield calculated on the basis of monomer was 1.6%, which could not be explained by statistical events considering the report of Wasserman in 1962 [3], although materials (supramolecular polymer chain vs. covalent chain) and conditions (e.g., concentration, solvent etc.) were totally different. Moreover, in rare cases, nano-[3]catenanes were also found. Surprisingly, when the poor solvent was changed from methylcyclohexane to less polar linear alkanes such as *n*-hexane and *n*-octane, the yield of the nano-catenane jumped to about 5–7%, accompanying the formation of up to nano-[5]catenane that resembled the Olympic symbol (Fig. 3.15b). This nano-[5]catenane was named "nanolympiadane" in homage to the Stoddart's "olympiadane" [41].

3.4 Mechanism of Nano-Catenane Formation

We anticipated that the hint of mechanism for the formation of nano-catenanes in the above high yields could be found from the aforementioned template-directed catenane synthesis developed by Sauvage and co-workers (Fig. 3.2a) [5]. The underlying principle in both the systems is quite similar, although the materials employed are different. Namely, if a precursor of circular objects exists near a cyclic compound, catenation would occur efficiently. The probability of nano-catenane formation in the course of supramolecular polymerization of monomer **DPN** should be much higher if nucleation more likely occurs on the surface of the preformed toroid. In particular, if nucleation occurs inside the toroid, the probability of catenation becomes comparable to the yield of ring-closing process (44%).

The phenomenon of nucleation of monomers on the surface of a preformed aggregate is called "secondary nucleation." It is also a crucial process in the aggregation of proteins that leads to the formation of amyloid fibrils, which are the cause of Alzheimer's and Parkinson's disease. In this regard, many of the rigid **SP**s reported so far are indeed bundled. For instance, the 1,4-naphthalene barbiturate afforded densely packed monolayer of cylindrical **SPs** with a uniform thickness of about 5.5 nm in MCH through supramolecular polymerization as observed by AFM [42]. Such close alignment of the **SPs** could be attributed to the secondary nucleation of monomers on the surface of preformed nanorods. Although bundling may occur during the drying process of the solution, as microscopic analysis is usually done in the dry state, we believe that a mechanism of the formation of new assembly on the surface of already formed **SP**s due to secondary nucleation should not be ignored in the future [43].

To prove the hypothesis "secondary nucleation on the toroid surface," we performed several experiments. As a direct evidence of the secondary nucleation, we performed a seeding experiment. Generally, the rate of the elongation process during supramolecular polymerization is significantly faster than that of the nucleation process. Thus, when fragments of supramolecular polymers are added as seeds to a monomeric solution wherein the spontaneous nucleation hardly occurs, the elongation process can occur at the termini of the seed [44]. **SP** seeds are usually prepared by fragmentation of supramolecular polymers by means of sonication or other protocols. What would happen if toroidal supramolecular polymers without the termini are used as seeds? If the nucleation process could be accelerated in the presence of such toroidal seeds, this would be a clear evidence of the secondary nucleation. However, there was a problem upon performing seeding experiments in our present system. As shown in Fig. 3.16, at the beginning of the elongation process of our **SP**s, the elongation pathway leading to closed toroids and that into opened random coils or helicoids competitively proceeds, and it was extremely difficult to completely control these pathways by the polymerization condition. Therefore, we attempted the purification of toroidal **SP**s based on the differences of stability and size of **SP**s as described below.

As mentioned in this chapter's sect. 2, thermal depolymerization of toroidal **SP**s hardly occurs because of the lack of termini [39, 40]. We thus initially prepared a mixture of toroidal and randomly coiled **SP**s by quenching the hot monomer solution (Fig. 3.16ii). Note that nano-catenanes rarely formed with this temperature protocol. At this stage, it was impossible to separate the two types of **SP**s because their sizes were not considerably different. However, when this quenched solution was heated to 90 °C, only the open-ended randomly coiled species depolymerized into monomer. Upon subsequently cooling the solution slowly to room temperature at a rate of 1 °C/min, the depolymerized monomers followed the pathway toward the helicoidal structure since the elongation process was thermodynamically controlled. It was

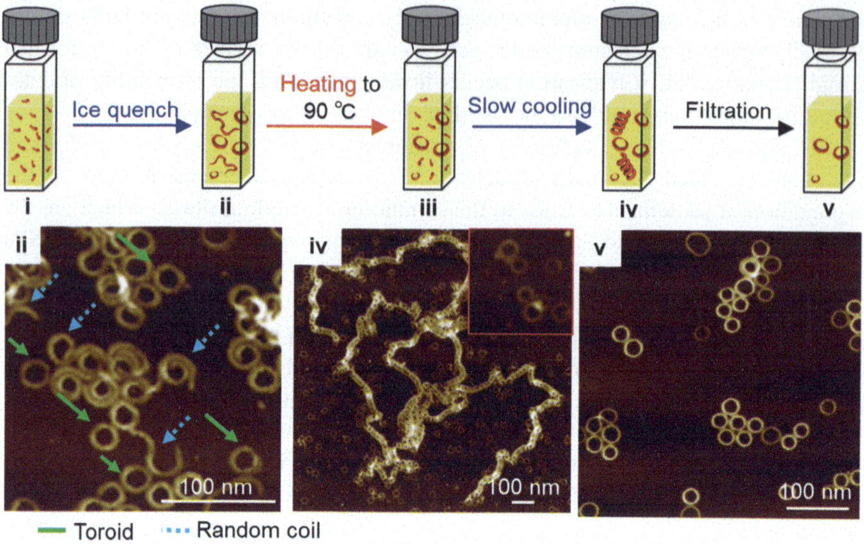

Fig. 3.16 Protocols for separating toroids from open-ended fibers and (**ii. iv. v**) the corresponding AFM images

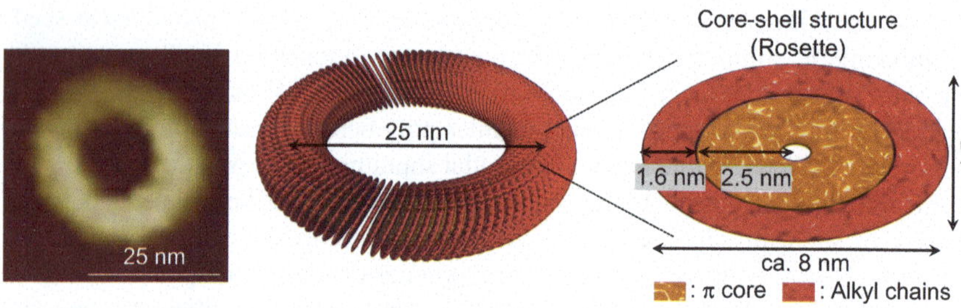

Fig. 3.17 Internal structure of a toroid elucidated by SAXS/SANS experiments

clearly shown by AFM that short randomly coiled **SP**s disappeared, while a large number of toroids and elongated helicoids with apparent lengths exceeding 1 μm were obtained (Fig. 3.16iv). By passing this solution through a membrane filter with a pore size of 200 nm, we succeeded in removing the large helicoids and accordingly obtained toroid **SP**s in the filtrate (Fig. 3.16v).

Small-angle neutron scattering (SANS) and small-angle X-ray scattering (SAXS) experiments confirmed the presence of toroids having size consistent with the AFM image, and their surface layer was covered with alkyl chains, proving that the van der Waals force is responsible for secondary nucleation of the monomers (Fig. 3.17).

The above isolated toroidal solution was used in a seeding experiment, wherein a hot monomer solution was cooled and kept at a constant temperature slightly below the nucleation temperature. When the solution was left as it was, the nucleation started after a lag time of ca. 5 min, as evident from the growth of the UV-Vis absorption band at 470 nm associated with the π–π stacking of the diphenyl naphthalene cores. On the other hand, when the toroid solution was added to the monomer solution after the temperature became constant, the nucleation started immediately without any lag phase. Accordingly, it was shown experimentally that the added ring drove the monomers to assemble on the ring surface thereby triggering the nucleation. Using a coarse-grained molecular dynamics simulation, we also computationally demonstrated that the oligomers of the assumed nucleated rosette can be stabilized at the ring surface.

"Scaling" is a method to investigate the contribution of elementary processes (primary nucleation, elongation, secondary nucleation, fragmentation, etc.) in protein aggregation [45–47]. In the scaling of protein aggregation established by Knowles et al., half-time (t_{50}) for the consumption of half of the total monomers by aggregation is plotted against various initial monomer concentrations (C_0). The slope of the plot which is called scaling exponent (γ) is related to the reaction order of the dominant mechanism associated with the generation of new aggregates. Therefore, we performed scaling for our system by following time-dependent changes of the UV-Vis absorption band at 470 nm at a constant temperature slightly below the nucleation temperature. We found that the γ value is consistent with monomer-dependent secondary nucleation processes. Furthermore, solvent dependence of secondary nucleation was shown by performing scaling experiments not only in cycloalkanes but also acyclic alkanes. Comparison of the obtained γ values in different solvents suggested considerably higher feasibility of secondary nucleation in acyclic alkanes compared to those in cyclic alkanes. This result was also consistent with a higher nano-catenane yield in acyclic alkanes than in cycloalkanes, as described above.

3.5 Nano-Polycatenanes

Based on the above results, we finally addressed the creation of nano-polycatenanes utilizing the nature of the secondary nucleation mechanism. Our method for obtaining nano-polycatenane contrasts with the polycatenane synthesis of Rowan et al. [14]. In their approach, metallosupramolecular polymers were pre-assembled by complexing cyclic and precursor molecules via Zn^{2+} coordination and then cyclized the precursor units by an olefin metathesis reaction to obtain polycatenanes (Fig. 3.18a). On the other hand, in the case of our **SP**s, such a pre-assembled structure is not possible because a toroid must be formed continuously inside preformed toroid. Based on this idea, we thought that the nano-catenanes could be elongated sequentially by injecting the monomer solution portionwise, rather than by injecting the monomer solution all at once to form nano-catenanes at the same time (Fig. 3.18b).

After attempting the injection speed, number of injections, and intervals, we finally optimized the condition: the monomer solution is injected into the poor

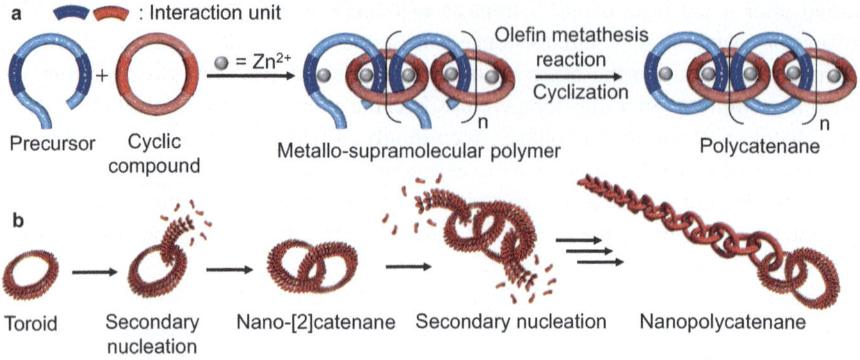

Fig. 3.18 (**a**) Rowan's strategy for the synthesis of polycatenanes. (**b**) Mechanism of the formation of our nano-catenane

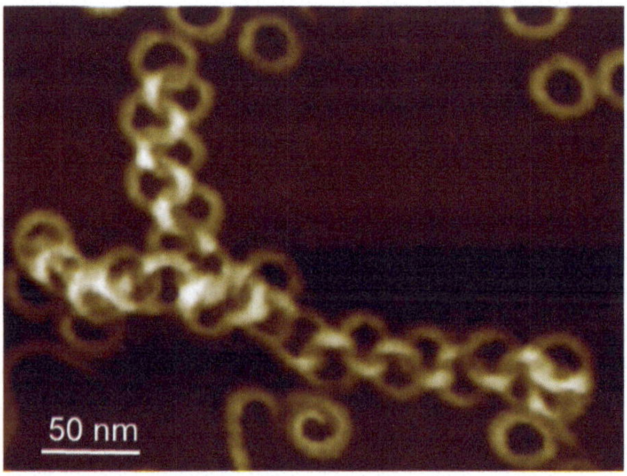

Fig. 3.19 AFM image of a nano-poly[22]catenane

solvent ten times at an interval of 1 second to ensure the formation of nano-polycatenanes. Figure 3.6 shows an AFM image of nano-poly[22]catenane that was the longest one we have found by this method (Fig. 3.19). Its length exceeds approximately 500 nm due to 20 linearly interlocked toroids. A branched structure is also visible along the chain, suggesting that three supramolecular polymer chains can pass through the internal void space of a single toroid.

We acquired a hundred of AFM images randomly, and the number of nano-[n] catenanes was counted as shown in Fig. 3.20. From the number of nano-catenanes, it is clearly evident that the number of catenanes increased by ten injections. The ratio of nano-[n]catenanes did not change much under both conditions, indicating that the sequential addition of monomers to the formed nano-[n]catenanes would lead to the formation of nano-[n + 1]catenanes.

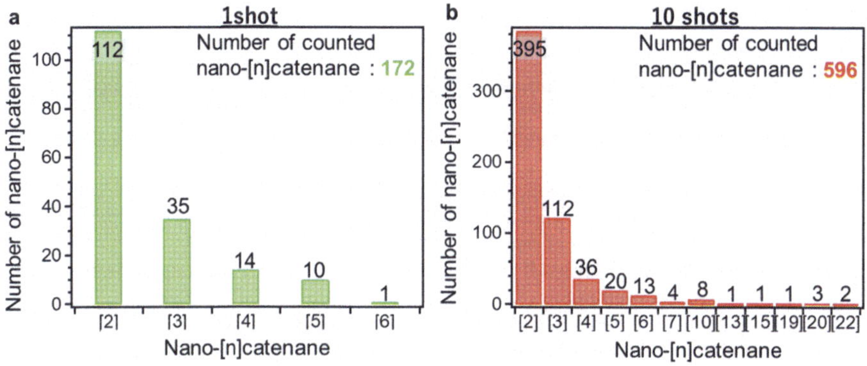

Fig. 3.20 Comparison of (**a**) 1-shot and (**b**) 10-shot injection

3.6 Conclusion

We have created diverse supramolecular polymer topologies by controlling not only elongation but also nucleation processes (Fig. 3.10a). The introduction of a new concept of "secondary nucleation" to the "monomers that polymerize into supramolecular polymer with intrinsic curvature" enabled creation of the first "self-assembled nano-polycatenanes" that could be formed spontaneously by simply injecting the monomers into a suitable solvent. Through a multifaceted approach, we have shown that this self-assembling phenomenon is due to the repetition of the elementary processes: nucleation → elongation → cyclization; secondary nucleation → elongation → cyclization; secondary nucleation → elongation → cyclization; and so on (Fig. 3.18b).

Although it is difficult to obtain average information on the structures of **SP**s by using AFM, it has allowed to visualize each polymer structure in detail. For example, proving the branching catenane structure is not easy by conventional method (e.g., size exclusion chromatography and DOSY), but it is quite easy with the AFM observation, as demonstrated in our study. Furthermore, AFM observations have already revealed several unprecedented self-assembled topologies, which will lead us to new research directions. In the near future, we plan to use high-speed AFM and other techniques to investigate the unusual motion of the mesoscopic topological assemblies and the unique properties of such assemblies. In addition, considering from broader perspective, secondary nucleation is expected to be a new tool in the future supramolecular polymer researches to induce unprecedented higher-order structures and sophisticated properties that cannot be obtained by conventional supramolecular polymerization driven only by primary, spontaneous nucleation.

References

1. Gil-Ramírez G, Leigh DA, Stephens AJ (2015) Catenanes: fifty years of molecular links. Angew Chem Int Ed 54:6110–6150
2. Wasserman E (1960) The preparation of interlocking rings: a catenane. J Am Chem Soc 82:4433–4434
3. Wasserman E (1962) Chemical topology. Sci Am 207:94–102
4. Brückner R (2019) Pioneering work on catenanes, rotaxanes, and a knotane in the university of Freiburg 1958–1988. Eur J Org Chem 2019:3289–3319
5. Dietrich-Buchecker CO, Sauvage JP, Kintzinger JP (1983) Une nouvelle famille de molecules: les metallo-catenanes. Tetrahedron Lett 24:5095–5098
6. Caballero A, Zapata F, White NG, Cost PJ, Félix V, Beer PD (2012) A halogen-bonding catenane for anion recognition and sensing. Angew Chem Int Ed 51:1876–1880
7. Gilday LC, Lang T, Caballero A, Costa PJ, Félix V, Beer PD (2013) A catenane assembled through a single charge-assisted halogen bond. Angew Chem Int Ed 52:4356–4360
8. Amabilino DB, Ashton PR, Reder AS, Spencer N, Stoddart JF (1994) Olympiadane. Angew Chem Int Ed 33:1286–1290
9. Fujita M, Ibukuro F, Hagihara H, Ogura K (1994) Quantitative self-assembly of a [2]catenane from two preformed molecular rings. Nature 367:720–723
10. Cougnon FBL, Ponnuswamy N, Jenkins NA, Pantos GD, Sanders JKM (2012) Structural parameters governing the dynamic combinatorial synthesis of catenanes in water. J Am Chem Soc 134:19129–19135
11. Li H, Zhang H, Lammer AD, Wang M, Li X, Lynch VM, Sessler JL (2015) Quantitative self-assembly of a purely organic three-dimensional catenane in water. Nat Chem 7:1003–1008
12. Segawa Y, Kuwayama M, Hijikata Y, Fushimi M, Nishihara T, Pirillo J, Shirasaki J, Kubota N, Itami K (2019) Topological molecular nanocarbons: all-benzene catenane and trefoil knot. Science 365:272–276
13. Segawa Y, Kuwayama M, Itami K (2020) Synthesis and structure of [9]cycloparaphenylene catenane: an all-benzene catenane consisting of small rings. Org Lett 22:1067–1070
14. Wu Q, Rauscher PM, Lang X, Wojtecki RJ, de Pablo JJ, Hore MJA, Rowan SJ (2017) Poly[n]catenanes: synthesis of molecular interlocked chains. Science 358:1434–1439
15. Rauscher PM, Schweizer KS, Rowan SJ, de Pablo JJ (2020) Thermodynamics and structure of poly[n]catenane melts. Macromolecules 53:3390–3408
16. Li G, Wang L, Wu L, Guo Z, Zhao J, Liu Y, Bai R, Yan X (2020) Woven polymer networks via the topological transformation of a [2]catenane. J Am Chem Soc 142:14343–14349
17. Hudson B, Vinograd J (1967) Catenated circular DNA molecules in HeLa cell mitochondria. Nature 216:647–652
18. Clayton DA, Vinograd J (1967) Circular dimer and catenate forms of mitochondrial DNA in human leukaemic leucocytes. Nature 216:652–657
19. Valero J, Pal N, Dhakal S, Walter NG, Famulok M (2018) A bio-hybrid DNA rotor–stator nanoengine that moves along predefined tracks. Nat Nanotechnol 13:496–503
20. Peil A, Zhan P, Liu N (2020) DNA origami catenanes templated by gold nanoparticles. Small 16:1905987
21. Yagai S, Goto Y, Karatsu T, Kitamura A, Kikkawa Y (2011) Catenation of self-assembled nanorings. Chem Eur J 17:13657–13660
22. De Greef TFA, Smulders MMJ, Wolffs M, Schenning APHJ, Sijbesma RP, Meijer EW (2009) Supramolecular polymerization. Chem Rev 109:5687–5754
23. Lin X, Suzuki M, Gushiken M, Yamauchi M, Karatsu T, Kizaki T, Tani Y, Nakayama K, Suzuki M, Yamada H, Kajitani T, Fukushima T, Kikkawa Y, Yagai S (2017) High-fidelity self-assembly pathways for hydrogen-bonding molecular semiconductors. Sci Rep 7:43098
24. Adhikari B, Lin X, Yamauchi M, Ouchi H, Aratsu K, Yagai S (2017) Hydrogen-bonded rosettes comprising π-conjugated systems as building blocks for functional one-dimensional assemblies. Chem Commun 53:9663–9683

25. Yagai S, Suzuki M, Lin X, Gushiken M, Noguchi T, Karatsu T, Kitamira A, Saeki A, Seki S, Kikkawa Y, Tani Y, Nakayama K (2014) Supramolecular engineering of oligothiophene nanorods without insulators: hierarchical association of rosettes and photovoltaic properties. Chem Eur J 20:16128–16137
26. Yamauchi M, Adhikari B, Prabhu DD, Lin X, Karatsu T, Ohba T, Shimizu N, Takagi H, Haruki R, Adachi S, Kajitani T, Fukushima T, Yagai S (2017) Supramolecular polymerization of supermacrocycles: effect of molecular conformations on kinetics and morphology. Chem Eur J 23:5270–5280
27. Hollamby MJ, Aratsu K, Pauw BR, Rogers SE, Smith AJ, Yamauchi M, Lin X, Yagai S (2016) Simultaneous SAXS and SANS analysis for the detection of toroidal supramolecular polymers composed of noncovalent supermacrocycles in solution. Angew Chem Int Ed 55:9890–9893
28. Yagai S, Usui M, Seki T, Murayama H, Kikkawa Y, Uemura S, Karatsu T, Kitamura A, Asano A, Seki S (2012) Supramolecularly engineered perylene bisimide assemblies exhibiting thermal transition from columnar to multilamellar structures. J Am Chem Soc 134:7983–7994
29. Yagai S, Kitamoto Y, Datta S, Adhikari B (2019) Supramolecular polymers capable of controlling their topology. Acc Chem Res 52:1325–1335
30. Yagai S, Goto Y, Lin X, Karatsu T, Kitamura A, Kuzuhara D, Yamada H, Kikkawa Y, Saeki A, Seki S (2012) Self-organization of hydrogen-bonding naphthalene chromophores into J-type nanorings and H-type nanorods: impact of regioisomerism. Angew Chem Int Ed 51:6643–6647
31. Aratsu K, Prabhu DD, Iwawaki H, Lin X, Yamauchi M, Karatsu T, Yagai S (2016) Self-sorting regioisomers through the hierarchical organization of hydrogen-bonded rosettes. Chem Commun 52:8211–8214
32. Adhikari B, Yamada Y, Yamauchi M, Wakita K, Lin X, Aratsu K, Ohba T, Karatsu T, Hollamby MJ, Shimizu N, Takagi H, Haruki R, Adachi S, Yagai S (2017) Light-induced unfolding and refolding of supramolecular polymer nanofibres. Nat Commun 8:15254
33. Prabhu DD, Aratsu K, Kitamoto Y, Ouchi H, Ohba T, Hollamby MJ, Shimizu N, Takagi H, Haruki R, Adachi S, Yagai S (2018) Self-folding of supramolecular polymers into bioinspired topology. Sci Adv 4:eaat8466
34. Kitamoto Y, Pan Z, Prabhu DD, Isobe A, Ohba T, Shimizu N, Takagi H, Haruki R, Adachi S, Yagai S (2019) One-shot preparation of topologically chimeric nanofibers via a gradient supramolecular copolymerization. Nat Commun 10:4578
35. Aizawa T, Aratsu K, Datta S, Mashimo T, Seki T, Kajitani T, Silly F, Yagai S (2020) Hydrogen bond-directed supramolecular polymorphism leading to soft and hard molecular ordering. Chem Commun 56:4280–4283
36. Aizawa T, Takahashi S, Isobe A, Datta S, Sotome H, Miyasaka H, Kajitani T, Yagai S (2020) Fluorescent supramolecular polymorphism driven by distinct hydrogen bonding lattice. Chem Lett 49:1009–1012
37. Aratsu K, Takeya R, Pauw BR, Hollamby MJ, Kitamoto Y, Shimizu N, Takagi H, Haruki R, Adachi S, Yagai S (2020) Supramolecular copolymerization driven by integrative self-sorting of hydrogen-bonded rosettes. Nat Commun 11:1623
38. Adhikari B, Aratsu K, Davis J, Yagai S (2019) Photoresponsive circular supramolecular polymers: a topological trap and photoinduced ring-opening elongation. Angew Chem Int Ed 58:3764–3768
39. Suzuki A, Aratsu K, Datta S, Shimizu N, Takagi H, Haruki R, Adachi S, Hollamby M, Silly F, Yagai S (2019) Topological impact on the kinetic stability of supramolecular polymers. J Am Chem Soc 141:13196–13202
40. Isobe A, Prabhu DD, Datta S, Aizawa T, Shiki Y (2020) Effect of an aromatic solvent on hydrogen-bond-directed supramolecular polymerization leading to distinct topologies. Chem Eur J 26:8997–9004
41. Datta S, Kato Y, Higashiharaguchi S, Aratsu K, Isobe A, Saito T, Prabhu DD, Kitamoto Y, Hollamby MJ, Smith AJ, Dalgliesh R, Mahmoudi N, Pesce L, Perego C, Pavan GM, Yagai S (2020) Self-assembled poly-catenanes from supramolecular toroidal building blocks. Nature 583:400–405

42. Yagai S, Yamauchi M, Kobayashi A, Karatsu T, Kitamura A, Ohba T, Kikkawa Y (2012) Control over hierarchy levels in the self-assembly of stackable nanotoroids. J Am Chem Soc 134:18205–18208
43. Sasaki N, Mabesoone MFJ, Kikkawa J, Fukui T, Shioya N, Shimoaka T, Hasegawa T, Takagi H, Haruki R, Shimizu N, Adachi S, Meijer EW, Takeuchi M, Sugiyasu K (2020) Supramolecular double-stranded Archimedean spirals and concentric toroids. Nat Commun 11:3578
44. Ogi S, Sugiyasu K, Manna S, Samitsu S, Takeuchi M (2014) Living supramolecular polymerization realized through a biomimetic approach. Nat Chem 6:188–195
45. Cohen SIA, Vendruscolo M, Dobson CM, Knowles TPJ (2012) From macroscopic measurements to microscopic mechanisms of protein aggregation. J Mol Biol 421:160–171
46. Meisl G, Kirkegaard JB, Arosio P, Michaels TCT, Vendruscolo M, Dobson CM, Linse S, Knowles TPJ (2016) Molecular mechanisms of protein aggregation from global fitting of kinetic models. Nat Protoc 11:252–272
47. Törnquist M, Michaels TCT, Sanagavarapu K, Yang X, Meisl G, Cohen SIA, Knowles TPJ, Linse S (2018) Secondary nucleation in amyloid formation. Chem Commun 54:8667–8684

Chapter 4
Molecular Architectonics Guide to the Fabrication of Self-Cleaning Materials

Mouli Konar and Thimmaiah Govindaraju

4.1 Introduction

Advances in science and technology have made the human life smooth and sophisticated. Use of luxury materials and products poses concerns of regular maintenance due to dust pollution and desperately seek to avoid the daily hazards and cost of cleaning chemical agents. This resulted in the rapid development of various technological platforms offering environmentally viable self-cleaning coatings. By definition, self-cleaning is an acquired property of a material or substance with self-aided cleaning ability attributed to its surface characteristics [1–3]. The term "self-cleaning" is closely related to the behavior of non-wettable surfaces. There are plenty of natural self-cleaning examples in nature including lotus leaves, rose petals, pitcher plant, mosquito eye, insect wings, lizard skin, water strider legs, shark skin, and fish scales [4–7]. All these beautiful and impeccable surfaces are manifested by nature to deter water and dust particles. These natural surfaces inspired researchers to mimic nature to create artificial self-cleaning surfaces. Such self-cleaning technologies have rapidly spread across wide range of materials and products [8]. The self-cleaning technologies are used indispensably starting from various essential commodities of daily life such as material coverings, cooking utensils, water-proof jackets, bags, sport shoes, window and door glasses, ceramic tiles, mirrors, mortars, concrete, and windscreen of motor vehicles to many classy items like pollutant-subsiding paints, solar panels, microfiltration membranes, water-harvesting membranes, and antibacterial layers among others [9]. The ever-growing demand of the waterproof and the dustproof materials forces researchers to scout for simple, cost-

M. Konar · T. Govindaraju (✉)
Bioorganic Chemistry Laboratory, New Chemistry Unit and School of Advanced Materials (SAMat), Jawaharlal Nehru Centre for Advanced Scientific Research, Bengaluru, Karnataka, India
e-mail: tgraju@jncasr.ac.in

T. Govindaraju, K. Ariga (eds.), *Molecular Architectonics and Nanoarchitectonics*, Nanostructure Science and Technology,
https://doi.org/10.1007/978-981-16-4189-3_4

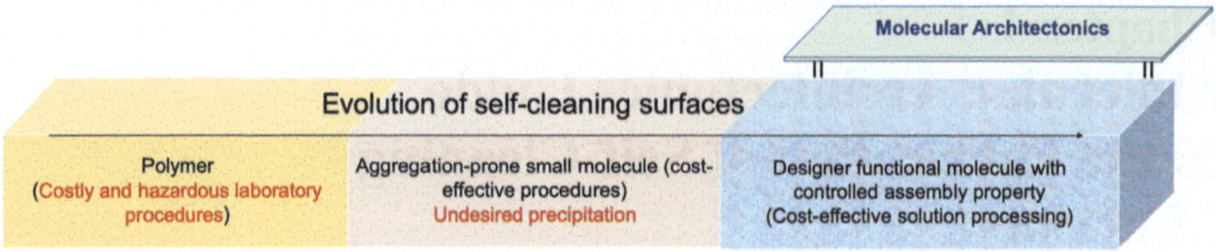

Fig. 4.1 Molecular architectonics-guided fabrication of cost-effective and solution processable self-cleaning material surfaces through the controlled molecular assembly of functional building block

effective, and robust fabrication techniques for the self-cleaning materials and surfaces. A number of strategies that demonstrate the fabrication of the self-cleaning surfaces via self-assembly processes, breath figure technique, solution deposition, chemical vapor deposition, and sol-gel methods are reported [10–12]. However, there is a growing interest in the fabrication of self-cleaning materials through bio-inspired approaches in order to maintain the ecological balance. At this juncture, researchers are challenged with conflict between the expectation and means of creating the self-cleaning surfaces. The majority of reported self-cleaning methods are based on polymer substrates that need costly laboratory techniques, and the fabrication procedures are not environmentally benign in many cases. Thus, adopting bio-inspired approaches is a better alternative overcoming these inherent limitations. In this context, small functional molecules with propensity for molecular assembly can compete with the polymer substrates, albeit they lack the control over their assembly process and result in unexpected crystallization or precipitation. There have been numerous efforts to overcome the limitations to develop a solution processable small molecule-based molecular assembly process to fabricate self-cleaning materials and surfaces. Contemporary developments in the field of molecular architectonics have the potential to fabricate self-cleaning surfaces through solution processable controlled molecular assembly of designer functional molecules (Fig. 4.1) [2, 13–16]. In brief, molecular architectonics is a state-of-the-art exercise to realize the controlled engineering of molecular assembly of functional small molecular building blocks [2, 17, 18]. The scheme of molecular architectonics involves designing functional building blocks and meticulously engineering their molecular assembly through controlled noncovalent interactions and skillful experimental conditions to generate nano-, micro-, and macroscale material architectures and surfaces [19–23]. Some recent works have demonstrated ways of fabricating the self-cleaning surfaces and coatings employing the molecular assembly of specially designed small molecules [2, 24].

Herein, we describe the concepts of molecular architectonics to fabricate definite hierarchical and functional architectures with required surface roughness and low-surface energy to exhibit self-cleaning property [18, 19]. In our work, we reported a molecular architectonics strategy to create hexagonal honeycomb microarrays which exhibited desired self-cleaning effect [25]. This work enlightened the potential of molecular architectonics to harvest self-cleaning coating using small functional molecule through simple solution processing technique. Success of this study motivated us to assemble the previously reported self-cleaning strategies through the controlled molecular assembly of small functional molecules. The theme of molecular architectonics is comparatively new within the scientific community, and some of the representative reports on the self-cleaning strategies fabricated through the controlled molecular assembly of small molecules are discussed under molecular architectonics-guided self-cleaning strategies. Thus, a summary of the progress in self-cleaning technologies developed based on the principles of molecular architectonics are presented in the present chapter. The preliminary sections depict the principle, parameters, and the mechanism of the self-cleaning effect. Later, we comprehensively discuss the guiding principles of molecular architectonics in fabricating the self-cleaning surfaces by considering few specific and selected examples from the reported literatures. Lastly, the chapter is concluded with a summary and future perspectives to produce technologically advanced and biologically and environmentally endurable self-cleaning products through molecular architectonics strategies.

4.2 Self-Cleaning Surfaces and Relevant Parameters

Self-cleaning is a surface property defined physically by the interaction of falling water droplets (oil sometimes) on the surface. The self-cleaning phenomena occur in nature following two extreme and mutually contradictory surface properties. In a different word, self-cleaning effect can be observed because of maximum or minimum interaction between the material surface and water droplets. The maximum interaction causes water droplets to spread over the surface and wets it completely. The deposited dust particles in this case are removed by the linear movement of water on the inclined surface. On the other hand, the surfaces having minimum interaction with water droplets show non-wettable characteristics. On such surfaces, the water droplets bead up and easily rolled off along with the dust particles [26]. The first type of self-cleaning phenomenon is observed with extremely hydrophilic or superhydrophilic surfaces and in the second case corresponds to extremely hydrophobic or superhydrophobic surfaces.

Therefore, artificial self-cleaning can be achieved by the creation of superhydrophilic or superhydrophobic material character on the surfaces. A widely used method to study such surface properties is the contact angle measurement at the solid (surface)-water interface. When the water contact angle (WCA) value reaches 10° or less, the associated hydrophilicity is termed as superhydrophilicity, and the surfaces with WCA value equal to or greater than 150° are considered superhydrophobic in nature [25]. For industrial applications, superhydrophobic surfaces are getting prioritized owing to their good biocompatibility, mechanical robustness, durability, and easy manufacturing procedures [27–31]. Apart from the contact angle measurement, the superhydrophobic self-cleaning surfaces take into account the smoothness of the movement of beading water droplets in a dircction

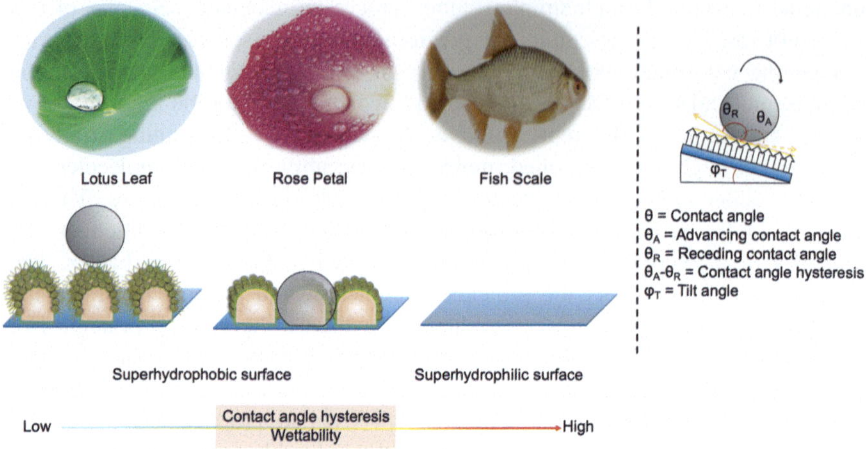

Fig. 4.2 The behavior of water droplet on superhydrophobic surfaces of low and high adhesive properties (lotus leaf and rose petal, respectively) and superhydrophilic surface. The right panel shows the self-cleaning parameters with respect to a water droplet

toward tilting and the degree of tilt surface. The latter is determined by a parameter called contact angle hysteresis (CAH). Theoretically, CAH is a mathematical difference between the advancing (θ_A) and the receding (θ_R) WCA values as shown in Fig. 4.2 [32]. The superiority of a superhydrophobic self-cleaning surface relies on the small values of the tilt angle (the threshold angle between the horizontal plane and the tilted plane) and CAH values of the tilted surface. Ideal superhydrophobicity-based self-cleaning surface will have nearly zero tilt angle and CAH value of $<5°$ [32]. There are other factors that control the self-cleaning property of a superhydrophobic surface, viz., surface energy, surface roughness, surface compositions, and surface adsorption or desorption abilities [33].

4.3 Theories of Superhydrophobic Property-Based Self-Cleaning Phenomena (Lotus Leaf Vs Rose Petal)

Ever since the non-wettable phenomenon of lotus leaf surface came into focus (1970s), research efforts are underway to study and understand the surface mechanism of the superhydrophobic property and its self-cleaning application. The development of various advanced microscopy techniques aided in unveiling the surface morphology of the lotus leaf and its special features of roughness. The self-cleaning effect of lotus leaf is popularly known as "lotus effect," and all the artificially made self-cleaning surfaces are described as mimics of lotus leaf [34]. It is documented after prolonged researches that the lotus effect originated from the hierarchical roughness of the lotus leaf surface, the base layer is composed of randomly grown up micropapillae structure with thickness in the range of 5 to 9 μm, and the upper

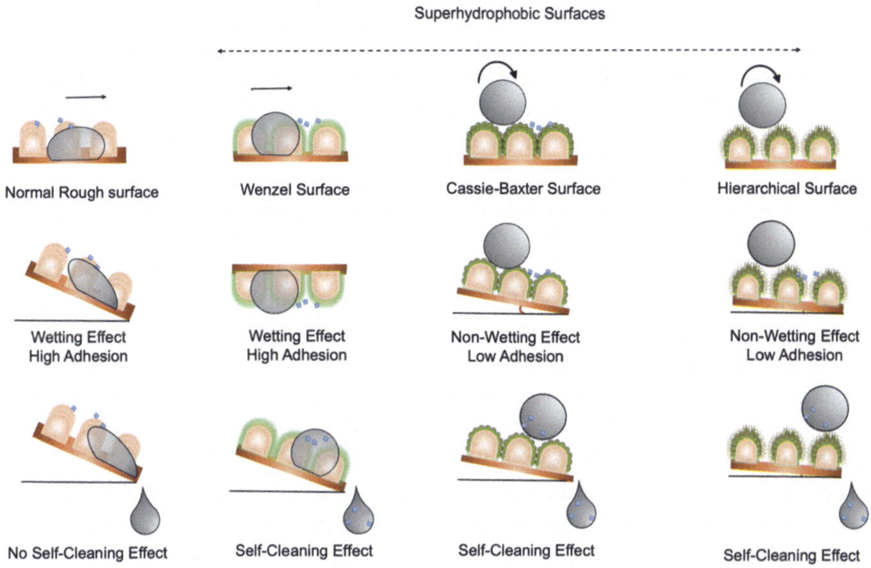

Fig. 4.3 Dynamics of water droplets on normal, Wenzel, Cassie–Baxter, and hierarchical surfaces

layer is made up with finely branched nano-sized wax tubules (diameter ~ 120 nm) placed on the micropapillae structure [35].

Later, researchers had noticed that the superhydrophobic surface of lotus leaf and the rose petal interacts differently with the water droplets. The beading droplets are rolled off on the lotus leaf surface due to low adhesion toward the surface, while beading droplets are pinned on to the rose petal surface due to strong adhesion. The rolling or pinning property of the beading droplets on the superhydrophobic surface is conventionally differentiated by the corresponding CAH values. In general, the low-adhesive superhydrophobic surfaces show lower CAH values (lesser than 5°) and high-adhesive surfaces exhibit higher values of CAH [36]. Detailed investigation on the lotus effect and the rose effect revealed that there is a significant difference in their wetting property, although both are superhydrophobic in nature. The reason for differential wetting property lies in the varying roughness textures of the two surfaces. To understand such differential superhydrophobic surface behavior, two mathematical models have been propounded, the Wenzel model (1936) and the Cassie–Baxter model (1944) (Fig. 4.3). The expressions of the Wenzel model (1) and the Cassie–Baxter model (2) are shown in the following equations [37, 38]:

$$\cos \theta w = R \cos \theta y \tag{1}$$

$$\cos \theta \mathrm{cb} = Fs(\cos \theta y + 1) - 1 \tag{2}$$

where θw and θy denote the apparent and the actual WCA values on the rough (Wenzel surface) and the smooth surface and θcb is the WCA value at Cassie–Baxter surface. The factors R and Fs are two ratio terms where R is the ratio of actual contact area to the projected contact area of the solid–liquid interface and *Fs* is the ratio of the apparent contact angle and the surface fraction of the solid, respectively. The Wenzel model considers that surface hydrophobicity is exclusively controlled by the surface roughness and attain superhydrophobicity for $\theta w >> \theta y$ ($\theta y > 90°$ and $R >> 1$) and a superhydrophilicity for $\theta w << \theta y$ ($\theta y < 90°$, $R >> 1$). The Wenzel model is applicable for homogeneously wettable rough surface where the water remains in contact with the surface as beaded (superhydrophobic) or spread (superhydrophilic) droplets and wets the covering area completely. For superhydrophobic surface obeying the Wenzel theory, the beading droplets are found to have high adhesion to the surface and thus pinned on it (θw is always less than 180°). The best example of the high adhesion Wenzel superhydrophobic surface is rose petal. The lotus effect can be explained by following the Cassie–Baxter model that considers heterogeneous or hierarchical rough surface in place of homogeneous rough surface. This model assumes distinct air layer at the interface of solid (surface) and water layers that hinder the wetting of the surface by beaded droplet which roll off the surface smoothly (θcb equals to 180°). In this case, the surface heterogenicity leads to spherical droplet in order to balance the equilibrium between the interface interactions of solid–air and air–liquid contacts [39]. Although majority of the artificial superhydrophobic surfaces exhibit lotus effect, the utilization of rose effect is found in fabricating anti-adhesive materials.

4.4 Molecular Architectonics-Guided Self-Cleaning Materials

Molecular architectonics guide the fabrication of well-defined nano-, micro-, and macroscale architectures through the assembly of designer molecular building blocks [30–35, 40, 41]. The meticulously designed functional building blocks can undergo self-assembly, co-assembly, or hetero-assembly through judiciously manipulated noncovalent interactions to form hierarchical architectures with desired shapes, sizes, and functionalities. Notably, the molecular architectonics of designer molecular building blocks generate nanoarchitectures with functional properties and applications [17, 18, 42–45]. The potential of molecular architectonics has been exploited to fabricate hierarchical surfaces for diverse applications across the areas of materials science to biomedicine [17, 18, 46]. The scheme of molecular architectonics offers huge scope for fabricating complex self-cleaning architectures with switchable and adjustable hydrophilic/hydrophobic character by modifying and tailoring the functionalities and assembly processes, respectively, of the building units [47]. Fabrication of efficient, mechanically robust, durable, and large-area coating of self-cleaning surfaces is associated with the understanding of the structure–property relationship of the building units to drive the controlled molecular assembly processes as desired. Generally, three steps are followed in the sequence to fabricate self-cleaning materials:

1. The core unit selection of the functional building unit, which is considered to be the important part in the entire process. In this step, suitability of the building unit to participate in the assembly process is ascertained.
2. The selection of appropriate auxiliaries for the functional core to transform it into desired functional building unit. This step offers huge scope to fine-tune the assembly properties of the functional building unit by judiciously conjugating the auxiliaries at proper positions of the core to achieve the control and desired assembly process. This step actually governs the superiority of self-cleaning property of the developed superhydrophobic surface by eliminating the formation of undesired material/surface architectures.
3. Judicious optimization of experimental parameters for the assembly of functional building block such as solvent medium, pH, ionic strength, time of assembly process, temperature, external radiation/stimuli, humidity, and selection of the casting substrate and its compositions is needed. This final step determines the outcome of the material/surface architecture with desired superhydrophobic and self-cleaning activities.

Overall, these three steps in the scheme of molecular architectonics allow the successful large-area fabrication of superhydrophobic and self-cleaning materials in a simple, cost-effective, and solution process (Fig. 4.4).

In this approach, derivatization of functional building units with bioactive or biomimetic auxiliaries can render the superhydrophobic surfaces biocompatible and non-cytotoxic property. Thus far, the self-cleaning materials fabricated, employing the scheme of molecular architectonics, are limited, possibly due to insufficient mechanical strength and durability. Nevertheless, there is enormous scope for the fabrication of robust and large area superhydrophobic films by the design of appropriate functional building block and adopting judicious experimental conditions. In the following section, we describe a few fabrication strategies adopted by the researcher to develop the self-cleaning materials by employing the principles of molecular architectonics.

(a) *Solution processable techniques*. Solution processable techniques refer to all molecular assembly process occurring in the solution phase. The derived molecular architectures can have a wide range of sizes and shapes in the nano-, micro-, and macroscale regimes. The bottom-up molecular assembly occurs through the meticulous manipulation of noncovalent interactions in three categories: (i) self-assembly, (ii) co-assembly or templated assembly, and (iii) hetero-assembly involving one, more than one, or different molecular building blocks, respectively. The molecular assembly process can be tweaked through the structure–property relationship study of the building units of same or different types to fabricate architectures ranging from 0D to 3D. Fabrication of superhydrophobic films or surfaces by molecular architectonics strategy involves the generation of the intended surface architecture through control assembly process that, in turn, renders hierarchy and roughness required to exhibit superhydrophobicity. In this context, self-assembled monolayer (SAM) is a widely used technique to fabricate one molecule thick molecular assembly layer on a chosen substrate either by

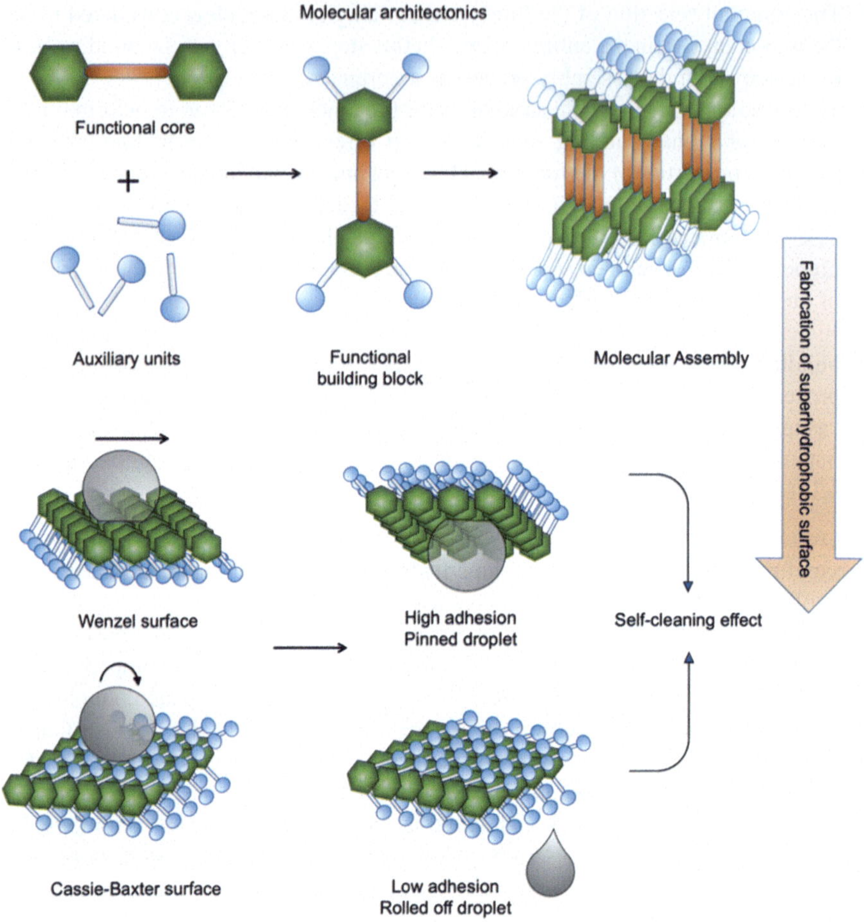

Fig. 4.4 Molecular architectonics and its utilization in fabricating Wenzel and Cassie–Baxter-type superhydrophobic and self-cleaning material surfaces

physisorption or chemisorption methods. To impart rough surface texture of low surface energy, the substrate on which the film to be prepared is coated with organic substance or polymer or metal or metal oxide through SAM technique.

(b) *Layer-by-layer (LBL) assembly technique*. The layer-by-layer (LBL) assembly is another interesting molecular assembly process in which layers of oppositely charged molecular materials are deposited as alternatively layers on a substrate to generate nano-/microscale hierarchical structures. In LBL assembly structure, the oppositely charged layers are electrostatically held together to form a definite architectural surface.

(c) *Breath figure technique (BFT)*. BFT is a popular bio-inspired technique for the fabrication of honeycomb-featured micro-arrays [48, 49]. Casting of polymer solution in volatile solvent on a substrate under humid conditions followed by

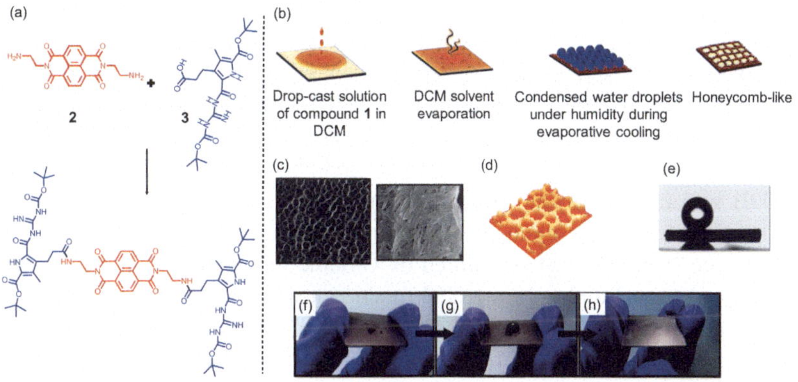

Fig. 4.5 (**a**) Design and preparation of functional building block **1** from the functional core **2** and auxiliary unit **3**. (**b**) Scheme representation shows the generation of hexagonal honeycomb-featured microarrays from the solution of **1** in dichloromethane (DCM) through BFT. (**c-e**) SEM and optical profile images of 3D film of **1** show nanobelt architectures constituting the walls of the microarrays and representative spherical droplet at its rolling state on the surface at a tilt angle of 3°. (**f-g**) Images of a gold-sputtered honeycomb-like microarray surface coated with 1H,1H,2H,2H-perfluorodecanethiol, which exhibit self-cleaning activity (of carbon dust). Reproduced with permission [2, 24]. Copyright,2020, 2012, Wiley-VCH

the rapid evaporation of solvent lead to the formation of honeycomb-like film (Fig. 4.5). In BFT, the well-organized hexagonal array architectures are formed due to the precipitation or assembly of polymer substrate around the uniformly distributed condensed water droplets during the evaporative cooling of volatile solvent. Subsequently, condensed water droplets evaporate leaving behind the hexagonally organized polymeric material. The porous BF film finds many potential applications in diverse fields [50, 51]. Recently, our group reported a fabrication of BF film with superhydrophobic self-cleaning property employing the concept of molecular architectonics [24]. In this unique BFT approach, small functional molecule was used in a simple and cost-effective solution processing to imprint the micro- or nanostructured honeycomb-like arrays.

4.5 Fabrication Superhydrophobic Self-Cleaning Surfaces by Molecular Architectonics

(a) *BFT-based fabrication strategy*. In recent years, BFT has become a popular method to fabricate rough and porous surfaces of typical honeycomb features with varying pore sizes from nanometer to macrometer [52–54]. Although BFT involves simple solution processing, its practice has been virtually restricted to polymers. Use of small molecule in BFT was not successful owing to their nonspecific and undesirable aggregation/crystallization during the evaporation of volatile solvent, which hamper the formation of well-defined surface architectures. Recently, we succeeded in overcoming the challenges to fabricate honeycomb-like hexagonal microstructures of excellent hydrophobicity through the molecular assembly of a designer small functional molecule by adopting BFT. The desired building block was designed and prepared by judiciously utilizing the knowledge of molecular architectonics. Naphthalene diimide (NDI) was chosen as the functional core which possesses intrinsic assembly properties due to its molecular planarity and high degree of π–π stacking. Moreover, NDI core can be tailored by variable functional auxiliaries to render both symmetrical and asymmetrical functional molecules for the fabrication of novel molecular and material architectures of different sizes and shapes [17, 55–57]. Because of the characteristic absorption, emission, and redox and semiconducting properties, micro-/nano-architectures of NDI derivatives have been exploited for optical, electronic, and biomedical applications [58, 59]. In this study, NDI was symmetrically functionalized with guanidinocarbonyl pyrrole (GCP) moieties to obtain the final building unit **1** (Fig. 4.5a) [24]. The solution of functional building block **1** in low-boiling chlorinated solvent dichloromethane (DCM) was drop casted on a silicon substrate. Evaporation of DCM leads to cooling effect and condensation of moisture into water droplets. DCM evaporation and enhanced hydrophobic effect in contact with condensed water droplets triggered the assembly **1** into nanobelt-like architectures around the water droplets (Fig. 4.5b). As the water droplets evaporate, the nanobelts stacked to form the sidewalls of the hexagonal well and thus produce closed-packed hierarchical rough microarrays over large area (Fig. 4.5c). The average pore (well) sizes of the microarrays were found to be 10–25 μm in diameter (internal), and pore wall thickness was approximately 10–15 μm (topographical) (Fig. 4.5d). However, a proportional alteration in the pore size was observed with the varying concentration of compound **1** as well as the relative humidity (RH). It was observed that the well diameter of the microarrays decreases with the increase in RH of the system and the pore walls became thicker at the higher concentrations. Overall, both the factors increased the degree of roughness of the resultant BF film. The surface hydrophobicity of the film significantly raised with increase in the WCA values from 63° to 134°. Further, BT film was coated with gold sputtering and independently functionalized with 11-mercaptoundecanoic acid (MSH), 1H,1H,2H,1-dodecanethiol (DSH), and 2H-perfluorodecanethiol (FSH). The honeycomb array surfaces of **1** with monolayer films of DSH and FSH showed superhydrophobic property corresponding to WCA values of 150° and 156°, respectively. We also demonstrated that the FSH-treated honeycomb array surface with high superhydrophobicity (156°) showed self-cleaning ability by complete removal of dust particles, as the water droplets roll off carrying the dust particles. The excellent self-cleaning property of the surface was further supported by the tilt angle and the CAH value of 3° and <1°, respectively (Fig. 4.5e-h). The emergence of BF microarrays of compound **1** in DCM solvent system could be attributed to the extensive intermolecular hydrogen bonding within the GCP moieties and also to the possible halogen-bonding interactions between the carbonyl functionalities of NDI core and the chlorinated DCM solvent molecules [24]. For the first time, we used BFT of small functional molecule to fabricate superhydrophobic and self-cleaning materials and surfaces. Besides the self-cleaning application, this molecular architectonics-based fabrication approach can be employed to fabricate decorated and colored hydrophobic surface formed through the noncovalent incorporation of the core with functional dyes like rhodamine B and perylene diimide. These decorated surfaces may find applications in advanced (bio)electronics and miniaturized bioanalytical separation techniques and assays.

(b) *Fabrication of small molecule assembly-based superhydrophobic surface.* The spontaneous and unique self-assembly propensity of biomolecules can play a key role in producing surface roughness with low surface energy. Among the biomolecules, the self-assembly property of short peptides has been exploited for the generation of diverse nano- and microstructures through various reductionistic approaches. In the scheme of molecular architectonics, amino acids and short peptides serve as auxiliaries to guide the assembly of core organic functional molecules. Depending on the number and the position of conjugated amino acids/peptides, the properties of the final assembly architectures in terms of size, shape, and chiroptical properties can be fine-tuned. Thus, peptide-based assembly can be considered as one of the reductionistic approaches for the construction of superhydrophobic materials. Diphenylalanine (Phe-Phe) is one such versatile dipeptide, and its self-assembly properties have been extensively studied. The self-assembly of Phe-Phe can be fine-tuned into all possible architectural dimensions from 0D nanoparticles to 3D nanofibrils, nanotubes, or nanogels under specific experimental conditions [60]. Lee et al. reported a strategy to fabricate thin film of superhydrophobic nature by solid-phase self-assembly process of Phe-Phe (**4**) [61]. When a drop of solution **4** in hexafluoroisopropanol (HFIP) dried under anhydrous condition, an amorphous thin peptide film was formed which on exposure to the vapors of aromatic amine solvents such as aniline and pentafluoroaniline at high temperature led to the growth of vertically aligned 2D nanowires with an average diameter of ~200 nm. The transformation of the peptide film from amorphous to well-aligned nanowire morphology introduced surface roughness under the treatment of amine vapors. The electron microscopy images indicated the formation of two distinct states (Fig. 4.6a-c). The aromatic amine treatment of rough peptide film improved the surface hydrophobicity with surface roughness. The hydrophobicity of the peptide film enhanced and became more prominent by the treatment with pentafluoroaniline vapor as compared to aniline. The contact angle measurement of the treated and the untreated peptide film revealed that the film formed under the pentafluoroaniline vapor exhibited superhydrophobic property with a WCA value of ~151°. Although the surface hydrophobicity of the aniline treated film was increased significantly in comparison to the untreated amorphous peptide film (WCA = ~37°), it was unable to show superhydrophobicity (WCA = ~128°) (Fig. 4.6d-f). To demonstrate the self-cleaning activity of the superhydrophobic pentafluoroaniline-treated peptide film, the vertically oriented nanowires were

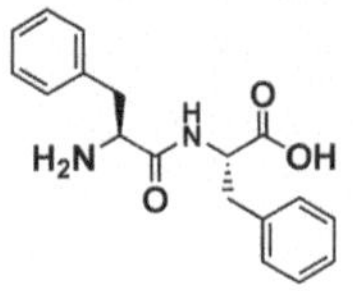

4

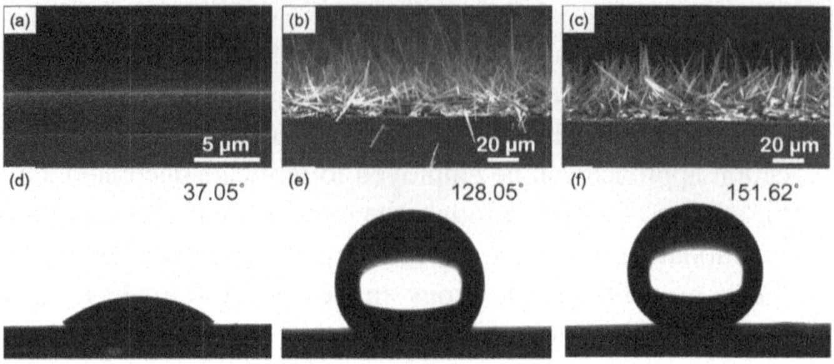

Fig. 4.6 (**a**) SEM image of the self-assembled films of diphenylalanine (Phe-Phe or FF) **4** before and after (**b**) aniline and (**c**) pentafluoroaniline treatment. (**d-f**) show the respective WCA values. Reproduced with permission [61]. Copyright 2009, Royal Society of Chemistry

further organized into hierarchical nano-/microstructures by the application of solvent evaporation-induced capillary force technique, and the resulting film showed a tilt angle value of <5°. The change in the morphology from the amorphous flat film to vertically grown nanowires was attributed to the solvent-assisted nucleation and further aggregation of the dipeptide molecules into nanowire structures under high temperature conditions. The authors have claimed that their approach to fabricate peptide self-assembly-based superhydrophobic materials may find diverse applications.

Apart from the aromatic dipeptide, synthetic aromatic molecules can form functional and well-defined nano-/microarchitectures. Liu and co-workers described hexabenzocoronene (HBC)-based strategy to fabricate superhydrophobic surface with self-cleaning property. A set of hexa-peri-hexabenzocoronene (HBC) derivatives (**5–7**) of different symmetry were prepared and shown to exhibit self-assembly property. HBC derivatives were subjected to molecular assembly at 200 °C, and their anti-corrosion activity on iron surface was studied [62]. Upon heating at 200 °C, HBC **5** was self-assembled and transformed into discotic columnar liquid crystal conformation (disc-shaped organization). Such transformation was followed by change in the surface morphology from fiber-like species with porous architectures to graphene-like architectures. Concurrently, the reorientation of the morphological features was also observed. In particular, HBC **5** displayed an end-on alignment of the fiber-like species at room temperature and face-on alignment of the graphene-like architecture generated at high temperature (Fig. 4.7a and c). Interestingly, the graphene-like architecture of HBC **5** exhibited high apparent WCA value (143.9°) with superior hydrophobicity (Fig. 4.7b and d). The hydrophobic graphene architecture of HBC **5** showed an outstanding anti-corrosion property which was far superior than other HBC derivatives (**6** and **7**) under similar treatment and conditions. The graphene architecture with enhanced hydrophobicity was formed due to high symmetry of the HBC **5** structure that facilitated heat-induced self-assembly process. The lack of similar hydrophobic property by the self-assembled structures of HBCs **6** and **7** is

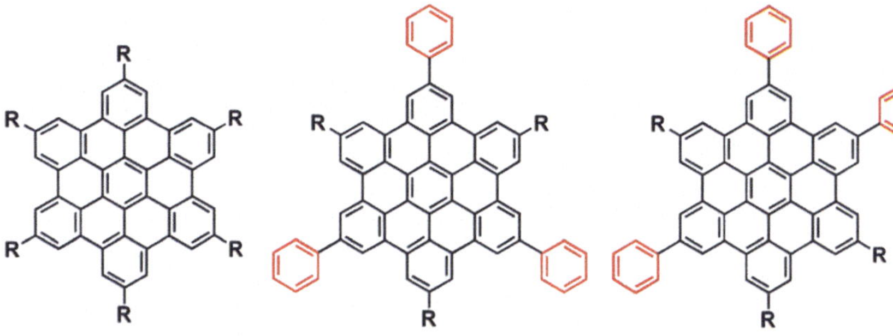

5 6 7

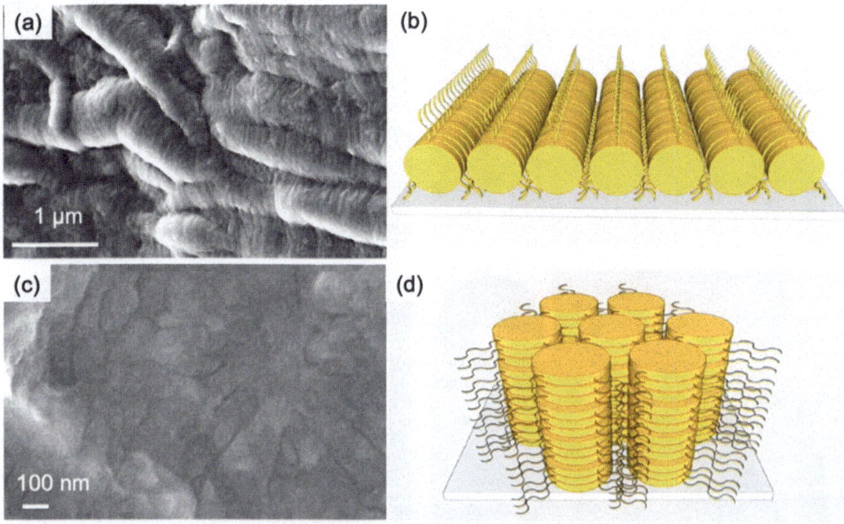

Fig. 4.7 Molecular structures of hexabenzocoronene (HBC) derivatives **5–7** (R = $C_{12}H_{25}$). SEM images of fiber-like assembly architecture of **5** (**a**) at room temperature and (**c**) after heating at 200 °C. (**b, d**) indicate "end-on" and "face-on" alignment of **5** in the assembled architectures before and after heating, respectively. Reproduced with permission [62]. Copyright 2018, Elsevier

possibly attributed to the low symmetry of their molecular structures. By appropriate fine-tuning of the molecular structure and the assembly conditions, HBC-based superhydrophobic self-cleaning coatings can be achieved.

(c) *Rose petal-like superhydrophobic and self-cleaning surfaces.* Rose petal-like self-cleaning surfaces are considerably rare in comparison to lotus leaf-like surfaces due to their limited applications. As expected, strategies for the fabrication of such surfaces by molecular architectonics are rare. In a recent report, Gao et al. described a synthetic strategy to fabricate superhydrophobic adhesion architecture using the molecular assembly of Y-shaped compound **8** from mixed solvent system. The compound **8** comprised of azobenzene groups linked to the

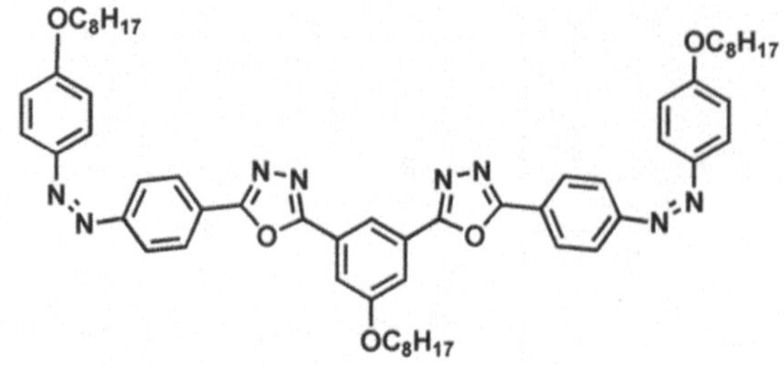

8

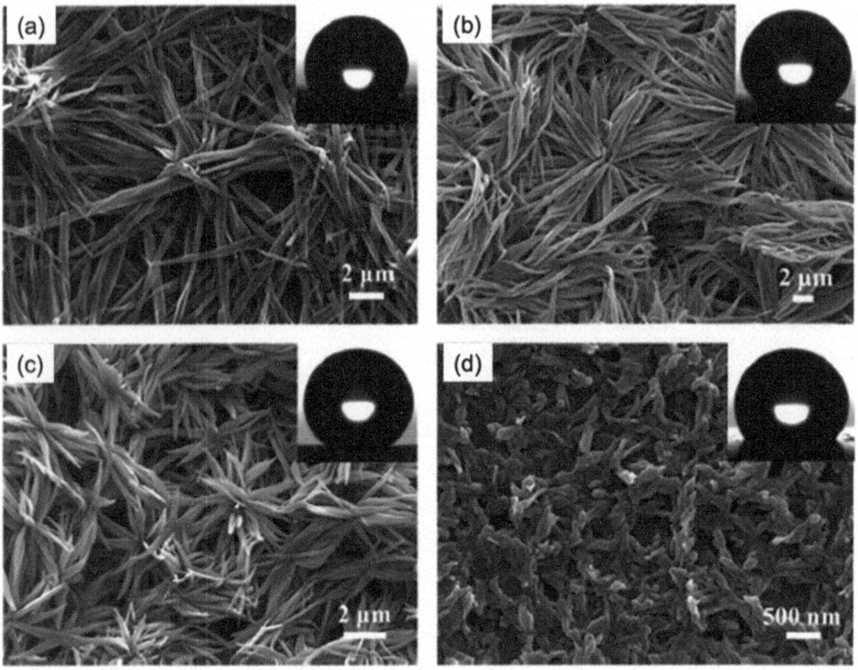

Fig. 4.8 SEM images of self-assembled microarchitectures of **8** in different volume ratios of the mixed solvent ($CHCl_3$:CH_3CN) system, (**a**) 90:10, (**b**) 50:50, (**c**) 40:60, and (**d**) 10:90. Reproduced with permission [63]. Copyright 2017, Royal Society of Chemistry

core phenolic ring at 3,5-positions through 1,3,4-oxadiazole linkers. Compound **8** is functionalized with three octyl chains as shown in Fig. 4.8 [63]. The azo compound **8** undergoes self-assembly in nonpolar acetonitrile-chloroform (CH_3CN/$CHCl_3$)-mixed solvent system. A careful modulation of the self-assembly of **8** by varying the ratio of the solvent composition of CH_3CN/$CHCl_3$ resulted in surface architectures with differential surface wettability and adhesion property. Specifically, self-assembly of **8** in 10% CH_3CN/$CHCl_3$ system formed elongated and flexible root-like fibrillar architectures of ~540 nm average thickness (Fig. 4.8a). In 50% and 60% volume ratios of

$CH_3CN/CHCl_3$, the fibrillar morphology changed to dense flower-like nanowire architectures of ~360 nm thickness. The nanoarchitectures formed in 50–60% $CH_3CN/CHCl_3$ system exhibit hierarchical roughness and contained abundant air trapping grooves. These trapped air between the interface of rough nano-structured surface and water droplets rendered it superhydrophobic with WCA value of 151.2° and sliding angle of 90° (Fig. 4.8b and c). Thus, fabricated superhydrophobic surface exhibited strong water adhesion property. The adhesion property of superhydrophobic surface also showed a dependency with the concentration of **8**. Further, a distinct dust-cleaning property of the superhydrophobic surface was observed with CAH value of less than 5°. However, at higher ratios of $CH_3CN/CHCl_3$ (more than 90%) system, the floral morphology of the superhydrophobic surface completely disappeared, and stick-like morphology was observed with comparatively low surface wettability (Fig. 4.8d). The authors have demonstrated that the superhydrophobic surface formed at 50% $CH_3CN/CHCl_3$ system was chemically stable under variable pH conditions (1–14) of the medium and thus can be utilized for many practical applications such as self-cleaning, selective transportation of droplets, and drop-based microdetection or microreaction.

4.6 Conclusions and Outlook

In conclusion, we have discussed the fabrication of superhydrophobic and self-cleaning materials under the scheme of molecular architectonics. The discussion focused the importance of programmed molecular assembly of consciously designed small functional building blocks to generate surface architectures with hierarchical roughness and low surface energy required for the superhydrophobic and self-cleaning surfaces. Thus, deeper understanding on the principles of molecular architectonics widens the scope of creating highly sophisticated self-cleaning materials.

Through the specific and selected examples, we have shown the advantage of custom design and engineering of the molecular assemblies of the functional building blocks meticulously synthesized by the conjugation of functional core and assembly-directing auxiliaries under subtle noncovalent interactions. Fine manipulation in the experimental parameters also guides the fabrication of the materials and surfaces with superhydrophobic property and self-cleaning effect under controlled noncovalent pathways. We believe that the fabrication of large area material and surface architectures with superhydrophobic property through simple, robust, and cost-effective solution processing by the scheme of molecular architectonics holds a great promise to deliver advanced self-cleaning technologies and decorative materials with (bio)electronics and bioanalytical assay applications.

The performance of multiscale architectures produced through the molecular architectonics scheme develops understanding on the structure–property relationship of the building components, viz., functional core and the auxiliary. For example, derivatization of the core structural units (naphthalene diimides, perylene diimides, pyrenes, anthracene, etc.) with appropriate auxiliaries obtained by thc minute structural mutation determines the outcome of the materials/surface architecture and the hydrophobic/hydrophilic/amphiphilic property. In this context, small biologically active auxiliaries can play a major role in imparting their inherent molecular recognition properties to guide the engineering of the molecular assemblies of functional core into hierarchical rough materials with superhydrophobic and self-cleaning properties. Similarly, meticulous manipulation of experimental parameters, viz., concentrations of the building units, solvent polarity, pH, temperatures, and other external additives or stimuli, also play key role in controlling the engineering of the molecular assemblies of the custom-designed molecules for the generation of functional superhydrophobic materials and surfaces. Suitably fabricated superhydrophobic materials and surfaces by employing the scheme of molecular architectonics can be supplemented with advanced catalytic, photocatalytic, optical, (bio)electronic, and (bio)sensing properties to cater to broad range of multipurpose applications.

References

1. Bhushan B, Jung YC (2011) Natural and biomimetic artificial surfaces for superhydrophobicity, self-cleaning, low adhesion, and drag reduction. Prog Mater Sci 56:1–108
2. Konar M, Roy B, Govindaraju T (2020) Molecular architectonics-guided fabrication of superhydrophobic and self-cleaning materials. Adv Mater Interfaces 7:2000246
3. Sarkar DK, Farzaneh M (2009) Superhydrophobic coatings with reduced ice adhesion. J Adhes Sci Technol 23:1215–1237
4. Wang J, Chen H, Sui T, Li A, Chen D (2009) Investigation on hydrophobicity of lotus leaf: experiment and theory. Plant Sci 176:687–695
5. Su B, Tian Y, Jiang L (2016) Bioinspired interfaces with superwettability: from materials to chemistry. J Am Chem Soc 138:1727–1748
6. Liu Y, Chen X, Xin JH (2008) Hydrophobic duck feathers and their simulation on textile substrates for water repellent treatment. Bioinspir Biomim 3:046007
7. Bhushan B (2009) Biomimetics: lessons from nature–an overview. Philos Trans A Math Phys Eng Sci 367:1445–1486
8. Howarter JA, Youngblood JP (2007) Self-cleaning and anti-fog surfaces via stimuli-responsive polymer brushes. Adv Mater 19:3838–3843
9. Liu K, Yao X, Jiang L (2010) Recent developments in bio-inspired special wettability. Chem Soc Rev 39:3240–3255
10. Liu M, Jiang L (2010) Switchable adhesion on liquid/solid interfaces. Adv Funct Mater 20:3753–3764
11. Long J, Fan P, Gong D, Jiang D, Zhang H, Li L, Zhong M (2015) Superhydrophobic surfaces fabricated by femtosecond laser with tunable water adhesion: from lotus leaf to rose petal. ACS Appl Mater Interfaces 7:9858–9865
12. Ju J, Xiao K, Yao X, Bai H, Jiang L (2013) Bioinspired conical copper wire with gradient wettability for continuous and efficient fog collection. Adv Mater 25:5937–5942
13. Avinash MB, Govindaraju T (2014) Nanoarchitectonics of biomolecular assemblies for functional applications. Nanoscale 6:13348–13369
14. Avinash MB, Samanta PK, Sandeepa KV, Pati SK, Govindaraju T (2013) Molecular architectonics of stereochemically constrained π-complementary functional modules. Eur J Org Chem 2013:5838–5847

15. Govindaraju T, Avinash MB (2012) Two-dimensional nanoarchitectonics: organic and hybrid materials. Nanoscale 4:6102–6117
16. Avinash MB, Swathi K, Narayan KS, Govindaraju T (2016) Molecular architectonics of naphthalenediimides for efficient structure–property correlation. ACS Appl Mater Interfaces 8:8678–8685
17. Avinash MB, Govindaraju T (2018) Architectonics: design of molecular architecture for functional applications. Acc Chem Res 51:414–426
18. Roy B, Govindaraju T (2019) Amino acids and peptides as functional components in arylenediimide-based molecular architectonics. Bull Chem Soc Jpn 92:1883–1901
19. Manchineella S, Govindaraju T (2017) Molecular self-assembly of cyclic dipeptide derivatives and their applications. ChemPlusChem 82:88–106
20. Watson GS, Cribb BW, Watson JA (2010) How micro/nanoarchitecture facilitates anti-wetting: an elegant hierarchical design on the termite wing. ACS Nano 4:129–136
21. Govindaraju T (ed) (2019) Templated DNA nanotechnology: functional DNA nanoarchitectonics, 1st edn. Jenny Stanford Publishing
22. Pandeeswar M, Senanayak SP, Govindaraju T (2016) Nanoarchitectonics of small molecule and DNA for ultrasensitive detection of mercury. ACS Appl Mater Interfaces 8:30362–30371
23. Manchineella S, Thrivikraman G, Khanum KK, Ramamurthy PC, Basu B, Govindaraju T (2016) Pigmented silk nanofibrous composite for skeletal muscle tissue engineering. Adv Healthcare Mater 5:1222–1232
24. Avinash MB, Verheggen E, Schmuck C, Govindaraju T (2012) Self-cleaning functional molecular materials. Angew Chem Int Ed 51:10324–10328
25. Koch K, Barthlott W (2009) Superhydrophobic and superhydrophilic plant surfaces: an inspiration for biomimetic materials. Philos Trans A Math Phys Eng Sci 367:1487–1509
26. Cassie ABD (1948) Contact angles. Discuss Faraday Soc 3:11–16
27. Ming W, Wu D, van Benthem R, de With G (2005) Superhydrophobic films from Raspberry-like particles. Nano Lett 5:2298–2301
28. Fürstner R, Barthlott W, Neinhuis C, Walzel P (2005) Wetting and self-cleaning properties of artificial superhydrophobic surfaces. Langmuir 21:956–961
29. Drelich J, Chibowski E, Meng DD, Terpilowski K (2011) Hydrophilic and superhydrophilic surfaces and materials. Soft Matter 7:9804–9828
30. Lim HS, Lee SG, Lee DH, Lee DY, Lee S, Cho K (2008) Superhydrophobic to superhydrophilic wetting transition with programmable ion-pairing interaction. Adv Mater 20:4438–4441
31. Xia F, Feng L, Wang S, Sun T, Song W, Jiang W, Jiang L (2006) Dual-responsive surfaces that switch between superhydrophilicity and superhydrophobicity. Adv Mater 18:432–436
32. Johnson RE, Dettre RH (1964) Contact angle hysteresis. Iii. Study of an idealized heterogeneous surface. J Phys Chem 68:1744–1750
33. Woodward JT, Gwin H, Schwartz DK (2000) Contact angles on surfaces with mesoscopic chemical heterogeneity. Langmuir 16:2957–2961
34. Barthlott W, Neinhuis C (1997) Purity of the sacred lotus, or escape from contamination in biological surfaces. Planta 202:1–8
35. Lafuma A, Quéré D (2003) Superhydrophobic states. Nat Mater 2:457–460
36. Bhushan B, Nosonovsky M (2010) The rose petal effect and the modes of superhydrophobicity. Philos Trans A Math Phys Eng Sci 368:4713–4728
37. Wenzel RN (1936) Resistance of solid surfaces to wetting by water. Ind Eng Chem 28:988–994
38. Cassie ABD, Baxter S (1944) Wettability of porous surfaces. Trans Faraday Soc 40:546–551
39. Baxter S, Cassie ABD (1945) 8—The water repellency of fabrics and a new water repellency test. J Text Inst 36:T67–T90
40. Pandeeswar M, Khare H, Ramakumar S, Govindaraju T (2015) Crystallographic insight-guided nanoarchitectonics and conductivity modulation of an n-type organic semiconductor through peptide conjugation. Chem Commun 51:8315–8318
41. Pandeeswar M, Khare H, Ramakumar S, Govindaraju T (2014) Biomimetic molecular organization of naphthalene diimide in the solid state: tunable (chiro-) optical, viscoelastic and nanoscale properties. RSC Adv 4:20154–20163

42. Ariga K, Lee MV, Mori T, Yu X-Y, Hill JP (2010) Two-dimensional nanoarchitectonics based on self-assembly. Adv Colloid Interf Sci 154:20–29
43. Ariga K, Leong DT, Mori T (2018) Nanoarchitectonics for hybrid and related materials for bio-oriented applications. Adv Funct Mater 28:1702905
44. Aono M, Ariga K (2016) The way to nanoarchitectonics and the way of nanoarchitectonics. Adv Mater 28:989–992
45. Ariga K, Nishikawa M, Mori T, Takeya J, Shrestha LK, Hill JP (2019) Self-assembly as a key player for materials nanoarchitectonics. Sci Technol Adv Mater 20:51–95
46. Avinash MB, Raut D, Mishra MK, Ramamurty U, Govindaraju T (2015) Bioinspired reductionistic peptide engineering for exceptional mechanical properties. Sci Rep 5:16070
47. Ganesh VA, Raut HK, Nair AS, Ramakrishna S (2011) A review on self-cleaning coatings. J Mater Chem 21:16304–16322
48. Widawski G, Rawiso M, François B (1994) Self-organized honeycomb morphology of star-polymer polystyrene films. Nature 369:387–389
49. Bunz UHF (2006) Breath figures as a dynamic templating method for polymers and nanomaterials. Adv Mater 18:973–989
50. Chiu Y-C, Kuo C-C, Lin C-J, Chen W-C (2011) Highly ordered luminescent microporous films prepared from crystalline conjugated rod-coil diblock copolymers of Pf-B-Psa and their superhydrophobic characteristics. Soft Matter 7:9350–9358
51. Zou J, Chen H, Chunder A, Yu Y, Huo Q, Zhai L (2008) Preparation of a superhydrophobic and conductive nanocomposite coating from a carbon-nanotube-conjugated block copolymer dispersion. Adv Mater 20:3337–3341
52. Zhang A, Bai H, Li L (2015) Breath figure: a nature-inspired preparation method for ordered porous films. Chem Rev 115:9801–9868
53. Ma H, Hao J (2011) Ordered patterns and structures via interfacial self-assembly: superlattices, honeycomb structures and coffee rings. Chem Soc Rev 40:5457–5471
54. Bai H, Du C, Zhang A, Li L (2013) Breath figure arrays: unconventional fabrications, functionalizations, and applications. Angew Chem Int Ed 52:12240–12255
55. Manchineella S, Prathyusha V, Priyakumar UD, Govindaraju T (2013) Solvent-induced helical assembly and reversible chiroptical switching of chiral cyclic-dipeptide-functionalized naphthalenediimides. Chem Eur J 19:16615–16624
56. Pandeeswar M, Govindaraju T (2013) Green-fluorescent naphthalene diimide: conducting layered hierarchical 2d nanosheets and reversible probe for detection of aromatic solvents. RSC Adv 3:11459–11462
57. Avinash MB, Govindaraju T (2012) Amino acid derivatized arylenediimides: a versatile modular approach for functional molecular materials. Adv Mater 24:3905–3922
58. Pandeeswar M, Senanayak SP, Narayan KS, Govindaraju T (2016) Multi-stimuli-responsive charge-transfer hydrogel for room-temperature organic ferroelectric thin-film devices. J Am Chem Soc 138:8259–8268
59. Avinash MB, Govindaraju T (2011) Engineering molecular organization of naphthalenediimides: large nanosheets with metallic conductivity and attoliter containers. Adv Funct Mater 21:3875–3882
60. Adler-Abramovich L, Gazit E (2014) The physical properties of supramolecular peptide assemblies: from building block association to technological applications. Chem Soc Rev 43:6881–6893
61. Lee JS, Ryu J, Park CB (2009) Bio-inspired fabrication of superhydrophobic surfaces through peptide self-assembly. Soft Matter 5:2717–2720
62. Liu Y-C, Huang W-J, Wu S-H, Lee M, Yeh J-M, Chen H-H (2018) Excellent superhydrophobic surface and anti-corrosion performance by nanostructure of discotic columnar liquid crystals. Corros Sci 138:1–7
63. Gao Q, He L, Li Y, Ran X, Guo L (2017) Controllable wettability and adhesion of superhydrophobic self-assembled surfaces based on a novel azobenzene derivative. RSC Adv 7:50403–50409

Chapter 5
Functional Discotic Liquid Crystals Through Molecular Self-Assembly: Toward Efficient Charge Transport Systems

Indu Bala, Joydip De, and Santanu Kumar Pal

5.1 Introduction

The development of sustainable energy resources to solve the increasing global energy crisis depends on the discovery of new economical and greener materials. In this respect, organic semiconductor materials provide opportunities to reduce production cost so that the economical electronic devices can be brought into the market [1]. Moreover, when compared with inorganic semiconductors, organic molecules offer the advantages of low-cost synthesis, inexpensive fabrication, and flexible and lightweight electronic products [1, 2]. Solubility remained one of the important issues for chemicals in industries. Organic materials can be easily deposited on thin films by cheaper solution-processable technique, unlike their inorganic counterparts. The formation of defect-free uniform thin films without the pinholes and grain boundaries is good for improving device efficiency and durability. In addition to processing sensitivity of thin-film morphology, the ability of materials to grow and nucleate in a constructive manner also plays a pivotal role and can be controlled by molecular design. Therefore, the molecular architecture, self-organizing features, and lastly the solution processability of the materials are important parameters for their efficacious applicability in electronic devices. In this regard, one class of materials is liquid crystals (LCs), in particular, discotic LCs (DLCs) [3–5], known to self-organize into columnar architecture due to supramolecular interactions among the central aromatic core. The self-assembly mechanism of LCs is a very striking feature, and large monodomain samples can be produced in a way that

Indu Bala and Joydip De contributed equally with all other contributors.

I. Bala · J. De · S. K. Pal (✉)
Department of Chemical Sciences, Indian Institute of Science Education and Research (IISER) Mohali, Mohali, India
e-mail: skpal@iisermohali.ac.in

T. Govindaraju, K. Ariga (eds.), *Molecular Architectonics and Nanoarchitectonics*, Nanostructure Science and Technology,
https://doi.org/10.1007/978-981-16-4189-3_5

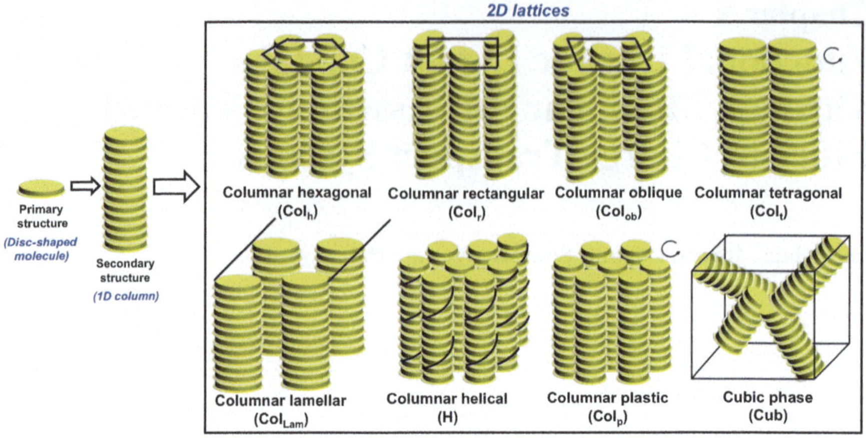

Fig. 5.1 Stacking of disc-shaped molecules into one-dimensional (1D) columns which can laterally self-organized into various two-dimensional (2D) lattices. (Redrawn from Ref. [5])

is otherwise not feasible for crystalline materials, and the LC nature of such films indicates the self-healing of defects by thermal annealing. Hence, the increasing interest in DLC materials arises from their inherent properties such as anisotropy in conduction, orientational control of the molecular order, ease of processability, and the self-healing of structural defects [3–6]. Most importantly, all these aspects can be tailored by careful molecular engineering, which is difficult in the case of single crystals or conjugated polymers.

DLCs, discovered by S. Chandrasekhar and co-workers in 1977, are known to form by the disc-shaped molecules consisting of a central aromatic core that is substituted by three or more flexible alkyl chains. DLCs usually form columnar phases by the self-assembly of disc-like molecules on top of each other decorated with peripheral alkyl chains and are the latest candidates in the field of organic electronics [3–5]. Based on the symmetry of the arrangement of columns, the degree of ordering within the column, and the orientation of the discs with respect to columnar axis, they are classified into a variety of phases, viz., columnar hexagonal (Col_h), columnar rectangular (Col_r), columnar oblique (Col_{ob}), columnar tetragonal (Col_t), columnar lamellar (Col_{Lam}), helical (H), columnar plastic (Col_p), and cubic (Cub) (Fig. 5.1) [5]. The Col_h and Col_r are the two most common mesophases exhibited by the disc-like molecules, while the other Col_{ob}, Col_t, Col_{Lam}, H, Col_p, and Cub phases are not so common. After the discovery of DLCs where the simplest benzene was used as a discotic core, many scientists started to explore larger aromatic cores to develop new discotic mesogens with the aim of increasing the orbital overlap. The increase in the orbital overlap is one of the prerequisites to achieve higher charge carrier mobilities which in turn critically determine the performance of organic semiconductor devices. In 1982, Simon and his group discussed the first idea of employing the DLCs as 1D conductors [7]. Then in 1989, Warman reported the first example of charge transport studies in columnar

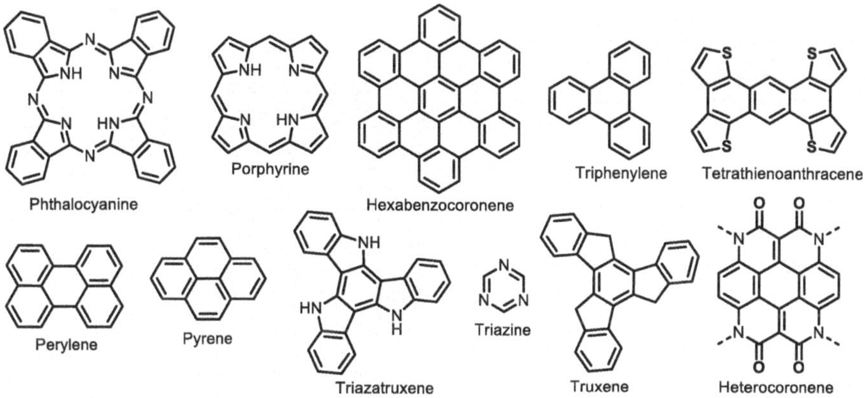

Fig. 5.2 Molecular structures of various discotic cores utilized to achieve DLC materials with efficient charge transport behavior

octa-*n*-alkoxy-substituted phthalocyanines by pulse-radiolysis time-resolved microwave conductivity (PR-TRMC) method [8]. In 1993, Adam et al. studied the charge transport in hexapentyloxytriphenylene by time-of-flight (TOF) method [9]. The same group reported a valuable example consisting of hexahexylthiotriphenylene and observed a high hole mobility of 0.1 $cm^2V^{-1}s^{-1}$ in the highly ordered H phase of DLCs [10]. After that, many groups initiated the charge transport measurements in DLC materials made up of several other different discotic cores by various techniques (vide infra) [11]. Figure 5.2 shows the various discotic cores that have been employed to develop functional DLCs with efficient charge transport properties.

5.2 Charge Transport in DLCs

In columnar phases of DLCs, the π-conjugated aromatic cores stack along the columnar direction where the stacking distance between two adjacent cores generally varies from ~3.4 to 3.7 Å (Fig. 5.3). The aromatic cores are responsible for the highest occupied molecular orbital (HOMO) and lowest occupied molecular orbital (LUMO) π-orbital overlap. Due to the finite dimensions of the conjugated DLC systems, the HOMO and LUMO energy levels are separated with the large energy gap. Thus, no intrinsic charge exists in the DLC materials. However, charges can be created in these systems by various means such as by injecting the charges, optical stimulation, or chemically doping the molecules with oxidizing or reducing species. In columnar DLCs, the charge carriers transport freely along the stacked direction while facing hindrance when moving along the columnar plane due to the presence of insulating alkyl chains. For instance, Arikainen and Boden's research groups studied the anisotropic nature of electron transport in triphenylene derivatives that showed the conductivity along the columnar direction is 1000 times greater than

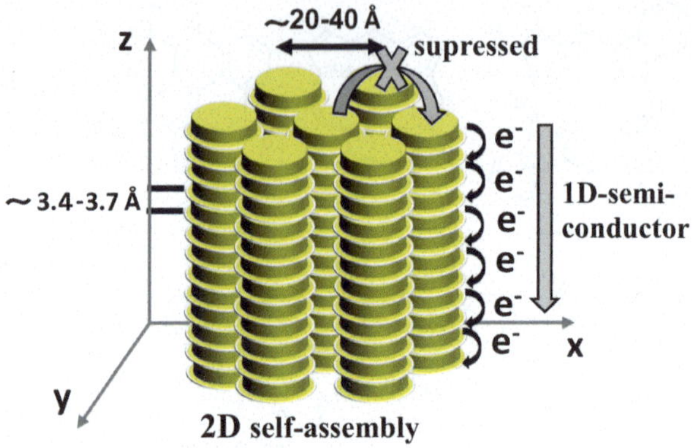

Fig. 5.3 2D columnar assembly of DLCs showing quasi-1D charge transport along the columnar stack

along the columnar plane [6, 12]. Thereby, columnar DLCs are also known as quasi-1D organic semiconductors (Fig. 5.3) [3–5].

Charge transport in DLCs can be understood by the phonon-assisted hopping formalism where the charges hop from one molecule to another with the rate described by semiclassical Marcus theory [13]. The more is the hopping rate, the faster will be the movement of charges that will ultimately lead to high mobility. As per the Marcus theory, the main parameters that govern the charge transfer rate, k_{et} (between the same molecules), are reorganization energy, λ, and the transfer integral, J, and are related to each other by the following equation [13]:

$$k_{et} = \frac{J^2}{\hbar}\sqrt{\frac{\pi}{\lambda k_B T}}\exp - \left[\frac{\lambda}{4k_B T}\right]$$

where $\hbar = \frac{h}{2\pi}$ and h is Planck's constant, k_B is Boltzmann's constant, and T is the temperature.

The transfer integral (J) defines the probability of electron transfer between the neighboring molecules (along the stack in columnar DLCs) and is directly proportional to the extent of molecular orbital overlap, thereby signifying the strength of interaction between the neighboring molecules. The orbital overlap not only depends upon the intermolecular distance d but also varies strongly with the relative orientation of the neighboring molecules. The larger value of J indicates higher mobility values. On the other hand, the reorganization energy λ does not depend upon the relative positions and orientation of the neighboring molecules. The reorganization energy λ reflects the occurrence of geometric changes to accommodate the charges (i.e., upon electron removal for hole mobility and electron addition for electron mobility). The smaller the value of λ, the higher the mobility value.

The efficiency of semiconducting material is measured in terms of charge carrier mobility parameter. For measuring the charge carrier mobility values of the DLC materials, various techniques, viz., PR-TRMC, TOF, space-charge-limited current (SCLC), and field-effect transistor (FET) method, have been employed in the literature [11, 14]. All these techniques demand different criteria to extract charge carrier mobility values. Among these techniques, TOF is the most widely used method due to its simple approximations usually taken into account for the calculation of the mobility values. This technique assumes material to be trap-free, and the charge carriers move under the constant electric field until it reaches to the opposite electrode and results in the production of current in the external circuit connected to the sample. The mobility of the sample extracted by TOF technique (μ_{TOF}) can be affected by many factors, e.g., if the sample under testing does not properly absorb the incident laser light, the exciton pair cannot be generated efficiently. In that case, the sample can be studied via charge injection techniques such as SCLC and FET. In SCLC technique, the mobility values (μ_{SCLC}) are generally affected if the charge injection is not proper (i.e., the charge injection barrier is more). Both SCLC and TOF mobilities of anisotropic DLC materials are highly dependent on the alignment of the molecules in the mobility measured cells. Thereby, the mobilities extracted from the SCLC and TOF techniques are the bulk mobilities of the sample that can be further improved by the proper alignment of the molecules. The homeotropic (or face-on) alignment improves the SCLC/TOF mobility values in comparison to the unaligned samples. In FET technique, the planar alignment (or edge-on) of discotic columnar molecules provides an efficient conducting channel for charge transport. However, in the PR-TRMC technique, the mobility values do not depend upon the alignment; therefore, this technique measures inherent mobilities of the material. The mobilities usually obtained by PR-TRMC are the upper mobility limits of the materials (i.e., $\Sigma\mu_{TRMC} \geq \mu_{TOF}$ or μ_{SCLC}) that can be achieved by SCLC and TOF technique under ideal conditions (i.e., for well-aligned monodomain sample). PR-TRMC give the sum of hole and electron charge carrier mobility [i.e., $\Sigma\mu_{TRMC} = \mu(-) + \mu(+)$], and thereby, hole and electron mobility values cannot be known separately by this technique. On the other hand, TOF, SCLC, and FET can measure both the hole and electron mobilities. In TOF and FET technique, hole and electron mobilities can be measured in the same cell by just reversing the polarity of the counter electrode or gate voltage, respectively, whereas, in SCLC technique, different device geometries, i.e., hole-only cell and electron-only cell, are needed.

5.2.1 *Charge Transport Studies in DLC Materials Based on Various Discotic Cores*

5.2.1.1 Phthalocyanine

Phthalocyanine is a planar symmetric aromatic moiety with a cavity inside it which is adequate to accommodate various metal ions by ligating with N-atoms of four iminoisoindoline units. Single crystals of phthalocyanine itself and its metallo-complexes were studied extensively for their interesting charge transport properties. Keeping this in mind, in 1989, Warman et al. [8] utilized peripherally alkoxy-substituted phthalocyanine derivatives (**Pc-1**, **Pc-2**, and **Pc-3**) (Fig. 5.4) for charge transport studies which were synthesized by van der Pol et al. [15]. These

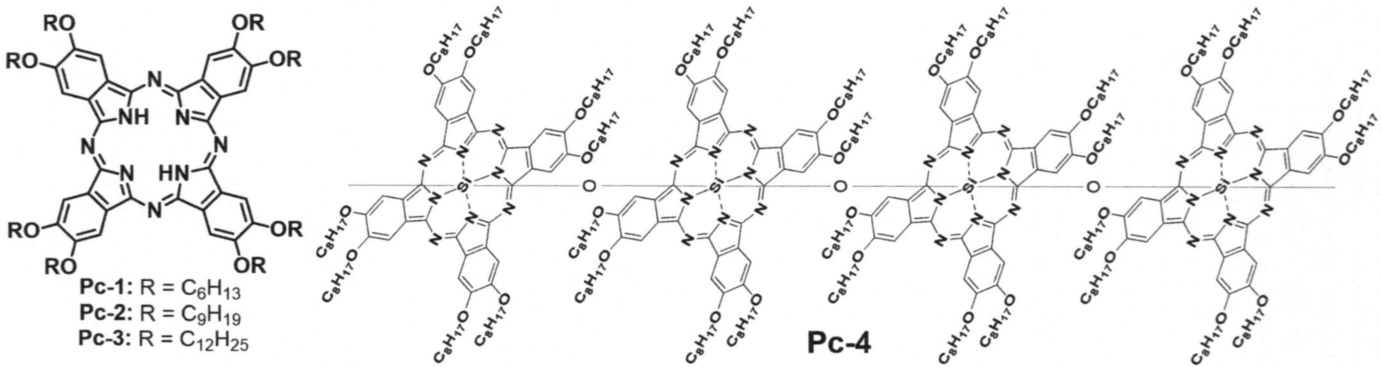

Fig. 5.4 Molecular structures of **Pc-1** to **Pc-4**. (Redrawn from Refs. [8 and 16])

Table 5.1 Phase behavior of **Pc-1** to **Pc-4** [8, 16]

Compound	Phase transition temperatures (in °C)^a
Pc-1	Cr 119 Col_h > 345† Iso
Pc-2	Cr 101 Col_h > 345† Iso
Pc-3	Cr 83 Col_h 309‡ Iso
Pc-4	Col_h > 345† Iso

^{a}Upon heating scan
† No isotropic (Iso) transition observed up to 345 °C
‡ Transition observed by polarized optical microscopy

disc-shaped phthalocyanine derivatives stacked co-facially on top of each other to form a columnar conduction channel for efficient charge transport properties.

All the three derivatives exhibited Col_h mesophase as depicted in Table 5.1. It was observed that all the phthalocyanine derivatives (**Pc-1** to **Pc-3**) showed the average isotropic mobility within the range of 0.4 to 40 $cm^2V^{-1}s^{-1}$ at room temperature (in crystalline (Cr) state) as measured by PR-TRMC technique. The mobility values decreased to approximately one order of magnitude in the transition of Cr to Col_h mesophase. This observation was explained in terms of extent of positional order in the materials, i.e., crystal has more positional order than that of LCs.

In 1992, Schouten et al. reported a comparative study of hexaalkoxy-substituted phthalocyanine (**Pc-3**) and a polymerized derivative (**Pc-4**) of its analogue (Fig. 5.4) in terms of intracolumnar conductivity along phthalocyanine stacks through PR-TRMC technique [16]. The polymerization of hexaalkoxy-substituted phthalocyanine was also performed by van der Pol et al. in 1990 [17], and it (**Pc-4**) showed room temperature Col_h mesophase which did not become isotropic up to 345 °C (Table 5.1). The advantage of polymerization of phthalocyanine was that its PR-TRMC mobility was measured in the order of 10^{-2} $cm^2V^{-1}s^{-1}$ for the first time in room temperature Col_h mesophase.

Until 2004, charge carrier mobility of phthalocyanine and its metal complexes were studied by PR-TRMC method, in which it is not possible to evaluate whether the observed mobility values are generated from hole or electron transport. In order to find out the individual contribution of hole and electron charge carriers in **Pc-5** (Fig. 5.5), TOF technique was employed [18]. **Pc-5** derivative exhibited columnar mesomorphism with the phase sequence mentioned in Fig. 5.5. TOF measurements led to the observation of non-dispersive transient photocurrent curves for both hole and electron transport. The high electron mobility of 2.60×10^{-3} $cm^2V^{-1}s^{-1}$ together with the high hole mobility of 2.99×10^{-3} $cm^2V^{-1}s^{-1}$ was observed at room temperature in the aligned semiconducting cell attained through supercooling of the ordered columnar mesophase. In addition, electric field-independent hole and electron mobilities were perceived for **Pc-5** derivative.

In 2004, Cherian et al. reported the fabrication of organic field-effect transistor (OFET) device utilizing a thin film of copper phthalocyanine-based DLC derivative (**Pc-6**) (Fig. 5.6a) [19]. They employed Langmuir-Blodgett technique to deposit the film of **Pc-6** on silicon substrates in bottom contact OFETs. The electrical

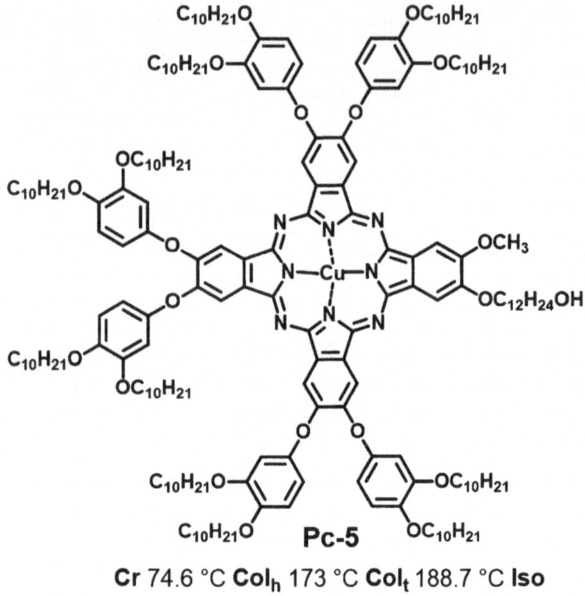

Fig. 5.5 Molecular structure of **Pc-5** along with its phase transition temperatures upon heating. (Redrawn from Ref. [18])

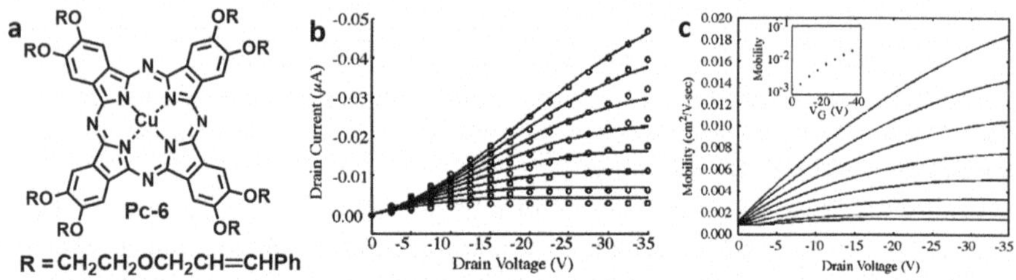

Fig. 5.6 (**a**) Molecular structure of **Pc-6**. (Redrawn from Ref. [19]) (**b**) Current-voltage (*I-V*) plots (fitted utilizing field-dependent mobility and contact barrier height of 0.415 eV) and (**c**) plot showing the variation in mobilities upon varying gate voltages in fabricated OFET device. The inset represents the variation of mobility as a function of gate voltage for drain-source voltage at −35 V. (Reproduced with the permission from Ref. [19])

measurements of resulting OFET devices were performed, and the device performances were evaluated by using a number of 2D simulations.

Among all models, only by using field-dependent mobility and contact barrier model, the experimental data were fitted very closely (Fig. 5.6b). The variation of extracted mobility values with varying gate voltages using this model is displayed in Fig. 5.6c and showed a high FET mobility of 0.018 $cm^2V^{-1}s^{-1}$.

In 2006, Deibel et al. reported the charge transport behavior of a phthalocyanine-based DLC derivative (**Pc-7**) by using three different methods (SCLC, TOF, and

Fig. 5.7 Molecular structure along with the thermal transitions of **Pc-7**. (Redrawn from Ref. [20])

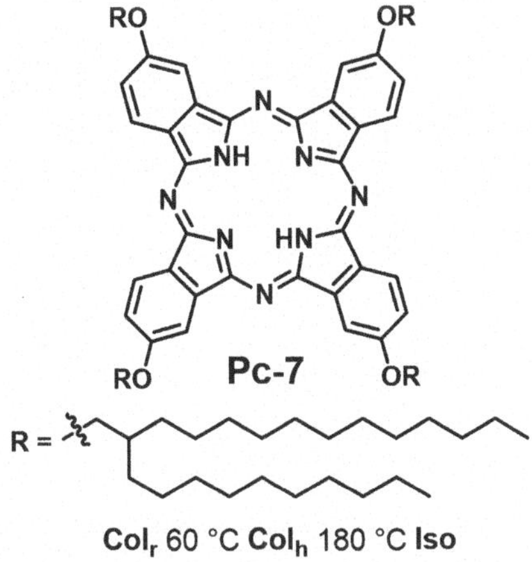

OFET) [20]. **Pc-7** derivative showed Col$_r$ and Col$_h$ mesophase at room temperature and higher temperature, respectively (Fig. 5.7). SCLC mobility measurement of **Pc-7** in two different device configurations was done, and better charge transport was observed in glass/ITO/DLC/ITO/glass than that of glass/ITO/OTS/DLC/OTS/ITO/glass (ITO is indium-tin oxide and OTS is octadecyltrichlorosilane). The reason is that in the first case the molecules were completely homeotropically aligned which provides unidirectional channels for 1D charge transport, whereas the latter one was partially homeotropic and not giving good channel for efficient flow of charge carriers. The extracted SCLC hole mobilities at room temperature and 141 °C are 7×10^{-4} cm^{2}V^{-1}s^{-1} and 4×10^{-3} cm^{2}V^{-1}s^{-1}, respectively. Further, mobility studies by TOF technique observed no signal for OTS-coated ITO cell, while the maximum hole mobility of 2.8×10^{-3} cm^{2}V^{-1}s^{-1} was obtained in glass/ITO/DLC/ITO/glass cell. On the other hand, the observation of no transistor signal in the case of ITO glass-based OFET was corroborated to the homeotropic alignment of **Pc-7** molecules which did not provide any channel for charge transport in OFETs. The charges are migrated from source to drain through semiconducting material in parallel to the substrate and, thereby, favor planar orientation of the molecules in OFETs. Further, another set of sample made by spin-coating the solution of **Pc-7** on SiO_2 (100 nm)/Si/Al substrate showed characteristic FET output with OFET mobility of the order of ~10^{-7} cm^{2}V^{-1}s^{-1}.

Imahori and co-workers reported segregated donor (D)-acceptor (A) columnar assembly in zinc phthalocyanine (as donor)- and fullerene (as acceptor)-based derivative **Pc-8** [21]. They prepared **Pc-8** along with the reference zinc phthalocyanine derivative **Pc-9** in order to have a comparative study (Fig. 5.8). The mobility of both compounds was measured by TOF as well as PR-TRMC technique. In TOF method, **Pc-8** exhibited high ambipolar charge transport with a hole mobility of 0.26 cm^{2}V^{-1}s^{-1} and electron mobility of 0.11 cm^{2}V^{-1}s^{-1}, whereas **Pc-9** showed only hole mobility of 2.8 cm^{2}V^{-1}s^{-1}. The PR-TRMC mobilities of **Pc-9** and **Pc-8** were measured as 3.0 cm^{2}V^{-1}s^{-1} and 0.52 cm^{2}V^{-1}s^{-1}, respectively. Thus, it is clear that in

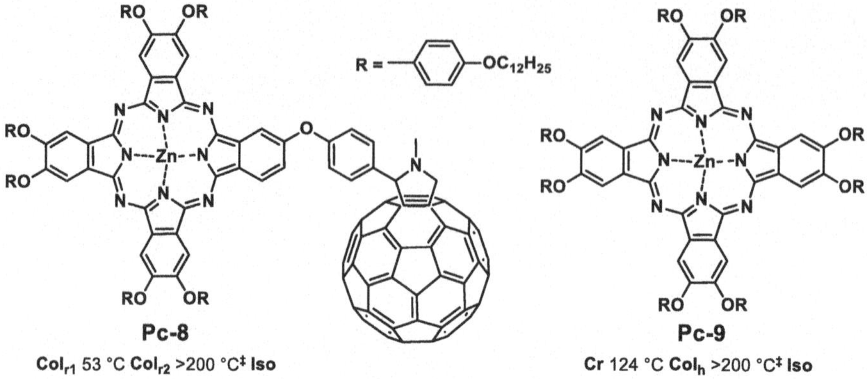

Fig. 5.8 Molecular structures of **Pc-8** and **Pc-9** along with their phase transition temperatures upon heating. ‡No isotropic transition was observed in both derivatives as above 200 °C compounds got decomposed. (Redrawn from Ref. [21])

PR-TRMC mobility, both the charges (hole and electron) contributed together and, hence, are responsible for the higher mobilities in comparison to the TOF mobilities. All the mobility values described above were measured at 25 °C after heating and cooling treatment.

5.2.1.2 Porphyrin

Porphyrin core is a highly conjugated macrocycle which is composed of four pyrrole rings that are joined by four inter-pyrrolic methine bridges. Porphyrins can be prepared through several methods, such as by the dimerization of dipyrromethanes and tetramerization of monopyrroles and from open-chain pyrrolic derivatives [5]. Many perylene derivatives are studied for charge transport. In 1991, Schouten et al. reported the PR-TRMC mobility of octakis-β-nonoxyethyl-substituted porphyrin (**Porphy-1**) and its zinc complex (**Porphy-2**) (Fig. 5.9a) [22]. Both the compounds were synthesized, and their mesomorphic behavior was studied by Gregg et al. in 1989 [23]. **Porphy-1** did not show any mesomorphic character, whereas **Porphy-2** exhibited Col$_h$ mesophase in the temperature range of 102–152 °C (Table 5.2). Both the compounds showed the columnar arrangement in Cr (**Porphy-1**) as well as in LC state (**Porphy-2**) where the molecules are tilted at an angle of 46° in the columns (Fig. 5.9b). It was observed that the inclusion of a central zinc atom to the porphyrin increases the mobility of the system.

Porphy-2 exhibited the mobility value greater than 0.6×10^{-2} cm^{2}V^{-1}s^{-1} and 2.6×10^{-2} cm^{2}V^{-1}s^{-1} in the Col$_h$ mesophase and Cr state, respectively. The decrease of mobility values in Col$_h$ phase than in Cr state is due to the presence of more order in the latter case as also observed in phthalocyanine derivatives.

Studies by Serrano and co-workers showed the mobility increment by 6–7 orders upon metalation in porphyrin derivatives compared to its metal-free analogue. A

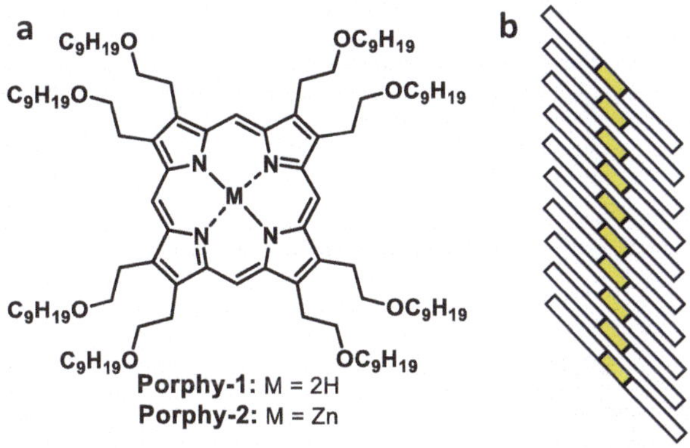

Fig. 5.9 (**a**) Molecular structure and (**b**) schematic of tilted columnar arrangement of **Porphy-1** and **Porphy-2**. (Redrawn from Ref. [22])

Table 5.2 Phase behavior of **Porphy-1** and **Porphy-2** [22]

Compound	Phase transition temperatures (in °C)[a]
Porphy-1	Cr 92 Iso
Porphy-2	Cr 102 Col_h 152 Iso

[a]Upon heating cycle

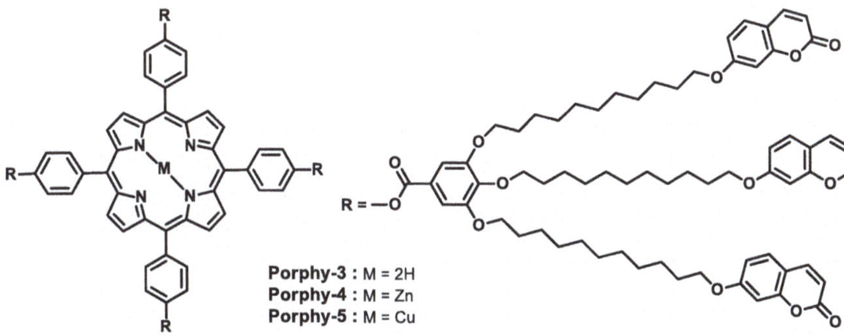

Fig. 5.10 Molecular structures of **Porphy-3**, **Porphy-4**, and **Porphy-5**. (Redrawn from Ref. [24])

new family of LC dendrimers based on porphyrin core attached with coumarin moiety at periphery through alkyl spacers was reported [24] where metal-free (**Porphy-3**) as well as Zn^{II}- and Cu^{II}-coordinated porphyrin derivatives (**Porphy-4** and **Porphy-5**) (Fig. 5.10) were prepared.

All the compounds exhibited room temperature discotic nematic (N_D) mesophase having a high tendency to align homeotropically, and upon metalation, the mesophase range increased in comparison to the metal-free analogue (Table 5.3). When used in SCLC devices, it was observed that **Porphy-3** showed the hole mobility of 1.2×10^{-7} $cm^2V^{-1}s^{-1}$, whereas its metal-coordinated analogues **Porphy-4** and **Porphy-5** exhibited the mobility of 1.1 $cm^2V^{-1}s^{-1}$ and 0.34 $cm^2V^{-1}s^{-1}$,

Table 5.3 Phase behavior and SCLC hole mobility ($\mu_{h,\,SCLC}$) of **Porphy-3** to **Porphy-5** [24]

Compound	Phase transition temperatures (in °C)[a]	$\mu_{h,\,SCLC}$ ($cm^2V^{-1}s^{-1}$)
Porphy-3	g 16 N_D 61 Iso	1.2×10^{-7}
Porphy-4	g 11 N_D 74 Iso	1.1
Porphy-5	g 17 N_D 73 Iso	0.34

[a]Upon second heating cycle
Abbreviation: *g* glass transition, N_D nematic discotic mesophase, *Iso* isotropic phase

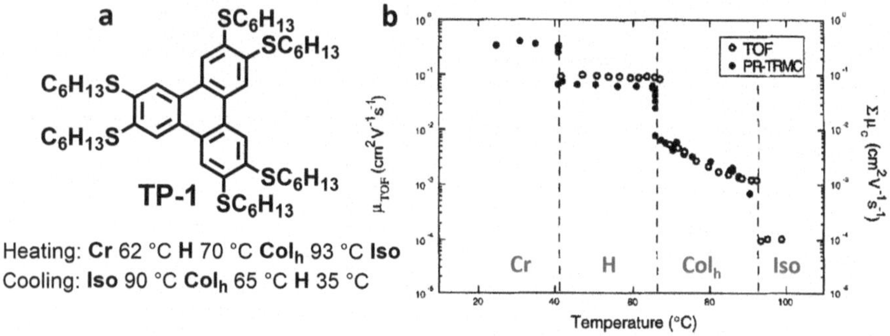

Fig. 5.11 (**a**) Molecular structure of hexahexylthiotriphenylene (**TP-1**). (Redrawn from Ref. [10]) (**b**) Temperature-dependent mobility of **TP-1** measured by TOF ($\mu_{h,\,TOF}$) (open circles) and PR-TRMC ($\Sigma\mu_{TRMC}$) (filled circles) technique upon cooling. (Reproduced with the permission from Ref. [27])

respectively (Table 5.3). This increment can be related to the effect of the inclusion of metal to the porphyrin derivative in accordance with better homeotropic alignment ability. Further, the same group studied the charge transport behavior in another class of LC dendrimers based on metalloporphyrin derivatives substituted with coumarin [25] and carbazole [26] units.

5.2.1.3 Triphenylene

Triphenylene derivatives are the most extensively studied discotic mesogens and considered as pillars of DLCs. Triphenylenes are widely used to induce stable columnar mesomorphism in many mesogenic or non-mesogenic components. The most significant charge transport study in triphenylene systems was reported in 1994 by Ringsdorf and Harrer group in hexahexylthiotriphenylene (**TP-1**) that self-assembled into a highly ordered columnar helical (H) phase of DLCs (Fig. 5.11a) [10]. This derivative showed a very high photoinduced hole charge carrier mobility ($\mu_{h,\,TOF}$) of the order of 0.1 $cm^2V^{-1}s^{-1}$ in the rarely observed H phase which was the highest hole mobility in DLCs at that time. The temperature-dependent photoinduced charge carrier mobility measurements in the heating sequence showed an abrupt change at the phase transitions and followed the mobility trend: H > Col_h > Iso (Fig. 5.11b). The increase in mobility values ongoing from isotropic to helical mesophase clearly showed how charge carrier mobility depends upon intracolumnar order. Interestingly, the mobilities in the phase sequence were also evaluated by PR-TRMC technique ($\Sigma\mu_{TRMC}$) [27]. The values observed in H phase by TOF method were 20% higher than obtained by PR-TRMC technique. These values are mutually supportive and indicated that the obtained mobility values are originating from the well-aligned monodomain sample and close to the 1D intracolumnar mobility. In addition to mutually supportive, the results are also complementary in terms that the mobilities in the isotropic phase are obtained by TOF while in a Cr state by PR-TRMC method. The mobility measurements in the Cr state by TOF method could not be possible due to the charge trapping in deep traps at grain boundaries which usually formed in crystalline state. On the other hand, the mobility in the isotropic phase (1×10^{-4} $cm^2V^{-1}s^{-1}$) was too small to be measured by PR-TRMC technique and, thereby, obtained by TOF. Therefore, with the use of both the techniques, the determination of mobility values was possible in all the four phases, i.e., Cr, helical, Col_h, and isotropic. From the combined results, the mobility ($\Sigma\mu_{TRMC}$) in the helical phase (40–66 °C) is 0.08 $cm^2V^{-1}s^{-1}$ and shows temperature independency in the helical phase. However, in the Col_h mesophase (67–93 °C), the mobility ($\Sigma\mu_{TRMC}$) increased from 1×10^{-3} $cm^2V^{-1}s^{-1}$ to 6×10^{-3} $cm^2V^{-1}s^{-1}$ on lowering the temperature, probably due to the thermally activated perturbations in the columnar structure. The mobility obtained by TOF method in the isotropic phase was 1×10^{-4} $cm^2V^{-1}s^{-1}$. After a decade, in 2005, Hanna group measured electron charge carrier mobility in the H phase (at 35 °C) of **TP-1** derivative, and an almost equivalent electron mobility of 0.08 $cm^2V^{-1}s^{-1}$ to that of the hole (0.1 $cm^2V^{-1}s^{-1}$) was obtained [28]. Charge transport studies of various hexaalkoxytriphenylenes (**TP-2** to **TP-7**) (Fig. 5.12) were studied by TOF method. From **TP-2** to **TP-7**, all compounds exhibited Col_h or ordered Col_h (Col_{ho}) mesophase provided that **TP-2** with *n*-butyloxy peripheral chain length displayed relatively more ordered columnar phase, i.e., columnar hexagonal plastic phase (Col_{hp}) (Table 5.4) [29]. The remarkable hole mobilities of 1 x 10^{-3} $cm^2V^{-1}s^{-1}$ were obtained for **TP-3** derivative with *n*-pentyloxy peripheral alkyl chain [9]. On the other hand, **TP-2** derivative exhibited intermediate mobility of 10^{-2} $cm^2V^{-1}s^{-1}$ in the Col_{hp} to that of H phase of hexahexylthiotriphenylene **TP-1** (10^{-1} $cm^2V^{-1}s^{-1}$) and Col_{ho} phase of **TP-3** (10^{-3}

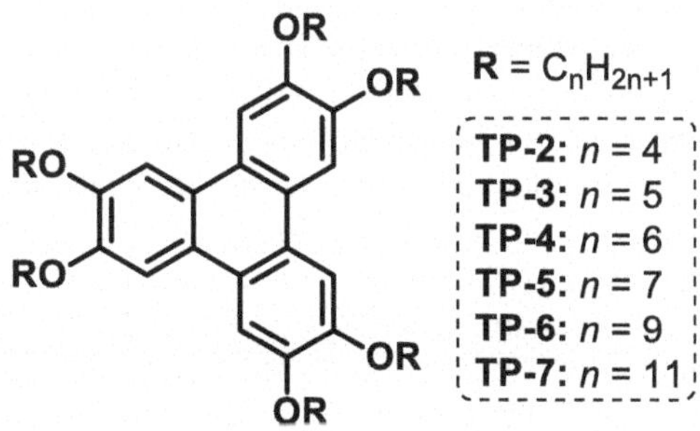

Fig. 5.12 Chemical structures of homologous series of hexaalkoxytriphenylene derivatives. (Redrawn from Refs. [29 to 32])

Table 5.4 Phase behavior of **TP-2** to **TP-7** [9, 29–31]

Compound	Phase transition temperatures (in °C)[a]	Compound	Phase transition temperatures (in °C)[a]
TP-2	Cr 90 Col_{hp} 147.3 Iso	**TP-5**	Cr 68.6 Col_{ho} 93.0 Iso
TP-3	Cr 69.0 Col_{ho} 122.0 Iso	**TP-6**	Cr 57.0 Col_{ho} 77.6 Iso
TP-4	Cr 70 Col_h 100 Iso	**TP-7**	Cr 54.0 Col_{ho} 66.0 Iso

[a]Upon heating scan

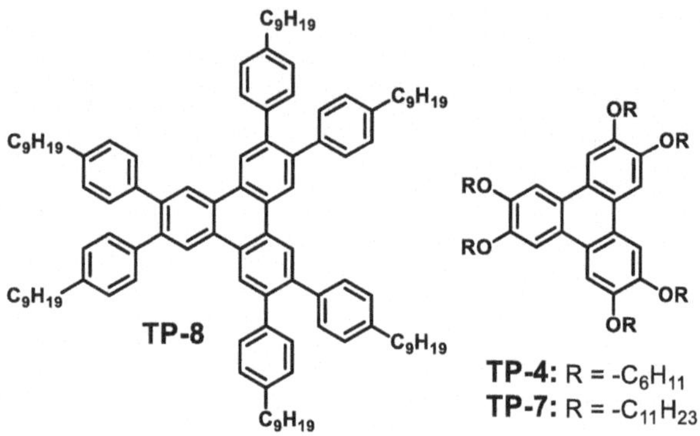

Fig. 5.13 Chemical structure of triphenylene derivative **TP-8**, **TP-4**, and **TP-7**. (Redrawn from Ref. [32])

$cm^2V^{-1}s^{-1}$) [29]. Boden et al. reported TOF mobilities of 10^{-4} $cm^2V^{-1}s^{-1}$ along the columnar stack of triphenylene derivative hexahexyloxytriphenylene (**TP-4**) forming Col_h phase (Fig. 5.12) [30]. The studies showed that the photocurrent decreased on increasing the peripheral chain length from **TP-3** to **TP-7** (for $n = 5$, 7, 9, and 11) in spite of exhibiting the similar intracolumnar distance [31]. However, the intercolumnar distances have increased with an increase in chain length which results in reducing the area covered by the photoconducting triphenylene cores, thereby explaining the exponential decrease in the photocurrent with increasing chain length.

In triphenylene systems, it had been shown that 1:1 binary mixture of hexaalkoxytriphenylenes **TP-4** and **TP-7** with the comparatively larger cored triphenylene derivative **TP-8** (Fig. 5.13) improved the mesomorphic behavior and formed structurally more ordered system in comparison to the individual compounds [32]. For instance, **TP-8** was Cr, while **TP-7** and **TP-4** displayed Col_h phase in a narrow temperature range (Table 5.5).

However, the binary mixtures, i.e., **TP-8**:**TP-4** and **TP-8**:**TP-7**, exhibited improved columnar behavior (Table 5.5) [33]. The formation of structurally more ordered columnar structures in the binary mixtures leads to their improved hole mobilities in comparison to the mixing component molecules. For example, the measured TOF mobility for hole carriers (Table 5.5) in the binary mixture **TP-8**:**TP-7** is 1.6×10^{-2} $cm^2V^{-1}s^{-1}$ (at 313 K), while the TOF mobility in the case of individual **TP-7** is 1.0×10^{-4} $cm^2V^{-1}s^{-1}$ (at 333 K). However, in the case of individual **TP-8**, mobility cannot be measured by the same technique due to their

Table 5.5 Phase behavior and TOF mobilities of **TP-4**, **TP-7**, and **TP-8** and their binary mixtures **TP-8:TP-4** and **TP-8:TP-7** [32, 33]

Compound	Phase transition temperatures (in °C)[a]	$\mu_{h,\ TOF}$ ($cm^2V^{-1}s^{-1}$)
TP-8	Cr 59 Iso	–
TP-4	Cr 70 Col_h 100 Iso	7.1×10^{-4}
TP-7	Cr 54 Col_h 66 Iso	1.0×10^{-4}
TP-8:TP-4	Cr 66 Col_h 155 Iso	1.4×10^{-2}
TP-8:TP-7	Cr 60 Col_h 170 Iso	1.6×10^{-2}

[a]Upon heating scan

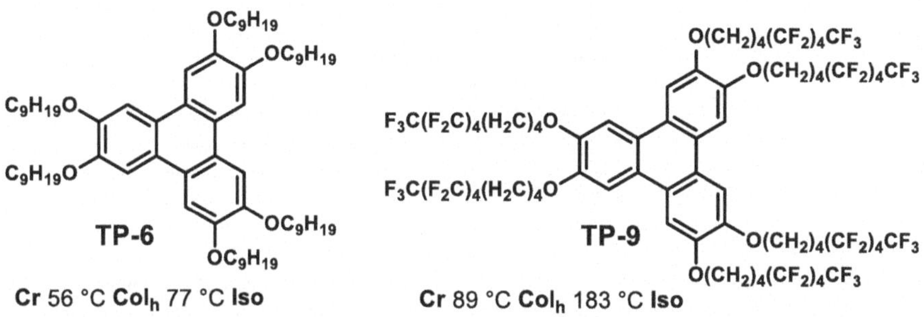

Fig. 5.14 Molecular structures of triphenylene derivatives **TP-6** and **TP-9** along with the phase transition temperatures upon heating. (Redrawn from Ref. [35])

non-mesomorphic nature. Similarly, in the case of another binary mixture, i.e., **TP-8**: **TP-4**, the obtained mobility value of 1.4×10^{-2} $cm^2V^{-1}s^{-1}$ (at 313 K) is two orders higher than the individual mesomorphic **TP-4** component (7.1×10^{-4} $cm^2V^{-1}s^{-1}$ at 353 K). The TOF as well as PR-TRMC mobility of these derivatives was also reported by another group and is in good agreement with the discussed results [34].

The influence of perfluoroalkylated chains (**TP-9**) in comparison to the non-fluorinated one (**TP-6**) on the mesomorphic as well as charge transport property of the triphenylene derivatives (Fig. 5.14) was studied by Shimizu group [35]. It has been found that on replacement of simple alkyl chains by perfluoroalkylated chains, however, mesomorphic range was found to be improved but the charge transport behavior did not show much improvement effect. Moreover, similar values of hole and electron mobilities of 3.7×10^{-4} $cm^2V^{-1}s^{-1}$ (at 180 °C) were observed for **TP-9** derivative.

The effect of doping on charge transport behavior of triphenylene-fused phenazine derivatives was studied by Kumar group where an optimized amount of oleylamine-capped CdS nanowires was dispersed in Col_h phase of phenazine-fused triphenylene derivative **TP-10** (Fig. 5.15) [36]. After dispersing CdS nanowires, there was not much impact on the phase transition temperatures (Table 5.6); however, XRD studies revealed that the core-core stacking distance in neat **TP-10** compound was 3.55 Å and decreased to 3.50 Å in the dispersed system (**TP-10/CdS**), suggesting the higher mobility in the latter case. But surprisingly, in

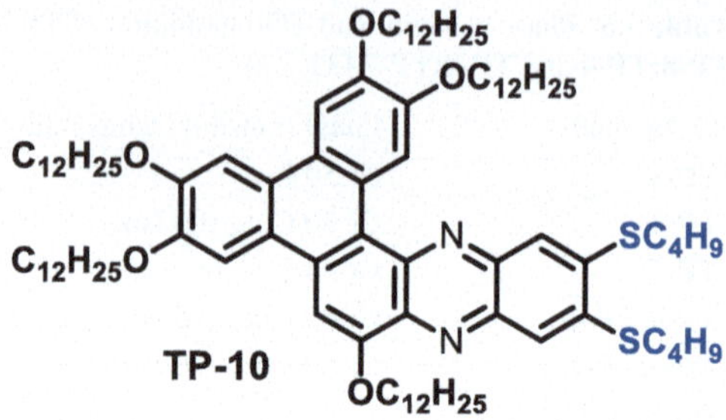

Fig. 5.15 Molecular structure of phenazine-fused triphenylene derivative **TP-10**. (Redrawn from Ref. [36])

Table 5.6 Phase behavior of **TP-10** and **TP-10/CdS** [36]

Compound	Phase transition temperatures (in °C)[a]
TP-10	Cr 55.1 Col_h 172 Iso Iso 170.3 Col_h 17.4 Cr
TP-10/CdS	Cr 37.1 Col_h 170.3 Iso Iso 167.3 Col_h 17.3 Cr

[a]Upon heating and cooling scan

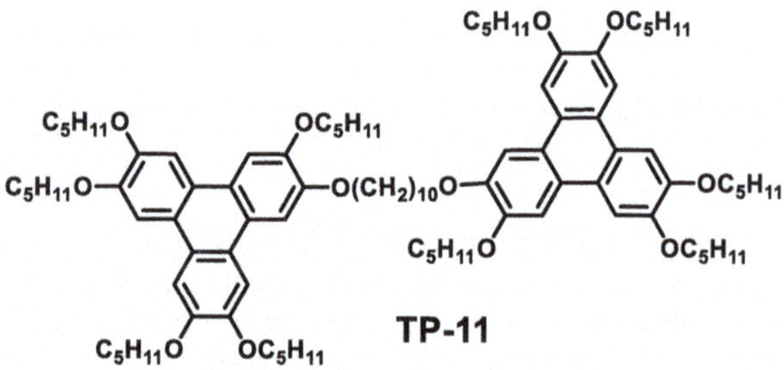

Fig. 5.16 Chemical structure of the "discotic twin" **TP-11** along with its phase transition temperatures upon heating (cooling) scan. (Redrawn from Ref. [37])

the dispersed system, comparatively larger transit time was observed that attributed to lower mobility value. The authors reasoned it to the probably decreased molecular ordering in the Col_h phase and formation of CdS charge traps in **TP-10** after dispersion.

In addition to monomeric triphenylene derivatives, oligomeric triphenylenes were also studied for their charge transport properties. Ringsdorf group explored the electronic charge carrier transport properties of "discotic twin" in 1995 [37]. Such kind of discotic dimers or oligomers can be considered as model compounds for discotic polymeric systems due to ease of purification, easy tailoring of mesophase temperature range simply by varying the spacer length, and possibility to freeze their discotic phase into a glassy state. In this work, discotic dimer (**TP-11**) of hexapentyloxytriphenylene (**TP-3**) is reported that exhibited a phase sequence as described in Fig. 5.16. The TOF measurements were performed in the phase sequence of **TP-11**. In the Col_h phase of the discotic twin, the same order mobility of 10^{-3} $cm^2V^{-1}s^{-1}$ was observed to that of the monomeric model compound **TP-3**.

This indicates that free rotation of the discotic mesogens that was possible in discotic monomer **TP-3** is impossible in the discotic twin, thereby neither promoting nor hindering the charge carrier transport. The lower mobility value was observed in the glassy state than in Col_h phase which was attributed to the "freezing out" of defects, while in the Col_h phase, the "dynamical healing" of defects reflected in larger value of mobility in the respective phase.

The mobility value variation over several orders of magnitude was observed in aligned versus non-aligned areas of isocyano-triphenylene palladium and platinum complexes (**TP-12** to **TP-15**) (Fig. 5.17) [38, 39]. The SCLC measurements of these samples were enormously dependent upon the thermal history of the samples. For instance, the obtained mobility values for the rapidly cooled samples from the isotropic liquid were in the range of 10^{-8}–10^{-9} $cm^2V^{-1}s^{-1}$ for all the complexes (Table 5.7). However, upon thermal annealing of the samples, homeotropic alignment was obtained, but not uniformly throughout the sample. Thereby, charge mobility values varied by several orders from the aligned to non-aligned areas, and surprisingly, the values were ten orders of magnitude higher in aligned than non-aligned areas. Due to the lack of alignment over the whole sample surface, the mobility values were scattered across the range from 10^{-3} to 10 $cm^2V^{-1}s^{-1}$ that are mentioned in Table 5.7.

Pal's group investigated the charge transport in electron donor (**TP-16**) and electron acceptor (2,4,7-trinitrofluorenone (**TNF**)) charge transfer complex of **TP-16** and **TNF** (i.e., **TP-16/TNF** complex) formed by mixing both the components in 1:2 ratio, respectively (Fig. 5.18a) [40]. The motive behind complexation was to

Fig. 5.17 Molecular structures of isocyano-triphenylene palladium/platinum complexes (**TP-12** to **TP-15**). (Redrawn from Refs. [39 and 40])

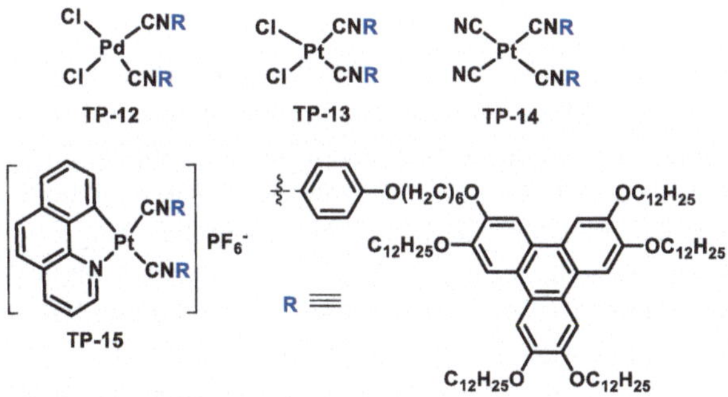

Table 5.7 Phase behavior and TOF mobility values of **TP-12** to **TP-15** [38, 39]

Compound	Phase transition temperatures (in °C)^a	$\mu_{h,\ TOF}$ ($cm^2V^{-1}s^{-1}$)	
		Aligned areas	Non-aligned areas
TP-12	Cr 34 Col_r 53 Iso	10^{-2}-10	10^{-8}-10^{-9}
TP-13	Cr 45 Col_r 79 Iso	10^{-2}-1	10^{-8}-10^{-9}
TP-14	Cr 35 Col_r 105 Iso	10^{-3}-10	$(3 \pm 2.8) \times 10^{-9}$
TP-15	Cr −9.0 Col_h 143 Iso	10^{-2}-1	$(5 \pm 2) \times 10^{-9}$

^{a}Upon heating scan

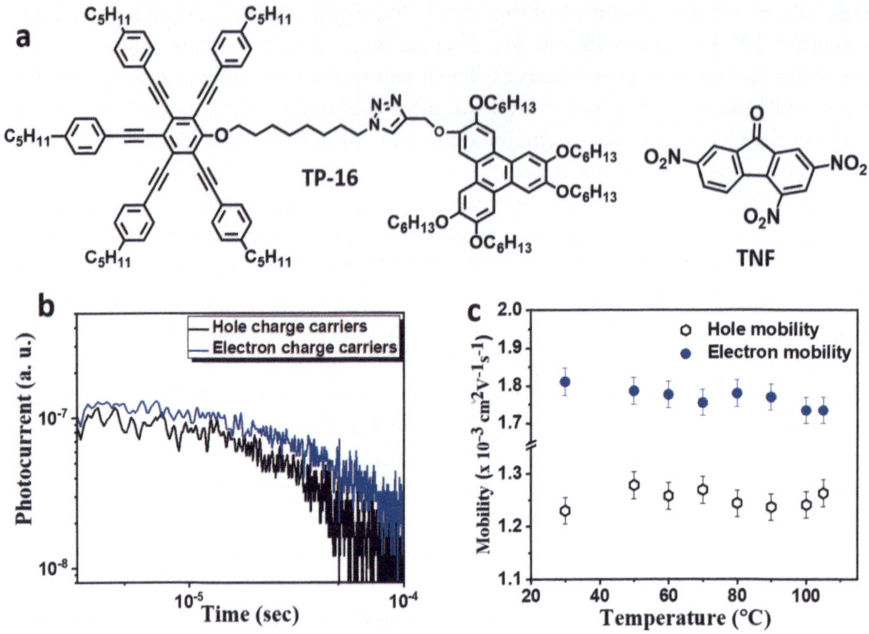

Fig. 5.18 (**a**) Molecular structure of triphenylene-pentaalkynylbenzene derivative (**TP-16**) and **TNF**. (**b**) Double logarithmic transient photocurrent curve obtained for positive (hole) and negative (electron) charge carriers of **TP-16/TNF** complex. (**c**) Temperature-independent mobility obtained for **TP-16/TNF** complex. (Redrawn and reproduced with the permission from Ref. [40])

stabilize the self-assembly behavior of **TP-16** derivative which was achieved by the strong complexation with its complementary electron-deficient **TNF** discotic component. The complex exhibited a broad mesophase range with the phase sequence: Col_h 132.8 °C Iso. As the complex consists of both electron-rich and electron-deficient moieties, therefore, the ambipolar charge mobility was measured at an optimized voltage of ± 10 V (Fig. 5.18b). The complex exhibited electron and hole mobilities of 1.78×10^{-3} $cm^2V^{-1}s^{-1}$ and 1.25×10^{-3} $cm^2V^{-1}s^{-1}$ at 30 °C, respectively, and was almost temperature invariant (Fig. 5.18c), revealing its potential application for ambipolar organic semiconducting material.

5.2.1.4 Coronene Family

Coronene is a polycyclic aromatic hydrocarbon which is composed of seven *peri*-fused benzene rings. The fusion of additional benzene rings to the periphery of coronene core results in several other larger aromatic cores, e.g., hexa-*peri*-hexabenzocoronene and so on. The modification to the *p*-type coronene core by the electron-withdrawing functionalities, e.g., imide, carboxaimides, and so on, changes the electronic properties of the core to *n*-type. Debije et al. reported several

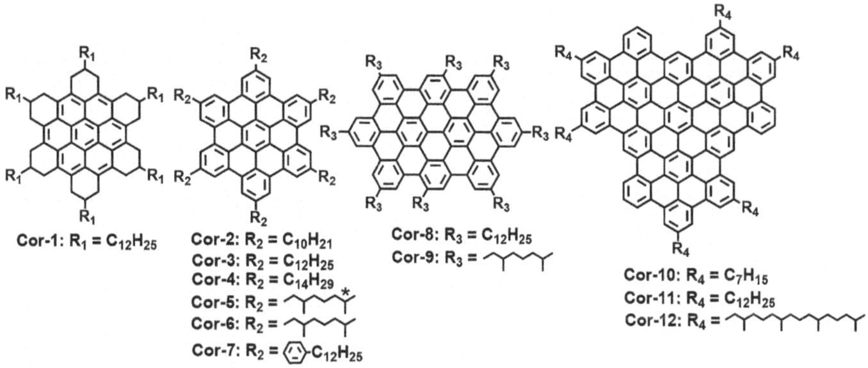

Fig. 5.19 Molecular structures of hexabenzocoronene derivatives **Cor-1** to **Cor-12**. (Redrawn from Ref. [41])

Table 5.8 Phase behavior and PR-TRMC mobility ($\Sigma\mu_{TRMC}$) of **Cor-1** to **Cor-12** [41]

Compound	Cr LC phase transition temperature[a]	$\Sigma\mu_{TRMC}$ ($cm^2V^{-1}s^{-1}$) at T (Cr LC) + 10 °C[b]	Compound	Cr LC phase transition temperature[a]	$\Sigma\mu_{TRMC}$ ($cm^2V^{-1}s^{-1}$) at T (Cr LC) + 10 °C[b]
Cor-1	Ca. 100 °C	0.1-0.2	**Cor-7**	RT LC	0.31[c]
Cor-2	124 °C	0.26	**Cor-8**	102 °C	0.26
Cor-3	105 °C	0.38	**Cor-9**	108 °C	0.29
Cor-4	109 °C	0.31	**Cor-10**	RT LC	0.23[c]
Cor-5	99 °C	0.26	**Cor-11**	RT LC	0.20[c]
Cor-6	81 °C	0.30	**Cor-12**	RT LC	0.08[c]

[a]Crystalline to liquid crystalline transition temperature
[b]Mobility values at about 10 °C above the transition temperatures on heating. RT LC = room temperature liquid crystalline
[c]Mobility values measured at 100 °C

larger aromatic core-based DLCs starting from coronene to hexabenzocoronene to extended coronene versions (Fig. 5.19) [41]. Their main aim was to validate the prediction that the mobilities should get higher and higher on increasing the size of the discotic core. The mobilities in the mesophases were extracted by using PR-TRMC technique. However, the mobility values do not follow any such trend when verified experimentally (Table 5.8), and this work finally concluded that "an increase in the size of the conjugated core does not necessarily ensure better transport properties." However, the obtained mobility values were very high for DLC-based organic semiconductors. Table 5.8 showed the mobility values for the **Cor-1** to **Cor-12** derivatives obtained at temperatures approximately 10 °C above the Cr to LC transitions, and for the room temperature LCs, the mobilities were measured at 100 °C. The temperature-dependent PR-TRMC mobility measurements for **Cor-1**, **Cor-3**, **Cor-8**, and **Cor-11** showed that the mobility values decreased

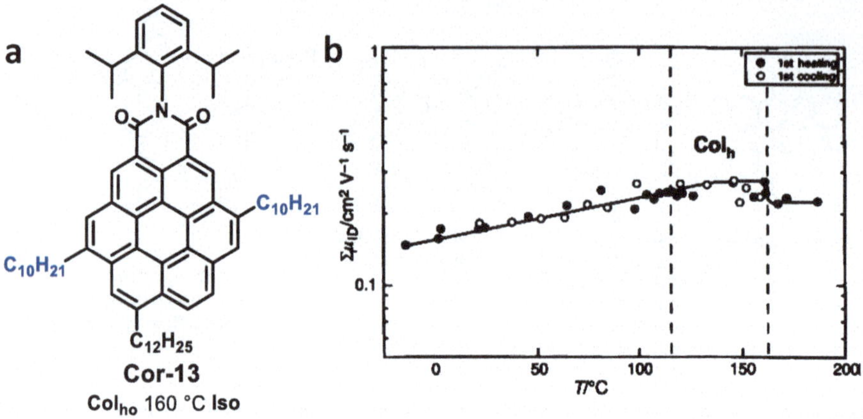

Fig. 5.20 (**a**) Molecular structure of coronenemonoimide derivative **Cor-13**. (Redrawn from Ref. [41]) along with its phase transition temperatures upon heating. (**b**) Temperature-dependent mobility curve of **Cor-13** derivative. (Reproduced with the permission from Ref. [42])

abruptly at crystalline to mesomorphic transition due to the increased motional freedom, resulting in decreased structural order within the columns in the mesophase. However, for **Cor-11**, the mobility value does not undergo significant change up to the maximum attainable limit of 200 °C in the PR-TRMC technique. Similarly, on cooling, the mobilities remain almost constant with an abrupt increase near to the mesophase to Cr temperatures.

Mullen group in 2001 reported the charge transport property of coronenemonoimide derivative **Cor-13** that showed room temperature columnar assembly with the phase sequence: Col_{ho} 160 °C Iso (Fig. 5.20a) [42]. The mobility derived from the PR-TRMC measurements was close to 0.2 $cm^2V^{-1}s^{-1}$ at room temperature and increased with slow pace upon raising temperature and reached maximum to 0.3 $cm^2V^{-1}s^{-1}$ just before entering into the isotropic phase at 160 °C. Interestingly, in the proposed isotropic phase of **Cor-13** derivative, the mobility value only dropped by 20% which is less than the usual decrement observed in isotropic phase of DLCs (Fig. 5.20b). This indicated that in the case of **Cor-13** derivative, the isotopic phase still intact the columnar order and more accurately the phase can be regarded as highly fluid mesophase rather isotropic phase.

Kuo et al. very recently reported hexabenzocoronene-based DLCs (**Cor-14** to **Cor-17**) displaying hexagonal phase (Fig. 5.21) [43]. In this report, hexabenzocoronene core is modified with the diacetylene chains, and the discotic columnar phase of LC is utilized here to align the diacetylene chains into a proper packing arrangement for the polymerization to occur. The polymerized columnar channels can lock the discotic system into the permanent columnar structure that may induce additional and more rigid conducting channels for the charge transport. Upon heating, all the compounds undergo polymerization, and for the **Cor-14**, **Cor-15**, and **Cor-16** compounds with dodecyl chains, the clearing temperatures did not appear before polymerization, but for **Cor-17** with shorter saturated alkyl

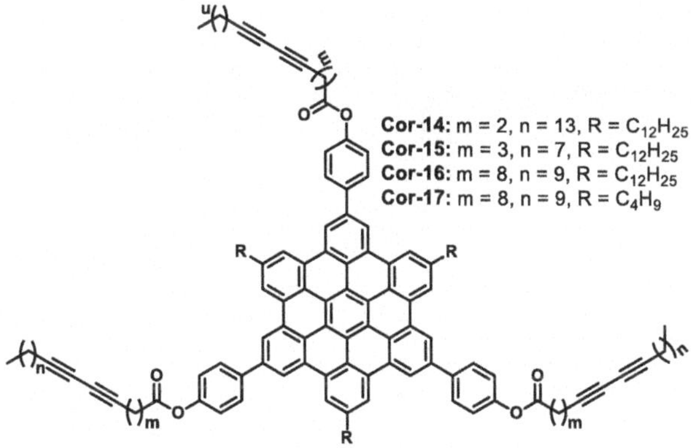

Fig. 5.21 Molecular structures of hexabenzocoronene derivatives (**Cor-14** to **Cor-17**). (Redrawn from Ref. [43])

Table 5.9 Phase behavior and PR-TRMC mobility ($\Sigma\mu_{TRMC}$) of **Cor-14** to **Cor-17** [43]

Compound	Phase transition temperatures (in °C)[a]	$\Sigma\mu_{TRMC}$ ($cm^2V^{-1}s^{-1}$)
Cor-14	Cr 31 Col_h 180 P	5.6×10^{-5}
Cor-15	Cr 26 Col_h 165 P	6.4×10^{-5}
Cor-16	Cr 47 Col_h 141 P	5.5×10^{-5}
Cor-17	Cr 33 Col_h 100 Iso 110 P	9.8×10^{-5}

[a]Upon heating cycle
Abbreviation: *P* polymer

chains entered into isotropic state before polymerization. After polymerization, the columnar assembly was retained. The PR-TRMC charge carrier mobilities were investigated in the polymerized derivatives, and the values are displayed in Table 5.9. The mobility in **Cor-17** is 1.5–1.8 times larger than the other three derivatives.

Donnio et al. reported fast charge carrier mobility in lamello-columnar oblique ($LamCol_{obl}$) mesophase of segregated donor-acceptor triad based on coronenediimide and triphenylene core units [44]. The X-ray diffraction (XRD) studies revealed that the donor and acceptor units self-sort and fully segregate into different columns with a regular alteration of rows of triphenylene and diimide units. **Cor-18** exhibited high clearing temperature with the phase sequence: $LamCol_{obl}$ 268 °C Iso (Fig. 5.22). The hole transport observed at 200 °C under the electric field of 5.0×10^4 Vcm^{-1} was 1.8×10^{-3} $cm^2V^{-1}s^{-1}$, and the electron transport under the same conditions was obtained to be 2.6×10^{-2} $cm^2V^{-1}s^{-1}$. These mobility values depend upon temperature and were found to be increased with increasing temperatures. For instance, the electron and hole mobilities gradually increased from 10^{-4} to 10^{-2} $cm^2V^{-1}s^{-1}$ and 10^{-5} to 10^{-3} $cm^2V^{-1}s^{-1}$, respectively, on raising the temperature from 30 to 210 °C.

Pal's group recently reported the ambipolar charge transport behavior in the heterocoronene derivatives (**Cor-19** to **Cor-21**) where the central heterocoronene core is connected to the wedge-shaped side groups (Fig. 5.23a, Table 5.10)

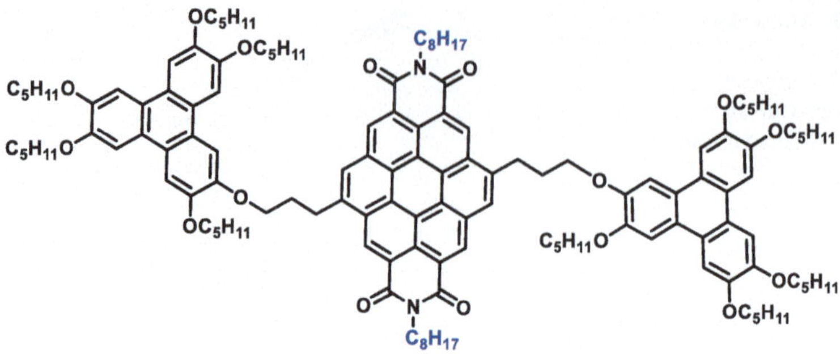

Fig. 5.22 Molecular structure of coronenediimide derivative **Cor-18** along with the phase transition temperatures upon heating. (Redrawn from Ref. [44])

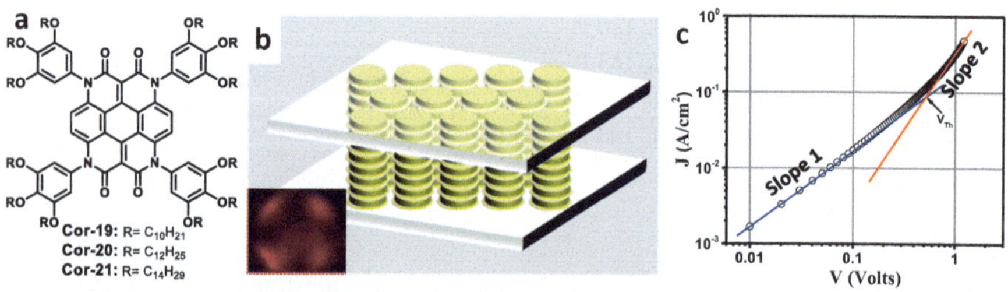

Fig. 5.23 (**a**) Molecular structures of heterocoronene derivatives **Cor-19** to **Cor-21**, (**b**) schematic diagram of homeotropic alignment in SCLC hole-only device of **Cor-19** (inset shows the corresponding conoscopic image), and (**c**) *J-V* curve in SCLC hole-only cell (thickness = 3.02 μm) of **Cor-19** (Slope 1 and Slope 2 represent the linear and SCLC regime, respectively). (Redrawn and reproduced with the permission from Ref. [45])

[45]. Three derivatives were reported by varying the length of alkyl tails connected to the wedge-shaped promesogenic groups. The derivative with the shortest decyloxy chain (R = -$C_{10}H_{21}$) exhibited Col$_r$ mesophase at lower temperatures and Col$_h$ at higher temperatures, while the other two derivatives with dodecyloxy (R = -$C_{12}H_{25}$) and tetradecyloxy (R = -$C_{14}H_{29}$) chains showed only Col$_h$ mesophase. Interestingly, all the derivatives were observed to be room temperature LCs. For **Cor-19**, the maximum value of hole mobility obtained by SCLC method was 8.84 $cm^2V^{-1}s^{-1}$ (at room temperature) as determined from the current-density voltage (*J-V*) curve shown in Fig. 5.23c. On the other hand, the highest electron mobility value of 3.59 $cm^2V^{-1}s^{-1}$ was perceived for **Cor-21** (Table 5.10). The high mobility values obtained in these heterocoronene derivatives were attributed to the high homeotropic alignment capability of these derivatives in the SCLC cells (Fig. 5.23b). Similarly, the observed difference in the mobility values for all the three derivatives is due to the extent of alignment obtained in the respective cells.

Table 5.10 Phase behavior and SCLC mobility (μ_{SCLC}) of **Cor-19** to **Cor-21** [45]

Compound	Phase transition temperatures (in °C)[a]	Hole mobility $\mu_{h,\ SCLC}$ $(cm^2V^{-1}s^{-1})^b$	Electron mobility $\mu_{e,\ SCLC}$ $(cm^2V^{-1}s^{-1})^b$
Cor-19	Col_r 64.7 Col_h 94.9 Iso	8.84	–
Cor-20	Col_h 73.7 Iso	0.51	0.36
Cor-21	Col_h 93.1 Iso	0.70	3.59

[a]Upon heating cycle
[b]Highest measured mobility values

The extent of alignment was measured by polarized optical microscopy (POM) and grazing-incidence small-angle X-ray scattering studies and justified the obtained mobility trend, i.e., alignment is higher in **Cor-19** > **Cor-20** > **Cor-21** and so is the mobility.

In 2005, Mullen and co-workers reported planar alignment of liquid crystalline hexabenzocoronene derivative (**Cor-3**) [46] by using a zone-casting technique on the poly(tetrafluoroethane) layer for the attainment of ordered semiconducting channels in the OFET [47]. High-resolution transmission electron microscopy image of the zone-cased film of **Cor-3** revealed homogeneous morphology in which molecules were arranged in highly ordered columnar fashion and columns were aligned along the direction of zone-casting. Upon employing this uniaxially aligned columnar layers of **Cor-3** in OFET, it exhibited FET mobility of 5×10^{-3} $cm^2V^{-1}s^{-1}$ with a high on/off ratio of 10^4 as determined from the output and transfer plots of the measured device. Further, Mullen and co-workers utilized magnetic field as an effective technique for the alignment of the thin film of hexabenzocoronene derivative (**Cor-7**) and studied charge transport by FET technique [48].

5.2.1.5 Perylene

In 2000, when phthalocyanines become famous as *p*-type discotic materials, at that time, the need was to explore their contemporary *n*-type materials. Struijk et al. reported the three N,N′-dialkyl-3,4,9,10-perylenetetracarboxyldiimides [49]. Out of the three synthesized derivatives, only **Pery-1** (Fig. 5.24a) was studied for charge

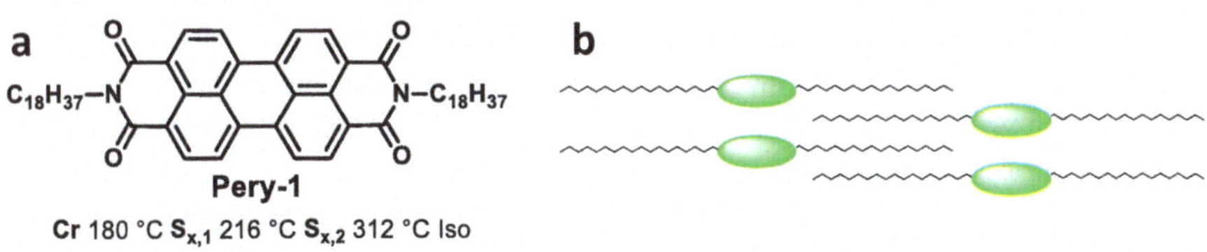

Fig. 5.24 (**a**) Chemical structure of **Pery-1** along with its phase transition temperatures (in °C obtained upon second heating, S_x denotes highly ordered LC phase) and (**b**) molecular packing of **Pery-1** derived from XRD studies. (Redrawn from Ref. [49])

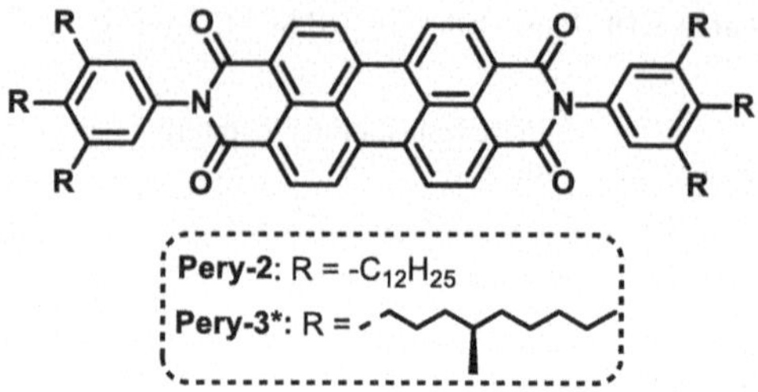

Fig. 5.25 Chemical structures of **Pery-2** and **Pery-3***. (Redrawn from Ref. [50])

Table 5.11 Phase behavior and PR-TRMC mobility ($\Sigma\mu_{TRMC}$) values of **Pery-2** and **Pery-3** [50]

Compound	Phase transition temperatures (in °C)	$\Sigma\mu_{TRMC}$ ($cm^2V^{-1}s^{-1}$)
Pery-2	Col_{hd} 300 Iso	0.0078
Pery-3*	Col_{ho} 349 Iso	0.030

transport measurements by PR-TRMC method. **Pery-1** substantiates smectic layer ordering, and within these layers, a 2D order was observed with cell parameter of 3.6 Å that originates from the π-π stacking of neighboring, co-facially stacked perylene cores. This assembly resulted in the formation of columnar stacks and showed the characteristics of both smectic and columnar discotic ordering (Fig. 5.24b). PR-TRMC measurements showed that **Pery-1** exhibited a charge carrier mobility of 0.11 $cm^2V^{-1}s^{-1}$ in its LC phase.

The chiral columnar mesophase can influence the charge transport properties as can be evident by the studies shown by Wurthner and co-workers in achiral (**Pery-2**) and chiral derivative (**Pery-3***) of perylene bisimide (Fig. 5.25) [50]. Interestingly, the achiral derivative **Pery-2** exhibited disordered hexagonal mesophase (Col_{hd}), while its chiral counterpart **Pery-3*** showed the existence of highly ordered hexagonal (Col_{ho}) mesophase (Table 5.11). This indicated that the intracolumnar order has improved upon the formation of chiral helical superstructures. The semiconducting properties explored by PR-TRMC technique revealed the mobility value of 0.0078 $cm^2V^{-1}s^{-1}$ in Col_{hd} phase of **Pery-2** at 110 °C and 0.030 $cm^2V^{-1}s^{-1}$ in Col_{ho} phase of **Pery-3*** (Table 5.11). The increase in intracolumnar order in **Pery-3*** is accounted for its higher charge carrier mobility in comparison to **Pery-2**.

Wang et al. in 2015 reported a LC derivative of perylene tetracarboxylic diester monoanhydride **Pery-4** (Fig. 5.26a) that self-assembled into a bundled stack columnar assembly [51]. The authors in this work deduced the detailed organization of the mesogenic molecules in the columnar π-stack from the experimentally obtained well-aligned 2D diffraction pattern which suggests the arrangement of 17 π-stacked molecules of **Pery-4** per column. Inside each π-stack, the mesogen rotates around its plane normal with respect to the neighboring mesogen, and the direction of rotation alternates along the stacking axis (Fig. 5.26b). Such kind of self-assembly is beneficial for high charge transport properties as it leads to the electronic coupling between the adjacent stacks which are otherwise restricted by the presence of either peripheral alkyl chains or hydrogen atoms. As a result, the reasonable high charge

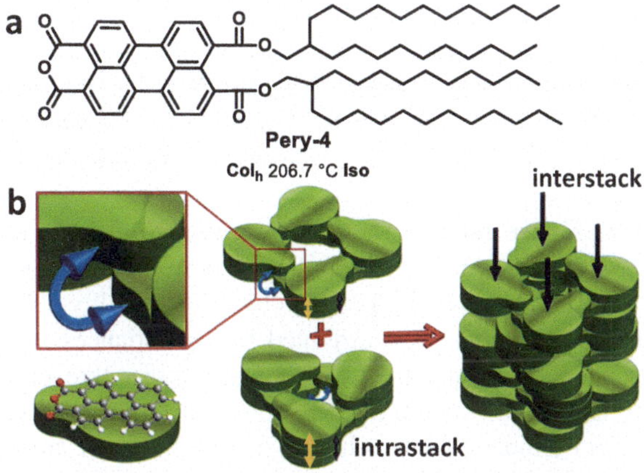

Fig. 5.26 (**a**) Chemical structure of **Pery-4**. (Redrawn from Ref. [51]) along with its phase transition temperature upon second heating. (**b**) Schematic of a single column packing showing the bundled stack organization of mesogens, charge, and energy transport directions. The inset shows a magnified drawing of inter-stack anhydride overlap. (Reproduced with the permission from Ref. [51])

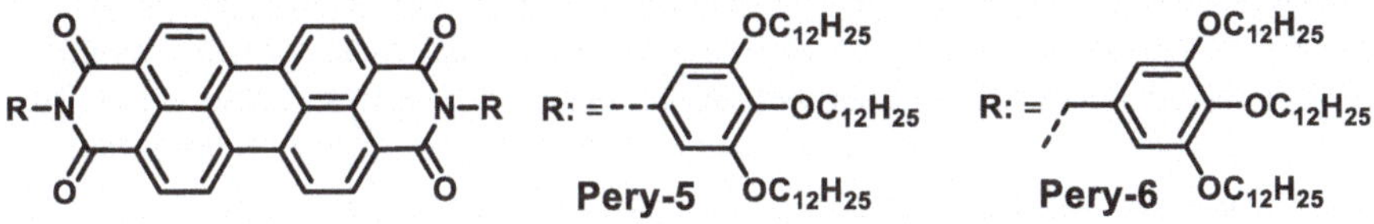

Fig. 5.27 Chemical structures of **Pery-5** and **Pery-6**. (Redrawn from Ref. [52])

Table 5.12 Phase behavior and SCLC mobility ($\mu_{e,\ SCLC}$) values of **Pery-5** and **Pery-6** [52]

Compound	Phase transition temperatures (in °C)	$\mu_{e,\ SCLC}$ ($cm^2V^{-1}s^{-1}$)
Pery-5	Cr −10 Col$_h$ 360 Iso	0.2 (in aligned sample) 10^{-4} (in unaligned sample)
Pery-6	Cr −10 Col$_h$ 226 Iso	1.3

carrier mobility of 5.41×10^{-2} $cm^2V^{-1}s^{-1}$ was reported for the **Pery-4** derivative by PR-TRMC technique.

Marder and group in 2005 reported two room temperature perylene bisimide-cored LCs **Pery-5** and **Pery-6** (Fig. 5.27) where the charge carrier mobility values were studied by SCLC technique [52]. The electron mobility of 0.2 $cm^2V^{-1}s^{-1}$ was reported for **Pery-5** (in aligned sample) and 1.3 $cm^2V^{-1}s^{-1}$ for **Pery-6** (Table 5.12). Both the derivatives differ in the sense that in **Pery-5**, the promesogenic tri (dodecyloxy)phenyl groups connected directly to the perylene bisimide moiety, but in **Pery-6**, the same groups are connected by methylene linker. **Pery-5** exhibited Col$_h$ mesophase from −10 °C to 360 °C with high isotropic transition, and **Pery-6** showed the Col$_h$ between −10 °C and 226 °C. In this work, the authors discussed clearly how for the same compound but two different samples of the same thickness (5 μm), the mobility values varied by several orders of magnitude. For instance, in the case of **Pery-5**, one sample exhibited 0.2 $cm^2V^{-1}s^{-1}$ mobility, while another sample showed mobility value less than 10^{-4} $cm^2V^{-1}s^{-1}$. This was explained by the fact that POM images of both the samples showed the formation of different types of domains.

The sample with higher mobility showed uniform region, while the sample with low mobility consisted of several small domains that might act as grain boundaries and hindered the free flow of charge, thereby less mobility. The morphology difference in both the samples was attributed to the variances in the processing conditions of the samples to which the LC materials are sensitive, unlike amorphous organic materials. On the other hand, **Pery-6** showed electron mobility as high as 1.3 $cm^2V^{-1}s^{-1}$ at ambient temperature which is higher than that of amorphous silicon. Similarly, Bechtold showed in his report how the homeotropic alignment led to 5 orders increase in mobility in comparison to the unaligned sample [53].

Thelakkat group reported perylene derivatives where the oligoethyleneglycol (OEG) chains induced LC phases (**Pery-8** and **Pery-9**), while the alkyl-substituted compound was Cr (**Pery-7**) (Fig. 5.28) [54]. The films prepared by doctor-blading technique from chloroform solutions of the perylene derivatives were further annealed for 5 min at 1–5 °C below isotropic temperatures in order to induce the better packing and morphology of the thin films. The SCLC charge carrier mobilities were extracted for the pristine as well as the annealed samples (Table 5.13). The results showed that annealing process does not have much impact on the improvement of mobility values in the case of Cr material **Pery-7**, while in the case of LC materials **Pery-8** and **Pery-9**, the mobility values increased by two orders of magnitude. The higher mobility values obtained upon annealing in LC materials are due to favorable structural reorganization of the molecules in the film, which is beneficial for charge transport. These findings indicate that the development of LC materials is significantly important for the organic electronics where the alignment of the molecules can be improved by thermal annealing and hence the charge transport can be enhanced. Moreover, LC materials also provide the advantage of self-healing from defect upon thermal annealing which otherwise can act as a trap for charge carriers and detrimental for charge transport. However, in Cr materials, it's difficult to control the orientation of molecules by thermal means due to their rigid nature, and thereby, domain boundaries can seriously limit their charge transport.

Sonoda group studied charge transport by TOF method in nanosegregated columnar phases of perylene tetracarboxylic bisimide derivatives bearing disiloxane [55, 56], trisiloxane [56, 57], and cyclotetrasiloxane [58] flexible units. These units are connected to perylene tetracarboxylic bisimide units through swallow-tail alkyl chains. The siloxane units are known for inducing high flexibility in the molecules, thereby responsible for inducing room temperature columnar mesophases. His group prepared several derivatives, some of them were found to be non-mesomorphic, and mobility measurements were majorly performed in mesomorphic samples. For instance, **Pery-10** to **Pery-13** all exhibit columnar mesomorphism at room temperature (Table 5.14). At periphery, the molecular structures of **Pery-10**, **Pery-11**, and

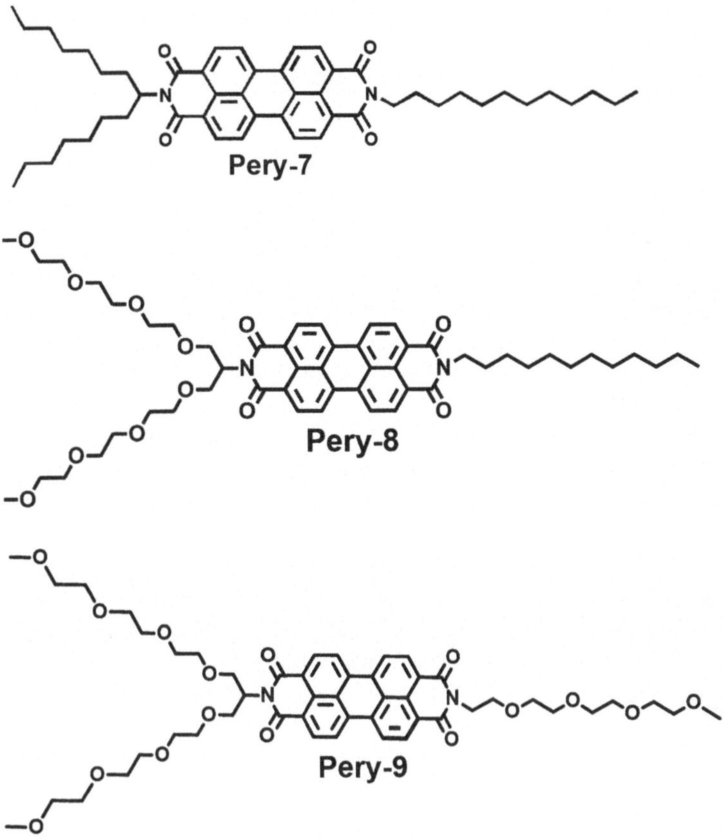

Fig. 5.28 Chemical structures of **Pery-7**, **Pery-8**, and **Pery-9**. (Redrawn from Ref. [54])

Table 5.13 Phase behavior and SCLC mobility ($\mu_{e,\ SCLC}$) values of **Pery-7** to **Pery-9** [54]

		$\mu_{e,\ SCLC}$ ($cm^2V^{-1}s^{-1}$)[b]	
Compound	Phase transition temperatures (in °C)[a]	Pristine films	Annealed films
Pery-7	$CrCol_{lam}$ 181.8 Iso	3×10^{-5}	8×10^{-5}
Pery-8	Col_h 136.4 Iso	2×10^{-5}	6×10^{-3}
Pery-9	Col_{lam} 164.0 Iso	6×10^{-5}	7×10^{-3}

[a] Upon second heating
[b] Average mobility values over varied active layer thickness
Abbreviation: $CrCol_{lam}$ Crystalline lamello-columnar

Pery-12 consisted of aliphatic oligosiloxane chains [56], while those of **Pery-13** bear cyclotetrasiloxane rings [58] (Fig. 5.29). The observed electron mobility trend in aliphatic oligosiloxane chains consisted derivatives was **Pery-10** > **Pery-11** > **Pery-12** [56].

Table 5.14 Phase behavior and SCLC electron mobility ($\mu_{e,\ SCLC}$) values of **Pery-10** to **Pery-13** [55–58]

Compound	Phase transition temperatures (in °C)[a]	$\mu_{e,\ SCLC}$ ($cm^2V^{-1}s^{-1}$)
Pery-10	−56.0 Col_{ro} 47.0 Col_{ho} 138.3 Iso	1.1×10^{-1} (at 37 °C)
Pery-11	−54.0 Col_{ho} 129.0 Iso	6.9×10^{-2} (at 97 °C)
Pery-12	−82.0 Col_{hd} 50.5 Iso	1×10^{-3} (at 30 °C)
Pery-13	−35.5 Col_{rd} 80.1 Iso	1×10^{-2} (at 25 °C)

[a]Upon heating scan

Pery-12 with trisiloxane bearing swallow chains exhibited disordered hexagonal phases and showed comparatively lower mobilities in comparison to the ordered columnar phases of **Pery-10** and **Pery-11** with disiloxane swallow alkyl tails. On the other hand, **Pery-13** bearing cyclotetrasiloxane flexible units with disordered Col_r mesophases exhibited mobility of the same order (i.e., 10^{-2} $cm^2V^{-1}s^{-1}$) to that of **Pery-11** (bearing disiloxane swallow tails) with ordered hexagonal mesophases. Interestingly, in all the four derivatives, TOF measurements revealed non-dispersive nature of the transient photocurrent curves clearly showing the transit time from which the mobility values could be easily estimated.

Pal and co-workers also reported the charge transport measurements in the ordered Col_r mesophase of perylene tetraester derivative **Pery-14** by TOF technique (Fig. 5.30) [59]. **Pery-14** exhibited room temperature ordered Col_r phase and melted into isotropic liquid at 115.5 °C. The mobility value was found to be 0.014 $cm^2V^{-1}s^{-1}$ in the mesophase. The perylene derivative with high electron mobility served as efficient green emitter in solution-processed green organic light-emitting diodes.

5.2.1.6 Pyrene

Pyrene is a planar discotic core consisting of four fused benzene rings and has been utilized to develop DLCs with efficient charge transport behavior. Kato et al. reported X-shaped pyrene-substituted oligothiophene derivatives **Pyr-1** and **Pyr-2** (Fig. 5.31a) where their photoconductive properties were studied by TOF method [60]. These X-shaped mesogens exhibited columnar mesomorphism with the occurrence of Col_h, Col_t, and Col_r mesophases over a wide temperature range (Table 5.15). The aim of attaching the oligothiophene segments to the pyrene core was to provide the extended conjugation which is beneficial for the better charge transport properties and contributed to the low band gap. The temperature variation of hole mobilities for **Pyr-1** and **Pyr-2** derivative are displayed in Fig. 5.31b. In the case of **Pyr-1**, the mobility in the Col_{h2} phase raised to 4.4×10^{-4} $cm^2V^{-1}s^{-1}$ (at 117 °C) than in the isotropic phase. This increase in the mobility value was corroborated to the formation of 1D π-π columnar stack. On further lowering the temperature from Col_{h2} to Col_{h1} to a glassy state, the mobility value slightly decreased.

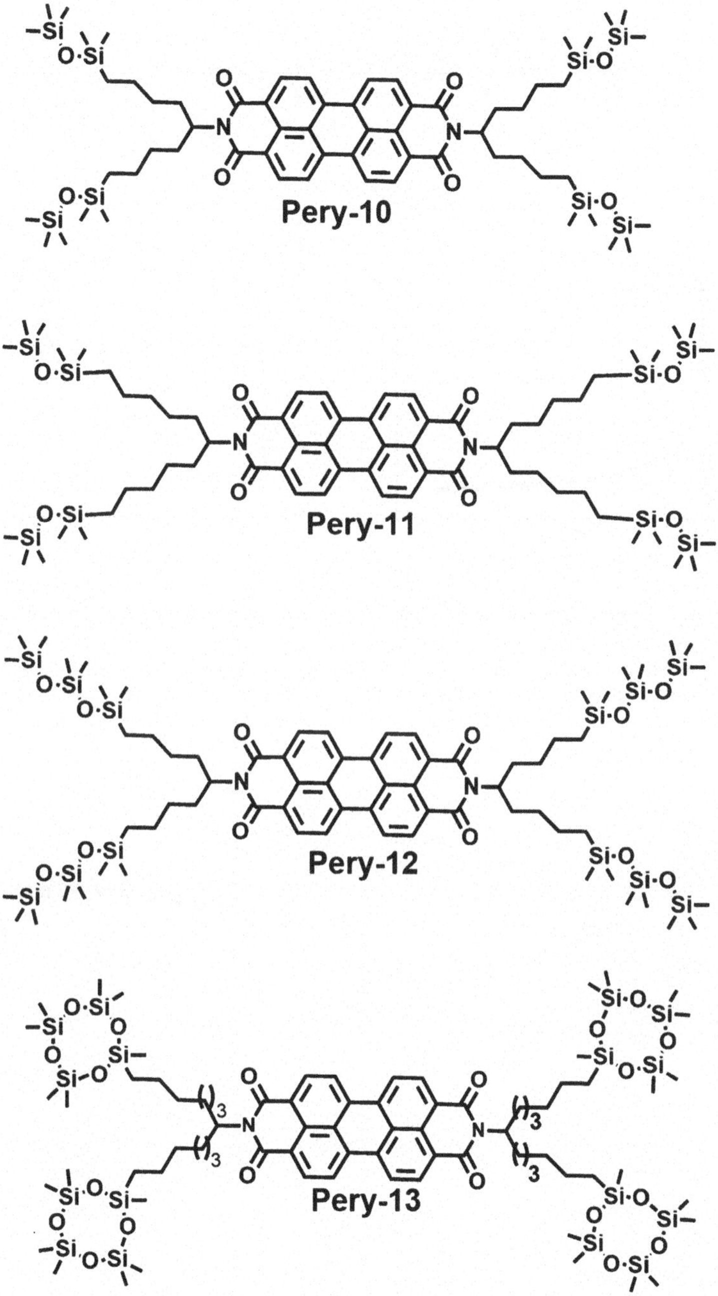

Fig. 5.29 Chemical structures of perylene tetracarboxylic bisimide derivatives **Pery-10**, **Pery-11**, **Pery-12**, and **Pery-13**. (Redrawn from Refs. [56–59])

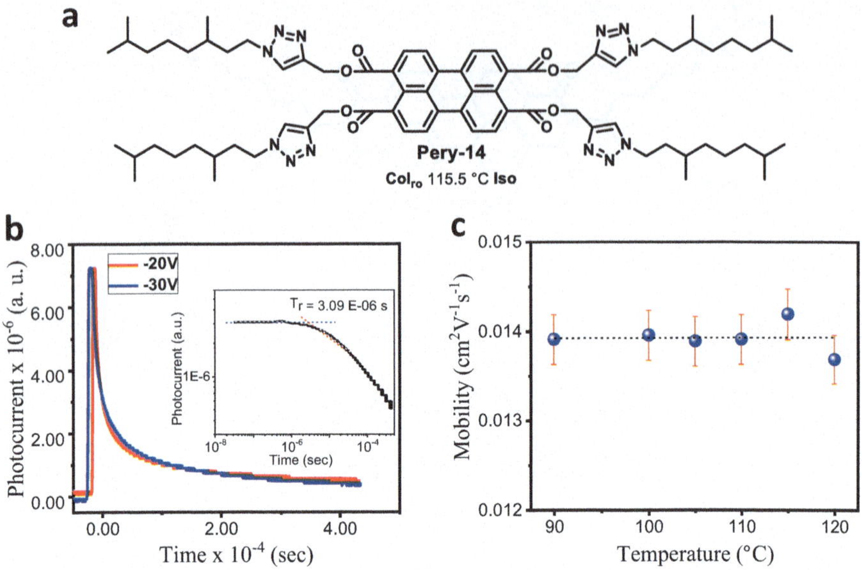

Fig. 5.30 (**a**) Chemical structure of perylene tetraester derivative **Pery-14** along with its phase transition temperature upon heating; (**b**) dispersive transient photocurrent curves of **Pery-14** at two different applied voltages, i.e., -20 V and – 30 V (the inset presents the log-log scale curve of dispersive transit photocurrent at −20 V to estimate transit time (T_r)); and (**c**) electron mobility versus temperature curve of **Pery-14**. (Redrawn and reproduced with the permission from Ref. [59])

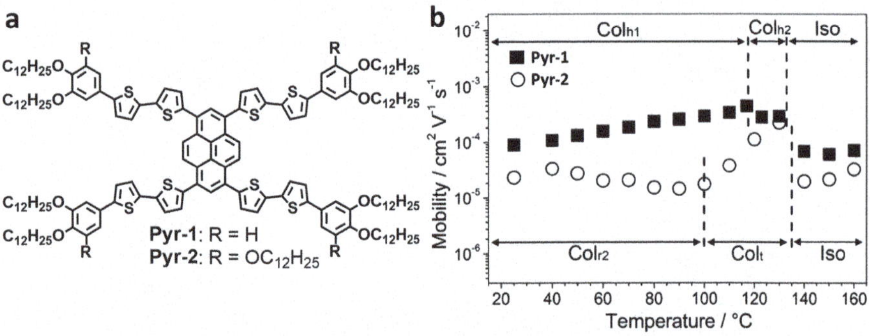

Fig. 5.31 (**a**) Molecular structures of **Pyr-1** and **Pyr-2**. (Redrawn from Ref. [60]) (**b**) Variation of hole mobilities with temperature in the case of **Pyr-1** and **Pyr-2**. (Reproduced with the permission from Ref. [60])

Table 5.15 Phase behavior and TOF mobility values of **Pyr-1** and **Pyr-2** [60]

Compound	Phase transition temperatures (in °C)[a]	$\mu_{h, TOF}$ (cm^{2}V^{-1}s^{-1})
Pyr-1	G 20 Col$_{h1}$ 117 Col$_{h2}$ 132 Iso	4.4×10^{-4} (at 117 °C)
Pyr-2	Col$_{r1}$ 10 Col$_{r2}$ 100 Col$_t$ 135 Iso	2.3×10^{-4} (at 130 °C)

[a]Upon heating scan

On the other hand, for **Pyr-2**, the maximum mobility value of 2.3 × 10^{-4} $cm^2V^{-1}s^{-1}$ (at 130 °C) was observed in the Col_t phase, and within this mesophase, the mobility values decreased by one order. However, in the Col_{r2} mesophase, no significant change was observed upon varying temperature.

5.2.1.7 Truxene Family

Truxene is a heptacyclic C_3 symmetric core composed of three fluorene units that shared a common benzene ring [5]. In truxene core, the methylene group of three fluorene units can be easily oxidized and form truxenone core. On the other hand, the replacement of methylenes with –NH moiety forming triazatruxene or triindole discotic cores. These cores are highly planar and are useful to produce functional DLC materials with good charge transport properties.

In 2008, Gomez-Lor and co-workers synthesized N-substituted truxene (i.e., triindole) surrounded with six decyloxy chains at the periphery (**Tx-1**, **Tx-2**, and **Tx-3**) (Fig. 5.32) [61]. All the three derivatives exhibited Cr state at room temperature and Col_h mesophase at higher temperatures (Table 5.16). The charge transport behavior of **Tx-1** to **Tx-3** was measured by SCLC method using ~5 μm cell. In Col_h mesophase, the hole mobilities of (4 ± 1) × 10^{-3} $cm^2V^{-1}s^{-1}$, (0.03 ± 0.01) $cm^2V^{-1}s^{-1}$, and (0.02 ± 0.005) $cm^2V^{-1}s^{-1}$ were obtained for **Tx-1**, **Tx-2**, and **Tx-3**, respectively (Table 5.16). The high mobilities in the mesophase were attributed to the high degree of homeotropic alignment in the macroscopic scale of columnar mesophases. Similarly, the same group reported several triindole-based

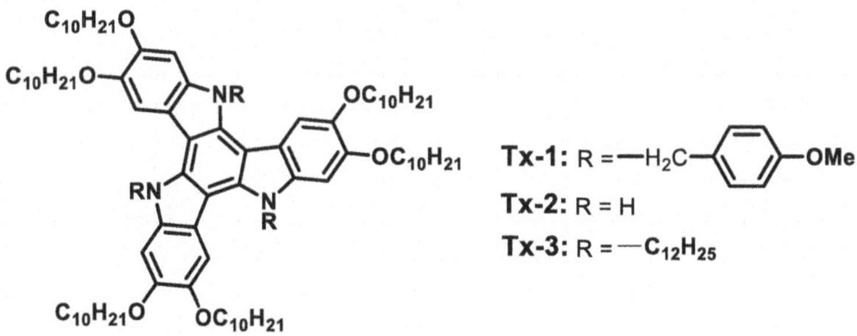

Fig. 5.32 Molecular structures of **Tx-1**, **Tx-2**, and **Tx-3**. (Redrawn from Ref. [61])

Table 5.16 Phase behavior and SCLC hole mobility ($\mu_{h, SCLC}$) of **Tx-1** to **Tx-3** [61]

Compound	Phase transition temperatures (in °C)[a]	$\mu_{h, SCLC}$ ($cm^2V^{-1}s^{-1}$)
Tx-1	Cr 84.7 Col_h 151.5 Iso	(4 ± 1) × 10^{-3}
Tx-2	Cr 76.0 Col_h 154.8 Iso	0.03 ± 0.01
Tx-3	Cr 38.9 Col_h 67.3 Iso	0.02 ± 0.005

[a]Upon first heating cycle

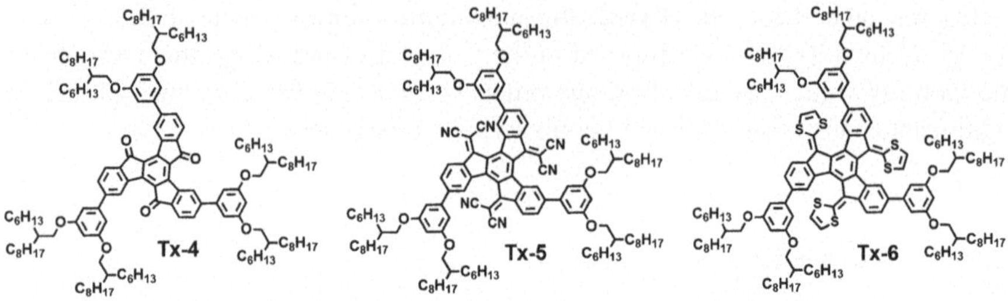

Fig. 5.33 Molecular structures of **Tx-4**, **Tx-5**, and **Tx-6**. (Redrawn from Ref. [64])

Table 5.17 Phase behavior and TOF mobilities of **Tx-4** to **Tx-6** [64]

		TOF mobility ($cm^2V^{-1}s^{-1}$)	
Compound	Phase transition temperatures (in °C)[a]	$\mu_{h, TOF}$	$\mu_{e, TOF}$
Tx-4	Col_h 106 Iso	–	3×10^{-4}
Tx-5	Col[b] 245 Iso	–	5×10^{-4}
Tx-6	Col[b] 161 Iso	5×10^{-4}	–

[a]Upon second heating scan
[b]Columnar mesophase

DLCs where the SCLC mobility was dependent on the extent of alignment of the sample on the macroscopic scale [62, 63].

Kato and co-workers reported substituted truxenone derivatives and showed how the electronic properties could be tuned and altered from *n*-type to *p*-type with the interplay of the substituents. For instance, **Tx-4** derivative was substituted with dicyanomethylene (**Tx-5**) and dithiafulvene (**Tx-6**) (Fig. 5.33) substituents in order to tune the electronic behavior [64]. All three compounds exhibited stable columnar LC mesophase (Table 5.17). It was observed that **Tx-5** and **Tx-6** had higher thermal range of mesophase than that of **Tx-4** which implied that upon inclusion of dicyanomethylene and dithiafulvene to the truxenone core stabilizes the columnar self-assembly. Further, the HOMO energy level (E_{HOMO}) of **Tx-6** (−4.96 eV) is much higher in comparison to **Tx-4** (−5.96 eV) and **Tx-5** (−6.13 eV), suggesting the high electron-donating capability of **Tx-6**. Thus, **Tx-6** and **Tx-5** were expected to show hole- and electron-transporting materials, respectively. Further, their charge transport properties were determined by TOF technique. **Tx-4** and **Tx-5** exhibited electron-transporting (*n*-type) behavior, whereas **Tx-6** showed hole-transporting (*p*-type) nature.

The hole mobility of **Tx-6** was determined to be 5×10^{-4} $cm^2V^{-1}s^{-1}$ at 100 °C (Table 5.17). It was also observed that with varying temperature from 60 to 120 °C in the columnar mesophase, the hole mobility of **Tx-6** remained almost constant. The electron mobility of **Tx-4** and **Tx-5** was calculated as 3×10^{-4} and 5×10^{-4} $cm^2V^{-1}s^{-1}$, respectively, in columnar mesophase at 100 °C (Table 5.17). In 2014, Hanna and co-workers synthesized a windmill-shaped truxene derivative attached with

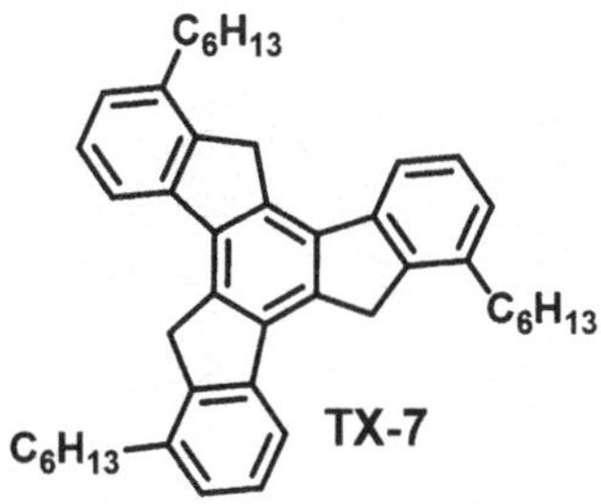

Cr 127 °C **Col$_r$** 144 °C **Col$_h$** 175 °C **Iso**

Fig. 5.34 Molecular structure of **Tx-7** along with the thermal transitions. (Redrawn from Ref. [65])

three alkyl chains at its periphery (**Tx-7**) [65]. **Tx-7** exhibited two types of columnar assembly (Col$_{ro}$ and Col$_{ho}$) in its mesophase range (Fig. 5.34). This is the first report where they showed ambipolar charge transport in truxene-based DLCs as measured by TOF method. Hole mobility of 0.2 $cm^2V^{-1}s^{-1}$ and electron mobility of 0.17 $cm^2V^{-1}s^{-1}$ was obtained for **Tx-7**. The temperature variation mobility studies of **Tx-7** showed that irrespective of the glassy phase and columnar phases, both electron and hole mobility values do not change while suddenly dropped in the isotropic phase. In addition to temperature independency, **Tx-7** displayed electric field-independent behavior, indicating its potential to be used as a potential ambipolar DLC material.

Later on, Gomez-Lor and co-worker studied the charge transport behavior in π-extended triazatruxene (i.e., triindole) derivative **Tx-8** attached with flexible dodecyl chains at the N-atoms and methyl group at the periphery (**Tx-8**) by two complementary techniques (Fig. 5.35) [66]. **Tx-8** showed Col$_h$ mesophase at room temperature with a broad range of temperature. The hole mobility of 0.65 $cm^2V^{-1}s^{-1}$ was estimated by SCLC technique in a diode-like configuration, whereas, in the solution-processed OFETs, **Tx-8** exhibited hole mobility of 1.04×10^{-4} $cm^2V^{-1}s^{-1}$. This difference of mobility values in two different methods was explained in terms of molecular alignment in the respective devices.

5.2.1.8 Thiophene

Thiophene is used as electron-rich ring and has been fused with several polycyclic aromatic cores to form various thiophene-fused aromatic cores. Donnio et al. recently investigated charge transport properties in thiophene-fused benzothienobenzothiophene-cored columnar mesogens [67]. Several such columnar mesogens were reported in this work by varying the peripheral alkyl chain lengths, while mobility studies were performed in one of the representative compounds **Thio-1** (Fig. 5.36) that possessed decyloxy alkyl chains at the periphery. **Thio-1** exhibited Cr phase below 87 °C and then Col$_h$ mesophase up to 260 °C upon cooling cycle. The non-dispersive transient photocurrent curves were observed for hole as well as

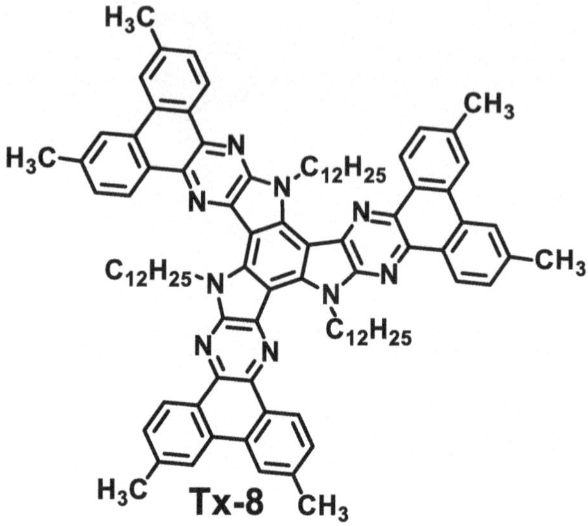

g 26 °C Col$_h$ 226 °C dec

Fig. 5.35 Molecular structure of **Tx-8** along with the thermal transitions. (Redrawn from Ref. [66])

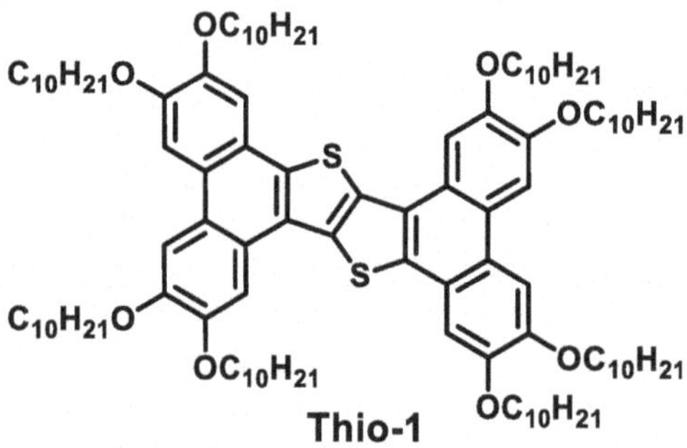

Cr 87 °C Col$_h$ 260 °C Iso

Fig. 5.36 Molecular structure of **Thio-1** along with the phase transition temperatures obtained upon cooling process. (Redrawn from Ref. [67])

electron charge carriers obtained in the hexagonal phase indicating the absence of charge traps and electrochemical stability of the derivative. However, the dispersive curves were obtained in the Cr range. The mobility studies by varying temperature showed that both hole and electron mobilities in the mesophase were higher than 10^{-3} $cm^2V^{-1}s^{-1}$ and were almost independent of temperature. The same group has explored many other fused thiophene derivatives for the charge transport studies [68].

Pal's group also explored the fused thiophene derivatives where the tetrathienoanthracene core was connected to the 3,4,5-trialkoxy phenyl linkers through the alkynyl spacers [69]. Herein, the alkynyl linkers act as twofold: one in extending the conjugation throughout the whole molecule and the other in providing the planarization to the overall molecular structure. The two derivatives **Thio-2** with decyloxy and **Thio-3** with dodecyloxy peripheral chain length (Fig. 5.37a) were

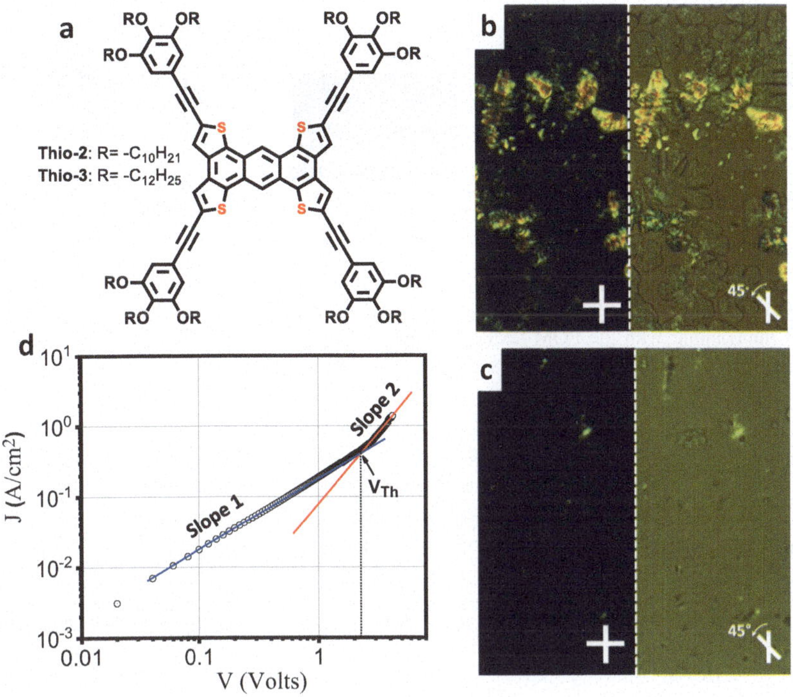

Fig. 5.37 (**a**) Molecular structures of tetrathienoanthracene derivatives **Thio-2** and **Thio-3** (Redrawn from Ref. [69]). POM images of SCLC mobility measured cell showing the extent of alignment observed in (**b**) **Thio-2** and (**c**) **Thio-3** derivative. (**d**) *J-V* curve for **Thio-3** consists of ideal ohmic (Slope 1) and SCLC regime (Slope 2). (Redrawn and reproduced with the permission from Ref. [69])

Table 5.18 Phase behavior and SCLC hole mobility ($\mu_{h,\ SCLC}$) values of **Thio-2** and **Thio-3** [69]

Compound	Phase transition temperatures (in °C)[a]	$\mu_{h,\ SCLC}$ ($cm^2V^{-1}s^{-1}$)[b]
Thio-2	Col_r 88.2 Col_h 181.2 Iso	$(4.13 \pm 0.69) \times 10^{-2}$
Thio-3	Col_r 78.7 Col_h 164.7 Iso	3.42 (4.22)[c] ± 0.86

[a]Upon heating scan. [b]Mobility values obtained as an average of five samples in different regions. [c]Parenthesis indicated the highest value obtained at room temperature

synthesized where the high hole mobility of 4.22 $cm^2V^{-1}s^{-1}$ was obtained for **Thio-3** (Table 5.18). However, the mobility value was in the order of 10^{-2} $cm^2V^{-1}s^{-1}$ in the case of **Thio-2** derivative (Table 5.18). The differences in the extent of homeotropically aligned samples in the respective SCLC cells (obtained by POM) (Fig. 5.37b, c) were attributed to the significant difference in the mobility values of these derivatives. A representative *J-V* curve is shown for **Thio-3** in Fig. 5.37d.

Chunyan Chi reported thiophene-fused tetracene diimide DLC material **Thio-4** (Fig. 5.38a) that assembled into lamellar structures (Fig. 5.38c) in the mesophase range as shown by DSC thermogram (Fig. 5.38b). **Thio-4** derivative showed

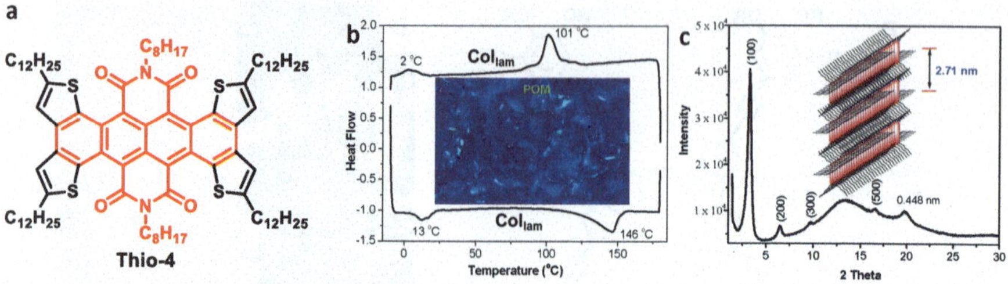

Fig. 5.38 (**a**) Molecular structure of thiophene-fused tetracene diimide derivative **Thio-4**. (Redrawn from Ref. [69]) (**b**) DSC trace obtained on heating and cooling cycle (recorded at 10 °C/min) along with the POM texture of **Thio-4**. and (**c**) lamellar self-assembly obtained for **Thio-4** as determined by XRD studies. (Reproduced with the permission from Ref. [70])

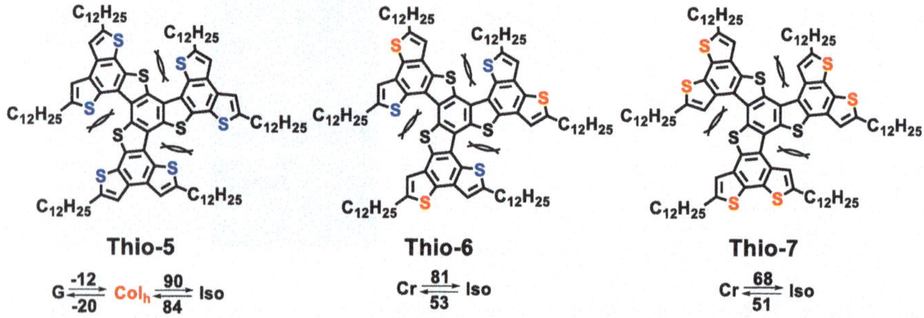

Fig. 5.39 Molecular structures of **Thio-5**, **Thio-6**, and **Thio-7**. (Redrawn from Ref. [71])

ambipolar behavior in thin-film FET [70]. **Thio-4** derivative exhibited low band gap of 1.52 eV with HOMO and LUMO energy level of −5.62 eV and − 4.10 eV, respectively. The energy levels are well corroborated with the ambipolar nature of the derivative. Solution-processed FET devices were fabricated for investigating the charge transport behavior of the **Thio-4** derivative. However, mainly *n*-type behavior was observed for **Thio-4** when the measurements were done under nitrogen atmosphere with electron mobility of 4.83×10^{-3} $cm^2V^{-1}s^{-1}$. But when the mobility measurements were performed in air atmosphere, the devices exhibited expected ambipolar character with hole mobility of 9.0×10^{-4} $cm^2V^{-1}s^{-1}$ and electron mobility of 7.4×10^{-4} $cm^2V^{-1}s^{-1}$.

Aida in 2013 reported propeller-shaped fused oligothiophene derivatives **Thio-5**, **Thio-6**, and **Thio-7** (Fig. 5.39) [71]. These derivatives differ in the position of sulfur (S) atom in the thiophene-fused derivatives as can be seen clearly from their molecular structures. With these structural features, the electronic mobilities of the LC derivative **Thio-5** were evaluated by TOF method which showed the significant amount of balanced ambipolar charge transport with hole and electron mobilities of 0.02 $cm^2V^{-1}s^{-1}$ and 0.03 $cm^2V^{-1}s^{-1}$, respectively (Table 5.19). Besides, the authors investigated the charge transport properties by PR-TRMC method in the LC phase of

Table 5.19 Phase behavior and mobility values of **Thio-5** to **Thio-7** [71]

Compound	Phase transition temperatures (in °C)[a]	Mobility ($cm^2V^{-1}s^{-1}$) μ_{TOF}	μ_{TRMC}
Thio-5	G – 12 Col_h 90 Iso	$\mu_{h,\ TOF} = 0.02$ $\mu_{e,\ TOF} = 0.03$	0.18
Thio-6	Cr* 81 Iso	–	0.13
Thio-7	Cr* 68 Iso	–	0.05

[a]Upon heating scan
Abbreviation: *Cr** Soft crystalline

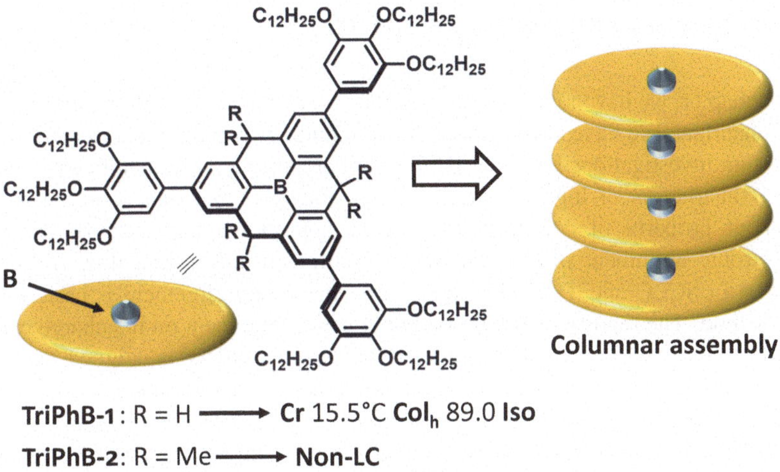

Fig. 5.40 Molecular structures and the columnar assembly of trigonally substituted mesogenic derivatives **TriphB-1** along with the phase transition temperatures (obtained upon heating) and **TriphB-2**. (Redrawn from Ref. [72])

Thio-5 and Cr state of **Thio-6** and **Thio-7**. As a result, 1D charge carrier mobilities of 0.18 $cm^2V^{-1}s^{-1}$, 0.13 $cm^2V^{-1}s^{-1}$, and 0.05 $cm^2V^{-1}s^{-1}$ were perceived for **Thio-5**, **Thio-6**, and **Thio-7**, respectively, in their corresponding phases (Table 5.19).

5.2.1.9 Triphenylborane

Kushida et al. studied the charge transport properties in a uniquely designed mesogenic system consisting of planarized triphenylborane core (Fig. 5.40) [72]. The vacant p orbital of the boron atom imparted these compounds with electron-accepting and Lewis acidic character. Thereby, the attachment of electron-rich 3,4,5-trialkoxyphenyl groups to the electron-deficient triphenylborane core has resulted in the formation of ambipolar structure that transport both holes and electrons.

The two reported derivatives differ in the central triphenylborane core in which the **TriPhB-2** derivative consisted of fully methylated core, and in **TriPhB-1**, the one corner of the core was substituted with hydrogen atoms. **TriphB-1** derivative was found to be columnar LC (Fig. 5.40), while the **TriphB-2** was non-LC. The ambipolar nature of the LC derivative was studied by the TOF method. The hole and electron mobilities calculated from the transit times obtained from their respective graphs were 3×10^{-5} $cm^2V^{-1}s^{-1}$ and approximately10^{-3} $cm^2V^{-1}s^{-1}$, respectively, and confirmed the ambipolar nature of the triphenylborane derivative.

5.3 Summary and Future Perspective

In this chapter, we have reviewed broadly the charge transport studies carried out in DLC materials. We have also covered the new findings from our laboratory. Above imprinting investigations based on DLC-based organic semiconductors have aimed at the construction of highly efficient charge transport materials with the use of alignment property of columnar LCs in an appropriate manner. For instance, homeotropic alignment of the discotic molecules ensures a free flow of charge transport in SCLC and TOF technique, while the planar alignment is preferred in FET method. The reports shed light on the effect of alignment on the charge carrier mobility values which usually improved by several orders in aligned samples. Besides alignment, other advantageous properties of DLCs include long-range self-assembling (order), self-healing (dynamics), ease of processing, and high solubility in organic solvents. However, most of the charge transport studies, especially by PR-TRMC, SCLC, and TOF techniques, were performed in the melt-processable samples and only FET technique majorly employing the solution-processable thin films. It has to be emphasized that the alignment of DLC materials by solution-processable is rather complicated and is still not well explored, and they are, in particular, crucial for the development of highly efficient DLC-based organic semiconductors in real devices. Therefore, there remain much to be done in this direction where the major focus of charge transport measurements of devices prepared from melt-processable technique should be shifted to the solution-processable thin films. This transition will be indeed helpful in further realizing the potential capabilities of the DLC materials in organic electronic devices.

Acknowledgment SKP acknowledges the SERB Project (CRG/2019/000901/OC) for financial support. I. Bala and J. De thank CSIR (09/947(0061)/2015-EMR-I) and CSIR (09/947(0220)/2019/EMR-I), respectively, for funding. We gratefully acknowledge our collaborator Dr. Upendra Kumar Pandey for contribution to the charge transport measurements.

References

1. Jacob MV (2014) Organic semiconductors: past, present and future. Electronics 3:594–597
2. Shirota Y, Kageyama H (2007) Charge carrier transporting molecular materials and their applications in devices. Chem Rev 107(4):953–1010
3. Kaafarani BR (2011) Discotic liquid crystals for opto-electronic applications. Chem Mater 23 (3):378–396
4. Wöhrle T, Wurzbach I, Kirres J, Kostidou A, Kapernaum N, Litterscheidt J, Haenle JC, Staffeld P, Baro A, Giesselmann F, Laschat S (2016) Discotic liquid crystals. Chem Rev 116 (3):1139–1241
5. Kumar S (2016) Chemistry of discotic liquid crystals: from monomers to polymers. CRC press
6. Boden N, Bushby RJ, Clements J (1993) Mechanism of quasi-one-dimensional electronic conductivity in discotic liquid crystals. J Chem Phys 98(7):5920–5931
7. Piechocki C, Simon J, Skoulios A, Guillon D, Weber P (1982) Annelides. 7. Discotic mesophases obtained from substituted metallophthalocyanines. Toward liquid crystalline one-dimensional conductors. J Am Chem Soc 104(19):5245–5247
8. Warman JM, de Haas MP, van der Pol JF, Drenth W (1989) Charge separation and recombination in pulse-irradiated columnar aggregates of peripherally octa-n-alkoxy-substituted phthalocyanines. Chem Phys Lett 164(6):581–586
9. Adam D, Closs F, Frey T, Funhoff D, Haarer D, Schuhmacher P, Siemensmeyer K (1993) Transient photoconductivity in a discotic liquid crystal. Phys Rev Lett 70(4):457
10. Adam D, Schuhmacher P, Simmerer J, Häussling L, Siemensmeyer K, Etzbachi KH, Ringsdorf H, Haarer D (1994) Fast photoconduction in the highly ordered columnar phase of a discotic liquid crystal. Nature 371(6493):141–143
11. Xue C, Li Q (2012) Self-organized semiconducting discotic liquid crystals for optoelectronic applications. In: Liquid crystals beyond displays: chemistry, physics, and applications, p 29
12. Arikainen EO, Boden N, Bushby RJ, Clements J, Movaghar B, Wood A (1995) Effects of side-chain length on the charge transport properties of discotic liquid crystals and their implications for the transport mechanism. J Mater Chem 5(12):2161–2165
13. Marcus RA (1993) Electron transfer reactions in chemistry. Theory Exp Rev Mod Phys 65 (3):599–610
14. Domercq B, Yu J, Kaafarani BR, Kondo T, Yoo S, Haddock JN, Barlow S, Marder SR, Kippelen B (2008) A comparative study of charge mobility measurements in a diamine and in a hexaazatrinaphthylene using different techniques. Mol Cryst Liq Cryst 481(1):80–93
15. Van der Pol JF, Neeleman E, Zwikker JW, Nolte RJM, Drenth W, Aerts J, Visser R, Picken SJ (1989) Homologous series of liquid-crystalline metal free and copper octa-n-alkoxyphthalocyanines. Liq Cryst 6(5):577–592
16. Schouten PG, Warman JM, De Haas MP, Van der Pol JF, Zwikker JW (1992) Radiation-induced conductivity in polymerized and nonpolymerized columnar aggregates of phthalocyanine. J Am Chem Soc 114(23):9028–9034
17. Van der Pol JF, Zwikker JW, Warman JM, De Haas MP (1990) Synthesis and transport properties of rigid poly (octasubstituted phthalocyaninatosiloxanes). Recueil des Travaux Chimiques des Pays-Bas 109(3):208–215
18. Fujikake H, Murashige T, Sugibayashi M, Ohta K (2004) Time-of-flight analysis of charge mobility in a Cu-phthalocyanine-based discotic liquid crystal semiconductor. Appl Phys Lett 85 (16):3474–3476
19. Cherian S, Donley C, Mathine D, LaRussa L, Xia W, Armstrong N (2004) Effects of field dependent mobility and contact barriers on liquid crystalline phthalocyanine organic transistors. J Appl Phys 96(10):5638–5643
20. Deibel C, Janssen D, Heremans P, De Cupere V, Geerts Y, Benkhedir ML, Adriaenssens GJ (2006) Charge transport properties of a metal-free phthalocyanine discotic liquid crystal. Org Electron 7(6):495–499

21. Hayashi H, Nihashi W, Umeyama T, Matano Y, Seki S, Shimizu Y, Imahori H (2011) Segregated donor–acceptor columns in liquid crystals that exhibit highly efficient ambipolar charge transport. J Am Chem Soc 133(28):10736–10739
22. Schouten PG, Warman JM, de Haas MP, Fox MA, Pan HL (1991) Charge migration in supramolecular stacks of peripherally substituted porphyrins. Nature 353(6346):736–737
23. Gregg BA, Fox MA, Bard AJ (1989) 2,3,7,8,12,13,17,18-Octakis (Beta.-hydroxyethyl) porphyrin (octaethanolporphyrin) and its liquid crystalline derivatives: synthesis and characterization. J Am Chem Soc 111(8):3024–3029
24. Concellón A, Marcos M, Romero P, Serrano JL, Termine R, Golemme A (2017) Not only columns: high hole mobility in a discotic nematic mesophase formed by metal-containing porphyrin-core dendrimers. Angew Chem 129(5):1279–1283
25. Concellón A, Termine R, Golemme A, Romero P, Marcos M, Serrano JL (2019) High hole mobility and light-harvesting in discotic nematic dendrimers prepared via 'click' chemistry. J Mater Chem C 7(10):2911–2918
26. Concellón A, Termine R, Golemme A, Romero P, Marcos M, Serrano JL (2020) Semiconducting and electropolymerizable liquid crystalline carbazole-containing porphyrin-core dendrimers. Organic Chem Front 7(15):2008–2015
27. van de Craats AM, Warman JM, de Haas MP, Adam D, Simmerer J, Haarer D, Schuhmacher P (1996) The mobility of charge carriers in all four phases of the columnar discotic material hexakis (hexylthio) triphenylene: combined TOF and PR-TRMC results. Adv Mater 8 (10):823–826
28. Iino H, Takayashiki Y, Hanna JI, Bushby RJ, Haarer D (2005) High electron mobility of 0.1 $cm^2V^{-1}s^{-1}$ in the highly ordered columnar phase of hexahexylthiotriphenylene. Appl Phys Lett 87(19):192105
29. Simmerer J, Glüsen B, Paulus W, Kettner A, Schuhmacher P, Adam D, Etzbach KH, Siemensmeyer K, Wendorff JH, Ringsdorf H, Haarer D (1996) Transient photoconductivity in a discotic hexagonal plastic crystal. Adv Mater 8(10):815–819
30. Boden N, Bushby RJ, Clements J, Movaghar B, Donovan KJ, Kreouzis T (1995) Mechanism of charge transport in discotic liquid crystals. Phys Rev B 52(18):13274
31. Bengs H, Closs F, Frey T, Funhoff D, Ringsdorf H, Siemensmeyer K (1993) Highly photoconductive discotic liquid crystals structure–property relations in the homologous series of hexa-alkoxytriphenylenes. Liq Cryst 15(5):565–574
32. Kreouzis T, Scott K, Donovan KJ, Boden N, Bushby RJ, Lozman OR, Liu Q (2000) Enhanced electronic transport properties in complementary binary discotic liquid crystal systems. Chem Phys 262(2-3):489–497
33. Arikainen EO, Boden N, Bushby RJ, Lozman OR, Vinter JG, Wood A (2000) Complimentary polytopic interactions. Angew Chem 112(13):2423–2426
34. Wegewijs BR, Siebbeles LDA, Boden N, Bushby RJ, Movaghar B, Lozman OR, Liu Q, Pecchia A, Mason LA (2002) Charge-carrier mobilities in binary mixtures of discotic triphenylene derivatives as a function of temperature. Phys Rev B 65(24):245112
35. Miyake Y, Fujii A, Ozaki M, Shimizu Y (2009) Carrier mobility of a columnar mesophase formed by a perfluoroalkylated triphenylene. Synth Met 159(9-10):875–879
36. Shah A, Duponchel B, Gowda A, Kumar S, Becuwe M, Davoisne C, Legrand C, Douali R, Singh DP (2020) Charge transport in phenazine-fused triphenylene discotic mesogens doped with CdS nanowires. New J Chem 44(35):14872–14878
37. Adam D, Schuhmacher P, Simmerer J, Häußling L, Paulus W, Siemensmeyer K, Etzbach KH, Ringsdorf H, Haarer D (1995) Photoconductivity in the columnar phases of a glassy discotic twin. Adv Mater 7(3):276–280
38. Chico R, de Domingo E, Dominguez C, Donnio B, Heinrich B, Termine R, Golemme A, Coco S, Espinet P (2017) High one-dimensional charge mobility in semiconducting columnar mesophases of isocyano-triphenylene metal complexes. Chem Mater 29(17):7587–7595
39. Tritto E, Chico R, Ortega J, Folcia CL, Etxebarria J, Coco S, Espinet P (2015) Synergistic π–π and Pt–Pt interactions in luminescent hybrid inorganic/organic dual columnar liquid crystals. J Mater Chem C 3(36):9385–9392

40. Bala I, Yang WY, Gupta SP, De J, Yadav RAK, Singh DP, Dubey DK, Jou JH, Douali R, Pal SK (2019) Room temperature discotic liquid crystalline triphenylene-pentaalkynylbenzene dyads as an emitter in blue OLEDs and their charge transfer complexes with ambipolar charge transport behaviour. J Mater Chem C 7(19):5724–5738
41. Debije MG, Piris J, de Haas MP, Warman JM, Tomović Ž, Simpson CD, Watson MD, Müllen K (2004) The optical and charge transport properties of discotic materials with large aromatic hydrocarbon cores. J Am Chem Soc 126(14):4641–4645
42. Rohr U, Kohl C, Müllen K, van de Craats A, Warman J (2001) Liquid crystalline coronene derivatives. J Mater Chem 11(7):1789–1799
43. Kuo YL, Tseng CY, Tseng CW, Chu KT, Liu YC, Chiang MH, Saeki A, Tao YT, Chen HH (2019) Polymerization of columnar mesogens tethered with diacetylenic side chains. ACS Applied Polymer Materials 2(2):248–255
44. Zhao KQ, An LL, Zhang XB, Yu WH, Hu P, Wang BQ, Xu J, Zeng QD, Monobe H, Shimizu Y, Heinrich B, Donnio B (2015) Highly segregated lamello-columnar mesophase organizations and fast charge carrier mobility in new discotic donor–acceptor triads. Chem Eur J 21(29):10379–10390
45. De J, Bala I, Gupta SP, Pandey UK, Pal SK (2019) High hole mobility and efficient Ambipolar charge transport in heterocoronene-based ordered columnar Discotics. J Am Chem Soc 141 (47):18799–18805
46. Herwig P, Kayser CW, Müllen K, Spiess HW (1996) Columnar mesophases of alkylated hexa-peri-hexabenzocoronenes with remarkably large phase widths. Adv Mater 8(6):510–513
47. Pisula W, Menon A, Stepputat M, Lieberwirth I, Kolb U, Tracz A, Sirringhaus H, Pakula T, Müllen K (2005) A zone-casting technique for device fabrication of field-effect transistors based on discotic hexa-peri-hexabenzocoronene. Adv Mater 17(6):684–689
48. Shklyarevskiy IO, Jonkheijm P, Stutzmann N, Wasserberg D, Wondergem HJ, Christianen PC, Schenning AP, de Leeuw DM, Tomović Ž, Wu J, Müllen K (2005) High anisotropy of the field-effect transistor mobility in magnetically aligned discotic liquid-crystalline semiconductors. J Am Chem Soc 127(46):16233–16237
49. Struijk CW, Sieval AB, Dakhorst JE, van Dijk M, Kimkes P, Koehorst RB, Donker H, Schaafsma TJ, Picken SJ, van de Craats AM, Warman JM (2000) Liquid crystalline perylene diimides: architecture and charge carrier mobilities. J Am Chem Soc 122(45):11057–11066
50. Dehm V, Chen Z, Baumeister U, Prins P, Siebbeles LD, Würthner F (2007) Helical growth of semiconducting columnar dye assemblies based on chiral perylene bisimides. Org Lett 9 (6):1085–1088
51. Wang B, Sun R, Günbaş DD, Zhang H, Grozema FC, Xiao K, Jin S (2015) A bundled-stack discotic columnar liquid crystalline phase with inter-stack electronic coupling. Chem Commun 51(59):11837–11840
52. An Z, Yu J, Jones SC, Barlow S, Yoo S, Domercq B, Prins P, Siebbeles LD, Kippelen B, Marder SR (2005) High electron mobility in room-temperature discotic liquid-crystalline perylene diimides. Adv Mater 17(21):2580–2583
53. Eccher J, Faria GC, Bock H, von Seggern H, Bechtold IH (2013) Order induced charge carrier mobility enhancement in columnar liquid crystal diodes. ACS Appl Mater Interfaces 5 (22):11935–11943
54. Muth MA, Gupta G, Wicklein A, Carrasco-Orozco M, Thurn-Albrecht T, Thelakkat M (2014) Crystalline vs liquid crystalline perylene bisimides: improved electron mobility via substituent alteration. J Phys Chem C 118(1):92–102
55. Funahashi M, Sonoda A (2012) High electron mobility in a columnar phase of liquid-crystalline perylene tetracarboxylic bisimide bearing oligosiloxane chains. J Mater Chem 22 (48):25190–25197
56. Funahashi M, Sonoda A (2014) Electron transport characteristics in nanosegregated columnar phases of perylene tetracarboxylic bisimide derivatives bearing oligosiloxane chains. Phys Chem Chem Phys 16(17):7754–7763

57. Funahashi M, Sonoda A (2012) Liquid–crystalline perylene tetracarboxylic acid bisimide bearing oligosiloxane chains with high electron mobility and solubility. Org Electron 13 (9):1633–1640
58. Funahashi M, Yamaoka M, Takenami K, Sonoda A (2013) Liquid-crystalline perylene tetracarboxylic bisimide derivatives bearing cyclotetrasiloxane moieties. J Mater Chem C 1 (47):7872–7878
59. Bala I, Singh N, Yadav RAK, De J, Gupta SP, Singh DP, Dubey DK, Jou JH, Douali R, Pal SK (2020) Room temperature perylene based columnar liquid crystals as solid-state fluorescent emitters in solution-processable organic light-emitting diodes. J Mater Chem C 8 (36):12485–12494
60. Gan KP, Yoshio M, Kato T (2016) Columnar liquid-crystalline assemblies of X-shaped pyrene–oligothiophene conjugates: photoconductivities and mechanochromic functions. J Mater Chem C 4(22):5073–5080
61. Talarico M, Termine R, Garcia-Frutos EM, Omenat A, Serrano JL, Gomez-Lor B, Golemme A (2008) New electrode-friendly triindole columnar phases with high hole mobility. Chem Mater 20(21):6589–6591
62. García-Frutos EM, Pandey UK, Termine R, Omenat A, Barberá J, Serrano JL, Golemme A, Gómez-Lor B (2011) High charge mobility in discotic liquid-crystalline triindoles: just a Core business? Angew Chem 123(32):7537–7540
63. Benito-Hernández A, Pandey UK, Cavero E, Termine R, García-Frutos EM, Serrano JL, Golemme A, Gómez-Lor B (2013) High hole mobility in triindole-based columnar phases: removing the bottleneck of homogeneous macroscopic orientation. Chem Mater 25(2):117–121
64. Isoda K, Yasuda T, Kato T (2009) Truxene-based columnar liquid crystals: self-assembled structures and electro-active properties. Chemistry–An Asian J 4(10):1619–1625
65. Liu X, Usui T, Hanna J (2014) A windmill-shaped discotic columnar liquid crystal with fast ambipolar charge carrier transport. Chem Mater 26(19):5437–5440
66. Ruiz C, Pandey UK, Termine R, García-Frutos EM, Lopez-Espejo G, Ortiz RP, Huang W, Marks TJ, Facchetti A, Ruiz Delgado MC, Golemme A, Gómez-Lor B (2016) Mobility versus alignment of a semiconducting π-extended discotic liquid-crystalline triindole. ACS Appl Mater Interfaces 8(40):26964–26971
67. Liu CX, Wang H, Du JQ, Zhao KQ, Hu P, Wang BQ, Monobe H, Heinrich B, Donnio B (2018) Molecular design of benzothienobenzothiophene-cored columnar mesogens: facile synthesis, mesomorphism, and charge carrier mobility. J Mater Chem C 6(16):4471–4478
68. Zhao KC, Du JQ, Wang HF, Zhao KQ, Hu P, Wang BQ, Monobe H, Heinrich B, Donnio B (2019) Board-like fused-thiophene liquid crystals and their benzene analogs: facile synthesis, self-assembly, p-type Semiconductivity, and photoluminescence. Chemistry–An Asian J 14 (3):462–4705
69. Bala I, De J, Gupta SP, Singh H, Pandey UK, Pal SK (2020) High hole mobility in room temperature discotic liquid crystalline tetrathienoanthracenes. Chem Commun 56 (42):5629–5632
70. Ye Q, Chang J, Huang KW, Chi C (2011) Thiophene-fused tetracene diimide with low band gap and ambipolar behavior. Org Lett 13(22):5960–5963
71. Xiao Q, Sakurai T, Fukino T, Akaike K, Honsho Y, Saeki A, Seki S, Kato K, Takata M, Aida T (2013) Propeller-shaped fused oligothiophenes: a remarkable effect of the topology of sulfur atoms on columnar stacking. J Am Chem Soc 135(49):18268–18271
72. Kushida T, Shuto A, Yoshio M, Kato T, Yamaguchi S (2015) A planarized triphenylborane mesogen: discotic liquid crystals with ambipolar charge-carrier transport properties. Angew Chem Int Ed 54(23):6922–6925

Part III
Architectonics of Peptides

Chapter 6
Dopamine-Based Materials: Recent Advances in Synthesis Methods and Applications

Hong Li and Junbai Li

6.1 Introduction

Dopamine, a neurotransmitter of the catecholamine group, is synthesized both in the central and the periphery nervous system. Through binding to the G protein-coupled receptors, the dopaminergic system is of vital importance in neuromodulation, such as motor control, motivation, reward, maternal, cognitive function, and reproductive behaviors [1]. Recently, dopamine has become a star molecule in material science due to its versatility in novel material construction. Inspired by the adhesive proteins of the mussel, Messersmith et al. reported a facile and robust method for a universal surface coating by using polydopamine (PDA) chemistry [2]. In this method, dopamine can self-polymerize to form PDA in weak alkaline solution with the presence of oxygen and form a surface coating layer on nearly arbitrary material, such as metals, oxides, ceramics, and polymers, as well as Teflon. PDA has high chemical reactivity with thiol- and amino-terminated molecules via Michael addition and/or Schiff base reactions, and it can also chelate with a variety of metal ions. Therefore, PDA has emerged as one of the most adaptable and simple single-step, material-independent surface modification strategies. With eumelanin-like molecule structure, PDA possesses attractive properties including antioxidant activity, high photothermal conversion efficiency, fluorescence quenching ability, and strong

H. Li
College of Chemistry and Chemical Engineering, Xi'an Shiyou University, Xi'an, China

J. Li (✉)
Beijing National Laboratory for Molecular Sciences, CAS Key Lab of Colloid, Interface and Chemical Thermodynamics, Institute of Chemistry, Chinese Academy of Sciences, Beijing, China

University of Chinese Academy of Sciences, Beijing, China
e-mail: jbli@iccas.ac.cn

T. Govindaraju, K. Ariga (eds.), *Molecular Architectonics and Nanoarchitectonics*, Nanostructure Science and Technology,
https://doi.org/10.1007/978-981-16-4189-3_6

metal ion chelation, rendering its broad applications in the fields ranging from biomedicine, sensing, energy to environment. The successes of PDA-based material have also inspired the research on the co-assembly of dopamine with other functional molecules. It is found that some molecules can participate in the PDA formation process, while other molecules inhibit the oxidative self-polymerization of dopamine and form some novel microstructures through the co-assembly of the building blocks.

This chapter is devoted to a review of recent development in dopamine-based materials and their applications. First, an overview of the different methodologies for fabrication of dopamine-based materials is summarized. We outline various polydopamine-based materials constructed with different preparation strategies and novel materials prepared via the co-assembly strategy, including the polydopamine-assisted co-deposition of functional molecules and novel dopamine-based nanostructures. Thereafter, we focus on the recent advances of the emerging applications of dopamine-based materials in different fields, such as cancer theranostics, bioimaging, self-adhesive bioelectronics, and removal of heavy metal ions. Finally, we discuss the critical unsolved challenges in this field and some potential opportunities for future research.

6.2 Polydopamine-Based Materials

Since the introduction of the simple and robust preparation method of PDA, a rapid development on novel fabrication strategies, functionalization, and applications of PDA-based materials has been witnessed, even if there is a major debate about the molecular structure of PDA being either a polymer or a supramolecular aggregate. Generally, it is accepted that PDA is composed of a collection of oligomeric species cross-linked by noncovalent interactions. The overall structure and properties of PDA-based materials yielded from the oxidative self-polymerization of dopamine are very similar to those of natural eumelanin. In this section, we highlight recent developments in the preparation of different types of PDA-based materials including PDA nanoparticles (NPs), core/shell nanoparticles, microcapsules, films, and hydrogels.

6.2.1 Polydopamine Nanoparticles

Melanin-like PDA nanoparticles were firstly synthesized via neutralization of dopamine hydrochloride with NaOH, followed by spontaneous oxidation of dopamine in the presence of dissolved O_2 [3]. In this synthesis method, a 2 mg mL^{-1} of dopamine hydrochloride solution was mixed with 1 N NaOH at 50 °C under vigorous stirring. The colorless solution rapidly changed to pale yellow and then gradually turned to dark brown. After the 5 h reaction, the nanoparticles with a size of 84 ± 16 nm were

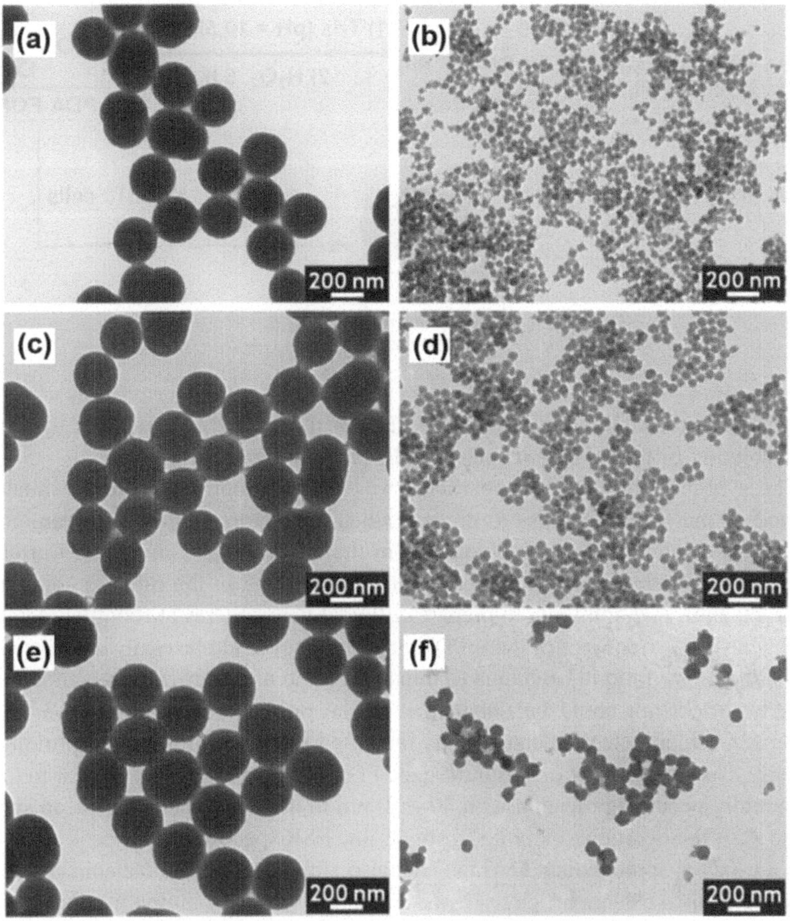

Fig. 6.1 TEM images of melanin-like nanoparticles synthesized with different synthetic parameters: the molar ratios of dopamine hydrochloride to NaOH of (**a**) 1:0.4 and (**b**) 1:1; the dopamine hydrochloride concentrations of (**c**) 4 and (**d**) 1.33 mg mL^{-1}; and the reaction temperatures at (**e**) 20 and (**f**) 70 °C. Reprinted with permission from [3] copyright (2011) American Chemical Society

collected by discarding large-sized aggregates via low-speed centrifugation. Moreover, the size of melanin-like nanoparticles could be modulated from <100 nm to hundreds of nanometers by varying the amount of NaOH, the temperature, and the concentration of dopamine hydrochloride (Fig. 6.1). For example, the particle size could decrease by elevating the reaction temperature from 20 to 70 °C (Fig. 6.1e, f). It was also demonstrated that the PDA nanoparticles with surface modification of thiol-terminated methoxy-poly(ethylene glycol) (mPEG-SH) exhibited good dispersion stability in biological media, such as phosphate buffer and fetal bovine serum. This synthesis approach was adopted by Amin et al. to obtain fluorescent

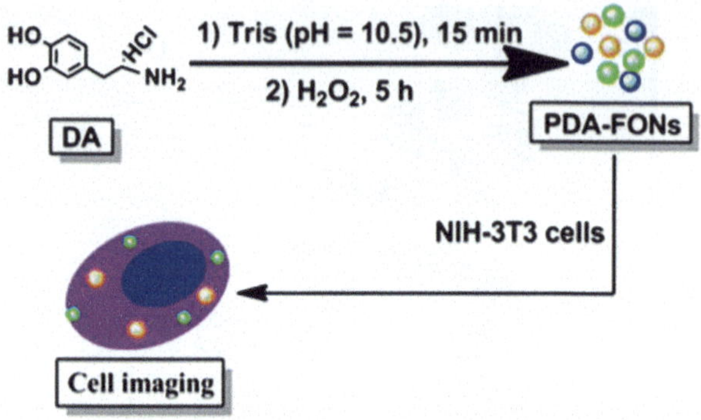

Fig. 6.2 Schematic illustration for preparation of PDA-FONs by oxidation of PDA and their application in cell imaging. Reprinted with permission from [7] copyright (2012) Royal Society of Chemistry

nanoparticles through adding rhodamine 123 or rhodamine B into either the dopamine solution or the mPEG-SH-capped PDA nanoparticles [4].

The synthesis of PDA nanoparticles was further improved with the analogy method of the Stöber process to prepare silica nanoparticles that added ammonia as a catalyst into the dopamine solution in the ethanol-water mixtures. Liu et al. explored the oxidative self-polymerization of dopamine in the mixture containing water, ethanol, and ammonia to prepare colloidal nanospheres at room temperature [5]. Interestingly, the sizes of these PDA nanoparticles could be easily modulated by tuning the molar ratio of ammonia to dopamine. The nanoparticles with an average diameter of 70 nm could be obtained with the ratio increased from 11.3 to 17. Moreover, the particle diameter slightly increased as the reaction time is prolonged. The size and morphology of as-synthesized PDA nanoparticles were similar to those of naturally occurring melanin with 40–150 nm in diameter and spherical in shape. These PDA nanoparticles exhibited a single-line ESR spectrum of the peak observed with a g-factor approaching 2, which was also similar to natural melanin. Notably, these melanin-like nanoparticles showed a broad absorption ranging from ultraviolet (UV) to near-infrared (NIR) wavelengths, making them the promising candidate for photothermal cancer therapy. In addition, as-prepared nanoparticles had good stability both in water and in 10% blood serum solution, greatly guaranteeing their potential biomedical application.

Generally, PDA is considered as a fluorescence quencher via Forster resonance energy transfer (FRET) and/or photoinduced electron transfer (PET) mechanisms, which may affect the fluorescence property of the loading dye molecules [6]. Zhang et al. first constructed fluorescent PDA with excellent biocompatibility [7]. The fluorescent PDA nanoparticles could be synthesized by the oxidation of PDA using the green oxidant of H_2O_2. Dopamine was first self-polymerized in the Tris buffer (pH 10.5) for 15 min to form PDA. Then, the concentrated H_2O_2 solution (30% w/w) was added and further reacted for another 5 h to obtain the PDA-FONs (Fig. 6.2). TEM and AFM images confirmed that the wormlike nanoparticles were produced with a size of several to tens of nanometers. Interestingly, the PDA-FONs showed an obvious fluorescent emission peak at 494 nm when excited by a 400 nm light. Moreover, the emission peaks would move to higher wavelength as the excitation wavelengths changed from 360 to 500 nm, which might be attributed to the wide size distribution of PDA-FONs caused by the uncontrollable polymerization of dopamine and oxidation of PDA.

6.2.2 Core/Shell Nanoparticles

The already formed nanoparticles can be further coated with thin PDA films. This concept has been applied to prepare fruitful PDA-based core/shell nanoparticles. A variety of nanoparticles, such as mesoporous silica nanoparticles (MSNs) [8, 9], polymeric nanoparticles [10, 11], graphene oxide (GO) [12, 13], and upconversion nanoparticles [14, 15], were utilized to be coated with PDA layer. Au nanoparticles are highly intriguing materials for anticancer applications due to the tunable NIR localized surface plasmon resonance, unique optical properties, good biocompatibility, and ease of synthesis. For the surface modification, gold-based nanomaterials, such as Au nanoparticles (Au NPs) [16, 17], Au nanorods (Au NRs) [18, 19], and Au nanostars (Au NSs) [20, 21], have been decorated with PDA layer. In an interesting work, Kumar et al. reported the oxidative nanopeeling chemistry of PDA for the synthesis of plasmonic nanobranched structures on the surface of PDA-coated spherical Au NPs [22]. Citrate-stabilized Au NPs were treated with dopamine in 10 mM Tris·HCl buffer (pH 8.5) at room temperature for 4 h to form 5 nm PDA-coated Au NPs, followed by the treatment of $HAuCl_4$ (5 mM), PVP (5% w/v), and hydroxyl amine (50 mM) for 5 min under shaking (Fig. 6.3a). The solution gradually changed from red color to blue and kept stable for several weeks. The oxidative peeling of the PDA layer on Au core and budding petal structures from the disrupted PDA layer could be seen from the TEM images (Fig. 6.3b-d). Finally, the branched Au core-petal nanoparticles with plasmonic petals on spherical Au cores could be obtained. Fe_3O_4 nanoparticles are of great interest for applications ranging from magnetic resonance imaging, hyperthermia, separation, purification to catalysis. However, Fe_3O_4 nanoparticles are easy to aggregate in water solution due to their insoluble outer layer, which is an important obstacle for their broad applications. Thus, PDA layer is also used to modify the Fe_3O_4 nanoparticles. Zhang et al. reported the core/shell magnetic molecularly imprinted polymers (MIPs) based on the PDA-coated Fe_3O_4 nanoparticles for protein recognition [23]. Fe_3O_4 nanoparticles were synthesized by the solvothermal method in a bisolvent containing ethylene glycol and diethylene glycol with the presence of PAA. PDA was further coated onto these Fe_3O_4 nanoparticles through the oxidative self-polymerization of dopamine in Tris buffer while embedding the template protein in the PDA shell. Moreover, the thickness of imprinted PDA shells could be controlled by varying the mass ratio of Fe_3O_4@PAA to dopamine from 1:0.25, 1:1 to 1: 4 to obtain the PDA thickness of 15 nm, 30 nm, and 40 nm, respectively. After washing the proteins embedded in nanohybrid with 5% SDS-HAc solution, the Fe_3O_4@Lyz-MIPs NPs with imprinted recognition sites in the PDA shell were acquired.

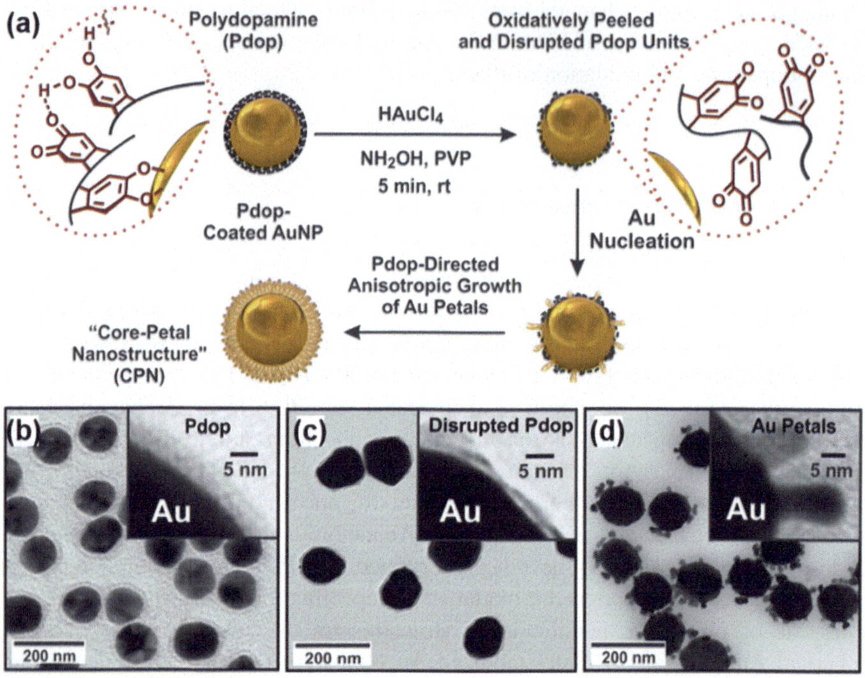

Fig. 6.3 (**a**) Schematic illustration of synthesis of plasmonic Au core-petal nanoparticles through the oxidative nanopeeling chemistry of PDA on Au NPs. TEM images of (**b**) PDA-Au NPs and (**c**) PDA-Au NPs immediately after adding $HAuCl_4$ and (d) PDA-Au NPs after adding $HAuCl_4$ for 1 min. Reprinted with permission from [22] copyright (2014) American Chemical Society

Carbon nanotubes (CNTs) have been extensively utilized as adsorbents to remove various environmental pollutants owing to their small size, large surface areas, and unique chemical structure. However, the adsorption performance of CNTs is severely limited by the poor dispersibility and lack of functional groups. Therefore, Zhang et al. proposed a novel method for preparation of amine-functionalized CNTs via the introduction of mussel-inspired chemistry [24]. In this approach, dopamine and CNTs were dispersed in Tris solution and stirred at room temperature for 8 h. And then, polyethylene polyamine with amine groups was further conjugated with PDA-coated CNTs via Michael addition reaction. CNTs-PDA and polyethylene polyamine were added into 1 M Tris solution for another 8 h to form the amine functionalized CNTs. With a similar process, PDA could also be applied to the coating of Fe_7S_8 nanorods to produce the Fe_7S_8@PDA nanorods, subsequently followed by the calcination at 600 °C for 2 h in an N_2 atmosphere to obtain Fe_7S_8@C nanorods [25].

6.2.3 Microcapsules

Hollow microcapsules have attracted much interest since they were first proposed by Möhwald's group in 1998 [26]. Layer-by-layer (LbL) assembly technique is the most frequently used method to prepare the hollow microcapsules, which renders them with controlled size, composition, and wall thickness [27]. Up to now, a wide variety of building blocks are available for LbL assembly, including polysaccharides, proteins, inorganic particles, micelles, virus particles, etc. Correspondingly, the driving forces for LbL assembly have also extended from conventional electrostatic interactions to hydrogen bonds, metal coordination, covalent bonds, charge transfer, and so on. However, the multistep procedure makes the LbL method labor-intensive and time-consuming, which dramatically restricts their application extension. To develop a convenient preparing method for hollow microcapsules, Caruso and co-workers developed a promising alternative method based on the oxidative self-polymerization of dopamine onto template surfaces. Their procedure involved the assembly of PDA thin films by the spontaneous oxidative polymerization of dopamine onto silica particles, followed by removal of the template particles with the HF solution to form the capsules. Such a PDA assembly process could be applied to nonporous silica particles with a range of sizes (0.5, 1, 3, 5μm in diameter) and nanoporous silica particles (SGX 1000, 5μm diameter; 20 nm pore size). It was found that the PDA capsule shell that assembled on monodisperse silica particles with a solid core and mesoporous shell (SC/MS) had a thickness of 45 ± 5 nm, approximately 2 times thicker than the shell thickness of PDA (~19 nm) deposited on nonporous silica particles. Smaller capsules (~210 nm) were shown to be prepared by using the template of SC/MS particles with a diameter of ~220 ± 10 nm. To avoid the potential toxicity induced by foreign PDA capsules in vivo, PDA capsules were expected to integrate fluorescence so that the biodistribution and pharmacokinetics of these nanocarriers could be monitored. To address this issue, they also prepared fluorescent PDA capsules via the oxidative polymerization of dopamine and subsequent H_2O_2 oxidation [28]. PDA microcapsules were also synthesized via polymerization of dopamine on silica templates and subsequent core removal. The resulting PDA capsules were further treated with dopamine and H_2O_2, leading to the formation of fluorescent PDA capsules (Fig. 6.4). The size of the fluorescent PDA capsules could be tuned by using different-sized templates (1.1 and 2.5μm). Moreover, the fluorescence property of these capsules could be modulated with the pH of buffer solutions and reaction time for the second dopamine polymerization step and the subsequent oxidation with H_2O_2. Aside from silica particles, polystyrene (PS) spheres could also play as the sacrificial templates for the synthesis of PDA microcapsules by removal of the PS cores using a THF solution [29].

For preparation of microcapsules, the removal of SiO_2 or PS templates requires harsh chemical reagents, such as organic solvents and acids, which might hinder their further application in biomedicine field. To develop a mild PDA microcapsule preparation method, we employed $MnCO_3$ particles that could dissolve in an EDTA

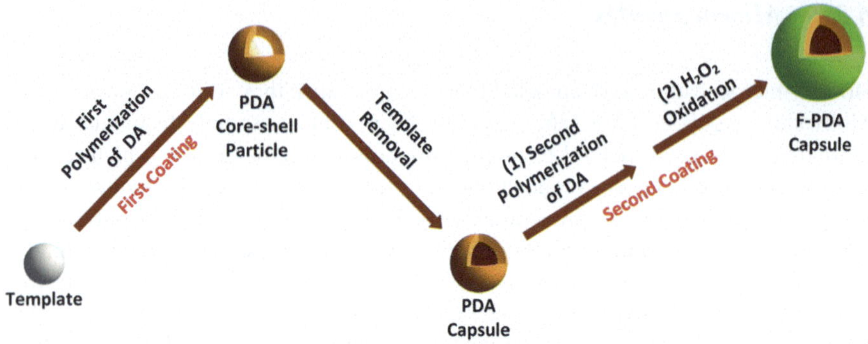

Fig. 6.4 Schematic illustration for synthesis of fluorescent PDA microcapsules. Reprinted with permission from [28] copyright (2014) American Chemical Society

solution as the template [30]. $MnCO_3$ particles with spherical shape and an average diameter of 4.5μm were synthesized by co-precipitation of $MnSO_4$ and NH_4HCO_3 solutions. To prepare the PDA capsules, $MnCO_3$ particles were exposed to dopamine solution (1 mg mL^{-1}) in Tris-HCl buffer (10 mM, pH 8.5). Dopamine would spontaneously self-polymerize onto the surface of $MnCO_3$ templates to form the PDA layer. After removal of $MnCO_3$ particles with EDTA solution, the hollow PDA microcapsules with structural integrity were obtained. Moreover, the concentration of dopamine, the pH of Tris-HCl buffer, and the deposition time could affect the shell thickness of microcapsules. Interestingly, the fluorescent property of PDA capsules labeled with FITC would be beneficial for monitoring their safety and efficacy as drug carriers. Notably, we showed that the PDA shells could also be coated onto insulin particles and exhibited pH-responsive drug release behavior, promising for oral administration of diabetic patients (Fig. 6.5).

Besides abovementioned hard templates, soft templates that can be easily removed under mild conditions are also utilized to fabricate PDA capsules. For example, Xu et al. took pristine alkane droplets of oil-in-water emulsions that stabilized solely by OH^- ions in the absence of any surfactant as the template for the formation of hollow capsules [31]. In this process, 0.2 mL of alkane, such as hexadecane, hexane, octane, or 1-octadecene, and 0.6 mL of the aqueous solution of NaOH (0.02 M) were consecutively added to 9.8 mL of water to produce pristine oil-in-water emulsions. Subsequently, 3 or 5 mg of dopamine was added to the as-prepared pristine emulsions with the pH value adjusted to 8.2 and then shaken for reaction at ambient conditions. Dopamine could selectively oxidize and self-polymerize on the OH^- ion-rich surfaces of the pristine oil droplets to produce a PDA capsule. SEM confirmed that PDA small particles with sizes of about 260 nm were formed after 1 h reaction, and the amount of PDA particles increased with the reaction time. In contrast to pristine alkane droplets, oil-in-water emulsions stabilized by ionic surfactants had rather weak interfacial basicity, which could not initiate the self-polymerization of dopamine on the surfaces of the oil droplets, thus leading to formation of macroporous structures. This study clearly implicated

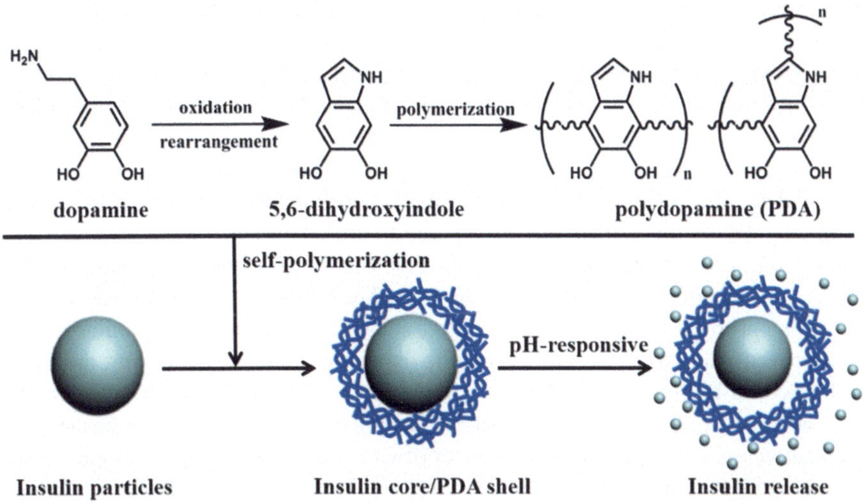

Fig. 6.5 Schematic illustration for PDA-coated insulin delivery system and the pH-responsive insulin release. Reprinted with permission from [30] copyright (2017) Elsevier

that the difference of the template effect between surfactant-free and surfactant-stabilized oil-in-water emulsions arose from their distinction in interfacial chemistry.

In a similar concept of the preparation of PDA microcapsules, curcumin cylindrical crystals could also act as the template to fabricate hollow PDA nanotubes [32]. In such scenario, the curcumin solid templates were prepared with phase separation upon addition of the water to an ethanol/acetone mixture (v/v 1:1) containing the curcumin and dopamine mixture. The solvent exchange could induce the precipitation of curcumin crystals, since the solubility of curcumin is quite low in water. Then, Tris-HCl buffer solution was added to the obtained mixture to reach the final concentration of 9 mM. The solution was kept in reaction for 24 h, where dopamine spontaneously oxidized and polymerized onto the surface of cylindrical crystals to form a PDA layer. Finally, hollow PDA nanotubes were obtained by dissolving the curcumin cores with ethanol. This approach has also been adapted to produce PDA nanotubes by using ZnO nanorods as the template cores [33].

6.2.4 *Films*

The unique adhesive capacity of PDA to be deposited as a thin layer onto virtually all types, shapes, and sizes of inorganic and/or organic surfaces via a simple dip coating procedure makes it distinguished from other strategies for surface modification. PDA layers employ multifold conjugating mechanisms, such as catechol-metal coordinations, $\pi - \pi$ interactions, hydrogen bonds, electrostatic interactions, and covalent reactions (e.g., Schiff base bonding or Michael addition), depending on the chemical characteristics of the material surface. Before the introduction of PDA coating, there are three dominant approaches in surface modification field: self-assembled monolayer (SAM) [34], layer-by-layer assembly [35], and LB films. In SAM, alkanethiol-functionalized molecules bind with noble metal surfaces through metal-thiolate bonds to form ordered monolayers, which therefore requires highly matching surface-adsorbate chemistry. In contrast, polymers used in LbL methods are often restricted by the interaction types, and the deposition process involving several coating cycles seems tremendous. Compared to aforementioned methods, the PDA coating is more simple and controllable that can be modulated by varying pH, concentration, buffer, and deposition time. Although the oxygen diffusion barrier might be a problem that affects the oxidative self-polymerization of dopamine, it can be improved via mild oscillation during the deposition process. Moreover, the PDA coating could be stable in solutions with a relative wide pH range [36]. Thanks to these advantages, the interest in the surface engineering with PDA layer has surged in recent years, resulting in a large number of works devoting in this hot topic. PDA is employed to functionalize metal oxides, noble metals, ceramics, and various polymers, such as polytetrafluoroethylene, polydimethylsiloxane, polystyrene, poly(lactic-co-glycolic) acid, polycarbonate, and poly-(caprolactone).

Superhydrophobic surfaces have aroused significant attention due to their extraordinary surface properties. Large effort has been made to fabricate such surfaces with applications from antireflectivity, self-cleaning, anti-fouling, and microfluidic valves. However, it remains a challenge to tailor the superhydrophobic surface properties. In this regard, Lee and co-workers developed a robust strategy to modify superhydrophobic surfaces that could be integrated with soft-lithographic techniques [37]. In their approach, anodic aluminum oxide (AAO) membranes were taken as templates for further gas-phase deposition of fluorosilane. Then, fluorosilane modified AAO superhydrophobic surfaces were immersed into the dopamine solution to reverse the wetting ability to hydrophilic. After coating with PDA for 18 h, the contact angle decreased from $(158.5 \pm 2.8)°$ to $(37.3 \pm 2.6)°$. Furthermore, such one-step PDA functionalization method was integrated with micromolding in capillaries techniques. By immersing the alkaline dopamine solution (2 mg mL^{-1}) into the microchannels for 18 h, the superhydrophobic (unmodified AAO)-hydrophilic (PDA) alternating micropatterns with 50 mm lines could be formed. Similar to the mechanism of water collection by a desert beetle, water could be concentrated and rolled specifically down along the PDA-coated area of modified superhydrophobic surface.

Mussel-inspired coating technique provides a promising avenue for preparation of nanofiltration membranes (NFMs). Mussel-inspired PDA layers are widely used as an interlayer or surface-modifying layer on nanofiltration membranes. However, the pure PDA coating could not produce the thin and dense skin layer for nanofiltration membranes, since large PDA aggregates are usually formed on the substrate surface, resulting in low rejection and low flux during nanofiltration process. In this regard, Lv et al. developed a PDA/poly(ethylene imine) (PEI) co-deposition strategy to construct the nanofiltration membranes with high efficiency [38]. They first employed the co-deposition of PDA and PEI to generate a

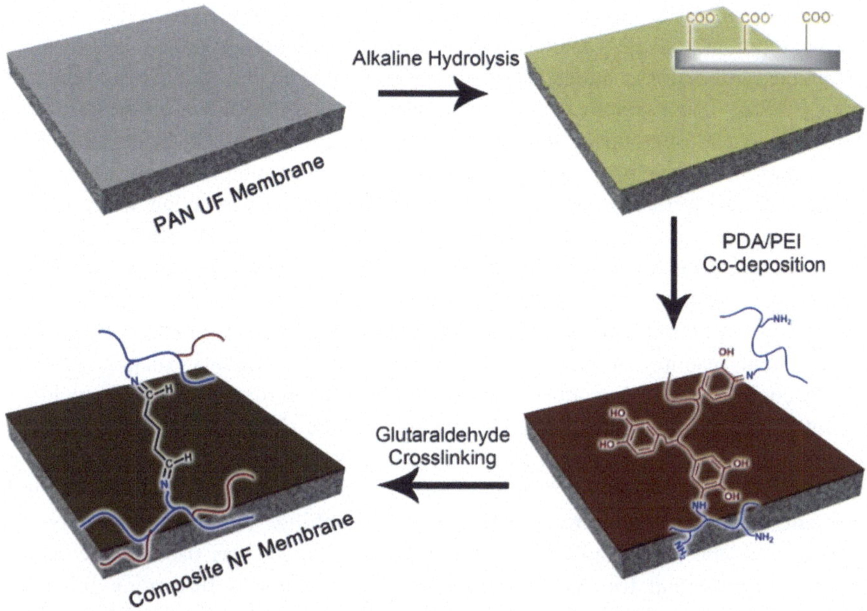

Fig. 6.6 Schematic illustration for preparation process of the composite nanofiltration membranes. Reprinted with permission from [38] copyright (2015) Elsevier

uniform and smooth composite layer on the hydrolyzed polyacrylonitrile ultrafiltration membrane substrate. Then, glutaraldehyde (GA) cross-linking was conducted to promote the cross-linking of PDA/PEI composite (Fig. 6.6). As-prepared NFMs were slightly positively charged as confirmed by Zeta potential measurements, thus resulting in a salt rejection sequence of $MgCl_2$, $CaCl_2$, $MgSO_4$, Na_2SO_4, NaCl at pH 5.5. The authors further incorporated SiO_2 nanoparticles onto the NFMs to enhance the mechanical strength, promote the structural stability, and improve the compaction resistance for long-term filtration process [39].

6.2.5 *Hydrogels*

Hydrogels, 3D cross-linked polymer networks dispersed in water, have attracted increasing attention owing to their outstanding mechanical and chemical properties similar with biological tissues. Inspired by mussel adhesion property, dopamine that contains both catechol and amine groups was first discovered to show one-step, material-independent surface chemistry in 2007 [2]. Since then, mussel-inspired chemistry has been considered a promising strategy to endow the hydrogel with excellent self-adhesiveness and modifiability. Although the mechanism for excellent adhesion property of PDA to material surfaces is still under active discussion and investigation, it is generally accepted that PDA could adhere to both organic and inorganic surfaces through the synergism of covalent and noncovalent interactions. Among them, noncovalent interactions, such as coordination, hydrogen bonding, π–π stacking, and cation-π interactions, are based on the catechol groups and benzene rings of PDA. These noncovalent interactions are highly dependent on the oxidative state of the catechol groups. Thus, the adhesive strength is significantly dependent on the oxidation of catechol groups to semiquinone and/or quinone groups. Covalent interactions are primarily resulted from the quinone groups reacting with nucleophilic groups on material surfaces, including $-NH_2$ and $-$ SH, via Michael addition and/or Schiff base bonding. Therefore, the redox balance between catechol and quinone groups in PDA is of critical importance for their universal adhesion.

Oxidant could accelerate the adhesiveness of mussel-inspired hydrogels through the covalent reaction of quinone. In the presence of oxidants, such as $NaIO_4$, the catechol groups of dopamine can be oxidized to form reactive quinone, yielding the intermolecular cross-linking and reacting with the -NH_2 or -SH groups on the material surfaces. For example, Liu et al. developed an injectable nanocomposite adhesive by combining PDA-functionalized poly(ethylene glycol) (PEG) and Laponite [40]. The four-armed PEG, the hydrophilic, biocompatible polymer approved by the Food and Drug Administration (FDA), was used as the major structural component for the adhesive hydrogel, which was further modified by dopamine. The catechol group in dopamine could be oxidized by $NaIO_4$ to form quinone capable of intermolecular cross-linking. Meanwhile, the quinone groups exhibited strong interfacial binding to biological tissue surfaces by reacting with -NH_2 or -SH groups, thus rendering the hydrogel with strong adhesiveness to tissues. Moreover, Laponite was incorporated into the gel through the noncovalent interactions with catechol groups. The authors demonstrated that the addition of 2 wt % Laponite could effectively enhance the bulk mechanical and adhesive properties of the hydrogels.

Metal ion-induced hydrogels are formed via coordination. Incorporation of metal ions can render PDA hydrogel with tunable and enhanced mechanical properties because catechol groups can effectively coordinate with metal ions forming the reversible cross-linking network in the hydrogel. Fe(III) is widely employed to initiate hydrogel gelation since Fe(III) exhibits pH-responsive coordination interaction with catechol-containing molecules. Jia et al. reported a novel dual self-catalytic system composed of metal ions and catechol-based molecules for efficient gelation of tough, conductive, transparent, and self-healing hydrogels at low temperature [41]. The system was comprised of the catechol-containing molecules, such as dopamine, and transition metal ions, both of which could activate ammonium persulfate to generate free radicals, thus triggering the free-radical polymerization of various vinyl monomers to yield hydrogels. On the one hand, the transition metal ions and reductive catechol-containing molecules could form redox pairs. The reduced metal ion acted as a good electron donor which could participate in single-electron transfer reactions. On the other hand, the oxidized products, quinone and semiquinone, together with catechol groups could also form a catechol-based catalytic system as an electron donor (Fig. 6.7). For the self-catalytic system developed with Fe(III) and dopamine, the radicals produced by the activation of

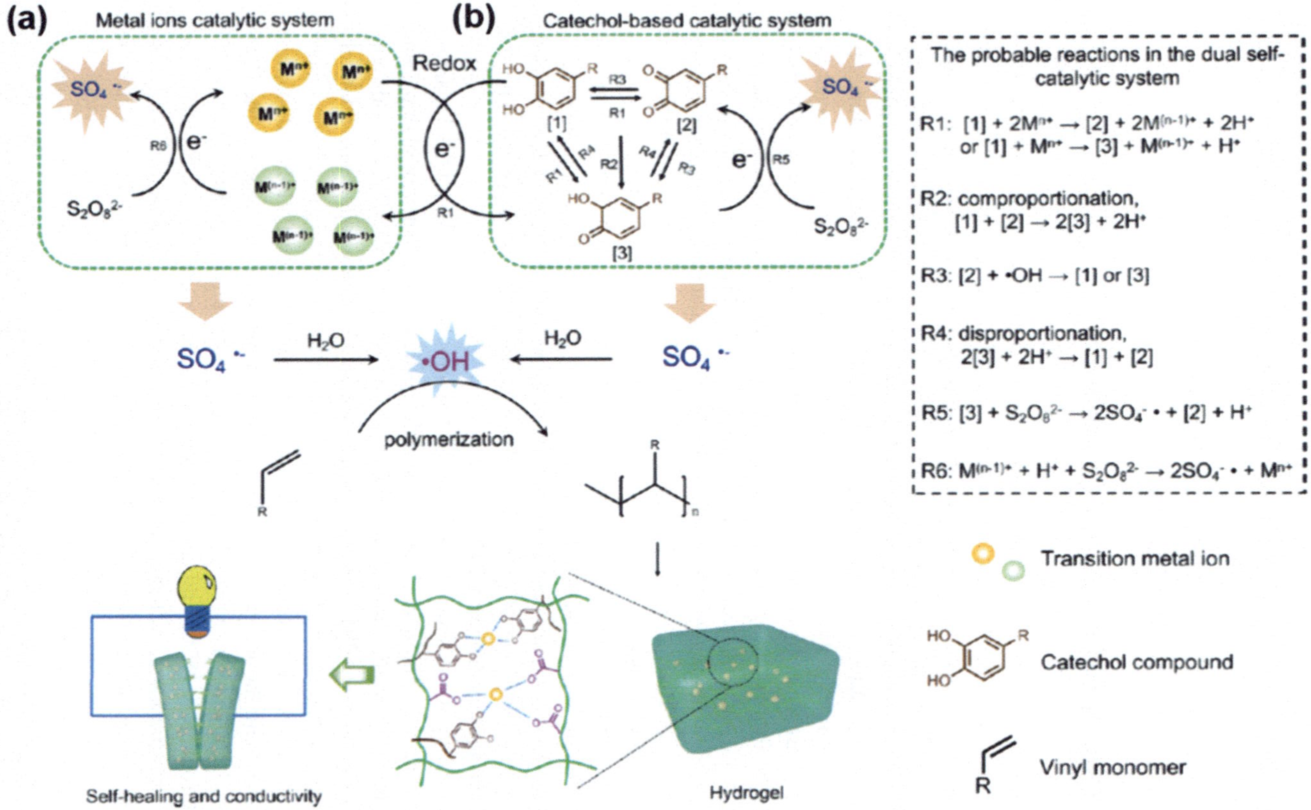

Fig. 6.7 Schematic illustration for a dual self-catalytic system triggering the polymerization of hydrogels and the possible reactions: (**a**) metal ion-based catalytic reaction, (**b**) catechol-based catalytic reaction. Schematic illustration for synthesis of fluorescent PDA microcapsules. Reprinted with permission from [41] copyright (2019) American Chemical Society

ammonium persulfate could rapidly trigger self-gelation of the hydrogel at low temperatures (6 °C) within 5 s.

6.3 Dopamine-Based Materials Prepared via the Co-assembly Strategy

6.3.1 Polydopamine-Assisted Co-deposition

For diverse applications, pure PDA coatings cannot meet all the requirements. Thus, PDA usually acts as the linker to incorporate other functional components onto the membranes/nanomaterials. In most case, two strategies are employed for dopamine-mediated surface functionalization: premodification and postfunctionalization [42]. For the premodification, dopamine is primarily grafted onto polymers, followed by deposition onto the surfaces. Elaborated synthesis and sometimes the protection of phenolic hydroxyl groups in dopamine are required, which make this process complicated. In comparison, postfunctionalization process seems to be simpler. The desired components are conjugated to the PDA-modified surface via chemical or physical interactions between them. Besides these two methods, there is another convenient strategy for functionalizing PDA materials with other components termed as PDA-assisted co-deposition. Compared to the abovementioned strategies, PDA-assisted co-deposition can realize the functionalization of PDA materials simultaneously in the deposition process, greatly simplifying the modification procedure. Currently, a variety of molecules, such as polymers, proteins [23], lipids [43], DNA [44], doxorubicin [45], Alcian Blue [46], C_3N_4 [47], $CaCO_3$ [48], hydroxyapatite [49], and Au [50], have been successfully co-deposited onto the material surfaces mediated by PDA.

A variety of polymers have been incorporated into the layers during the dopamine deposition process. Zhang et al. first studied the co-deposition behavior of dopamine and nonionic polymers including PEG, poly(vinyl alcohol) (PVA), and poly (N-vinylpyrrolidone) (PVP) onto silica substrates [51]. They found that relatively weak hydrogen bonds could facilitate the co-deposition of PDA and polymers, such as PEG and PVA. However, strong hydrogen bonds between PVP and dopamine might overwhelm the self-polymerization of dopamine itself and suppress the PDA formation. Following this principle, polyhydroxy compounds including dextran [52] and hyaluronic acid [53] could also be loaded with dopamine in a single deposition process. Furthermore, Zhang et al. showed the co-deposition of poly (N-isopropylacrylamide) (pNiPAAm) and PDA both on planar substrates and on colloidal substrates [54]. Different sizes of templates were utilized to yield the capsules. Larger template particles could produce capsules with thicker mixed polymer membrane and different permeability based on the type of pNiPAAm with a sequence of PDA > PDA/pNiPAAm-NH_2 > PDA/pNiPAAm-HB capsules. However, the capsules prepared with smaller template particles showed a reverse permeability sequence of PDA < PDA/pNiPAAm-NH_2 < PDA/pNiPAAm-HB capsules.

Aside from polymers, biomacromolecules, such as proteins [23], lipids [43], and DNA [44], and small molecules, such as doxorubicin [45] and Alcian Blue [46], were also reported as second components in polydopamine-assisted co-deposition to fabricate functional materials. Some proteins, such as human serum albumin (HSA) and bovine pancreatic alkaline phosphatase, could accelerate the production rate of PDA in solution [55]. Moreover, HSA could reduce the diameter of PDA nanoparticles with 24 h of oxidation in the presence of 50 mM Tris buffer (pH = 8.5). Surprisingly, the presence of HSA slowed down the growth of hybrid films on solid surfaces. Some other proteins, such as hen egg white lysozyme and bovine a-lactalbumin, could not control the size of the PDA aggregates and did not inhibit the deposition of PDA films on solid surfaces. Thus, the mechanism about the influence of proteins on the formation of PDA is worthy of deep discussion from the fundamental point of view.

Inorganic materials including C_3N_4 [47], hydroxyapatite [49], and Au [50] could also be co-deposited in the formation process of PDA to produce organic-inorganic hybrid materials. For example, Dong et al. reported the one-pot, dopamine-assisted biomineralization method using a gas diffusion procedure to synthesize $CaCO_3$-PDA hollow nanoparticles [48]. In the enclosed chamber, the CO_2 and NH_3 gases were produced by the natural decomposition of NH_4HCO_3, which would continuously diffuse into the ethanol solution with Ca^{2+} ions and dopamine, facilitating the formation of $CaCO_3$. At the same time, dopamine could not only speed up the growth of $CaCO_3$ but also oxidatively self-polymerize to form the PDA in such alkaline environment, yielding the $CaCO_3$-PDA composite nanoparticles (Fig. 6.8a). During this process, the solution would change from colorless to blue, attributing to the coordination between Ca^{2+} and PDA. Moreover, the morphology of as-synthesized $CaCO_3$-PDA nanoparticles could be easily modulated by varying the ratios of dopamine and calcium chloride (Fig. 6.8b).

6.3.2 Novel Dopamine-Based Nanostructures

Inspired by the mussel adhesive proteins, a variety of dopamine-based materials are constructed based on the oxidative self-polymerization formation of PDA for broad applications, ranging from photothermal therapy, bioimaging, tissue engineering, self-adhesive bioelectronics, catalysis to wastewater treatment. Among them, the modulation of the morphology of PDA nanostructures, such as the shape and the size, is of pivotal importance, which has been devoted too much effort. Recently, there are several emerging researches focusing on the modulation of dopamine-based co-assembly with other functional molecules to explore novel nanostructures with distinctive properties. Some molecules participate in the PDA formation via covalent interactions, while some other molecules suppress the oxidative self-polymerization of dopamine through the supramolecular interactions.

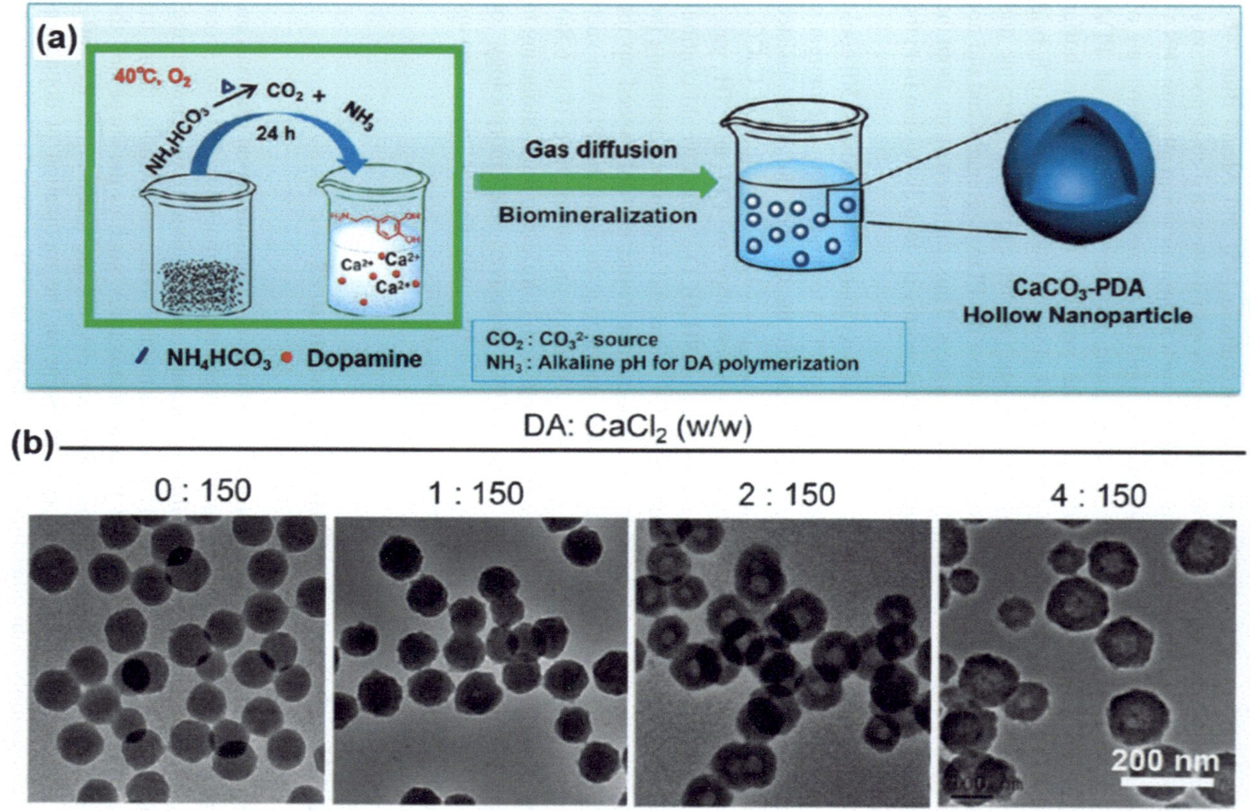

Fig. 6.8 (**a**) Schematic illustration for synthesis of $CaCO_3$-PDA nanoparticles. (**b**) TEM images of $CaCO_3$ and $CaCO_3$-PDA nanoparticles prepared with different ratios of dopamine and $CaCl_2$. Reprinted with permission from [48] copyright (2018) American Chemical Society

Small molecules of carboxylic acid including folic acid [56] and caffeic acid [57] were found to have a control over the morphology and size of PDA assemblies. For example, Yu et al. reported curled PDA nanofibers formed with the presence of folic acid [56]. Folic acid (0.15 mg mL^{-1}) and dopamine (0.3 mg mL^{-1}) were mixed at 60 °C for 1 day, followed by the addition of Tris buffer (10 mM, pH = 8.8) to trigger the self-polymerization of dopamine. The yield of nanoparticles increased with their morphology changed from belts to curled nanofibers as the oxidation reaction time varied from 1 to 30 h. Moreover, nanofibers were preferentially formed in the dark environment protecting the folic acid from photodegradation. The authors claimed that supramolecular interactions between folic acid and protomolecules of PDA, such as π-π stacking and hydrogen bonding, played a vital role for the formation of these nanofibers. Besides folic acid, Kohri et al. prepared raspberry-like particles through oxidative self-polymerization of dopamine with the assistance of a carboxylic acid-containing compound, such as caffeic acid, protocatechuic acid, trans-cinnamic acid, and benzoic acid [57]. The carboxylic acid group was the key moiety for the generation of raspberry-like particles on polystyrene particles. The protomolecules of PDA became larger due to the interactions with carboxylic acid via hydrogen bonding, resulting in the yield of PDA nodule grown on the polystyrene particles from vertical direction. Nevertheless, further studies evaluating the interaction mechanism between carboxylic acid and dopamine molecules would be needed from the state-of-the-art view.

Polyoxometalates (POMs) are a class of metal-oxide nanoclusters facilitating the preparation of organic-inorganic hybrid materials with diverse electronic, magnetic, photochemical, and catalytic properties. POMs, such as Keggin-type phosphotungstic acid [58] and Weakley-type polyoxometalate $Na_9[EuW_{10}O_{36}]\cdot 32H_2O$ (EuW_{10}) [59], could not only inhibit the self-polymerization of dopamine in alkaline environment but also assemble with dopamine to form novel hierarchical nanostructures. We first proposed the preparation of flowerlike 3D hierarchical nanostructures through the co-assembly of dopamine and phosphotungstic acid. These hybrid nanoflowers had a size from 2 to 6.5μm and were composed of smooth nanopetals with 20 nm in thickness and 300–500 nm in width, as confirmed by the SEM images. The ratio of two building blocks, the concentrations, and the pH of the initial Tris-HCl solution played critical roles in controlling the morphology and size of such nanostructures. For the formation mechanism, there might be a two-stage growth process deduced from temporal evolution experiments [58]. Moreover, we demonstrated that these hierarchical nanostructures could be used for the in situ synthesis of nanostructured Ag with the catechol groups acting as the reductant [60]. Similar to the hierarchical nanostructures assembled with dopamine and phosphotungstic acid, Zhang et al. showed the ionic co-assembly of EuW_{10} and dopamine in aqueous solution to produce similar flowerlike nanostructure [59]. After forming EuW_{10}/DA nanoflowers, the fluorescence of EuW_{10} would be quenched due to the blocking of the hopping of d_1 electron in EuW_{10}. It is no doubt that these studies will open a new door for fabrication of hierarchical nanostructures by combination of bioactive molecules and flexible inorganic polyoxoanions.

Besides the supramolecular interactions, covalent interactions are also employed to interact with dopamine through participating in the PDA formation or inhibiting the PDA formation. Dopamine monomers can self-polymerize under weak alkaline environment to form PDA generally considered as light quenching agents. To eliminate the π-π stacking of PDA oligomers, amine- or thiol-containing organic molecules are used in the polymerization process of dopamine via Schiff base or Michael addition reaction to participate in PDA formation [61]. In some cases, the extra organic molecules could effectively decrease the stacking interaction, thus generating auto-fluorescence. For example, polyamines are quite different from other polyelectrolytes when co-deposited with dopamine. Hong et al. proposed the air/water interfacial formation of freestanding Janus microfilm of PDA films in the presence of poly(ethylene imine) [62]. Single PDA could form a thin film at the air/water interface, which was not mechanically robust and freestanding film. After the addition of PEI (Mw = 750 kDa), the PEI acted as a polymeric template enabling freestanding formation of hybrid films through cross-linking with the catechol moieties. The film was Janus-like morphology with two distinct sides of the homogeneous and flat top layer and porous bottom layer, as confirmed by the SEM and energy-dispersive X-ray spectroscopy (EDS). In the following work, Liu et al. developed fluorescent PDA nanoparticles via the co-polymerization of dopamine and PEI [63, 64]. PEI could not only provide the alkaline environment to enhance dopamine polymerization but also react with dopamine via Michael addition and/or Schiff base reaction. It was found that the PDA-PEI nanoparticles exhibited strong green fluorescence, and the emission wavelength was independent of the excitation wavelength. But the fluorescence intensity would be changed correspondingly when the nanoparticles were excited by different wavelengths. Moreover, as PEI concentration increased, the size of the resulting nanoparticles became smaller with stronger fluorescence intensity. The authors hypothesized that PEI reduced the aggregation of PDA, and thus weakened the intermolecular coupling effect, resulting in fluorescence enhancement compared to that of pure PDA.

Considering that dopamine possesses catechol group and amino group in its structure, it can easily react with the cross-linking agents, such as genipin and glutaraldehyde. We first reported the dopamine-based nanoparticles covalently assembled with dopamine and genipin [65]. Typically, dopamine and genipin were dissolved in ultrapure water and then mixed under vigorous stirring. The pH of the mixture was adjusted to 7.5 with 0.1 M NaOH solution. It could be seen that the colorless solution immediately changed into yellow and then became darker and darker until a dark blue suspension collected after 24 h reaction. The resulting products were monodispersed nanospheres with the size varied from 130 to 360 nm by changing the molar ratio and the concentration of dopamine and genipin (Fig. 6.9). Interestingly, as-prepared nanospheres exhibited intrinsic photosensitive properties that they could convert O_2 to singlet oxygen ($^1O_2$), as confirmed by the chemical trapping probe 9,10-anthracenediyl-bis(methylene) dimalonic acid (ABDA). Moreover, the catechol group on the nanospheres could enable further loading of anticancer drug bortezomib via a reversible catechol-boronate linkage, which facilitated pH-responsive drug release in an acidic tumor microenvironment.

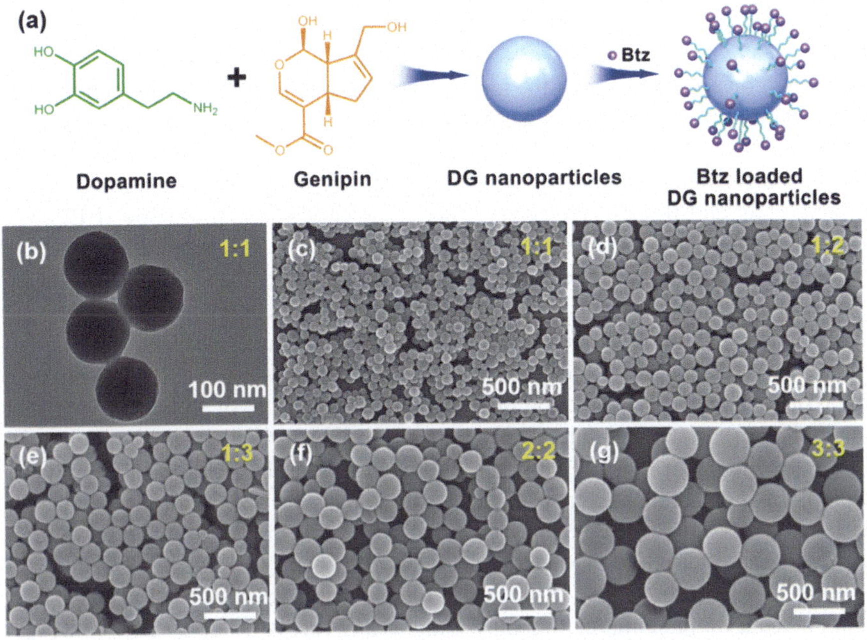

Fig. 6.9 (**a**) Schematic illustration of the assembly of dopamine and genipin into nanoparticles and the loading of anticancer drug bortezomib. (**b**) TEM image and (**c-g**) SEM images of the assembled nanoparticles obtained with varied molar ratios and concentrations of dopamine and genipin. Reprinted with permission from [65] copyright (2019) Royal Society of Chemistry

Similar to boronic ester bonds, Schiff base bonds are also pH-sensitive under acidic environment and have been widely used in the construction of acid-activable drug delivery systems for anticancer therapy. Thus, we developed a covalent assembly strategy by introducing Schiff base bonds into the assembly of dopamine and glutaraldehyde [66]. Schiff base bonds endowed the dopamine-based nanoparticles with autofluorescence and pH-dependent degradation properties, allowing the facile monitoring and controlled drug release in acidic environment.

6.4 Applications of Dopamine-Based Materials

Dopamine-based materials have attracted increasing attention due to their excellent physicochemical properties, such as good biocompatibility, reducibility, fluorescence quenching ability, and high photothermal conversion capability. With the capacity to embed functional molecules into dopamine-based materials, they have shown broad applications in different fields, such as cancer theranostics, bioimaging, self-adhesive bioelectronics, and removal of heavy metal ions.

6.4.1 Cancer Theranostics

Cancer is considered as one of the most severe diseases jeopardizing human beings' health worldwide. Currently, clinical therapies are limited to surgery, radiotherapy, and chemotherapy. However, these treatments exert effects on the tumor tissues while also destroying the immune system of patients, causing strong side effects on normal cells and tissues. To address this issue, photothermal therapy (PTT) has attracted much attention due to the high selectivity and low side effects. In this treatment, PTT agents gathered in the region of tumor tissues could kill cells by converting light to heat with high efficiency under the localized irradiation of NIR laser. Liu el al. first proposed PDA nanospheres could play as an efficient photothermal therapeutic agent [5]. PDA nanospheres were synthesized through the oxidative self-polymerization of dopamine in a mixture of water, ethanol, and ammonia at room temperature. The size of nanospheres could be easily modulated by tuning the molar ratio of ammonia to dopamine and the reaction time. Encouragingly, as-prepared PDA nanospheres exhibited strong NIR absorption and high photothermal conversion efficiency, and the temperature of the nanospheres solution increased rapidly by 33.6 °C under 808 nm laser (2 W cm^{-2}) irradiation for 500 s. In vivo experiments confirmed that the 4 T1 tumors treated with PDA nanospheres were ablated by photothermal therapy without any regrowth for 10 days of observation. Furthermore, PDA nanospheres could react with molecules terminated with thiol and amino groups, facilitating the design of multifunctional nanoplatforms for simultaneous cancer diagnosis and therapy.

PDA also possesses strong coordination capability with metal ions, such as iron, gadolinium, and manganese, which could be further used as the contrast agent for magnetic resonance imaging (MRI). For example, Chen et al. demonstrated the conjugation of PDA with Fe^{3+} and subsequent modification of benzene-1,3,5-tricarboxylic acid (H_3btc) [67]. Through alternative deposition of $FeCl_3$ and H_3btc onto the surface of PDA nanoparticles, biodegradable coordination polymer was obtained (Fig. 6.10a). The size of as-synthesized nanocomplex could be tuned from 131 ± 9.03 nm to 144 ± 8.74 nm with three growth cycles. Magnetic resonance imaging demonstrated that the nanocomplex could act as a T_1/T_2 dual-mode contrast agent with high longitudinal relaxivity (r_1 = 7.524 mM^{-1} s^{-1}) and transverse relaxivity (r_2 = 45.92 mM^{-1} s^{-1}) (Fig. 6.10b, c). Thanks to the good photothermal conversion efficiency of PDA, the nanocomplex loading with doxorubicin (DOX) could not only produce heat to kill the tumor cells but also facilitate the chemotherapy through NIR- triggered release of DOX, resulting in synergistic cancer treatment effect both in vitro and in vivo (Fig. 6.10d-f).

Activable nanoplatforms that improve the therapeutic efficiency while reducing side effects are highly attractive for precision cancer treatment. $CaCO_3$-PDA hollow nanoparticles were synthesized with one-pot, dopamine-assisted biomineralization method that oxidized a Ca^{2+} salt-dopamine mixture in the atmosphere of O_2 and the gas bubbled with NH_4HCO_3 [48]. As-obtained nanoparticles were subsequently modified with PEG to enhance the physiological stability. Intriguingly,

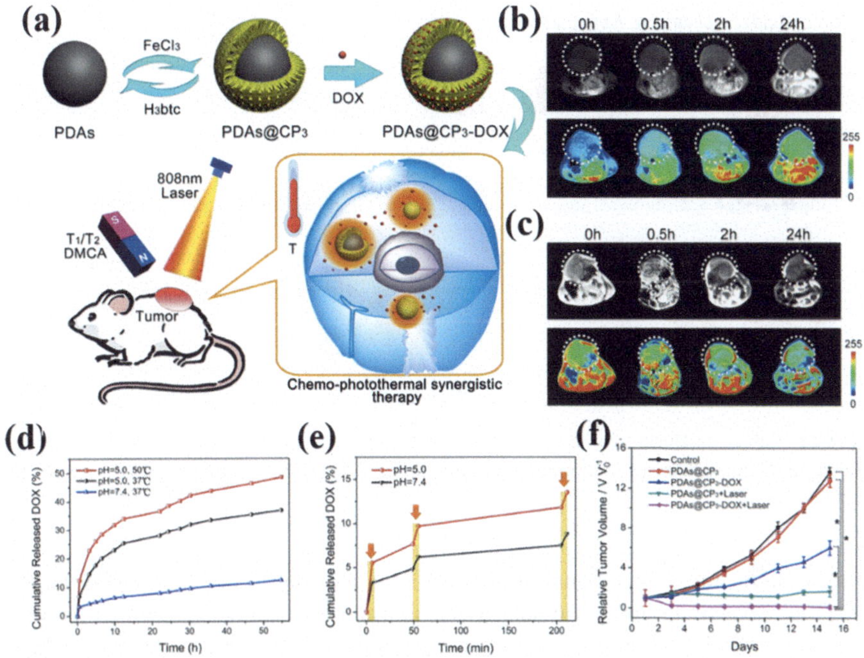

Fig. 6.10 (**a**) Schematic illustration for preparation and application of PDAs@CP_3-DOX. (**b**) T_1 and (**c**) T_2 MR images of mice with intravenous injection of PDAs@CP_3. (**d**) The temperature- and pH-dependent and (**e**) NIR-triggered release of DOX from PDAs@CP_3-DOX. (**f**) Tumor growth curves of the mice under varied treatments. Reprinted with permission from [67] copyright (2016) Elsevier

photosensitizer chlorin e6 (Ce6) loaded onto the nanocomposites was quenched by the strong absorption of PDA under neutral pH and activated to recover the fluorescence and singlet oxygen generation ability under reduced pH due to the degradation of the nanocomposites. Moreover, the nanocomposites could also facilitate the photoacoustic (PA) imaging and T_1-weighted MR imaging by the chelation with Mn^{2+}. Therefore, enhanced in vivo antitumor photodynamic therapy (PDT) was realized with these nanocomposites in the acidic environment of tumor under the guidance of multimodal imaging, while they were quenched at normal physiological pH showing scarce side effect to normal tissues (Fig. 6.11).

6.4.2 *Bioimaging*

Biological optical imaging is widely employed to analyze the characteristics and morphology of specific regions of the cell, tissue, or organism, which can be divided into fluorescence imaging, bioluminescence imaging, optical tomography, photoacoustic imaging, etc. Among them, fluorescence bioimaging, based on the

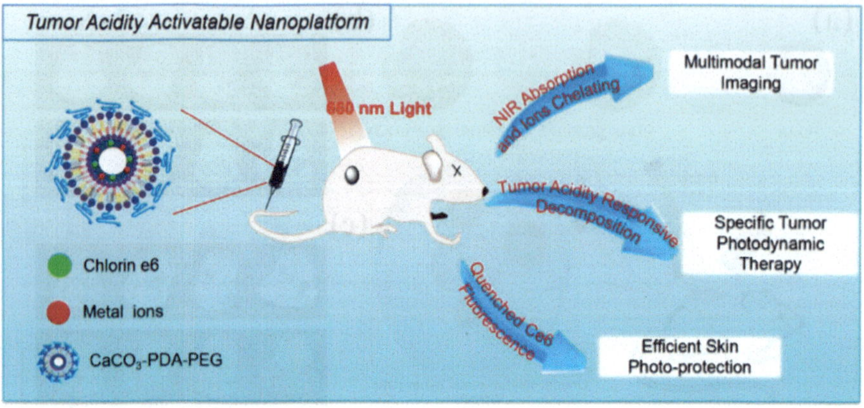

Fig. 6.11 Schematic illustration of Ce6@$CaCO_3$-PDA-PEG nanocomplex as multimodal imaging-guided PDT nanoplatform. Reprinted with permission from [48] copyright (2018) American Chemical Society

acquisition of fluorescent signal of biological matter using specific probes or labels with high sensitivity and good selectivity, is the most commonly used technique. Recently, PDA-based nanomaterials have provided an attractive pathway for fluorescence bioimaging. Zhang et al. first proposed the preparation of fluorescent organic nanoparticles (FONs) based on the direct oxidation of PDA nanoparticles in the hydrogen peroxide (H_2O_2) solution (Fig. 6.12a) [7]. Notably, these PDA FONs showed the special photoluminescence emission spectra that the emission peaks could move to higher wavelengths as the excitation increased from 360 to 500 nm. The NIH-3 T3 cells incubated with PDA FONs presented green and green-yellow fluorescence when excited by 405 and 458 nm laser, respectively (Fig. 6.12b, c). Moreover, the cell cytotoxicity experiments demonstrated that PDA FONs did show any obvious cytotoxicity at a concentration up to 160μg mL^{-1}. Following this work, Chen et al. prepared fluorescent PDA capsules with controlled size and morphology by using sacrificial templates and subsequent oxidation with hydrogen peroxide [28]. MTT cell viability assays suggested that as-prepared PDA capsules could be internalized into HeLa cells through endocytosis while showing negligible cytotoxicity. Therefore, the fluorescent PDA capsules were capable for tracking the materials in biological systems.

Surface-enhanced Raman scattering (SERS) imaging has attracted much attention because it can provide molecular fingerprint spectra at single-molecule level. Compared with fluorescence imaging technique, SERS imaging may avoid the disadvantages of photobleaching and autofluorescence. Sun et al. prepared the versatile SERS probes by using PDA layer [68]. SERS active tags could be co-deposited onto the surface of Au NPs during the formation process of PDA. With this one-pot synthesis method, three SERS probes were constructed for SERS imaging of prostate cancer cells. As confirmed by the SERS images, three tumor-associated antigens on the prostate cancer cells could be distinguished from each other. In another interesting

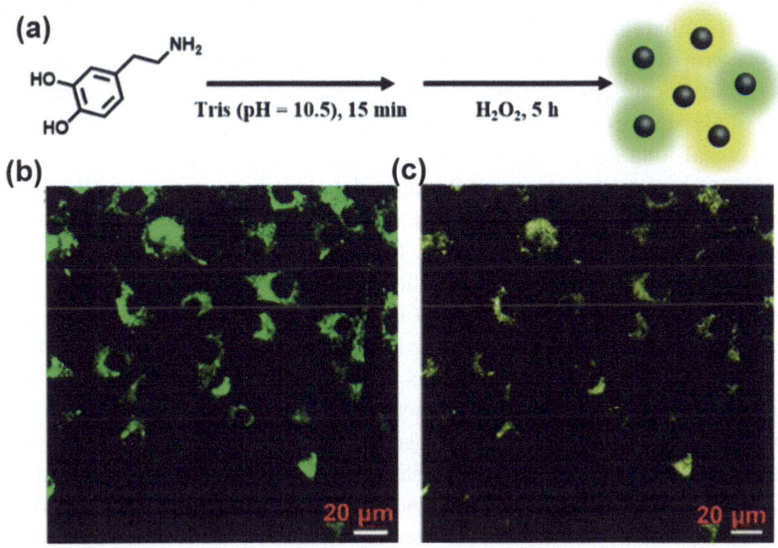

Fig. 6.12 (**a**) Schematic illustration of the preparation of PDA FONs. Confocal laser scanning microscopy images of NIH-3 T3 cells excited by the laser with wavelengths of (**b**) 405 nm and (**c**) 458 nm. Reprinted with permission from [7] copyright (2012) Royal Society of Chemistry

work, Ju et al. designed and prepared a dual-mode nanoprobe for SERS imaging and T_1-weighted MRI with PDA-coated gold nanoparticles [16, 69]. Fe^{3+}-complexed PDA nanoparticles showed high relaxivity value for T_1-weighted MRI. In vitro MRI images of MDA-MB 231 cells treated with these nanocomposites presented the concentration-dependent signal enhancement effect. Moreover, Raman scattering of PDA layer located at 1596 and 1415 cm^{-1} could be amplified by the gold substrate. Thus, the MDA-MB 231 cells could also be imaged with relatively high Raman signals.

6.4.3 Self-Adhesive Bioelectronics

Wearable bioelectronics have attracted a great deal of attention due to their promising potential in personalized healthcare. Most bioelectronics depend on external aids to attach to the human body, resulting in unstable performance in practical application. To solve this problem, self-adhesive bioelectronics with reliable and conformal contact with tissue and fidelity in the signal detection are proposed. Nevertheless, it is remaining a big challenge for materials to realize long-term stable self-adhesion to soft and wet biological tissues. Fortunately, mussel-inspired hydrogels shed new light on the preparation of self-adhesive bioelectronics. Moreover, the modifiability of PDA provides a unique avenue for integrating multiple functional properties to all-in-one bioelectronic devices, which presents great expectation for healthcare applications [70]. For instance, Han et al. prepared a highly tough polydopamine-polyacrylamide (PDA-PAM) hydrogel that exhibited super toughness and long-term adhesiveness [71]. In a two-step hydrogel synthesis, the dopamine monomers were first polymerized to produce precursor PDA in the alkaline environment; then, acrylamide monomers were added to the PDA solution to form the PDA-PAM hydrogel with the aid of the initiator and cross-linker. Inspired by the natural mussel adhesive mechanism, that is, mussels secrete reductive cysteine-rich proteins to maintain a high concentration of catechol groups for high adhesiveness, the authors prevented overoxidation of PDA by initiating the polymerization of acrylamide monomers in the PDA precursor solution to isolate unreacted catechol groups from the oxygen environment. With enough free catechol groups in as-prepared PDA-PAM hydrogel, it showed several excellent properties, such as super stretchability and repeatable adhesion ability. The hydrogel could be extended to a large strain of 3300% under a fracture energy of 2400 J m^{-2}. In addition, the covalent interactions between the catechol groups on the PDA and the amino groups on the PAM could enhance the toughness of the PDA-PAM network.

Transparent bioelectronics are of great importance for see-through surgical operations. The ideal bioelectronics for see-through surgical operations should satisfy these requirements, such as excellent transparency, stable conductivity, self-adhesive properties, and good biocompatibility. Ionic hydrogels are one type of transparent hydrogels, which are produced by trapping ions in or adding ionic liquid to the hydrogel network to ensure both conductivity and transparency. Noncovalent interactions including coordination interactions, hydrogen bonding, host-guest interactions, and π-π stacking are usually employed to trap the ions within the hydrogel networks. Recently, Jia et al. reported a dual self-catalytic system based on a catechol-metal ion complex that could catalyze the formation of tough, conductive, and self-healing hydrogels [41]. Such a system was comprised of the catechol-containing molecules, such as dopamine, tannic acid, and tea polyphenol, and transition metal ions (e.g., Fe^{3+}, Co^{2+}, Ni^{3+}, and Zr^{3+}). Both of the reduced metal ion and the catechol groups with their oxidized products could activate ammonium persulfate (APS) to generate radicals. These radicals further triggered the free-radical polymerization of various vinyl monomers, such as acrylic acid and acrylamide, thus producing multifunctional hydrogels with trapped metal ions through catechol-based coordination. The light transmittance of as-prepared hydrogel was high with 91% at a wavelength of 650 nm, resulting in a clear observation of a flower covered by the hydrogel (Fig. 6.13a). The conductivity was ~38 S m^{-1}, and the maximum tensile strain of the hydrogels was 2600% (Fig. 6.13b). Moreover, the superior light transmission and electrical conductivity enabled their applications in sensors for detection of electrocardiogram (ECG) and electromyography (EMG) signals in human body (Fig. 6.13d).

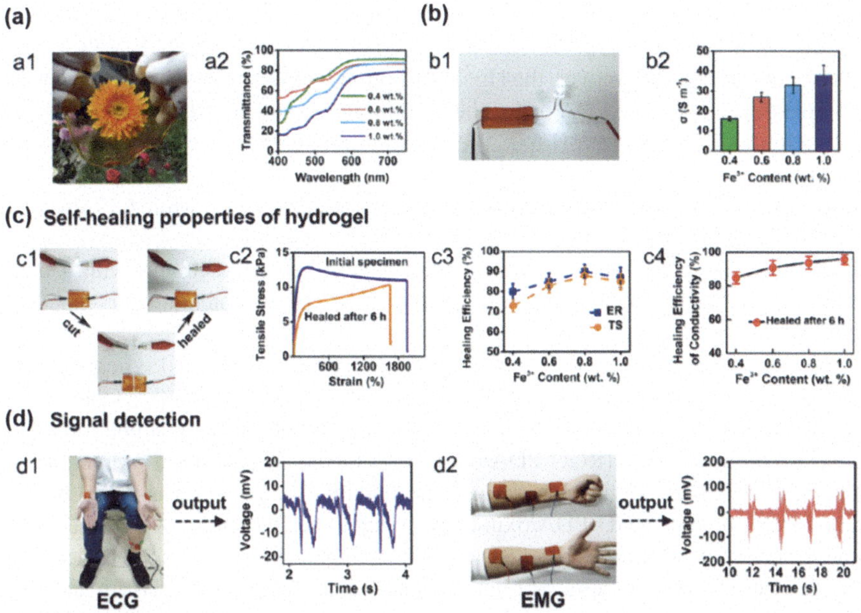

Fig. 6.13 Transparent, conductive, and self-healing hydrogels fabricated with a dual-catalytic system composed of the metal ions and catechol-containing molecules. (**a**) Transmittance of the hydrogel. (**b**) Conductivity of the hydrogel. (**c**) Self-healing properties of the hydrogel. (**d**) Signal detection with hydrogel to monitor ECG and EMG signals. Reprinted with permission from [41] copyright (2019) American Chemical Society

6.4.4 Removal of Heavy Metal Ions

With the fast development of modern technology, water pollution has become a global concern due to their overall environmental toxicity. The heavy metal ions are one of main sources of water contamination, since they are not biodegradable and tend to accumulate in living organisms. Long-time uptake of contaminated water with heavy metals even at a low concentration may cause cardiovascular, immunological, pulmonary, neurological, and endocrine disorders and cancers. Thus, much attention has been focused on the removal of heavy metal ions from contaminated water through various methods, such as ion-exchange, adsorption, chemical precipitation, membrane filtration, coagulation-flocculation, and electrochemical methods [72]. Among them, adsorption of the heavy metals is one of the most widely employed methods due to its simplicity, low cost, and environmental friendliness. Considering the existence of abundant functional groups including catechol groups, amine groups, ketone groups, and aromatic moieties, PDA could directly bind with heavy metal ions for waste watertreatment. For example, Zhang et al. showed the preparation of PDA nanospheres in the presence of Tris-buffer mixed with isopropyl alcohol, ethylene glycol, or ethanol [73]. The adsorption equilibrium could be reached rapidly within 4 h by using the resultant nanospheres towards the removal of Hg^{2+}. The maximum adsorption capacities of Hg^{2+} increased as the temperature elevated with 1861.72 mg/g, 2037.22 mg/g, and 2076.81 mg/g at 298 K, 313 K, and 328 K, respectively. Moreover, the Hg^{2+} adsorption by PDA nanospheres was found to be pH dependent. The PDA nanospheres exhibited complete adsorption at pH 1.0 and optimum adsorption at pH 6.0. The PDA nanospheres showed highly selective adsorption towards Hg^{2+}, compared with the ions of Cd^{2+}, Zn^{2+}, Cu^{2+}, and Pb^{2+}. NO_3^-, K^+, or Ca^{2+} had scarce effects on the adsorption of Hg^{2+}, while Cl^- strongly impended the adsorption process. After five times of adsorption-desorption recycles, no obvious change of the structure of PDA nanospheres was found with a high adsorption capacity maintained, suggesting their great potential for practical applications.

Besides the direct application in adsorption, PDA can also be used synergistically with other components to exert their adsorption function. Recently, Qian et al. reported a one-step method to synthesize salicylaldoxime-/polydopamine-modified reduced graphene oxide (RGO-PDA/oxime) via the co-deposition of oxime during dopamine polymerization process [74]. The maximum adsorption amount of uranium for as-prepared RGO-PDA/oxime was found to be 1049 mg/g, which was 3–16 times larger than those of the reported single PDA or oxime-modified nanostructures. The authors attributed outstanding adsorption performance of RGO-PDA/oxime complex to the electrostatic repulsion between GO and salicylaldoxime and the effective synergism between PDA and oxime molecules. Furthermore, the presence of co-existing ions even at high concentrations had a slight influence on the uranium adsorption. The absorbent was still stable after four times of adsorption-desorption cycles. In addition, the magnetic materials are widely used for adsorption due to its ease of collection and reuse. Zhu et al. prepared magnetic porous nitrogen-doped carbon materials containing well-dispersed active Fe nanocrystals for highly efficient U(VI) adsorption and reduction [75]. The porous nitrogen-doped carbon materials were synthesized via the co-deposition of Fe^{3+} during the PDA formation process in Tris solution, followed by in situ post-carbonization at 700 °C (Fig. 6.14a). As-prepared carbon materials were nanospheres with many macroporous concaves and ultrahigh surface area (343.02 m^2/g), resulting in high-efficient removal performance of U(VI) with adsorption capacity of 232.54 mg/g. The macroscopic adsorption experiments confirmed that the U(VI) adsorption kinetics and isotherms onto Fe/N-C-700 agreed well with the pseudo-second-order kinetic model and Langmuir model, respectively (Fig. 6.14b). After six adsorption recycling tests, U(VI) removal onto Fe/N-C-700 could remain high efficiency. All these characteristics made magnetic porous nitrogen-doped carbon nanospheres good candidate for U(VI)-contaminated wastewater treatment (Fig. 6.14d).

Ultrafiltration (UF) membrane is considered as a potential technology for municipal drinking water and wastewater treatment owing to the compactness, relatively low energy consumption, easy automation, and the ability to remove turbidity, microorganisms, and natural organic matters. Nevertheless, most of the UF membranes separate the contaminants based on the size exclusion effect, which hinders their use for heavy metal ion removal. Recently, adsorptive membrane, combining

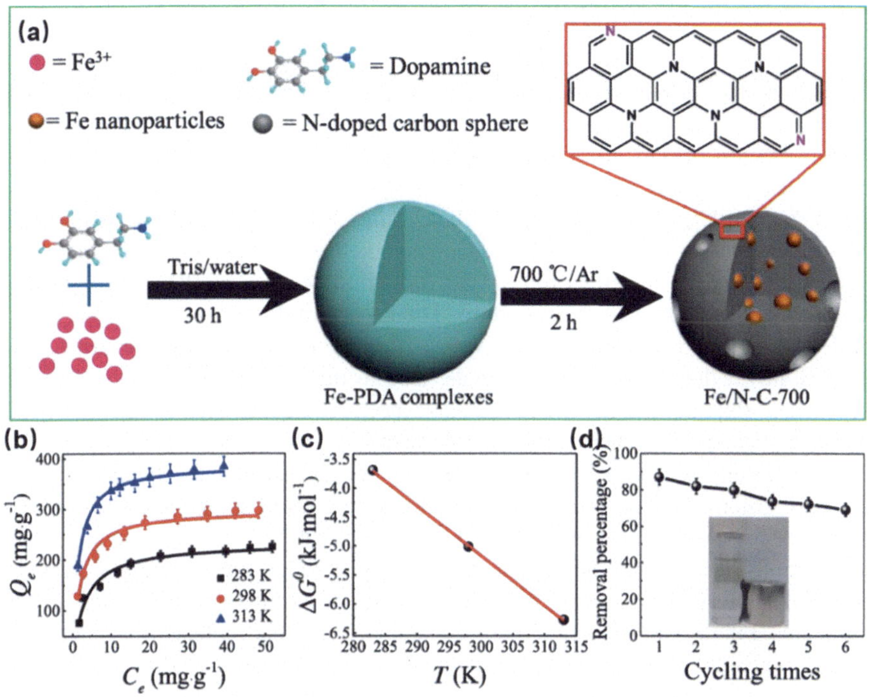

Fig. 6.14 (**a**) Schematic illustration for synthesis of magnetic porous nitrogen-doped carbon materials. (**b**) Adsorption isotherms, (**c**) plot of ΔG^0 versus T, and (**d**) recycle property of the U (VI) removal with the porous nitrogen-doped carbon materials. The inset in (**d**) is a digital photograph of the magnetic separation after six recycling tests. Reprinted with permission from [75] copyright (2018) Elsevier

the advantages of membrane technology and adsorption process, has provided an alternative solution for the removal of heavy metals from aqueous solution with UF membranes. Due to the gentle experiment conditions, high modification efficiency, and abundant functional groups, mussel-inspired PDA coatings are utilized as a universal method for preparation of adsorptive membrane. For instance, Fang et al. developed a new strategy for self-deposition of PDA nanoparticles onto internal pores of polymeric membranes to fabricate ultrafiltration-adsorption membranes [76]. Especially, circular filtrating was employed to control the self-polymerization of dopamine solution onto the cross-section of asymmetric polyethersulfone (PES) UF membrane from reverse direction. The PDA modification of internal pores endowed such ultrafiltration-adsorption membranes with enhanced adsorption capacity for heavy metals, and the static adsorption amounts for Pb^{2+}, Cd^{2+}, and Cu^{2+} were 20.23 mg/g, 17.01 mg/g, and 10.42 mg/g, respectively. In addition, the ultrafiltration-adsorption membrane with an area of 1 m^2 could purify 1096 L of 100 ppb Pb^{2+} and 50 mg/L BSA containing water to meet the requirement of maximum contaminant level (MCL) for drinking water, showing great potential for practical applications in the decontamination of municipal drinking water.

6.5 Summary and Outlook

This chapter has described recent advances in the fabrication of dopamine-based materials for diverse applications. Since the discovery of oxidative self-polymerization of dopamine for preparation of PDA materials, different strategies have been employed to construct fruitful polydopamine-based materials, such as PDA nanoparticles, core/shell nanoparticles, microcapsules, films, and hydrogels. Moreover, one-pot PDA-assisted co-deposition method could incorporate functional molecules into the materials during the formation process of PDA, thus greatly simplifying the modification procedure. The co-assembly of dopamine with other molecules through covalent or noncovalent interactions is an alternative strategy to prepare novel dopamine-based materials. Due to their intriguing physicochemical properties, dopamine-based materials have been widely utilized in the fields of biomedicine, energy, sensing, and environment.

In the future, deep understanding of the exact polymerization mechanism and the detailed structures of the components of PDA will be of immediate interest and relevance for development of this field. Further exploration is necessary to find other applications of dopamine-based materials in various fields, while it highly depends on the discovery of new physicochemical properties and the construction of novel structures. Finally, the co-assembly of dopamine with other molecules is an emerging approach that devotes to be paid more efforts for fabrication of dopamine-based nanoplatform. Combination of dopamine with functional components will create new opportunities for developing novel materials, properties, and functions. Dopamine-based materials will become a completely new solution to many issues in material science.

Acknowledgments The authors thank the National Natural Science Foundation of China (Nos. 21703169 and 21972084), the Scientific Research Plan of Shaanxi Province of China (No. 2021KJXX-39), and the Young Talent Fund of University Association for Science and Technology in Shaanxi Province of China (No. 20190605).

References

1. Klein MO, Battagello DS, Cardoso AR et al (2019) Dopamine: functions, signaling, and association with neurological diseases. Cell Mol Neurobiol 39(1):31–59
2. Lee H, Dellatore SM, Miller WM et al (2007) Mussel-inspired surface chemistry for multifunctional coatings. Science 318(5849):426–430
3. Ju K-Y, Lee Y, Lee S et al (2011) Bioinspired polymerization of dopamine to generate melanin-like nanoparticles having an excellent free-radical-scavenging property. Biomacromolecules 12 (3):625–632
4. Amin DR, Sugnaux C, Lau KHA et al (2017) Size control and fluorescence labeling of polydopamine melanin-mimetic nanoparticles for intracellular imaging. Biomimetics 2(3):17
5. Liu YL, Ai KL, Liu JH et al (2013) Dopamine-melanin colloidal nanospheres: an efficient near-infrared photothermal therapeutic agent for in vivo cancer therapy. Adv Mater 25(9):1353–1359

6. Qiang W, Li W, Li X et al (2014) Bioinspired polydopamine nanospheres: a superquencher for fluorescence sensing of biomolecules. Chem Sci 5(8):3018–3024
7. Zhang X, Wang S, Xu L et al (2012c) Biocompatible polydopamine fluorescent organic nanoparticles: facile preparation and cell imaging. Nanoscale 4(18):5581–5584
8. Hu J, Zhang X, Wen Z et al (2016) Asn-Gly-Arg-modified polydopamine-coated nanoparticles for dual-targeting therapy of brain glioma in rats. Oncotarget 7(45):73681–73696
9. Zheng Q, Lin T, Wu H et al (2014) Mussel-inspired polydopamine coated mesoporous silica nanoparticles as pH-sensitive nanocarriers for controlled release. Int J Pharm 463(1):22–26
10. Xiong W, Peng L, Chen H et al (2015) Surface modification of MPEG-b-PCL-based nanoparticles via oxidative self-polymerization of dopamine for malignant melanoma therapy. Int J Nanomedicine 10:2985–2996
11. Xu G, Yu X, Zhang J et al (2016) Robust aptamer-polydopamine-functionalized m-PLGA-TPGS nanoparticles for targeted delivery of docetaxel and enhanced cervical cancer therapy. Int J Nanomedicine 11:2953–2965
12. Lin Q, Huang X, Tang J et al (2013) Environmentally friendly, one-pot synthesis of folic acid-decorated graphene oxide-based drug delivery system. J Nanopart Res 15(12):2144
13. Sharker SM, Kang EB, Shin C-I et al (2016) Near-infrared-active and pH-responsive fluorescent polymer-integrated hybrid graphene oxide nanoparticles for the detection and treatment of cancer. J Appl Polym Sci 133(32):43791
14. Liu F, He X, Lei Z et al (2015a) Facile preparation of doxorubicin-loaded upconversion@polydopamine nanoplatforms for simultaneous in vivo multimodality imaging and chemophotothermal synergistic therapy. Adv Health Mater 4(4):559–568
15. Liu T, Li S, Liu Y et al (2016) Mn-complex modified $NaDyF_4$:Yb@$NaLuF_4$:Yb, Er@polydopamine core-shell nanocomposites for multifunctional imaging-guided photothermal therapy. J Mater Chem B 4(15):2697–2705
16. Ju K-Y, Lee S, Pyo J et al (2015) Bio-inspired development of a dual-mode nanoprobe for MRI and Raman imaging. Small 11(1):84–89
17. Li C, Liu Z, Yao P (2016a) Gold nanoparticles coated with a polydopamine layer and dextran brush surface for diagnosis and highly efficient photothermal therapy of tumors. RSC Adv 6 (39):33083–33091
18. Wang S, Zhao X, Wang S et al (2016) Biologically inspired polydopamine capped gold nanorods for drug delivery and light-mediated cancer therapy. ACS Appl Mater Interf 8 (37):24368–24384
19. Zhang L, Su H, Cai J et al (2016) A multifunctional platform for tumor angiogenesis-targeted chemo-thermal therapy using polydopamine-coated gold nanorods. ACS Nano 10 (11):10404–10417
20. Du B, Gu X, Zhao W et al (2016) Hybrid of gold nanostar and indocyanine green for targeted imaging-guided diagnosis and phototherapy using low-density laser irradiation. J Mater Chem B 4(35):5842–5849
21. Li D, Zhang Y, Wen S et al (2016b) Construction of polydopamine-coated gold nanostars for CT imaging and enhanced photothermal therapy of tumors: an innovative theranostic strategy. J Mater Chem B 4(23):4216–4226
22. Kumar A, Kumar S, Rhim W-K et al (2014) Oxidative nanopeeling chemistry-based synthesis and photodynamic and photothermal therapeutic applications of plasmonic core-petal nanostructures. J Am Chem Soc 136(46):16317–16325
23. Zhang M, Zhang X, He X et al (2012) A self-assembled polydopamine film on the surface of magnetic nanoparticles for specific capture of protein. Nanoscale 4(10):3141–3147
24. Zhang X, Huang Q, Liu M et al (2015) Preparation of amine functionalized carbon nanotubes via a bioinspired strategy and their application in Cu^{2+} removal. Appl Surf Sci 343:19–27
25. Zhang Q, Liao J, Liao M et al (2019) One-dimensional Fe_7S_8@C nanorods as anode materials for high-rate and long-life lithium-ion batteries. Appl Surf Sci 473:799–806
26. Donath E, Sukhorukov GB, Caruso F et al (1998) Novel hollow polymer shells by colloid-templated assembly of polyelectrolytes. Angew Chem Int Ed 37(16):2201–2205

27. Jia Y, Li J (2019) Molecular assemblies of biomimetic microcapsules. Langmuir 35 (26):8557–8564
28. Chen X, Yan Y, Muellner M et al (2014) Engineering fluorescent poly(dopamine) capsules. Langmuir 30(10):2921–2925
29. Cheng F-F, Zhang J-J, Xu F et al (2013) pH-sensitive polydopamine nanocapsules for cell imaging and drug delivery based on folate receptor targeting. J Biomed Nanotechnol 9 (7):1155–1163
30. Li H, Jia Y, Feng X et al (2017) Facile fabrication of robust polydopamine microcapsules for insulin delivery. J Colloid Interf Sci 487:12–19
31. Xu H, Liu X, Wang D (2011) Interfacial basicity-guided formation of polydopamine hollow capsules in pristine o/w emulsions-toward understanding of emulsion template roles. Chem Mater 23(23):5105–5110
32. Xue J, Zheng W, Wang L et al (2016) Scalable fabrication of polydopamine nanotubes based on curcumin crystals. ACS Biomater Sci Eng 2(4):489–493
33. Yan D, Xu P, Xiang Q et al (2016) Polydopamine nanotubes: bio-inspired synthesis, formaldehyde sensing properties and thermodynamic investigation. J Mater Chem A 4(9):3487–3493
34. Nuzzo RG, Allara DL (1983) Adsorption of bifunctional organic disulfides on gold surfaces. J Am Chem Soc 105(13):4481–4483
35. Decher G, Hong JD (1991) Buildup of ultrathin multilayer films by a self-assembly process, 1 consecutive adsorption of anionic and cationic bipolar amphiphiles on charged surfaces. In: Makromolekulare Chemie. Macromolecular symposia, Wiley online library, pp 321–327
36. Wei H, Ren J, Han B et al (2013) Stability of polydopamine and poly(DOPA) melanin-like films on the surface of polymer membranes under strongly acidic and alkaline conditions. Colloids Surf B Biointerfaces 110:22–28
37. Kang SM, You I, Cho WK et al (2010) One-step modification of superhydrophobic surfaces by a mussel-inspired polymer coating. Angew Chem Int Ed 49(49):9401–9404
38. Lv Y, Yang H-C, Liang H-Q et al (2015) Nanofiltration membranes via co-deposition of polydopamine/polyethylenimine followed by cross-linking. J Membr Sci 476:50–58
39. Lv Y, Du Y, Qiu W-Z et al (2017) Nanocomposite membranes via the codeposition of polydopamine/polyethylenimine with silica nanoparticles for enhanced mechanical strength and high water permeability. ACS Appl Mater Interf 9(3):2966–2972
40. Liu Y, Meng H, Konst S et al (2014b) Injectable dopamine-modified poly(ethylene glycol) nanocomposite hydrogel with enhanced adhesive property and bioactivity. ACS Appl Mater Interf 6(19):16982–16992
41. Jia Z, Zeng Y, Tang P et al (2019) Conductive, tough, transparent, and self-healing hydrogels based on catechol-metal ion dual self-catalysis. Chem Mater 31(15):5625–5632
42. Yang HC, Waldman RZ, Wu MB et al (2018) Dopamine: just the right medicine for membranes. Adv Funct Mater 8(8):1705327
43. Lynge ME, Teo BM, Laursen MB et al (2013) Cargo delivery to adhering myoblast cells from liposome-containing poly(dopamine) composite coatings. Biomater Sci 1(11):1181–1192
44. Ding T, Xing Y, Wang Z et al (2019) Structural complementarity from DNA for directing two-dimensional polydopamine nanomaterials with biomedical applications. Nanoscale Horiz 4 (3):652–657
45. Wang Y, Wu Y, Li K et al (2019b) Ultralong circulating lollipop-like nanoparticles assembled with gossypol, doxorubicin, and polydopamine via π–π stacking for synergistic tumor therapy. Adv Funct Mater 29(1):1805582
46. Ponzio F, Bour J, Ball V (2015) Composite films of polydopamine-alcian blue for colored coating with new physical properties. J Colloid Interf Sci 459:29–35
47. Wang H, Lin Q, Yin L et al (2019a) Biomimetic design of hollow flower-like g-C_3N_4@ PDA organic framework nanospheres for realizing an efficient photoreactivity. Small 15 (16):1900011

48. Dong Z, Feng L, Hao Y et al (2018) Synthesis of hollow biomineralized $CaCO_3$-polydopamine nanoparticles for multimodal imaging-guided cancer photodynamic therapy with reduced skin photosensitivity. J Am Chem Soc 140(6):2165–2178
49. Chien CY, Liu TY, Kuo WH et al (2013) Dopamine-assisted immobilization of hydroxyapatite nanoparticles and RGD peptides to improve the osteoconductivity of titanium. J Biomed Mater Res A 101(3):740–747
50. Jiao L, Xu W, Yan H et al (2019) A dopamine-induced au hydrogel nanozyme for enhanced biomimetic catalysis. Chem Commun 55(66):9865–9868
51. Zhang Y, Thingholm B, Goldie KN et al (2012d) Assembly of poly(dopamine) films mixed with a nonionic polymer. Langmuir 28(51):17585–17592
52. Liu Y, Chang C-P, Sun T (2014a) Dopamine-assisted deposition of dextran for nonfouling applications. Langmuir 30(11):3118–3126
53. Huang R, Liu X, Ye H et al (2015) Conjugation of hyaluronic acid onto surfaces via the interfacial polymerization of dopamine to prevent protein adsorption. Langmuir 31 (44):12061–12070
54. Zhang Y, Teo BM, Goldie KN et al (2014b) Poly(N-isopropylacrylamide)/poly (dopamine) capsules. Langmuir 30(19):5592–5598
55. Chassepot A, Ball V (2014) Human serum albumin and other proteins as templating agents for the synthesis of nanosized dopamine-eumelanin. J Colloid Interf Sci 414:97–102
56. Yu X, Fan H, Wang L et al (2014) Formation of polydopamine nanofibers with the aid of folic acid. Angew Chem Int Ed 53(46):12600–12604
57. Kohri M, Nannichi Y, Kohma H et al (2014) Size control of polydopamine nodules formed on polystyrene particles during dopamine polymerization with carboxylic acid-containing compounds for the fabrication of raspberry-like particles. Colloids Surf A Physicochem Eng Asp 449:114–120
58. Li H, Jia Y, Wang A et al (2014) Self-assembly of hierarchical nanostructures from dopamine and polyoxometalate for oral drug delivery. Chem Eur J 20(2):499–504
59. Zhang H, Guo L-Y, Jiao J et al (2017a) Ionic self-assembly of polyoxometalate-dopamine hybrid nanoflowers with excellent catalytic activity for dyes. ACS Sustain Chem Eng 5 (2):1358–1367
60. Li H, Yan Y, Gu X et al (2018) Organic-inorganic hybrid based on co-assembly of polyoxometalate and dopamine for synthesis of nanostructured ag. Colloids Surf A Physicochem Eng Asp 538:513–518
61. Tang L, Mo S, Liu SG et al (2018) Preparation of bright fluorescent polydopamine-glutathione nanoparticles and their application for sensing of hydrogen peroxide and glucose. Sensor Actuat B Chem 259:467–474
62. Hong S, Schaber CF, Dening K et al (2014) Air/water interfacial formation of freestanding, stimuli-responsive, self-healing catecholamine Janus-faced microfilms. Adv Mater 26 (45):7581–7587
63. Liu M, Ji J, Zhang X et al (2015b) Self-polymerization of dopamine and polyethyleneimine: novel fluorescent organic nanoprobes for biological imaging applications. J Mater Chem B 3 (17):3476–3482
64. Zhao C, Zuo F, Liao Z et al (2015) Mussel-inspired one-pot synthesis of a fluorescent and water-soluble polydopamine-polyethyleneimine copolymer. Macromol Rapid Commun 36 (10):909–915
65. Li H, Zhao Y, Jia Y et al (2019) Covalently assembled dopamine nanoparticle as an intrinsic photosensitizer and pH-responsive nanocarrier for potential application of anticancer therapy. Chem Commun 55:15057–15060
66. Li H, Zhao Y, Jia Y et al (2020) pH-responsive dopamine-based nanoparticles assembled via Schiff base bonds for synergistic anticancer therapy. Chem Commun 56:13347–13350
67. Chen Y, Ai K, Liu J et al (2016) Polydopamine-based coordination nanocomplex for T_1/T_2 dual mode magnetic resonance imaging-guided chemo-photothermal synergistic therapy. Biomaterials 77:198–206

68. Sun C, Zhang L, Zhang R et al (2015) Facilely synthesized polydopamine encapsulated surface-enhanced raman scattering (SRES) probes for multiplex tumor associated cell surface antigen detection using SRES imaging. RSC Adv 5(88):72369–72372
69. Ju K-Y, Lee JW, Im GH et al (2013) Bio-inspired, melanin-like nanoparticles as a highly efficient contrast agent for T_1-weighted magnetic resonance imaging. Biomacromolecules 14 (10):3491–3497
70. Xie C, Wang X, He H et al (2020) Mussel-inspired hydrogels for self-adhesive bioelectronics. Adv Funct Mater 30(25):1909954
71. Han L, Yan L, Wang K et al (2017) Tough, self-healable and tissue-adhesive hydrogel with tunable multifunctionality. NPG Asia Mater 9(4):e372
72. Zhang X, Huang Q, Deng F et al (2017b) Mussel-inspired fabrication of functional materials and their environmental applications: progress and prospects. Appl Mater Today 7:222–238
73. Zhang X, Jia X, Zhang G et al (2014a) Efficient removal and highly selective adsorption of Hg^{2+} by polydopamine nanospheres with total recycle capacity. Appl Surf Sci 314:166–173
74. Qian Y, Yuan Y, Wang H et al (2018) Highly efficient uranium adsorption by salicylaldoxime/polydopamine graphene oxide nanocomposites. J Mater Chem A 6(48):24676–24685
75. Zhu K, Chen C, Xu M et al (2018) In situ carbothermal reduction synthesis of Fe nanocrystals embedded into N-doped carbon nanospheres for highly efficient U(VI) adsorption and reduction. Chem Eng J 331:395–405
76. Fang X, Li J, Li X et al (2017) Internal pore decoration with polydopamine nanoparticle on polymeric ultrafiltration membrane for enhanced heavy metal removal. Chem Eng J 314:38–49

Chapter 7
Peptide-Based Nanoarchitectonics: Self-Assembly and Biological Applications

Luyang Zhao and Xuehai Yan

7.1 Introduction

Self-assembly is ubiquitous in life. It bridges single molecules to the delicate architecture with fascinating functionalities. For example, chlorophylls as the pigment of photosynthesis self-assemble into large supramolecular structures in chlorosomes to enable efficient light harvesting and energy transfer [1]. The surficial shell of calicivirus is constituted by 90 copies of capsid protein dimer molecules [2]. The self-assembly of these biomolecules is based on precise control of intermolecular interactions, which mainly include electrostatic interaction, hydrogen-bonding interaction, hydrophobic interaction, and Van der Waals interaction. Inspired by nature, artificial functional supramolecular systems, namely, architectonics, undergo fast development and become an important approach toward future biological and biomimetic materials. The major differences between natural and artificial self-assembly systems are that natural systems self-assembled in a directional and hierarchical way with complicated functions, while artificial ones are relatively simple in both chemical structure and functionality [3, 4]. One of the

L. Zhao
State Key Laboratory of Biochemical Engineering, Institute of Process Engineering, Chinese Academy of Sciences, Beijing, P. R. China

X. Yan (✉)
State Key Laboratory of Biochemical Engineering, Institute of Process Engineering, Chinese Academy of Sciences, Beijing, P. R. China

School of Chemical Engineering, University of Chinese Academy of Sciences, Beijing, P. R. China

Center for Mesoscience, Institute of Process Engineering, Chinese Academy of Sciences, Beijing, P. R. China
e-mail: yanxh@ipe.ac.cn

T. Govindaraju, K. Ariga (eds.), *Molecular Architectonics and Nanoarchitectonics*, Nanostructure Science and Technology,
https://doi.org/10.1007/978-981-16-4189-3_7

challenges in this field is how to employ simple molecules to construct complex architecture with desired functions.

Peptides are a class of natural products that widely distributed in nature. They are generally composed of less than 50 amino acids and have specific secondary structures. Due to these properties, many peptides are facile to serve as building blocks and self-assemble to various supramolecular architectures like nanotubes, nanofibrils, nanovesicles, nanobelts, and hydrogels with many interesting properties [5–9]. Here, the peptide assemblies are contributed by the diverse amino acids which provide abundant intermolecular interactions. In addition, the interactions between peptides and other functional biomolecules further afford multicomponent architectures with a broader range of applications [10–13]. Consequently, peptide self-assembly has become a fascinating research field for future development of novel nanoarchitectonics.

The topic of this chapter is peptide-modulated nanoarchitectonics, including both the supramolecular mechanisms and diverse biological applications. As the basis of peptide-based nanoarchitectonics, how the peptide and other species self-assemble is crucial. Among various biological applications of peptides, tumor imaging, and phototherapy, as well as biomimetic photosynthetic systems are highlighted in this chapter. We intend this chapter to provide a comprehensive resource on the peptide self-assembly science, and to show great promise for future exploration of novel peptide-based biomaterials.

7.2 Self-Assembly Mechanisms

Peptides may exhibit diverse supramolecular morphologies like nanostrands, nanofibrils, nanoparticles, nanosheets, and so forth. The formation of these morphologies relies on the intermolecular interactions of the building blocks [14].

The regulation of peptide intermolecular interactions could be realized by either designing molecular structures or controlling the self-assembly process [15–18]. Diphenylalanine is a simple dipeptide showing excellent self-assembly properties. The ability to construct multiple nanoarchitectures of the diphenylalanine results from its unique molecular structure, where the phenyl moieties provide hydrophobic or, more specifically, π-π interactions, and the charged amino and carboxylic groups provide electrostatic and hydrogen-bonding interactions. These multiple interactions are synergistically driving the build-block molecules to construct a definite supramolecular structure under certain environmental conditions. A simple derivative of diphenylalanine, 9-fluorenylmethyloxycarbonyl-*L*-phenylalanyl-*L*-phenylalanine (Fmoc-FF), was further studied to represent how the nanoarchitecture was regulated by external environmental conditions (Fig. 7.1) [19]. In acidic solution, Fmoc-FF was neutral in charge and tends to form fibrous hydrogel which is similar to a β-sheet supramolecular structure. Comparably, by tuning the solution from acidic to basic, the charge of Fmoc-FF was converted from neutral to negative due to the carboxylic group. The synergistic effect of electrostatic

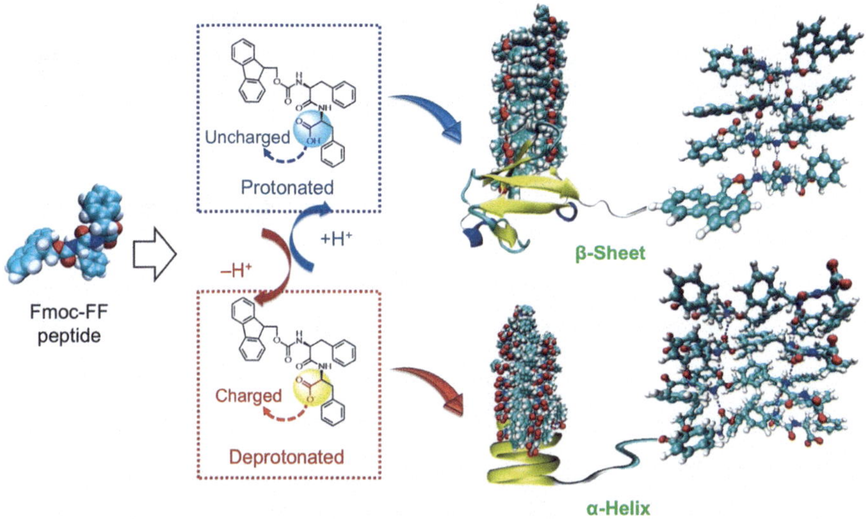

Fig. 7.1 Mechanism of pH-regulated self-assembly morphologies of Fmoc-FF. Reprinted with permission [19]. Copyright 2018, Wiley-VCH

interaction and the hydrophobic interaction was that the Fmoc-FF self-assembled to a helical structure.

In addition to the peptide themselves, peptide may be coassembled with other functional species, forming complex or even hierarchical supramolecular nanoarchitectures. 5,10,15,20-Tetrakis(4-sulfonatophenyl)porphyrin (TPPS) is a widely applied light-harvesting dye. Aqueous mixture of TPPS and diphenylalanine (FF) at acidic conditions underwent a two-step coassembly process [20]. The TPPS was firstly self-assembled to oligomeric *J*-aggregates, which was then followed by interacting with FF and resulted in porous multichambered microspheres. The self-assembly of TPPS was driven by the strong π-π interactions, while the interaction between TPPS and FF was dominated by both electrostatic interaction and π-π interaction. More interestingly, coassembly of TPPS and dilysine (KK) could afford hierarchical fiber bundles [21]. Different from the assembly process of TPPS/FF system, the TPPS and KK were coassembled to nanorods at the beginning, which was then followed by self-orienting growth to afford fiber bundles. Here KK was more hydrophilic than FF with abundant hydrogen-bonding interactions, thereby leading to a more ordered supramolecular structure than FF did.

In many cases, the self-assembly processes of a certain building block were not unique but affected by kinetic conditions, thereby representing assembly pathway complexity [22, 23]. A model compound, diphenylalanine-conjugated zinc (II) phthalocyanine (PF), could self-assemble to nanoparticles by rapid mixing water with its concentrated DMSO solution (Fig. 7.2a) [24]. Comparably, raising the initial DMSO/water volume ratio and elongating the PF assembly time to several days resulted in helical nanofibrils (Fig. 7.2b). The nanofibrils further showed

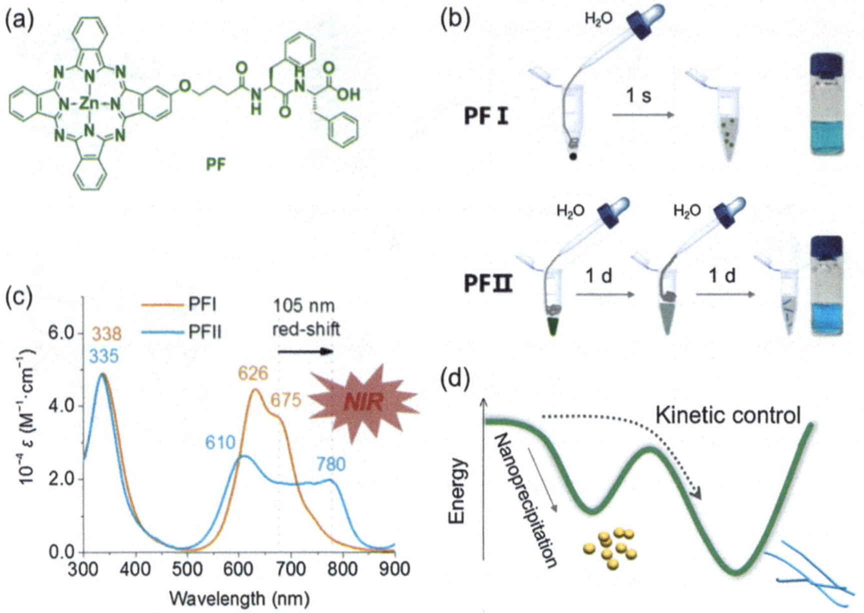

Fig. 7.2 Kinetic control of phthalocyanine-peptide conjugate self-assembly enabling different nanoarchitectures with superlarge redshifted absorption. (**a**) The chemical structure of phthalocyanine-peptide conjugate. (**b**) Schematic illustration of rapid precipitation (PFI) and kinetic assembly (PFII). (**c**) Absorption spectra of the nanoparticle (PFI) and the nanofibrils (PFII). (**d**) The energy landscape of the two nanoarchitectures. Reprinted with permission [24]. Copyright 2019, Chinese Chemical Society

superlarge redshifted absorption of 105 nm in comparison with both the monomeric PF solution and PF nanoparticles, showing a near-infrared absorption (Fig. 7.2c). Here, the formation of PF nanoparticle was an entropy-dominated process with relatively high thermodynamic energy and random molecular orientation due to extremely short assembly time, while the formation of PF nanofibrils was an enthalpy-dominated process and reached thermodynamic equilibrium (Fig. 7.2d). The extra addition of DMSO reduced the free energy barrier between the two nanoarchitectures, so that the formation of nanofibrils is enabled. The significances of kinetic assembly lie on two aspects: Firstly, it provides more flexible method to regulate the supramolecular structure, and secondly, the kinetically afforded assemblies may exhibit more attractive properties (in this case, superlarge redshifted absorption) that are otherwise cannot be obtained. Since peptides have diverse functional groups and flexible chemical structure, they are naturally excellent kinetic assembly candidates.

It is more interesting that kinetic assembly may be coupled with covalent conjugation reaction, resulting in a chemical evolution process. A typical example is tyrosine oxidation, which leads to the formation of melanin or cross-linked proteins based on dityrosine. The chemical evolution of tyrosine could also be

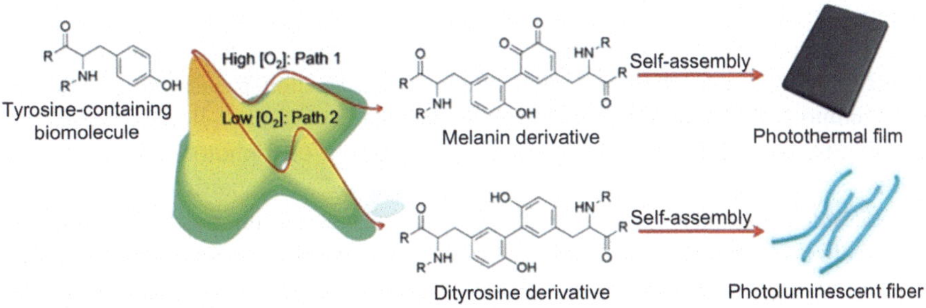

Fig. 7.3 Schematic diagram of selective chemical evolution from tyrosine to melanin or dityrosine-based functional materials under the control of oxygen concentration during assembly. Reprinted with permission [25]. Copyright 2019, Wiley-VCH

artificially tuned by oxygen concentration (Fig. 7.3) [25]. A simple tyrosine derivative, *N*-carbobenzyloxy-*L*-tyrosine (ZY), was irradiated by UV light (254 nm) for 12 h under oxygen-rich condition and resulted in two-dimensional microfilms with visible and near-infrared light absorptions. The microfilms were probably formed by the combination of chemical reaction which converted tyrosine to melanin and self-assembly which was driven by mainly π–π stacking and hydrophobic interactions. Comparably, light-irradiated ZY under oxygen-deficient condition afforded luminescent dityrosine nanofibrils. The nanofibrils were self-assembled under π-π stacking and hydrogen-bonding interactions. The critical oxygen/argon centration value between the two chemical evolution pathways was further determined to be 18.6% (v/v). Therefore, reaction-participated self-assembly process provides more regulating methods for functional materials.

7.3 Tumor Imaging and Phototherapeutic Biomaterials

Peptides are versatile natural species for biological purposes like tumor therapy, antibacterial, as well as tissue engineering. The combination of biomedicine and nanotechnology has paved the way from supramolecular materials to nanomedicines with optimized targeting efficiency, pharmacokinetics, and biocompatibility during treatment [26, 27]. Peptide-based phototherapeutic biomaterials are distinguished from conventional liposome drug delivery systems on the aspect that the peptide is one of the building blocks involved in the assembly which contributes to the stability and drug loading efficiency of the nanoarchitectures. Consequently, peptide-modulated phototherapeutic nanomaterials enable more precise preparation and higher complexity than conventional ones [28].

Phototherapy mainly includes photodynamic therapy and photothermal therapy. Photodynamic therapy utilizes the singlet oxygen or reactive oxygen species that is converted by the photosensitizers from light-harvesting, and photothermal therapy employs the photothermal conversion to generate heat so that lead to the ablation of tumors [29, 30]. To achieve this phototherapeutic efficacy, a photosensitizing agent may be assembled either individually or combined with other functional species to afford integrated nanoplatform. Interestingly, a simple diphenylalanine derivative, (*S*)-2-amino-*N*-((*S*)-1-amino-1-oxo-3-phenylpropan-2-yl)-3-phenylpropanamide (H-Phe-Phe-NH_2, DPA), could template the assembly of a photosensitizing compound, chlorin e6 (Ce6), affording well-ordered nanoparticles [31]. The nanoparticles showed high colloid stability due to inherent intermolecular interactions including π-stacking and electrostatic interaction. The nanoparticles were also responsive to pH and enzymes, which contribute to the controlled release of photosensitizers in tumor microenvironment. Compared with free Ce6, the DPA/Ce6 nanoparticles exhibited fourfold enhancement of photocytotoxicity and completed tumor ablation efficacy.

In addition to the binary biomedical nanoplatform, peptides may also integrate multiple functional species and enable simultaneous improvement of therapeutic performance and reduction of side effect. Inspired by the multicomponent self-organization in life, we proposed coassembling platform of metal ion-participated nanoarchitectonics. The combination of zinc(II) cation, Ce6, and fluorenylmethoxycarbonyl-*L*-histidine afforded uniform nanoparticles (Fig. 7.4a). Their self-assembly process was driven by not only hydrophobic and electrostatic interactions but also coordination interaction between the imidazole group and zinc ion [32]. The coordination is a relatively strong intermolecular interaction and sensitive to environmental conditions, thereby enabling stimuli-responsive nanoparticle disassembly. The trinary nanoagents were demonstrated to be stable during blood circulation while undergoing efficient disassembly and drug release in tumor environmental conditions (pH 6.5–7.2 and high GSH concentration level). Compared to those unencapsulated photosensitizers, such metal ion-participated nanodrugs could prolong the blood circulation life, enhance the tumor accumulation, and improve the antitumor effect (Fig. 7.4b, c).

The combination of peptide and metal ion assembly is a versatile protocol for innovation of biomedical materials [33–35]. For example, the assembly of biliverdin, manganese(II) ion, and a histidine derivative resulted in endogenous near-infrared photothermal nanoagent for cancer theranostics, where biliverdin served as the near-infrared photothermal agent, manganese(II) ion as the magnetic resonance imaging (MRI) agent, and histidine derivative as the assembly template [36]. The nanoagents showed high selectivity in tumor accumulation, efficient photothermal tumor ablation, as well as multimodel contrast for imaging techniques with high biocompatibility. The assembly of curcumin, zinc(II) ion, and a histidine derivative (9-fluorenylmethyloxycarbonyl-*L*-histidine, Fmoc-His-OH) resulted in well-defined and spherical nanoparticles with tunable sizes [37]. The coordination-included assembly was proved practical to overcome rapid degradation and poor tissue absorption of the biomaterial.

Efficient control of the morphology and function of nanoplatforms could also be realized by covalently conjugating peptides to the photosensitizers [38, 39]. Porphyrin is a good photothermal agent, while simple self-assembly of individual porphyrins just provides precipitates. Comparably, covalent reaction of diphenylalanine to

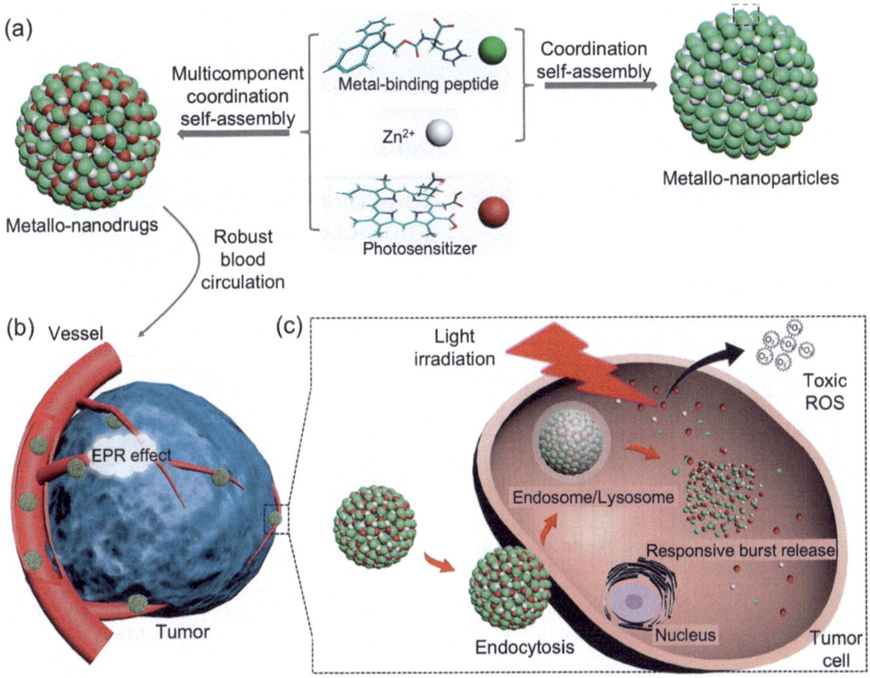

Fig. 7.4 Supramolecular coordination assembly of zinc(II) cation, Ce6, and peptide derivative for efficient photodynamic therapy. (**a**) Construction of the trinary nanoarchitectures. (**b**) Tumor accumulation of the nanoparticles via blood circulation. (**c**) Cellular internalization of the nanoparticles in a stimuli-responsive manner enabling efficient photodynamic therapy. Reprinted with permission [32]. Copyright 2018, American Chemical Society

the porphyrin moiety afforded a peptide-porphyrin conjugate, which acted as the self-assembly building block toward the self-assembled nanoplatform and photothermal agents (Fig. 7.5) [40]. The nanodots exhibited excellent colloid stability, which originated from the strong π-π stacking and hydrophobic interactions. Due to the π-π stacking, the intrinsic fluorescence of porphyrin was quenched and in turn showed efficient photothermal conversion as high as 54.7%, which was comparable to many other photothermal agents. The nanodots showed much improved tumor accumulation and long circulation in vivo, and rendered temperature ascending to as high as 58.1 °C in tumor site, leading to complete tumor ablation.

Sometimes the covalent conjugation may be coupled with self-assembly. Namely, the covalent reaction and self-assembly are simultaneously progressed. Such “covalent assembly” strategy provides an alternative way forward for efficient tumor therapeutic nanoarchitectonic agents. Genipin is a typical iridoid presenting in gardenia fruits. Genipin and the peptide DPA are intrinsically colorless nonpigmented biomolecules [41]. Intriguingly, covalent assembly of genipin and DPA (keeping in aqueous solution at 70 °C for 24 h) afforded dark blue nanoparticles with a diameter of 170 ± 54 nm. The nanoparticle size could be

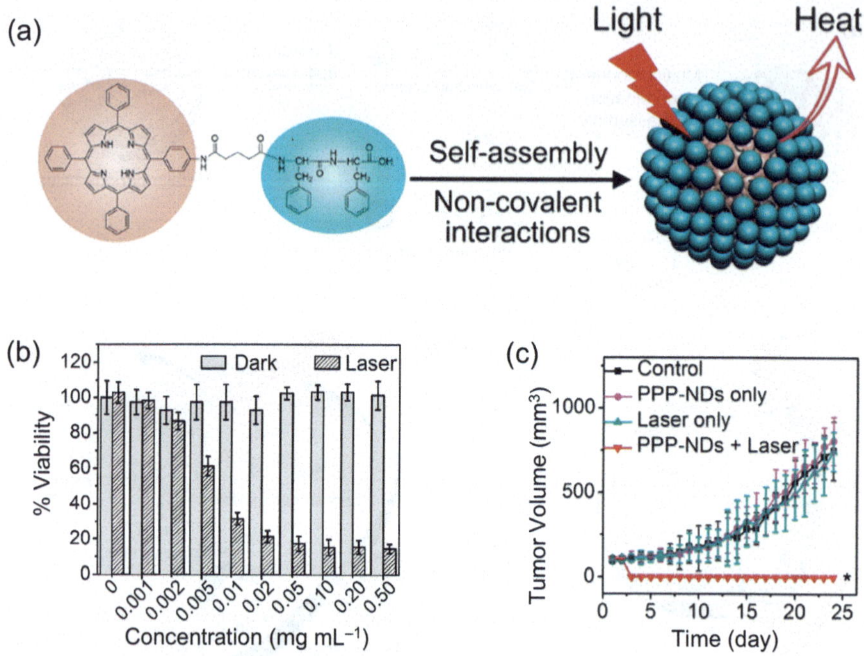

Fig. 7.5 Self-assembly of a peptide-porphyrin conjugate affording photothermal nanodots. (**a**) Molecular structure of the peptide-porphyrin conjugate and schematic self-assembly protocol. (**b**) Cell viability under different nanoparticle concentrations and light/dark condition. (**c**) In vivo tumor photothermal therapeutic effect. Reprinted with permission [40]. Copyright 2017 American Chemical Society

facilely tuned by the pH, stoichiometry, as well as temperature. Here, both the nucleophilic reaction between genipin and DPA and the noncovalent interactions like π-π and hydrogen-bonding interactions contributed to the nanoparticle formation. The covalent assembly showed two advantages: Firstly, the fabrication procedure was easy and facile to regulation; and secondly, the resulted nanoparticles were robust and stable. These merits gave rise to efficient tumor accumulation and consequently selective tumor ablation with good biocompatibility.

7.4 Biomimetic Photosynthetic Architectures

Sunlight is the most important energy source fueling the evolution of life. The conversion of solar energy to chemical energy was realized by the delicate photosynthetic systems rooted in bacteria and plants. Mimicking natural photosynthetic systems is significant for solar energy utilization. The challenge remains on fabricating highly efficient photosynthetic systems with biomimetic structure and functions using simple methods.

Recently, we proved that coassembly of simple peptide and porphyrin, namely, 9-fluorenylmethoxycarbonyl-*L*-lysine (Fmoc-K) and 5,10,15,20-tetrakis(4-sulfonatophenyl) porphyrin, could lead to urchin-like light-harvesting antenna nanoarchitectonics [42]. Fmoc-K served as the template to control the pigment assembly. Based on the binary fibers, platinum (Pt) were grown in situ on the fiber surface, resulting in a light-harvesting antenna. In the antenna, TPPS absorbed light and converted from ground state to excited state, and then transferred the electron to Pt. The electron stored on Pt sequentially reduced proton to dihydrogen. Replacing Fmoc-K by dilysine (KK) led to similar photosynthetic nanoarchitectonic systems which mimicked the functions of green sulfur bacteria [43]. Based on the nanoarchitectonic light-harvesting antenna, more complex system, for example, a hierarchical architecture involving cysteine, TPPS, alcohol dehydrogenase (ADH), and platinum nanodots, could be prepared to mimic natural organelles. The hybrid assemblies coupled photocatalytic hydrogen evolution and enzymatic nicotinamide adenine dinucleotide (NADH) generation, which mimicked the light and dark reactions of chloroplasts [44]. These studies provide a simple strategy for the construction of biomimetic photosystem for sustainable production of green energy.

Enzyme mimics have attracted broad research interest due to their extensive applications. The challenges of mimicking photocatalytic enzyme lie on the integration of designed host molecules into three-dimensional architectures. We explored that Fmoc-His-OH could induce the assembly of phthalocyanine to nanovesicles with tunable size and membrane thickness (Fig. 7.6) [45]. The nanoarchitecture improved the photosensitivity and stability of phthalocyanine, thereby exhibiting superior and sustainable photocatalysis in the conversion from dopamine to leucodopaminechrome. The catalytic site, phthalocyanine, converted the solar energy to produce singlet oxygen via photoinduced energy transfer process. The generated singlet oxygen was reacted with dopamine and resulted leucodopaminechrome. The phthalocyanine-containing nanovesicles well mimicked the natural photooxidase in a sense that the photosensitizer was activated by the coassembled membrane structure which was different from the covalent strategy that direct bound photosensitizer to the hydrophobic area of lipoprotein in photooxidase. This work also inspires the preparation of functional reinforcement materials with tunable structure using multicomponent collaborative self-assembly strategy.

7.5 Conclusions and Perspective

In summary, we highlighted recent works regarding peptide-based self-assembling nanoarchitectonics. Self-assembly of individual peptides could afford diverse morphological architectures, and coassembly of peptide and biological species could further provide significant nanoplatforms and functional materials where the peptide served as assembly templates. Furtherly, various thermodynamic or kinetic strategies

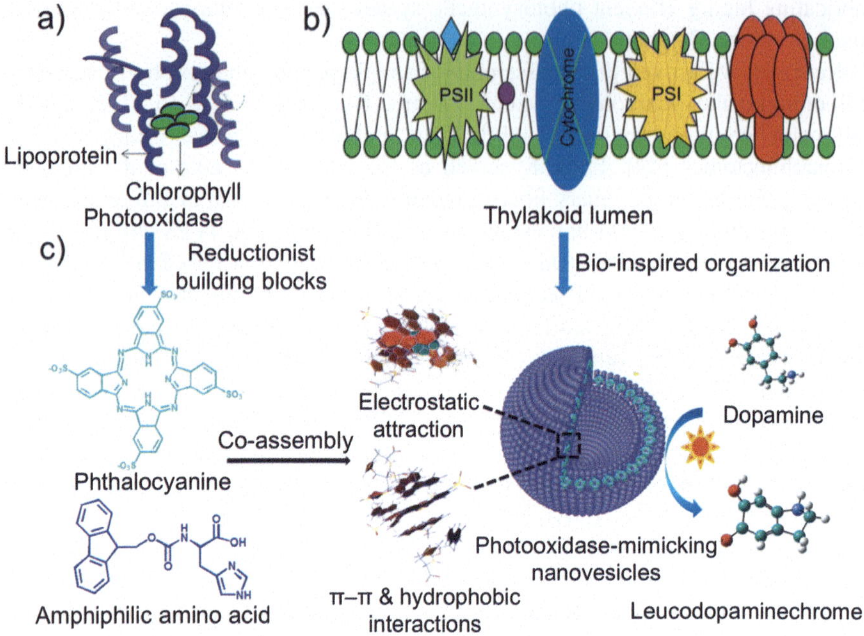

Fig. 7.6 Schematic illustration of the biomimetic photocatalytic nanovesicle system. (**a**) Natural photooxidase of lipoprotein and chlorophyll binary system. (**b**) Artificial enzyme mimics of natural thylakoid membrane structure of host cytochrome. (**c**) Chemical structure of phthalocyanine and amino acid derivatives and their coassembly toward nanovesicles enabling the conversion of dopamine into leucodopaminechrome. Reprinted with permission [45]. Copyright 2019, Wiley-VCH

were explored to precisely regulate the assembly process, affording complicated and hierarchical nanoplatforms. Such peptide-based self-assembling systems showed great potential in various biological applications. Through coassembly with phototherapeutic sensitizers, the peptide served as assembly templates and significantly improved the phototherapeutic efficiency of the photosensitizer, which thereby lead to enhanced tumor ablation efficacy. Covalent conjugation of peptide and the photosensitizer before assembly also resulted in successful tumor phototherapeutic biomaterials. Peptide-modulated biomimetic photosystems mainly adopted a coassembly strategy with light-harvesting species. The resulted nanosystem enabled efficient coupling between the photoinduced redox process and biochemical reactions to provide energetic species and thereby represented as integrated biomimetic photosystems.

It is worth noting that the key factors to regulate the assembly process, the supramolecular structure, as well as the biomaterial functionalities are the noncovalent interactions. However, the strategies of regulating noncovalent interactions are still qualitative, which limited precise fabrication of novel peptide-based nanoarchitectonics with concrete supramolecular structures. Therefore, one of the great challenges remaining in this field is how to control the noncovalent interactions in a quantitative way. In addition, the noncovalent interactions include several types, which synergistically drive the assembly process. How to recognize the contribution of each kind of noncovalent interactions and then to rationally regulate the synergistic effect toward more complex and multifunctional peptide-based materials are also of huge significance. It is quite promising that solving these problems will greatly help the design of novel peptide-based materials with more precise structure, more attractive properties, and more abundant functionalities.

Acknowledgment This work was financially supported by the National Natural Science Foundation of China (Project No. 21703252), the Innovation Research Community Science Fund (No. 21821005), the National Natural Science Fund BRICS STI Framework Programme (No. 51861145304), the Key Research Program of Frontier Sciences of Chinese Academy of Sciences (CAS, Grant No. QYZDB-SSW-JSC034), and the CAS President's International Fellowship Initiative (2018VEA0005).

References

1. Orf GS, Blankenship RE (2013) Chlorosome antenna complexes from green photosynthetic bacteria. Photosynth Res 116:315–331
2. Prasad BVV, Hardy ME, Rossmannn MG, Estesz MK (1999) X-ray crystallographic structure of the norwalk virus capsid. Science 286:287–290
3. Ariga K, Li J (2016) Nanoarchitectonics for advanced materials: strategy beyond nanotechnology. Adv Mater 28:987–988
4. Lutz J-F, Lehn J-M, Meijer EW, Matyjaszewski K (2016) From precision polymers to complex materials and systems. Nat Rev Mater 1:16024
5. Zou Q, Chang R, Xing R, Yuan C, Yan X (2020) Injectable self-assembled bola-dipeptide hydrogels for sustained photodynamic prodrug delivery and enhanced tumor therapy. J Control Release 319:344–351
6. Ji W, Xue B, Bera S, Guerin S, Liu Y, Yuan H, Li Q, Yuan C, Shimon LJW, Ma Q et al (2020) Tunable mechanical and optoelectronic properties of organic cocrystals by unexpected stacking transformation from H- to J- to X-aggregation. ACS Nano 14:10704–10715
7. Sun B, Tao K, Jia Y, Yan X, Zou Q, Gazit E, Li J (2019) Photoactive properties of supramolecular assembled short peptides. Chem Soc Rev 48:4387–4400
8. Li J, Wang A, Ren P, Yan X, Bai S (2019) One-step co-assembly method to fabricate photosensitive peptide nanoparticles for two-photon photodynamic therapy. Chem Commun 55:3191–3194
9. Yuan C, Ji W, Xing R, Li J, Gazit E, Yan X (2019) Hierarchically oriented organization in supramolecular peptide crystals. Nat Rev Chem 3:567–588
10. Song J, Xing R, Jiao T, Peng Q, Yuan C, Möhwald H, Yan X (2018) Crystalline dipeptide nanobelts based on solid-solid phase transformation self-assembly and their polarization imaging of cells. ACS Appl Mater Interfaces 10:2368–2376
11. Li J, Wang A, Zhao L, Dong Q, Wang M, Xu H, Yan X, Bai S (2018) Self-assembly of monomeric hydrophobic photosensitizers with short peptides forming photodynamic nanoparticles with real-time tracking property and without the need of release in vivo. ACS Appl Mater Interfaces 10:28420–28427
12. Ji W, Yuan C, Chakraborty P, Gilead S, Yan X (2019) Stoichiometry-controlled secondary structure transition of amyloid-derived supramolecular dipeptide co-assemblies. Commun Chem 2:65

13. Zhang H, Liu K, Li S, Xin X, Yuan S, Ma G, Yan X (2018) Self-assembled minimalist multifunctional theranostic nanoplatform for magnetic resonance imaging-guided tumor photodynamic therapy. ACS Nano 12:8266–8276
14. Wang J, Zou Q, Yan X (2017) Peptide supramolecular self-assembly: structural precise regulation and functionalization. Acta Chim Sin 75:933–942
15. Yan X, Cui Y, He Q, Wang K, Li J (2008) Organogels bbased on self-assembly of diphenylalanine peptide and their application to immobilize quantum dots. Chem Mater 20:1522–1526
16. Yan X, Zhu P, Li J (2010) Self-assembly and application of diphenylalanine-based nanostructures. Chem Soc Rev 39:1877–1890
17. Lee J-H, Heo K, Schulz-Schönhagen K, Lee JH, Desai MS, Jin H-E, Lee S-W (2018) Diphenylalanine peptide nanotube energy harvesters. ACS Nano 12:8138–8144
18. Guo C, Luo Y, Zhou R, Wei G (2012) Probing the self-assembly mechanism of diphenylalanine-based peptide nanovesicles and nanotubes. ACS Nano 6:3907–3918
19. Xing R, Yuan C, Li S, Song J, Li J, Yan X (2018) Charge-induced secondary structure transformation of amyloid-derived dipeptide assemblies from β-sheet to α-helix. Angew Chem Int Ed 57:1537–1542
20. Zou Q, Zhang L, Yan X, Wang A, Ma G, Li J, Möhwald H, Mann S (2014) Multifunctional porous microspheres based on peptide-porphyrin hierarchical co-assembly. Angew Chem Int Ed 53:2366–2370
21. Liu K, Xing R, Chen C, Shen G, Yan L, Zou Q, Ma G, Möhwald H, Yan X (2015) Peptide-induced hierarchical long-range order and photocatalytic activity of porphyrin assemblies. Angew Chem Int Ed 54:500–505
22. Wehner M, Röhr MIS, Bühler M, Stepanenko V, Wagner W, Würthner F (2019) Supramolecular polymorphism in one-dimensional self-assembly by kinetic pathway control. J Am Chem Soc 151:6092–6107
23. Deng J, Walther A (2020) Pathway complexity in fuel-driven DNA nanostructures with autonomous reconfiguration of multiple dynamic steady states. J Am Chem Soc 142:685–689
24. Zhao L, Li S, Liu Y, Xing R, Yan X (2019) Kinetically controlled self-assembly of phthalocyanine – peptide conjugate nanofibrils enabling superlarge redshifted absorption. CCS Chem 1:173–180
25. Ren X, Zou Q, Yuan C, Chang R, Xing R, Yan X (2019) The dominant role of oxygen in modulating the chemical evolution pathways of tyrosine in peptides: Dityrosine or melanin. Angew Chem Int Ed 58:5872–5876
26. Zhao L, Liu Y, Chang R, Xing R, Yan X (2019) Supramolecular photothermal nanomaterials as an emerging paradigm toward precision cancer therapy. Adv Funct Mater 29:1806877
27. Xing R, Liu Y, Yan X (2019) Self-assembled injectable biomolecular hydrogels towards phototherapy. Nanoscale 11:22182–22195
28. Li S, Zhang W, Xue H, Yan X (2020) Tumor microenvironment-oriented adaptive nanodrugs based on peptide self-assembly. Chem Sci 11:8644–8656
29. Abbas M, Zou Q, Li S, Yan X (2017) Self-assembled peptide- and protein-based nanomaterials for antitumor photodynamic and photothermal therapy. Adv Mater 29:1605021
30. Zhao L, Liu Y, Xing R, Yan X (2020) Supramolecular photothermal effects: a promising mechanism for efficient thermal conversion. Angew Chem Int Ed 59:3793–3801
31. Liu K, Xing R, Zou Q, Ma G, Möhwald H, Yan X (2016) Simple peptide-tuned self-assembly of photosensitizers towards anticancer photodynamic therapy. Angew Chem Int Ed 55:3036–3039
32. Li S, Zou Q, Li Y, Yuan C, Xing R, Yan X (2018) Smart peptide-based supramolecular photodynamic metallo-nanodrugs designed by multicomponent coordination self-assembly. J Am Chem Soc 140:10794–10802
33. Zhang H, Kang L, Zou Q, Xin X, Yan X (2019) Coordination-assembled supramolecular nanoplatforms: structural modulation and theranostic applications. Curr Opin Biotechnol 58:45–52

34. Cao M, Xing R, Chang R, Wang Y, Yan X (2019) Peptide-coordination self-assembly for the precise design of theranostic nanodrugs. Coord Chem Rev 397:14–27
35. Zou Q, Yan X (2018) Amino acid coordinated self-assembly. Chem Eur J 24:755–761
36. Xing R, Zou Q, Yuan C, Zhao L, Chang R, Yan X (2019) Self-assembling endogenous biliverdin as a versatile near-infrared photothermal nanoagent for cancer theranostics. Adv Mater 31:1900822
37. Li Y, Zou Q, Yuan C, Li S, Xing R, Yan X (2018) Antitumor therapy amino acid coordination driven self-assembly for enhancing both the biological stability and tumor accumulation of curcumin. Angew Chem Int Ed 57:17084–17088
38. Li S, Zhao L, Chang R, Xing R, Yan X (2019) Spatiotemporally coupled photoactivity of phthalocyanine-peptide conjugate self-assemblies for adaptive tumor theranostics. Chem – A Eur J 25:13429–13435
39. Li S, Liu Y, Xing R, Yan X (2019) Covalently assembled dipeptide nanoparticles with adjustable fluorescence emission for multicolor bioimaging. Chem Bio Chem 20:555–560
40. Zou Q, Abbas M, Zhao L, Li S, Shen G, Yan X (2017) Biological photothermal nanodots based on self-assembly of peptide-porphyrin conjugates for antitumor therapy. J Am Chem Soc 139:1921–1927
41. Liu Y, Shen G, Zhao L, Zou Q, Jiao T, Yan X (2019) Robust photothermal nanodrugs based on covalent assembly of nonpigmented biomolecules for antitumor therapy. ACS Appl Mater Interfaces 11:41898–41905
42. Liu K, Abass M, Zou Q, Yan X (2017) Self-assembly of biomimetic light-harvesting complexes capable of hydrogen evolution. Green Energy Environ 2:58–63
43. Yan X (2017) Bio-inspired photosystem for green energy. Green Energy Environ 2:66
44. Liu K, Yuan C, Zou Q, Xie Z, Yan X (2017) Self-assembled zinc/cystine-based chloroplast mimics capable of photoenzymatic reactions for sustainable fuel synthesis. Angew Chem Int Ed 56:7876–7880
45. Han J, Liu K, Chang R, Zhao L, Yan X (2019) Photooxidase-mimicking nanovesicles with superior photocatalytic activity and stability based on amphiphilic amino acid and phthalocyanine co-assembly. Angew Chem Int Ed 58:2000–2004

Chapter 8
Peptide Cross-β Nanoarchitectures: Characterizing Self-Assembly Mechanisms, Structure, and Physicochemical Properties

Christopher W. Jones, Hannah E. Distaffen, and Bradley L. Nilsson

8.1 Introduction

Peptide and protein self-assembly into β-sheet fibrils is a characteristic of both amyloid misfolding disorders and functional biomaterials [1]. Alzheimer's disease, Huntington's disease, and Parkinson's disease are prominent examples of protein misfolding disorders that are characterized by the aberrant self-assembly of proteins or peptides into amyloid fibrils [2]. These fibrils and/or their metastable oligomeric folding intermediates have been shown to be cytotoxic to cells and tissues and are thought to play an important role in the etiology of these amyloid diseases. This relevance to disease has prompted intense scrutiny of the structure and function of amyloid fibrils, including efforts to understand the mechanisms of self-assembly through thermodynamic and kinetic models, the development of tools to more accurately determine the molecular architecture of these materials, and interrogation of the emergent properties of these assemblies and the related prefibrillar aggregates [3, 4]. These efforts have led to discoveries that amyloid fibrils are also found in functional biological materials in which the assembly of the fibrils is carefully regulated [5, 6]. These discoveries have been coupled with efforts to design and engineer novel biomaterials inspired by self-assembled amyloid fibrils for applications in biomedicine and materials [7].

Amyloid fibrils adopt a common quaternary structure that has been termed the cross-β fold. Cross-β fibrils are β-sheet assemblies of peptides and proteins (Fig. 8.1) [8]. These one-dimensional assemblies are unbranched fibrils that are typically ~10 nanometers in width and up to tens or hundreds of micrometers in length. The constituent peptides or proteins adopt parallel or antiparallel β-sheet configurations

C. W. Jones · H. E. Distaffen · B. L. Nilsson (✉)
Department of Chemistry, University of Rochester, Rochester, New York, USA
e-mail: bradley.nilsson@rochester.edu

T. Govindaraju, K. Ariga (eds.), *Molecular Architectonics and Nanoarchitectonics*, Nanostructure Science and Technology,
https://doi.org/10.1007/978-981-16-4189-3_8

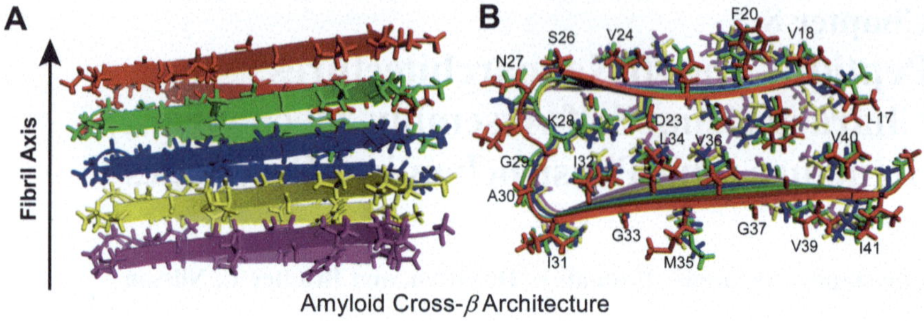

Fig. 8.1 Illustration of a representative structure of amyloid fibrils of the amyloid-β 1–42 peptide (Aβ42). The N-terminal fragment comprising residues 1–16 is not shown. (**a**) Five molecules of Aβ42 (shown in red, green, blue, yellow, and magenta) stacked in an amyloid cross-β fibril with the fibril axis indicated. An extensive intermolecular hydrogen bond network composed of the amide groups, which are oriented parallel to the fibril axis, is formed between neighboring peptides. (**b**) A view of the Aβ42 fibril segment from panel A that has been rotated 90°. This view illustrates the β-sheet-rich structure of these amyloid fibrils, which feature a parallel alignment with a sheet-loop-sheet structure of the constituent peptides. The alignment of the side chain groups perpendicular to the fibril axis becomes readily apparent in this view. This structural model of a segment of Aβ42 fibrils was reproduced from the Protein Data Bank structure reported in Lührs, T.; Ritter, C.; Adrian, M.; Riek-Lohner, D.; Bohrmann, B.; Döbeli, H.; Schubert, D.; Riek, R. *Proc. Natl. Acad. Sci. USA* **2005**, *102*, 17,342–17,347 (PDB ID Code 2BEG, DOI: https://doi.org/10.2210/pdb2BEG/pdb)

in which there is an intermolecular hydrogen bond network between the amide backbone entities of neighboring peptides that stabilizes the assembled system. The amide and carbonyl groups are oriented parallel to the fibril axis, which enables the formation of this hydrogen bond network. The side chain groups of the constituent peptides are perpendicular to the fibril axis (Fig. 8.1b), hence the "cross-β" designation. Side chain groups further stabilize the self-assembled fibrils through hydrophobic, Van der Waals, aromatic π-π, and charge-charge interactions. The fibril-forming proteins associated with the known amyloid disorders have no significant sequence or length homology, which has given rise to the hypothesis that all proteins and peptides are capable of forming amyloid fibrils under appropriate concentration and solvent conditions of pH and ionic strength. A great variety of protein and peptide sequences have been demonstrated to form amyloid fibrils, and these fibrils share the common features of the cross-β structure even while the specific structural elements (parallel versus antiparallel β-sheet, length of β-sheet element, the presence/absence of loop structures, etc.) of a given fibril are unique to that sequence. All amyloid fibrils exhibit an X-ray scattering pattern with scattering intensities of 4.8 Å and ~ 10 Å, corresponding to the backbone peptide-peptide distances between hydrogen bonded β-sheets and the intersheet distance between laminated β-sheets, respectively [9]. In addition, amyloid fibrils commonly bind histological staining agents, including Congo red, and exhibit a characteristic red-green birefringence when viewed under polarized light [10, 11].

Researchers have committed significant effort to understanding cross-β amyloid fibrils and related biomaterials [9]. Cross-β fibrils have proven to be challenging subjects of study. This is due, in part, to the physical characteristics of cross-β fibrils, which are often insoluble and can be found as part of complex heterogeneous mixtures of subtly different fibril morphologies as well as prefibrillar aggregates. The inherent cytotoxicity of these aggregates has presented challenges to understanding their functional properties. Engineered cross-β biomaterials must be nontoxic, which has placed increased urgency on the need to understand the interactions of cross-β fibrils and their prefibrillar aggregates with cells. Structural understanding has been complicated by the non-crystalline nature of most cross-β fibrils as well as the large molecular mass of these aggregates, which make high-resolution crystallography and solution-state NMR problematic for structural determination. Thus, the development of methods for the characterization of cross-β assemblies has been critical to provide deeper insight into these materials [12, 13]. The objective of this chapter is to present an introduction to current methods used to characterize cross-β fibrils. These methods are divided into those used for interrogation of the self-assembly mechanisms (kinetics and thermodynamics), molecular structure, and the emergent properties of cross-β fibrils.

8.2 Mechanisms of Cross-β Self-Assembly

8.2.1 General Mechanistic Considerations

The self-assembly of amyloidogenic peptides into cross-β fibrils typically proceeds in a three-step mechanism (Fig. 8.2) [14]. The first step, nucleation, is often referred to as the “lag phase.” Nucleation is characterized by an associative collapse of monomeric peptides into oligomeric and prefibrillar aggregates [14]. These oligomers give rise to protofibrils, which have the characteristic β-sheet secondary structure of amyloidogenic peptide assemblies. The elongation phase is the second step of amyloid formation. In the elongation phase, addition of monomeric peptide to protofibril results in maturation of the fibril through a lengthening process. Finally, the third phase is marked by a saturation plateau where the monomers in solution have reached equilibrium with the fibrils. The kinetics of amyloid formation typically display a sigmoidal shaped curve in thioflavin T (ThT) assays (discussed in Sect. 2.2). The nucleation phase can be bypassed by “seeding,” where preformed fibrils are added to monomer solutions acting as a template for immediate elongation without the characteristic lag phase, indicating that nucleation is a kinetically slow step in cross-β self-assembly [15]. The aggregation process occurs over a variable time range that is dependent on both peptide/protein identity and environment. Secondary structural conformation changes can occur on the order of milliseconds, while the formation of fibrillar aggregates that are observable to the naked eye can occur on the order of days [4]. Characterization of the kinetic and thermodynamic parameters of cross-β assembly is critical in building mechanistic models of self-assembly [16–18]. Methods that are used to interrogate the kinetics and thermodynamics of cross-β assembly are outlined in the following subsections.

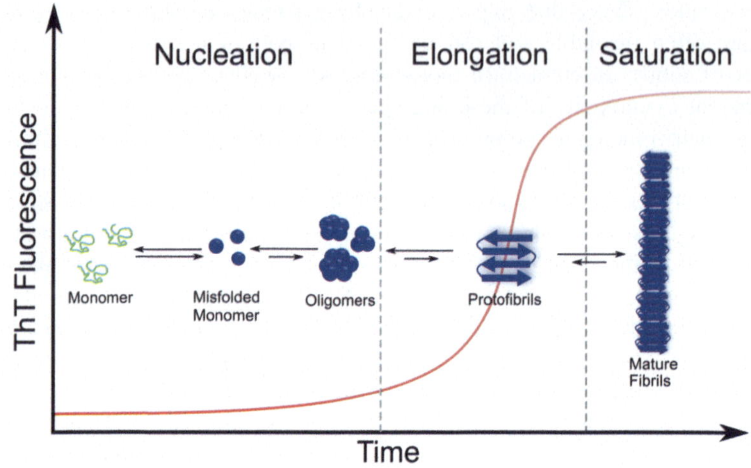

Fig. 8.2 A kinetic depiction of the self-assembly of cross-β fibrils showing the initial aggregation into prefibrillar oligomeric aggregates followed by conversion to one-dimensional amyloid fibrils. The initial nucleation of peptides and proteins into oligomers occurs during the "lag phase" which is followed by the exponential elongation of protofibrils into mature cross-β fibrils. These types of kinetics experiments are often conducted using thioflavin T (ThT) as a fluorescent reporter of fibril formation (see Sect. 2.2)

Fig. 8.3 Chemical structure of thioflavin T (ThT)

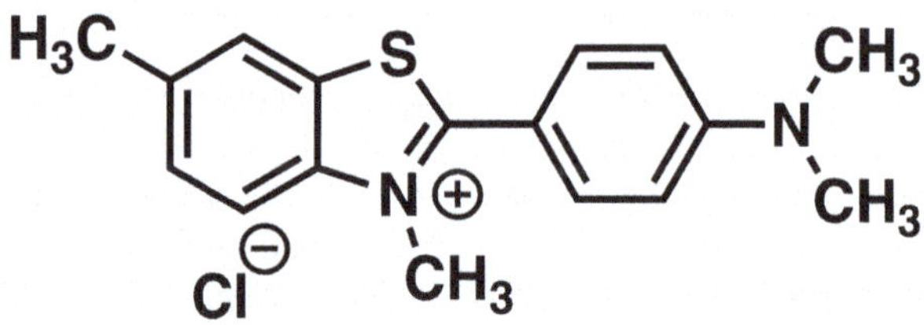

8.2.2 Fluorescent Reporters of Cross-β Assembly, Including ThT

The kinetics of cross-β amyloid self-assembly are commonly characterized by a thioflavin T (ThT) fluorescence assay [19]. ThT (Fig. 8.3) is a fluorescent dye that has been used for histological staining. ThT has also been useful in studying amyloid assembly since it exhibits a shift in the fluorescence emission spectrum from 510 nm to 480 nm upon binding to β-sheet rich structures, producing an observable change that can be used to conveniently monitor fibril formation as a function of time (Fig. 8.2) [19]. To monitor fibrillization, either the self-assembling peptide is incubated with ThT and the fluorescence signal is read continuously or aliquots of the fibril forming solution are taken at fixed time points and diluted into a ThT-containing solution, and the fluorescence of the resulting solution is recorded. Assuming the concentration of ThT is constant, the fluorescence signal intensity at 480 nm is directly related to the fibril concentration. ThT does not typically bind to prefibrillar oligomeric aggregates; thus this assay reports directly on the degree of fibril formation. As shown in Fig. 8.2, ThT binding assays typically exhibit a sigmoidal curve shape that represents the initial nucleation step as a lag phase, followed by the exponential elongation phase and reaching a final saturation of ThT fluorescence as peptide/protein monomer and fibril reach equilibrium.

There are some caveats to the use of ThT fluorescence to monitor amyloid self-assembly. First, ThT fluorescence can be induced by binding to molecules other than peptide/protein β-sheets, including DNA or cyclodextrin. ThT fluorescence is thus most useful for in vitro experiments where all components of the mixture can be carefully controlled and quantified. Additionally, ThT fluorescence varies depending on peptide identity and solution pH, ionic strength, and viscosity. This complicates the use of ThT fluorescence for the comparative quantification of absolute fibril concentrations for peptides and proteins with different sequences. Despite these limitations, ThT fluorescence is accepted as a robust tool for the characterization of cross-β self-assembly. While ThT is the most commonly used reporter for fibril self-assembly, others have been developed that have also found utility for the kinetic analysis of self-assembly processes [11].

8.2.3 Turbidity

Turbidity, a measurement of the amount of light scattered by a given material, is another common method for probing the kinetics of cross-β fibril formation [15]. In this technique, samples are analyzed by UV-Vis spectroscopy, and the degree to which light is scattered by fibrils and other aggregates in solution is quantified [20]. Typically, samples are analyzed at 350–450 nm, where the fibrils do not absorb, so that any changes in transmittance can be attributed to light scattering effects [16]. The main advantages of turbidity measurements for interrogating peptide self-assembly kinetics include simplicity of application, since sample modification or the addition of external reagents is not required, and the high reproducibility of the measurements. However, the sensitivity of this method is low and scattering from the largest insoluble particles dominates the data, which fails to provide a complete picture of all aggregate species in solution. Despite these drawbacks, turbidity measurements are frequently used to characterize the self-assembly kinetics of amyloid systems [21–23].

8.2.4 Infrared Spectroscopy

Fourier-transform infrared spectroscopy (FTIR) has also been used to follow the kinetics of cross-β self-assembly by monitoring the development of the β-sheet structure that is characteristic of these materials as a function of time [24]. Peptides display a prominent amide I band between 1600 and 1700 cm^{-1} that results from stretching vibrations of the carbonyl group in the peptide backbone [25]. Peptides and proteins lacking a distinct secondary structure have a broad amide I band around 1650 cm^{-1}, while proteins with significant β-sheet secondary structure, such as cross-β fibrils, display a narrower peak from 1620 to 1625 cm^{-1}. It is important to note that the water bending absorption band overlaps with the amide I band, which requires that experiments interrogating peptide/protein secondary structure be carried out in D_2O or at a very high peptide concentration. The kinetics of cross-β assembly can be characterized by monitoring the formation of β-sheet structure in the amide I IR region as a function of time. In one illustrative example, the aggregation of $A\beta_{1-28}$ upon acidification was studied to model how Aβ acts in the acidic lysosomes of cells [26]. In this study, the kinetics of $A\beta_{1-28}$ aggregation under these conditions was followed by observing an increase in the intensity of the β-sheet band at ~1620 cm^{-1} accompanied by a decrease in intensity of the random coil band around 1651 cm^{-1}, indicating a loss of random coil structure and transition to β-sheets upon acidification of the environment.

8.2.5 Circular Dichroism (CD) Spectroscopy

Circular dichroism (CD) is a spectroscopic method commonly used to characterize the secondary structure of peptides and proteins that has also been used to monitor cross-β self-assembly [19]. CD spectra report the differential absorption of left- and right-handed circularly polarized light by chiral molecules [25]. Peptides that lack a clear secondary structure present a very weak CD signal, while β-sheets, α-helices, and random coils present distinct CD spectra that have been used to monitor the formation of β-sheet-rich cross-β structures over time. β-sheets are characterized by a negative CD band at 220 nm. While ThT fluorescence reports selectively on cross-β fibrils without providing information about many prefibrillar aggregates, CD spectra report on formation of β-sheet structure generally in both prefibrillar and fibrillar aggregates [27]. In one example, CD spectroscopy was used to monitor the formation of Aβ fibrils as a function of time in the presence of Cu^{2+} in low pH environments, where it has been shown that copper ions inhibit β-sheet formation and higher temperatures increase aggregation rates for fibers incubated in the absence of copper ions [27].

8.2.6 Dynamic Light Scattering (DLS)

Dynamic light scattering (DLS) is another technique that is used to characterize cross-β self-assembly mechanisms by measuring the size of particle aggregates in solution [16]. DLS spectroscopy measures the correlation function of light scattered by particles in solution or suspension which can subsequently be used to determine the translational diffusion coefficient of molecules undergoing Brownian motion to determine the size of particles present [28]. DLS techniques assume that particles are spherical. DLS can therefore be used to monitor amyloid formation as a function of increasing particle size and is especially useful to characterize the early events of self-assembly which includes the formation of nonfibrillar aggregates. Unfortunately, DLS signals are often dominated by the largest particles in solution, making it a low-resolution technique that is often incapable of monitoring monomers or oligomers directly once fibrils begin to appear. Despite these drawbacks, DLS has been useful to interrogate early events in cross-β amyloid aggregation. In one example, DLS was used to measure the kinetics of aggregation of the amyloid-β 40 peptide (Aβ40) at slightly acidic pH values similar to those that would be found in endosomes during cellular update of this peptide [29]. In solutions of Aβ40 under these conditions, DLS was used to detect several early phases of aggregation: an initial rapid burst phase followed by slower phase 1 and phase 2 association phases. In the burst phase, large aggregates with an average hydrodynamic radius (R_h) of 84 nm to 196 nm were observed by DLS in the first 12 minutes of incubation; these aggregates were shown to be amorphous in morphology by correlating atomic force microscopy. The association phases gave rise to smaller spherical aggregates with R_h values that were ~ 36 nm in diameter. This work clearly demonstrates the utility of DLS methods to give insight into early aggregation events in cross-β, specifically showing an initial hydrophobic collapse of peptides into amorphous aggregates followed by reorganization into ordered oligomeric aggregates prior to fibril formation in the case of Aβ40 under these conditions [29].

8.2.7 Transmission Electron Microscopy (TEM), Atomic Force Microscopy (AFM), and High-Speed AFM (HS-AFM)

A limitation of many of the spectroscopic methods that have been used to study cross-β self-assembly is that they monitor the aggregates that dominate at any given time in the self-assembly process and often fail to capture the complexity of heterogeneous mixtures [30]. They give information about the species that are present in the highest concentration, while species in lower concentrations are overlooked. Microscopic imaging methods, including transmission electron microscopy (TEM) and atomic force microscopy (AFM), have the potential to reflect more accurately the diversity of aggregate species that exist at any given point during self-assembly [30]. This enables deeper mechanistic insight into the heterogeneous structures found in the course of cross-β assembly. TEM and AFM are most often used to structurally characterize the morphology of the aggregates that are formed by peptide self-assembly. Recent advances in high-speed AFM (HS-AFM) have enabled the use of this technique to follow self-assembly over time. For example, HS-AFM has been used to rapidly acquire consecutive images of the elongation of

Aβ42 fibrils, enabling the simultaneous observation of structural morphology and elongation kinetics in compiled video files [31, 32].

8.2.8 Sedimentation Analysis

The thermodynamics of peptide cross-β self-assembly can be quantified using sedimentation analysis [33]. Sedimentation analysis exploits the selective sedimentation of self-assembled aggregates, both cross-β fibrils and prefibrillar aggregates, when subjected to high-speed ultracentrifugation. Under appropriate centrifugal forces, all aggregate structures can be sedimented, leaving monomeric peptides in solution. Quantification of monomeric peptide by high-performance liquid chromatography (HPLC) correlation to a standard curve can be used to extrapolate comparative free energies ($\Delta\Delta G^\circ$) of self-assembly for specific peptide systems. The concentration of monomer in solution at equilibrium, the critical concentration (C_r), can be related to the equilibrium association constant (K_a) for the addition of a molecule of monomeric peptide to the end of a fibril, which is dependent on both the identity of the peptide and the assembly environment. The association constant, K_a, is the reciprocal of the critical concentration, $1/C_r$. The free energy of assembly for that peptide, ΔG°, can be determined from K_a using the standard expression $\Delta G^\circ = -RT(\ln K_a)$. In an application of this method, Nilsson and coworkers have used sedimentation analysis to characterize the relative thermodynamics of nine variants of the amyloid-β 16–22 fragment (Aβ(16–22)) [34, 35]. In the wild-type peptide, residues 19 and 20 are Phe. Variant peptides were prepared in which Phe 19 and 20 were changed to residues of differing hydrophobic and aromatic character in order to understand the possible role of aromatic effects on the thermodynamics of Aβ(16–22) self-assembly. Sedimentation analysis of each of the variant peptides provided $\Delta\Delta G^\circ$ comparisons with the wild-type peptide, giving insight into the thermodynamic role of the Phe side chain groups in promoting self-assembly.

Sedimentation analysis using a range of centrifugal forces has also facilitated insight into the nature of the prefibrillar oligomer species that precede cross-β fibril self-assembly [36]. Prefibrillar oligomers are of interest from both a mechanistic and functional perspective: these oligomers are believed to be the major toxins in many amyloid disorders [37]. Oligomers tend to be soluble aggregates of varying molecular weight. Oligomer species in solution can be separated based on size using a range of centrifugal conditions. Low-speed centrifugation for short time periods can be used to selectively sediment insoluble fibrils and larger aggregates. High-speed centrifugation for longer periods of time can separate smaller aggregate species. Optical density, fluorescence, or related analytical methods can be used to determine sedimentation coefficients for aggregates of varying sizes that can then be used to obtain accurate molecular weights for these aggregates. There are two main techniques in analytical ultracentrifugation: sedimentation equilibrium and sedimentation velocity [38]. Sedimentation equilibrium experiments use short solution columns and low centrifugal speeds to bring the system to equilibrium. This technique has been used to characterize the subunits that form fibrils; however, it is not useful for characterizing fibrils themselves as they are too large to form equilibrium gradients and simply sediment to the vessel floor. Sedimentation velocity is thus used to study the larger fibrils by using hydrodynamic theory to extrapolate the size and shape of molecules moving through a strong centrifugal field [39]. The insight these techniques provide on the size and range of aggregates that exist during cross-β self-assembly is a critical element to the mechanistic puzzle for these processes.

8.2.9 Electrospray Ionization-Ion Mobility-Mass Spectrometry (ESI-IMS-MS)

Electrospray ionization-ion mobility-mass spectrometry (ESI-IMS-MS) has also been used for characterizing the heterogeneous aggregates generated during cross-β fibrillization to gain understanding of the nature of prefibrillar oligomer species [40]. This technique separates ions based on their shape and/or charge and is therefore capable of separating multiple conformations and/or aggregates of a peptide that have the same m/z ratio. This is significant since many proteins undergo initial misfolding prior to aggregation and the initial oligomeric aggregates are of varying molecular weight and conformation. Prefibrillar oligomers also rapidly interconvert, complicating identification. ESI-IMS-MS has been successfully used to detect distinct conformers of cross-β-forming proteins as well as early aggregation intermediates. In one illustrative example, ESI-IMS-MS analysis of the islet amyloid polypeptide (IAPP) revealed the early formation of β-hairpin conformations, which were hypothesized to be precursors to oligomer aggregation [41, 42]. These and other studies exploiting this technique have provided critical glimpses of elusive early events in cross-β assembly [40].

8.2.10 Quartz Crystal Microbalance (QCM) Analysis

Quartz crystal microbalance (QCM) methods have been used to characterize the enthalpic and entropic contributions to the overall free energy barriers of fibril elongation [43]. QCM techniques monitor the elongation of fibril seeds that have been deposited on the sensor surface by detecting the change in frequency of the quartz crystal resonator as the hydrodynamic mass of the aggregates increases. This takes advantage of "seeded" elongation in which the addition of fibrils to solutions of monomeric peptide results in immediate elongation of fibrils by monomer without any nucleation phase. This rate of accelerated seeded addition of monomer to fibril corresponds to diffusional motion over a single free energy barrier without any intermediate species. Thus, the rate of change of the hydrodynamic mass of fibrils in

QCM experiments is proportional to the rate of fibril elongation. QCM analysis of ten amyloid systems at variable temperature was performed by Buell et al. to derive enthalpic ($\Delta H^{\neq}$) and entropic ($\Delta S^{\neq}$) contributions to the free energy transition state barriers for elongation of each of these systems [43]. It was found that the enthalpic contributions to these barriers were unfavorable while the entropic contributions were favorable and that these values correspond to peptide/protein characteristics of size and initial folded structure. QCM analysis has been an insightful tool for examining the specific energetic barriers that contribute to the kinetics of amyloid elongation.

8.2.11 Surface Plasmon Resonance (SPR)

Surface plasmon resonance (SPR) is another biosensor technique that has been used in the analysis of cross-β self-assembly [44]. In SPR analysis of amyloid self-assembly, fibril seed structures have been immobilized to the metal sensor surface, and changes to this surface as monomer is flowed over can be monitored to interrogate fibril elongation as a function of monomer adsorption to the immobilized fibril seed. This analysis facilitates characterization of the kinetics of seeded fibril elongation. Monomeric amyloid peptide has also been immobilized to the sensor surface for the analysis of nucleation kinetics of unseeded aggregation processes [44]. In one example of the use of SPR to assess the kinetics and thermodynamics of fibril extension and dissociation reactions, Aβ fibrils were immobilized on the sensor [45]. The rate of extension by addition of monomer to the fibril ends was measured by flowing monomeric Aβ over the immobilized fibrils. The rate of monomer dissociation from the fibril ends was measured by flowing buffered water over the immobilized fibrils. The rates of the association and dissociation reactions were both found to be proportional to the concentration of Aβ and the initial density of fibers immobilized on the surface of the sensor, consistent with a first-order kinetic model. These experiments also enabled determinations of critical monomer concentrations at equilibrium, which were used to characterize the monomer-fibril association constant, K. SPR was thus demonstrated to be a useful tool for the interrogation of both kinetic and thermodynamic parameters for important cross-β self-assembly processes.

8.2.12 Isothermal Titration Calorimetry (ITC) and Differential Scanning Calorimetry (DSC)

Isothermal titration calorimetry (ITC) has been used to interrogate the thermodynamics of cross-β peptide and protein self-assembly [46]. ITC can be used to characterize enthalpic and entropic contributions to molecular processes by precisely measuring the heat released to or absorbed from the environment during exothermic and endothermic processes, respectively. These measurements can be used to extrapolate enthalpic and entropic contributions to chemical processes, including peptide self-assembly. Swanekamp et al. have used ITC to characterize the relative enthalpies of the self-assembly of pleated β-sheet nanofibrils and coassembly of enantiomeric peptides into rippled β-sheet nanofibrils [47]. L-Ac-$(FKFE)_2$-NH_2 has been shown to self-assemble into pleated β-sheet nanoribbons, while equimolar mixtures of L- and D-Ac-$(FKFE)_2$-NH_2 coassembled into rippled β-sheet fibrils with an alternating arrangement of L/D peptides in the fibrils. ITC was used to compare the enthalpies of these reactions by enforcing coassembly through alteration of the charge of the peptides. Specifically, the L-Ac-$(FEFE)_2$-NH_2 peptide and L and D-Ac-$(FKFK)_2$-NH_2 peptides were prepared; at neutral pH these peptides do not effectively self-assemble due to charge repulsion, but they readily coassemble with complementary charged partners. The ITC experiments were conducted by titrating solutions of either L or D-Ac-$(FKFK)_2$-NH_2 peptides into a solution of the L-Ac-$(FEFE)_2$-NH_2 peptide and measuring the heat of reaction. The enthalpy for both the L/L and L/D mixtures was favorable, with the L/D mixture showing an enthalpic advantage over the L/L mixture. ITC provided critical insight into the thermodynamic parameters that govern the respective formation of rippled and pleated β-sheet assembly.

Differential scanning calorimetry (DSC) is another tool that has been used to study the thermodynamics of amyloid fibril formation [48]. In this technique, the peptide sample is loaded into the sample cell, and an equal volume of solvent is loaded into the reference cell. As both cells are heated at the same rate, the additional release or absorbance of heat from the sample cell is quantified and compensated for by a feedback heater, giving information about the heat capacity difference between the two cells. DSC analysis of cross-β amyloid materials has given interesting and variable results that depend on both the peptide/protein that comprises the amyloid fibrils and the conditions. Some fibrils (N47A Spc-SH3 domain and WL fibrils of β_2-microglobulin (β_2-m)) exhibit cooperative thermal transitions that are characteristic of fibril melting. Other fibrils (β_2-m, HEWL, and Aβ) show transformation of the fibrils due to presumed transient inter-fibrillar association during heating. These variable outcomes have made DSC a useful method for interrogation of the complex thermodynamics at play in the formation and stability of self-assembled cross-β fibrils.

8.2.13 *In Silico Simulations*

Computer modeling and simulation provide critical tools to complement experimental approaches for understanding the fundamental physicochemical parameters that underlie cross-β self-assembly processes [49]. These efforts include the development of computational algorithms to predict the propensity of peptides and proteins to undergo cross-β assembly based on amino acid sequence [50–53]. These models are based on both phenomenological observations and fundamental physicochemical properties, including hydrophobicity and β-sheet propensity. Simulation approaches have also been harnessed to understand early conformational changes in cross-β assembly as well as the key mechanistic events that occur during nucleation and elongation. Challenges associated with the application of classical molecular dynamics simulations models to peptide/protein self-assembly include managing the high atom count of cross-β systems and achieving meaningful simulations that encompass useful time scales [54]. The use of coarse-grained models, which retain only the necessary information about shape and average interactions, has helped to bridge this gap. For example, coarse-grained computer simulations have accurately replicated the two-step mechanism that includes initial formation of prefibrillar oligomers which convert into β-sheet-rich aggregates that can seed elongation [55, 56]. A detailed treatment of the use of computational simulations to gain insight into cross-β self-assembly processes is beyond the scope of this review. However, the experimental complexities of amyloid self-assembly processes have made in silico approaches of critical importance to understanding the fundamental principles that govern cross-β formation.

8.3 Structural Characterization of Cross-β Nanomaterials

8.3.1 Introduction

Structure and function are indelibly connected in biological systems, making interrogation of the structure of cross-β peptide nanomaterials a subject of intense interest. Understanding of the cross-β fold has progressed from the initial recognition that amyloid aggregates are ordered, as evidenced by their Congo red binding properties, to the elucidation of atom-level packing models of cross-β systems [57–59]. Techniques that are used to probe the architecture of self-assembled cross-β fibrils and their prefibrillar predecessors range from bulk scale structure analysis by microscopic imaging methods to spectroscopic methods that can provide increasing levels of structural understanding. Spectroscopic methods have further enabled the identification of structural information ranging from secondary structure motifs found in aggregates to β-sheet packing alignment to all-atom models of the aggregates. Structure determination of cross-β amyloid presents significant challenges. Cross-β fibrils tend to be non-crystalline, which complicates the use of high-resolution X-ray diffraction methods. In addition, the high molecular weight of these aggregates likewise complicates the use of solution-phase NMR methods to determine structure. Finally, metastable prefibrillar aggregates that give rise to cross-β structures tend to exist in complex heterogeneous mixtures, which has made interrogation of oligomer structures exceedingly challenging. Innovation in solid-state NMR and electron microscopy have had a paradigm-shifting impact on structural understanding of cross-β nanomaterials. This section will introduce experimental methods that are commonly used to characterize the hierarchical structure of assembled cross-β aggregates, beginning with low-resolution techniques and proceeding to high-resolution methods.

8.3.2 Circular Dichroism

Circular dichroism (CD) spectroscopy (see Sect. 2.5) is a chiral UV method that provides information about the secondary structure of peptides and proteins [60]. CD spectroscopy can differentiate between secondary structures in chiral protein and peptide systems based on the variable absorption patterns of left-handed and right-handed circularly polarized light. The three-dimensional shape of an asymmetric peptide or protein will differentially absorb left- or right-handed polarized light; the difference of the two spectra provides a characteristic CD spectrum that can assist in the identification of α-helical, β-sheet, and disordered structures (Fig. 8.4). CD spectra of β-sheet motifs exhibit a minimum between 215 and 222 nm [19]. CD is an excellent technique to rapidly determine the secondary structure of assembled peptides and to monitor structural changes in different environmental conditions. CD spectroscopy is a low-resolution technique; however, cross-β systems will exhibit distinct β-sheet character in CD spectra. While it can confirm that self-assembled fibrils are rich in β-sheet secondary structure, CD techniques alone cannot be used as

Fig. 8.4 Representative CD spectra for α-helical, β-sheet, collagen triple helix, and disordered peptide structures. Adapted with permission from Springer Nature, Greenfield N (2006) Using circular dichroism spectra to estimate protein secondary structure. Nat Protoc 1 (6):2876–2890. https://doi.org/10.1038/nprot.2006.202 (reference [60]), Copyright 2007, Nature Publishing Group

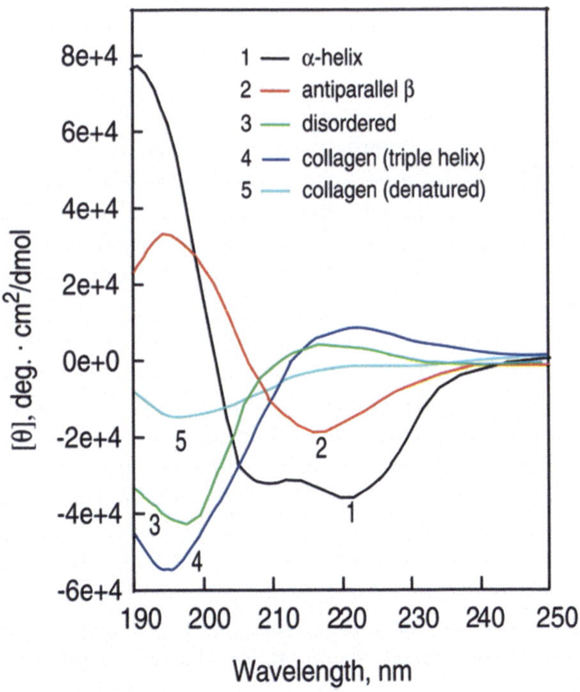

absolute confirmation of a cross-β arrangement in self-assembled nanomaterials. Additional techniques, including TEM or AFM microscopy, must be used in conjunction with CD spectroscopy to provide confirmation of cross-β architecture. CD spectroscopy also has significant drawbacks specifically for the analysis of cross-β assemblies. Many amyloid materials precipitate upon assembly and are high in molecular weight; both of these characteristics can lead to complicated light scattering effects in CD analyses. In spite of these limitations, CD still has an important place among the spectroscopic tools used to assist in the structural characterization of cross-β amyloid assemblies.

8.3.3 Vibrational Spectroscopy

8.3.3.1 Infrared (IR) Spectroscopy

Fourier transform infrared spectroscopy (FTIR) is a vibrational spectroscopy technique that is used to confirm the secondary structure of cross-β aggregates from characteristic molecular vibrations in the amide backbone [61]. Molecular vibrational modes in IR spectroscopy are excited by irradiation with infrared light. IR vibrational transitions in the amide I region, from ~1600 to 1800 cm^{-1}, are distinct for peptides and proteins that adopt α-helical, β-sheet, and disordered structures. IR spectroscopy can be conveniently applied to both solution-phase and solid-phase samples, a significant advantage over CD spectroscopy for the routine analysis of β-sheet assemblies. Cross-β fibrils typically display an amide I stretch between 1615 and 1630 cm^{-1}, consistent with β-sheet secondary structure (Fig. 8.5a). In addition, empirical observations suggest that the parallel or antiparallel orientation of peptide strands in cross-β fibrils can be determined using IR spectroscopy [62]. Antiparallel strands have been shown to display a weak transition between 1670 and 1680 cm^{-1}. This band transition is often not observed in parallel β-sheet structures. Standard IR spectroscopy is similar to CD spectroscopy in that it is a low-resolution spectroscopic method that is primarily useful for corroborating that fibrillar assemblies are rich in β-sheet secondary structure.

Isotope-edited FTIR spectroscopy (IE-IR) is a technique that provides additional information about strand alignment within β-sheet assemblies [64, 65]. IE-IR requires the incorporation of stable-isotope 1-^{13}C carbonyl labels at selected amino acids in cross-β self-assembling peptides. Incorporating an isotopically labeled carbonyl results in a redshift in vibrational frequency due to the lower vibrational frequency of the heavier ^{13}C isotope. This will lead to band separation of the amide I stretch due to the presence of the labeled carbonyl atom (see Fig. 8.5a for examples of this line splitting). When two isotopically labeled carbonyls are organized directly cross-strand from each other, these carbonyls undergo vibrational coupling, leading to changes in the line splitting patterns relative to uncoupled carbonyl groups (Fig. 8.5a). Strategic incorporation of 1-^{13}C labels can thus be used to characterize

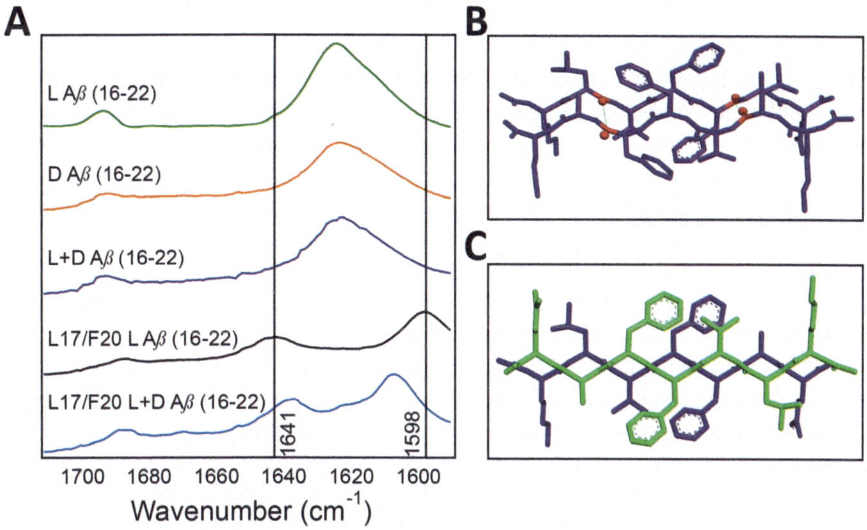

Fig. 8.5 (**a**) Fourier transform infrared spectra overlays of self-assembled L- and D-Aβ(16–22), coassembled L/D-Aβ(16–22), 1-^{13}C labeled L-Aβ(16–22) (labels at Leu 17 and Phe 20), and coassembled 1-^{13}C labeled L-Aβ(16–22) with unlabeled D-Aβ(16–22). These spectra illustrate typical β-sheet amide I signatures for cross-β fibrils (green, orange, and purple lines). In addition, coupling effects due to selective incorporation of 1-^{13}C carbonyl labels are shown (black and blue lines). (**b**) A structural model for β-sheet strand alignment in putative pleated β-sheets of self-assembled L-Aβ(16–22) with 1-^{13}C labeled positions highlighted in red. The close proximity of the 1-^{13}C labels across strand leads to coupling as evidenced by the line splitting of the major amide I stretch at ~1620 cm^{-1}. (**c**) A proposed structural model of β-sheet alignment in coassembled L/D-Aβ (16–22) rippled β-sheets. Reproduced with permission from Urban JM, Ho J, Piester G, Fu R, Nilsson BL (2019) Rippled β-Sheet Formation by an Amyloid-β Fragment Indicates Expanded Scope of Sequence Space for Enantiomeric β-Sheet Peptide Coassembly. Molecules 24:1983. https://doi.org/10.3390/molecules24101983 (reference [63]) Copyright 2019 the authors, some rights reserved; exclusive licensee MDPI. Distributed under a Creative Commons Attribution License 4.0 (CC BY) https://creativecommons.org/licenses/by/4.0/

strand registry within the cross-β packing structure, a significant enhancement in resolution of structural information over standard FTIR spectroscopy.

In a recent example, IE-IR spectroscopy was applied to the characterization of cross-β systems derived from enantiomers of the Aβ(16–22) peptide and mixtures thereof to compare self-assembly of single enantiomer solutions into pleated β-sheet fibrils and coassembly of equimolar mixtures of L- and D-peptides into two-component rippled β-sheet fibrils [63]. L-Aβ(16–22) was prepared with 1-^{13}C labels incorporated at Leu17 and Phe21, which are expected to be directly aligned in pleated β-sheet assemblies of this peptide (Fig. 8.5b). The IR spectra of the unlabeled L- and D-Aβ(16–22) peptides exhibited typical antiparallel β-sheet spectra with an amide I stretch at ~1620 cm^{-1} and a weak stretch at ~1690 cm^{-1} (Fig. 8.5a). The IE-IR spectra of the Leu17/Phe20 1-^{13}C labeled L-Aβ(16–22) peptide displayed a separated amide I band with absorbances at 1598 cm^{-1} and 1641 cm^{-1}, consistent with a cross-strand coupling interaction between the labeled carbonyl groups. This labeled peptide was then mixed with an equimolar amount of unlabeled D-Aβ (16–22). Mixtures of enantiomeric β-sheet peptides have been predicted to coassemble into so-called "rippled" β-sheet assemblies with alternating arrangements of the L- and D-peptides [66]. In this case, the mixture of the unlabeled D-Aβ(16–22) peptide with the Leu17/Phe20 1-^{13}C labeled L-Aβ(16–22) peptide should provide a strand orientation in which the alternating L/D pattern interferes with the cross-strand coupling of the labeled carbonyl groups (Fig. 8.5c). Indeed, the splitting pattern was altered in this mixture of peptides, with the IR spectrum showing unique absorbances at 1638 cm^{-1} and 1608 cm^{-1}, consistent with coassembly into L/D patterned rippled β-sheet assembly.

Two-dimensional infrared (2D-IR) spectroscopy is a further refinement of IR analysis that has been used to characterize cross-β assembly [67]. A beneficial aspect of 2D IR spectroscopy is the nonlinearity of the signal strength. Signals in 1D IR spectroscopy are scaled linearly in relation to the concentration and molar absorptivity of the analyte molecule, whereas 2D IR signals are scaled linearly in relation to concentration and quadratically in relation to molar absorptivity. This difference provides sharper spectra and greater sensitivity to secondary structure, and inaccurate baselines from weak absorbers are eliminated. Additional information provided by 2D-IR spectroscopy in the analysis of protein secondary structure includes anharmonic shifts, 2D line shapes, lifetimes, and vibrational dynamics. In a recent application, Zanni and coworkers have leveraged 2D-IR techniques to structurally identify cataract-causing cross-β aggregates [68]. The identification of cross-β in the eye lens using FTIR has been challenging. However, by analyzing 2D-IR spectral features, including diagonal frequency, anharmonic shift, and shifted cross-peak frequencies, Zanni and coworkers were able to differentiate between native β-sheets in juvenile lenses that do not contain cataracts and amyloid cross-β structures in 65-year-old lenses that do contain cataracts. Thus, 2D-IR has the potential to be leveraged in situ for the identification of amyloid in tissues. As with FTIR, 2D-IR provides a higher level of structural detail in cross-β assemblies, enabling the elucidation of strand alignment of peptides within the cross-β fibrils.

8.3.3.2 Raman Spectroscopy

Raman spectroscopy is an additional form of vibrational spectroscopy used to interrogate the structure of self-assembled cross-β materials [69]. These techniques include normal Raman (NR), deep UV Raman (DUVRR), surface enhanced Raman, and tip-enhanced Raman spectroscopy (TERS), which rely on the Raman effect to obtain structural information. The Raman effect is based on a vibrational frequency difference between the incident and scattered light when a molecule is irradiated with electromagnetic radiation. The spectrum created by Raman spectroscopy is influenced by vibrational modes from the peptide backbone, and aromatic and non-aromatic amino acid side chains, which provides structural information. These techniques can be used to determine the core structure of fibrils, the local environment around aromatic amino acids, the surface organization of amyloid fibrils, and nanoscale structural organization of fibrils. The amide I band from 1600 to 1700 cm^{-1}, the amide II band at ~1550 cm^{-1}, and the amide III band from 1200 to 1340 cm^{-1} provide information on secondary structure, hydrogen bonding interactions, and aromatic and sulfur side chain environment in the various forms of Raman spectroscopy.

Recently, TERS was used to analyze the differences in structure of assemblies of wild-type Aβ42, the L34T Aβ42 variant that has been shown to form less toxic fibrils, and the G37C Aβ42 variant that gives rise to highly toxic oligomers [70]. TERS exploits atomic force microscopy methodology to obtain structural information of cross-β materials at the single fibril level. In this study, TERS was used to determine that wild-type Aβ42 and the L34T variant assembled into parallel β-sheets while the assemblies formed by the G37C Aβ42 variant were organized into antiparallel β-sheets. This is an interesting demonstration of Raman capability that shows the enhanced range of analysis of Raman compared to closely related IR techniques. Similar to IR methods, Raman spectroscopy can rapidly provide higher-level structural analysis of cross-β materials but falls short of providing atomic-level structural detail.

8.4 Solid-State NMR (SSNMR)

NMR spectroscopy is an emerging tool to provide detailed structural constraints of self-assembled cross-β systems that can enable the development of molecular models of higher resolution and accuracy [71]. There are several NMR techniques that have been used to help characterize cross-β peptide assemblies. Solution-state NMR experiments have provided structural insight relevant to early events in self-assembly, including structural transitions and early interactions in the self-assembly process [72]. However, solution-state NMR is of limited use in the structural analysis of high molecular weight peptide and protein cross-β aggregates. The high molecular weight and frequent insolubility of cross-β assemblies result in significant line-broadening effects due to anisotropy, which makes standard solution-state NMR unsuitable for the acquisition of correlative data for the construction of higher-resolution cross-β structural models.

Solid-state NMR (SSNMR) spectroscopy has emerged as a critical higher-resolution tool for the resolution characterization of self-assembled cross-β nanomaterials [72, 73]. Magic angle spinning SSNMR techniques are well-suited for the structural interrogation of non-crystalline solids. SSNMR methods exploit pulse sequences that enable the accurate measurement of site-specific intermolecular distances within cross-β assemblies, often between amino acids that are selectively labeled with stable isotopes (including ^{13}C and ^{15}N). These experimental distance measurements have been effectively used as constraints to build accurate high-resolution structural models of cross-β fibrils [74]. SSNMR spectroscopy studies have been critical in the establishment of the fundamental principles of amyloid

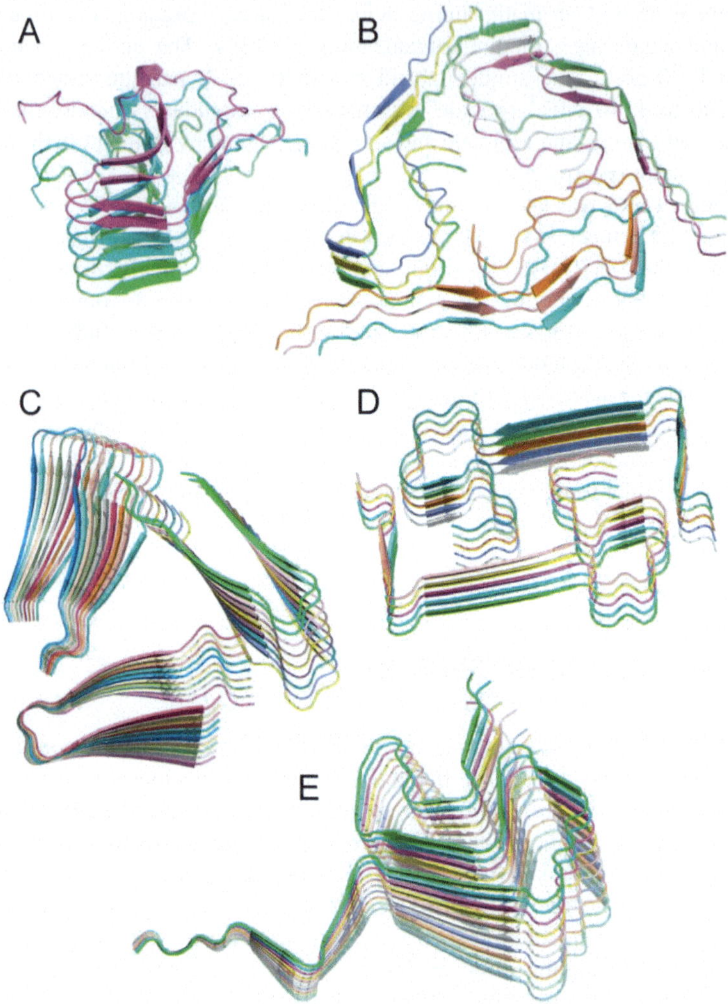

Fig. 8.6 Structural models of cross-β amyloid fibrils developed using correlative solid-state NMR methods. (**a**) HET-s(218–289) prion fibrils, (**b**) Aβ1–40 fibrils in Alzheimer's disease brain tissue, (**c**) Aβ1–40 D23N "Iowa" mutant fibrils, (**d**) Aβ1–40 E22D "Osaka" mutant fibrils; E. Human a-synuclein fibrils. Reprinted from Loquet A, El Mammeri N, Stanek J, Berbon M, Bardiaux B., Pintacuda G., Habenstein B. (2018) 3D structure determination of amyloid fibrils using solid-state NMR spectroscopy. Methods 138–139:26–38. https://doi.org/10.1016/j.ymeth.2018.03.014 (reference [74]), Copyright 2018, with permission from Elsevier

structure. Figure 8.6 depicts several cross-β fibril structures that have been solved using SSNMR. Early pioneering work by Tycko and others [75, 76] was critical in the development of SSNMR for cross-β structure analysis. In the last two decades, SSNMR has become more broadly applied in this field.

8.5 Diffraction Techniques

X-ray diffraction (XRD) and related fiber diffraction techniques have been used in the structural analysis of cross-β aggregates for over 50 years [77]. X-ray fiber diffraction of cross-β fibrils reveals meridional reflections at 4.7 Å indicative of the β-strand structural repeat parallel to the fibril axis and equatorial reflections at ~10 Å indicative of β-sheet spacing perpendicular to the fibril axis. This characteristic scattering pattern is a fundamental diagnostic confirmation for cross-β assemblies. Wide-angle X-ray scattering (WAXS) is now complemented by small-angle X-ray scattering (SAXS) and small-angle neutron scattering (SANS), which have been used to reveal higher-order structural features of self-assembled cross-β fibrils [78]. For example, Lynn and coworkers exploited small-angle scattering in the characterization of unique cross-β assemblies formed by the Aβ(16–22) peptide under acidic conditions in mixed water/acetonitrile solvents [79]. Under these assembly conditions, Aβ(16–22) was found to assemble into cross-β nanotubes with tube diameters of ~82 nm reported by electron microscopy measurements. In contrast, SANS and SAXS measurements revealed a more accurate diameter of 52 nm. The higher accuracy of the small-angle scattering experiments is due to the ability to perform these experiments in situ without the need to process the samples. The wider apparent diameter of the tubes in microscopic images is due to collapse of the tubes after deposition and drying on surfaces. The scattering data in these experiments also revealed that the inner and outer walls of the hollow nanotubes had a thickness of 4 nm. Thus, wide and small angle diffraction experiments have proven useful in the determination of higher-order structural elements of cross-β assemblies.

X-ray crystallography has also been applied as a high-resolution technique to visualize the atomic-level packing architecture of crystallized cross-β assemblies [80]. High-resolution crystallography requires crystalline samples, presenting a significant barrier for the use of this methodology in the examination of assembled cross-β fibrils, which tend to be non-crystalline. Nonetheless, some cross-β systems have been shown to form microcrystals that are closely related to their fibrillar polymorphs [59]. In an early example, Eisenberg and coworkers successfully crystallized a fragment of Sup35, GNNQQNY, and were able to use focused X-rays from a synchrotron source to elucidate the packing structure of this amyloid system [81]. In another recent example, Eisenberg and Bowers utilized X-ray microdiffraction crystallography to visualize the molecular packing structures of two amyloid pentapeptides derived from [Leu5]-enkephalin and determined the factors that caused differences in the mode of cross-β assembly for these short peptides [82].

Nowick and coworkers have designed a strategy to template amyloid-forming peptides as macrocyclic β-hairpins, facilitating self-assembly into discrete crystals that have been instructive in understanding the structure of cross-β systems [83]. In an example of this strategy, Nowick and coworkers engineered a derived macrocyclic β-hairpin peptide derived from the Aβ 16–36 fragment to facilitate crystallization (Fig. 8.7a and b) [84]. Crystallization was aided by connection of the N- and C-termini of Aβ(16–36) with a δ-linked ornithine turn unit and incorporation of a cross-strand disulfide bond that replaced Ala21 and Ile31, which stabilized the β-hairpin. In addition, an N-methyl group on Gly33 was introduced to prevent uncontrolled aggregation, further facilitating crystallization. High-resolution X-ray diffraction crystallography revealed the atomic structure of these aggregates, giving insight into possible structures relevant to amyloid prefibrillar oligomers. These methods have been applied to a number of other systems with great success [83]. Thus, insightful engineering of self-assembling peptides along with advances in X-ray diffraction methodology has provided important structural understanding of both cross-β fibrils and their prefibrillar aggregates.

8.6 Electron Microscopy

Electron microscopy (EM) and atomic force microscopy (AFM) have long been employed in the structural analysis of cross-β amyloid aggregates [85, 32]. Negative stain transmission electron microscopy (TEM) and cryo-EM are fundamental techniques used to observe the physical morphology and geometric parameters of peptide assemblies, including length, width, pitch, etc. Although negative stain TEM provides nanometer-scale images of cross-β materials, it cannot provide information on atomic-scale structure. TEM requires staining with heavy metal salts such as uranyl acetate for increased resolution, and samples need to be deposited onto a surface for imaging. These events can alter the structure of the peptide assembly, sometimes providing an inaccurate representation of the aggregate. AFM often visualizes cross-β materials in a dried state deposited onto a surface, making this technique vulnerable to the same limitations as negative stain TEM. AFM is also possible in solution conditions, which has enabled real-time visualization of nanofiber formation (Sect. 2.7). When imaging low-contrast amyloid fibrils, scanning transmission electron microscopy (STEM) is a powerful method to visualize amyloid structure [86]. STEM is able to directly measure the mass-per-length ratio of individual fibers providing substantial information that can assist in creating a molecular model.

Cryo-EM overcomes many of the disadvantages of other microscopy methods for the analysis of cross-β fibrils [87]. TEM, STEM, and AFM require sample fixation onto flat surfaces and staining, both of which can alter the native structure of an assembly. However, cryo-EM thermally fixates the sample in a thin layer of its original solvent, eliminating problems with sample alteration common to other microscopy methods. In addition, recent advances in cryo-EM data acquisition and processing provide data of sufficient resolution to enable reconstruction of high-resolution packing architectures of cross-β aggregates [88]. Cryo-EM model reconstruction is done from single fibril aggregate analysis at long length scales compared to SSNMR methods, which rely on highly local distance constraints to build models. Cryo-EM has become a transformative tool for structural interrogation of

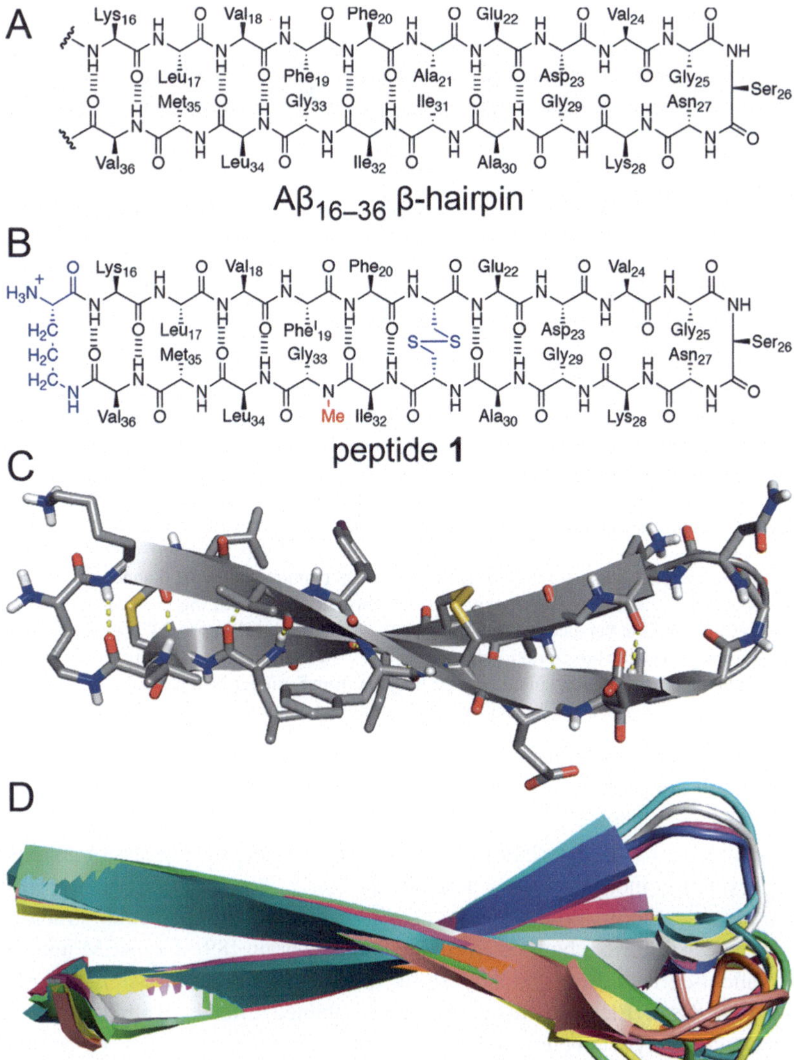

Fig. 8.7 (**a**) Chemical structure of a putative hairpin structure formed by the $A\beta_{16-36}$ fragment. (**b**) Chemical structure of peptide **1**, a hairpin of the $A\beta_{16-36}$ fragment stabilized by peptide macrocyclization by amide bond formation between the δ-amine of Lys 16 and the C-terminus of Val 36. (**c**) X-ray crystallographic structure of a representative β-hairpin monomer formed by peptide **1** (PDB 6WXM). (**d**) An overlay of the 11 peptide **1** β-hairpins in the asymmetric unit of the crystallographic structure that illustrates the conformational regularity of each hairpin in the crystal and shows the variability in the loop region. Reprinted with permission from Kreutzer AG, Samdin TD, Guaglianone G, Spencer RK, Nowick JS (2020) X-ray Crystallography Reveals Parallel and Antiparallel β-Sheet Dimers of a β-Hairpin Derived from $A\beta_{16-36}$ that Assemble to Form Different Tetramers. ACS Chem Neurosci 11 (15):2340–2347. https://doi.org/10.1021/acschemneuro.0cc00290 (reference [84]), Copyright 2020 American Chemical Society

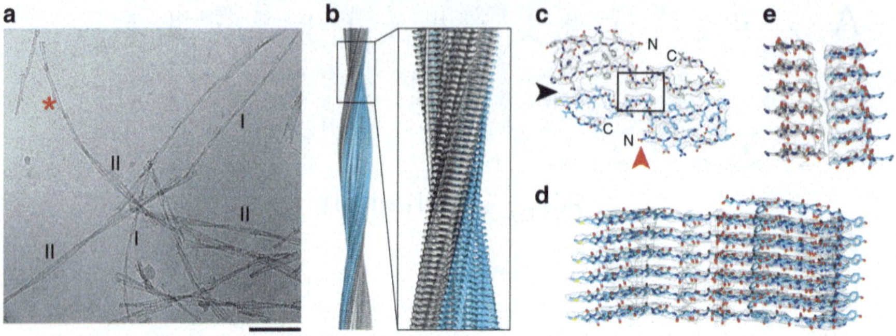

Fig. 8.8 Illustration of the use of cryo-EM techniques to develop models of fibril morphology of Aβ fibrils isolated from Alzheimer's brain tissue. (**a**) A cryo-EM image that depicts two distinct fibril types found in Alzheimer's brain tissue (scale bar is 50 nm). Fibrils of each morphology are labeled with I and II. The asterisk shows a fibril in which a morphology I-like fibril is emerging from a fibril segment that is of morphology II. (**b-d**) These panels show 3D views of the reconstructed 3D map of fibrils of morphology I at 4.4 Å resolution. Fibrils are composed of two β-sheet stacks, which are shown in gray and blue in these structures. (**b**) A side view of the reconstructed 3D map. **c** A cross-sectional view of one molecular layer of the fibril superimposed with the molecular model, illustrating the folded structure of each constituent peptide within the β-sheet assemblies. (**d**) A side view depiction of a six-layer peptide stack superimposed with the molecular model. This stack is viewed from the perspective indicated by the red arrowhead in panel C. (**e**) A side view of six molecular layers of the boxed region from panel C, viewed from the perspective indicated by the black arrowhead in panel C. Reprinted with permission from Springer Nature, Kollmer M, Close W, Funk L, Rasmussen J, Bsoul A, Schierhorn A, Schmidt M, Sigurdson CJ, Jucker M, Fändrich M (2019) Cryo-EM structure and polymorphism of Aβ amyloid fibrils purified from Alzheimer's brain tissue. Nat Commun 10 (1):4760. https://doi.org/10.1038/s41467-019-12683-8 (reference [89]), Copyright 2019, The Authors

self-assembled cross-β materials, dramatically accelerating the process of generating atomic resolution models of these assemblies.

In a recent example, cryo-EM was used to determine the molecular structure of polymorphic Aβ fibrils obtained from Alzheimer's brain [89]. Cryo-EM images of these patient-derived fibrils clearly show the polymorphic characteristics of Aβ fibrils (Fig. 8.8a). Electron diffraction data obtained directly from these images revealed symmetry elements and enabled the construction of a high-resolution three-dimensional structure consistent with this data to a resolution of 4.5 Å between the reconstructed map and the model. The constructed map shows two distinct stacks of β-sheets (Fig. 8.8b-e). The model provided atomic-level resolution and refined structural information enabling most of the structural characteristics of brain-derived Aβ fibrils and their polymorphs to be determined. Cryo-EM has the potential to surpass traditional microscopic and spectroscopic techniques with ongoing advances in methodology and as cryo-EM availability increases.

8.7 Emergent Physicochemical Properties of Cross-β Nanomaterials

Cross-β nanomaterials are of great interest due to their emergent physicochemical properties. In biological cross-β amyloid, the physicochemical properties of primary interest are the modes of action that lead to cellular dysfunction. Understanding these modes of action is manifested in efforts to correlate the cytotoxicity of cross-β fibrils and the associated prefibrillar aggregates with the structure of the aggregates. These efforts are as diverse as the diseases that are associated with cross-β amyloid and are specific to the disease states. As such, a detailed discussion of the various methods applied to interrogating the dysfunctional effects of cross-β aggregates is beyond the scope of this chapter. We will instead briefly discuss cross-β structures as biomaterials and provide an introduction to efforts to understand these materials both in terms of their inherent materials properties and their emergent biochemical functions.

Functional cross-β amyloid also possesses interesting physicochemical properties [5, 6]. For example, curli fibrils are components of the complex extracellular matrices produced by *Enterobacteriaceae*, including *Escherichia coli* [90]. Curli fibrils are assembled under carefully regulated conditions and mediate adhesion to surfaces, cell aggregation, biofilm formation, and host cell adhesion and invasion. Each of these functional emergent biochemical properties has been the subject of intense study. Again, a detailed discussion of the large body of work that has resulted in elucidation of the emergent physicochemical properties of functional amyloid is beyond the scope of this work. However, this class of amyloid can be used to illustrate the interrogation of cross-β structures using material techniques. As an example, the tensile strength of the interactions between curli fibrils and host fibronectin networks has been probed using AFM cantilever manipulation [91]. Correlation of the fundamental physicochemical properties of cross-β structures with emergent function is now a common mode of interrogation for these types of materials, with the applied methods depending on the function of the material that is being assessed.

Cross-β fibrils that form hydrogel networks are an illustrative example of a system for which the physicochemical properties are of primary significance. Cross-β systems that are engineered to sequester hydrophobic side chain group to the fibril interior while exposing hydrophilic side chain functionality to solvent self-assemble rapidly and at high concentrations entangle to form an emergent hydrogel network [92]. Many of these hydrogel-forming peptides have been designed based on understanding of the structure and mechanism of cross-β self-assembly [93]. The viscoelasticity, or "stiffness" of a hydrogel, is critical to its function. Oscillatory rheology is a critical method used in the interrogation of the emergent viscoelastic properties of these types of hydrogels [94]. These hydrogels have found use in diverse applications that include tissue engineering, regenerative medicine, and drug delivery. Each of these applications has an attendant set of requisite emergent properties that are characterized during the design process. The vast scope of techniques used to assess these properties depends on the property that is being interrogated. We will forego a detailed discussion of these varying techniques.

Finally, cross-β assemblies have been broadly used as scaffolds for the multivalent display of functional chemical groups [95]. These biomaterials have been used for tissue engineering, drug delivery, self-adjuvanting vaccine, antimicrobial, catalytic, and optical applications. The emergent properties that are interrogated for these materials depend on the specific application. For example, a material for tissue engineering may require the multivalent display of a cell-signaling motif on a cross-β scaffold. The relevant biochemical properties that would be assessed would include cellular adhesion and perhaps detection of specific ligand-receptor interactions between the cross-β display and the cells of interest. Detailed discussion of the many properties that are interrogated in these types of cross-β materials is beyond the scope of this chapter. The broad range of applications that cross-β materials have been designed to address illustrates that characterization of these materials must go far beyond typical structural and mechanistic analysis of cross-β assemblies. Cross-β assemblies are an emerging class of next-generation biomaterial for applications across virtually all fields of materials science. As such, it is expected that an ever-increasing range of techniques will be used to assess these materials.

8.8 Conclusion

Peptide and protein cross-β nanomaterials are the object of increasing fascination and study. The relevance of these supramolecular assemblies to both amyloid disease etiology and evolved biological materials motivates much of this interest. In the last several decades, there has also been an explosion in the development of engineering biomaterials based on peptide and protein cross-β assemblies. Designed cross-β systems have found wide-ranging use as biocompatible materials for regenerative medicine, drug delivery, wound healing, and as self-adjuvanting scaffolds for epitope presentation in vaccines. This broad relevance of cross-β architecture to both medical and materials science has increased the urgency to understand the relationships between peptide and protein sequence and self-assembly properties and to understand the correlations between aggregate structure, mechanism of assembly, and emergent functional properties. There have been transformative advances in understanding the mechanisms of cross-β assembly, in the available methods for the structural characterization of cross-β materials, and in the exploitation of these materials for diverse applications. These advances have dramatically quickened discoveries of the structural and functional properties of cross-β aggregates. The next decade will undoubtedly bring additional innovation, which will lead to transformative acceleration in the study of cross-β structure and the design of next-generation cross-β materials.

Acknowledgements The preparation of this chapter was supported by the National Science Foundation (CHE-1904528).

References

1. Ke PC, Zhou R, Serpell LC, Riek R, Knowles TPJ, Lashuel HA, Gazit E, Hamley IW, Davis TP, Fändrich M, Otzen DE, Chapman MR, Dobson CM, Eisenberg DS, Mezzenga R (2020) Half a century of amyloids: past, present and future. Chem Soc Rev 49(15):5473–5509
2. Harrison RS, Sharpe PC, Singh Y, Fairlie DP (2007) Amyloid peptides and proteins in review. Rev Physiol Biochem Pharmacol 159:1–77
3. Cawood EE, Karamanos TK, Wilson AJ, Radford SE (2021) Visualizing and trapping transient oligomers in amyloid assembly pathways. Biophys Chem 268:106505
4. Almeida ZL, Brito RMM (2020) Structure and aggregation mechanisms in amyloids. Molecules 25(5):1195
5. Levkovich SA, Gazit E, Laor Bar-Yosef D (2021) Two decades of studying functional amyloids in microorganisms. Trends Microbiol 29(3):251–265
6. Sergeeva AV, Galkin AP (2020) Functional amyloids of eukaryotes: criteria, classification, and biological significance. Curr Genet 66(5):849–866
7. Abdelrahman S, Alghrably M, Lachowicz JI, Emwas A-H, Hauser CAE, Jaremko M (2020) "What Doesn't kill you makes you stronger": future applications of amyloid aggregates in biomedicine. Molecules 25(22):5245
8. Kajava AV, Squire JM, Parry DAD (2006) β-Structures in fibrous proteins. Adv Protein Chem 73:1–15
9. Balasco N, Diaferia C, Morelli G, Vitagliano L, Accardo A (2021) Amyloid-like aggregation in diseases and biomaterials: osmosis of structural information. Front Bioeng Biotechnol 9(130)
10. Westermark GT, Johnson KH, Westermark P (1999) Staining methods for identification of amyloid in tissue. Methods Enzymol 309:3–25
11. Aliyan A, Cook NP, Martí AA (2019) Interrogating amyloid aggregates using fluorescent probes. Chem Rev 119(23):11819–11856
12. Castelletto V, Hamley IW (2018) Methods to characterize the nanostructure and molecular Organization of Amphiphilic Peptide Assemblies. Methods Mol Biol 1777:3–21
13. Vadukul DM, Al-Hilaly YK, Serpell LC (2019) Methods for structural analysis of amyloid fibrils in Misfolding diseases. Methods Mol Biol 1873:109–122
14. Sahoo BR, Cox SJ, Ramamoorthy A (2020) High-resolution probing of early events in amyloid-β aggregation related to Alzheimer's disease. Chem Commun 56(34):4627–4639
15. Zhao R, So M, Maat H, Ray NJ, Arisaka F, Goto Y, Carver JA, Hall D (2016) Measurement of amyloid formation by turbidity assay—seeing through the cloud. Biophys Rev 8(4):445–471
16. Kumar EK, Haque N, Prabhu NP (2017) Kinetics of protein fibril formation: methods and mechanisms. Int J Biol Macromol 100:3–10
17. Wetzel R (2006) Kinetics and thermodynamics of amyloid fibril assembly. Acc Chem Res 39 (9):671–679
18. Wang J, Liu K, Xing R, Yan X (2016) Peptide self-assembly: thermodynamics and kinetics. Chem Soc Rev 45(20):5589–5604
19. Gade Malmos K, Blancas-Mejia LM, Weber B, Buchner J, Ramirez-Alvarado M, Naiki H, Otzen D (2017) ThT 101: a primer on the use of thioflavin T to investigate amyloid formation. Amyloid 24(1):1–16
20. Pignataro MF, Herrera MG, Dodero VI (2020) Evaluation of peptide/protein self-assembly and aggregation by spectroscopic methods. Molecules 25(20):4854
21. Barykin EP, Petrushanko IY, Kozin SA, Telegin GB, Chernov AS, Lopina OD, Radko SP, Mitkevich VA, Makarov AA (2018) Phosphorylation of the amyloid-Beta peptide inhibits zinc-dependent aggregation, prevents Na, K-ATPase inhibition, and reduces cerebral plaque deposition. Front Mol Neurosci 11(302)
22. Profit AA, Vedad J, Desamero RZB (2017) Peptide conjugates of benzene carboxylic acids as agonists and antagonists of amylin aggregation. Bioconjug Chem 28(2):666–677

23. Khalili Samani E, Mofid MR, Malakoutikhah M (2020) The effect of terminal groups and halogenation of KLVFF peptide on its activity as an inhibitor of β-amyloid aggregation. J Pept Sci 26(2):e3227
24. Martial B, Lefèvre T, Auger M (2018) Understanding amyloid fibril formation using protein fragments: structural investigations via vibrational spectroscopy and solid-state NMR. Biophys Rev 10(4):1133–1149
25. Cristóvão JS, Henriques BJ, Gomes CM (2019) Biophysical and spectroscopic methods for monitoring protein Misfolding and amyloid aggregation. Methods Mol Biol 1873:3–18
26. Perálvarez-Marín A, Barth A, Gräslund A (2008) Time-resolved infrared spectroscopy of pH-induced aggregation of the Alzheimer Aβ1–28 peptide. J Mol Biol 379(3):589–596
27. Bin Y, Li X, He Y, Chen S, Xiang J (2013) Amyloid-β peptide (1–42) aggregation induced by copper ions under acidic conditions. Acta Biochim Biophys Sin 45(7):570–577
28. Radko SP, Khmeleva SA, Suprun EV, Kozin SA, Bodoev NV, Makarov AA, Archakov AI, Shumyantseva VV (2015) Physico-chemical methods for studying amyloid-[beta] aggregation. Biochem Moscow Suppl Ser B 9(3):258–274
29. Gorman PM, Yip CM, Fraser PE, Chakrabartty A (2003) Alternate aggregation pathways of the Alzheimer β-amyloid peptide: Aβ association kinetics at endosomal pH. J Mol Biol 325 (4):743–757
30. Ruggeri FS, Habchi J, Cerreta A, Dietler G (2016) AFM-based single molecule techniques: unraveling the amyloid pathogenic species. Curr Pharm Des 22(26):3950–3970
31. Watanabe-Nakayama T, Ono K (2018) High-speed atomic force microscopy of individual Amyloidogenic protein assemblies. Methods Mol Biol 1814:201–212
32. Watanabe-Nakayama T, Ono K, Itami M, Takahashi R, Teplow DB, Yamada M (2016) High-speed atomic force microscopy reveals structural dynamics of amyloid $\beta_{1\text{-}42}$ aggregates. Proc Natl Acad Sci 113(21):5835–5840
33. O'Nuallain B, Thakur AK, Williams AD, Bhattacharyya AM, Chen S, Thiagarajan G, Wetzel R (2006) Kinetics and thermodynamics of amyloid assembly using a high-performance liquid chromatography–based sedimentation assay. Methods Enzymol 413:34–74
34. Senguen FT, Lee NR, Gu X, Ryan DM, Doran TM, Anderson EA, Nilsson BL (2011) Probing aromatic, hydrophobic, and steric effects on the self-assembly of an amyloid-β fragment peptide. Mol BioSyst 7(2):486–496
35. Senguen FT, Doran TM, Anderson EA, Nilsson BL (2011) Clarifying the influence of core amino acid hydrophobicity, secondary structure propensity, and molecular volume on amyloid-β 16–22 self-assembly. Mol BioSyst 7(2):497–510
36. Mok Y-F, Howlett GJ, Griffin MDW (2015) Sedimentation velocity analysis of the size distribution of amyloid oligomers and fibrils. Methods Enzymol 562:241–256
37. Lee SJC, Nam E, Lee HJ, Savelieff MG, Lim MH (2017) Towards an understanding of amyloid-β oligomers: characterization, toxicity mechanisms, and inhibitors. Chem Soc Rev 46(2):310–323
38. Pham CLL, Mok Y-F, Howlett GJ (2011) Sedimentation velocity analysis of amyloid fibrils. Methods Mol Biol 752:179–196
39. Cole JL, Lary JW, Moody T, Laue TM (2008) Analytical ultracentrifugation: sedimentation velocity and sedimentation equilibrium. Methods Cell Biol 84:143–179
40. Woods LA, Radford SE, Ashcroft AE (2013) Advances in ion mobility spectrometry–mass spectrometry reveal key insights into amyloid assembly. Biochim Biophys Acta 1834 (6):1257–1268
41. Dupuis NF, Wu C, Shea J-E, Bowers MT (2009) Human islet amyloid polypeptide monomers form ordered β-hairpins: a possible direct Amyloidogenic precursor. J Am Chem Soc 131 (51):18283–18292
42. Dupuis NF, Wu C, Shea J-E, Bowers MT (2011) The amyloid formation mechanism in human IAPP: dimers have β-Strand monomer–monomer interfaces. J Am Chem Soc 133 (19):7240–7243

43. Buell AK, Dhulesia A, White DA, Knowles TPJ, Dobson CM, Welland ME (2012) Detailed analysis of the energy barriers for amyloid fibril growth. Angew Chem Int Ed 51 (21):5247–5251
44. White DA, Buell AK, Dobson CM, Welland ME, Knowles TPJ (2009) Biosensor-based label-free assays of amyloid growth. FEBS Lett 583(16):2587–2592
45. Hasegawa K, Ono K, Yamada M, Naiki H (2002) Kinetic modeling and determination of reaction constants of Alzheimer's β-amyloid fibril extension and dissociation using surface Plasmon resonance. Biochemistry 41(46):13489–13498
46. Kabiri M, Unsworth LD (2014) Application of isothermal titration calorimetry for characterizing thermodynamic parameters of biomolecular interactions: peptide self-assembly and protein adsorption case studies. Biomacromolecules 15(10):3463–3473
47. Swanekamp RJ, DiMaio JTM, Bowerman CJ, Nilsson BL (2012) Coassembly of enantiomeric amphipathic peptides into amyloid-inspired rippled β-sheet fibrils. J Am Chem Soc 134 (12):5556–5559
48. Sasahara K, Goto Y (2013) Application and use of differential scanning calorimetry in studies of thermal fluctuation associated with amyloid fibril formation. Biophys Rev 5(3):259–269
49. Ilie IM, Caflisch A (2019) Simulation studies of Amyloidogenic polypeptides and their aggregates. Chem Rev 119(12):6956–6993
50. Caflisch A (2006) Computational models for the prediction of polypeptide aggregation propensity. Curr Opin Chem Biol 10(5):437–444
51. Ahmed AB, Kajava AV (2013) Breaking the amyloidogenicity code: methods to predict amyloids from amino acid sequence. FEBS Lett 587(8):1089–1095
52. Huang C, Ghanati E, Schmit JD (2018) Theory of sequence effects in amyloid aggregation. J Phys Chem B 122(21):5567–5578
53. Wojciechowski JW, Kotulska M (2020) PATH–prediction of Amyloidogenicity by threading and machine learning. Sci Rep 10(1):7721
54. Michaels TCT, Šarić A, Habchi J, Chia S, Meisl G, Vendruscolo M, Dobson CM, Knowles TPJ (2018) Chemical kinetics for bridging molecular mechanisms and macroscopic measurements of amyloid fibril formation. Annu Rev Phys Chem 69(1):273–298
55. Šarić A, Chebaro YC, Knowles TPJ, Frenkel D (2014) Crucial role of nonspecific interactions in amyloid nucleation. Proc Natl Acad Sci 111(50):17869–17874
56. Nguyen P, Derreumaux P (2014) Understanding amyloid fibril nucleation and Aβ oligomer/drug interactions from computer simulations. Acc Chem Res 47(2):603–611
57. Li D, Liu C (2020) Structural diversity of amyloid fibrils and advances in their structure determination. Biochemistry 59(5):639–646
58. Iadanza MG, Jackson MP, Hewitt EW, Ranson NA, Radford SE (2018) A new era for understanding amyloid structures and disease. Nat Rev Mol Cell Biol 19(12):755–773
59. Eisenberg DS, Sawaya MR (2017) Structural studies of amyloid proteins at the molecular level. Annu Rev Biochem 86(1):69–95
60. Greenfield NJ (2006) Using circular dichroism spectra to estimate protein secondary structure. Nat Protoc 1(6):2876–2890
61. Li H, Lantz R, Du D (2019) Vibrational approach to the dynamics and structure of protein amyloids. Molecules 24(1)
62. Lomont JP, Ostrander JS, Ho J-J, Petti MK, Zanni MT (2017) Not all β-sheets are the same: amyloid infrared spectra, transition dipole strengths, and couplings investigated by 2D IR spectroscopy. J Phys Chem B 121(38):8935–8945
63. Urban JM, Ho J, Piester G, Fu R, Nilsson BL (2019) Rippled β-sheet formation by an amyloid-β fragment indicates expanded scope of sequence space for enantiomeric β-sheet peptide Coassembly. Molecules 24(10)
64. Hiramatsu H, Kitagawa T (2005) FT-IR approaches on amyloid fibril structure. Biochim Biophys Acta 1753(1):100–107
65. Decatur SM (2006) Elucidation of residue-level structure and dynamics of polypeptides via isotope-edited infrared spectroscopy. Acc Chem Res 39(3):169–175

66. Pauling L, Corey RB (1953) Two rippled-sheet configurations of polypeptide chains, and a note about the pleated sheets. Proc Natl Acad Sci 39(4):253
67. Moran SD, Zanni MT (2014) How to get insight into amyloid structure and formation from infrared spectroscopy. J Phys Chem Lett 5(11):1984–1993
68. Alperstein AM, Ostrander JS, TO Z, Zanni MT (2019) Amyloid found in human cataracts with two-dimensional infrared spectroscopy. Proc Natl Acad Sci 116(14):6602
69. Kurouski D, Van Duyne RP, Lednev IK (2015) Exploring the structure and formation mechanism of amyloid fibrils by Raman spectroscopy: a review. Analyst 140(15):4967–4980
70. Bonhommeau S, Talaga D, Hunel J, Cullin C, Lecomte S (2017) Tip-enhanced Raman spectroscopy to distinguish toxic oligomers from Aβ1–42 fibrils at the nanometer scale. Angew Chem Int Ed 56(7):1771–1774
71. Tycko R (2011) Solid-state NMR studies of amyloid fibril structure. Annu Rev Phys Chem 62 (1):279–299
72. Karamanos TK, Kalverda AP, Thompson GS, Radford SE (2015) Mechanisms of amyloid formation revealed by solution NMR. Prog Nucl Magn Reson Spectrosc 88-89:86–104
73. Jaroniec CP (2019) Two decades of progress in structural and dynamic studies of amyloids by solid-state NMR. J Magn Reson 306:42–47
74. Loquet A, El Mammeri N, Stanek J, Berbon M, Bardiaux B, Pintacuda G, Habenstein B (2018) 3D structure determination of amyloid fibrils using solid-state NMR spectroscopy. Methods 138-139:26–38
75. Balbach JJ, Ishii Y, Antzutkin ON, Leapman RD, Rizzo NW, Dyda F, Reed J, Tycko R (2000) Amyloid fibril formation by Aβ16-22, a seven-residue fragment of the Alzheimer's β-amyloid peptide, and structural characterization by solid state NMR. Biochemistry 39(45):13748–13759
76. Potapov A, Yau W-M, Ghirlando R, Thurber KR, Tycko R (2015) Successive stages of amyloid-β self-assembly characterized by solid-state nuclear magnetic resonance with dynamic nuclear polarization. J Am Chem Soc 137(25):8294–8307
77. Serpell LC, Fraser PE, Sunde M (1999) X-Ray fiber diffraction of amyloid fibrils. Methods Enzymol 309:526–536
78. Ricci C, Spinozzi F, Mariani P, Ortore MG (2016) Protein Amyloidogenesis investigated by small angle scattering. Curr Pharm Des 22(26):3937–3949
79. Lu K, Jacob J, Thiyagarajan P, Conticello VP, Lynn DG (2003) Exploiting amyloid fibril lamination for nanotube self-assembly. J Am Chem Soc 125(21):6391–6393
80. Toyama BH, Weissman JS (2011) Amyloid structure: conformational diversity and consequences. Annu Rev Biochem 80(1):557–585
81. Balbirnie M, Grothe R, Eisenberg DS (2001) An amyloid-forming peptide from the yeast prion Sup35 reveals a dehydrated beta-sheet structure for amyloid. Proc Natl Acad Sci U S A 98 (5):2375–2380
82. Do TD, LaPointe NE, Sangwan S, Teplow DB, Feinstein SC, Sawaya MR, Eisenberg DS, Bowers MT (2014) Factors that drive peptide assembly from native to amyloid structures: experimental and theoretical analysis of [Leu-5]-Enkephalin mutants. J Phys Chem B 118 (26):7247–7256
83. Kreutzer AG, Nowick JS (2018) Elucidating the structures of amyloid oligomers with macrocyclic β-hairpin peptides: insights into Alzheimer's disease and other amyloid diseases. Acc Chem Res 51(3):706–718
84. Kreutzer AG, Samdin TD, Guaglianone G, Spencer RK, Nowick JS (2020) X-ray crystallography reveals parallel and antiparallel β-sheet dimers of a β-hairpin derived from Aβ16–36 that assemble to form different tetramers. ACS Chem Neurosci 11(15):2340–2347
85. Gras SL, Waddington LJ, Goldie KN (2011) Transmission electron microscopy of amyloid fibrils. Methods Mol Biol 752:197–214
86. Goldsbury C, Baxa U, Simon MN, Steven AC, Engel A, Wall JS, Aebi U, Müller SA (2011) Amyloid structure and assembly: insights from scanning transmission electron microscopy. J Struct Biol 173(1):1–13

87. Newcomb CJ, Moyer TJ, Lee SS, Stupp SI (2012) Advances in cryogenic transmission electron microscopy for the characterization of dynamic self-assembling nanostructures. Curr Opin Colloid Interface Sci 17(6):350–359
88. Fitzpatrick AWP, Saibil HR (2019) Cryo-EM of amyloid fibrils and cellular aggregates. Curr Opin Struct Biol 58:34–42
89. Kollmer M, Close W, Funk L, Rasmussen J, Bsoul A, Schierhorn A, Schmidt M, Sigurdson CJ, Jucker M, Fändrich M (2019) Cryo-EM structure and polymorphism of Aβ amyloid fibrils purified from Alzheimer's brain tissue. Nat Commun 10(1):4760
90. Barnhart MM, Chapman MR (2006) Curli biogenesis and function. Annu Rev Microbiol 60 (1):131–147
91. Oh YJ, Hubauer-Brenner M, Gruber HJ, Cui Y, Traxler L, Siligan C, Park S, Hinterdorfer P (2016) Curli mediate bacterial adhesion to fibronectin via tensile multiple bonds. Sci Rep 6 (1):33909
92. Bowerman CJ, Nilsson BL (2012) Review self-assembly of amphipathic β-sheet peptides: insights and applications. Pept Sci 98(3):169–184
93. Fu K, Wu H, Su Z (2021) Self-assembling peptide-based hydrogels: fabrication, properties, and applications. Biotechnol Adv 49:107752
94. Yan C, Pochan DJ (2010) Rheological properties of peptide-based hydrogels for biomedical and other applications. Chem Soc Rev 39(9):3528–3540
95. Distaffen HE, Jones CW, Abraham BL, Nilsson BL (2021) Multivalent display of chemical signals on self-assembled peptide scaffolds. Pept Sci 113(2):e24224

Chapter 9
Function-Inspired Design of Molecular Hydrogels: Paradigm-Shifting Biomaterials for *Biomedical* Applications

Ashish Dhayani, Preethem Srinath, Sujanthi Easwara Prasad, Afsana Naaz, Mukesh Dhanka, Sanjeeb Kalita, and Praveen Kumar Vemula

9.1 Introduction

Formation of gel from molecules of natural and synthetic origin is well documented. When gels are formed by physical interactions such as surface forces and weak interactions, the resulting gels are thermo-reversible. Some gels are formed by chemical crosslinking, for example, polyacrylamide, which are not thermo-reversible. Polymeric hydrogels have been used, but they show variability when used for biomedical applications. Factors such as molecular weight difference, degree of substitution, degree of crosslinking, degradability, and biocompatibility can all show variation depending on the route of synthesis. Thus, small-molecular hydrogels or low-molecular-weight gelators (LMWG) came to be known, which could address all the limitations associated with polymeric hydrogels. Non-polymeric (molecular) hydrogels are formed because of the self-assembly of small amphiphilic molecules through non-covalent interactions such as hydrogen bonding, π-π interactions, electrostatic interactions, hydrophobic interactions, and van der Waals forces (Fig. 9.1).

A. Dhayani · S. E. Prasad
Institute for Stem Cell Science and Regenerative Medicine (inStem), Bengaluru, Karnataka, India

School of Chemical and Biotechnology, SASTRA University, Thanjavur, Tamil Nadu, India
e-mail: ashishdd@instem.res.in

P. Srinath · A. Naaz · M. Dhanka · S. Kalita · P. K. Vemula (✉)
Institute for Stem Cell Science and Regenerative Medicine (inStem), Bengaluru, Karnataka, India
e-mail: praveenv@instem.res.in

T. Govindaraju, K. Ariga (eds.), *Molecular Architectonics and Nanoarchitectonics*, Nanostructure Science and Technology,
https://doi.org/10.1007/978-981-16-4189-3_9

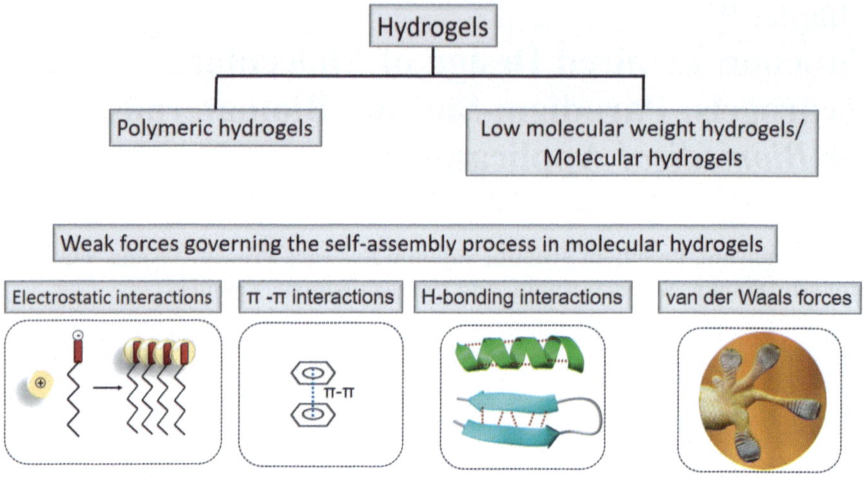

Fig. 9.1 Molecular interactions underpinning self-assembly of low-molecular-weight amphiphiles. Hydrogels can be broadly classified into polymeric and molecular hydrogels. Weak interactions govern the supramolecular self-assembly of amphiphiles in molecular hydrogels. These include electrostatic interactions, π-π stacking, H-bonding, and van der Waals forces

Molecular hydrogels typically constitute of >90% water, which makes them highly biocompatible. These small amphiphiles are tunable and can be designed as stimuli-responsive materials. Further, the morphology and strength resemble that of extracellular matrix, and they may also serve as drug delivery depots, in vivo. They can be derived from bio-based or bio-benign sources and should ideally be cleared from the body without generating a strong immune response.

Self-assembled molecular hydrogels have been used in biomedical engineering fields as cargo delivery biomaterials, cell retention and cell growth matrices, stimuli-responsive drug delivery for immunotherapy in various diseases, tissue adhesives and self-healing biomaterials, supramolecular prodrug-based systems, antimicrobial wound healing processes, in situ forming complexes, implantable biomaterials, and so on (Fig. 9.2). The limitations of polymeric systems include differences in chemistry, degree of substitution, difficult degradation and clearance. However, many of these limitations can be eliminated using small molecular amphiphiles as biomaterials, which makes them attractive for adopting for biomedical applications.

9.2 Molecular Hydrogels from Self-Assembling Peptides (SAPs)

Self-assembling peptides (SAPs) have been used for several biomedical applications, including tissue-healing sealants, stimuli-responsive anticancer drug delivery, wound healing, sustained-release vehicles and low-immune response-generating

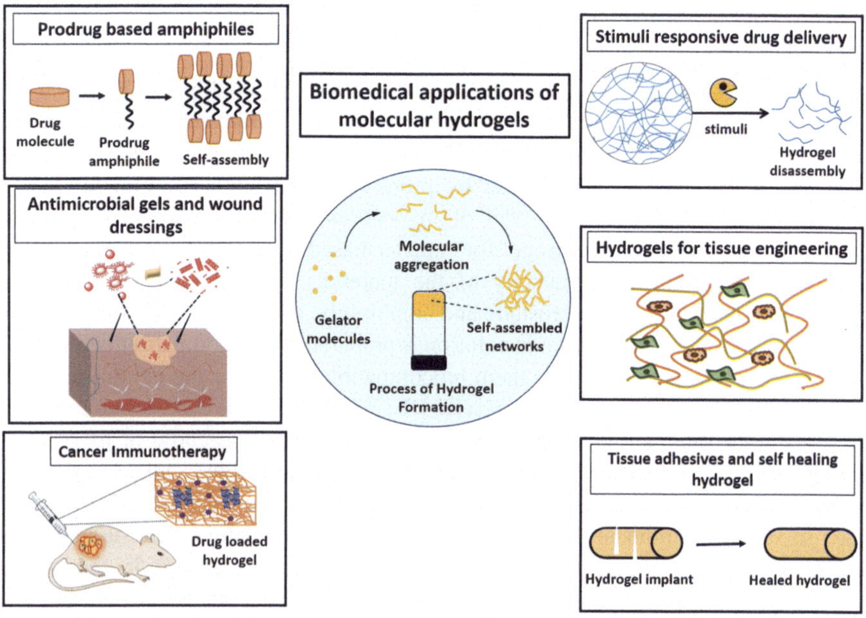

Fig. 9.2 Biomedical applications of molecular hydrogels. Self-assembled, small-molecular-weight hydrogelators could be used for various applications ranging from self-healing gels to tissue adhesives, tissue-engineering scaffolds, sustained and stimuli-responsive drug depots, prodrugs, antimicrobial and wound dressings, and cancer immunotherapy

materials, and antibacterial networks. Even though peptides are large molecules, their physical self-assembly can give rise to supramolecular networks and secondary structures and hence can be classified as "molecular gels." SAPs form an interesting class of molecular gels and are studied for biomedical applications because of various attractive properties like their resemblance to ECM, possessing pore size <500 nm making it useful for the capillary formation and endothelial cell adhesions, biocompatibility, biodegradability, insubstantial batch-to-batch variations, and feasibility toward sequence variations [1]. Secondary structures formed by beta sheet, alpha helix, triple helix, ELP-like, or amphiphile structures of engineered peptides were used in developing various self-assembled hydrogels. For introducing sol-gel transitions, triggers such as changing ionic strength, temperature, and pH are used [2]. Some prominent series of self-assembling peptides discovered are Ac-(AEAEAKAKAEAEAKAK)-NH_2 (EAK16), Ac-(KLDLKLDLKLDL)-NH_2 (KLD12), Ac-(RADARADARADARADA)-NH_2 (RAD16-I), and Ac-(RARADADARARADADA)-NH_2 (RAD16-II). Due to alternating hydrophilic and hydrophobic amino acids, this peptide forms β-sheets [3].

9.2.1 Self-Healing SAPs for Cardiovascular Disease

An injectable, structurally dynamic hydrogel based on SAP amphiphiles is ideal for a noninvasive method of biomaterial applications in cardiovascular disease. A re-healable viscoelastic hydrogel was formed by the linearization of the otherwise sterically constrained cyclic peptide by proteolytic activation resulting in SAPs [4]. This was tested in a rat model of myocardial infarction. One of the major bottlenecks in developing hydrogel for catheter-based injections for cardiovascular diseases is the high viscoelasticity of the material. Ordered nanostructures are formed by SAPs through electrostatic and amphiphilic interactions. A cyclic peptide progelator which is free-flowing in solution is prepared. Hemocompatibility and low viscosity of these peptides make them biocompatible and allow the gel to flow into the catheter. A repeat sequence of $(KLDL)_3$, known to be non-immunogenic, non-hemolytic, and antimicrobial, is used as a gellable core because of its β-sheet forming nature from the alternating lysine and arginine residue and hydrophobic interactions arising from the leucine residue. An analog sequence of $(KFDF)_3$ was used to study the implication of secondary structure on gelation. The study determined that altering the sequence resulted only in a modest change in secondary structure and viscoelastic property, which did not elicit major changes in the capacity to self-assemble. Functionalized $(KLDL)_3$ and $(KFDF)_3$ both experienced a switch in secondary structure to form a parallel β-sheet from unfunctionalized antiparallel sheets without compromising their fibrillar network formation. This property confirmed that the SAPs could be modified by adding a drug moiety or fluorescent tags without affecting the gelation property. Therefore, a substrate recognition sequence for myocardial infarction-related proteases that are endogenously produced was added. Introducing steric constraints by cyclization provided a free-flowing solution at neutral pH. A proof-of-concept study showed enzyme-triggered gelation when a mixture of cyclic peptide and cyclic peptide with fluorescent tag was used. The mixture was amenable for injection through a catheter and showed in situ gelation in the rat ischemia-reperfusion model. Further, the histological and fluorescence-based study revealed that there was no increase in the inflammatory response in the infarct region.

9.2.2 SAP-Based Molecular Hydrogels in Accelerated Wound Healing

Wound healing is one of the major concerns in the healthcare system. Wound healing is a highly orchestrated process which comprises hemostasis, inflammation, proliferation, and remodeling. Improper functioning of any of these can affect the rate of wound healing. Hydrogels have been a major candidate therapy for wound healing as they provide a moist environment that aids in reepithelialization and wound closure. An ultrashort peptide sequence tailored with a hydrophobic tail and a hydrophilic head group comprising sequence LIVAGKC (LK_6C) was specifically designed to address wound healing requirement as wound dressing [3]. The major advantage of peptide-based self-assembled systems is their tenability to single amino acid sequence modification for optimization of self-assembly. The cysteine end in the design of a short peptide sequence helps in crosslinking to provide increased strength and impart shape fidelity to the gel, whereas modifying the cationic head group (K) into an anionic D (LD_6C) or neutral S (LS_6C) interfered with the gelation ability. It was concluded that stiffer gels are more transparent because of less phase separation or precipitation. Shortening the length of peptide to tetramers also formed gels with varying properties. After studying the effect of single amino acid modifications in gelation property, LK_6C was chosen for its lower gelation concentration and stiffness for application in cutaneous wound healing. Functionalization of LK_6C with integrin-binding peptide, CRGD formed spontaneous gel with a fibrous network. Crosslinking by disulfide linkage resulted in more pronounced β-turn characteristics which imparted structure fidelity for a prolonged duration in water, when compared to non-crosslinked gels. Magnusson-Kligman maximization test on guinea pigs showed low or weak allergenic potential of the crosslinked gel. These gels were used in a mouse model of surgically induced full-thickness excisional wound. Topical application of this gel combined with growth medium was evaluated as a wound dressing. When applied to an excisional wound experimental model, the gel elicited reepithelialization with the appearance of stratum corneum and stratum granulosum, thus demonstrating its wound healing potential.

9.2.3 *Hydrogels to Regulate Immune Response Toward the Implant*

An innate immune response is often observed with implantation of biomaterials used for various applications as a part of the healing process. Delivering an immunosuppressant drug such as dexamethasone (Dex) can be used to reduce post-implant-related inflammatory response. Combining peptide amphiphiles (PA) with other receptor targeting ligands, biomaterials, or proteins with a conserved self-assembled moiety have been developed. A self-assembled peptide amphiphile (PA) system comprising a short peptide sequence and a hydrophobic tail (C_{16}-$V_2A_2E_2K$-Dex) was designed with dexamethasone attached via a covalent hydrazone linkage to control drug release [5]. PA-Dex system was used to form a robust nanofiber gel using charge screening with $CaCl_2$. While a physical mixture of gel containing C_{16}-$V_2A_2E_2$ with Dex showed burst release of 50% drug within 5 days, C_{16}-$V_2A_2E_2K$-Dex resulted in a prolonged sustained release of Dex of about 40% over 32 days, thus demonstrating potential as a delivery agent for Dex release, in vivo. A cell surrogate system using polystyrene microparticles was injected with a control system of PA in the left flank, and a PA-Dex system with calcium-based gelation was injected in the right flank in SKH1-E mice to study its localized immune suppression ability. The study showed that the PA-Dex system led to significantly reduced ROS formation which was confirmed using luminescence-based live animal imaging corroborated by histopathological examination for immune cell infiltration.

A peptide-based amphiphile has shown potential in nitric oxide delivery to treat neointimal hyperplasia that occurs during arterial interventions [6]. The peptide amphiphile has a hydrophobic segment made of fatty acids and a peptide hydrophilic segment that can bind to negatively charged biopolymers in an aqueous environment, leading to self-assembly due to aggregation of hydrophobic segments. Heparin, a negatively charged polysaccharide, was included to enhance nanofiber growth and formation of networks. Local delivery of this gel in vivo showed a reduction in intimal hyperplasia by release of nitric oxide attached in the form of diazeniumdiolates.

Several peptide-based hydrogel systems have also been tested for antibacterial activity. Peptide mimics of anthranilamide have antibacterial property. A library of anthranilamide-based short peptides designed by varying acyl group, capping moiety, or incorporating them in bolaamphiphiles or C3 symmetrical structures, were studied. Derivatives such as N-benzoyl capping and a para-substituted bolaamphiphile exhibited antibacterial activity against *Staphylococcus aureus*, the most common cause of skin and soft tissue infections [7]. Similarly, a zinc ion-binding peptide sequence Nap-GFFYGGGHGRGD formed by self-assembled hydrogelation induced by Zn^{2+} binding showed antibacterial efficacy against the enteropathogenic *Escherichia coli* [8].

9.3 Prodrug-Based Self-Assembled Hydrogels

Most of the drugs that target specific diseases are limited from reaching their therapeutic potential because of systemic toxicity and limited bioavailability. Delivering the drug in an inactive form termed as prodrug is one of the strategies used in improving the pharmacokinetics of the drug [9, 10]. Prodrugs that are designed to incorporate components that self-assemble to form molecular gels have been extensively studied. These self-assembled prodrugs are designed to overcome the limitations of drug molecules by altering their solubility, enhancing drug loading capacity, and avoiding the use of harmful excipients. Self-assembled prodrugs also are amenable for modification and can be tailored for enzyme hydrolysis as well as other stimuli depending on the disease symptoms [11]. Many self-assembled prodrug-based hydrogels are in developmental stage with detailed in vitro proof of concepts studied. Even though extensive in vitro characterization has shown better drug loading and reduced burst release, the in vivo efficacy and further clinical feasibility remain to be evaluated. Prodrug-based hydrogels can be divided into two main classes. Prodrug-based reversible hydrogels are the ones where the incorporation of pro-moiety inhibits the function of the parent drug. Irreversible hydrogels are the ones when the prodrug maintains its bioactivity even upon the addition of the pro-moiety. Both classes of prodrugs have been described in detail, and we direct our readers to the cited references. The first set of reversible prodrug-based hydrogels are a class of acetaminophen prodrugs and were described by Vemula et al. [12]. Irreversible prodrug-based hydrogels studied in vitro include ibuprofen [13], amygdalin [14], and paclitaxel [15] which were formed by the incorporation of fatty acid, peptide, or sugar moiety that aids in self-assembly process. The pro-moiety added here helps in self-assembly without interfering the bioactivity of the drug. Other examples which form irreversible self-assembled prodrug-based hydrogels include paclitaxel [16], vancomycin [17], and glucosamine [18].

9.4 Stimuli-Guided Self-Assembly and Disassembly (Disease-Responsive Disassembly) of Small Molecules

Design of disease-responsive drug delivery systems is an exciting area in biomaterials for biomedicine and nanotechnology. Such systems enable titration of dose and timed drug release in response to physiological cues such as enzymes, pH, temperature, and redox reactions. A vast number of examples can be found in literature for such systems. Here, we describe two examples of enzyme-responsive systems studied by our group and how they have been used for various inflammatory conditions such as vascularized composite allotransplantation (VCA), inflammatory bowel disease (IBD), and rheumatoid arthritis (RA) in various animal models using various small molecular amphiphiles. We further discuss other stimuli-responsive systems that have been designed for various other applications such as cancer therapy.

9.4.1 Enzyme-Responsive Hydrogels for Delivery of Immunosuppressants in Vascularized Composite Allotransplantation (VCA) and Arthritis

Gajanayake et al. described an amphiphilic system using triglycerol monostearate (TGMS or TG-18) for the delivery of immunosuppressants to prevent allograft rejection [19]. TGMS is an amphiphile that is approved by the Food and Drug Administration (FDA) as a "generally recognized as safe" (GRAS) agent. It can self-assemble to form hydrogels as well as degrade in the presence of esterases and proteases which are overexpressed during immune activation in graft rejection. TGMS loaded with immunosuppressant, tacrolimus, could prevent graft rejection in a rodent VCA model for >100 days compared to 33 days in control animals given with free drug, showing the importance of the enzyme-responsive drug delivery systems. The risk-to-benefit ratio is high in conventional therapy owing to a plethora of side effects—these are completely overcome using a localized drug delivery approach [20]. The hydrogel-based drug delivery system can be monitored

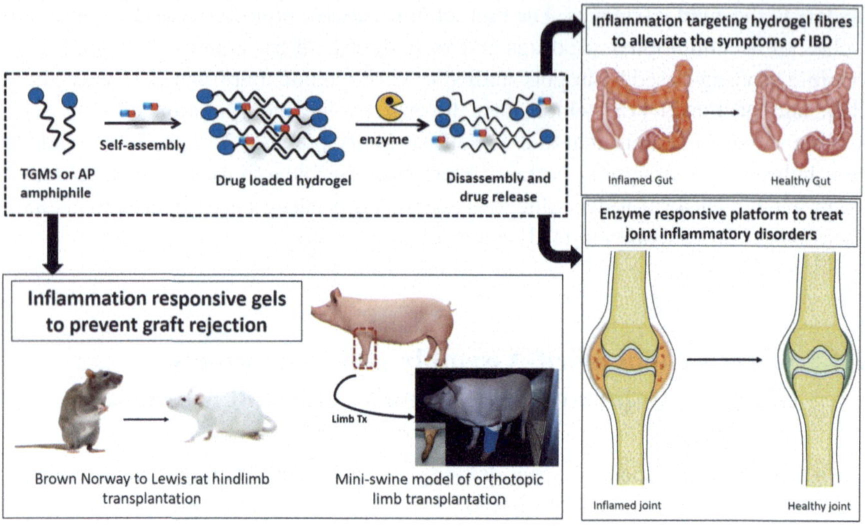

Fig. 9.3 Disease-responsive drug delivery platforms for inflammatory diseases. Triglycerol monostearate (TGMS) and ascorbyl palmitate (AP)-based amphiphiles can self-assemble in the presence of drug molecules leading to drug encapsulation. The gels can be degraded by various inflammatory enzymes triggered either due to rejection episodes during transplantation or in inflamed gut as well as in inflamed joint leading to release of the drug to alleviate the disease symptoms

noninvasively using a near-infrared dye (NIRD) [21]. TGMS-TAC-based system performed with similar efficacy in a porcine model of limb VCA [22].

The same TGMS-based system was used for the delivery of triamcinolone acetonide (TA) in response to flare-ups during inflammatory arthritis (IA) [23]. Multiple biomaterial-based strategies have been developed using liposomes, micelles, solid lipid nanoparticles, and hydrogels. However, most of the reports show sustained release of drugs as well as passive release which do not account for the disease severity during arthritis which show flare-ups as well as low disease activity. The TGMS hydrogels loaded with TA could selectively release the drug depending on the concentration of inflammatory enzymes as shown by in vitro studies using lipase and MMPs. Moreover, the platform could also show severity-dependent release in the arthritic animal as shown by the dye release in vivo and the lowering of the disease severity score when compared with equivalent dose of free drug. However, even though there is no burst release of drug, there is a small amount of baseline release of the drug in response to ECM enzymes such as MMPs which might not necessarily be overexpressed during IA. The small burst release can be due to passive diffusion or surface-encapsulated drug. Also, the biomaterial showed good biocompatibility. Thus, the enzyme-responsive system represents a novel platform to treat inflammatory diseases since these are autonomous systems and should see the light of clinical translation (Fig. 9.3).

9.4.2 Ascorbyl Palmitate (AP or AP-16) Hydrogel Fibers for Charge-Dependent Localization, Adherence, and Enzyme-Responsive Drug Delivery in Inflammatory Bowel Disease (IBD) and Rheumatoid Arthritis

Zhang et al. reported an inflammation-targeting fibrous hydrogel made of ascorbyl palmitate, which could self-assemble and encapsulate anti-inflammatory agent, dexamethasone, to treat ulcerative colitis (UC) [24]. The ascorbyl palmitate monomers owing to their negative charge could selectively bind to inflamed intestinal epithelium because of accumulation of positively charged antimicrobial peptides on the epithelial surface. This design helped to deliver the drug locally to the inflamed tissue and significantly alleviate the symptoms of IBD in two different mouse models when the hydrogel fibers were given as enemas. Since AP is a GRAS agent, it did not lead to any toxic effects. Such biomaterials represent a novel way to deliver drugs to sites of inflammation.

The same hydrogelator, ascorbyl palmitate, was earlier used by Vemula et al. for the delivery of anti-inflammatory drugs in rheumatoid arthritis (RA) [25]. The authors showed conclusively the self-assembly and dye loading, along with disassembly in the presence of enzymes such as MMP-2 and 9 which are overexpressed during RA. Synovial fluid collected from RA patient joints showed significant release of the encapsulated dye which was not observed in the synovial fluid collected from healthy joints. Also, the gels were stable for over 10 days and could sustain mechanical stress without any burst release. Thus, such hydrogels could act as viable platforms for multiple proteolytic diseases as shown by treatment of multiple diseases and should be studied in better details for their use in humans (Fig. 9.3).

9.4.3 Stimuli-Responsive Molecular Hydrogels for Cancer Immunotherapy

Cancer is currently one of the leading causes of death, globally [26]. *Advancement in cancer treatment modalities,* including chemotherapy, immunotherapy, and hormonal therapy, has been beneficial to fight cancer. Administration of traditional anticancer drugs requires multiple injections and high dosage, leading to many systemic side effects, off-target accumulation, toxicity, and low patient compliance. Hydrogels as drug delivery systems have come to our rescue by improving pharmacokinetics and biodistribution of anticancer drugs. However, conventional hydrogels exhibit minimal changes in their properties in response to the surrounding environment. Their rigid shapes allow them to be used only in limited ways [27–29]. Large pore size and excellent water absorption capacity of hydrogels make it challenging to control the release kinetics of the drug entrapped, often resulting in burst release, limiting their efficacy in cancer treatment. Stimuli-responsive hydrogels have grabbed attention due to their unique ability to undergo morphological changes in response to external and internal stimuli like pH, temperature, ultrasound, light, ionic strength, or enzymes, etc. Stimuli-responsive hydrogels can deliver drugs to tumor sites with minimal leakage of the drug in the blood circulation. The control over permeability, conformation, swelling, and degradation of these hydrogels can be monitored, thus ensuring localized drug release at the tumor sites, reducing the drug dosage, and providing long term therapeutic benefits. However, hydrogels that respond to external cues can pose translational limitations in drug delivery, such as the requirement of bulky and costly equipment, the need for trained personnel, and sometimes the inability of the stimulus to reach deeper tissues and organs. Harnessing the cancer-specific microenvironment or other physiological cues as triggers for degradation of hydrogels and drug release is clinically more assuring. In this regard, molecular hydrogels have emerged as suitable alternatives to polymeric materials. They can be designed for controlled release of anticancer drugs by incorporating an enzyme cleavable bond in the gelator. Molecular hydrogels provide good biocompatibility and biodegradability. In this section, we will highlight the use of stimuli-responsive molecular hydrogels in cancer therapy.

Molecular hydrogels are formed by supramolecular self-assembly of small molecules where non-covalent interactions such as H-bonds, electrostatic interactions, hydrophobic interactions, etc. control the mechanical properties of the hydrogel [30]. The release of entrapped anticancer drugs can be monitored by controlling the degradation of the hydrogels.

In the last decade, *enzyme-responsive self-assembled hydrogels* have attracted attention. In one study, Yang et al. developed a molecular hydrogelator of folic acid (FA)-taxol conjugates by disulfide bond reduction triggered by glutathione (GSH) [31]. FA was attached to taxol through peptide sequences of GFKE, GKE, KE, and K. The shorter peptide linker K was provided the advantage of easy synthesis and improved yield. It had the highest loading capacity of taxol and showed slowest release of the drug. FA-taxol conjugates could efficiently inhibit tumor growth in 14 days in a mouse tumor model. The FA-taxol conjugates could sustain drug release through ester bond hydrolysis. A single dose of intratumor administration of FA-taxol conjugates was more efficient in inhibiting tumor growth than four times the intravenous dose of clinically used taxol. Molecular hydrogels of therapeutic agents provide a sustained release of the drug without burst release. Many such molecular hydrogels of taxol and other anticancer agents have been judiciously designed and reported [16, 32–34]. In another study in 2013, Xu and coworkers developed supramolecular nanofibers/hydrogels of D-peptides via dephosphorylation by alkaline phosphatase (ALP) [35]. They attached taxol to D-peptide-based hydrogelators, which self-assembled after enzymatic dephosphorylation for intratumoral chemotherapy. Acquired drug resistance of cancer cells is one of the major challenges of chemotherapy and an important reason behind the failure of drugs in clinical trials. To address this limitation and minimize drug resistance, the same group in 2016 used a similar technique to design a molecule for subcellular targeting in cancer cells [36]. The molecule comprised the tetrapeptide self-assembling backbone Phe-Phe-Tyr-Lys (FFYK) phosphorylated on tyrosine and conjugated with a fluorophore (4-nitro-2,1,3-benzoxadiazole (NBD) at the N-terminal (NBD-FFpYK) to report intracellular molecular self-assembly. Triphenyl phosphonium (TPP), a redox modulator to target mitochondria, was attached to the ε-amine of the lysine residue on the NBD-FFpYK to form the precursor NBD-FFpYK-TPP. Dephosphorylation of tyrosine phosphate (pY) triggered by ALP resulted in the formation of nanoscale assemblies via non-covalent interaction. TPP conjugates selectively targeted cancer cells because cancer cells overexpress ALP. They selectively killed human osteosarcoma cells (Saos2) while being ineffective in killing normal cells (HS5). The TPP nanoscale assemblies were formed on the surface of cancer cells and were engulfed by endocytosis. Inside the cancer cells, the TPP conjugates disrupted the integrity of mitochondria resulting in the release of cytochrome c thus killing the cells.

pH-sensitive molecular hydrogels that respond to pH changes and ionic strength also make promising candidates for drug delivery in cancer therapy. Aerobic glycolysis releases lactic acid in tumor cells. It makes the tumor microenvironment acidic compared to normal tissues [37]. The tumor tissue's endosomal pH is between 4.5 and 6.5 [38]. The changes in surrounding pH control the self-assembly of pH-sensitive molecular hydrogels loaded with anticancer drugs. Liang et al., in 2013, prepared acid-responsive self-assembled micelles from amphiphilic peptides loaded with doxorubicin [39]. They studied how a change in the external pH affected the drug release. The peptide had KKGRGDS and VVVVVV sequence in the hydrophilic head and hydrophobic tail, respectively. In neutral and basic pH, the hydrogen bonds and the hydrophobic interactions aided the peptide to self-assemble into micelles. In an acidic medium, the micelle dissociated because of electrostatic repulsion between charged lysine residues of the peptide. As a result, the DOX-loaded hydrogels showed sustained release at neutral pH (pH = 7.0) and fast release in acidic pH (pH = 5) of the tumor cells. The amphiphilic peptide hydrogel developed is a promising drug delivery system for targeted delivery of anticancer drugs and their rapid release because of pH change.

Photoresponsive molecular hydrogels in cancer therapy have also been a subject of great interest because of the inherent advantage of light as a stimulus. Manipulation by light is noninvasive and allows remote and easy control of hydrogels with minimum byproducts. The irradiation dosage and the photoreactions can be precisely controlled in both time and space by modulating the intensity of the light source and exposure time. In one interesting study, Yata et al. designed a novel composite-type DNA hydrogel having photothermal and immunostimulatory properties for cancer therapy [40]. The hydrogel had a hexapod-like structured DNA (hexapodna) with CpG sequences and gold nanoparticles. Gold nanospheres (AuNS) and gold nanorods (AuNR) were used as photosensitizers. The high affinity of gold nanoparticles AuNS and AuNR toward DNA allows them to directly load DNA on their surfaces. The DNA hydrogels were prepared by self-assembly of gold nanostructured DNAs using oligodeoxynucleotides (ODNs) containing CpG sequences. AuNS-ODN(cg) was prepared by the absorption of ODN(cg) and a CpG ODN with oligoadenine sequence onto the surface of AuNS. Similarly AuNS-ODN(cg) was prepared by using ODN(cg)-oligoA and GpC ODN. Here, (cg) and (gc) signify CpG or GpC sequence, respectively. Hexapodnas (hPODNA(cg) and hPODNA(gc)) were prepared separately by annealing equimolar mixtures of ODNs where half of the ODNs were complementary to the other half in the mixture. Finally, AuNR and AuNS hydrogels were prepared by mixing hPODNAs with AuNRODN and AuNSODN, respectively. hPODNAs and AuNRODN or AuNSODN had complementary DNA ends to facilitate DNA hybridization and the formation of these hydrogels. 532 nm laser irradiation triggered the release of DNA and AuNS from AuNS hydrogels (cg). Similar irradiation of AuNS-DNA hydrogel (cg) also induced the release of TNF-α and IL-6 from mouse macrophage-like, TLR9-positive RAW264.7 cells. They used AuNR DNA hydrogel(cg), with absorption peaks around 650–900 nm for mice studies to avoid the harmful effects of 532 nm laser. AuNR DNA hydrogel(cg) inhibited the tumor growth in EG7-OVA tumor-bearing mice and enhanced their survival. To the best of our knowledge, this is the first successful attempt to develop self-assembled hydrogels for photothermal cancer immunotherapy. Recently in 2020, a group of researchers have used supramolecular strategy to design injectable self-assembled bola peptide (a di-FF derivative, DFF) hydrogels for sustained and localized delivery of photodynamic prodrug (5-aminolevulinic acid (5-ALA)) [41]. L-Phe-L-Phe (FF) is a widely used peptide for self-assembly. The prodrug-encapsulated hydrogels were biocompatible and inhibited nontargeted leakage of the prodrug. As compared to intra-tumoral injection of 5-ALA, 5-ALA-loaded bola peptide hydrogels enhanced the formation of active photosensitizers from the inactive prodrug when the tumor sites were irradiated with 635 nm radiation.

Stimuli-responsive molecular hydrogels, because of their minimal invasiveness, have emerged out as an effective drug delivery system for localized and sustained delivery of anticancer drugs and prodrugs, highlighting their potential in clinical translation. However, complex and dynamic physiological conditions make it challenging to understand their in vivo assembly/degradation kinetics. To promote the use of smart responsive hydrogels in clinical studies, there is an urgent need to develop novel approaches to study the process of self-assembly, the resultant structure, their biological effects, and their degradation kinetics.

9.5 In Situ Forming Gels

The in situ forming injectable gels are premixed sol state of therapeutic molecules (drug, protein, nanoparticles, cells) and low-molecular-weight gelling compounds, which form a hydrogel upon in vivo injection because of chemical or physical stimuli. Subsequently, after sol to gel transition in the in vivo condition, the therapeutic molecules can release out from the matrix. Several low-molecular-weight hydrogelators (LMWHs)-based in situ forming injectable hydrogels have been widely employed in biomedical engineering for therapeutics delivery, antimicrobials, regenerative medicine, and tissue-engineering applications. LMWHs include steroids, fatty acids, sugars, urea, amino acid, nucleoside, and guanosine-derived amphiphilic molecules [42–44]. (Fig. 9.4).

In Situ Forming Molecular Gels for Chemotherapy Localized intra-tumoral sustained delivery of anticancer drugs via in situ forming molecular gels not only helps to reduce the associated systemic drug toxicity but also improves solubility, increases bioavailability, and thereby increases their efficacy [45]. In a study, Singh and co-workers developed doxorubicin (DOX)-entrapped hydrazide L-alanine (ALA-HYD) stable nano-assembled injectable gel (DOX-GEL) for localized treatment of tumors [46]. To assess anticancer efficacy, DOX-GEL was subcutaneously injected near the tumor site in the mice. DOX-GEL could effectively decrease tumor load in mice compared to intravenously injected DOX and free DOX injected at the local tumor site. Zhang and his group synthesized paclitaxel (PTX)- based anticancer gel (PTX-GEL) in which the drug self-assembled to form gel in situ. The in vivo anticancer efficacy of PTX-GEL was evaluated in the 4T1 tumor mouse model showing effective inhibition of cancer metastasis [47].

Zhi et al. studied a natural triterpenoid (*Poria cocos* (Schw.) Wolf and *Liquidambar formosana*)-based bioactive self-assembled natural product gel (NPG) loaded with doxorubicin for anticancer therapy (DOX-3/NPG). The system could undergo in situ gelation after local injection. In vivo studies on the 4T1 solid tumor containing BALB/c mice showed that the animal group treated with DOX-3/NPG had 97.6% inhibition. On the 14th day of treatment, tumor growth was completely inhibited. The strong anti-tumor activity of the gel was not only because of its ability to induce apoptosis but also due to the strong anti-inflammatory property as evaluated by prostaglandin E2 (PGE2) in the mice blood [48]. Maity et al. synthesized a biodegradable self-assembled injectable gel of aromatic diamine conjugated bile acid (LCA-Gly-Gly) and explored it for in vivo anticancer drug therapy in BALB/c mice. In vivo results for triple drugs such as doxorubicin, combretastatin A4, and dexamethasone-loaded gel (TRI-Gel) showed a significant reduction in tumor growth kinetics in Lewis lung carcinoma-bearing BALB/c mice and confirmed the high anticancer efficacy of TRI-Gel formulation [49].

In Situ Forming Molecular Gels for Tissue Engineering and Regeneration The localized in situ forming injectable hydrogels or scaffolds in tissue regeneration have been widely used to improve the therapeutic efficacy of cells and growth factors/ biological signaling molecules by providing the sustained delivery at the target site [50, 51].

In Situ Forming Molecular Gels for Bone and Cartilage Regeneration Development of a biomaterial with mechanical properties similar to that of the targeted tissues is crucial for tissue regeneration. Kim et al. synthesized a self-assembled in situ forming molecular hydrogel of the collagen mimetic peptides (CMPs), which showed the structural conformation similarity to collagen-like triple helices. Further, to improve the mechanical properties, poly(L-lactide-co-caprolactone) (PLCL)-based scaffolds were loaded in nano-assembled peptide-based bioactive hydrogels. To evaluate in vivo regeneration potential, rabbit bone

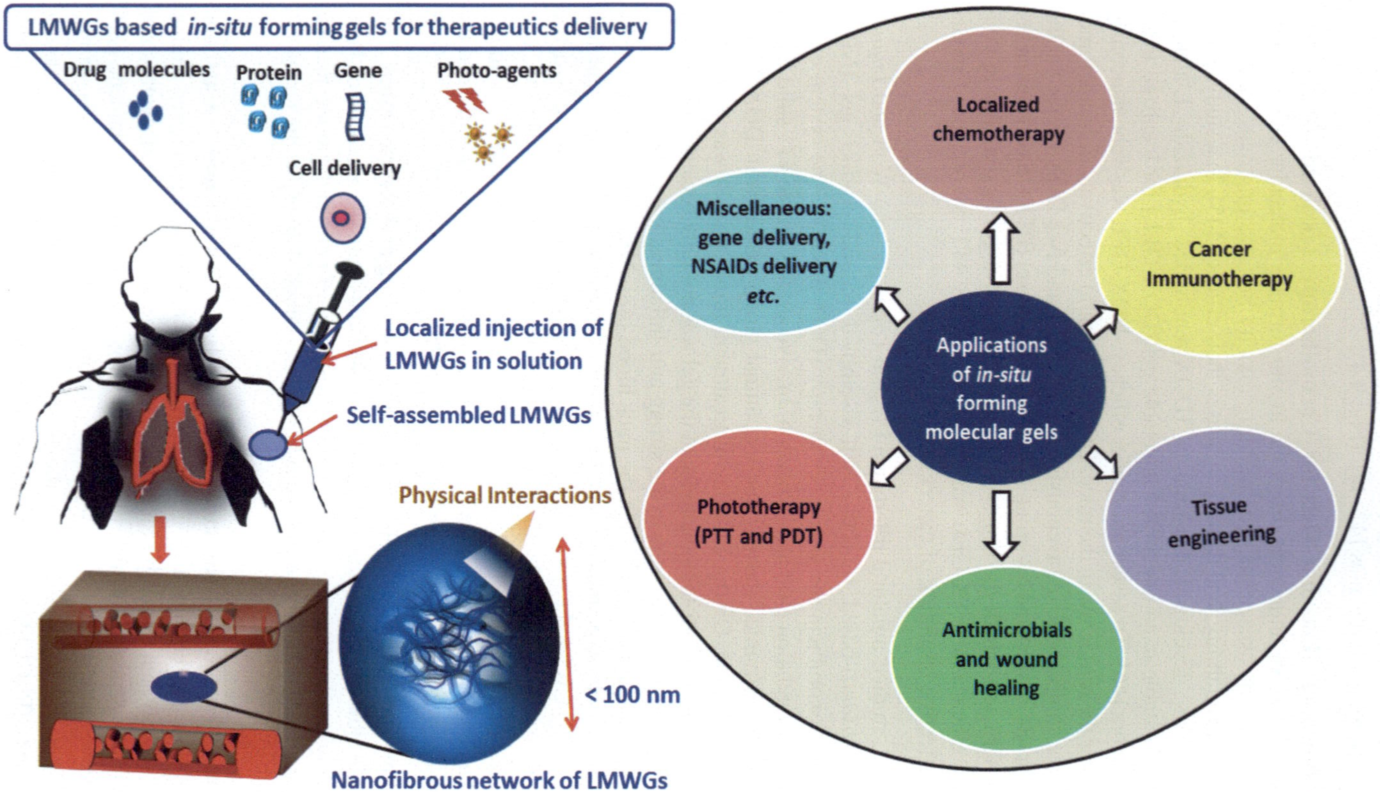

Fig. 9.4 Schematic showing injectable in situ forming molecular hydrogels and their biomedical applications. Gelators or formulation precursors form the self-assembled nanofibrous gel depot at the localized site in the body which can further be used for the desired application

marrow stromal cells (rBMSCs) loaded into PLCL-peptide hydrogel (rBMSC/PLCL-peptide complexes) were injected into the subcutaneous dorsal portion of nude mice. The in vivo regeneration results showed that in situ chondrogenesis and cartilaginous tissue formation occurred after 5 weeks of hydrogel composite injection [52].

In Situ Forming Molecular Gels for Dental Regeneration To solve the problem associated with degenerated dental pulp, a consequence of incurable dental caries in nondeciduous teeth, several in situ forming hydrogels were developed for localized applications. Dental pulp stem cells encapsulated in dentinogenic peptide hydrogel with a beta-sheet nanofiber structure were developed by Nguyen et al. [53]. The cell-gel construct provided a suitable environment for the proliferation of autologous cells and also showed dentinogenic potential in vivo. Galler et al. loaded dental pulp stem cells to an enzyme-cleavable nano-assembled peptide hydrogel with important cell adhesive features. Further, the cell-gel construct was co-encapsulated with different growth factors such as fibroblast growth factor, basic transforming growth factor b1, and vascular endothelial growth factor for quick recovery of teeth pulp tissues. Subcutaneous implantation of hydrogel system in dentin cylinders in immunocompromised mice revealed a faster vascularization in soft connective tissue, which confirmed the promising potential of developed hydrogel for regenerative endodontics application.

In Situ Forming Molecular Gels for Cardiac Regeneration Cardiomyocyte necrosis and apoptosis because of coronary artery disease can cause damage to the myocardium. Myocardium cannot regenerate, leading to heart failure and a substantial number of deaths worldwide. Using biomaterials in regenerative therapies can repair the myocardium damage and help regenerate cardiomyocytes [54]. Bastings et al. reported a supramolecular pH-responsive hydrogel of ureidopyrimidinone (UPy) for local catheter injection in infarcted myocardium. The formulation was liquid due to low viscosity at pH 8.5, which can be injected via a 1-m-long catheter. The formulation readily converted into gel form when it came in contact with the tissue interface (pH 7.4). In vivo studies in large pig animal models revealed that growth factor (hepatocyte growth factor and insulin-like growth factor-1)-loaded UPy hydrogel after trans-endocardial injection at the local site released the growth factors in a sustained manner and reduced the infarct scar [55]. Thus, the pH-responsive injectable UPy hydrogel formulation showed the in vivo regenerative potential in a large animal model.

In Situ Forming Molecular Gel for Delivery of Antimicrobials and Wound Healing For the better management of a chronic bacterial skin wound, various LMWGs-based injectable in situ forming gels have been employed for multiple therapeutics delivery at the localized wound site such as antimicrobials, growth factors, active molecules, and cells. The in situ forming molecular gels not only fill the infected skin at the wound site but also avoid scar formation and avert infection-associated complications [56]. Zhao et al. studied a peptide-based self-assembled nanogel for skin wound healing. In a study, they reported

Fmoc-FFGGRGD peptide hydrogel encapsulating resveratrol (polyphenol anti-inflammatory compound) for wound dressing (Pep/RES) applications. In vivo wound healing data confirmed that Pep/RES hydrogel accelerates wound healing rate, reduces inflammation, and prevents scar progress [57].

In Situ Forming Molecular Gels for Phototherapies (Photodynamic and Photothermal Therapy) Photodynamic therapy (PDT) and photothermal therapy (PTT) are highly effective therapeutic modalities in biomedical engineering, owing to their high specificity and noninvasive characteristics. PTT employs hyperpyrexia while PDT employs the reactive oxygen species (ROS) against malignant tumor and bacterial infection. ROS and hyperpyrexia are generated at the localized site through irradiation of photo-agents with laser. However, low bioavailability and high toxicity associated with photo-agents are responsible for their limited use in phototherapy. Molecular injectable gels are a potential delivery vehicle for the localized sustained delivery of photosensitive drugs to effectively eradicate the tumor and bacterial infections [58, 59]. Abbas et al. synthesized an injectable Fmoc-FF/Chlornin e6 (Ce6) hydrogel, which was constructed by encapsulation of Ce6 into positively charged poly-L-lysine (PLL) and negatively charged Fmoc-FF electrostatic inclusion. The injectable system turned sol during injection. Upon injection, gel formation occurred in situ, which showed a sustained localized release of Ce6 photosensitive drug at the localized intra-tumoral injection site. Such injection efficiently ablated the tumor upon laser illumination. In another study by Zhang and co-workers, a hybrid biomaterial was designed by combining an Fmoc-FF peptide based in situ forming hydrogel with fullerene derivative nanoparticles (C60-PTC, avg. diameter 6 nm), to treat localized bacterial infections in wounds through effective phototherapy. In vivo results in mice showed potent antibacterial efficacy against *S. aureus* after irradiation with NIR light (808 nm) without any detectable toxicity [60].

9.5.1 Other Applications: LMWHs for Gene Therapy and Delivery of NSAIDs

Zhen et al. for the first time reported an in situ forming molecular organogel of soybean oil and methyl (S)-2,5-ditetradecanamidopentanoate (MDP) gelator for the controlled delivery of candesartan cilexetil (CC) for effective management of hypertension. The in situ forming gel formulation released the drug in a controlled manner in vivo and maintained the plasma candesartan cilexetil concentration for over 10 days [61]. Raymond et al. synthesized a self-assembled in situ forming supramolecular hydrogel of cationic derivatives of phenylalanine (Phe) for the localized sustained delivery of diclofenac sodium, which is one of the most effective nonsteroidal anti-inflammatory drugs (NSAIDs) for joint pain and inflammation management. To synthesize the phenylalanine derivative-based hydrogelators, the N-fluorenylmethoxycarbonyl phenylalanine (Fmoc-Phe) compound was chemically modified by using the diaminopropane (DAP) molecule at the carboxylic end to synthesized Fmoc-Phe-DAP conjugate, which could form a self-assembled hydrogel upon addition of an appropriate amount of NaCl. The presence of diclofenac caused the self-assembly of Fmoc-Phe-DAP molecules. Fmoc-F5-Phe-DAP/diclofenac hydrogel injected into the ankle joint of mice could efficiently alleviate pain and provide a sustained in vivo drug release for 2 weeks [62]. Li et al. reported an injectable nucleoside-based in situ forming hydrogel for in vivo protein molecule delivery. The supramolecular hydrogel of N4-octanoyl-2-deoxytidine nucleoside hydrogelators was used to encapsulate proteins and characterize their toxicity and drug delivery in vivo. The in vivo studies highlighted high biocompatibility and controlled delivery of protein [63].

9.6 Tissue-Engineering Scaffolds for Regenerative Medicine

Small molecules and oligos described thus far have several characteristics that warrant their consideration as cell-delivery vehicles and tissue-engineering scaffolds. The materials are biocompatible, shear-thinning (injectable, amenable to bioprinting), and self-healing, undergo self-assembly in physiologically relevant conditions (aqueous media, 37C, neutral pH), possess cell-binding motifs (peptide hydrogels), can be tuned for sustained delivery of relevant biomolecules (growth factors, drugs), and have low immunogenicity.

From a "bioinspired" perspective, molecular self-assembly of relatively low-molecular-weight biomolecules is a ubiquitous phenomenon in biogenesis, maintenance, and remodeling of natural extracellular matrix. Typical participants include laminin, fibronectin, tenascin, thrombospondin, and proteoglycans, which associate with a "base" polymeric network of high-molecular-weight proteins such as collagen I to form biologically functional ECM. Research efforts to identify the most critical functional motifs for directed cell adhesion, migration, proliferation, and differentiation are on. With structural modifications based on principles of self-assembly, such an approach could replace natural ECM as a scaffolding material for tissue engineering and move toward a more reproducible "defined ECM" platform for developing high-throughput cell-based organoids for in vitro and therapeutic applications.

Molecular hydrogels may even have properties that make them superior to polymer hydrogels in some ways. Self-assembly is primarily driven by physical crosslinking using weak interactions, avoiding the use of potentially cytotoxic crosslinking agents. With molecular hydrogels, it is relatively easy to reverse gelation by altering the environment (easy to elute cells after culture, get rid of the scaffold). A modular approach to fine-tune mechanical properties by altering concentrations and molar ratios of gelators is also achievable. Many times, these gels are optically clear, thus making visual examination of encapsulated tissues and microscopy easier. Such an approach enables "xeno-free" scaffolding biomaterials and paves the way for the development of "defined ECM substitutes."

However, molecular hydrogels also have drawbacks compared to polymer hydrogels. With the most relevant class of molecular gelators for TE, peptides are synthesized in solid phase, which is expensive and difficult to scale up. Also, functionalization by planar stacking molecules such as Fmoc is almost always required to achieve robust gelation, and this also requires equipment and proficiency in organic synthesis.

Exploring the applicability of molecular hydrogels for tissue engineering and regenerative medicine appears to be a relatively new topic, with a bulk of relevant studies published only over the last decade.

In one study [64], Loo et al. in 2019 screened a series of hexamer self-assembling peptides for their compatibility with iPSCs to support proliferation, maintenance of pluripotency, and differentiation protocols. Peptides containing aspartic acid, serine, and lysine were synthesized and evaluated. Of these, one peptide was identified—lysine containing Ac-ILVAGK-NH2 was found to maintain H1, H9, and iPSCs for up to 30 passages when maintained in mTESR1 medium (comparable to well-established Matrigel cultures) and supported differentiation into ectoderm, mesoderm, and endoderm lineages when appropriate media cues were provided. This peptide underwent gelation within 5 minutes of incubation at 37C to form optically clear gels, suitable for nondestructive monitoring by fluorescence- and absorbance-based modalities. Encapsulated colonies of pluripotent stem cells were passaged by digestion with TrypLE Express, followed by vigorous pipetting and dilution to obtain a single-cell suspension; these cells were re-encapsulated for the next passage. The same group demonstrates in another study [65] that the hydrogel is also amenable to extrusion-based bioprinting. Here, the hydrogel bioink was printed into a 5x PBS bath to create noodle-like structures, extruded into predetermined paths, and deposited into droplet arrays. Bioink was combined with H1 ESCs or MSCs and directed to differentiate into intestinal/skin organoids (H1) or adipocytes (MSC). The researchers acknowledge that the peptide hydrogel does not maintain pluripotency or direct differentiation by its merit but rather supports cells due to its entangled fiber-like architecture, pointing out that media-derived cues are essential to drive such processes. However, the researchers also state that further decoration of the peptides with more functional moieties by click chemistry could lead to more precise guidance of cells toward a pluripotent or lineage-specific differentiated phenotype.

Another study [66] by Tong et al. in 2018 demonstrates similar applicability of non-peptide tripodal squaramide molecules with specific alkyl side chain length in supporting the generation of human iPSC spheroids that could be eluted from self-assembled gels with minimal stress and high viability, with retention of pluripotent markers. The researchers highlight that, due to lack of adhesive peptide motifs, single iPSCs develop into compact spheroids, devoid of cell-hydrogel receptor-ligand interaction.

In another study [67], Godbe et al. in 2020 have generated hydrogels from a peptide amphiphile, achieved tunable stiffness (G') by doping the hydrogel with oligolysines of various chain lengths, and identified the most suited stiffness that promotes survival of dopaminergic neurons. Interestingly, the other mechanical properties such as shear strain response, flow behavior, and stress-relaxation remained similar, and only the storage modulus varied. The researchers compare this series of stiffness to that found in the brain, wherein different parts have different stiffness and claim that such an approach could selectively enrich for specific neuronal populations and preserve target phenotype within injectable gels that could have applications in cell-based therapies for neurodegenerative diseases.

Another similar implementation has been described in a series of studies [68, 69] by Alkapa et al. since 2016. Fmoc-diphenylalanine and Fmoc-serine peptides are combined in a 1:1 ratio, and gels are formed at different concentrations to obtain a wide range of physiologically relevant stiffnesses (G' ranging from 0.1 kPa to 32 kPa). It was shown that a self-assembled nanofibrous structure emerged in the presence of calcium-containing cell culture media, and the fibers themselves had a core-shell morphology, with Fmoc-diphenylalanine forming the core fiber and Fmoc-serine forming a shell around it. The most striking results of these studies are that a 13 kPa gel developed in this way promoted cultured pericytes to differentiate into chondrocyte-like phenotype (shown by gene expression, metabolomics, and ECM production) while avoiding the default program of hypertrophy, characteristic of endochondral ossification rather than the formation of cartilage. The researchers also highlight and demonstrate improved ease of performing metabolomic analysis and identifying small molecule mediators of directing stem cell fate, compared to crosslinked biopolymer-based hydrogels. Small molecules that may be capable of guiding pericytes to chondrogenic phenotype were also identified and tested.

Another promising area of applicability of molecular hydrogels to regenerative medicine is to use them as carriers for therapeutic stem cells in vivo. A study by Takeuchi et al. [70] in 2016 demonstrates the applicability of RADA16 (a commercially available self-assembling peptide) in delivering periodontal ligament cells to the surgical site of gum surgery in rats and shows enhanced healing compared to using Matrigel for the same purpose. In another study [71], Li et al. in 2018 demonstrate the applicability of folic acid-modified tripeptides, further linked to a dipeptide by disulfide linkage (FA-FFF-ss-EE, FA-VVV-ss-EE, and FA-AAA-ss-EE) for encapsulating iPSCs and forming an injectable cell therapy. This iPSC-laden hydrogel was injected into myocardial infarct sites in mice. The procedure was shown to have improved cell retention at the injection site and elicited superior healing response and cardiac output compared to injection of iPSCs alone. Along the same lines, Huang et al. in 2019 [72] demonstrate the use of a similar molecule—naphthalene functionalized with peptides (Nap-GFFYK-ss-EE) in delivering MSCs to improve clinical outcomes in a mouse hind limb ischemia model.

In this way, molecular hydrogels find their application in tissue engineering and regenerative medicine, with enormous scope for future work. Many peptide motifs used are "bioinspired" in that they are truncated versions of naturally occurring proteins or they are discovered serendipitously. Extensive theoretical simulation and a directed molecular docking approach can enable identification of novel peptide sequences that support physiologically relevant gelation and cell adhesion. Such hydrogel platforms can serve as high-performance tissue-engineering scaffolds and cell-delivery vehicles. As mentioned earlier in this chapter, efforts are also underway to develop fully defined ECM substitutes using a modular approach. Mimics of individual ECM components and common naturally derived materials have been developed by targeted design, such as type I collagen [73], fibroin [74], elastin [74], laminin [75], and fibronectin [76]. However, as the studies themselves outline, these systems are far from perfect in biomimicry. For example, the type I collagen mimic mentioned [73] forms amyloid structures as opposed to fibrous morphology characteristic of natural collagen I. Thus, much potential exists for improving upon the current status in this area of research.

The authors would like to direct the reader to informative contemporary reviews [77–81] that provide the fundamentals and advanced concepts into the applicability of molecular hydrogels for regenerative medicine.

9.7 Future Perspectives

A large number of functional outcomes have been achieved using molecular hydrogels across the areas of drug delivery and tissue engineering. Self-assembly is a highly complex phenomenon, and there is much potential to expand upon the current status. Utilization of artificial intelligence and molecular docking studies to generate theoretical constructs could accelerate discovery of functional molecular hydrogels and expand their clinical applications. Large-scale safety and efficacy studies too are required to translate the latent potential of this exciting and extremely promising biomaterial platform for clinical applications.

9.8 Conclusions

A wealth of preclinical data has been presented in this chapter that warrants the application of molecular hydrogels as a highly versatile drug delivery platform, capable of encapsulating a wide variety of cargo and smart delivery characteristics. Efforts have also been made to provide the reader with a diverse set of example studies that have harnessed a plethora of structural characteristics to achieve functional outcomes. While fewer in number, studies do show that molecular hydrogels may also apply to cell and tissue culture, possessing some significantly advantageous characteristics over traditional biopolymer-based scaffolds, warranting their further investigation in tissue engineering. The authors hope this chapter will serve as a primer for the reader to grasp the intricacies of molecular hydrogel design, applications, and potential.

Acknowledgments A. Dhayani thanks University Grant Commission for Senior Research Fellowship. MKD thanks bridging postdoctoral fellowship, CORE Grant from DBT-inStem. The authors have no other relevant affiliations or financial involvement with any organization or entity with a financial interest in or financial conflict with the subject matter or materials discussed in the manuscript apart from those disclosed.

References

1. Yuan X, He B, Lv Z, Luo S (2014) Fabrication of self-assembling peptide nanofiber hydrogels for myocardial repair. RSC Adv 4:53801–53811
2. Chen J, Zou X (2019) Self-assemble peptide biomaterials and their biomedical applications. Bioact Mater 4:120–131
3. Seow WY, Salgado G, Lane EB, Hauser CAE (2016) Transparent crosslinked ultrashort peptide hydrogel dressing with high shape-fidelity accelerates healing of full-thickness excision wounds. Sci Rep 6:1–12
4. Carlini AS et al (2019) Enzyme-responsive progelator cyclic peptides for minimally invasive delivery to the heart post-myocardial infarction. Nat Commun 10:1–14
5. Webber MJ, Matson JB, Tamboli VK, Stupp SI (2012) Controlled release of dexamethasone from peptide nanofiber gels to modulate inflammatory response. Biomaterials 33:6823–6832
6. Kapadia MR et al (2008) Nitric oxide and nanotechnology: a novel approach to inhibit neointimal hyperplasia. J Vasc Surg 47:173–182
7. Aldilla VR et al (2020) Anthranilamide-based short peptides self-assembled hydrogels as antibacterial agents. Sci Rep 10:1–12
8. Xu C, Cai Y, Ren C, Gao J, Hao J (2015) Zinc-triggered hydrogelation of self-assembled small molecules to inhibit bacterial growth. Sci Rep 5:1–7
9. Ettmayer P, Amidon GL, Clement B, Testa B (2004) Lessons learned from marketed and investigational prodrugs. J Med Chem 47:2393–2404
10. Rautio J et al (2008) Prodrugs: design and clinical applications. Nat Rev Drug Discov 7:255–270
11. Vemula PK et al (2013) Prodrugs as self-assembled hydrogels: a new paradigm for biomaterials. Curr Opin Biotechnol 24:1174–1182
12. Vemula PK, Cruikshank GA, Karp JM, John G (2009) Self-assembled prodrugs: an enzymatically triggered drug-delivery platform. Biomaterials 30:383–393
13. Bhuniya S, Seo YJ, Kim BH (2006) (S)-(+)-Ibuprofen-based hydrogelators: an approach toward anti-inflammatory drug delivery. Tetrahedron Lett 47:7153–7156
14. Vemula PK, Li J, John G (2006) Enzyme catalysis: tool to make and break amygdalin hydrogelators from renewable resources: a delivery model for hydrophobic drugs. J Am Chem Soc 128:8932–8938
15. Nicolaou KC, Guy RK, Pitsinos EN, Wrasidlo W (1994) A water-soluble prodrug of Taxol with self-assembling properties. Angew Chemie Int Ed English 33:1583–1587
16. Gao Y et al (2009) Enzyme-instructed molecular self-assembly confers nanofibers and a supramolecular hydrogel of taxol derivative. J Am Chem Soc 131:13576–13577
17. Xing B et al (2002) Hydrophobic interaction and hydrogen bonding cooperatively confer a vancomycin hydrogel: a potential candidate for biomaterials. J Am Chem Soc 124:14846–14847
18. Yang Z et al (2007) D-glucosamine-based supramolecular hydrogels to improve wound healing. Chem Commun 8:843–845
19. Gajanayake T et al (2014) A single localized dose of enzyme-responsive hydrogel improves long-term survival of a vascularized composite allograft. Sci Transl Med 6:249ra110

20. Dzhonova DV et al (2018) Local injections of tacrolimus-loaded hydrogel reduce systemic immunosuppression-related toxicity in vascularized composite allotransplantation. Transplantation
21. Dzhonova D et al (2018) Local release of tacrolimus from hydrogel- based drug delivery system is controlled by inflammatory enzymes in vivo and can be monitored non-invasively using in vivo imaging. PLoS One 13:1–16
22. Anton Fries C et al (2019) Graft-implanted, enzyme responsive, tacrolimus-eluting hydrogel enables long-term survival of orthotopic porcine limb vascularized composite allografts: a proof of concept study. PLoS One 14:1–15
23. Joshi N et al (2018) Towards an arthritis flare-responsive drug delivery system. Nat Commun 9
24. Zhang S et al (2015) An inflammation-targeting hydrogel for local drug delivery in inflammatory bowel disease. Sci Transl Med 7:1275
25. Vemula PK et al (2011) On-demand drug delivery from self-assembled nanofibrous gels: a new approach for treatment of proteolytic disease. J Biomed Mater Res–Part A 97(A):103–110
26. Jemal A, Bray F, Ferlay J (2011) Global cancer statistics: 2011. CA Cancer J Clin 61:69–90
27. Yu L, Ding J (2008) Injectable hydrogels as unique biomedical materials. Chem Soc Rev 37:1473–1481
28. Tan H, Chu CR, Payne KA, Marra KG (2009) Injectable in situ forming biodegradable chitosan-hyaluronic acid based hydrogels for cartilage tissue engineering. Biomaterials 30:2499–2506
29. Li Y, Rodrigues J, Tomás H (2012) Injectable and biodegradable hydrogels: gelation, biodegradation and biomedical applications. Chem Soc Rev 41:2193–2221
30. Du X, Zhou J, Shi J, Xu B (2015) Supramolecular Hydrogelators and hydrogels: from soft matter to molecular biomaterials. Chem Rev 115:13165–13307
31. Yang C et al (2013) Disulfide bond reduction-triggered molecular hydrogels of folic acid-Taxol conjugates. Org Biomol Chem 11:6946–6951
32. Wang H et al (2012) The inhibition of tumor growth and metastasis by self-assembled nanofibers of taxol. Biomaterials 33:5848–5853
33. Ren C et al (2014) Gemcitabine induced supramolecular hydrogelations of aldehyde-containing short peptides. RSC Adv 4:34729–34732
34. Wang H et al (2011) Self-assembled nanospheres as a novel delivery system for taxol: a molecular hydrogel with nanosphere morphology. Chem Commun 47:4439–4441
35. Li J et al (2013) Dephosphorylation of d -peptide derivatives to form biofunctional, supramolecular nanofibers/hydrogels and their potential applications for intracellular imaging and intratumoral chemotherapy. J Am Chem Soc 135:9907–9914
36. Wang H et al (2016) Integrating enzymatic self-assembly and mitochondria targeting for selectively killing cancer cells without acquired drug resistance. J Am Chem Soc 138:16046–16055
37. Kenney RM, Boyce MW, Whitman NA, Kromhout BP, Lockett MR (2018) A pH-sensing Optode for mapping spatiotemporal gradients in 3D paper-based cell cultures. Anal Chem 90:2376–2383
38. Daglar B, Ozgur E, Corman ME, Uzun L, Demirel GB (2014) Polymeric nanocarriers for expected nanomedicine: current challenges and future prospects. RSC Adv 4:48639–48659
39. Liang J, Wu WL, Xu XD, Zhuo RX, Zhang XZ (2014) PH responsive micelle self-assembled from a new amphiphilic peptide as anti-tumor drug carrier. Colloids Surfaces B Biointerfaces 114:398–403
40. Yata T et al (2017) DNA nanotechnology-based composite-type gold nanoparticle-immunostimulatory DNA hydrogel for tumor photothermal immunotherapy. Biomaterials 146:136–145
41. Zou Q, Chang R, Xing R, Yuan C, Yan X (2020) Injectable self-assembled bola-dipeptide hydrogels for sustained photodynamic prodrug delivery and enhanced tumor therapy. J Control Release 319:344–351

42. Weiss RG, Terech P (2006) Molecular gels: materials with self-assembled fibrillar networks. Molecular Gels: Materials with Self-Assembled Fibrillar Networks
43. Smith DK (2008) Molecular gels–nanostructured soft materials. Organic Nanostructures
44. Truong WT, Su Y, Meijer JT, Thordarson P, Braet F (2011) Self-assembled gels for biomedical applications. Chem–An Asian J 6:30–42
45. Kumar S, Bajaj A (2020) Advances in self-assembled injectable hydrogels for cancer therapy. Biomater Sci 8:2055–2073
46. Singh M et al (2014) Injectable small molecule hydrogel as a potential nanocarrier for localized and sustained in vivo delivery of doxorubicin. Nanoscale 6:12849–12855
47. Zhang K, Zhou L, Chen F, Chen Y, Luo X (2019) Injectable gel self-assembled by paclitaxel itself for in situ inhibition of tumor growth. J Control Release 315:197–205
48. Zhi K, Wang J, Zhao H, Yang X (2020) Self-assembled small molecule natural product gel for drug delivery: a breakthrough in new application of small molecule natural products. Acta Pharm Sin B 10:913–927
49. Maity M, Maitra U (2017) Supramolecular gels from conjugates of bile acids and amino acids and their applications. European J Org Chem 2017:1713–1720
50. Yang JA, Yeom J, Hwang BW, Hoffman AS, Hahn SK (2014) In situ-forming injectable hydrogels for regenerative medicine. Prog Polym Sci 39:1973–1986
51. Koutsopoulos S (2016) Self-assembling peptide nanofiber hydrogels in tissue engineering and regenerative medicine: Progress, design guidelines, and applications. J Biomed Mater Res–Part A 104:1002–1016
52. Kim JE, Kim SH, Jung Y (2015) In situ chondrogenic differentiation of bone marrow stromal cells in bioactive self-assembled peptide gels. J Biosci Bioeng 120:91–98
53. Nguyen PK et al (2018) Self-assembly of a Dentinogenic peptide hydrogel. ACS Omega 3:5980–5987
54. Hsieh PCH, Davis ME, Gannon J, MacGillivray C, Lee RT (2006) Controlled delivery of PDGF-BB for myocardial protection using injectable self-assembling peptide nanofibers. J Clin Invest 116:237–248
55. Bastings MMC et al (2014) A fast pH-switchable and self-healing supramolecular hydrogel carrier for guided, local catheter injection in the infarcted myocardium. Adv Healthc Mater 3:70–78
56. Elnaggar YSR, Talaat SM, Bahey-El-Din M, Abdallah OY (2016) Novel lecithin-integrated liquid crystalline nanogels for enhanced cutaneous targeting of terconazole: development, in vitro and in vivo studies. Int J Nanomedicine 11:5531–5547
57. Zhao CC et al (2019) Resveratrol-loaded peptide-hydrogels inhibit scar formation in wound healing through suppressing inflammation. Regen Biomater 7:99–107
58. Xing R, Liu Y, Zou Q, Yan X (2019) Self-assembled injectable biomolecular hydrogels towards phototherapy. Nanoscale 11:22182–22195
59. Abbas M, Zou Q, Li S, Yan X (2017) Self-assembled peptide- and protein-based nanomaterials for antitumor photodynamic and Photothermal therapy. Adv Mater 29
60. Zhang Y et al (2018) An injectable dipeptide-fullerene supramolecular hydrogel for photodynamic antibacterial therapy. J Mater Chem B 6:7335–7342
61. Raymond DM et al (2019) Low-molecular-weight supramolecular hydrogels for sustained and localized in vivo drug delivery. ACS Appl. Bio Mater. 2:2116–2124
62. Li Z et al (2016) Self-assembled drug delivery system based on low-molecular-weight bis-amide organogelator: synthesis, properties and in vivo evaluation. Drug Deliv 23:3168–3178
63. Faidra Angelerou MG et al (2020) Mechanistic investigations into the encapsulation and release of small molecules and proteins from a supramolecular nucleoside gel in vitro and in vivo. J Control Release 317:118–129
64. Loo Y et al (2019) A chemically well-defined, self-assembling 3D substrate for long-term culture of human pluripotent stem cells. ACS Appl Bio Mater 2:1406–1412

65. Loo Y et al (2015) Peptide bioink: self-assembling Nanofibrous scaffolds for three-dimensional Organotypic cultures. Nano Lett 15:6919–6925
66. Tong C et al (2018) Squaramide-based supramolecular materials for three-dimensional cell culture of human induced pluripotent stem cells and their derivatives. Biomacromolecules 19:1091–1099
67. Godbe JM et al (2020) Gelator length precisely tunes supramolecular hydrogel stiffness and neuronal phenotype in 3D culture. ACS Biomater Sci Eng 6:1196–1207
68. Alakpa EV et al (2016) Tunable supramolecular hydrogels for selection of lineage-guiding metabolites in stem cell cultures. Chem 1:298–319
69. Alakpa EV et al (2017) Improving cartilage phenotype from differentiated pericytes in tunable peptide hydrogels. Sci Rep 7:1–11
70. Takeuchi T et al (2016) Enhanced healing of surgical periodontal defects in rats following application of a self-assembling peptide nanofibre hydrogel. J Clin Periodontol 43:279–288
71. Li H et al (2018) Folic acid derived hydrogel enhances the survival and promotes therapeutic efficacy of iPS cells for acute myocardial infarction. ACS Appl Mater Interfaces 10:24459–24468
72. Huang A et al (2019) Self-assembled GFFYK peptide hydrogel enhances the therapeutic efficacy of mesenchymal stem cells in a mouse hindlimb ischemia model. Acta Biomater 85:94–105
73. Pal VK, Jain R, Roy S (2020) Tuning the supramolecular structure and function of collagen mimetic ionic complementary peptides via electrostatic interactions. Langmuir 36:1003–1013
74. Ibáñez-Fonseca A et al (2020) Influence of the thermodynamic and kinetic control of self-assembly on the microstructure evolution of silk-elastin-like Recombinamer hydrogels. Small 16
75. Barros D, Amaral IF, Pêgo AP (2020) Laminin-inspired cell-instructive microenvironments for neural stem cells. Biomacromolecules 21:276–293
76. Lash JW, Linask KK, Yamada KM (1987) Synthetic peptides that mimic the adhesive recognition signal of fibronectin: differential effects on cell-cell and cell-substratum adhesion in embryonic chick cells. Dev Biol 123:411–420
77. Teng L, Chen Y, Jia YG, Ren L (2019) Supramolecular and dynamic covalent hydrogel scaffolds: from gelation chemistry to enhanced cell retention and cartilage regeneration. J Mater Chem B 7:6705–6736
78. Gaspar VM, Lavrador P, Borges J, Oliveira MB, Mano JF (2020) Advanced bottom-up engineering of living architectures. Adv Mater 32
79. Nicolas J et al (2020) 3D extracellular matrix mimics: fundamental concepts and role of materials chemistry to influence stem cell fate. Biomacromolecules 21:1968–1994
80. Gavel PK, Kumar N, Parmar HS, Das AK (2020) Evaluation of a peptide-based Coassembled Nanofibrous and thixotropic hydrogel for dermal wound healing. ACS Appl Bio Mater 3:3326–3336
81. Das AK, Gavel PK (2020) Low molecular weight self-assembling peptide-based materials for cell culture, antimicrobial, anti-inflammatory, wound healing, anticancer, drug delivery, bioimaging and 3D bioprinting applications. Soft Matter 16:10065–10095

Chapter 10
Smart Peptide Assembly Architectures to Mimic Biology's Adaptive Properties and Applications

Debasish Haldar

10.1 Introduction

During the last couple of decades, significant progress has been made in developing molecular architectonics and nanoarchitectonics with advanced functionalities [1–3]. There are two approaches for the fabrication of molecular architectonics and nanoarchitectonics, namely, the "top-down approach" and "bottom-up approach. These molecular architectonics may have specific interfaces or structures that may show explicit properties as synergistic effect. The "bottom-up" approach [4], where molecule-by-molecule packing provides novel supramolecular nanoarchitectonics [5, 6], requires a thorough understanding of each molecular building blocks and their physical properties, chemical structures, and crystal packing propensities [7, 8]. The nanoarchitected materials are becoming increasingly popular for fabricating various devices with biology's adaptive properties and applications. The development of molecular architectonics and nanoarchitectonics by correlation of structural information with the molecular recognition process of the chemical constituents and the function of bulk materials is highly interesting [9–11]. The self-assembly of the building blocks occurred through weak and noncovalent interactions such as hydrogen bonding, π-π stacking, ionic interactions, and van der Wall interactions that glued the building blocks [12]. So, rationally designed nanoarchitectonics consisting of different, complementary building blocks can evolve into advance functional materials with unique property combinations [13]. However, to enable cost-effective synthesis and superior performance, further improvement of current technologies is strongly needed.

D. Haldar (✉)
Department of Chemical Sciences, Center for Advance Functional Materials and Center for Climate and Environmental Studies, Indian Institute of Science Education and Research Kolkata, Mohanpur, West Bengal, India
e-mail: deba_h76@iiserkol.ac.in

T. Govindaraju, K. Ariga (eds.), *Molecular Architectonics and Nanoarchitectonics*, Nanostructure Science and Technology,
https://doi.org/10.1007/978-981-16-4189-3_10

10.2 Different Nanoarchitectonics

Different nanostructures including nanofibrils, nanotubes, nanotapes, nanovesicles, and micelles from short designer peptide are very well known as peptides that have great tendency to form hydrogen bonding, hydrophobic interaction, π-π interaction, or electrostatic interaction in molecular levels (Fig. 10.1). The combinations of all these forces make it difficult to understand the behavior of self-assembling peptides. Peptides that have a hydrophilic head and a hydrophobic tail part will likely to self-assemble into spherical micelles or vesicles. By elongation of the micelles or vesicles, it transform into fibers or tubes respectively. Peptides with a tendency to form β-sheet tend to assemble into flat structures such as tapes or ribbons. However, when the concentration of the peptides increases, these tapes and ribbons will stack on each other and form more tightly packed fibers.

10.2.1 Micelles

Spherical micelles can be designed when building blocks with a hydrophilic head and a hydrophobic tail self-assemble into a supramolecular core-shell structure (Fig. 10.1). Depending on various factors like temperature, pH, concentration, and interactions between the peptides, the size and shape of the micelle can vary. The micelle can encapsulate water-soluble and water-insoluble model drugs and thus can be used as a novel and biocompatible drug delivery tool.

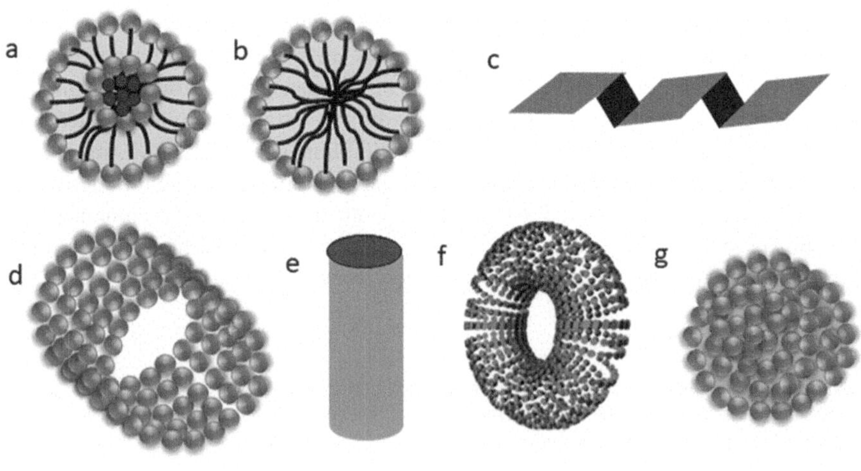

Fig. 10.1 Schematic presentation of (**a**) vesicle, (**b**) micelle, (**c**) ribbons, (**d**) nanotube, (**e**) nanofibers, (**f**) toroid, and (**g**) nanospheres

10.2.2 Vesicles

In vesicle, the hydrophilic part points out to both the outer and inner surfaces. It is not completely clear how the vesicles are being formed exactly from peptides (Fig. 10.1). It is thought that the peptide monomers first create a bilayer by connecting tail to tail and then convert into a spherical or tubelike shape structure due to hydrophobic interactions. This peptide nanovesicle can be used as a new class of molecular scaffolds for organizing metal and inorganic nanocrystals to fabricate devices at the nanometer scale. The self-assembled peptide nanovesicle can be used to encapsulate different molecules like vitamin B12, riboflavin, bioactive peptides, and small protein molecule.

10.2.3 Fibers

Fibers can be considered as elongated micelle (Fig. 10.1). However, fibers are not spherical like micelle; fibers are like cylindrical. So, it is obvious that the peptides which can form micelle have a tendency to form fibers. It was reported that peptide amphiphile self-assembled to nanofibers which is further used for the slow release and delivery of antisense oligonucleotide drug. Peptides that can form β-sheet also have a tendency to form fibers.

10.2.4 Tubes

Tubes can be considered as elongated vesicles (Fig. 10.1). Generally, the same peptides that are capable of forming vesicles can also form tubes. The structure is open-ended and has a hollow inside, and the monomers that are linked together by their hydrophobic tail parts and their hydrophilic head parts are directed outwards. If the tubes are like a ring, it is known as toroid.

10.2.5 Tapes and Ribbons

The amino acid sequence that has a tendency to form β-sheets can also form tape-like structures in higher concentration (Fig. 10.1). These tapes were several hundred nanometer long consisting of β-sheets in twisted shape. It was also known very well that most tapes form a gel-like substance. Ribbons can be developed from two stacked tapes.

10.2.6 Nanospheres

Self-assembled peptide *nanospheres* are the spherical particles which have the size between 10 and 200 nm in diameter (Fig. 10.1). Peptide *nanospheres* are colloidal systems that entrap therapeutic agents (drugs) within a colloidal matrix and are used as drug formulations or delivery vehicles. Depending on the external stimuli, the nanospheres can release the drug sustainably.

10.3 Self-Assembly Amino Acids to Nanoarchitectonics

We have developed eggshell-like nanoarchitectonics as drug delivery systems from methyl 6-amino-2-oxo-2H-thiochromene-3-carboxylate [14]. X-ray crystallography reveals that the ε-amino ester adopts planer structure and self-associates to form four-member-ring-like porous architecture through multiple N-H···O and C-H···S intermolecular hydrogen bonding interactions (Fig. 10.2a). At higher-order packing, the ε-amino ester forms an antiparallel sheetlike structure by intermolecular hydrogen bonding as well as face-to-face π-π stacking interactions (Fig. 10.2b). Field emission scanning electron microscopy (FE-SEM) revealed that the ε-amino ester exhibits eggshell-like porous nanovesicle morphology (Fig. 10.2c). These eggshell-like nanovesicles have been used as potential carrier for the bacteriostatic antibiotic sulfamethoxazole. The spectroscopic studies as well as the growth inhibition of *E. coli* exhibit that the ε-amino ester-sulfamethoxazole formulation leads to pH-responsive sustained release of the drug.

We have also developed different nanoarchitectonics by varying the amino acid side chains exploiting 4-biphenyl carboxylic acid and L-alanine, L-leucine, and L-phenylalanine [15]. The compound containing L-alanine exhibits polydispersed microsphere morphology. From Fig. 10.3a, the diameter of the microspheres is ca. 300 nm. However, for L-leucine-containing compound, several-micrometer-long sheetlike morphology has appeared (Fig. 10.3b). From FE-SEM the compound with L-phenylalanine exhibits nanorod morphology (Fig. 10.3c). The diameter of the nanorod is ca. 400 nm and several micrometers long. Hence, irrespective of N-terminal 4-biphenylcarboxy protection, the α-amino acids exhibit diverse self-assembly and morphology.

The single crystal X-ray diffraction analysis shows that compounds containing alanine form supramolecular parallel sheetlike structure through intermolecular hydrogen bonding interactions and π-π stacking interactions, whereas compound containing leucine forms supramolecular antiparallel sheetlike structure through intermolecular hydrogen bonding interactions. The compound containing phenylalanine self-assembles through intermolecular C=O. . . .HN hydrogen bonds to form tubular assembly. The inner diameter of the tubes is ca 4Å [15]. Irrespective of N-terminal 4-biphenylcarboxy protection, the minor change in the amino acid side chain has significant effect on self-assembly and nanoarchitectonics.

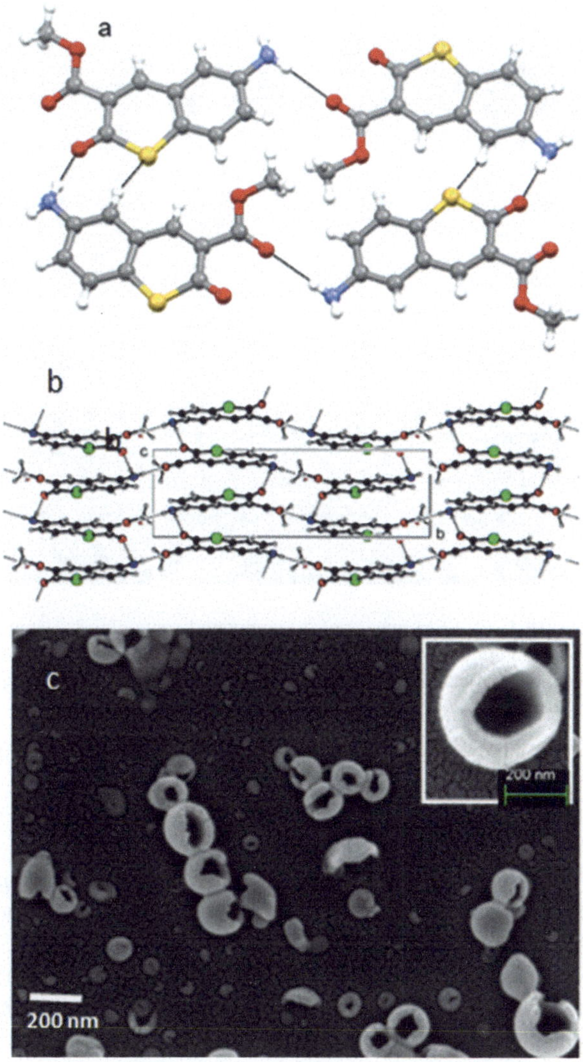

Fig. 10.2 (**a**) Supramolecular four-member-ring-like porous structures. Hydrogen bonds are shown as dotted line. (Carbon atoms are black, nitrogen blue, oxygen red, and sulfur yellow.) (**b**) The supramolecular sheetlike structure (carbon atoms are black, nitrogen blue, oxygen red, and sulfur green). (**c**) The FE-SEM image showing eggshell-like nanovesicle morphology. (Adapted from [14])

We have also synthesized 4-biphenylcarboxy-protected serine and tyrosine, and their crystal packing-induced diverse optical properties in solid state were investigated [16]. The L-serine-containing compound exhibits thin flake-like nanoarchitectonics and green-gold birefringence under polarized light (Fig. 10.4a,b). But the L-containing compound does not show any birefringence by POM (Fig. 10.4c, d). This indicates that not only the chromophoric system but also the crystal packing pattern is highly important for this diverse optical behavior. To explore further, the solid-state UV-Vis spectroscopy and fluorescence spectroscopy were performed. The compound containing L-serine crystals show absorbance

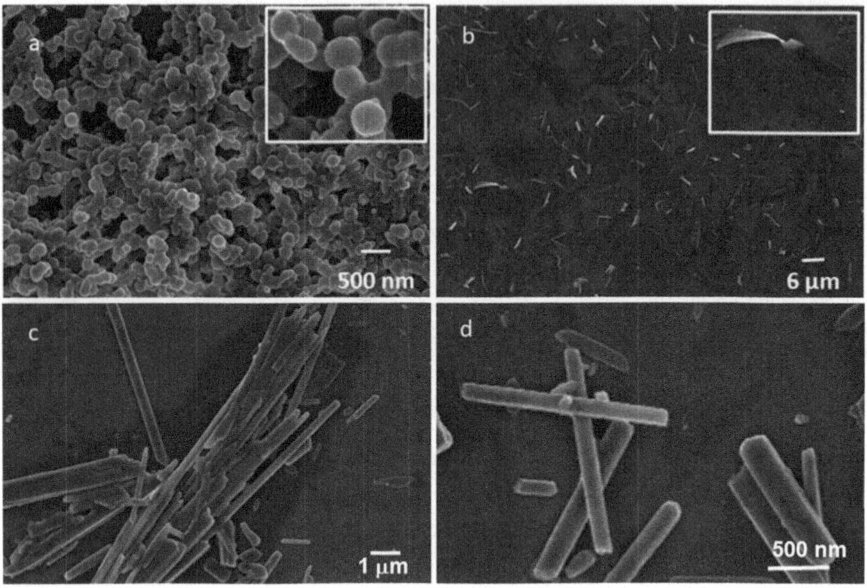

Fig. 10.3 FE-SEM images of (**a**) compound containing L-alanine showing polydispersed microsphere morphology, (**b**) compound containing L-leucine showing sheetlike morphology, (**c**) and (**d**) compound containing L-phenylalanine showing nanorod-like morphology. (Adapted from [15])

peak at 355 nm and band around 500–700 nm (Fig. 10.4e) and emission band at 525 nm on excitation at 270 nm (Fig. 10.4f). However, the crystals of L-tyrosine-containing compound do not show any absorbance peak at 355 nm and band around 500–700 nm (Fig. 10.4a) and emission at 525 nm on excitation at 270 nm (Fig. 10.4b). This indicates that irrespective of the presence of same biphenyl chromophore, the fluorescence properties of the compounds depend on the crystal packing pattern of the molecules.

The crystal structure shows that irrespective of the π-π stacking interactions between biphenyl moieties, the hydrogen bond between amide C=O and L-serine side chain HO directs the supramolecular assembly to a parallel sheetlike structure. But, the L-tyrosine-containing compound forms a supramolecular strand through intermolecular hydrogen bonds between ester C=O and L-tyrosine side chain HO. Though the biphenyl moieties are in a parallel arrangement in the strand, there are no face-to-face π-π stacking interactions. In packing diagram, two supramolecular strands are perpendicular to each other. There is only a single-point interaction between two supramolecular strands [16].

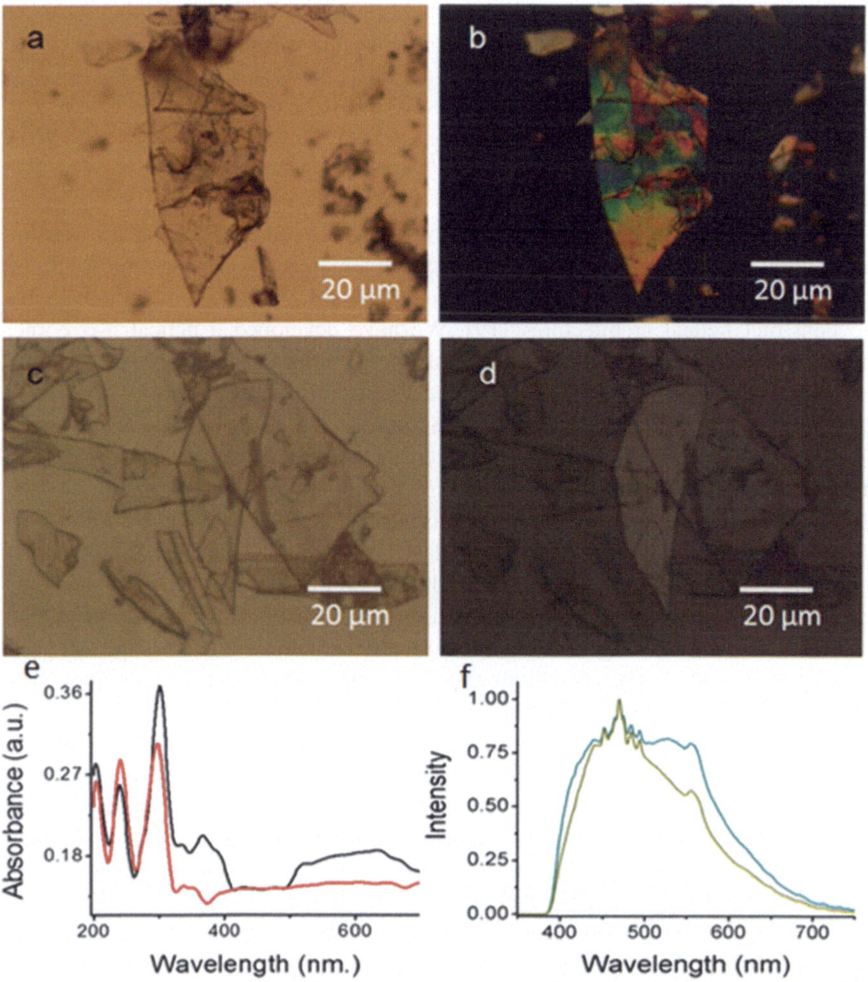

Fig. 10.4 POM images of compound containing L-serine in normal light (**a**) and under polarized light (**b**) POM images of compound containing L-tyrosine in normal light (**c**) and under polarized light (**d**) Solid-state (**e**) UV-Vis spectra of compounds containing L-serine (black) and containing L-tyrosine (red) and (**f**) emission spectra of compounds containing L-serine (blue) and compounds containing L-tyrosine (green). Excitation at 270 nm. (Adapted from [16])

10.4 Peptide Self-Assembly to Nanoarchitectonics

10.4.1 Supramolecular Helices

Supramolecular helices Nanoarchitectonics: peptide self-assembly from different small peptide backbone through self-organization are very common in biological system. Previously a number of different approaches have been pursued to create various types of supramolecular helical structures from peptides containing coded or noncoded amino acids [17]. Three types of helices have been reported.

10.4.2 Single-Stranded Supramolecular Helix

There are remarkable progresses in the de novo design of helical architectures using the natural α-amino acids [18]. Even, noncoded amino acids like α-aminoisobutyric acid, α,β-dehydropheylalanine, β-amino acids, etc. were also used in peptide chains to construct helical architectures [19]. We have developed a helical α-α corner mimetic that self-assembled to form rodlike supramolecular structure which bends and closes end to end like a cyclization reaction to form uniform toroids [20]. Each peptide fragment containing L-leucine, α-aminoisobutyric acid (Aib), and L-tyrosine forms rigid 3_{10} helical structures stabilized by multiple intramolecular N-H. . . .O hydrogen bonds. Two 3_{10} helices are connected by the spacer 3-aminomethyl-benzylamine and maintain an angular distance of 120° and therefore mimic the α-α corner motif of protein supersecondary structure. The individual α-α corner subunits are themselves regularly interlinked through multiple water-mediated intermolecular hydrogen-bonding interactions to form the rodlike supramolecular structure and toroids (Fig. 10.5a). The formation of supramolecular structure has been proven with X-ray crystallography and other spectroscopic techniques. The cyclization of supramolecular structure and toroid formation were studied by optical microscope, afm and fe-sem experiments (Fig. 10.5c,d). Despite other assignment such as exfoliation of graphene from graphite, the compound exhibits significant memory to produce the toroids finally [20]. The water-mediated supramolecular helix has also been reported by the same group where water molecules interlinked two peptide units through intermolecular H-bonding to form single-stranded helical structure [21]. Li and his coworker reported two bolaamphiphiles based on D- or L-glutamic acid which formed supramolecular helical aggregates [22]. Parthasarathy and his coworkers have also presented a series of tripeptides that produce extended helical structures with intervening water molecules between two consecutive peptide molecules [23]. Single-stranded helical structures from several peptides were demonstrated by Balaram and his coworkers [24]. Recently, Balaram's group has reported an 18-membered synthetic peptide that forms single-stranded helical architecture [25].

10.4.3 Double-Stranded Supramolecular Helix

A wide range of double helical structures and functions from relatively a small number of building blocks such as 4 nucleobases and 20 amino acids are abundant in nature [26]. DNA, the hereditary material in humans and almost all other organisms, have double helical structure. The designing of oligomers that mimic double helical

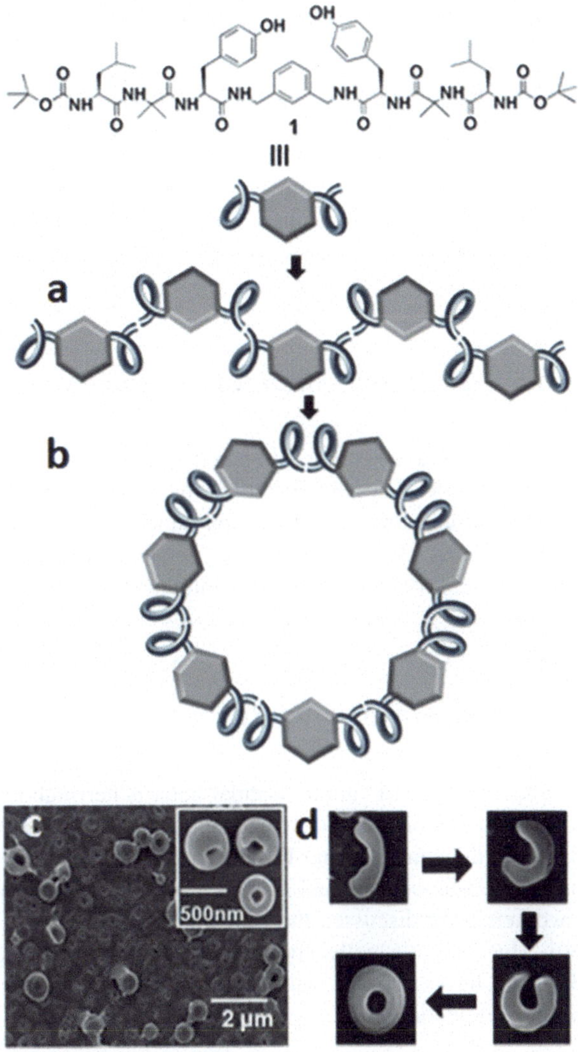

Fig. 10.5 The peptide mimetic and the schematic presentation of (**a**) rodlike supramolecular structure from peptidomimetic building blocks; (**b**) ring-like supramolecular polymer. (**c**) The FE-SEM images exhibiting toroid structure. Inset: the pierced holes of toroids. (**d**) The mechanism of toroid formation by self-assembly of peptide mimetic via rodlike structure and croissant-like structure. (Adapted from [20])

structures like DNA is very interesting due to their structural versatility, biocompatibility, robustness, and a relative experimental simplicity. Many oligomers have been reported to construct double helical structures in literature [27]. But double helix from small peptides is very rare. Benedetti et al. have demonstrated double-stranded β-helix from a D,L-octapeptide from its crystal structure in 1979 [28]. Görbitz has reported left-handed double helix from hydrophobic dipeptides [29]. The formation of a double helix from dipeptide containing N-terminal β-amino acids was reported by Banerjee and his coworkers [30]. We have also reported the construction of double helix from terminally protected tripeptides [31] and also from terminally protected capped γ-peptides [32].

10.4.4 Triple-Stranded Supramolecular Helix

Triple helical structures are very common in collagen and DNA. Different approaches have been made to mimic the triple helical structures including metal-based collagen-forming building units, chiral diamine/diolin literature. But the formation of triple helix from small peptide units is very rare in literature. In 2005, Banerjee and his coworkers have reported the formation of collagen-like triple-helix structure from a water-soluble, tetrapeptide from its crystal structure [33].

10.4.5 Quadruple-Stranded Supramolecular Helix

Jiang, Huc, and coworkers prepared quinoxaline carboxamide-based 4-mer and 8-mer foldamers. Crystal structure of the 4-mer revealed that two molecules stacked in a head-to-tail dimer and two such dimers were further self-assembled to form a quadruple helix [34]. In contrast, the octameric oligoamide formed only the double helix, and no further self-assembly into quadruple helix could be attained.

10.4.6 Herringbone Helix

Another rare type of supramolecular helix is herringbone helix. The monomers are here self-assembled to form herringbone gear-like architecture. Lin et al. have discussed the construction of 2D herringbone-like supramolecular helical architecture from zinc-coordinated complex from a helical motif [35]. Meng and his coworkers have discussed the formation of herringbone helix from metal complex [36]. Huc and his colleagues have reported herringbone helix from foldamers [37]. We have also constructed herringbone-like supramolecular structure from synthetic dipeptide [38].

10.4.6.1 Supramolecular β-Sheets

The constructions of supramolecular β-sheet from different short peptides are very important for their potential application in bioscience and material science. Enormous efforts were given since the last 30 years to design and synthesize peptide models to construct β-sheet structures having special importance in science.

10.4.6.2 β-Sheet from Cyclic Peptide Foldamers

Self-assembly of cyclic peptides into supramolecular β-sheet structures has been widely studied and very interesting due to their implications in nanotechnology, mainly in the construction of peptide nanotubes. Ghadiri et al. have designed and synthesized open-ended nanotubes by ring staking of cyclic D,L-peptides as subunits which on further self-assembly gives supramolecular β-sheet structures [39]. The same group also has also demonstrated the formation of cylinders from some cyclic peptides through supramolecular β-sheet structures. Granja et al. have reported the preparation of self-assembling peptide tubelets composed of 32-membered rings with alternating α-amino acids and cis-3-aminocyclohexanecarboxylic acids [40]. These molecules are self-assembled through supramolecular β-sheet structures to form tubelets. Nanotubes are also reported from enantiomeric pair of cyclic peptides containing alternative D, L-amino acids and supramolecular β-sheet forming cyclic peptides by alternating α- and β-amino acids, all β-amino acids, and alternating D- and L-α-amino acids. Saha and coworkers reported vesicle formation of a cyclic peptide by sheetlike aggregation which can be transformed to nanotube by addition of H^+ ion [41].

10.4.6.3 β-Sheet from Acyclic Peptide Foldamers

Construction of supramolecular β-sheet structures from design acyclic peptides is very difficult due to their high propensity to form high molecular weight β-sheet quaternary structure. Different approaches have been made to construct supramolecular β-sheet structure by incorporating different rigid moieties into peptide backbone is convensitional. Kemp and his coworkers have designed and constructed an antiparallel β-sheet model from peptide containing rigid epindolidone [42].

Banerjee and coworkers have also established that a more rigid N-substituted 3-amino phenyl acetic acid residue and α-aminoisobutyric acid residue containing pseudopeptide adopt almost an extended backbone conformation which self-assembles to form an antiparallel β-sheet structure in solid state in an another issue [43]. Banerjee and coworkers have also inserted conformationally constrained α-aminoisobutyric acid (Aib) into peptide backbone to form antiparallel β-sheet columns through intermolecular hydrogen bonds and, on further self-assembly of the individual columns stacked via van der Waals interactions, form a highly ordered supramolecular cross β-sheet structure. Recently we have reported a dipeptide containing a *p*-nitrophenylalanine moiety which adopts supramolecular antiparallel β-sheet structure using hydrogen bonding, as well as π-π stacking interactions, in the solid state, whereas other dipeptide containing a *p*-nitrophenylacetic acid moiety adopts a parallel β-sheet structure [44].

The insertion of ω-amino acids like β-alanine, δ-aminovaleric acid, γ-aminobutyric, acid and ε-aminocaproic acid may help to provide extended nature to the peptide backbone. Pramanik and his coworkers have reported the formation of β-sheet structure from model dipeptides using β-alanine and γ-aminobutyric acid as flexible part. We also reported antiparallel β-sheet formation by a dipeptide containing 2-(aminomethyl)benzoic acid and α-aminoisobutyric acid (Aib) [45]. A short peptide containing β-alanine can also adopt a β-turn conformation and form higher-order β-sheet ladder structure through intermolecular hydrogen bonds in solid state.

10.4.7 Factors on Self-Assembly of Folded Peptides

Understanding the mechanism and control self-assembly of peptide is very important to regulate the formation of nano- or microstructures, formulate peptide-drug complexes, and further control the complex size and shape for different applications. There are different internal and external factors that influence the peptide self-assembly. Peptide sequence, concentration, pH of the solution, ionic strength, solvent polarity, mechanical force, presence of denaturizing agents, metal coordination, sonication, and surface and its properties are some factors that control the self-assembly of peptide.

10.4.8 Effect of Amino Acid Sequence

The peptide sequence plays a very important role on self-assembly of peptides. The type, number, and arrangement of amino acid determine the secondary structure as well as the nanostructure of peptide. Hai Xu et al. have reported a series of cationic peptide surfactants A*m*K ($m = 3$, 6, and 9), which show different types of self-assembled nanostructures. They have demonstrated that membrane sheets are formed from A_3K, long nanofibers (wormlike micelles) are formed from A_6K, and short nanorods are formed from A_9K [46]. [f]The difference in charge distribution due to the arrangement of amino acids can change peptide self-assembly pattern. Hong and his coworkers reported that EAK16-I and EAK16-II form fibril-like nanostructures, where EAK16-IV forms globular aggregates at neutral pH [46]. Gazit and coworkers have reported the transition of nanotube to hollow nanocage structure by simple introduction of a thiol group at the terminal position of the dipeptide [47]. We have prepared a series of tyrosine-modified analogues of diphenylalanine and found inhibition of fibril formation [48].

10.4.9 Effect of Concentration

The concentration also controls the self-assembling behavior of peptide moiety. The secondary structures of peptide and hence nanostructure differ depending on concentration. Li and his coworkers reported that dipeptide NH_3-Phe-Phe-$CONH_2$

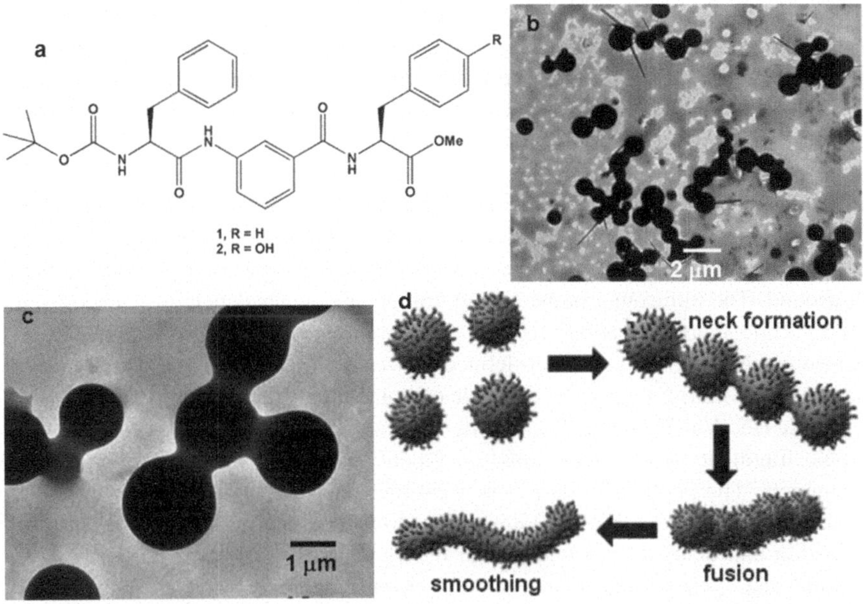

Fig. 10.6 (**a**) Chemical structure of tripeptides, (**b**) and (**c**) TEM image showing the transition from sphere-to-rod-like microvesicles. (**d**) Schematic representation of the possible mechanisms leading to sphere-to-rod transition of microvesicles. (Adapted from [50])

showed tubular nanostructure at a concentration of 8 mg/mL, while at a lower concentration (1 mg/mL), the peptide self-assembled to form vesicular nanostructure [49]. The EAK16-I can change its nanostructure from globular structure to well-defined fibrillar structure on increasing concentration. We have reported the construction of hollow peptide microspheres and the mechanism of transition of microspheres to rodlike vesicles at low concentration [50] (Fig. 10.6). The tripeptides Boc-Phe-Maba-Phe-OMe and Boc-Phe-Maba-Tyr-OMe, each of them containing a rigid m-aminobenzoic acid (Maba) template at the central position, form microspheres at a concentration of 1.6 mM in methanol. At low concentration, these vesicular structures are fused through neck formation, and this leads to sphere-to-rod transition of vesicles. Sizes of these microspheres increase with increasing concentration. We have successfully characterized this transition by fluorescence spectroscopy, DLS, and electron microscopic study. The scanning electron microscopy clearly shows that these are hollow spheres.

10.4.10 Effect of Sonication

Sonication energy is another external stimuli that regulate the self-assembly behavior of peptides. Ultrasound can produce morphologies markedly different to those encountered under silent conditions, thereby representing a valuable route to synthesis new materials. Wang et al. reported that N-Fmoc-octylglycine produces a precipitate consisting of unbranched nanowires on heating and subsequent cooling. Use of ultrasound as stimulus switches to intermolecular hydrogen bonds that facilitate the formation of interconnected nanofibers and hence gelation [51]. A H-bonding gelator, N-lauryl-L-glutamic acid di-n-butylamide, which forms separate spherulites in solution, can be transformed into a 3D fibrillar network on sonication. The Fmoc-protected dipeptide produces precipitates from hot homogeneous alkane solutions on cooling. But it forms well-defined fiber network by application of ultrasound. The ultrasound induces morphological changes depending on exposure times. While short sonication (10 s) favors formation of sheetlike particles, longer exposures (20–60 s) lead to a well-defined network of fibers. We have also discussed the change of morphology of a tripeptide on sonication [52]. A terminally protected tripeptide Boc-Val-Phe-Phe-OMe having sequence similarity with the central hydrophobic fragment 18–20 of the amyloid β-peptide $A\beta^{42}$ forms sonication-induced organogels. The peptide forms a very weak gel only in 1,2-dichlorobenzene after heating, cooling, and aging for 3 days. But ultrasound energy induces instant fibril formation and gelation in a wide range of organic solvents starting from hexane, cyclohexane, petrol, and kerosene to benzene, toluene, xylene, and ethanol (Fig. 10.7). CD, FT-IR, NMR, and wide-angle X-ray scattering (WAXS) studies of the peptide exhibit distinct structural changes before and after sonication. Field emission scanning electron microscopy (FE-SEM) of the xerogels reveals a nanofibrillar morphology, which is obtained by the sonication-induced self-assembly of the gelator (Fig. 10.7).

Recently Ulijn and his coworkers demonstrated that ultrasound energy may also be used to achieve transient reorganization of supramolecular nanostructures, which revert back to the original state when sound is switched off [53]. They also demonstrated an in situ ultrasonic approach to influence self-assembly across the supramolecular to micron length scales, showing enhancement of supramolecular interactions, chirality, and orientation, which depends on the peptide sequence and solvent environment.

10.4.11 Effect of Spacer

Chemists and material scientists are interested in complex biological systems and morphologies for fabrication of new materials. Bioinspired design of short peptide conjugates produces various self-assembled structures, depending on their structural diversity. Spacer plays a key role in controlling the morphology of chemical compounds. Verma and his coworkers reported Phe-Phe aromatic dipeptide which is an interesting candidate for designing a variety of molecular scaffolds. They described the synthesis of two symmetrical peptide conjugates by coupling with C2 and C3 symmetric functional linkers, which spontaneously generate different self-assembled ultra-structures [54].

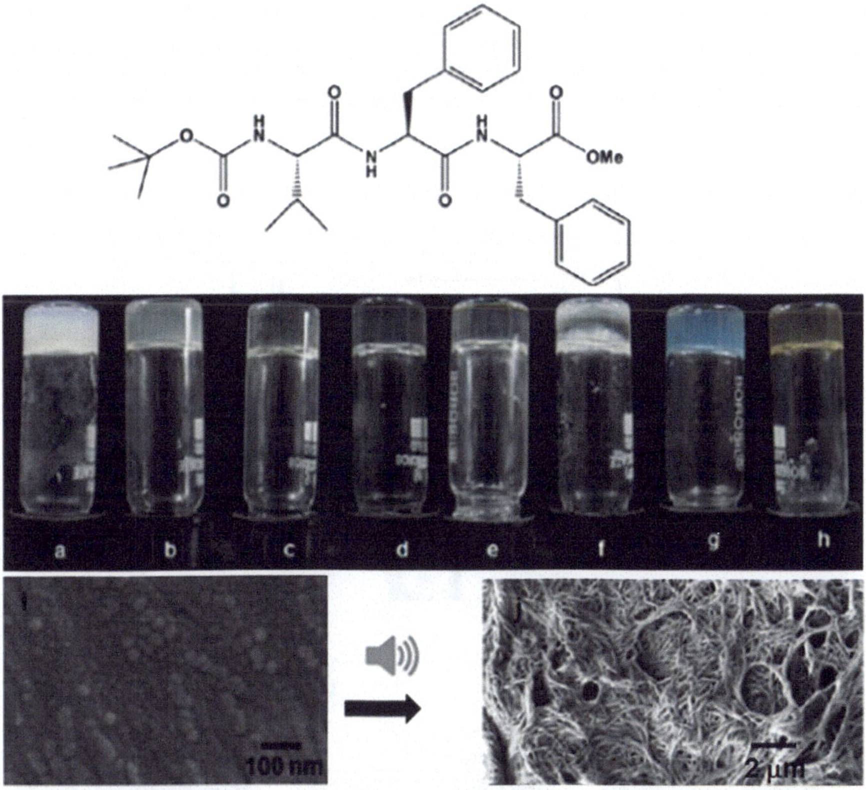

Fig. 10.7 Chemical structures of tripeptide Boc-Val-Phe-Phe-OMe. Gels are formed after being subjected to sonication in (**a**) hexane, (**b**) benzene, (**c**) *o*-xylene, (**d**) *m*-xylene, (**e**) *p*-xylene, (**f**) toluene, (**g**) kerosene, and (**h**) petrol. (**i**) FE-SEM images of nanospheres from as-synthesized peptide. (**j**) FE-SEM images of entangled networks of peptide nanofibers from toluene after 40 s sonication. (Adapted from [52])

10.5 Effect of pH

In biological systems, the solution pH is an important environmental factor affecting protein and peptide structures. The pH change will influence the ionic state of the charged residues as well as the net charge of peptides and proteins. The globular nanostructures occur at pH between 6.5 and 7.5, while nanofiber networks appear at pH beyond this range. Menzenski et al. also reported a series of new peptide bolaamphiphiles containing bis(N-a-amino-phenylalanine) dicarboxylate moieties which transform their nanostructure on changing the pH of the medium [55]. Recently, we have reported acid-responsive assembly of benzoyloxycarbonyl-L-phenylalanine to fibrils as well as urease-assisted disassembly of the said fibrils [56]. Under acidic pH, the compound formed O-H O=C hydrogen-bonded dimer. The dimers are further self-assembled by intermolecular N-H O=C hydrogen bonds and π-π stacking interactions to form fibrils with high mechanical properties, from such a simple molecule. However, the self-assembly process is dynamic. Hence, the in situ generated NH_3 increased the pH uniformly and led to homogeneous disassembly of the fibrils (Fig. 10.8). Schmuck and his group have reported pH-controlled formation of a stable β-sheet and amyloid-like fibers from an

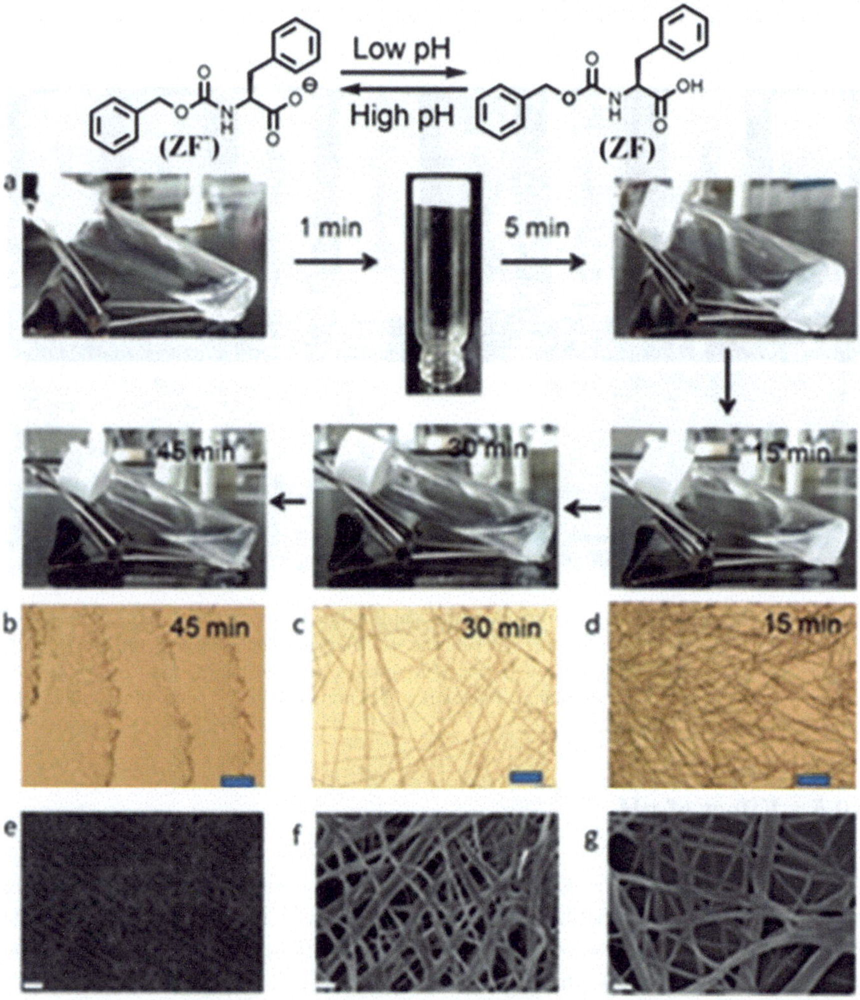

Fig. 10.8 Fig. 10.4 (**a**) Transient gel formation by benzoyloxycarbonyl-L-phenylalanine [**ZF**] in the presence of urease. [**ZF**] 2 mg/ mL, [urease] 0.6 mg/mL, [urea] 0.06 M. (**b**) POM image after 45 min of urease activity showing absence of fibers. (**c**) POM image after 30 min of urease activity showing decrease of fiber density. (**d**) POM image after 15 min of urease activity showing dense fiber network. (**e**) FE-SEM image of sol showing absence of fiber or aggregated structure. (**f**) FE-SEM image of gel by ZF in the presence of urease. [**ZF**] 2 mg/ mL, [urease] 0.6 mg/mL, [urea] 0.06 M. (**g**) Fe-SEM image of pH-responsive [**ZF**] gel. (Adapted from [56])

amphiphilic peptide composed of guanidiniocarbonyl pyrrole (GCP) groups [57]. Matsui and his coworker reported *N,N*-Di-*p*-benzoic acid dodecane diacid diamide self-assemble to form microspheres at pH 8 and nanotubes at pH 7 in citric acid/sodium hydroxide solutions [58].

10.6 Effect of Solvent

Peptide amphiphiles are known to form a variety of distinctive self-assembled nanostructures depending upon the solvent conditions also. Since the discovery of dipeptide self-assembly, Phe-Phe (FF)-based dipeptides have been widely investigated in a variety of fields. Kim and coworkers reported the controlled self-assembly of diphenylalanine to form nanotubes and nanowires in the aqueous phase [59]; however, FF formed organogels in non-hydrogen bonding solvent like chloroform or toluene. The effect of solvent in peptide self-assembly is well manifested by V. Harmandaris et al. who have shown that by manipulating different solvent compositions and different ratios of solvents, dipeptide Val-Ala can undergo different kinds of self-assembled architectures [60]. Recently our group also reported solvent-dependent appearances on nanoarchitectonics (Fig. 10.9) [61]. The hexagonal polydispersed micro-tubes of Boc-protected diphenylalanine (FF) obtained from methanol solution and micro-rods of FF appeared from aqueous solution (Fig. 10.9). Moreover, through the laser-assisted self-assembly of diphenylalanine, we obtained stable annular ring microstructures which exhibit waveguiding properties. In addition, we also obtain self-assembled micro-rod structures in the absence of external stimuli. Interestingly, the self-assembled structures display prominent spectrally asymmetric Fano resonances in the waveguiding response to broadband light (Fig. 10.9). We also demonstrate the potential of such microstructures as sensors from the changes in their Fano resonance profiles and waveguiding response on exposure to Congo red dye.

10.7 Effect of Other Stimulus

Other stimuli like temperature, metal coordination in peptide backbone, and salt effect sometimes may play crucial role for the formation of various types of self-assembled micro- and nanostructure of peptides. Temperature is a powerful factor in varying hydrogen-bonding and hydrophobic interactions. Elevation of temperatures sometimes may cause weakening or breaking of intermolecular hydrogen bonds; however, it increases the hydrophobic effect. Salt ions may decrease in the hydrogen bonding of peptide and solvents by capturing the solvent molecules through solvation. Moreover, salt effects, including pH, type of salt ion, and ionic strength of the solution, also have a great influence on electrostatic interactions. Kim and Ihee

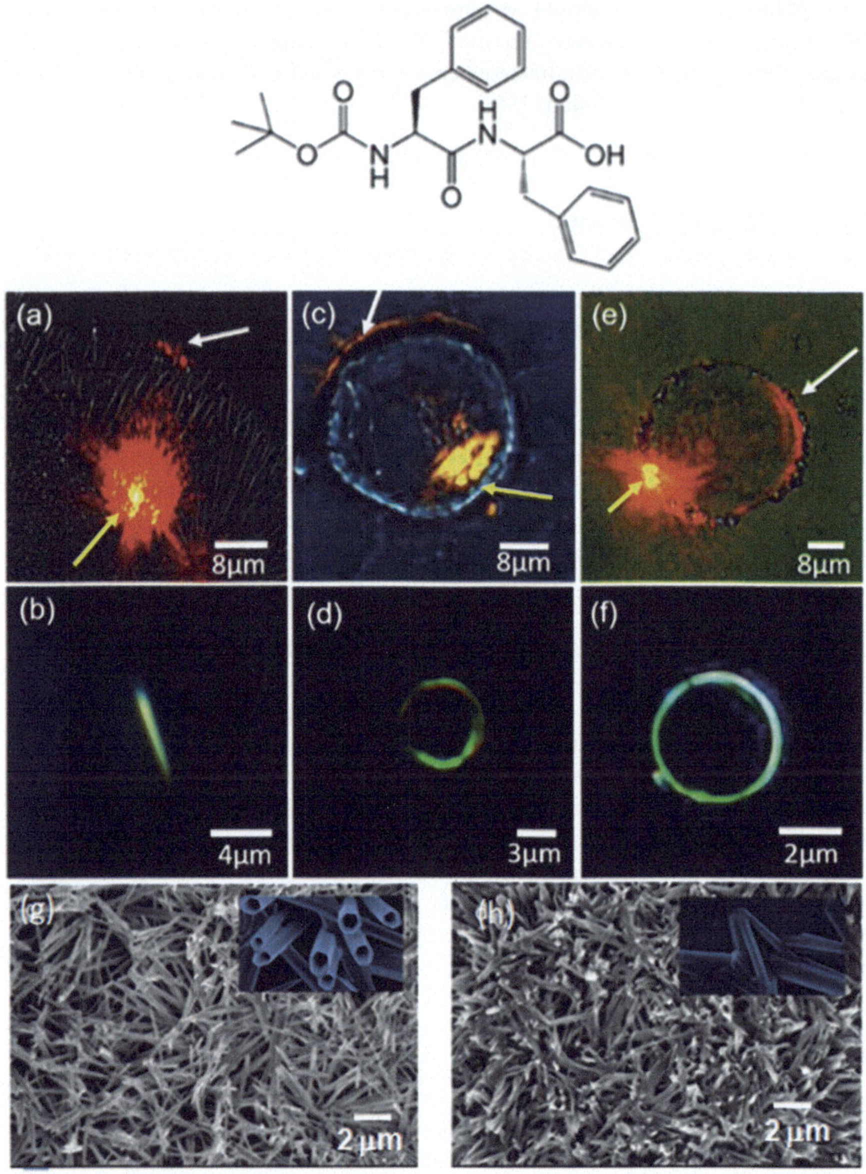

Fig. 10.9 (**a**) Bright field image of a radial cluster of FF rods acting as waveguides when excited by a 671 nm laser; (**b**) a single FF rod waveguide for broadband excitation in a dark field configuration; (c) micro-bubble-based FF ring waveguide for 671 nm laser (bright field) and (**d**) white light (dark field), respectively; (**e**) FF ring formed due to melting and subsequent self-assembly, acting as a waveguide for 671 nm laser (bright field) and (**f**) white light (dark field), respectively. The dark field images do not represent the true color of light that is waveguided since there exists a filter in front of the camera attached to the microscope. (**g**) The hexagonal polydispersed micro-tubes of FF obtained from methanol solution and (**h**) micro-rods of FF as appeared from aqueous solution. (Adapted from [61])

showed the effect of various kinetic factors on thermodynamic interactions in their work [59]. They showed that hydrogen bonding between water and diphenylalanine (FF) molecules played a role in the assembly and induced a structural transition between nanowire and nanotube by changing the concentrations, temperature, ions, and sonication. One can see that all these factors mainly affect the hydrogen bonding, e.g., temperature and sonication decrease hydrogen bonding, concentration adjusts the ratio of hydrogen bonding between each molecule and solvent molecules, and ions can reduce the free solvent and can weaken the hydrogen bonding between peptides and solvent molecules. Thus, the compromise and balance of hydrogen bonds influenced by these kinetic factors enable structural regulation of different self-assembled nanostructures.

Self-assembled peptide-based smart materials can efficiently encapsulate and release various types of drugs in controlled fashion depending upon the environment and stimuli applied on it. Ulijn and coworkers have described the enzyme-responsive on-demand release of proteins by using charged peptide-based actuators which could be matched to the protein payload [62]. They also reported the use of a polypeptide containing an albumin-binding domain linked to a potential and diabetic peptide for sustainable drug release through sequential cleavage by native plasma proteases. Recently Pramanik and coworkers have reported the design, synthesis, and self-assembly of a leucine-enkephalin analogue in different nanostructures and used the nanovesicles as potassium ion-responsive drug delivery vehicles [63]. Pramanik and coworkers also reported the solvent-assisted self-assembly of two modified tripeptides (Boc-Leu-Aib-Val-Maba-OMe and Boc-Phe-Aib-Leu-Maba-OMe) into multilayer vesicles, tubes, various porous structures, and organogels and the application of the multilayer vesicles as a potential drug (curcumin) delivery vehicle [63]. Our group has reported the formation of mesoporous vesicles from supramolecular helical pentapeptide Boc-Leu-Aib-Phe-Phe-Aib-OMe which are capable of encapsulating and releasing the antibiotic drug sulfamethoxazole under acidic pH (Fig. 10.10) [64]. The peptide forms a supramolecular helical column through intra- and intermolecular hydrogen bonding interactions and an interdigitated helical bundle structure in the solid state. In methanol solution, the peptide forms mesoporous vesicles, where the diameters of the vesicles vary with the concentration in direct proportion. The most important property of these mesoporous vesicular structures is the encapsulation of a potent bacteriostatic antibiotic, sulfamethoxazole. Moreover, at pH 6.2, the drug-loaded vesicles can effectively release the encapsulated drug slowly [64].

10.8 Conclusion

In conclusion, this chapter has discussed the field of fabrication of molecular architectonics and nanoarchitectonics by self-assembly of amino acids and peptides. It is clear that there is significant interest and activity in the development of

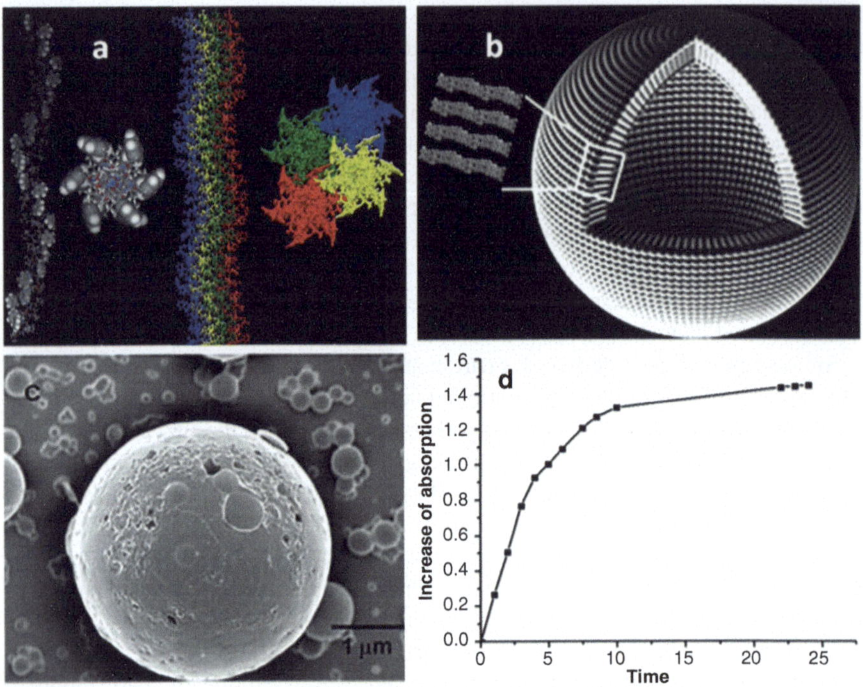

Fig. 10.10 (**a**) Helical bundle like packing of the pentapeptide in solid state, (**b**) schematic presentation of microvesicle formation from supramolecular helical bundles, (**c**) FE-SEM images of mesoporous vesicles, (**d**) Sulfamethoxazole release profile from peptide vesicles at pH 6.2. (Adapted from [64])

molecular architectonics and nanoarchitectonics by environment-friendly manner. In particular, the key advantages of self-assembly of amino acids and peptides as building blocks for the synthesis of molecular architectonics and nanoarchitectonics with high biocompatibility and biodegradability mean that these nanomaterials have potential to solve some pressing environmental issues. The noncovalent interactions including ionic, hydrogen bonding, hydrophobic, π-π stacking, and van der Waals interactions between the peptide building blocks are very important to determine the thermodynamic stable structure and self-assembly pattern. The compounds used in this discussion are new amino acids and peptides containing coded or noncoded amino acids. A number of examples of high-impact research have been presented in this article. However, the rational design of molecular architectonics and nanoarchitectonics capable of bioactivities is still challenging. The concepts, strategies, and examples described in this chapter will be useful in the future development of molecular architectonics and nanoarchitectonics with environmental remediation.

References

1. Shimizu T, Ding W, Kameta N (2020) Soft-matter nanotubes: a platform for diverse functions and applications. Chem Rev 120:2347
2. Ariga K, Ji Q, Mori T, Naito M, Yamauchi Y, Abe H, Hill JP (2013) Enzyme nanoarchitectonics: organization and device application. Chem Soc Rev 42:6322
3. Giussi JM, Cortez ML, Marmisolle WA, Azzaroni O (2019) Practical use of polymer brushes in sustainable energy applications: interfacial nanoarchitectonics for high-efficiency devices. Chem Soc Rev 48:814
4. Cai J, Ruffieux P, Jaafar R, Bieri M, Braun T, Blankenburg S, Muoth M, Seitsonen AP, Saleh M, Feng X, Müllen K, Fasel R (2010) Atomically precise bottom-up fabrication of graphene nanoribbons. Nature 466:470
5. Yang L, Tan X, Wang Z, Zhang X (2015) Supramolecular polymers: historical development, preparation, characterization, and functions, Chem Rev 115:7196
6. Würthner F, Saha-Möller CR, Fimmel B, Ogi S, Leowanawat P, Schmidt D (2016) Perylene bisimide dye assemblies as archetype functional supramolecular materials. Chem Rev 116:962
7. Du X, Zhou J, Shi J, Xu B (2015) Supramolecular hydrogelators and hydrogels: from soft matter to molecular biomaterials, Chem Rev 115:13165
8. Krieg E, Bastings MMC, Besenius P, Rybtchinski B (2016) Supramolecular polymers in aqueous media. Chem Rev 116:2414
9. Canrinus TR, Cerpentier FJR, Feringa BL, Browne WR (2017) Remarkable solvent isotope dependence on gelation strength in low molecular weight hydro-gelators. Chem Commun 53:1719
10. Gunnlaugsson T (2016) Accessible self-assembly, Nat Chem 8:6
11. Martínez-Calvo M, Kotova O, Möbius ME, Bell AP, McCabe T, Boland JJ, Gunnlaugsson T (2015) Healable luminescent self-assembly supramolecular metallogels possessing lanthanide (eu/tb) dependent rheological and morphological properties. J Am Chem Soc 137:1983
12. Hect S, Huc I (2007) Foldamers: structure, properties and applications. Wiley-VCH, Weinheim
13. Shimizu T, Masuda M, Minamikawa H (2005) Supramolecular nanotube architectures based on amphiphilic molecules. Chem Rev 105:1401
14. Debnath M, Sasmal S, Haldar D (2017) Fabrication of egg shell-like nanovesicles from a thiocoumarin-based ε-amino ester: a potential carrier. J Mat Chem B 5:5450
15. Sasmal S, Nandi SK, Kumar S, Haldar D (2019) Atomic-level insight of self-assembled nanorods from 4-biphenylcarboxy protected l-phenylalanine. Chemistry Select 4:11172
16. Sasmal S, Podder D, Debnath M, Nandi SK, Haldar D (2019) Assembly-induced diverse optical property of 4-biphenylcarboxy-protected serine and tyrosine. Chemistry Select 4:10302
17. Seebach D, Gardiner J (2008) β-Peptidic peptidomimetics. Acc Chem Res 41:1366
18. Horne WS, Price JL, Keck JL, Gellman SH (2007) Helix bundle quaternary structure from α/β-peptide foldamers. J Am Chem Soc 129:4178
19. Haldar D, Banerjee A, Drew MGB, Das AK, Banerjee A (2003) First crystallographic signature of an acyclic peptidenanorod: molecular mechanism of nanorod formation by a self-assembled tetrapeptide. Chem Commun:1406
20. Podder D, Bera S, Debnath M, Das T, Haldar D (2017) Formation of toroids by self-assembly of an α–α corner mimetic: supramolecular cyclization. J Mater Chem B 5:7583
21. Guha S, Drew MGB, Banerjee A (2006) Formation of a one-dimensional helical alignment of water molecules within a water-mediated supramolecular helix using molecular self-assembly of a water-soluble short pseudopeptide. Tetrahedron Lett 47:7951
22. Wang H, Zhu W, Li J, Tian T, Lan Y, Gao N, Wang C, Zhang M, Faul CFJ, Li G (2015) Helically structured metal–organic frameworks fabricated by using supramolecular assemblies as templates. Chem Sci 6:1910
23. Parthasarathy R, Chaturvedi S (1990) Design of crystalline helices of short oligopeptides as a possible model for nucleation of alpha-helix: role of water molecules in stabilizing helices. Proc Natl Acad Sci U S A 87:871

24. Aravinda S, Shamala N, Bandyopadhyay A, Balaram P (2003) Probing the role of the C–H⊙⊙⊙ o hydrogen bond stabilized polypeptide chain reversal at the c-terminus of designed peptide helices. Structural characterization of three decapeptides. J Am Chem Soc 125:15065
25. Vasudev PG, Chatterjee S, Shamala N, Balaram P (2011) Structural chemistry of peptides containing backbone expanded amino acid residues: conformational features of β, γ, and hybrid peptides. Chem Rev 111:657
26. Haldar D, Schmuck C (2009) Metal-free double helices from abiotic backbones. Chem Soc Rev 38:363
27. Goto H, Furusho Y, Yashima E (2007) Double helical oligoresorcinols specifically recognize oligosaccharides via heteroduplex formation through noncovalent interactions in water. J Am Chem Soc 129:9168
28. Benedetti E, Blasio BD, Pedone C, Lorenzi GP, Tomasic L, Gramlich V (1979) A double-stranded *β*-helix with antiparallel chains in a crystalline oligo-L–D-peptide. Nature 282:630
29. Görbitz CH (2007) Microporous organic materials from hydrophobic dipeptides. Chem–Eur J 13:1022
30. Guha S, Drew MGB, Banerjee A (2007) A new molecular scaffold for the formation of supramolecular peptide double helices: the crystallographic insight. Org Lett 9:1347
31. Jana P, Maity S, Maity SK, Haldar D (2011) A new peptide motif in the formation of supramolecular double helices. Chem Commun 47:2092
32. Maity SK, Maity S, Jana P, Haldar D (2012) Supramolecular double helix from capped γ-peptide. Chem Commun 48:711
33. Das AK, Haldar D, Hegde RP, Shamala N, Banerjee A (2005) X-Ray crystallographic signature of supramolecular triple helix formation from a water soluble synthetic tetrapeptide. Chem Commun 1836
34. Gan Q, Bao C, Kauffmann B, Grelard A, Xiang J, Liu S, Huc I, Jiang H (2008) Quadruple and double helices of 8-fluoroquinoline oligoamides. Angew Chem Int Ed 47:1715
35. Su Y, Zang S, Meng Q, Lu Z, Zhu H, Wen L, Lin J (2007) A novel 2D herringbone-like zinc coordination polymer built from helical motif: hydrothermal synthesis, structure and properties. Ing Chem Commun 10:74
36. Wen L-L, Dang D-B, Duan C-Y, Li Y-Z, Tian Z-F, Meng Q-J (2005) 1D helix, 2D brick-wall and herringbone, and 3D interpenetration d10 metal-organic framework structures assembled from pyridine-2,6-dicarboxylic acid N-oxide. Inorg Chem 44:7161
37. Delsuc N, Godde F, Kauffmann B, Leger J-M, Huc I (2007) The herringbone helix: a noncanonical folding in aromatic-aliphatic peptides. J Am Chem Soc 129:11348
38. Maity S, Jana P, Maity SK, Haldar D (2012) Conformational heterogeneity, self-assembly, and gas adsorption studies of isomeric hybrid peptides. Crys Growth Des 12:422
39. Bong DT, Clark TD, Granja JR, Ghadiri MR (2001) Organische nanoröhren durch selbstorganisation. Angew Chem 113:1016
40. Amorín M, Castedo L, Granja JR (2005) Self-assembled peptide tubelets with 7 Å pores. Chem Eur J 11:6543
41. Mitra A, Panda DK, Corson LJ, Saha S (2013) Controllable self-assembly of amphiphilic macrocycles into closed-shell and open-shell vesicles, nanotubes, and fibers. Chem Commun 49:4601
42. Kemp DS, Bowen BR, Muendel CC (1990) Synthesis and conformational analysis of epindolidione-derived peptide models for.beta.-sheet formation. J Org Chem 55:4650
43. Banerjee A, Maji SK, Drew MGB, Haldar D, Banerjee A (2003) An amyloid-like fibril forming antiparallel supramolecular β-sheet from a synthetic tripeptide: a crystallographic signature. Tetrahedron Lett 44:6741
44. Maity SK, Kumar R, Ambast DKS, Pal B, Haldar D (2012) J Mater Chem 22:22198
45. Maity SK, Bera S, Paikar A, Pramanik A, Haldar D (2013) Self-assembly and nonlinear optical properties of a synthetic dipeptide. Cryst Eng Comm 15:5860
46. Hong Y, Legge RL, Zhang S, Chen P (2003) Effect of amino acid sequence and ph on nanofiber formation of self-assembling peptides EAK16-II and EAK16-IV. Biomacromolecules 4:1433

47. Reches M, Gazit E (2004) Formation of closed-cage nanostructures by self-assembly of aromatic dipeptides. Nano Lett 4:581
48. Bera S, Jana P, Maity SK, Haldar D (2014) Inhibition of fibril formation by tyrosine modification of diphenylalanine: crystallographic insights. Cryst Growth Des 14:1032–1038
49. Yan X, He Q, Wang K, Duan L, Cui Y, Li J (2007) Transition of cationic dipeptide nanotubes into vesicles and oligonucleotide delivery. Angew Chem Int Ed 2007(46):2431
50. Maity S, Jana P, Maity SK, Haldar D (2011) Fabrication of hollow self-assembled peptide microvesicles and transition from sphere-to-rod structure. Langmuir 27:3835
51. Wang Y, Zang C, Fu H, Li X, Sheng X, Zhao Y, Xiao D, Ma Y, Ma JS, Yao J (2008) Switch from intra- to intermolecular h-bonds by ultrasound: induced gelation and distinct nanoscale morphologies. Langmuir 24:7635
52. Maity S, Kumar P, Haldar D (2011) Sonication-induced instant amyloid-like fibril formation and organogelation by a tripeptide. Soft Matter 7:5239
53. Pappas CG, Mutasa T, Frederix PWJM, Fleming S, Bai S, Debnath S, Kelly SM, Gachaganb A, Ulijn RV (2015) Transient supramolecular reconfiguration of peptide nanostructures using ultrasound. Mater Horiz 2:198
54. Ghosh S, Adler-Abramovich L, Gazit E, Verma S (2013) Spacer driven morphological twist in Phe-Phe dipeptide conjugates. Tetrahedron 69:2004
55. Menzenski MZ, Banerjee IA (2007) Self-assembly of supramolecular nanostructures from phenylalanine derived bolaamphiphiles. New J Chem 31:1674
56. Mondal S, Podder D, Nandi SK, Chowdhury SR, Haldar D (2020) Acid-responsive fibrillation and urease-assisted defibrillation of phenylalanine: a transient supramolecular hydrogel. Soft Matter 16:10115
57. Jana P, Ehlers M, Zellermann E, Samanta K, Schmuck C (2016) pH-Controlled formation of a stable β-sheet and amyloid-like fibers from an amphiphilic peptide: the importance of a tailor-made binding motif for secondary structure formation. Angew Chem Int Ed 55:15287
58. Matsui H, Holtman C (2002) Organic nanotube bridge fabrication by controlling molecular self-assembly processes between spherical and tubular formations. Nano Lett 2:887
59. Kim J, Han TH, Kim YI, Park JS, Choi J, Churchill DG, Kim SO, Ihee H (2010) Role of water in directing diphenylalanine assembly into nanotubes and nanowires. Adv Mater 22:583
60. Erdogan H, Babur E, Yilmaz M, Candas E, Gordesel M, Dede Y, Oren EE, Demirel GB, Ozturk MK, Yavuz MS, Demirel G (2015) Morphological versatility in the self-assembly of Val-Ala and Ala-Val dipeptides. Langmuir 31:7337
61. Tiwari R, Maji K, Ajmal SG, Ghosh N, Haldar D, Banerjee A (2020) Laser assisted self-assembly of diphenylalanine: emergence of robust waveguiding properties and Fano resonances. J Mater Chem C 8:9663
62. McDonald TO, Qu H, Saunders BR, Ulijn RV (2009) Branched peptide actuators for enzyme responsive hydrogel particles. Soft Matter 5:1728
63. Koley P, Gayen A, Drew MGB, Mukhopadhyay C, Pramanik A (2012) Design and self-assembly of a leucine-enkephalin analogue in different nanostructures: application of nanovesicles. Small 8:984
64. Maity S, Jana P, Maity SK, Haldar D (2011) Mesoporous vesicles from supramolecular helical peptide as drug carrier. Soft Matter 7:10174

Part IV
Architectonics of Nucleic Acids

Chapter 11
Bio-inspired Functional DNA Architectures

Eugen Stulz

11.1 Introduction

DNA is well-known as bearer of the genetic code. Since its structure elucidation around seven decades ago by Watson, Crick, Wilkins and Franklin [1–3], much has been learnt about its detailed structure, function and genetic coding. In the field of DNA chemistry, solid-phase synthesis (SPS) of DNA certainly is one of the most influential developments of the last century [4–6] as it allows to synthesise DNA in any desired sequence and in lengths of up to hundreds of bases in the very best case. This has advanced the field of DNA research substantially. In addition, classic organic synthesis has allowed the introduction of a very large diversity of modifications in the DNA in a sequence-specific manner, which have initially been targeted at altering the biological function of DNA.

In recent years, DNA has become a very attractive scaffold in supramolecular chemistry, where DNA is taken out of its biological role and serves as both stick-and-glue molecule to assemble novel functional structures with nanometre precision [7]. By attaching a variety of functional groups to the DNA, either at the periphery or internally, we are now able to design novel molecules with a range of applications, from electronics back to biological system manipulation and medicinal chemistry. Bio-inspired functional architectures are becoming a mature field from where inspiration can be drawn to advance research in many ways.

The fundamental principles to work with DNA are relatively simple: DNA likes to form a double-stranded system (dsDNA) with the well-known helical B-DNA structure (Fig. 11.1). The complementarity is governed by the Watson–Crick base

E. Stulz (✉)
School of Chemistry and Institute for Life Sciences, University of Southampton, Southampton, UK
e-mail: est@soton.ac.uk

T. Govindaraju, K. Ariga (eds.), *Molecular Architectonics and Nanoarchitectonics*, Nanostructure Science and Technology,
https://doi.org/10.1007/978-981-16-4189-3_11

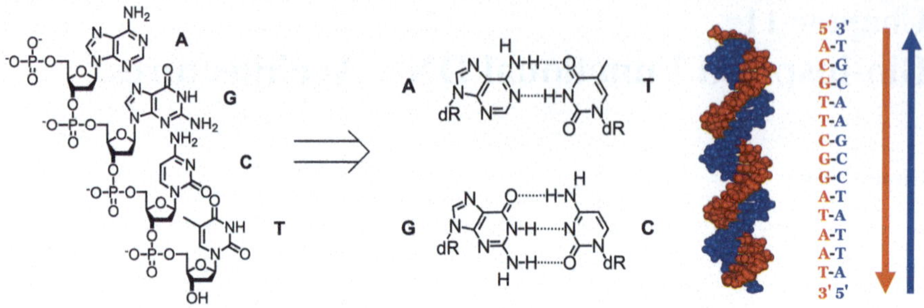

Fig. 11.1 Primary structure of DNA (DNA sequence), showing the basic nucleotide building blocks adenosine (A), guanosine (G), cytidine (C) and thymidine (T). The complementary base pairing (A–T and G–C) is responsible for the selective recognition of the DNA strands, leading to an antiparallel arrangement of the dsDNA with its well-known B-DNA structure

pairing (A–T and G–C), and the strands are aligned in an antiparallel manner. Duplexes from about seven to eight base pairs (bp) are stable under ambient conditions, and the dsDNA has a persistence length of around 40–50 nm (stick-like). Single-stranded DNA (ssDNA), on the other hand, is rather flexible in structure, but as it binds well to the complementary strand, it can serve as a glue-like molecule to hold different strands together. The combination of both properties now gives rise to a Lego-like building block system, which is very well explored in the field of DNA bio-nanotechnology and very impressively in the field of DNA origami. Adding substituents through classical organic synthesis enhances the repertoire of available functionalities greatly, and the size of the chemical space around modified DNA is almost endless. Today, the field likes to be referred to as *DNA bio-nanoarchitectonics*. Since the beginning of using DNA as architectural material, the field has taken its own pace and a huge amount of research has been published [8].

This book chapter is not intended to give a comprehensive account, which would mean a book in its own right, but rather an overview of different aspects with selected examples, which are hopefully inspirational to the reader. It will focus on the use of dsDNA which has been modified covalently with a variety of functional molecules. It will leave out all the systems that have been achieved by exploring non-covalent interactions, non-canonical DNA structures and most of the more complex higher-hierarchy 3D constructs.

11.2 Modification Strategies

Before we look at the formation of DNA-based architectures, it is worth noting how different strategies to modify DNA are being used. The introduction of modifications on DNA can be achieved through various methods, which very much depend on the final aim.

Most convenient sites available for modification include the 3′- and 5′-terminal positions, the 2′-position on the ribose and the nucleobase itself. For the latter, the most convenient site to use is the 5′-position of thymidine (more precisely 2-′-deoxyuridine, dU), because the corresponding 5-iodo-dU (5-I-dU) is readily available and modified through *Sonogashira* coupling. In addition, the nucleobase itself can be substituted with designer molecules (artificial nucleobases, base surrogates), or the entire nucleotide can be replaced with moieties that mimic the function and structure of the DNA. One should not forget the many commercially available modifiers for either internal or terminal modifications; the chemistries to introduce these are, however, basically the same, but the synthesis can be outsourced should this be desirable (the reader is referred to the many commercial DNA synthesis companies).

The general way to obtain DNA is via solid-phase synthesis, which is relatively fast, reliable, convenient and generally high yielding for standard DNA, though the yields vary greatly with additional functionalities and in particular with the nature, the quality and the reactivity of the modifier. The general synthesis includes the addition of a phosphoramidite nucleotide to the growing chain on the solid phase, followed by oxidation of the P(III) to P(V), and capping of unreacted growing DNA chain. Deprotection with strong acid is again followed by coupling, and the final product is cleaved and fully deprotected in concentrated ammonia. There are numerous fine-tuning variations available with regard to protecting groups and mild deprotecting strategies. Most intriguing is the fact that any molecule which has two hydroxyl groups that can be transformed into a DMT-protected phosphoramidite can be used in SPS and combined with DNA at will, whether it resembles a nucleoside or not.

One should also note that DNA synthesis is rather wasteful, and 90% of the nucleotide building blocks are discarded because a large excess (tenfold) is used to ensure highest possible coupling close to 100%. In the best case, the excess nucleotide can be reduced to about fourfold. This should be taken into account when preparing the phosphoramidite building blocks of precious material. In addition, the functionalities need to survive strongly acidic and basic, as well as oxidising conditions, which limits the functional groups that can directly be introduced during DNA synthesis. However, post-synthetic modifications are an alternative method and certainly worth taking into account; this would most conveniently be done using amino or carboxy-modifiers and amide coupling with the desired functionality, or alkyne–azide click chemistry.

The different strategies have been reviewed in the past and will not be covered again here in more detail [7, 9, 10]. However, as an overview the strategies are summarised in Fig. 11.2; the modifications mentioned in this chapter will have been introduced using one of these strategies, or modifications thereof, which can be found in the original literature. In short, there is a wide range of strategies available to suit the needs, and which one is best will have to be decided on a case-by-case basis and depends on the application as well as on the modifiers. The use of modified DNA, either home-made or commercially available, is indispensable in the design

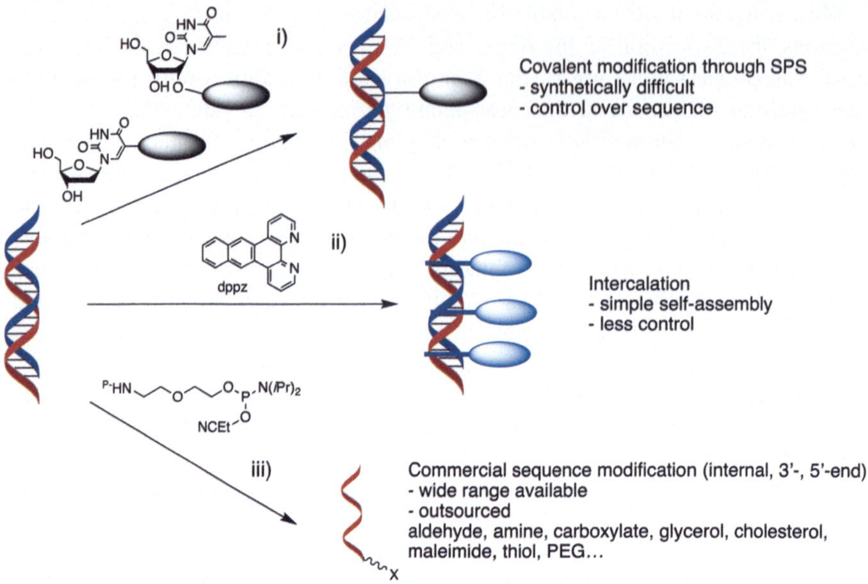

Fig. 11.2 General strategies to incorporate functional molecule into DNA. (i) Covalent attachment of self-made modifiers in programmed sequence. (ii) Self-assembly of intercalators, alternative to external binding through electrostatic interactions. (iii) Outsourcing of the DNA synthesis by ordering commercially available modifications. If direct attachment is not feasible during SPS, the "simple" amino or carboxylate functionalities can be used for post-synthetic modification via amide coupling, either internally or at the periphery

and synthesis of DNA nanostructures, and these days hardly any nano-architecture is being reported that does not contain a modified DNA strand of some sort.

11.3 DNA Duplexes with External Modifications

The development of synthetic procedures to attach modifiers to the nucleotides, which—after SPS—will be protruding out of the duplex DNA, has greatly influenced the development of dsDNA-based functional molecules. Most of these aim for applications in electronics, as the use of electronically active molecules is attractive on one side and relatively straightforward on the other side. The effort that has gone into the initial systems should not be underestimated, though.

The modifications are usually based on aromatic systems which show electronic properties such as strong absorbance and emission (chromophores), or electrochemical response. The main players in the field comprise polyaromatics (PAs, pyrene, anthracene), dimides, porphyrins, fullerenes but also metal complexes based on phenanthroline, bipyridine (bipy), terpyridine (terpy) or metallocenes

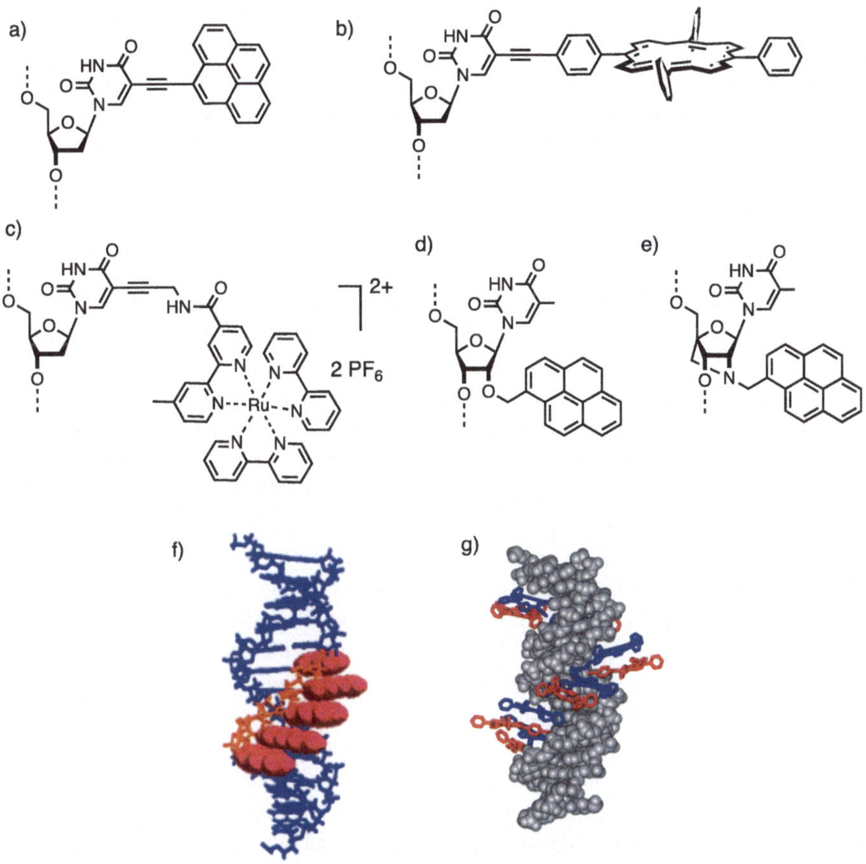

Fig. 11.3 Selection of modifiers attached to (**a**)–(**c**) either the nucleobase, predominantly the 5′-position of deoxy-uridine, or (**d**)–(**e**) the 2′-position of the ribose, and two early examples of multichromophore arrays with (**f**) pyrene and (**g**) porphyrin. (**a**) Pyrene–nucleoside by *Wagenknecht* [11], (**b**) porphyrin–nucleoside by *Stulz* [12], (**c**) ruthenium–bipy–nucleoside by *Grinstaff* [13], (**d**) pyrene–2′-ribose nucleoside by *Nakamura* and *Yamana* [14], (**e**) pyrene–2′-LNA by *Wengel* [15], (**f**) representation of the pyrene stack along DNA by *Wagenknecht* (reprinted by permission of John Wiley & Sons, Inc. from Ref. [16]) and (**g**) modelled structure of a multiporphyrin–DNA by *Stulz* [17]

(predominantly ferrocene); Fig. 11.3 shows a selection of representative examples. The aromatic systems behave to some extent similar yet are complementary. They are mostly fluorescent which is dependent on their environment, they are hydrophobic which leads to association/aggregation, and they show large induced chiroptical properties which can be detected by circular dichroism (CD) spectroscopy.

The general strategy of attaching the modifications either to the 5′-position of T via *Sonogashira* coupling or to the 2′-position of a ribose nucleoside is most suitable; modifications that are attached to this position will be placed outside of the base-pair region but within the major groove, thus giving relatively little disturbance to the dsDNA structure. This, however, holds only true for either a small number of modifications or substituents that are small in size; large substituents such as porphyrins lead to a significant decrease in thermodynamic stability (see below) [12]. The hydrophobic nature of these substituents also renders the handling of DNA, which contains large number of modifications, rather cumbersome if the length of the DNA is not long enough to ensure good water solubility. The substituents tend to associate, leading to multiple DNA strands being bound together. This may as well have a positive aspect as it can be regarded as a "molecular glue" [18].

The initial systems were aimed at exploring the possibilities to use the DNA as a scaffold for making helical stacks of chromophores. Multiple attachments of small (pyrene) modifications by *Wagenknecht* [16] and large (porphyrin) modifications by *Berova* [19, 20] and *Stulz* [12] were incorporated into DNA. This work pioneered the field of helical multichromophore arrays on DNA (Fig. 11.3). Attachment of multiple chromophores (and substituents in general) leads to an array of molecules, whose exact composition is programmable through the DNA sequence [21]. This has led to the first examples of truly programmable supramolecular systems with multiple chromophores using automated synthesis. Starting with just a few modifications to explore the basic concept, high numbers of up to 5 consecutive modifications with pyrene [22] or 12 porphyrins [12] show that there is in principle no limitation from a synthetic point of view. However, the thermodynamic duplex stability can be greatly compromised, which is measured in the change of duplex melting temperature (the temperature T_m where 50% of the DNA is denatured); the ΔT_m can be as large as −7 °C per modification. It was quickly realised that a zipper-like arrangement, where the substituents are placed on the complementary strands at every other nucleobase, is a highly beneficiary approach: in the annealed state, the interlocked stacking of the hydrophobic moieties leads to a greatly enhanced thermodynamic stability [23]. In extreme cases, the thermodynamic stability increase ΔT_m can be as high as +40 °C.

To date, many different multi-chromophore systems based on the DNA scaffolding approach have been created [11]. The thus constructed arrays show high absorbance and emission and can be used for the formation of systems for light harvesting and energy transfer [24]. Examples include assemblies of covalent pyrene modification (dsDNA, G-quadruplex) and complexes with $Ru(II)(bpy)_3$ and Zn–tetramethyl pyridinium porphyrin (ZnTMPyP) for photochemical upconversion [25, 26]; fullerene–pyrene–Nile-red conjugates for DNA-based solar cells [27]; pyrene-functionalised locked nucleic acids (LNA) for diagnostics and nanotechnology [28]; and pyrene–perylene FRET pairs for DNA analysis [29]. Porphyrins have been used extensively, for example, in conjugation with a black-hole quencher for energy transfer [30], in membrane–anchored assemblies for energy and electron transfer [31–33], in porphyrin dimers for photoresponsive electrodes [34] and in programmable conformational regulation of dimers [35, 36]. Phthalocyanines are rarely used because of their solubility issues, but it was shown that a dodeca-trifluoroethoxy-substituted phthalocyanine–DNA remains monomeric in solution [37]. A somewhat different system, as it comprises commercially available fluorophores, was shown by *Buckhout-White* and *Spillmann*, who investigated a large number of increasingly complex assembly designs including linear, bifurcated, Holliday junction, stars and dendrimers involving up to five different dyes, thus assembling programmable FRET-based photonic networks using designer DNA scaffolds [38]. This should highlight that not only custom-made chromophore building blocks are efficient, but the repertoire in commercially available fluorophores should be explored, too.

Circular dichroism (CD) spectroscopy is a very versatile method when it comes to analysing the global structure of DNA. While chromophores normally are achiral and do not show any CD signals, attachment to the chiral DNA renders any modification also chiral (induced chirality), thus giving rise to a specific CD signature. If more than one chromophore is attached to the DNA, then an exciton coupling between the dipoles of the chromophores will arise, provided they are in close enough proximity. To investigate and make use of this phenomenon, CD spectroscopy with porphyrin-modified DNA has been pioneered by *Balaz* and *Berova*: two porphyrins attached to DNA will give a Soret band CD exciton couplet signal around 420 nm, which is outside the DNA signal below 300 nm. Thus, the two spectral regions can be used in conjunction to obtain information on structure and conformational changes. With this, two porphyrins attached to both ends of dsDNA were used to detect the salt-induced B- to Z-DNA transition (Fig. 11.4) [39–41]. For covalent porphyrins which are attached in the major groove of the DNA with short linkers, the direct face-to-face contact of the porphyrins was used by *Stulz* et al. to obtain an insight in the relative orientation of multiple chromophore assemblies [23]. Here, both the nature of the linker and the metallation state of the porphyrin had a strong impact on the induced CD signal and exciton coupling [17], but also helped to confirm the presence of intermolecular interactions (DNA–DNA association through porphyrin–porphyrin interactions) [42].

The use of specific substituents can be used to determine distances within the DNA. Since the B-DNA is well defined with a distance of 0.332 nm between base pairs, the DNA actually acts as a sub-nanometre molecular ruler. Different techniques have been applied to this end, in particular exciton-coupled CD spectroscopy using stilbene derivatives [43] or porphyrins [44]; the latter was used to demonstrate intercalation with base-pair resolution. Alternative methods to CD spectroscopy include photoacoustic tomography (PAT) which showed distance-dependent photoacoustic signal enhancement [45], or EPR spectroscopy using paramagnetic metal complexes [46]. The EPR spectroscopy was based on Cu(II) metallated porphyrins and did not only reveal the intramolecular porphyrin–porphyrin distance, but was also used to demonstrate that in multiporphyrin–DNA systems three to four porphyrin–DNA strands associate (but not precipitate), which confirmed earlier SAXS measurements [17]. On a larger scale upwards of 10 nm, which is usually the limit of FRET, the use of nanoparticles has shown to be versatile, leading to nanoplasmonic rulers. Noteworthy early examples include the coupling of single pairs of gold and silver nanoparticles, which was used to study the kinetics of single DNA hybridisation events [47], and a single Au nanoparticle tethered to dsDNA which showed changes in plasmonic resonance depending on the length of the DNA

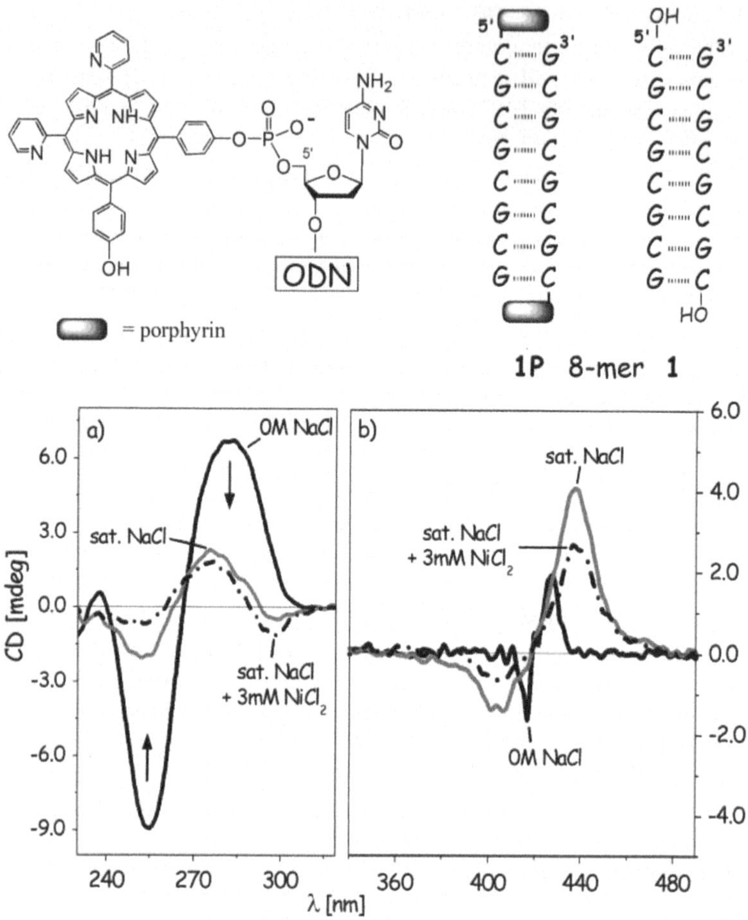

Fig. 11.4 End-of-DNA modification with porphyrin which acts as a helical reported for detecting the B- to Z-DNA transition using CD spectroscopy. Reproduced from Ref. [39] with permission from the Royal Society of Chemistry

[48]. This latter system did show sub-nanometre resolution, despite its relative size of 20 nm nanoparticles and >50 bp DNA, where the change in plasmon resonance wavelength was about 1.24 nm per base pair and serves as a DNA footprinting platform and mapping system for specific protein–DNA binding.

As should become obvious, the possibilities to modify the DNA at the periphery are endless: basically, any molecule, from small chromophores to nanoparticles or even proteins, can be attached in a programmable manner. The specific interactions between the attached modifiers are governed by the underlying DNA sequence, and "simply" by changing the sequence of the DNA through programming of the synthesiser allows for a fine-tuning of the overall functional molecule. However, the synthetic effort is not negligible as multiple nonstandard phosphoramidites may need to be prepared simultaneously.

11.4 DNA Duplexes with Internal Modifications

The internal region of the dsDNA, i.e. the base-pair region, is particularly attractive for the insertion of modifications which will be stacked on top of each other (H-aggregate), in contrast to major groove located modifications which will be stacked in a slipped fashion (J-aggregate). These, however, will need to fulfil certain requirements: they need to be aromatic, should not have large substituents and be in the range of the base-pair area (ca 0.9×0.5 nm^2). Many of the aromatic DNA modifiers that have been described in Sect. 11.5 have been inserted into the DNA as base surrogates, as well as metal coordination systems. An early example involves the replacement of the base pair with a porphyrin, where the porphyrin was directly embedded inside the duplex region and replaced the entire nucleoside [49]. Alternatively, replacing only the base part to yield a porphyrin-C–nucleoside retained its fluorescence within the DNA, where it strongly intercalates, and showed enhanced duplex stability when placed at the end of the dsDNA [50]. Inside the DNA, however, a destabilisation of 8–10 °C was observed. Consequently, also multiporphyrin stacks inside the DNA double helix were synthesised, which results in the formation of H-aggregates showing fluorescence quenching along with intense exciton-coupled CD effects [51].

Pyrenes again show a predominant role in the formation of internal stacks of chromophores. A self-organising system composed of two oligopyrene strands that leads to the formation of an interstrand helical stack embedded in a double-stranded DNA was reported, which contains up to 14 consecutive achiral pyrene building blocks flanked by 5 to 9 base-pair duplexes on either side (Fig. 11.5a) [52]. The flanking DNA region thus efficiently induces a helical superstructure in the pyrene stack. Similarly, up to six perylene-3,4:9,10-tetracarboxylic acid bisimides (PBs) were incorporated in the middle of an 18-mer DNA duplex [54]. The stacking of the

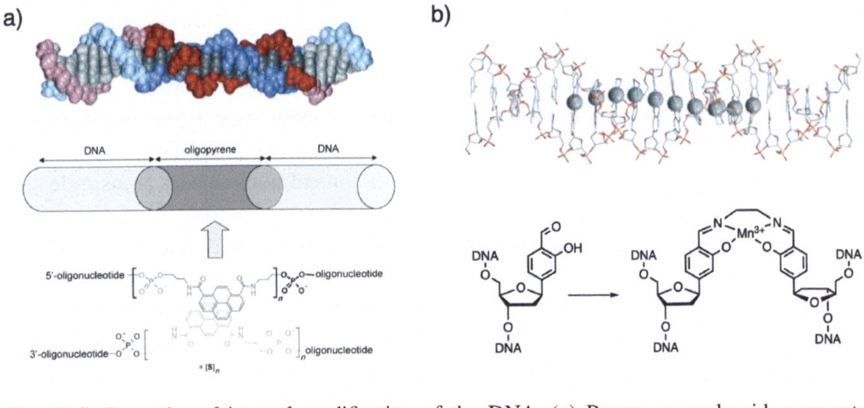

Fig. 11.5 Examples of internal modification of the DNA. (**a**) Pyrene as nucleoside surrogate leading to the self-assembly of a 14-mer pyrene stack (reprinted by permission of John Wiley & Sons, Inc. from Ref. [52]) and (**b**) formation of a metallic wire comprised of ten transition metals within the base-pairing region (reprinted by permission of John Wiley & Sons, Inc. from Ref. [53])

PB chromophores was dependent on the presence or absence of thymidines opposite the PBs, and a zipper-like arrangement was proposed. Abasic sites in the DNA were further used as both binding pockets and reactive sites for the DNA-templated synthesis of perylene diimide (PDI) stacks [55]; a PDI–diamine binds inside the DNA pocket and is consequently covalently attached to the abasic site with the use of uracil DNA glycosylase, followed by reduction of the formed Schiff base with $NaBH_3CN$. This shows elegantly how SPS with modified nucleosides can be circumvented, and post-synthetic modification using a DNA-templated approach leads to the desired chromophore assembly.

As was mentioned earlier in this book chapter, aromatic substituents on DNA tend to associate, which can be used as a "molecular glue". This is particularly true when the modifiers are attached to the end of the DNA, where significant end-to-end stacking can occur, leading to extended DNA stacks if both ends are modified. This was shown with porphyrins [41] and PDI [18], and the structure of a complex involving pyrene and phenanthrene was solved crystallographically [56].

Based on these initial studies, a multitude of further internal multichromophore systems were developed [11, 21, 57] including tetrathiafulvalene (TTF) and TTF–pyrene hybrid systems [58], PDI stacks [59], self-assembled pyrene foldamers [60], PDI–porphyrin hybrid structures [61], and perylene–pyrene hybrid systems to mimic quantum dots [62]. Many of the systems show efficient energy transfer, for example, the perylene–pyrene combination demonstrated efficient FRET and could be used to discriminate single-base changes within DNA [29]. The energy transfer properties of chromophores were elegantly used by *Häner* et al. to create a light-harvesting complex, which was composed of a π-stacked multichromophore array in a DNA three-way junction [63]. The modular approach, combining a phenanthrene antenna and acceptors like pyrene (exciplex formation), PDI (quencher) or cyanine dye (emitter), demonstrates the use of DNA for a successful combination strategy. The same group also used DNA to construct well-defined heterochromophoric stacks; in this system electronic coupling among non-adjacent chromophores of the same type leads to the co-existence of PDI and pyrene H-aggregates in hybrids composed of alternating chromophore stacks [64].

Very different systems are composed of metal complexes within the DNA. For example, a macrocyclic cyclidene metal complex as nucleotide replacement results in a new family of metal-functionalised DNA derivatives [65]. They are able to distinguish electrochemically between all four canonical nucleobases at a single site when redox-active metals are used, such as Cu(II). A related system comprises terpyridine and diphenyl-phenanthroline ligands as nucleotide replacements, which can bind different metals selectively and recruit Fe(II), Co(II), Cu(I) or Cu(II), depending on the combination of the ligands in the complementary strands [66]. On the other hand, a major effort was made to study both metal complexes and nucleobase–metal ion interactions as artificial base pairs, leading to multimetallic wires in the centre of the DNA [67, 68]. The first example is based on copper(II)-mediated pyridine-2,6-dicarboxylate (Dipic) and pyridine (Py) metallo-base pair (first generation), and pyridine-2,6-dicarboxamide (Dipam) and a pyridine (Py) Cu(II) base pair (second generation) [69]. A salen-based metal–base pair leads to an increase of the melting temperature of >40 °C, which was believed to originate from a combination of reversible cross-linking of the strands and axial coordination of the metal with heteroatoms of the neighbouring base pairs [70, 71]. This system was then used to form an impressive controlled stacking of ten transition metal ions inside the dsDNA (Fig. 11.5b) [53]. Also the formation of metal complexes with natural bases, with or without ligands as base surrogates or modified nucleobases, has subsequently been explored to form various stable metallic wires within the DNA duplex. These include the formation of antiferromagnetically coupled and stacked Cu(II)-salen complexes [72], a dinuclear Ag(I)-mediated base pair with ethenoadenine and thymidine [73], Ag(I) base pairs with 7-deazaguanine and cytosine [74], a glycol nucleic acid analogue containing phenanthroline to selectively bind Ag(I) with cytosine and Hg(II) with thymidine [75], a phenanthroline-derived nucleobase surrogate to form a stable Hg(II) complex with T [76] and even up to four Fe(III) ions within a DNA triple strand [77]. A 5-carboxyuracil (caU) derivative was found to not only form a hydrogen-bonded base pair with itself, but also selectively with all four natural bases, in particular, the base pairs caU–Cu(II)–caU, caU–Hg(II)–T, caU–Ag(I)–C, caU–Cu(II)–G and caU–Cu(II)–A [78]. This could be used to create designer metal wires within the DNA duplex.

The internal region of the DNA thus allows for quite different types of stacks of molecules than the exterior of the DNA. While many of the aromatic modifiers can be used either way, bulky modifiers will not work inside the DNA. On the other hand, well-defined metal wires where the metal ions are in direct contact cannot be achieved in the grooves of the DNA. Potential applications have already been demonstrated in view of energy transfer, and the metallic systems hold promise for the creation of efficient conducting wires. However, this field has still great potential for exploration, and we will certainly see more exciting results in the near future.

11.5 Higher-Order DNA Architectures

The concept of designing higher-order DNA nanostructures is—in principle—relatively simple: complementary DNA sequences will anneal to the duplex dsDNA, whereas non-complementary sequences will not; mismatches will lower the thermodynamic stability and allow to discriminate between match and mismatch in the base pairing. Sequences of 10 to 20 base pairs (bp) will usually hybridise spontaneously at ambient temperature, but more complex architectures will require a denaturing–annealing process in order to disrupt potential secondary structures and unwanted duplex stretches; slow cooling ensures that predominantly the desired duplexes are formed. Having DNA strands with partial complementarity will allow for DNA strands to selectively hybridise, eventually folding into the desired 3D architecture (Fig. 11.6).

Pioneering work is certainly accredited to *Seeman* and *Turberfield*, who have created some of the first DNA-based self-assembled structures. Both *Seeman's*

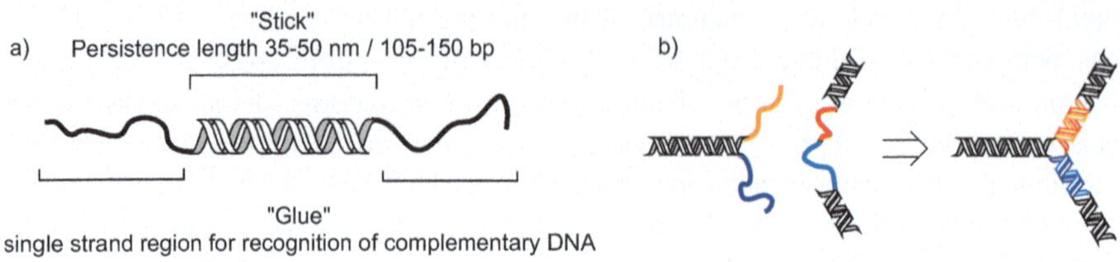

Fig. 11.6 (**a**) Concept of DNA as stick-and-glue molecule. (**b**) While dsDNA is rather a rigid stick, ssDNA overhangs act as glue to assemble other complementary sequences that may carry further self-assembled DNA systems

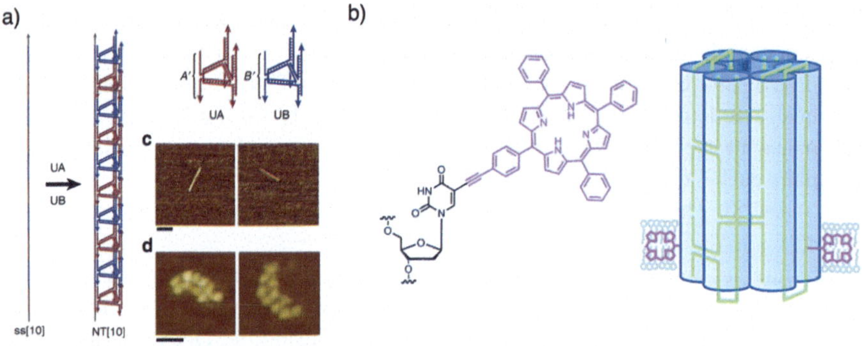

Fig. 11.7 Self-assembled DNA structures that can exhibit specifically designed functions. (**a**) Single-stranded DNA-guided nanotube assembly (Reprinted by permission from Springer Nature [90]) and (**b**) a porphyrin-modified DNA allow a self-assembled DNA bundle to stably insert into a lipid bilayer creating a nanopore (reprinted by permission of John Wiley & Sons, Inc. from Ref. [91])

octahedron [79] and *Turberfield's* tetrahedron [80, 81] have been paving the way to more complex DNA structures. Since then, a wide range of arrays [82], such as grids and lattices on surfaces [83, 84], linked frayed wires [85], nanoscale patterns through folded DNA [86], tube structures [87, 88] and bipyramids [89] have been assembled successfully. *Sleiman* has taken the systems forward to introduce, for example, sequential growth of long DNA strands (Fig. 11.7a) [90]. By drawing inspiration from SPS, they described a DNA assembly method which combines sequential addition and ligation to yield monodisperse ssDNA of up to 1 kb. This can then be used to create specific patterns where specifically placed unique sites within a repeating motif are sufficient, and greatly minimises the number of components used. Self-assembled DNA systems that show potential in medicinal applications have also been reported, most notably a DNA membrane nanopore (Fig. 11.7b) [91, 92]. Several DNA strands assemble to form a six-helix bundle that, when containing hydrophobic modifications like a belt, will insert into lipid bilayers. Either 78 short alkyl chains [92] or 2 porphyrins [91] will be sufficient to stably insert the nanopore and create a pore that allows for the flux of ions. The DNA pore shows cytotoxicity which could be explored in medicinal chemistry [93]. It is worthy to note that a DNA pore does not necessarily have to have several DNA duplexes bundled together, and the smallest DNA-based membrane nanopore was actually constructed from a six-porphyrin dsDNA. The inserted duplex creates a vortex in the membrane which also allows for an ion flux, most likely along the charged phosphate diester backbone of the DNA [94].

Extending the recognition pattern of DNA beyond the standard two base pairs greatly increases the available structural motifs. An example is the use of melamine, which mimics the hydrogen bonding pattern of adenine but on three sides [95]. It is therefore capable of binding multiple thymines, but in a complex that was structurally solved by X-ray crystallography, two poly-T strands wrap around a central helical column of melamine in an antiparallel manner. This complex can be used as a tool for DNA nanotechnology and incorporated into two-dimensional grids and hybrid DNA small-molecule polymers.

The introduction of the toehold concept, in which a single-strand overhang is recognised by the complementary strand and subsequently peeled off the structure, has made possible that reversible structural changes can be realised [96]. A pioneering system by *Yurke et al.*, which is termed "molecular tweezers", can be termed the first DNA-fuelled molecular machine uniquely made of DNA (Fig. 11.8a) [97]. Their "machine" intriguingly uses DNA for both structural material and fuel. In this case, three strands hybridise into a pair of tweezers, which can be opened and closed by the addition of auxiliary strands (the fuel). In this and subsequent examples [96, 100], each catalytic cycle produces one dsDNA as waste, which eventually will start to interfere with the reactions.

Taking a quantum leap forward was certainly the conception of the DNA origami concept (though it should correctly be termed ori-DNA) by *Rothemund* [86]. The use of a long template strand (e.g. 7.2 kb genome of phage M13mp18) in the form of ssDNA is stitched together with the help of hundreds of short synthetic staple strands. This seemingly simple yet very intriguing concept allows the formation of almost any three-dimensional nanostructure through careful design of the system. Nowadays, programs like caDNAno, which is probably the most widely used, are indispensable in the design of 3D-DNA origami structures due to the complexity of the strand combinations and overlaps, which quickly becomes unmanageable otherwise. One main issue to consider is the helical turn of DNA, which is 10.5 bp. Periodic crossovers in origami design for flat structures are separated by odd numbers of half-turns, with 16 bp considered as 1.5 turns. This leads to a twist density of 10.67 bp per turn (16 bp/1.5 turns), which does not reflect the 10.5 bp/helical turn and gives a slight underwinding of all DNA in B-form DNA [101]. The solution to this problem is to occasionally use only 15 bp in a crossover.

Early examples of DNA origami structures include bottles and ball-shaped structures [102], but ever more elaborate constructs are being reported, including tubes [103] and self-assembled DNA crystals (Fig. 11.8b) [98]. The folding pathway of complex DNA structures has been studied as well, and key findings are that the assembly of DNA origami is highly cooperative, that reversible bond formation is important in recovering from transient misfoldings and that early formation of long-

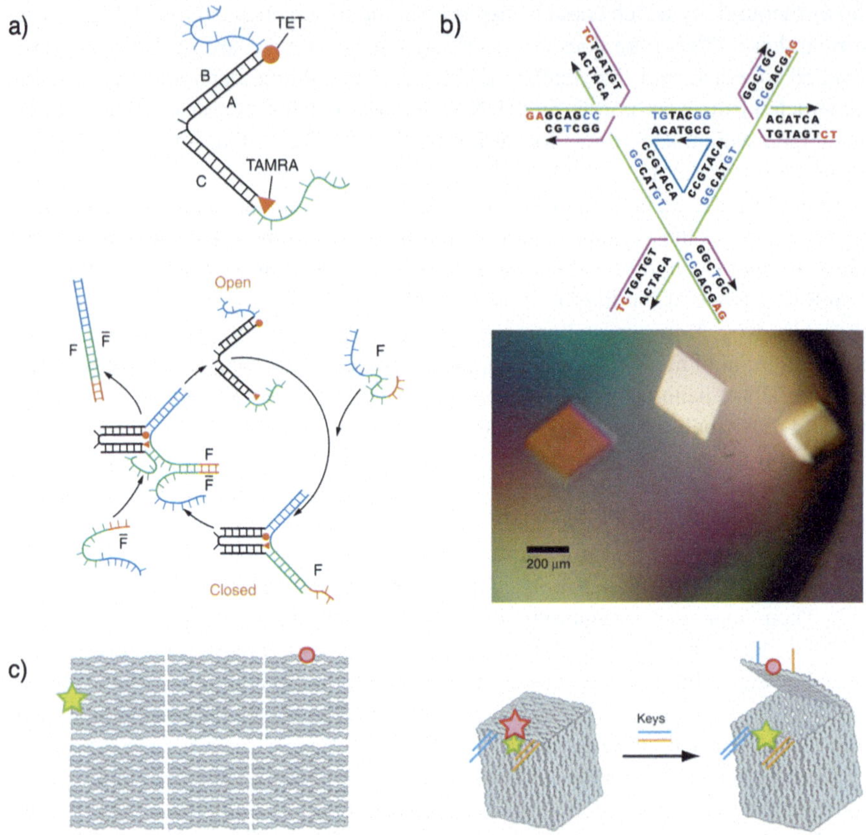

Fig. 11.8 Molecular machines, self-assembled crystals and DNA origami boxes made of DNA. (**a**) A DNA-fuelled molecular machine made of DNA displaying opening and closing upon sequential addition of appropriate DNA strands (Reprinted by permission from Springer Nature [97]). (**b**) Rational design of a self-assembled DNA crystal (Reprinted by permission from Springer Nature [98]). (**c**) A DNA origami-based self-assembled box where the lid can be opened and closed with the use of DNA keys (Reprinted by permission from Springer Nature [99])

range connections can effectively enforce particular folds [104]. Again, the toehold concept has found its way to introduce structural changes. An example is the DNA origami-based cube, where the lid is held in a lock position through duplex DNA formation between the lid and the box (Fig. 11.8c) [99]. Complementary DNA acts as the key to open the lid, and the box can release its cargo. Removal of the key through even longer complementary DNA will again close the lid and lock the box shut.

The field has been reviewed extensively [8, 105]; some recent examples include topics on therapeutic and biomedicine [106–112], drug delivery [113–115], sensing [116–120], bio-interface [121–127], computing and electronics [128–130], functional materials including robotics and machines [131–142] and general origami studies [143–154]. Surely more reviews in this direction will be published constantly, and it is well worth keeping an eye on the advancements in the field.

11.6 Conclusions and Outlook

In this book chapter, we have looked at the formation of bio-inspired functional DNA architectures, although we have only covered a very small section of the entire supramolecular space that is available; as one would say, we scratched the surface. DNA is far more than just the bearer of the genetic material in all living organisms: it is a most versatile construction material that should not be underestimated. On one hand, its simplicity and predictability in structure, limited number of natural building blocks and tendency to form a duplex with the complementary strand without further obvious function (apart from storing the genetic code, of course) would suggest it is a boring molecule. However, the advent of reliable synthetic procedures, the use of synthetic chemistry and clever design to introduce modifications, the programmability of the system and the stability of the DNA have opened the field of DNA-based bio-nanoarchitectonics. One could easily write several further chapters covering the use of other non-canonical DNA structures such as G-quadruplexes, i-motifs and triple helices, self-assembly of amphiphilic DNA structures and the self-assembly of modified DNA strands to higher hierarchies of 3D structures (in particular DNA origami).

Where is the field heading? As with every research, it is unpredictable what will be discovered in the next 5–10 years, and the field may take completely different turns that are not foreseeable, and this is very likely to happen. In the short term, most of these systems will find real-world applications in drug delivery and medicinal applications; indeed, many of these are already in clinical trial or even in use, for example, in antisense strategy and gene therapy. Likewise, biotechnology and diagnostics have made quantum leaps due to modified DNA. Whether functionalised DNA will once form the basis of computers and other electronic systems or artificial photosynthesis remains to be seen, but efforts in this direction are being made with impressive results. The versatility of the use of modified DNA allows to bring in ideas from all other disciplines, including medicine, biology, engineering, electronics or physics. The ultimate outcome of future research directions is only limited by the researcher's imagination.

References

1. Watson JD, Crick FHC (1953) Molecular structure of nucleic acids: a structure for deoxyribose nucleic acid. Nature 171:737–738
2. Franklin RE, Gosling RG (1953) Molecular configuration in sodium thymonucleate. Nature 171:740–741

3. Wilkins MHF, Stokes AR, Wilson HR (1953) Molecular structure of nucleic acids: molecular structure of deoxypentose nucleic acids. Nature 171:738–740
4. Sinha ND, Biernat J, McManus J, Koster H (1984) Polymer support oligonucleotide synthesis XVIII: use of beta-cyanoethyl-N,N-dialkylamino-/N-morpholino phosphoramidite of deoxynucleosides for the synthesis of DNA fragments simplifying deprotection and isolation of the final product. Nucl Acids Res 12:4539–4557
5. Beaucage SL, Caruthers MH (1981) Deoxynucleoside phosphoramidites - a new class of key intermediates for deoxypolynucleotide synthesis. Tetrahedron Lett 22:1859–1862
6. Matteucci MD, Caruthers MH (1981) Synthesis of deoxyoligonucleotides on a polymer support. J Am Chem Soc 103:3185–3191
7. Bandy TJ, Brewer A, Burns JR, Marth G, Nguyen T, Stulz E (2011) DNA as supramolecular scaffold for functional molecules: progress in DNA nanotechnology. Chem Soc Rev 40:138–148
8. Stulz E (2012) DNA architectonics: towards the next generation of bio-inspired materials. Chem Eur J 18:4456–4469
9. Leumann CJ (2002) DNA analogues: from supramolecular principles to biological properties. Bioorg Med Chem 10:841–854
10. Stulz E, Clever G, Shionoya M, Mao CD (2011) DNA in a modern world. Chem Soc Rev 40:5633–5635
11. Varghese R, Wagenknecht HA (2009) DNA as a supramolecular framework for the helical arrangements of chromophores: towards photoactive DNA-based nanomaterials. Chem Commun 19:2615–2624
12. Fendt LA, Bouamaied I, Thöni S, Amiot N, Stulz E (2007) DNA as supramolecular scaffold for porphyrin arrays on the nanometer scale. J Am Chem Soc 129:15319–15329
13. Khan SI, Beilstein AE, Smith GD, Sykora M, Grinstaff MW (1999) Synthesis and excited-state properties of a novel ruthenium nucleoside: 5-[Ru(bpy)(2)(4-m-4'-pa-bpy)](2+)-2'-deoxyuridine. Inorg Chem 38:2411–2415
14. Nakamura M, Shimomura Y, Ohtoshi Y, Sasa K, Hayashi H, Nakano H, Yamana K (2007) Pyrene aromatic arrays on RNA duplexes as helical templates. Org Biomol Chem 5:1945–1951
15. Hrdlicka PJ, Babu BR, Sorensen MD, Wengel J (2004) Interstrand communication between 2'-N-(pyren-1-yl)methyl-2'-amino-LNA monomers in nucleic acid duplexes: directional control and signalling of full complementarity. Chem Commun 2004:1478–1479
16. Mayer-Enthart E, Wagenknecht H-A (2006) Structure-sensitive and self-assembled helical pyrene array based on DNA architecture. Angew Chem Int Ed 45:3372–3375
17. Brewer A, Siligardi G, Neylon C, Stulz E (2011) Introducing structural flexibility into porphyrin-DNA zipper arrays. Org Biomol Chem 9:777–782
18. Baumstark D, Wagenknecht HA (2008) Perylene bisimide dimers as fluorescent “glue” for DNA and for base-mismatch detection. Angew Chem Int Ed Engl 47:2612–2614
19. Mammana A, Asakawa T, Bitsch-Jensen K, Wolfe A, Chaturantabut S, Otani Y, Li XX, Li ZM, Nakanishi K, Balaz M, Ellestad GA, Berova N (2008) Synthesis and characterization of water-soluble free-base, zinc and copper porphyrin-oligonucleotide conjugates. Bioorg Med Chem 16:6544–6551
20. Balaz M, Holmes AE, Benedetti M, Proni G, Berova N (2005) Porphyrin substituted phosphoramidites: new building blocks for porphyrin-oligonucleotide syntheses. Bioorg Med Chem 13:2413–2421
21. Kashida H, Asanuma H (2012) Preparation of supramolecular chromophoric assemblies using a DNA duplex. Phys Chem Chem Phys 14:7196–7204
22. Barbaric J, Wagenknecht H-A (2006) DNA as a supramolecular scaffold for the helical arrangement of a stack of 1-ethynylpyrene chromophores. Org Biomol Chem 4:2088–2090
23. Nguyen T, Brewer A, Stulz E (2009) Duplex stabilization and energy transfer in zipper porphyrin-DNA. Angew Chem Int Ed 48:1974–1977

24. Albinsson B, Hannestad JK, Borjesson K (2012) Functionalized DNA nanostructures for light harvesting and charge separation. Coord Chem Rev 256:2399–2413
25. Doluca O, Withers JM, Loo TS, Edwards PJB, Gonzalez C, Filichev VV (2015) Interdependence of pyrene interactions and tetramolecular G4-DNA assembly. Org Biomol Chem 13:3742–3748
26. Mutsamwira S, Ainscough EW, Partridge AC, Derrick PJ, Filichev VV (2016) G-Quadruplex supramolecular assemblies in photochemical upconversion. Chem Eur J 22:10376–10381
27. Ensslen P, Gartner S, Glaser K, Colsmann A, Wagenknecht HA (2016) A DNA-fullerene conjugate as a template for supramolecular chromophore assemblies: towards DNA-based solar cells. Angew Chem Int Ed 55:1904–1908
28. Østergaard ME, Hrdlicka PJ (2011) Pyrene-functionalized oligonucleotides and locked nucleic acids (LNAs): tools for fundamental research, diagnostics, and nanotechnology. Chem Soc Rev 40:5771–5788
29. Kashida H, Takatsu T, Sekiguchi K, Asanuma H (2010) An efficient fluorescence resonance energy transfer (FRET) between pyrene and perylene assembled in a DNA duplex and its potential for discriminating single-base changes. Chem Eur J 16:2479–2486
30. Ohya Y, Hashimoto N, Jo S, Nohori T, Yoshikuni T, Ouchi T, Tamiaki H (2009) Synthesis of oligo-DNA containing hydrophilic porphyrin in the main chain, and its energy transfer behaviour in duplex state. Supramol Chem 21:301–309
31. Börjesson K, Tumpane J, Ljungdahl T, Wilhelmsson LM, Norden B, Brown T, Martensson J, Albinsson B (2009) Membrane-anchored DNA assembly for energy and electron transfer. J Am Chem Soc 131:2831–2839
32. Borjesson K, Wiberg J, El-Sagheer AH, Ljungdahl T, Martensson J, Brown T, Norden B, Albinsson B (2010) Functionalized nanostructures: redox-active porphyrin anchors for supramolecular DNA assemblies. ACS Nano 4:5037–5046
33. Börjesson K, Woller JG, Parsa E, Martensson J, Albinsson B (2012) A bioinspired self assembled dimeric porphyrin pocket that binds electron accepting ligands. Chem Commun 48:1793–1795
34. Takada T, Iwaki T, Nakamura M, Yamana K (2017) Photoresponsive electrodes modified with DNA duplexes possessing a porphyrin dimer. Chem Eur J 23:18258–18263
35. Endo M, Fujitsuka M, Majima T (2008) Diastereochemically controlled porphyrin dimer formation on a DNA duplex scaffold. J Org Chem 73:1106–1112
36. Endo M, Fujitsuka M, Majima T (2008) Programmable conformational regulation of porphyrin dimers on geometric scaffold of duplex DNA. Tetrahedron 64:1839–1846
37. Reddy MR, Shibata N, Kondo Y, Nakamura S, Toru T (2006) Design, synthesis, and spectroscopic investigation of zinc dodecakis (trifluoroethoxy) phthalocyanines conjugated with deoxyribonucleosides. Angew Chem Int Ed 45:8163–8166
38. Buckhout-White S, Spillmann CM, Algar WR, Khachatrian A, Melinger JS, Goldman ER, Ancona MG, Medintz IL (2014) Assembling programmable FRET-based photonic networks using designer DNA scaffolds. Nat Commun
39. Balaz M, Li BC, Steinkruger JD, Ellestad GA, Nakanishi K, Berova N (2006) Porphyrins conjugated to DNA as CD reporters of the salt-induced B to Z-DNA transition. Org Biomol Chem 4:1865–1867
40. Balaz M, Bitsch-Jensen K, Mammana A, Ellestad GA, Nakanishi K, Berova N (2007) Porphyrins as spectroscopic sensors for conformational studies of DNA. Pure Appl Chem 79:801–809
41. Mammana A, Pescitelli G, Asakawa T, Jockusch S, Petrovic AG, Monaco RR, Purrello R, Turro NJ, Nakanishi K, Ellestad GA, Balaz M, Berova N (2009) Role of environmental factors on the structure and spectroscopic response of 5'-DNA-porphyrin conjugates caused by changes in the porphyrin-porphyrin interactions. Chem Eur J 15:11853–11866
42. Singleton DG, Hussain R, Siligardi G, Kumar P, Hrdlicka PJ, Berova N, Stulz E (2016) Increased duplex stabilization in porphyrin-LNA zipper arrays with structure dependent exciton coupling. Org Biomol Chem 14:149–157

43. Lewis FD, Zhang LG, Liu XY, Zuo XB, Tiede DM, Long H, Schatz GC (2005) DNA as helical ruler: exciton-coupled circular dichroism in DNA conjugates. J Am Chem Soc 127:14445–14453
44. Burns JR, Wood JW, Stulz E (2020) A porphyrin-DNA chiroptical molecular ruler with base pair resolution. Front Chem 8:8
45. Joseph J, Baumann KN, Koehler P, Zuehlsdorff TJ, Cole DJ, Weber J, Bohndiek SE, Hernandez-Ainsa S (2017) Distance dependent photoacoustics revealed through DNA nanostructures. Nanoscale 9:16193–16199
46. Nguyen T, Hakansson P, Edge R, Collison D, Goodman BA, Burns JR, Stulz E (2014) EPR based distance measurement in Cu-porphyrin-DNA. New J Chem 38:5254–5259
47. Sonnichsen C, Reinhard BM, Liphardt J, Alivisatos AP (2005) A molecular ruler based on plasmon coupling of single gold and silver nanoparticles. Nat Biotechnol 23:741–745
48. Liu GL, Yin YD, Kunchakarra S, Mukherjee B, Gerion D, Jett SD, Bear DG, Gray JW, Alivisatos AP, Lee LP, Chen FQF (2006) A nanoplasmonic molecular ruler for measuring nuclease activity and DNA footprinting. Nat Nanotechnol 1:47–52
49. Berlin K, Jain RK, Simon MD, Richert C (1998) A porphyrin embedded in DNA. J Org Chem 63:1527–1535
50. Morales-Rojas H, Kool ET (2002) A porphyrin C-nucleoside incorporated into DNA. Org Lett 4:4377–4380
51. Vybornyi M, Nussbaumer AL, Langenegger SM, Häner R (2014) Assembling multiporphyrin stacks inside the DNA double Helix. Bioconjug Chem 25:1785–1793
52. Malinovskii VL, Samain F, Häner R (2007) Helical arrangement of interstrand stacked pyrenes in a DNA framework. Angew Chem Int Ed 46:4464–4467
53. Clever GH, Carell T (2007) Controlled stacking of 10 transition-metal ions inside a DNA duplex. Angew Chem Int Ed 46:250–253
54. Baumstark D, Wagenknecht HA (2008) Fluorescent hydrophobic zippers inside duplex DNA: interstrand stacking of perylene-3,4: 9,10-tetracarboxylic acid bisimides as artificial DNA base dyes. Chem Eur J 14:6640–6645
55. Takada T, Ido M, Ashida A, Nakamura M, Yamana K (2016) DNA-templated synthesis of perylenediimide stacks utilizing Abasic sites as binding pockets and reactive sites. Chembiochem 17:2230–2233
56. Probst M, Aeschimann W, Chau TTH, Langenegger SM, Stocker A, Haner R (2016) Structural insight into DNA-assembled oligochromophores: crystallographic analysis of pyrene- and phenanthrene-modified DNA in complex with BpuJI endonuclease. Nucleic Acids Res 44:7079–7089
57. Malinovskii VL, Wenger D, Häner R (2010) Nucleic acid-guided assembly of aromatic chromophores. Chem Soc Rev 39:410–422
58. Bouquin N, Malinovskii VL, Guegano X, Liu SX, Decurtins S, Häner R (2008) TTF-modified DNA. Chem Eur J 14:5732–5736
59. Bouquin N, Malinovskii VL, Haner R (2008) Highly efficient quenching of excimer fluorescence by perylene diimide in DNA. Chem Commun 7:1974–1976
60. Häner R, Samain F, Malinovskii VL (2009) DNA-assisted self-assembly of pyrene Foldamers. Chem Eur J 15:5701–5708
61. Xiang YG, Zhang QY, Li ZB, Chen H (2017) Role of electrostatic complementarity between perylenediimide and porphyrin in highly stabilized GNA. Mater Sci Eng C Mater Biol Appl 70:1156–1162
62. Kashida H, Sekiguchi K, Liang X, Asanuma H (2010) Accumulation of fluorophores into DNA duplexes to mimic the properties of quantum dots. J Am Chem Soc 132:6223–6230
63. Probst M, Langenegger SM, Häner R (2014) A modular LHC built on the DNA three-way junction. Chem Commun 50:159–161
64. Winiger CB, Langenegger SM, Calzaferri G, Haner R (2015) Formation of two homochromophoric H-aggregates in DNA-assembled alternating dye stacks. Angew Chem Int Ed 54:3643–3647

65. Duprey J, Carr-Smith J, Horswell SL, Kowalski J, Tucker JHR (2016) Macrocyclic metal complex-DNA conjugates for electrochemical sensing of single nucleobase changes in DNA. J Am Chem Soc 138:746–749
66. Yang H, Rys AZ, McLaughlin CK, Sleiman HF (2009) Templated ligand environments for the selective incorporation of different metals into DNA. Angew Chem Int Ed 48:9919–9923
67. Clever GH, Kaul C, Carell T (2007) DNA-metal base pairs. Angew Chem Int Ed 46:6226–6236
68. Clever GH, Shionoya M (2010) Metal-base pairing in DNA. Coord Chem Rev 254:2391–2402
69. Zimmermann N, Meggers E, Schultz PG (2004) A second-generation copper(II)-mediated metallo-DNA-base pair. Bioorg Chem 32:13–25
70. Clever GH, Polborn K, Carell T (2005) A highly DNA-duplex-stabilizing metal-Salen Base pair. Angew Chem Int Ed 44:7204–7208
71. Clever GH, Soltl Y, Burks H, Spahl W, Carell T (2006) Metal-salen-base-pair complexes inside DNA: complexation overrides sequence information. Chem Eur J 12:8708–8718
72. Clever GH, Reitmeier SJ, Carell T, Schiemann O (2010) Antiferromagnetic coupling of stacked Cu-II-Salen complexes in DNA. Angew Chem Int Ed 49:4927–4929
73. Mandal S, Hebenbrock M, Muller J (2018) A dinuclear silver(I)-mediated base pair in DNA formed from 1,N-6-ethenoadenine and thymine. Inorg Chim Acta 472:229–233
74. Mendez-Arriaga JM, Maldonado CR, Dobado JA, Galindo MA (2018) Silver(I)-Mediated Base pairs in DNA sequences containing 7-Deazaguanine/cytosine: towards DNA with entirely metallated Watson-Crick base pairs. Chem Eur J 24:4583–4589
75. Jash B, Muller J (2018) Concomitant site-specific incorporation of silver(I) and mercury (II) ions into a DNA duplex. Chem Eur J 24:10636–10640
76. Jash B, Muller J (2020) Stable Hg(II)-mediated base pairs with a phenanthroline-derived nucleobase surrogate in antiparallel-stranded DNA. J Biol Inorg Chem 25:647–654
77. Takezawa Y, Maeda W, Tanaka K, Shionoya M (2009) Discrete self-assembly of iron(III) ions inside triple-stranded artificial DNA. Angew Chem Int Ed 48:1081–1084
78. Takezawa Y, Suzuki A, Nakaya M, Nishiyama K, Shionoya M (2020) Metal-dependent dna base pairing of 5-carboxyuracil with itself and all four canonical nucleobases. J Am Chem Soc 142(52):21640–22164
79. Zhang YW, Seeman NC (1994) Construction of a DNA-truncated octahedron. J Am Chem Soc 116:1661–1669
80. Goodman RP, Schaap IAT, Tardin CF, Erben CM, Berry RM, Schmidt CF, Turberfield AJ (2005) Rapid chiral assembly of rigid DNA building blocks for molecular nanofabrication. Science 310:1661–1665
81. Goodman RP, Berry RM, Turberfield AJ (2004) The single-step synthesis of a DNA tetrahedron. Chem Commun 12:1372–1373
82. Pistol C, Dwyer C (2007) Scalable, low-cost, hierarchical assembly of programmable DNA nanostructures. Nanotechnology
83. Yan H, Park SH, Finkelstein G, Reif JH, LaBean TH (2003) DNA-templated self-assembly of protein arrays and highly conductive nanowires. Science 301:1882–1884
84. Park SH, Pistol C, Ahn SJ, Reif JH, Lebeck AR, Dwyer C, LaBean TH (2006) Finite-size, fully addressable DNA tile lattices formed by hierarchical assembly procedures. Angew Chem Int Ed 45:735–739
85. Batalia MA, Protozanova E, MacGregor RB Jr, Erie DA (2002) Self-assembly of frayed wires and frayed-wire networks: nanoconstruction with multistranded DNA. Nano Lett 2:269–274
86. Rothemund PWK (2006) Folding DNA to create nanoscale shapes and patterns. Nature 440:297–302
87. O'Neill P, Rothemund PW, Kumar A, Fygenson DK (2006) Sturdier DNA nanotubes via ligation. Nano Lett 6:1379–1383
88. Endo M, Seeman NC, Majima T (2005) DNA tube structures controlled by a four-way-branched DNA connector. Angew Chem Int Ed 44:6074–6077

89. Erben CM, Goodman RP, Turberfield AJ (2007) A self-assembled DNA bipyramid. J Am Chem Soc 129:6992–6993
90. Hamblin GD, Rahbani JF, Sleiman HF (2015) Sequential growth of long DNA strands with user-defined patterns for nanostructures and scaffolds. Nat Commun 6:8
91. Burns JR, Gopfrich K, Wood JW, Thacker VV, Stulz E, Keyser UF, Howorka S (2013) Lipid-bilayer-spanning DNA nanopores with a bifunctional porphyrin anchor. Angew Chem Int Ed 52:12069–12072
92. Burns JR, Stulz E, Howorka S (2013) Self-assembled DNA nanopores that span lipid bilayers. Nano Lett 13:2351–2356
93. Burns JR, Al-Juffali N, Janes SM, Howorka S (2014) Membrane-spanning DNA nanopores with cytotoxic effect. Angew Chem-Int Ed 53:12466–12470
94. Göpfrich K, Li C-Y, Mames I, Bhamidimarri SP, Ricci M, Yoo J, Mames A, Ohmann A, Winterhalter M, Stulz E, Aksimentiev A, Keyser UF (2016) Ion channels made from a single membrane-spanning DNA Duplex. Nano Lett 16:4665–4669
95. Li Q, Zhao JM, Liu LF, Jonchhe S, Rizzuto FJ, Mandal S, He HW, Wei SS, Sleiman HF, Mao HB, Mao CD (2020) A poly(thymine)-melamine duplex for the assembly of DNA nanomaterials. Nat Mater 19:1012
96. Green SJ, Lubrich D, Turberfield AJ (2006) DNA hairpins: fuel for autonomous DNA devices. Biophys J 91:2966–2975
97. Yurke B, Turberfield AJ, Mills AP, Simmel FC, Neumann JL (2000) A DNA-fuelled molecular machine made of DNA. Nature 406:605–608
98. Zheng JP, Birktoft JJ, Chen Y, Wang T, Sha RJ, Constantinou PE, Ginell SL, Mao CD, Seeman NC (2009) From molecular to macroscopic via the rational design of a self-assembled 3D DNA crystal. Nature 461:74–77
99. Andersen ES, Dong M, Nielsen MM, Jahn K, Subramani R, Mamdouh W, Golas MM, Sander B, Stark H, Oliveira CLP, Pedersen JS, Birkedal V, Besenbacher F, Gothelf KV, Kjems J (2009) Self-assembly of a nanoscale DNA box with a controllable lid. Nature 459:73–76
100. Turberfield AJ, Mitchell JC, Yurke B, Mills AP, Blakey MI, Simmel FC (2003) DNA fuel for free-running nanomachines. Phys Rev Lett 90
101. Li Z, Liu MH, Wang L, Nangreave J, Yan H, Liu Y (2010) Molecular behavior of DNA origami in higher-order self-assembly. J Am Chem Soc 132:13545–13552
102. Han DR, Pal S, Nangreave J, Deng ZT, Liu Y, Yan H (2011) DNA origami with complex curvatures in three-dimensional space. Science 332:342–346
103. Lo PK, Karam P, Aldaye FA, McLaughlin CK, Hamblin GD, Cosa G, Sleiman HF (2010) Loading and selective release of cargo in DNA nanotubes with longitudinal variation. Nat Chem 2:319–328
104. Dunn KE, Dannenberg F, Ouldridge TE, Kwiatkowska M, Turberfield AJ, Bath J (2015) Guiding the folding pathway of DNA origami. Nature 525:82–86
105. Hernandez-Ainsa S, Keyser UF (2014) DNA origami nanopores: developments, challenges and perspectives. Nanoscale 6:14121–14132
106. Keller A, Linko V (2020) Challenges and perspectives of DNA nanostructures in biomedicine. Angew Chem Int Ed 59:15818–15833
107. Sun QY, Han Y, Yang YM, de La Fuente JM, Cui DX, Wang XQ (2020) Application of DNA nanostructures in cancer therapy. Appl Mater Today 21:15
108. Wang H, Luo D, Wang H, Wang F, Liu X (2020) Construction of smart stimuli-responsive DNA Nanostructures for biomedical applications. Chem Eur J 16
109. Jahanban-Esfahlan A, Seidi K, Jaymand M, Schmidt TL, Majdi H, Javaheri T, Jahanban-Esfahlan R, Zare P (2019) Dynamic DNA nanostructures in biomedicine: beauty, utility and limits. J Control Release 315:166–185
110. Huang XG, Blum NT, Lin J, Shi JJ, Zhang C, Huang P (2021) Chemotherapeutic drug-DNA hybrid nanostructures for anti-tumor therapy. Mater Horizons 8:78–101

111. Yoon Y, Lee C, Kim SJ (2021) Structural DNA nanotechnology towards therapeutic applications. J Korean Phys Soc 8
112. Zhang XY, Gong CC, Akakuru OU, Su ZQ, Wu AG, Wei G (2019) The design and biomedical applications of self-assembled two-dimensional organic biomaterials. Chem Soc Rev 48:5564–5595
113. Hu QQ, Li H, Wang LH, Gu HZ, Fan CH (2019) DNA nanotechnology-enabled drug delivery systems. Chem Rev 119:6459–6506
114. Lu XH, Liu JB, Wu XH, Ding BQ (2019) Multifunctional DNA origami Nanoplatforms for drug delivery. Chem Asian J 14:2193–2202
115. Xu F, Xia Q, Wang PF (2020) Rationally designed DNA nanostructures for drug delivery. Front Chem 8:13
116. Sameiyan E, Bagheri E, Ramezani M, Alibolandi M, Abnous K, Taghdisi SM (2019) DNA origami-based aptasensors. Biosens Bioelectron 143:12
117. Xiao MS, Lai W, Man TT, Chang BB, Li L, Chandrasekaran AR, Pei H (2019) Rationally engineered nucleic acid architectures for biosensing applications. Chem Rev 119:11631–11717
118. Loretan M, Domljanovic I, Lakatos M, Ruegg C, Acuna GP (2020) DNA origami as emerging technology for the engineering of fluorescent and plasmonic-based biosensors. Materials 13:24
119. Fu XY, Peng FQ, Lee J, Yang Q, Zhang F, Xiong MY, Kong GZ, Meng HM, Ke GL, Zhang XB (2020) Aptamer-functionalized DNA nanostructures for biological applications. Top Curr Chem 378:43
120. Kogikoski S, Paschoalino WJ, Cantelli L, Silva W, Kubota LT (2019) Electrochemical sensing based on DNA nanotechnology. TrAC Trends Anal Chem 118:597–605
121. Feng L, Li J, Sun J, Wang L, Fan C, Shen J (2021) Recent advances of DNA nanostructure-based cell membrane engineering. Adv Healthcare Mater
122. Liu XG, Zhao Y, Liu P, Wang LH, Lin JP, Fan CH (2019) Biomimetic DNA nanotubes: nanoscale channel design and applications. Angew Chem Int Ed 58:8996–9011
123. Stephanopoulos N (2020) Hybrid nanostructures from the self-assembly of proteins and DNA. Chem 6:364–405
124. Dong YC, Mao YD (2019) DNA origami as scaffolds for self-assembly of lipids and proteins. Chembiochem 20:2422–2431
125. Mishra S, Feng YH, Endo M, Sugiyama H (2020) Advances in DNA origami-cell interfaces. Chembiochem 21:33–44
126. Li X, Yang DL, Shen LY, Xu F, Wang PF (2020) Programmable assembly of DNA-protein hybrid structures. Chem Res Chin Univ 36:211–218
127. Bae W, Kocabey S, Liedl T (2019) DNA nanostructures in vitro, in vivo and on membranes. Nano Today 26:98–107
128. Yin J, Wang JK, Niu RJ, Ren SK, Wang DX, Chao J (2020) DNA nanotechnology-based biocomputing. Chem Res Chin Univ 36:219–226
129. Wang JK, Yin J, Niu RJ, Ren SK, Chao J (2020) DNA computing and DNA nanotechnology. J Electron Inf Technol 42:1313–1325
130. Vittala SK, Han D (2020) DNA-guided assemblies toward nanoelectronic applications. ACS Appl Biol Mater 3:2702–2722
131. Nummelin S, Shen BX, Piskunen P, Liu Q, Kostiainen MA, Linko V (2020) Robotic DNA nanostructures. ACS Synth Biol 9:1923–1940
132. Li H, Hao YY, Wang F, Wang LH, Liu G (2020) DNA nanostructures in the study of molecular interactions. Acta Polym Sin 51:728–737
133. Hannewald N, Winterwerber P, Zechel S, Ng DYW, Hager MD, Weil T, Schubert US (2020) DNA origami meets polymers: a powerful tool for the design of defined nanostructures. Angew Chem Int Ed
134. Ghosh D, Datta LP, Govindaraju T (2020) Molecular architectonics of DNA for functional nanoarchitectures. Beilstein J Nanotechnol 11:124–140

135. Lombardo D, Calandra P, Pasqua L, Magazu S (2020) Self-assembly of organic nanomaterials and biomaterials: the bottom-up approach for functional nanostructures formation and advanced applications. Materials 13:42
136. Johnson JA, Dehankar A, Robbins A, Kabtiyal P, Jergens E, Lee KH, Johnston-Halperin E, Poirier M, Castro CE, Winter JO (2019) The path towards functional nanoparticle-DNA origami composites. Mater Sci Eng R Rep 138:153–209
137. Shin SW, Yuk JS, Chun SH, Lim YT, Um SH (2020) Hybrid material of structural DNA with inorganic compound: synthesis, applications, and perspective. Nano Converg 7:12
138. DeLuca M, Shi Z, Castro CE, Arya G (2020) Dynamic DNA nanotechnology: toward functional nanoscale devices. Nanoscale Horiz 5:182–201
139. Zhang YW, Pan V, Li X, Yang XQ, Li HF, Wang PF, Ke YG (2019) Dynamic DNA structures. Small 15:9
140. Ma NN, Minevich B, Liu JL, Ji M, Tian Y, Gang O (2020) Directional assembly of nanoparticles by DNA shapes: towards designed architectures and functionality. Top Curr Chem 378:34
141. Heuer-Jungemann A, Liedl T (2019) From DNA tiles to functional DNA materials. Trends Chem 1:799–814
142. Yang Y, Zhang R, Fan CH (2020) Shaping functional materials with DNA frameworks. Trends Chem 2:137–147
143. Chen YH, Sun W, Yang CY, Zhu Z (2020) Scaling up DNA self-assembly. ACS Appl Bio Mater 3:2805–2815
144. Platnich CM, Rizzuto FJ, Cosa G, Sleiman HF (2020) Single-molecule methods in structural DNA nanotechnology. Chem Soc Rev 49:4220–4233
145. Kekic T, Barisic I (2020) In silico modelling of DNA nanostructures. Comp Struct Biotechnol J 18:1191–1201
146. Bush J, Singh S, Vargas M, Oktay E, Hu CH, Veneziano R (2020) Synthesis of DNA origami scaffolds: current and emerging strategies. Molecules 25:18
147. Dunn KE (2020) The business of DNA nanotechnology: commercialization of origami and other technologies. Molecules 25:11
148. Jaekel A, Lill P, Whitelam S, Sacca B (2020) Insights into the structure and energy of DNA nanoassemblies. Molecules 25:25
149. Piskunen P, Nummelin S, Shen BX, Kostiainen MA, Linko V (2020) Increasing complexity in wireframe DNA nanostructures. Molecules 25:17
150. Zhao Y, Dai XP, Wang F, Zhang XL, Fan CH, Liu XG (2019) Nanofabrication based on DNA nanotechnology. Nano Today 26:123–148
151. Yan XH, Huang SJ, Wang Y, Tang YY, Tian Y (2020) Bottom-up self-assembly based on DNA nanotechnology. Nano 10:14
152. Fan SS, Wang DF, Kenaan A, Cheng J, Cui DX, Song J (2019) Create nanoscale patterns with DNA Origami. Small 15:12
153. Dong JY, Zhou C, Wang QB (2020) Towards active self-assembly through DNA nanotechnology. Top Curr Chem 378:25
154. Scalvini B, Sheikhhassani V, Woodard J, Aupi J, Dame RT, Jerala R, Mashaghi A (2020) Topology of folded molecular chains: from single biomolecules to engineered origami. Trends Chem 2:609–622

Chapter 12
Functional Molecule-Templated DNA Molecular Architectonics

Lakshmi Priya Datta, Debasis Ghosh, and Thimmaiah Govindaraju

12.1 Introduction

DNA is the molecule of life responsible for information storage and transmission from one generation to another through its molecular assembly modulation. The intriguing and built-in information for molecular recognition to modulate the assembly-disassembly process allowed researchers to exploit DNA as a molecular building block to construct novel material architectures through predictable and programmable nucleobases (weak noncovalent interactions) driven by hybridization process [1, 2]. Single-stranded DNA (ssDNA) sequences are employed as building blocks to generate higher-ordered DNA assembly architectures in the celebrated area of DNA nanotechnology and DNA origami [3]. In classical DNA nanotechnology, ssDNA strands are used as "smart glues" to form geometrical motifs of branched structures at nanoscale dimensions. The structural DNA nanotechnology harvests the formation of double-stranded DNA (dsDNA) bound together by canonical Watson-Crick (WC) hydrogen bonding interactions with nanoscale structural characteristics, viz., thickness of 2 nm and 40–50 nm persistence length [4, 5]. These molecular and nanoscale structural characteristics of DNA are of significant interest to researchers to construct material architectures.

The competence of DNA strands to form noncovalent interactions driven by predictable structures has opened a plethora of opportunities to engineer the molecular interactions for designing co-assembly and assembly-based novel hybrid architectures. Further, the novel hybrid DNA architectures can be accomplished by employing meticulously designed external functional molecules as templates within

L. P. Datta · D. Ghosh · T. Govindaraju (✉)
Bioorganic Chemistry Laboratory, New Chemistry Unit, and School of Advanced Materials (SAMAt), Jawaharlal Nehru Centre for Advanced Scientific Research, Bengaluru, Karnataka, India
e-mail: tgraju@jncasr.ac.in

T. Govindaraju, K. Ariga (eds.), *Molecular Architectonics and Nanoarchitectonics*, Nanostructure Science and Technology,
https://doi.org/10.1007/978-981-16-4189-3_12

the natural molecular systems to orchestrate the noncovalent interactions with precision. In essence, the arbitrary concepts of fabrication of material and molecular architectures are fast shifting toward the advanced and precision design to gain structure and functional superiority for technological applications. The principles of molecular architectonics, a state-of-the-art innovation referring to custom design and engineering of molecular assembly to construct advanced molecular and material architectures with defined shape, size, and properties, are invaluable in the design and construction of small-molecule-templated DNA architectures [6–11]. The scheme of molecular architectonics offers to manipulate complex yet universal noncovalent interactions at the molecular level to create higher-order molecular assemblies and co-assemblies of modular building units supplemented with chemical or biomolecular auxiliaries. Biomolecules are fundamental to the process of evolution and exemplify extraordinary specificity and fidelity in terms of molecular recognition in a myriad of living systems. Notably, controlling the noncovalent interactions at the molecular level in synthetic molecular systems is a challenging task. The art of translating nature's specificity to achieve controlled molecular organization of synthetic molecular systems can possibly be accomplished by using functional auxiliaries. The molecular building blocks of life, viz., amino acids, peptides, nucleobases, sugars, and lipids with built-in information for molecular recognition, can serve as auxiliaries to drive the controlled molecular organization of functional synthetic molecules to well-defined molecular and material architectures [6, 7, 12, 13]. In the scheme of molecular architectonics, biomolecular auxiliary-driven molecular recognition plays a key role in guiding the controlled organization of modular units to construct of 0D, 1D, 2D, and 3D architectures with functional properties and applications in the domains of health, energy, and environment [6–11]. In particular, the biomolecules or biomimetic auxiliaries aid the assembly of functional modular units to construction of well-defined molecular and nanoarchitectures [12, 14–16].

The molecular architectonics offer a paradigm shift from the classical DNA nanotechnology to exploit the material property of DNA through mutual assembly with molecular templates [17]. The canonical and noncanonical hydrogen bonding interactions including the aromatic π-π and electrostatic interactions drive the formation of functional molecule-templated DNA architectures (Fig. 12.1) [6, 7, 9, 10, 12, 15, 18]. The scheme of molecular architectonics guides the establishment of small functional molecule (SFM)-templated DNA nanotechnology to construct functional DNA nanoarchitectures [6, 17, 19–21]. The templated DNA nanotechnology integrates the functional properties of small molecules and versatility of DNA to overcome the limitations of classical DNA nanotechnology for practical applications [17, 19]. Short oligonucleotide sequences (dB_n, where B = nucleobase and $n < 40$–50) are easy to synthesize or procure commercially and hence considered as ideal candidates to design robust and cost-effective functional nanoarchitectures [21]. The assembly of SFMs through hydrophobic and aromatic π-π interactions supports the templating platforms for dB_n. Such molecular templating platform and translation of molecular assembly characteristics lead to the creation of molecular and nanoarchitectures of hybrid SFM-DNA ensembles.

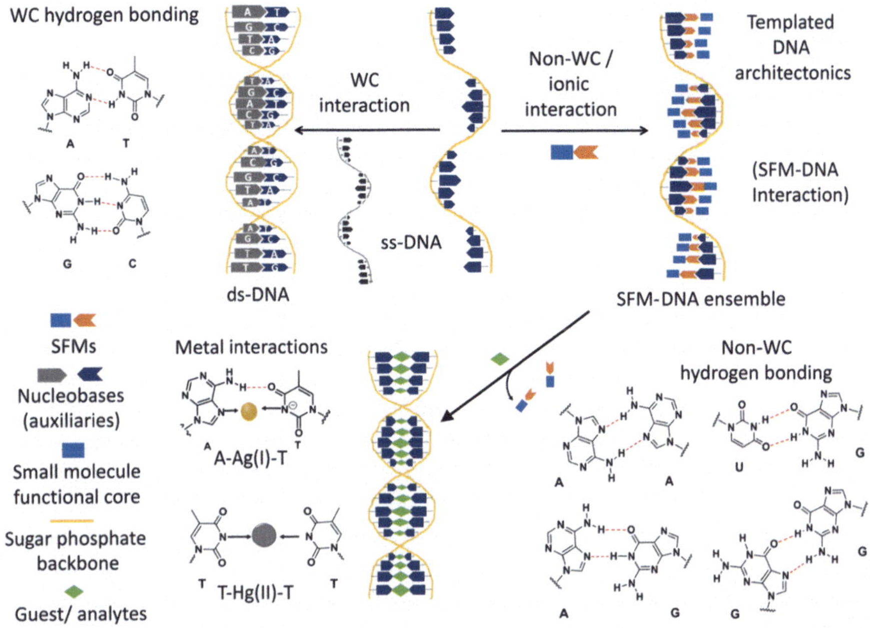

Fig. 12.1 Schematic outline of canonical (WC) and noncanonical (non-WC) hydrogen bonding and nucleobase-metal ion interactions in DNA assembly and small functional molecule (SFM)-templated DNA ensembles

Investigation of the assembly mechanisms at molecular level is important in understanding the templated co-assembly systems. The bottom-up fabrication methodology, integration of SFMs, use of short oligonucleotides, reproducibility, and overall cost-effectiveness guarantee the practical utility of templated DNA nanotechnology. In this chapter, we present the design of mutually templated SFM-DNA molecular and nanoarchitectures constructed through canonical/Watson-Crick (WC) and non-canonical/non-Watson-Crick (non-WC) hydrogen bonding interactions of nucleobases. The nucleobases serve as auxiliaries to aid the tailorability of SFM-templated DNA ensembles employing the principles of molecular architectonics. Further, meticulous use of various dynamic noncovalent interactions such as hydrogen bonding and hydrophobic, aromatic π-π, ionic, and metal-guided interactions for the construction of SFM-DNA architectures are discussed (Fig. 12.1). Overall, recent contributions in the establishment of SFM-DNA molecular and materials architectonics are highlighted with a range of applications in materials science and biomedicine with practical treatment of the subject from the viewpoint of the current progress and future directions.

12.1.1 SFM Toolbox

In the scheme of molecular architectonics, we propose the creation pool of meticulously designed SFMs to support the construction of tailored-made templated DNA ensembles with desired functional properties and applications [6, 10, 17, 20]. DNA and its assembly through reversible hybridization have been shown to play a vital role in the construction of higher-ordered architectures [21, 22]. The templated DNA nanoarchitectonics presented in this chapter is broadly classified into two categories: (i) SFM governs the assembly process through noncovalent hydrogen bonding and hydrophobic or electrostatic interactions and (ii) SFM-metal or nanoparticle interactions help to stabilize the templated interactions. The suitably functionalized polycyclic aromatic chromophores and metallocycles have been extensively designed and studied as SFMs to support the bottom-up templated assembly through aromatic π-π, hydrogen bonding and metal coordination interactions to create the well-defined hybrid ensembles of ssDNA [23–25]. The π-conjugated systems are particularly advantageous as the functional core in the design of SFMs, as they offer directionality for hydrogen bonding and hydrophobic, electrostatic and metal-mediated interactions. In an attempt to design SFMs to template ssDNA, π-conjugated functional cores functionalized with nucleobases have been developed in our laboratory [7, 8, 26, 27]. In particular, naphthalenediimide (NDI) and perylenediimide (PDI) cores with nucleobase auxiliaries are designed as SFMs to mediate the mutual templating of short oligonucleotides through canonical or noncanonical hydrogen bonding, metal-base pair, and ionic interactions (Fig. 12.2) [22, 28, 29, 36].

12.1.2 Templated DNA Architectures

The templated molecular and nanoarchitectonics of DNA employs SFMs and nanomaterials as templates to guide the assembly of DNA into well-defined architectures. The molecular-level interactions of SFMs consisting of different synthetic organic cores contribute to optical and electronic properties which are essential to realize the practical applications of templated DNA architectures in imaging, biosensing, and drug delivery, among others. Nanoparticles exhibit unique high surface-to-volume ratio and optical, electronic, plasmonic, and magnetic properties different from their bulk materials making them special candidates for the templated organization of DNA to generate diverse assembly architectures with applications ranging from sensor devices, imaging, and gene delivery agents. For clarity of presentation and understanding, various templated interactions are classified into different categories.

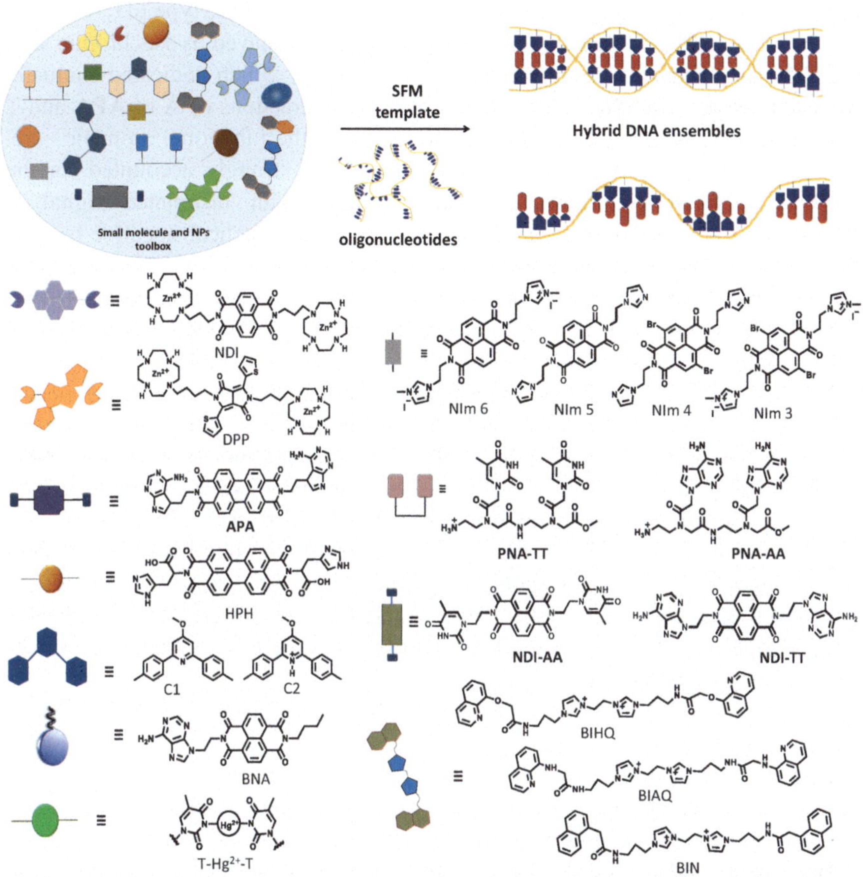

Fig. 12.2 Small functional molecules (SFMs) as a toolbox to choose a suitable molecular template to construct SFM-DNA molecular and nanoarchitectures. Molecular structures of selected SFMs to construct templated DNA architectures [22, 28–35]

12.1.2.1 SFM-Templated DNA Architectonics Driven by Canonical Hydrogen Bonding Interactions

We developed methodologies for the construction of SFM-templated DNA nanoarchitectures through the scheme of molecular architectonics to cater to applications ranging from imaging, neurotoxic metal sensing to drug delivery. Envisaging the π-stacking interaction and propensity for helical organization, planar organic dyes tethered with nucleobases were designed as small-molecule templates. These nucleobase-functionalized SFMs are anticipated to undergo helical assembly and mutually templated ssDNA (oligonucleotide) through canonical or noncanonical hydrogen bonding interactions to produce chiral hybrid ensembles. The symmetrically functionalized PDI with adenine (A) units (**APA**) templates ssDNA through canonical A-T base pairing to form hybrid DNA ensemble of the type [dT_n-(**APA**)$_n$-dT_n, $n = 10$ or 20] [22]. [dT_n-(**APA**)$_n$-dT_n] ensemble displayed hypsochromic shift in UV-Vis spectrum as compared to **APA** assembly, which indicated mutually templated organization through hydrogen bonding interactions of A of **APA** and T of deoxyoligothymidine (dT_n) with reorganization in the functional aromatic core. Similarly, circular dichroism (CD) spectroscopy data strongly accounted for the presence of mutually imprinted chiral assembly with bisignated signal at 400–600 nm corresponding to the PDI chromophore and helical negative band at 250–300 nm corresponding to nucleobase absorption (Fig. 12.3a). These studies have confirmed the significant contribution of canonical hydrogen bonding (A-T pairing) interactions and suitably adjusted aromatic π-π stacking interactions of perylene core to the templated co-assembly formation. Remarkably, the atomic force microscopy (AFM) study confirmed the formation of double-zipper helical assembly of [dT_n-(**APA**)$_n$-dT_n] with left-handed helical (M-helix) signature and lateral thickness of ~5 nm (Fig. 12.3a). The mutual templating between **APA** and dA_n or dG_n was found to occur through noncanonical hydrogen bonding (A/A and A/G, respectively) interactions to form P- and M-helical architectures. The hybrid ensembles [dB_n-(**APA**)$_n$-dB_n] are formed mainly through hydrogen bonding interactions and serve as pH-responsive systems. The double-zipper assemblies are formed at neutral conditions and collapse in acidic conditions (<pH 6), and hence, these DNA ensembles can be used as pH-responsive delivery systems for small-molecule or gene-based drugs.

In another study, NDI functionalized with alkyl (butyl) chain and A (BNA) was used as SFM to construct templated chiral and conducting hybrid DNA ensembles with dT_n through canonical A-T base pairing interaction supported by π-π stacking of the NDI core, van der Waals, and hydrophobic interactions of alkyl chain [29]. The NDI-based SFM (BNA) was chosen for its semiconducting property and the BNA-templated DNA ensemble was shown to exhibit inherent conductivity. The higher-ordered hierarchical ensembles of BNA and dT_n ($n = 6, 10, 20$) were constructed based on canonical A-T (A of BNA and T of dT_n) base pairing supported by aromatic π-stacking of NDI core and cooperative hydrophobic interaction of butyl chain (Fig. 12.3b). Remarkably, [BNA_n-dT_n] formed 2D nanosheets and the lateral dimensions of the sheets were found to be dependent on the dT_n sequence length. The conductivity of [BNA_n-dT_n] material architectures was measured by constructing a field-effect transistor (FET) device (Fig. 12.3b). BNA alone self-assembles to form 1D tape and the electrical conductivity differs greatly between 2D nanosheet and 1D tape architectures of [BNA_n-dT_n] and BNA, respectively. The transport characteristics were accurately evaluated by analyzing and comparing the doping levels of BNA and [BNA_n-dT_n], which was found to be 1×10^9 for 1D tapes of BNA and 1×10^{11} for 2D nanosheets of [BNA_n-dT_n]. These transport characteristics are attributed to favorable π-π stacking of BNA in the templated DNA nanoarchitecture. A significant and predictable variation in the conducting property of [BNA_n-dT_n] ensemble was observed as a function of dT_n length, which is attributed to enhanced crystalline molecular ordering as revealed by the powder X-ray diffraction (PXRD) data. Such conducting nanoarchitectures of DNA are

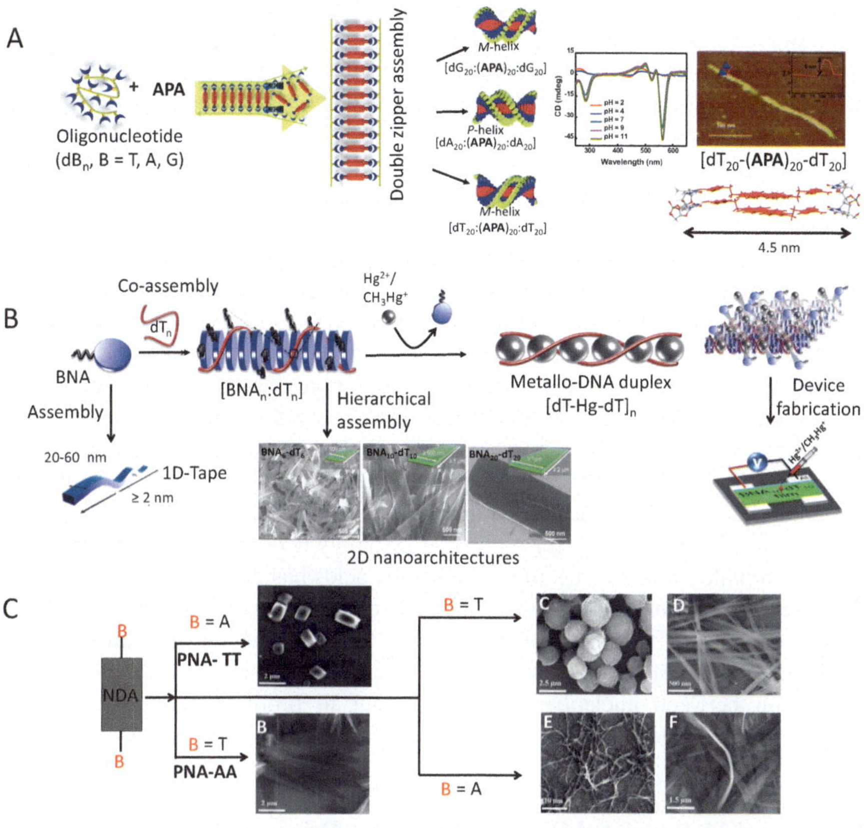

Fig. 12.3 (**a**) Schematic representation of the mutually templated double helical zipper assembly of **APA** and dB_n (T, A, or G; $n = 10, 20$) through canonical and noncanonical hydrogen bonding interactions. pH-dependent CD measurement confirmed dissociation of the ensemble architecture in acidic conditions. Also shown are the P- and M-type double-zipper helical assemblies. AFM image of [dT_{20}-(**APA**)$_{20}$-dT_{20}] and its height profile data (5 nm, inset); typical thickness (4.5 nm). (**b**) 2D nanoarchitectures (nanosheets) were formed in a mutually templated assembly of BNA and dT_n. The lateral dimension of [BNA_n-dT_n] of 2D nanosheets was directly proportional to the chain length of the dT_n. Also shown is the mercury-mediated displacement of BNA from the [BNA_n-dT_n] ensemble and subsequent formation of the metallo-DNA duplex [dT-Hg-dT]n. Schematic representation of FET device fabricated using [BNA_{10}-dT_{10}] for the detection of Hg^{2+} or CH_3Hg^+. (**c**) Molecular structures of nucleobase-tethered NDI molecules (NDI-AA and NDI-TT) and their self-assembly, and templated co-assembly with **PNA** dimer clamps. (**a**) Adapted with permission from Ref. [22], Royal Society of Chemistry (RSC). (**b**) Adapted with permission from Ref. [29], American Chemical Society. (**c**) Reproduced by permission of the Royal Society of Chemistry (RSC) on behalf of the Centre National de la Recherche Scientifique (CNRS) and the RSC

expected to find potential applications in the field of bio-optoelectronics. The material architectures of [BNA_n-dT_n] were used for the chiroptical and conductivity-based ultrasensitive detection of neurotoxic mercury at sub-nanomolar concentration (vide supra).

Iazawa and coworkers have demonstrated an interesting design strategy of developing helical nanofibers utilizing the A-T base pairing concept [37]. Anthracene core was used in the design of thymidylic acid-appended 2,6-bis[5-(3′-thymidylic acid) pentyloxy]anthracene derivative (**1**) as SFM and its self-assembly behavior was studied. The noncovalent interaction of **1** with oligodeoxyadenylic acid (dA_{20}) was investigated. The binary assembly between **1** and dA_{20} via complementary base pairing association resulted in the formation of helical J-aggregates with diameter 5.1 nm. Similarly, Shimizu group has reported thymidylic acid-conjugated oligo(*p*-phenylenevinylene) ligand as SFM and studied single component as well as binary molecular assembly with dA_{20} [38]. The studies involving variable stoichiometric ratios of ligand-appended thymine/adenine (1:1 and 2:1) have confirmed the formation of right-handed helical stacks having diameters of 6.4- and 5.1-nm, respectively, depending on the residual stoichiometries. The bottom-up fabrication of nanoscale architectures from molecular building units utilizing the noncovalent interactions is a challenging yet interesting task. The robust molecular recognition properties of DNA facilitate the templating effect induced as well as multiple-recognition-governed synthesis of higher-ordered functional molecular assembly architectures. In this context, the molecular assembly between thymidine-appended bola amphiphiles with a series of oligoadenylic acids has been evaluated [39]. The nanofibers generated from the binary molecular assembly with equimolar stoichiometric adenine and thymine residues resulted in the formation of hydrogel upon incubation for few days. The molecular-level control and established complementary association between nucleobases are important to promote harmonized three-dimensional templated DNA molecular assemblies with elegant functional features.

12.1.2.2 SFM-Templated DNA Architectonics Driven by Noncanonical Hydrogen Bonding Interactions

The exploitation of noncanonical (non-WC base pairing) interactions to fabricate SFM-templated DNA molecular and material architectures is the subject of intense research. This exercise is a particularly challenging task and involves suitably designed nucleobase-functionalized SFMs, which facilitates non-WC base pairing interaction between noncomplementary nucleobases. Utilization of non-WC base pairing interaction in the construction of SFM-templated DNA architectures has been introduced from our laboratory. In addition to canonical interactions with dT_n (vide supra), **APA** is capable of interacting with oligoadenine (dA_n) and oligoguanidine (dG_n) sequences through noncanonical hydrogen bonding interactions attributed to the special property of A to form non-WC-type (A/A and A/G) base pairing interactions [17, 22]. The spectroscopy and microscopy studies have revealed successful construction of hybrid ensemble of the type [dA_n-(**APA**)$_n$-dA_n] and [dG_n-(**APA**)$_n$-dG_n] between **APA** and dA_n/dG_n, respectively, through non-WC interactions. A similar concept was attempted previously with NDI analogue, which was not successful [28]. This investigation revealed that the selection of SFM is crucial in the templated organization of ssDNA. Utilization of synthetic functional cores with nucleobase as versatile hydrogen bonding sites is an attractive strategy to generate well-defined architectures. In an exploratory work with nucleobase-appended NDI derivatives, peptide nucleic acid (**PNA**) dimers were used as molecular clamps (**PNA-TT** and **PNA-AA**) to derive molecular assemblies exploiting complementary hydrogen bonding and aromatic interactions [28]. The noncanonical hydrogen bonding interactions between adenine and thymine units along with aromatic interactions exerted by the NDI core facilitate the adenine-NDI-adenine (**NDA-AA**) to organize into nanoribbons via solution processing and the mutual templating with **PNA-TT** promote the formation of 2D microstructures. Alternatively, the complementary **PNA-AA** templating with thymine-NDI-thymine (**NDA-TT**) conjugates leads to the generation of porous spheres, tapes, and petal-like 2D sheets. The ^{1}H NMR and 2D-NOESY spectra gave insight into the contribution of hydrogen bonding to develop new sets of spatial Hoogsteen base pairing interactions between two complementary nucleobases. The detailed analysis highlights the development of a bio-inspired strategy to modulate the molecular assembly of NDI-chromophores via nucleobase auxiliary. The molecular recognition from nucleobases resulted in a variety of structural evolution encompassing the presence of nine-membered hydrogen-bonded structure instead of eight-membered structure present in WC pairs.

Cyanuric acid (CA) has been explored as a potential SFM to construct the templated DNA assembly in water by reprogramming the nucleobase interactions [40]. CA consisting of three thymine-like faces interacts with polyadenine units forming noncanonical and orthogonal assembly. The hybridized structure of CA-dA_{15} forms fiber-like architectures through the hydrogen bonding interactions that form "hexameric rosette"-like cascade. This work is an excellent example to demonstrate the ability of SFMs to induce noncovalent molecular assembly of nucleic acids in water ensuing the production of functional architectures. The chemical modification of naphthalene moiety to generate a series of derivatives for the templated nanofabrication is a highly innovative strategy. The naphthalene cores chemically functionalized with nucleobases to control their interactions with noncomplementary nucleobases have been evaluated. In an interesting study, the diaminopurine-appended naphthalene derivative was chosen to interact with dT_n to form pH-triggered templated assembly [41]. Interestingly, the right-handed helicity of the templated assembly was converted to left-handed prototype upon changing the solution pH from neutral to acidic region. The detailed analysis revealed that the protonation of naphthalene derivative plays a key role in "helix-reversal" without affecting the base pairing interactions between nucleobases. Overall, specific examples discussed above emphasize the fact that noncanonical hydrogen binding interactions play a vital role in the templated assembly of SFMs and DNA to construct the functional molecular and material architectures.

12.1.2.3 SFM-Templated DNA Architectonics Driven by Ionic Interactions

In nature, the importance of electrostatic interactions is next only to hydrogen bonding interactions to maintain the higher-ordered structural integrity of biomacromolecules and their complex architectures. Accordingly, ionic interactions are expected to play a profound role in stabilizing the templated organizations of SFMs and DNA. Armitage group reported the ionic interaction mediated by co-assembly between organic dyes and ssDNA [42]. Cyanine dyes are valuable optical probes used as sensitizers in photographic application. The two-dimensional orientation of the dye aggregates across the planar surface of DNA resulted in the formation of J-aggregates or H-aggregates [43]. The self-association of dye molecules is a complicated process and depends on several external parameters, like dye and DNA conformations, pH, polarity of medium, and temperature [44]. Modulation of dye aggregation was achieved by employing dimerization of the benzothiazole-functionalized cyanine dye across the minor groove of dsDNA [42]. The dimerization-induced templating allowed the cooperative binding of dye molecules and formation of assembled structures. Subsequently to this work, a tricationic cyanine derivative was reported to transform to helical J-aggregates in the presence of dsDNA structure with a high cooperativity effect [45]. In the presence of physical or chemical stimuli like temperature and ionic strength, the J-aggregates interconvert to H-aggregates. This process was used as a model system to achieve the electrostatic interactions driven by structural conversion of templated DNA systems. A benzothiazole-functionalized cyanine dye was used in the DNA-templated organization to create donor-bridge-acceptor (D-B-A) energy transfer system (Fig. 12.4a) [46]. In this design strategy, Alexa Fluor 350, 30 bp (9.7 nm) DNA-templated cyanine aggregate, and Alexa Fluor 555 were chosen as the donor, molecular bridge, and acceptor, respectively. The energy transfer efficiency of the dye-aggregated molecular wire was found to be 96% without the loss of any excitation energy. The energy transfer in the dye-templated organization of up to 32 nm length was determined, and the loss of energy per nanometer was about <1%, suggesting its potential use in photonics. Lee and coworkers reported a pyridinium derivative (C1) which in acidic pH undergoes protonation (C2) and serves as a coating material to wrap DNA molecules via electrostatic interactions (between cationic pyridine and anionic DNA) to generate virus-mimic filamentous morphology [30]. The C2-coated DNA exhibits right-handed helical conformation at physiological pH, while transformation to left-handed conformation was observed in acidic pH conditions (Fig. 12.4b). The pH-dependent helical inversion was investigated *in cellulo* conditions, which suggest the ionic interactions are important noncovalent interactions to construct SFM-templated DNA architectonics.

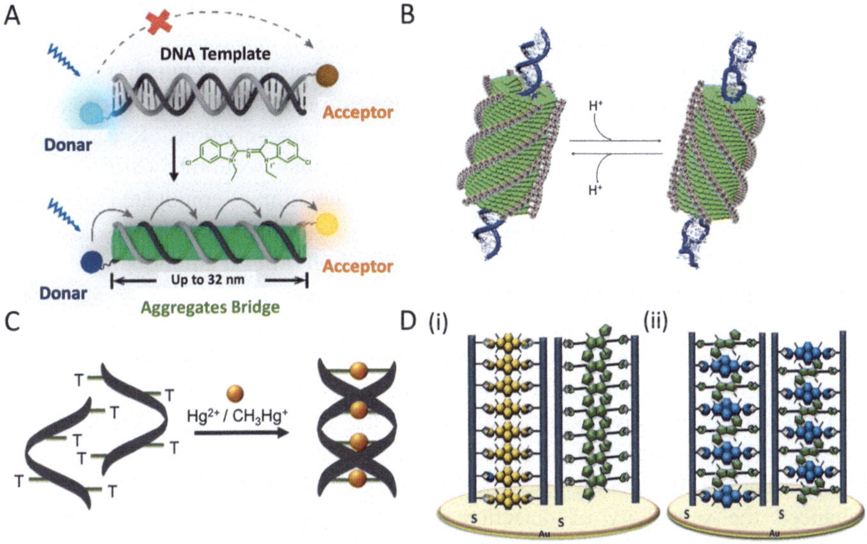

Fig. 12.4 (**a**) Schematic of DNA-templated cyanine dye aggregates with donor-bridge-acceptor (D-B-A) three-component energy transfer construct. Alexa Fluor 350 as a donor and Alexa Fluor 555 as an acceptor are conjugated onto two ends of a poly(A)$_{24}$/poly(T)$_{24}$ dsDNA template. (**b**) Synthetic coat assembly construct of bipyridine and methylpyridinium derivatives, respectively. Collective motion in the helicity inversion of DNA and the synthetic coat assembly triggered by the pH change. (**c**) Construction of metallo-DNA duplex by T-Hg(II)-T interaction followed 2:1 molar ratio of T and Hg(II). (**d**) (i) Schematic presentation of the ssDNA-templated discrete (DPP-dT$_{40}$) and (NDI-dT$_{40}$) multichromophoric arrays via co-immobilizing donor-acceptor units on a gold electrode and (ii) random assembly of co-immobilized stack of aromatic donor and acceptor systems (DPP-NDI-dT$_{40}$) used for the design of nanoarchitectures. (**a**) Adapted with permission from Ref. [46] American Chemical Society. (**b**) Adapted with permission from Ref. [30], Nature group. (**d**) Adapted with permission from Ref. [47], Wiley & Sons

12.1.2.4 SFM-Templated DNA Architectonics Driven by Metal-Base Pair Interactions

The interaction of nucleobases with various metal ions is considered as alternative to WC interactions to create functional co-assemblies. The metal ion-mediated base pairing interactions have the potential to contribute in the scheme of molecular architectonics in facilitating templated co-assembly of SFM and DNA to produce functional molecular and nanoarchitectures.

The contribution of specific metal ions to stabilize DNA duplexes via metal-mediated base pairing interactions has been investigated by Tanaka and Ono groups (Fig. 12.4c) [31]. Interaction of mercury (Hg) with thymine results in the formation of stable [thymine-mercury-thymine: T-Hg-T] structures. From the structural perspective, metal-base pair interactions are excellent replacement for WC interactions in the native DNA structure. Further, the complexation of nucleobases with metal ions can be strategically exploited to create functional DNA architectures with potential application in metal ion sensing. The detection of ultra-low concentration of neurotoxic Hg in water is highly desirable for the protection of human health and environment. Despite numerous efforts, the conventional analytical methods are not capable of sensing mercury in water at ultra-low concentrations. Recently, we developed a fluorescence and surface-enhanced Raman scattering (SERS)-based mercury sensing platform using histidine (thymine mimic)-conjugated perylene diimide (PDI), which allowed us to achieve the highest ever reported selectivity and sensitivity for Hg with the limit of detection (LOD) of 60 attomolar (aM, 0.01 parts-per-quadrillion (ppq)) in water [32].

Taking advantage of the strong and selective [T-Hg-T] metal-base pair interaction, DNA-based biosensors were developed for the detection of Hg [48, 49]. However, [T-Hg-T] interaction-based ultrasensitive and selective detection platform for mercury remained elusive possibly due to the use of DNA sequences with few T units. Surprisingly, use of dT_n sequences for the detection of Hg through the [T-Hg-T] base pairing interactions was not attempted, which is anticipated to have enhanced the sensitivity of detection. We conceptualized the construction of SFM-templated ssDNA (dT_n) ensemble through the scheme of molecular architectonics for the chiroptical and conductometry-based ultrasensitive detection of Hg through [T-Hg-T] base pairing interactions [29]. In the presence of Hg, the displacement of SFM from the templated ssDNA assembly (SFM-dT_n) leads to the formation of stable metallo-DNA duplex $[dT\text{-}Hg\text{-}dT]_n$ duplex. In contrast to the aptamer-based design possessing few T units, the usage of oligothymidine (dT_n) sequence in our strategy allowed to fabricate an ultrasensitive detection platform.

Organic semiconductor BNA consisting of a butyl group to enforce hydrophobicity and chiral/opto-electronically active NDI core and adenine (A) as functional auxiliary was designed as SFM (Fig. 12.2). The mutually templated hybrid ensemble $[BNA_{10}\text{-}dT_{10}]$ was constructed as left-handed helical assembly architecture, as evident from the induced circular dichroism (ICD) signature. Addition of organic or inorganic Hg instantly triggers the dissociation of BNA from the ensemble which resulted in the structural transformation of 2D nanoarchitectures to 1D architectures; these architectural changes were also reflected in quantitative changes of ~20% and 40% observed in the CD intensities at 5 and 10 nM of Hg. Next, we fabricated a field-effect transistor (FET) device using $[BNA_{10}\text{-}dT_{10}]$ material architectures (Fig. 12.3). The favorable interaction of Hg with dT_n leads to preferential displacement of BNA from the $[BNA_{10}\text{-}dT_{10}]$ ensemble. Interestingly, the rate of BNA displacement is proportional to the rate of Hg addition, and distinctive changes in the conductivity were recorded for Hg with concentration as low as 0.1 nM. This conductometry-based Hg sensing platform is exceptionally sensitive in comparison to optical or spectroscopy-based sensing platforms. In particular, the meticulously designed molecular and nanoarchitectonics-guided mutually templated assembly of organic semiconductor BNA and oligonucleotide dT_n provided a facile technology toward the development of Hg sensor with sub-nanomolar sensitivity and selectivity, which exploited the metal ion-assisted T-Hg-T interaction. In a significantly improved and sophisticated design, imidazole was used as a T-surrogate to complex with Hg.

Zn(II) macrocyclic tetra-amine (cyclen) complex was reported with selective nucleoside recognition property in aqueous solution [50]. The complex binds selectively to the thymine (T) via coordination of Zn (II) center to the imide nitrogen atom of the nucleobase and hydrogen bonding interaction between the carbonyl group of T and NH group of cyclen complex. The aromatic chromophore tethered Zn (II)-biscyclen system shown to form the mutually templated bicomponent co-assembly with nucleobases as charge transport system [33]. The Zn (II)-biscyclen complex-functionalized NDI serves as a recognition unit for dT_n. The mutually templated assembly between NDI-tethered Zn (II)-biscyclen and dT_n forms a multi-chromophoric array on the gold substrate, which resulted in the generation of photocurrent due to high electrical conductivity through the stack of NDI aromatic core. In a subsequent report, two different zinc tethered platforms, Zn(II)-bis (cyclen)-NDI-dT_{40} and Zn(II)-bis(cyclen)-DPP-dT_{40}, were developed, which behaved as donor-acceptor heterojunction system (Fig. 12.4d) [47]. In the heterojunction system, DPP (diketopyrrolopyrrole) acts as donor and NDI core as acceptor unit. The SFM-DNA-based heterojunction multichromophoric assembly was found to be aligned vertically on the gold substrate that promotes exothermic charge separation-mediated photocurrent generation instead of ground state charge transfer (CT) complexation between NDI and DPP (Fig. 12.4d(i)). Notably, the random organization of DPP-NDI-dT_{40} restricts the photocurrent generation due to the ground state CT complexation (Fig. 12.4d(ii)). These thoughtful designs have outlined the important role of metal complexes to derive SFM coordinated mutually templated assembly fabricating hybrid DNA molecular assembly architectures with potential applications.

12.2 Nanoparticle-Templated DNA Architectonics

The interaction of nucleobases with nanoscale material templates holds great promise in the development of templated DNA nanoarchitectures. DNA with its virtue of molecular recognition is capable of interacting with metal and non-metal nanoparticles. The classical DNA nanotechnology gave birth to the novel concept of DNA origami which has been successfully integrated in the field of nanoarchitectonics. Following the breakthrough works on DNA-templated assembly of gold (Au) nanoparticles from the Mirkin [51] and Alivisatos [52] groups, the nanoparticle-supported construction of DNA nanoarchitectures has gained considerable attention in the research community. The thermo-responsiveness of DNA has been effectively exploited for the nanoparticle co-assembly to fine-tune their optical properties. In particular, the pioneering works pave the way for spatiotemporal control of nanoparticle assembly using DNA templates. This concept has been further extended to design DNA responsive nanoparticle systems. In a typical example, gold nanoparticles are functionalized with thiolated ssDNA consisting of two stretches of cytosine-rich domains (forming i-motif under acidic conditions) [53]. In high pH conditions, i-motif DNA transforms into random coil structure, which makes the gold nanoparticles remain in a non-aggregated state due to electrostatic repulsions. Under low pH condition, the ssDNA assembly triggers the generation of interparticle i-motif conformation that triggers the nanoparticle aggregation. pH-induced aggregation-disaggregation process is evaluated by distinct changes in gold nanoparticle solution color and useful in the design of pH-responsive DNA machines and devices. Au nanoparticle-templated DNA conjugates were developed as a heavy metal sensor, where two different types of Au nanoparticles (probe A and probe B) were thiol functionalized with complementary DNA sequences (5′ HS-C_{10}-A_{10}-T-A_{10} 3′; probe B, 5′ HS-C_{10}-T_{10}-T-T_{10} 3′) except with a single T-T mismatch (Fig. 12.5a) [54]. These ssDNA-modified gold nanoparticle clusters form stable aggregation due to complementary association and exhibit a low melting temperature. Introduction of Hg within the system triggers a significant enhancement in the melting temperature due to coordination of Hg with the T-T mismatch center. This enzyme-free colorimetric method in principle can be used for the detection of heavy metals in water.

Keyser et al. reported the use of DNA origami nanosheets for the accurate positioning of Au nanoparticles to promote the plasmonic interactions between two particles [58]. The gold nanoparticles were arranged within the gap of 3.3 ± 1 nm of the origami sheets, and the SERS confirmed high local field enhancement with potential application in the field of biosensing. Kuzyk and Liu groups have reported the formation of three-dimensional DNA nanoarchitectures by assimilating Au nanorod within the DNA origami architecture [59]. In this design, DNA acts as an engineered template to organize Au nanoparticles in a three-dimensional orientation as well as serves as fuel molecules to adapt the plasmonic architecture in different conformational states. The three-dimensional reconfigurable plasmonic nanoarchitectures exhibit DNA-triggered conformational changes at nanoscale interface and serve as optical reporters. Interaction with individual biomolecules alters the plasmonic intersection angles, which can be detected by CD spectra. In the DNA-printed nanoparticle technology, unique 3D molecular transfer property was integrated within Au nanoparticle-wrapped DNA icosahedron geometry [55]. DNA icosahedron cage was developed as a carrier system to transport the patterned DNA strands to the nanoparticle surface. As shown in Fig. 12.5b, the molecular transfer of the patterned DNA strands is divided into two different steps. In first step, two different half icosahedrons consisting of dithiolated 10-nucleotide-long polythymine sequence are brought together to form a compact icosahedron cage (I-cage), and Au nanoparticle is encapsulated within the caged structure. The dithiolated overhangs from the DNA strands form covalent conjugation with the Au nanoparticle via Au-S bond. In the next step, the I-cage structure was removed, and the DNA pattern remained on the AuNP surface.

The customized design approach to print molecular information from the template external shell to the core of the nanostructure is a highly appealing strategy and extends the application of DNA nanotechnology to nanolithography. Such tailored design is a highly promising example of nanoparticle-templated DNA architecture

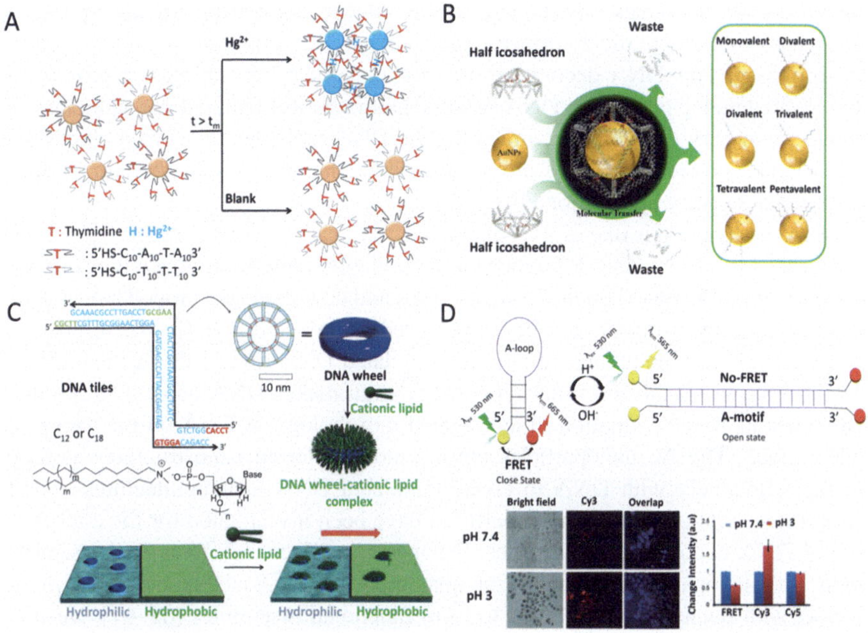

Fig. 12.5 (**a**) Schematic representation of DNA-Au nanoparticle-based colorimetric detection of Hg^{2+}, two types of Au nanoparticles tethered with complementary DNA sequences have been decorated with a single thymine-thymine (T-T) mismatch, and interaction with Hg^{2+} results in the entrapment of the metal ion, aggregation of nanoparticles, and changes in the thermal and optical properties. (**b**) Schematic overview of the 3D transfer of DNA patterns onto AuNP via the DNA icosahedron cage. Various DNA patterns have been designed by modulating the structural association of DNA. (**c**) Schematic representation of complexation of DNA wheel and cationic lipid which leads to the transfer of the DNA wheel to a hydrophobic surface induced by lipid complexation. (**d**) pH-triggered transformation of the LMB (closed state) to A-motif (open state) sensor upon excitation at 530 nm. The epifluorescence microscopy images depicting LMB sensor-treated HeLa cells at two different pHs. The images were captured at 500 ms exposure duration, scale = 40 mm. The bar diagram displays the changes in fluorescence intensity of LMB in HeLa cells at pH 7 and 3, upon excitation at 512 nm. (**a**) Adapted with permission from Ref. [54], Wiley & Sons. (**b**) Adapted with permission from Ref. [55], American Chemical Society. (**c**) Reproduced from Ref. [56] with permission from the PCCP Owner Societies, Royal Society of Chemistry. (**d**) Adapted with permission from Ref. [57], Royal Society of Chemistry

with possible application in the area of sensing and opto-electronics. The covalent interactions between DNA and Au nanoparticles typically involve the thiolated chemistry (Au-S bond) between thiol-functionalized oligonucleotide strands and AuNPs. Nonetheless, the procedure is quite challenging to fine-tune the conformations of the surface-tethered DNA strands and modulate their hybridization efficiency. Fan et al. reported the use of polyadenine as a replacement of the thiol strategy to construct DNA-Au nanoparticle conjugates [60]. The polyadenine sequences are used as anchoring block to adsorb over the nanoparticle surface, and interestingly, the tethered recognition sequences maintain an upright conformation across the surface to promote further hybridization. This non-thiolation-based functionalization strategy of Au nanoparticle prevents the nonspecific absorption of DNA onto the nanoparticle surface and facilitates the sequence-mediated upright conformation. This DNA decoration of Au nanoparticle fabrication is a promising strategy to develop plasmonic DNA sensors. Dithiothreitol-tethered Au nanoclusters form a "raspberry-like" architecture along the DNA molecule to act as a biomolecular shield to protect the DNA from nuclease degradation and holds potential utility in the gene delivery applications [61].

DNA methylation is one of the epigenetic modifications considered as a potential biomarker for cancer. The differences in methylation pattern influence the solution behavior of DNA. Based on this concept, Trau and coworkers developed a diagnosis platform for cancer detection without the need of any sensor [62]. The affinity of methylated, hyper-methylated, and non-methylated DNA toward the Au nanoparticle substrate was analyzed. The hypermethylation makes DNA more hydrophobic which results in the formation of aggregated morphology of DNA on the nanoparticle surface. The Au nanoparticles are the most preferred nanostructures as they interact effectively with DNA to create templated DNA nanoarchitectures. Apart from Au, several other metal nanoparticles have been investigated for the construction of DNA-nanoparticle ensembles. For instance, silver (Ag) nanoparticles, platinum (Pt) nanoparticles, and several quantum dots (QDs) have been shown to interact with nucleotides and DNA to form functional assemblies [63–67]. Despite considerable attempts, the optimization of conjugation strategies to prepare robust DNA-(Ag/Pt/QD) nanoparticle ensembles is yet to be achieved. For instance, the thiol linkages are not effective for conjugation of DNA with Pt nanoparticle, while phosphorothioate linkages are found to be useful for the surface functionalization of Ag nanoparticles [68, 69]. These observations and importance of DNA-nanoparticle conjugation highlight the vast opportunity available for researchers in the area of nanoparticle-guided construction of DNA architectures.

12.3 Biomolecule-Templated DNA Architectonics

The inherent molecular recognition and predictable hybridization make DNA an exceptional and robust material building block. In living systems, the cellular DNA in the nucleus is present in the condensed state wrapped around the histone proteins. The protein-bound DNA is the brilliant example of nature's templated architectonics, and these dynamic architectures are fundamental to information storage and transmission and serve as molecular machinery of life. Nature-inspired biomolecule-templated DNA architectonics has the potential to create well-defined functional architectures. In living systems, enzymes play a pivotal role in the regulation of metabolism. The cascades of enzymes are interconnected through several chemical transformations to allow for signal transduction. Niemeyer and coworkers reported a series of studies on the generation of DNA-templated synthetic nano-factories for the perfect positioning of multienzyme cascades with nanoscale precision for the regulation of cellular metabolism [70]. The covalently conjugated DNA-enzyme adducts serve as versatile molecular adaptors with suitable recognition motifs to interact with complementary and noncomplementary nucleobases. In a subsequent study, nucleotide-modified enzyme systems were introduced to perform the site-selective hybridization with complementary oligonucleotide sequences [71]. This methodology generated higher-ordered assemblies of 2D and 3D nanoarchitectures with strong enzyme immobilization efficiency.

Lipids have been used as a chemical fuel to translocate wheel-shaped DNA tiles from a hydrophilic to hydrophobic surface [56]. Yonamine et al. have shown that DNA wheel positioned on hydrophilic surface stay fixed, which upon addition of a cationic lipid undergoes co-assembly to form hydrophobic complexed ensemble. During the complexation time, the DNA wheels translocate through a patterned 2D substrate toward the hydrophobic region. DNA moves in a flip-flop fashion on the substrate without any alteration to its structure (Fig. 12.5c). This study demonstrated the utility of lipid to template the organization of DNA wheels, which cause significant effect on the localization and locomotion of DNA on the patterned substrates and possibly find applications in microlithography and gene delivery. Proteins are the most important biomacromolecules in both biological and materials interfaces. The spatiotemporal control over protein interactions can be accomplished by the templated assembly of DNA molecules. The protein of interest is covalently conjugated with both ssDNA and dsDNA. The ssDNA is flexible in shape, while the dsDNA is rigid with persistence length of ~50 nm. The structural differences of ssDNA and dsDNA can control the intermolecular distance between the two protein molecules. This concept resulted in the creation of DNA-tweezers that can be switched on and off by modulating fuel and antifuel strands [72]. In the closed conformation, the two DNA ligands are in a cooperative association to hold thrombin. In the stem loop conformation, the open state of the tweezer triggers the dissociation of ligands followed by the release of thrombin. In the context of DNA tweezers, DNA devices or nanomachines are promising contribution of templated DNA nanotechnology. In a seminal work by Seeman et al., the double crossover tiles of DNA were designed as switchable motors to construct molecular device [73].

A substantial progress has been made in the design of templated DNA architectures as DNA devices and machines [21]. We have constructed a labeled molecular beacon (LMB)-based DNA device with FRET pair, which is responsive to pH conditions [57]. The nanomachine is made up of 24 nucleobases and forms a hairpin-like molecular beacon (MB) structure under normal physiological pH (closed state). The 5′- and 3′-ends of the duplex stem structure are functionalized with FRET pair dyes, Cy3 (donor, $\lambda_{em} = 565$ nm) and Cy5 (acceptor, $\lambda_{ex} = 650$ nm) (Fig. 12.5d). Under closed conformation, the close proximity between the two dyes results in FRET response. On changing the pH to acidic (5–3) conditions, protonation of the A bases triggers the transformation of closed state (MB) to A-motif conformation (open state), which is stabilized by reverse Hoogsteen hydrogen bonding and ionic interactions causing disruption of FRET response. The sensing of acidic pH (5–3) within a synthetic vesicle (cell mimic) using LMB was performed by the confocal scanning fluorescence microscopy. *In cellulo* study revealed excellent cellular uptake of the LMB device in HeLa cells without the requirement of any vectors and efficient sensing of low intracellular pH. The lowering of pH is a pathological hallmark in several life-threatening disease conditions including cancer. Remarkably, our design approach has made intracellular delivery of the DNA device and sensing of acidic pH (5–3) with a low step size of 0.2–0.3 possible. Further, we have performed the live epifluorescence imaging of the HeLa cells encapsulating the LMB sensors at two different pH conditions (pH 7.4 and 3) (Fig. 12.5d). The cell images captured at pH 3 exhibit higher loss of FRET signals as compared to the images acquired at pH 7.4 validating the efficient sensing of low pH by the LMB device. Further the plate reader-based FRET studies were performed by changing the solution pH from pH 7.4 to 3 upon addition of HCl. It was observed that upon exciting at 512 nm, the cells show increased Cy3 intensity and a reduction in the FRET signal between Cy3 and Cy5, which further confirmed the ability of the DNA device to sense acidic pH.

Similarly, in vivo uptake and clearance of DNA nanomachine by the lysosomal degradation was assessed in multicellular organism *C. elegans* [74]. DNA duplex nanostructure device consisting of two ssDNAs exhibits half-life of 8 h, while lowering of ssDNA domains increases the half-life up to 11 h. Interestingly, the DNA architecture shaped into compact icosahedron without the presence of any free DNA ends is resistant to lysosomal degradation for a period of 24 h, which reiterates that specifically designed DNA architecture is decisive to maintain the structure-activity relationship [75].

12.3.1 Threading Intercalator-Guided DNA Architectonics

Small molecules interact with DNA through two major modes of binding: i) intercalation and minor groove binding [76–79]. The intercalators are generally flat aromatic ligands that bind to DNA by inserting between the nucleobases, while groove binding molecules interact mostly with the minor groove owing to its narrow structural features that specifically accommodate certain ligands. Intercalator and minor groove binding ligands interact reversibly with DNA and are exploited in a range of diagnostic and therapeutic applications. The major drawbacks of both intercalator and groove binding ligands are their limited base pair or sequence specificity and inability to recognize long DNA stretches to achieve improved specificity and selectivity. A new class of small molecules termed as threading intercalators is capable of binding to DNA through both intercalation and groove binding modes [80]. The threading intercalators bind to DNA through intercalation and threading through both minor and major grooves to recognize long stretches of DNA. Thus, threading intercalators are promising candidates for the sequence-selective targeting of DNA molecules for diagnostic and therapeutic applications. In fact, threading intercalators exhibit strong affinity and slow kinetics desirable for DNA-targeted diagnostics and therapeutics [81]. The threading intercalation of small-molecule binding causes significant deformation in the DNA conformation, which is the preliminary rate-limiting step leading to slow association and dissociation kinetics of the complex. We reported simple and versatile threading intercalators as prospective DNA molecular adaptors. In particular, a series of imidazolyl-naphthalenediimide (Im-NDI) conjugates were designed as modular units to interact with DNA [34]. The 1-(3-aminopropyl) imidazole was conjugated on either side of NDI core as imide substituent to generate NIm 1–8 by varying three structural parameters, viz., linker length (between NDI core and Im), heavy atom substitution at the NDI core, and charge on the imidazole ring. NDI moiety was considered in the design due to its intercalation property, and the imidazole was incorporated as a minor/major groove binding unit. The characteristic intercalation and groove binding structural features make Im-NDI systems as potential threading intercalators to guide the DNA molecular architectures. Based on the analysis, NIm 6 was found to be the best threading intercalator. Detailed theoretical studies demonstrated the optimum mode of threading intercalation of NIm 6 with DNA leading to the formation of NIm6/DNA molecular complex. Further studies revealed the biological significance of the NDI-imidazole conjugates, as assessed by the cytotoxicity against HeLa cell lines. NIm 6 showed significant anticancer activity due to its high affinity binding with DNA. Further, the DNA topoisomerase I (Top1) unwinding assay demonstrated the potential Top1 inhibitory effect of NIm 6.

The condensation-decondensation of DNA is an essential biological process for gene regulation and chromosomal maintenance. In this context, the molecular architectonics regulating the condensation-decondensation cycle is an interesting event to study, as the conformation and functional features of DNA change between the condensed and decondensed states. In an advanced approach, the concept of molecular architectonics has been adapted successfully to modulate the condensation of DNA by a group of threading intercalators, endogenous factor-driven decondensation, and effective transfection in cells [35]. The thorough investigation of DNA condensation events is worth studying to meet the challenges associated with the gene delivery. We have developed a series of threading intercalators bis-(imidazolium-naphthyl) (BIN), bis-(imidazolium-hydroxy-quinolinyl) (BIHQ), and bis-(imidazolium-amino-quinolinyl) (BIAQ) as potential transfection agents for the targeted delivery of DNA to cells. The interactions of the threading intercalators with DNA trigger the formation of nanoarchitectures, termed as DNA nanocondensates (Fig. 12.6a). The controlled DNA condensation through specific interactions (threading intercalation) avoiding nonspecific electrostatic interactions was achieved by designer ligands (BIN, BIHQ, and BIAQ) with bis-imidazolium (groove binding) core and aromatic (naphthyl or quinolinyl) moieties (intercalation) connected by amide or ether linkages (see Fig. 12.2). The imidazolium groups facilitate threading of the ligand through the grooves of target DNA via specific interactions of heteroatoms, while the aromatic moieties bind through intercalation mode. The quinolinyl moiety upon protonation in acidic pH conditions of endolysosomal pathway serves as fluorescence reporter for the intracellular tracking of nanocondensates. As shown in Fig. 12.6b, the free DNA shows thread-like morphology, while condensation triggered by the threading intercalator BIHQ led to the formation of spherical nanocondensates (Fig. 12.6). In an interesting strategy, we evaluated the role of specific endogenous metal ions (Mg^{2+}, Na^{+}, K^{+}, Ca^{2+}, and

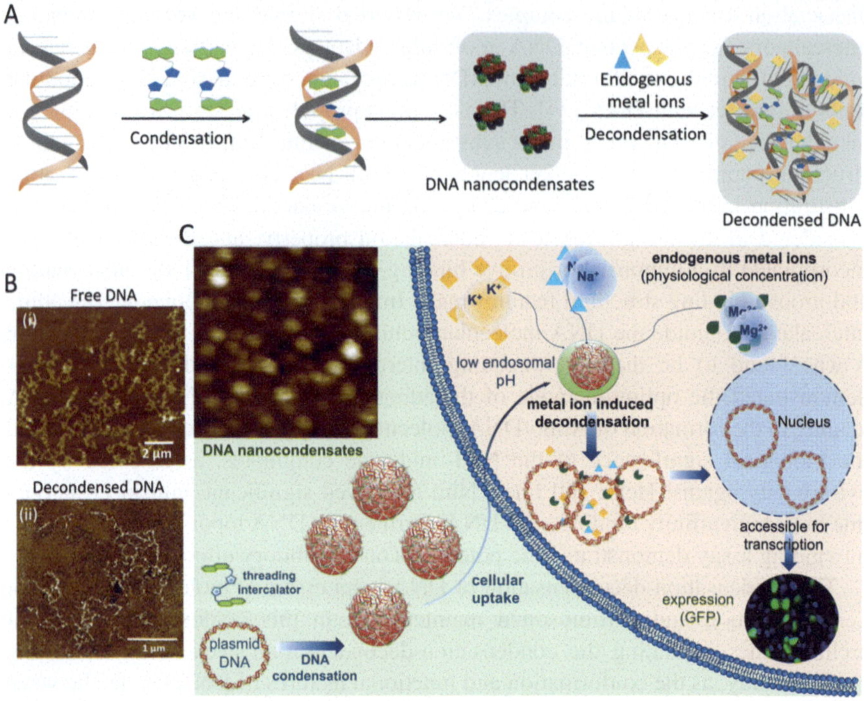

Fig. 12.6 (**a**) Schematic representation of threading intercalator-guided formation of DNA nanocondensates and decondensation of nanocondensates in the presence of endogenous metal ions. (**b**) Characterization of DNA and decondensed DNA by AFM. (i) AFM images of free pDNA and (ii) Mg^{2+}-induced decondensation of DNA nanocondensates. (**c**) Schematic representation of cellular uptake, fluorescence-based tracking of ligand/DNA nanocondensates, and metal ion-induced decondensation and transfection; the inset depicts the formation of nanocondensates by the interaction of DNA with threading intercalator BIHQ. Figure is adapted with permission from Ref. [35], American Chemical Society

Zn^{2+}) for the decondensation of DNA nanocondensates. Notably, the decondensation of nanocondensates by the endogenous metal ions was investigated at their physiological concentrations. The AFM data in Fig. 12.6b show the Mg^{2+}-induced decondensation of nanocondensates to fine thread-like morphology, which indeed suggests effective decondensation of nanocondensates by endogenous metal ion. The detailed analysis on the uptake of DNA nanocondensates into cells followed by their decondensation and transfection ability provided imperative mechanistic understanding over the condensation-decondensation-mediated DNA transfection (Fig. 12.6c). This minimalistic approach is a proof of concept in the design of SFM with considerable impact in the area of threading intercalator-guided molecular architectures. Most importantly, this simple yet conceptual design strategy could be a compelling example in the field of stimuli-responsive assembly and disassembly of nucleic acids triggering formation of nano-, micro-, and macroarchitectures [82, 83].

12.4 Conclusions and Future Perspectives

We presented the design of SFM-templated DNA molecular and material architectures employing the principles of molecular architectonics. From the viewpoint of practical applications, templated DNA nanotechnology or functional DNA nanoarchitectonics overcomes the limitations of classical DNA nanotechnology. In this contemporary approach, functional DNA architectures can be designed and constructed using SFMs as templates through simple and cost-effective protocols. The canonical and noncanonical hydrogen bonding interactions of nucleobases are supported by the interplay of aromatic π-π, metal coordination, and electrostatic interactions to construct templated co-assemblies of SFMs and ssDNA (short oligonucleotides) to produce novel DNA molecular and nanoarchitectures with unique functional properties and applications.

The trivial challenge associated with templated nucleic acid (DNA/RNA) nanotechnology is the selection or design of appropriate SFM with desired functional core and auxiliary. The organic molecules with known opto-electronic or redox properties can serve as functional core, while the basic biomolecules and their mimics with built-in molecular recognition properties are employed as functional auxiliaries to guide the construction of SFM-DNA/RNA hybrid ensembles and their material architectures through the scheme of molecular architectonics. The future of SFM-templated DNA architectonics lies in achieving complete control on the molecular organization through programmable noncovalent interactions (advanced molecular architectonics strategies) via mutually templated assembly and co-assembly of SFMs and DNA/RNA/mimics to produce structurally (0D, 1D, 2D, and 3D) and functionally well-defined molecular and material architectures with intriguing properties to cater to applications in the diverse areas of materials science and biology.

Acknowledgments We thank Prof. C. N. R. Rao FRS for his constant support and encouragement, JNCASR, Nano mission (SR/NM/TP-25/2016), SwarnaJayanthi Fellowship Grant (DST/SJF/CSA-02/2015-2016), DST, Government of India, and Sheikh Saqr Laboratory (SSL), ICMS-JNCASR, for financial support.

References

1. Seeman NC (1998) DNA nanotechnology: novel DNA constructions. Annu Rev Biophys Biomol Struct 27:225–248
2. Seeman NC (1982) Nucleic acid junctions and lattices. J Theor Biol 99:237–247
3. Rothemund PWK (2006) Folding DNA to create nanoscale shapes and patterns. Nature 440:297–302
4. Blackburn GM, Gait MJ, Loakes D, Williams DM, Egli M, Flavell A et al (2006) Nucleic acids in chemistry and biology. Royal Society of Chemistry, Cambridge, UK
5. Seeman NC (2003) DNA in a material world. Nature 421:427–431

6. Avinash MB, Govindaraju T (2018) Architectonics: design of molecular architecture for functional applications. Acc Chem Res 51:414–426
7. Avinash MB, Swathi K, Narayan KS, Govindaraju T (2016) Molecular architectonics of naphthalenediimides for efficient structure–property correlation. ACS Appl Mater Interfaces 8:8678–8685
8. Avinash MB, Govindaraju T (2011) Engineering molecular organization of naphthalenediimides: large nanosheets with metallic conductivity and attoliter containers. Adv Funct Mater 21:3875–3882
9. Avinash MB, Govindaraju T (2012) Amino acid derivatized arylenediimides: a versatile modular approach for functional molecular materials. Adv Mater 24:3905–3922
10. Roy B, Govindaraju T (2019) Amino acids and peptides as functional components in arylenediimide-based molecular architectonics. Bull Chem Soc Jpn 92:1883–1901
11. Konar M, Roy B, Govindaraju T (2020) Molecular architectonics-guided fabrication of superhydrophobic and self-cleaning materials. Adv Mater Interfaces 7:2000246
12. Avinash MB, Govindaraju T (2014) Nanoarchitectonics of biomolecular assemblies for functional applications. Nanoscale 6:13348–13369
13. Avinash MB, Samanta PK, Sandeepa KV, Pati SK, Govindaraju T (2013) Molecular architectonics of stereochemically constrained π-complementary functional modules. Eur J Org Chem 2013:5838–5847
14. Komiyama M, Yoshimoto K, Sisido M, Ariga K (2017) Chemistry can make strict and fuzzy controls for bio-systems: DNA nanoarchitectonics and cell-macromolecular nanoarchitectonics. Bull Chem Soc Jpn 90:967–1004
15. Govindaraju T, Avinash MB (2012) Two-dimensional nanoarchitectonics: organic and hybrid materials. Nanoscale 4:6102–6117
16. Pandeeswar M, Khare H, Ramakumar S, Govindaraju T (2015) Crystallographic insight-guided nanoarchitectonics and conductivity modulation of an n-type organic semiconductor through peptide conjugation. Chem Commun (Camb) 51:8315–8318
17. Ghosh D, Datta LP, Govindaraju T (2020) Molecular architectonics of DNA for functional nanoarchitectures. Beilstein J Nanotechnol 11:124–140
18. Ariga K, Li J, Fei J, Ji Q, Hill JP (2016) Nanoarchitectonics for dynamic functional materials from atomic-/molecular-level manipulation to macroscopic action. Adv Mater 28:1251–1286
19. Govindaraju T, Templated DNA (2019) Nanotechnology: functional DNA nanoarchitectonics. Pan Stanford Publishing, New York
20. Roy B, Ghosh D, Govindaraju T (2019) Functional molecule–templated DNA nanoarchitectures. In: Govindaraju T (ed) Templated DNA nanotechnology: functional DNA nanoarchitectonics, 1st edn. Pan Stanford Publishing, New York, pp 69–106
21. Roy B, Ramesh M, Govindaraju T (2019) DNA-based nanoswitches and devices. In: Govindaraju T (ed) Templated DNA nanotechnology: functional DNA nanoarchitectonics, 1st edn. Pan Stanford Publishing, New York, pp 365–408
22. Narayanaswamy N, Suresh G, Priyakumar UD, Govindaraju T (2015) Double zipper helical assembly of deoxyoligonucleotides: mutual templating and chiral imprinting to form hybrid DNA ensembles. Chem Commun 51:5493–5496
23. Pandeeswar M, Govindaraju T (2016) Engineering molecular self-assembly of perylene diimide through pH-responsive chiroptical switching. Mol Syst Des Eng 1:202–207
24. Avinash MB, Govindaraju T (2013) Extremely slow dynamics of an abiotic helical assembly: unusual relevance to the secondary structure of proteins. J Phys Chem Lett 4:583–588
25. Stulz E (2017) Nanoarchitectonics with porphyrin functionalized DNA. Acc Chem Res 50:823–831
26. Pandeeswar M, Khare H, Ramakumar S, Govindaraju T (2014) Biomimetic molecular organization of naphthalene diimide in the solid state: tunable (chiro-) optical, viscoelastic and nanoscale properties. RSC Adv 4:20154–20163

27. Pandeeswar M, Govindaraju T (2014) Bioinspired nanoarchitectonics of naphthalene diimide to access 2D sheets of tunable size, shape, and optoelectronic properties. J Inorg Organomet Polymers Mater 25:293–300
28. Narayanaswamy N, Avinash MB, Govindaraju T (2013) Exploring hydrogen bonding and weak aromatic interactions induced assembly of adenine and thymine functionalised naphthalenediimides. New J Chem 37:1302–1306
29. Pandeeswar M, Senanayak SP, Govindaraju T (2016) Nanoarchitectonics of small molecule and DNA for ultrasensitive detection of mercury. ACS Appl Mater Interfaces 8:30362–30371
30. Kim Y, Li H, He Y, Chen X, Ma X, Lee M (2017) Collective helicity switching of a DNA–coat assembly. Nat Nanotechnol 12:551–556
31. Miyake Y, Togashi H, Tashiro M, Yamaguchi H, Oda S, Kudo M et al (2006) MercuryII-mediated formation of thymine-HgII-Thymine Base pairs in DNA duplexes. J Am Chem Soc 128:2172–2173
32. Makam P, Shilpa R, Kandjani AE, Periasamy SR, Sabri YM, Madhu C et al (2018) SERS and fluorescence-based ultrasensitive detection of mercury in water. Biosens Bioelectron 100:556–564
33. Nakamura M, Okaue T, Takada T, Yamana K (2012) DNA-templated assembly of naphthalenediimide arrays. Chem Eur J 18:196–201
34. Suseela YV, Das S, Pati SK, Govindaraju T (2016) Imidazolyl-naphthalenediimide-based threading intercalators of DNA. Chembiochem 17:2162–2171
35. Pratihar S, Suseela YV, Govindaraju T (2020) Threading intercalator-induced nanocondensates and role of endogenous metal ions in decondensation for DNA delivery. ACS Appl Bio Mater 3:6979–6991
36. Narayanaswamy N, Unnikrishnan M, Gupta M, Govindaraju T (2015) Fluorescence reporting of G-quadruplex structures and modulating their DNAzyme activity using polyethylenimine–pyrene conjugate. Bioorg Med Chem Lett 25:2395–2400
37. Iwaura R, Ohnishi-Kameyama M, Iizawa T (2009) Construction of helical J-aggregates self-assembled from a Thymidylic acid appended anthracene dye and DNA as a template. Chem Eur J 15:3729–3735
38. Iwaura R, Hoeben FJM, Masuda M, Schenning APHJ, Meijer EW, Shimizu T (2006) Molecular-level helical stack of a nucleotide-appended oligo(p-phenylenevinylene) directed by supramolecular self-assembly with a complementary oligonucleotide as a template. J Am Chem Soc 128:13298–13304
39. Iwaura R, Yoshida K, Masuda M, Ohnishi-Kameyama M, Yoshida M, Shimizu T (2003) Oligonucleotide-templated self-assembly of nucleotide Bolaamphiphiles: DNA-like nanofibers edged by a double-helical arrangement of A–T Base pairs. Angew Chem Int Ed 42:1009–1012
40. Avakyan N, Greschner AA, Aldaye F, Serpell CJ, Toader V, Petitjean A et al (2016) Reprogramming the assembly of unmodified DNA with a small molecule. Nat Chem 8:368–376
41. Janssen PGA, Ruiz-Carretero A, González-Rodríguez D, Meijer EW, Schenning APHJ (2009) pH-switchable helicity of DNA-templated assemblies. Angew Chem Int Ed 48:8103–8106
42. Seifert JL, Connor RE, Kushon SA, Wang M, Armitage BA (1999) Spontaneous assembly of helical cyanine dye aggregates on DNA nanotemplates. J Am Chem Soc 121:2987–2995
43. Harrison WJ, Mateer DL, Tiddy GJT (1996) Liquid-crystalline J-aggregates formed by aqueous ionic cyanine dyes. J Phys Chem 100:2310–2321
44. Herz AH (1977) Aggregation of sensitizing dyes in solution and their adsorption onto silver halides. Adv Colloid Interf Sci 8:237–298
45. Wang M, Silva GL, Armitage BA (2000) DNA-templated formation of a helical cyanine dye J-aggregate. J Am Chem Soc 122:9977–9986
46. Zhou X, Mandal S, Jiang S, Lin S, Yang J, Liu Y et al (2019) Efficient long-range, directional energy transfer through DNA-templated dye aggregates. J Am Chem Soc 141:8473–8481
47. Nakamura M, Tsuto K, Jomura A, Takada T, Yamana K (2015) Donor–acceptor heterojunction configurations based on DNA–multichromophore arrays. Chem Eur J 21:11788–11792

48. Liu C-W, Tsai T-C, Osawa M, Chang H-C, Yang R-J (2018) Aptamer-based sensor for quantitative detection of mercury(II) ions by attenuated total reflection surface enhanced infrared absorption spectroscopy. Anal Chim Acta 1033:137–147
49. Zheng Y, Wang X, He S, Gao Z, Di Y, Lu K et al (2019) Aptamer-DNA concatamer-quantum dots based electrochemical biosensing strategy for green and ultrasensitive detection of tumor cells via mercury-free anodic stripping voltammetry. Biosens Bioelectron 126:261–268
50. Shionoya M, Kimura E, Shiro M (1993) A new ternary zinc(II) complex with [12]aneN4 (=1,4,7,10-tetraazacyclododecane) and AZT (=3′-azido-3′-deoxythymidine). Highly selective recognition of thymidine and its related nucleosides by a zinc(II) macrocyclic tetraamine complex with novel complementary associations. J Am Chem Soc 115:6730–6737
51. Mirkin CA, Letsinger RL, Mucic RC, Storhoff JJ (1996) A DNA-based method for rationally assembling nanoparticles into macroscopic materials. Nature 382:607–609
52. Alivisatos AP, Johnsson KP, Peng X, Wilson TE, Loweth CJ, Bruchez MP et al (1996) Organization of 'nanocrystal molecules' using DNA. Nature 382:609–611
53. Wang W, Liu H, Liu D, Xu Y, Yang ZD (2007) Use of the interparticle i-motif for the controlled assembly of gold nanoparticles. Langmuir 23:11956–11959
54. Lee J-S, Han MS, Mirkin CA (2007) Colorimetric detection of mercuric ion (Hg2+) in aqueous media using DNA-functionalized gold nanoparticles. Angew Chem Int Ed 46:4093–4096
55. Xie N, Liu S, Fang H, Yang Y, Quan K, Li J et al (2019) Three-dimensional molecular transfer from DNA nanocages to inner gold nanoparticle surfaces. ACS Nano 13:4174–4182
56. Yonamine Y, Cervantes-Salguero K, Nakanishi W, Kawamata I, Minami K, Komatsu H et al (2015) In situ 2D-extraction of DNA wheels by 3D through-solution transport. PCCP 17:32122–32125
57. Narayanaswamy N, Nair RR, Suseela YV, Saini DK, Govindaraju T (2016) A molecular beacon-based DNA switch for reversible pH sensing in vesicles and live cells. Chem Commun 52:8741–8744
58. Thacker VV, Herrmann LO, Sigle DO, Zhang T, Liedl T, Baumberg JJ et al (2014) DNA origami based assembly of gold nanoparticle dimers for surface-enhanced Raman scattering. Nat Commun 5:3448
59. Kuzyk A, Schreiber R, Zhang H, Govorov AO, Liedl T, Liu N (2014) Reconfigurable 3D plasmonic metamolecules. Nat Mater 13:862–866
60. Pei H, Li F, Wan Y, Wei M, Liu H, Su Y et al (2012) Designed diblock oligonucleotide for the synthesis of spatially isolated and highly hybridizable functionalization of DNA–gold nanoparticle nanoconjugates. J Am Chem Soc 134:11876–11879
61. Zhang L, Li Z, Jing J, Ding H, Chen Z (2019) DTT–Au NCs interact with DNA to form raspberry-like particles. Part Part Syst Charac 36:1800517
62. Sina AAI, Carrascosa LG, Liang Z, Grewal YS, Wardiana A, Shiddiky MJA et al (2018) Epigenetically reprogrammed methylation landscape drives the DNA self-assembly and serves as a universal cancer biomarker. Nat Commun 9:4915
63. Datta LP, Chatterjee A, Acharya K, De P, Das M (2017) Enzyme responsive nucleotide functionalized silver nanoparticles with effective antimicrobial and anticancer activity. New J Chem 41:1538–1548
64. Thompson DG, Enright A, Faulds K, Smith WE, Graham D (2008) Ultrasensitive DNA detection using oligonucleotide−silver nanoparticle conjugates. Anal Chem 80:2805–2810
65. Gehrke H, Pelka J, Hartinger CG, Blank H, Bleimund F, Schneider R et al (2011) Platinum nanoparticles and their cellular uptake and DNA platination at non-cytotoxic concentrations. Arch Toxicol 85:799–812
66. Zhu N, Chang Z, He P, Fang Y (2005) Electrochemical DNA biosensors based on platinum nanoparticles combined carbon nanotubes. Anal Chim Acta 545:21–26
67. Srivastava S, Samanta B, Arumugam P, Han G, Rotello VM (2007) DNA-mediated assembly of iron platinum (FePt) nanoparticles. J Mater Chem 17:52–55
68. Wang L, Feng Y, Yang Z, He Y-M, Fan Q-H, Liu D (2012) Reversibly controlled morphology transformation of an amphiphilic DNA–dendron hybrid. Chem Commun 48:3715–3717

69. Zheng Y, Li Y, Deng Z (2012) Silver nanoparticle–DNA bionanoconjugates bearing a discrete number of DNA ligands. Chem Commun 48:6160–6162
70. Niemeyer CM (2001) Bioorganic applications of semisynthetic DNA–protein conjugates. Chem Eur J 7:3188–3195
71. Niemeyer CM, Koehler J, Wuerdemann C (2002) DNA-directed assembly of bienzymic complexes from in vivo biotinylated NAD(P)H:FMN oxidoreductase and luciferase. Chembiochem 3:242–245
72. Liu D, Balasubramanian S (2003) A proton-fuelled DNA nanomachine. Angew Chem Int Ed 42:5734–5736
73. Li X, Yang X, Qi J, Seeman NC (1996) Antiparallel DNA double crossover molecules as components for nanoconstruction. J Am Chem Soc 118:6131–6140
74. Surana S, Bhatia D, Krishnan Y (2013) A method to study in vivo stability of DNA nanostructures. Methods 64:94–100
75. Walsh AS, Yin H, Erben CM, Wood MJA, Turberfield AJ (2011) DNA cage delivery to mammalian cells. ACS Nano 5:5427–5432
76. Suseela YV, Narayanaswamy N, Pratihar S, Govindaraju T (2018) Far-red fluorescent probes for canonical and non-canonical nucleic acid structures: current progress and future implications. Chem Soc Rev 47:1098–1131
77. Narayanaswamy N, Kumar M, Das S, Sharma R, Samanta PK, Pati SK et al (2014) A Thiazole Coumarin (TC) turn-on fluorescence probe for AT-base pair detection and multipurpose applications in different biological systems. Sci Rep 4:6476
78. Narayanaswamy N, Das S, Samanta PK, Banu K, Sharma GP, Mondal N et al (2015) Sequence-specific recognition of DNA minor groove by an NIR-fluorescence switch-on probe and its potential applications. Nucleic Acids Res 43:8651–8663
79. Holman GG, Zewail-Foote M, Smith AR, Johnson KA, Iverson BL (2011) A sequence-specific threading tetra-intercalator with an extremely slow dissociation rate constant. Nat Chem 3:875–881
80. Takagi M, Yokoyama H, Takenaka S, Yokoyama M, Kondo H (1998) Poly-intercalators carrying threading intercalator moieties as novel DNA targeting ligands. J Incl Phenom Mol Recognit Chem 32:375–383
81. Almaqwashi AA, Andersson J, Lincoln P, Rouzina I, Westerlund F, Williams MC (2016) DNA intercalation optimized by two-step molecular lock mechanism. Sci Rep 6:37993
82. Ariga K, Aono M (2019) Advanced supramolecular nanoarchitectonics. Elsevier Science, Amsterdam
83. Baillet J, Desvergnes V, Hamoud A, Latxague L, Barthélémy P (2018) Lipid and nucleic acid chemistries: combining the best of both worlds to construct advanced biomaterials. Adv Mater 30:1705078

Chapter 13
Architectures of Nucleolipid Assemblies and Their Applications

Manisha B. Walunj, Swagata Dutta, and Seergazhi G. Srivatsan

13.1 Introduction

The ways in which biopolymers fold, self-assemble and function in nature have constantly inspired our minds to develop simpler synthetic models that would essentially mimic or even show superior properties compared to the native systems. This has been possible due to the architectonic qualities of biomolecules like nucleic acids, proteins and sugars, which use diverse combinations of chemical interactions to form complex but well-defined architectures with unique functions. In particular, nucleic acids have a clear advantage because of their exquisite recognition and self-assembling properties. Initially, most efforts were curiosity driven and focused on constructing programmable self-assembled architectures of different sizes and shapes [1, 2], but more recently, nucleic acid architectures are built for defined applications, which range from technology to medicine [3–5]. While nucleic acid architectonics has gathered immense interest, the major technological challenge is scalability and fabrication, which is largely limited by batch-to-batch reproducibility in the self-assembling process and yields [6]. Alternatively, nucleic acid building blocks, namely, nucleobases, nucleosides and nucleotides, serve as excellent frameworks for the bottom-up construction of various hierarchical architectures [7, 8]. An obvious advantage is that these scaffolds can be conveniently derivatized with self-assembling features without compromising the intrinsic recognition and metal ion binding properties of the parent building blocks (Fig. 13.1). Design of such hybrid systems has again taken cue from naturally occurring nucleoside-lipid conjugates,

M. B. Walunj · S. Dutta · S. G. Srivatsan (✉)
Department of Chemistry, Indian Institute of Science Education and Research (IISER), Pune, India
e-mail: srivatsan@iiserpune.ac.in

T. Govindaraju, K. Ariga (eds.), *Molecular Architectonics and Nanoarchitectonics*, Nanostructure Science and Technology,
https://doi.org/10.1007/978-981-16-4189-3_13

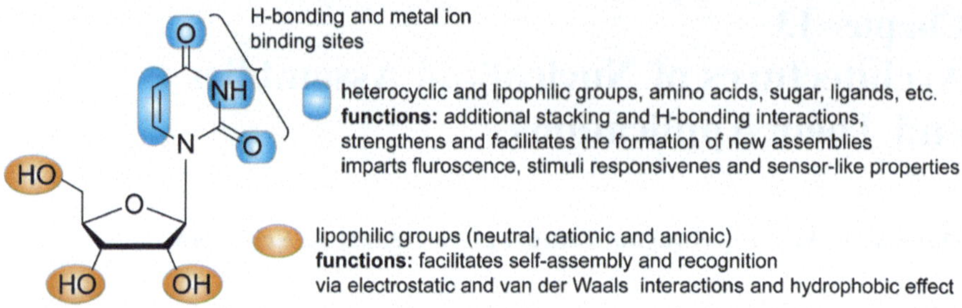

Fig. 13.1 A representative example of a nucleoside showing potential sites for chemical derivatization

which though have been primarily used in therapeutic applications rather than as supramolecular materials [9, 10].

Most strategies to build synthons use synthetic combinations of nucleic acid components, lipids, sugars and peptides that invoke basic supramolecular principles along with inherent recognition features to form architectures with new functions and applications [11, 12]. A simple conjugate of a nucleobase, nucleoside and nucleotide with a lipophilic component (e.g. fatty acid) is called a nucleolipid. Systems containing additional features obtained by modifying nucleobase and ribose sugar residues are called hybrid nucleolipids. The modifications are typically chosen to (i) augment H-bonding, stacking and van der Waals interactions, hydrophobic effect and metal ion chelation and (ii) impart sensor-like features, multi-stimuli responsiveness and optoelectronic properties. Many such tailor-made nucleolipid synthons self-assemble into interesting structures, which include fibres, vesicles, microspheres and nanotubes. These systems have found practical applications as biocompatible drug and gene delivery vehicles, injectable gels, responsive materials and sensors [8, 13].

In this chapter, we review the recent advances in the design and applications of supramolecular nucleolipid assemblies in the field of medicine and material sciences. First, we discuss in detail on various design principles used in constructing nucleolipid synthons and their self-assembled architectures. In the second part, we provide an elaborate discussion on the recent applications of nucleolipid assemblies in biomedical and materials research. Emphasis is laid on delivery tools, injectable gels, tissue engineering scaffolds, sensors and environment remedial systems. Assemblies made of non-lipid modified nucleosides or nucleotides are not discussed here, and hence, the readers may refer to recent review articles available in literature on such systems [14, 15].

13.2 Architectonic Landscape of Nucleolipids

The ease with which the chemical space of nucleosides can be expanded by using simple modification strategies is the origin for the vast architectural landscape exhibited by nucleolipid assemblies. There are two regions where functionalities can be introduced in nucleosides to generate self-assembling building blocks – (i) ribose sugar and (ii) nucleobase. Typically, lipophilic moieties are attached to the sugar hydroxyls or sometimes to the nucleobase, and such modifications impart amphiphilicity. Depending on the charge of the lipophilic group, the supramolecular synthons are categorized as neutral, anionic, cationic or zwitterionic nucleolipids [16]. Further, functionalities that impart additional supramolecular interactions and controls, and components that impart optical, material, therapeutic and recognition properties are introduced to expand the architectural and functional space of nucleolipids (Fig. 13.2). In this section, we survey recent examples of hybrid nucleolipids and their self-assembly process, architectures and functional tuning by chemical modification strategies.

Much like other small-molecule supramolecular building blocks, nucleolipids also exhibit hierarchical self-assembling processes driven by combinations of noncovalent interactions, which depends on the chemical design of the nucleolipid itself and the physicochemical environment [8]. In terms of the chemical design, the following sets of modification strategies are followed. (i) Nucleobase/nucleoside/nucleotide and lipid moieties are conjugated via an ester, amide, ether or a triazole group, (ii) the lipid chain length and saturation-unsaturation ratio are varied, (iii) charge on the lipids is varied, and (iv) a wide range of chemical moieties are installed on the nucleobase. While most modifications are intended to retain the native base-pairing ability of nucleobases, there are instances wherein changes in design afford the formation of non-canonical nucleobase H-bonding interactions in the assembled state. Further, modification of sugar hydroxyl groups alters its puckering, and attachment of larger groups to nucleobases (e.g. guanine) alters the preference of *syn* versus *anti* glycosidic conformation of the nucleoside [17]. All these design aspects are manifested in the hierarchical assembling process, and when subjected to different physicochemical environments, often they produce amazingly different self-assembled structures and morphologies with interesting properties [9, 13, 18].

13.2.1 Design and Tuning of Nucleolipid Assemblies

Although predicting the self-assembling process and architecture is not straightforward, a systematic design of a family of nucleolipid synthons has shown ways to tune the architecture and properties of the self-assemblies. For convenience, we have categorized nucleolipids as non-ionic, ionic and nucleobase/sugar-modified hybrids, which include heterocycle-modified nucleolipids and glyconucleolipids.

13.2.2 Non-ionic Nucleolipids

Initially strategies mainly focused on building simple nucleolipid amphiphiles for packaging and delivering biomacromolecules, particularly nucleic acids. The groups

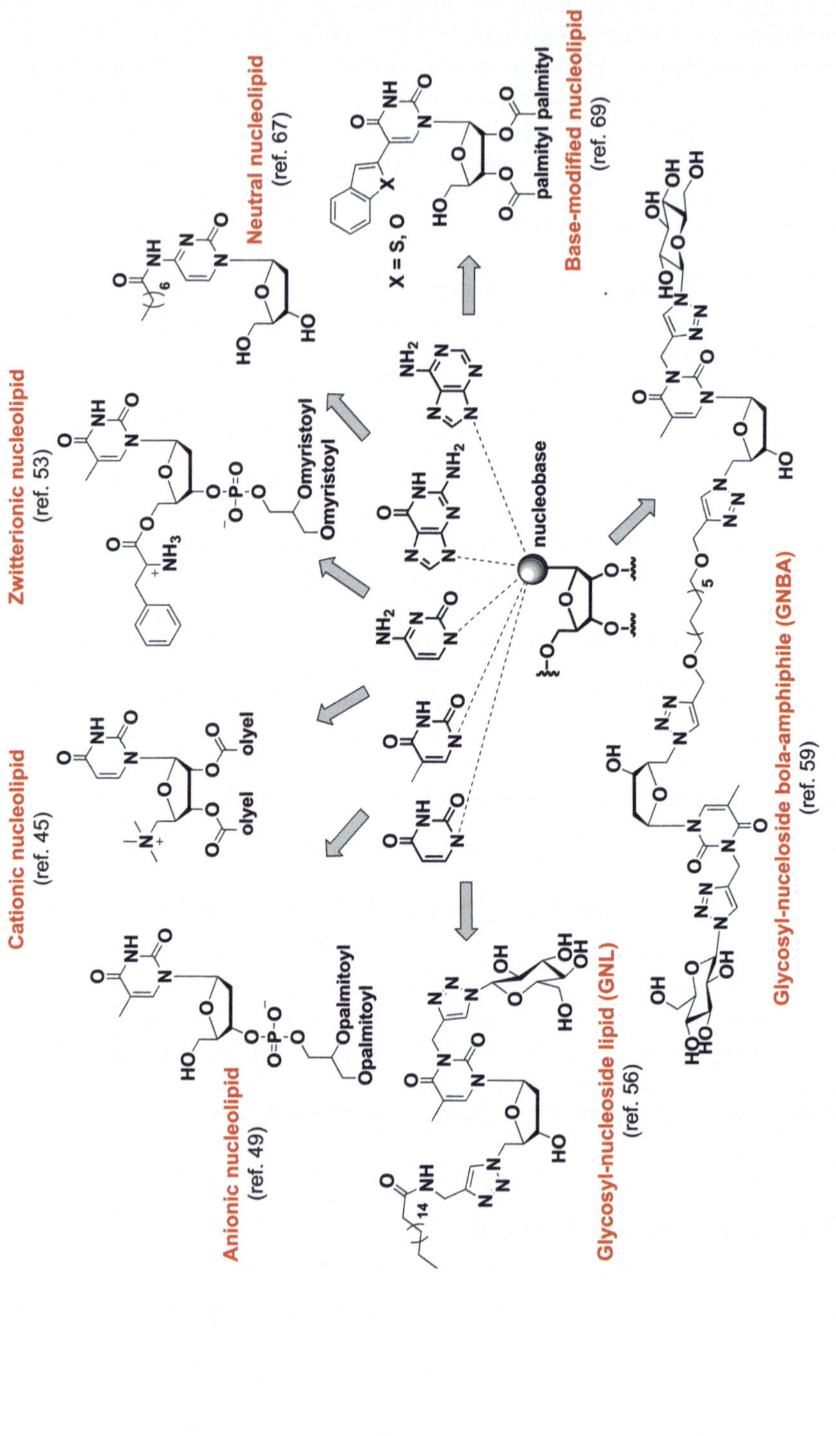

Fig. 13.2 Representative examples illustrating the chemical space of hybrid nucleolipids

of Grinstaff and Barthélémy synthesized a small series of sugar-modified uridine nucleolipids to study the role and interplay of nucleobase, alkyl chains and polyethylene glycol (PEG) in the formation of DNA-amphiphile assemblies [19]. The 5′-position was tethered to different PEGs and 2′ and 3′ positions were attached to long chain fatty acids through an ester linkage. PEG 5000-containing nucleolipid only formed micelles and also condensed plasmid DNA on mica surface, which was confirmed by atomic force microscopy (AFM) analysis. A shorter PEG chain and a non-nucleolipid version did not support the formation of micelles, suggesting the importance of hydrophilic-hydrophobic balance and nucleobase interaction in the assembling process. Drummond and coworkers studied the effect of the presence of unsaturation in the fatty acid chains [20]. For this purpose, oleoyl- and dioleoyl-substituted thymidine nucleolipids were synthesized and the self-assembling process was studied in the neat form and upon hydration. In the neat form, 3′-mono and 3′,5′-difatty acid-substituted thymidines both adopted a lamellar crystalline phase; however, 3′-phytanoylthymidine containing a saturated chain exhibited a fluid isotropic phase. While single-chain-containing (oleoyl and phytanoyl) nucleolipids upon hydration supported the formation of a lyotropic liquid-crystalline phase, 3′,5′-dioleoylthymidine failed to show such a behaviour.

Srivatsan and coworkers carried out a systematic study to investigate the effects of fatty acid substitution on the morphology and mechanism of self-assembling process of pyrimidine nucleolipids [21–23]. A series of 2′-deoxy and ribonucleolipids containing one or more fatty acid chains were synthesized, and the self-assembling process was studied in different solvents and conditions by using conventional supramolecular characterization techniques. Amazingly, these simple nucleolipids exhibited interesting and distinct self-organization processes, which depended on the fatty acid chain length, position of modification and physicochemical parameters. First, 3′-*O*-monofatty acid-substituted thymidines and 3′,5′-*O*-difatty acid-substituted thymidines containing different hydrocarbon chain lengths (12–18 carbon) were synthesized, and their self-assembling process was studied in different solvents (Fig. 13.3) [21]. 3′-*O*-Monosubstituted thymidines did not support gelation process in several of the tested organic solvents. Interestingly, organic dispersions of the nucleolipids in the presence of a small amount of water immediately formed opaque gels. 20% of water was required to induce the gelation process, and beyond this percentage, precipitation occurred. The critical gelation concentration (CGC) was found to depend on the fatty acid chain length. The lowest CGC was observed for stearoyl-modified thymidine and the highest was observed for shortest chain (dodecanoyl) containing nucleolipid. A detailed analysis of these gels by single crystal X-ray, powder X-ray diffraction (PXRD) and nuclear magnetic resonance (NMR) analyses revealed a combination of Watson-Crick base pairing between the nucleobases of different layers and H-bonding network involving O4′ and O5′-H atoms of sugar and water molecules to afford the hierarchical self-assembly. On the contrary, 5′-*O*-monofatty acid-substituted thymidines reported by Kim's group formed stable organogels in several nonpolar organic solvents without the assistance of water [24]. 3′,5′-*O*-Disubstituted thymidines formed stable organogels in *N,N*-dimethyl sulfoxide (DMSO) and acetonitrile upon heating-

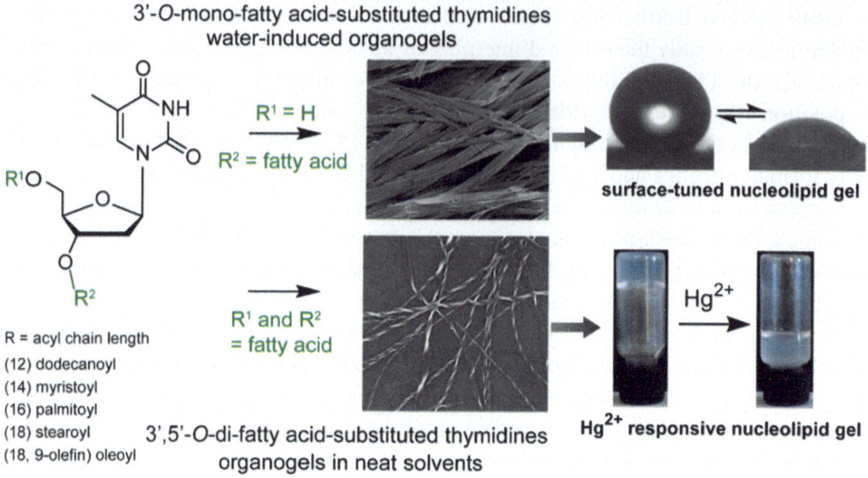

Fig. 13.3 Depending on the site of fatty acid substitution, thymidine nucleolipids showed different course of self-organization process to form gels. This figure is reproduced with the permission of the journal [21]

cooling/sonication steps. Rheology analysis indicated that a water-triggered gel of monofatty acid-substituted thymidine had noticeably higher viscoelastic character compared to an organogel of disubstituted thymidine containing the same chain length. All these observations were also manifested in their morphologies and the assembling process. SEM images of xerogels of monosubstituted thymidines exhibited the formation of highly entangled long-range sheets with 2–8μm in width and several μm in length, whereas 3′,5′-*O*-disubstituted thymidines formed long-range twisted tapes, which further substantiated the observed differences in self-assembling process and mechanical strength (Fig. 13.3).

Taking forward this study, Srivatsan and coworkers focused their attention on the self-assembling behaviour of 2′,3′-*O*-difatty acid-substituted purine and pyrimidine ribonucleosides (Fig. 13.4) [22]. Individually, the ribonucleolipids adopted distinctly different morphologies with uridine nucleolipids forming organogels and adenosine nucleolipids not supporting the gelation process. This was possible because the uridine nucleolipids aggregated to form highly entangled fibres and tapes (morphologies that stabilize gels), whereas adenosine analogs formed highly porous microspheres (Fig. 13.4a). Using X-ray and ^{1}H-NMR analyses, it was confirmed that the interactions that were present in the crystal packing were also present in the gel state. The structure of uridine nucleolipid containing myristoyl chains showed bifurcated H-bonding network formed by both canonical and non-canonical H-bonds (Fig. 13.4b). This network was further steadied by additional H-bonding interaction between C2 carbonyl oxygen atom and 5′-OH atoms of adjacent uridine. The sheets were formed by head-to-head and tail-to-tail interactions between nucleolipids, which was supported by interlayer hydrophobic contacts. The self-assembling process also depended on the length of the chains, wherein

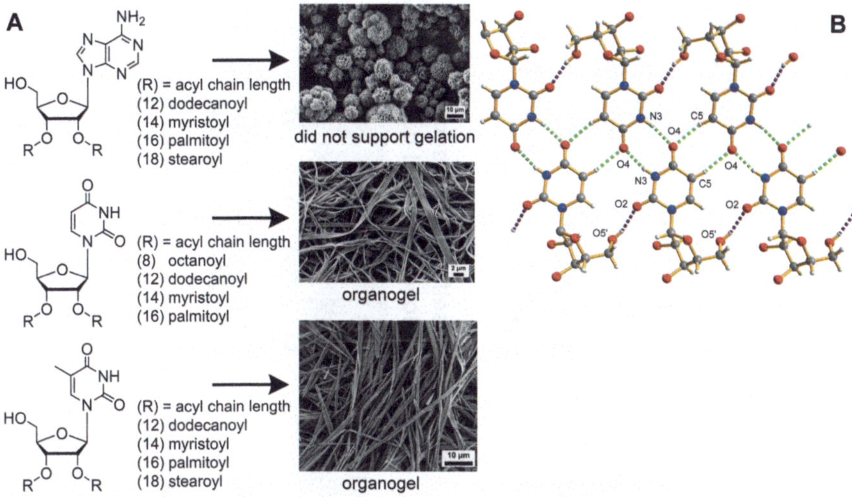

Fig. 13.4 (**a**) Chemical structures of purine and pyrimidine nucleolipids and their morphologies as studied by SEM. (**b**) Crystal packing showing the H-bonding network present in uridine nucleolipid containing myristoyl chains. SEM images and X-ray structure have been reproduced with permission of the journals [22, 25]

dodecanoyl chains, although displayed base pairing, did not show the complex H-bonding shown by the myristoyl-containing nucleolipids.

Interesting architectures were formed when different mixtures of these nucleolipids and complementary ribonucleosides were subjected to self-assembling process (Fig. 13.5). For example, when an uridine nucleolipid, which by itself forms fibre, was mixed with microsphere-forming adenosine nucleolipid of the same chain length, a self-sorted assembly formed resulting in the disruption of the gel [22]. The porous microspheres were covered by the fibrous network of uridine nucleolipid. NMR analysis and 3D rendered images obtained by using 3D X-ray microtomography confirmed the sorting phenomenon. Unprecedented growth pattern as a result of self-sorting was observed when uridine nucleolipids were mixed with adenosine. Here, uridine nucleolipids formed tapes first, which served as microscopic templates for the clustering of complementary adenosine. Such a pattern was not obtained upon mixing other nucleosides. The difference in self-sorting behaviour was attributed to the kinetics of self-organization process of individual supramolecular synthons. In contrast, mixtures of uridine nucleolipids containing different chain lengths formed constructive co-assemblies with significantly enhanced mechanical strength. This two-component assembling platform could be also implemented on other nucleosides, which could pave ways to construct unique higher-ordered architectures.

In another study, self-assembling properties of 2′,3′-*O*-difatty acid-modified ribothymidine and ribocytidine and 2′,3′,5′-*O*-triacylated ribothymidine nucleolipids were compared. While di-substituted ribothymidines formed stable organogels in

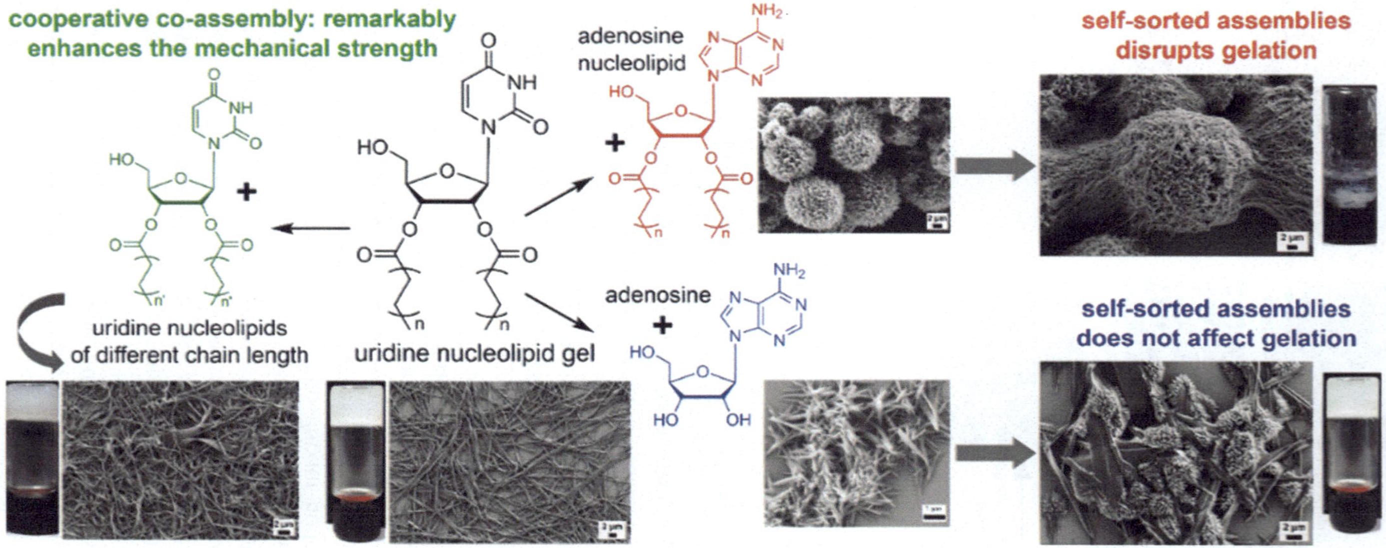

Fig. 13.5 Two-component systems made of differently self-assembling building blocks exhibited cooperative and disruptive self-sorting and co-assembling behaviour. This figure is reproduced by permission of The Royal Society of Chemistry (RSC) [22]

various polar and nonpolar solvents, cytidine and tri-substituted thymidines did not self-assemble to form gels (Fig. 13.4a) [25].

Nuthanakanti and Srivatsan used simple 5′-*O*-monofatty acid-substituted pyrimidine ribonucleolipids to obtain organogels and hydrogels by changing the chemical environment [23]. The fatty acid chains of two ribothymidine and uridine synthons interdigitated to form ordered lamellar aggregates, which further assembled to form a fibrous network, thereby gelling nonpolar organic solvents. However, upon partial hydrolysis in the presence of NaOH, heterotypic hydrogels composed of the nucleolipid and fatty acid carboxylate were formed. This did not happen in the presence of LiOH and KOH. However, addition of 0.05 M NaCl to nucleolipids containing KOH or LiOH resulted in fibres, suggesting the specific role of Na^+ ions in the self-organization process. The ability of the gel to interact with cations was used in trapping dye pollutants, which is discussed in applications section.

Instead of using fatty acids as the lipophilic part, researchers also used other biologically relevant hydrophobic moieties to construct nucleolipid supramolecular synthons. Rosemeyer and coworkers reported the synthesis of *N*-geranylated and farnesylated 2′-deoxyinosine, thymidine, uridine and 5-methyluridine derivatives, which represented the first example of synthetic nucleoterpene mimetic [26, 27]. The phosphoramidites of the nucleoside derivatives were incorporated into oligonucleotides, and their insertion into artificial lipid bilayers was evaluated by single-molecule fluorescence spectroscopy and microscopy techniques. While N(3)-geranyl thymidine inserted well into the bilayer, farnesylated nucleosides showed better stability towards perfusion. It should be noted that the self-assembling process at the nucleolipid level, though would be interesting, was not studied. Taking cue from this bioinspired system, Bourgaux and coworkers developed a series of squalene-modified nucleolipids and studied the relationship between the chemical structure and architecture of the self-assembled nanoparticles (Fig. 13.6a) [28]. Base- and sugar-modified squalene derivatives of purine and pyrimidine nucleosides were subjected to nanoprecipitation process in water, and thus the obtained nanoparticles were analysed by synchrotron small-angle X-ray scattering and cryo-TEM imaging. Even small differences in chemical structures resulted in significant differences in the self-assembly affording diverse architectures (Fig. 13.6a). The assembling process depended on the nucleobase and the position of squalene conjugation. For example, squalene grafted on the nucleobases produced sponge-like nanoparticles or unilamellar liposomes, whereas sugar-modified squalenes self-organized as lamellar phases. The authors claimed that this study could provide means to fine-tune the supramolecular organization of such squalene prodrugs and hence their pharmacological properties.

Arbuzova and coworkers came up with a nice combination of a two-component self-assembling system composed of cholesteryl nucleolipid, namely, cholesterylaminouridine (CholAU) and phosphatidylcholines (PCs, Fig. 13.6b) [29, 30]. The choice of using cholesteryl nucleolipid as a supramolecular synthon probably originated from an earlier report wherein an interplay of cholesterol-cholesterol and uracil-uracil interactions in uracil-conjugated cholesterol formed fibres gelating cyclohexane [31]. Aqueous suspensions of CholAU and

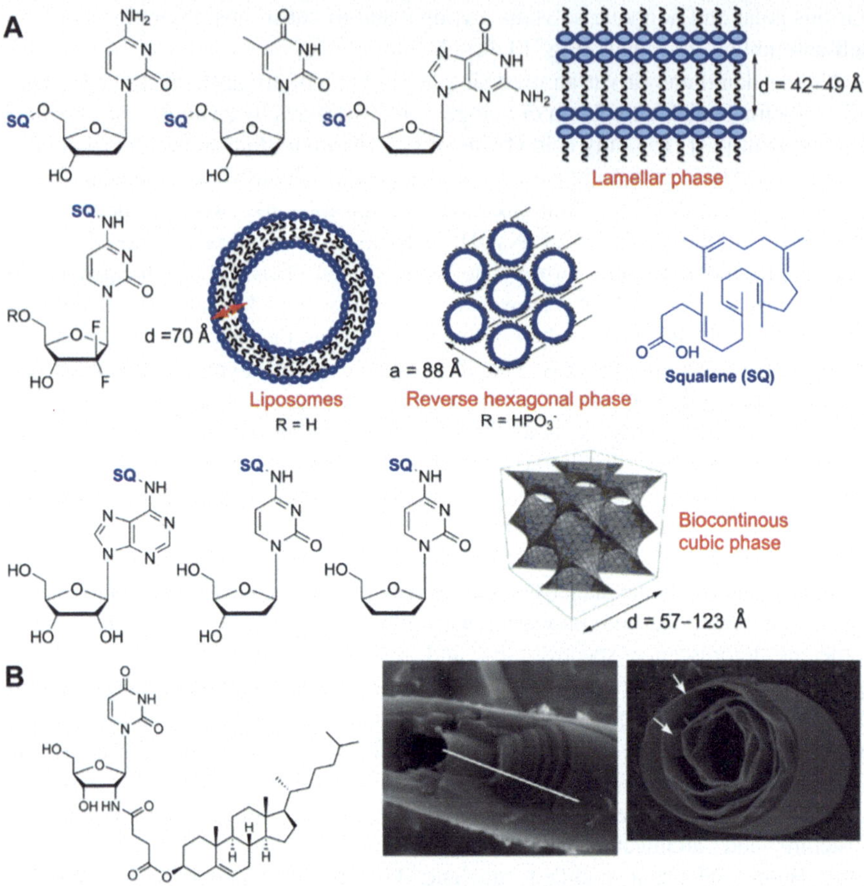

Fig. 13.6 (**a**) Architectures of squalene-modified nucleolipid assemblies. While squalene conjugated to nucleobases formed sponge-like nanoparticles or unilamellar liposomes, sugar-modified squalenes formed lamellar phases. (**b**) Self-assemblies of a cholesteryl nucleolipid CholAU in the presence of PC. Part of this figure is adapted with permission of the journals [28, 29]

Phosphatidylcholine (PC) (dioleoylphosphatidylcholine (DOPC) and dimyristoleoylphosphatidylcholine (DMoPC)), depending on the molar ratio of the components, self-assembled to form nano- and micro-tubules (Fig. 13.6b). Cryo-SEM images revealed that the nanotubules were densely packed, but micro-tubules were loosely rolled by lamellas [29]. This notion was further supported by AFM analysis, which indicated the presence of nm-sized CholAU-rich bilayers in nanotubes. In this co-assembly, it was identified that PCs serve as a platform to support the self-assembly of cholesteryl nucleolipid in aqueous medium.

Supramolecular assemblies generated using guanine and its derivatives require special mention as they can self-organize to form various architectures including G-ribbons, G-sheets and ion channels [32]. This is because the chemical space of

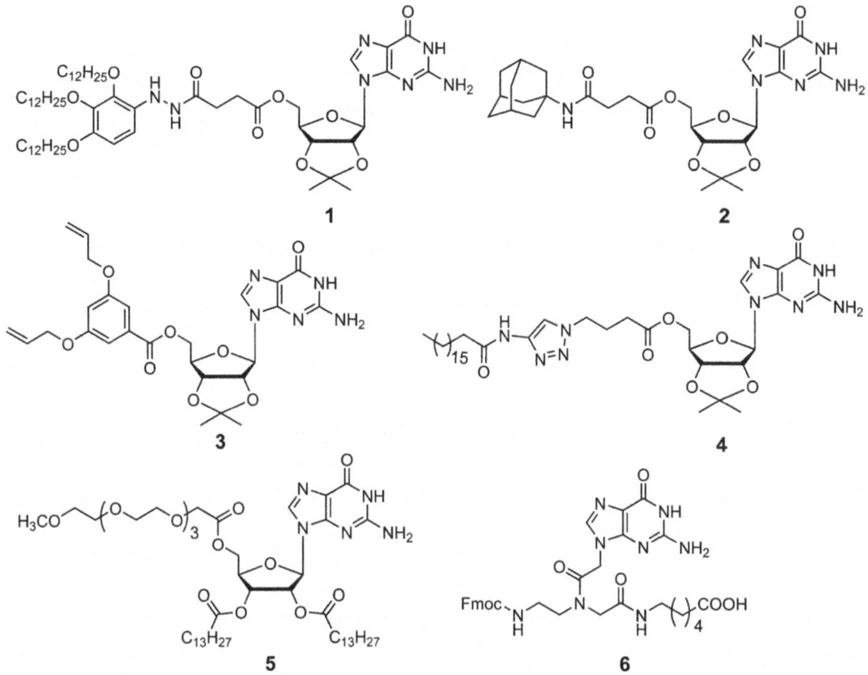

Fig. 13.7 Selected examples of guanine-based nucleolipids

guanine allows H-bonding through Watson-Crick and Hoogsteen faces, metal ion and stacking interactions. Notably, the formation of these structures depends on the concentration and the presence of different metals ions. For example, in the absence of metal ions, guanine adopts G-ribbons, which further assembles to form G-sheets. However, in the presence of Na^+ or K^+ ions, it forms cyclic G4-quartets, which stack one above the other to form fibres. At certain concentrations these assemblies eventually support hydrogel formation. In order to expand the repertoire of guanosine assemblies, lipophilic groups have been attached. Excellent reviews exclusively focused on guanine-based supramolecular assemblies have been recently published [32–34], and, hence, here we limit our discussion to some selected examples of lipophilic guanine derivatives as shown in Fig. 13.7.

2′,3′-*O*-isopropylideneguanosine derivatives (**1–4**) containing long alkyl chains were developed, and their self-assembling properties were studied in different solvents and in the presence of the complementary base (Fig. 13.7) [35–38]. In particular, lipophilic guanosine **1** formed a reversible organogel, wherein ribbon-like structures that stabilized the organogel in the absence of metal ions transformed into G-quartets upon addition of K^+ ions, which disassembled the gel [35]. Interestingly, when the cryptand [2.2.2], a K^+ chelator, was added, the G-quartets reverted to ribbon-like structures and gelled the solvent. On the other hand, at low concentrations adamantine-modified guanosine **2** adopted a G-quadruplex structure in acetonitrile without any templating ions [36]. However, at a higher concentration, the G-quadruplex transformed into a linear G-ribbon supporting the gel network. In another report, Montesarchio and coworkers reported acylated polyether functionalized guanosine derivative **5**, which formed stable organogels in polar solvents [39].

Davis and coworkers developed ionophores based on supramolecular assemblies formed by 5′-(3,5-bis(allyloxy)benzoyl)-2′,3′-isopropylidene guanosine (**3**) in the presence of potassium picrate, followed by crosslinking of neighbouring guanine units within the G-quadruplex framework using ruthenium-catalyzed ring-closing metathesis reaction [37]. The ionophore is integrated well into the phospholipid membrane and selectively transported Na^+ ions over the K^+ ions across the phospholipid bilayer membrane. The exterior of oligonucleotide G-quadruplexes is composed of phosphate backbone, and hence, insertion of such negatively charged structures in the lipid membrane is not straightforward. Dash and coworkers elegantly used a lipophilic guanosine (**4**) and G-quadruplex forming human telomeric repeat oligonucleotide to construct a synthetic ionophore [38]. The lipophilic guanosine was found to bind telomeric G-quadruplex in a 3:1 ratio with guanine base stacking over the tetrad and the along alkyl tail rapping around the groove. The complex nicely inserted itself into the lipid bilayer to form an ionophore, which was ably stabilized by the hydrophobic interaction between the hydrophobic chain and lipid layer. This assembly efficiently transported K^+ ions in a stimuli-responsive manner across CHO and K-562 cell membranes. In a recent report, Dash and coworkers synthesized lipophilic tert-butyldimethylsilyl (TBDMS) protected 8-bromoguanosine, which formed different nanostructures (rings, dumbbells and G-wires) depending on the incubation time. The nanostructure also formed discrete transmembrane ion channels and enabled the transport of K^+ ions [40].

Gazit and coworkers came up with an altogether new family of supramolecular synthons by using peptide nucleic acid building blocks. First, various combinations of di-PNA building blocks were prepared and evaluated for their ability to self-organize in different solvents and at different pH [41]. Only guanine-containing PNAs, namely, CG, GC and GG, formed ordered assemblies (long rods and spheroids) in alkaline solution. Interestingly, the assemblies exhibited voltage-gated electroluminescence and fluorescence emission in a wide range. In a similar approach, self-assemblies of Fmoc-protected guanine peptide nucleic acid (PNA) conjugated to a hydrophobic moiety were studied [42]. This synthon (**6**) formed a cyclic tetrameric structure as a result of head-to-tail H-bonds between the Fmoc group and guanine base of an adjacent molecule. Depending on the solvent and concentration, it formed diverse structures including porous microspheres and nanoribbons. Importantly, the assembly made of cyclic quartet arrangement displayed very high Young's modulus, suggesting potential applications in nanotechnology.

Xu and coworkers reported a biocompatible N-terminus-capped naphthalene nucleopeptide in which the nucleobases were attached to the peptide backbone [43]. This hybrid system formed hydrogels made of nanofibres, and the assembly interacted with ssDNA, plasmid DNA and hairpin DNA through complementary hydrogen bonding interactions. The assembly also assisted the transfection of DNA into live HeLa cells with low cytotoxicity. In a similar approach, a single thymine-capped nonapeptide was developed, which formed fibrillar supramolecular assemblies selectively in the presence of ATP [44]. This enabled the sequestering of ATP molecules in cancer cells, thereby reducing the efflux of doxorubicin and increasing its anti-cancer efficacy.

13.2.3 Ionic Nucleolipids

Unlike non-ionic nucleolipids, ionic nucleolipids containing cationic, anionic or zwitterionic lipid chains provide access to self-assembled architectures from aqueous solution. Many such designer-made ionic nucleolipids were mainly developed for biomedical applications, particularly as biocompatible delivery vehicles and tissue engineering materials. Grinstaff and coworkers reported the first example of cationic 2′,3′-dioleoyl derivative of uridine (DOTAU **7**) for gene delivery (Fig. 13.8a) [45]. Liposomes ranging from 60 nm to 15μm are formed in aqueous solution, which in the presence of calf thymus DNA (ctDNA) formed multilamellar vesicles having ~7 nm bilayers size. Taking forward the self-assembling property and ensuing utility of DOTAU, Barthélémy and coworkers came up with a novel nucleolipid (**8**), wherein dioleyl chains were attached to the 2′ and 3′ positions via a

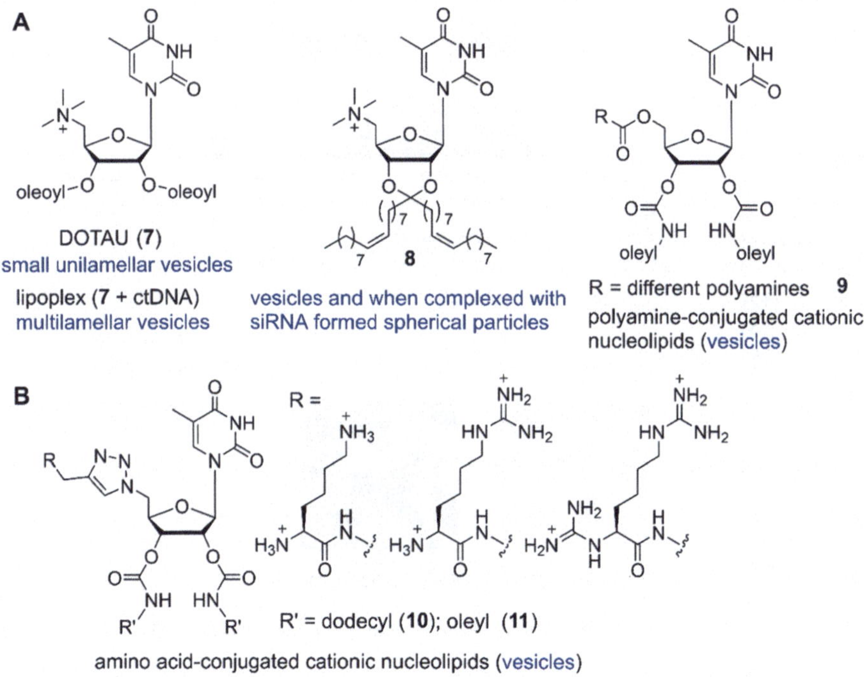

Fig. 13.8 Examples of nucleolipids containing fatty acids and cationic groups attached to the sugar residue

ketal group [46]. Dynamic light scattering (DLS) and transmission electron microscopy (TEM) studies indicated that the nucleolipid alone formed vesicles in aqueous buffer, and when complexed with siRNA formed spherical particles (~100 nm). Notably, the assembly was esterase resistant and helped in efficient transfection of siRNA into hepatic cell lines as compared to its diester analogue. During this time, Kim and coworkers reported cationic nucleolipids based on polyamines (**9**) and amino acids (**10, 11**), which formed vesicles and served as good anti-VEGF siRNA carriers (Fig. 13.8) [47, 48].

Barthélémy and coworkers developed an anionic nucleolipid (**12**) by using thymidine-3′-monophosphate as the polar head and 1,2 dipalmitoyl-sn-glycerol as the hydrophobic tail for in vitro delivery of nucleic acids [49]. Though examples of anionic nucleolipids for gene delivery are rare due to electrostatic repulsion with the cell membrane, here they formed lipoplexes with plasmid DNA in the presence of Ca^{2+} ions, which showed enhanced transfection efficacy in mammalian cell lines. This anionic nucleolipid has found applications in the delivery of cisplatin and in the preparation of injectable hydrogels, which are discussed in detail in the next section [50, 51]. Using a similar strategy, guanine (**13**)- and cytosine (**14**)-based anionic nucleolipids containing 3′-monophosphate nucleosides attached to 1,2-dipalmitoyl-sn-glycerol were synthesized and their assemblies in the presence and absence of Na^+ ions were studied (Fig. 13.9a) [52]. While cytosine analogs required NaCl to

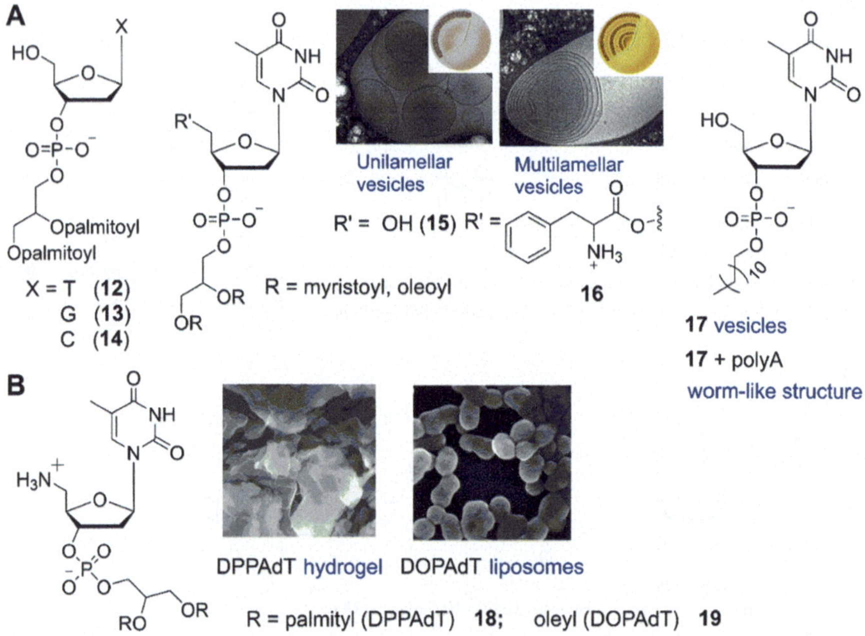

Fig. 13.9 (**a**) Examples of anionic nucleolipids and their assemblies. (**b**) Examples of zwitterionic nucleolipids and their assemblies. Microscopy images are reproduced with permission of the journals (ACS and RSC) [53, 55]

support the formation of hydrogels, guanosine derivatives formed gels even in the absence of the salt. In contrast, 1,2-myristoyl/oleoyl-sn-glycerol containing thymidine-3′-monophosphates (**15**) formed unilamellar vesicles. Interestingly, conjugation of phenylaniline at the 5′ position of these nucleolipids (**16**) resulted in the formation of a multilamellar assembly as a result of added stacking interaction provided by the aromatic amino acid (Fig. 13.9a) [53]. Zhang and coworkers synthesized anionic lipids based on 3′-*O*-alkyl nucleotides containing alkyl chain lengths of 8, 12 and 16 carbon atoms and studied their complexation with polyadenylic acid (polyA) [54]. TEM images of nucleolipid (**17**) changed dramatically after addition of polyA. Vesicle structures formed by the nucleolipid transformed into worm-like morphology in the presence of polyA.

Though zwitterionic nucleolipids provide access to different types of self-organized architectures, their applications are limited [16]. In a recent work, Yang and coworkers designed 5′-amino,3′-phospho nucleolipids containing 1,2-palmityl/oleyl-sn-glycerol moiety attached to the phosphate group [55]. SEM and TEM analysis of DPPAdT **18** containing saturated fatty acid chains showed the formation of an opaque hydrogel as a result of lamellar structure (Fig. 13.9b). In contrast, the nucleolipid containing unsaturated oleyl chains (DOPAdT **19**) formed spherical-like nanostructures in phosphate buffer. Further, circular dichroism (CD) and AFM studies showed that these 5′-aminonucleoside phospholipids interacted with polyA and calf thymus DNA through π-π stacking and H-bonding interactions.

13.2.4 Glycosyl-Based Nucleolipids

Natural polysaccharides specifically interact with polynucleotides via hydrogen bonding, Van der Waals forces and electrostatic interactions, and these findings prompted researchers to develop nucleolipid assemblies with sugar moieties for biomedical applications. Barthélémy and coworkers introduced triazole groups by attaching lipid and sugar moieties to the ribose ring and nucleoside, respectively, via double click chemistry [56]. Four glycosyl nucleolipids (GNLs) featuring β-D-glucopyranoside, thymidine and lipid moieties were synthesized. Clear gels were formed in water and chloroform for **20** and **21**, whereas **22** in water and **23** in chloroform formed opaque gels (Fig. 13.10a). Interestingly, control amphiphiles without nucleoside were not able to form gels in water or chloroform. This indicated that all three components glucopyranoside, nucleoside and lipid are essential for the gelation process. A combination of H-bonding between the sugar moieties as well as the amide moieties, π-π stacking between the bases and triazole moieties and additional hydrophobicity induced by the lipid moiety favoured the formation of fibres and nanotubes, which were confirmed by phase contrast microscopy and small-angle X-ray scattering (SAXS). Hydrogels made of gelator having an unsaturated oleylamide (**20**) formed entangled nanofibres. Interestingly, compact circular

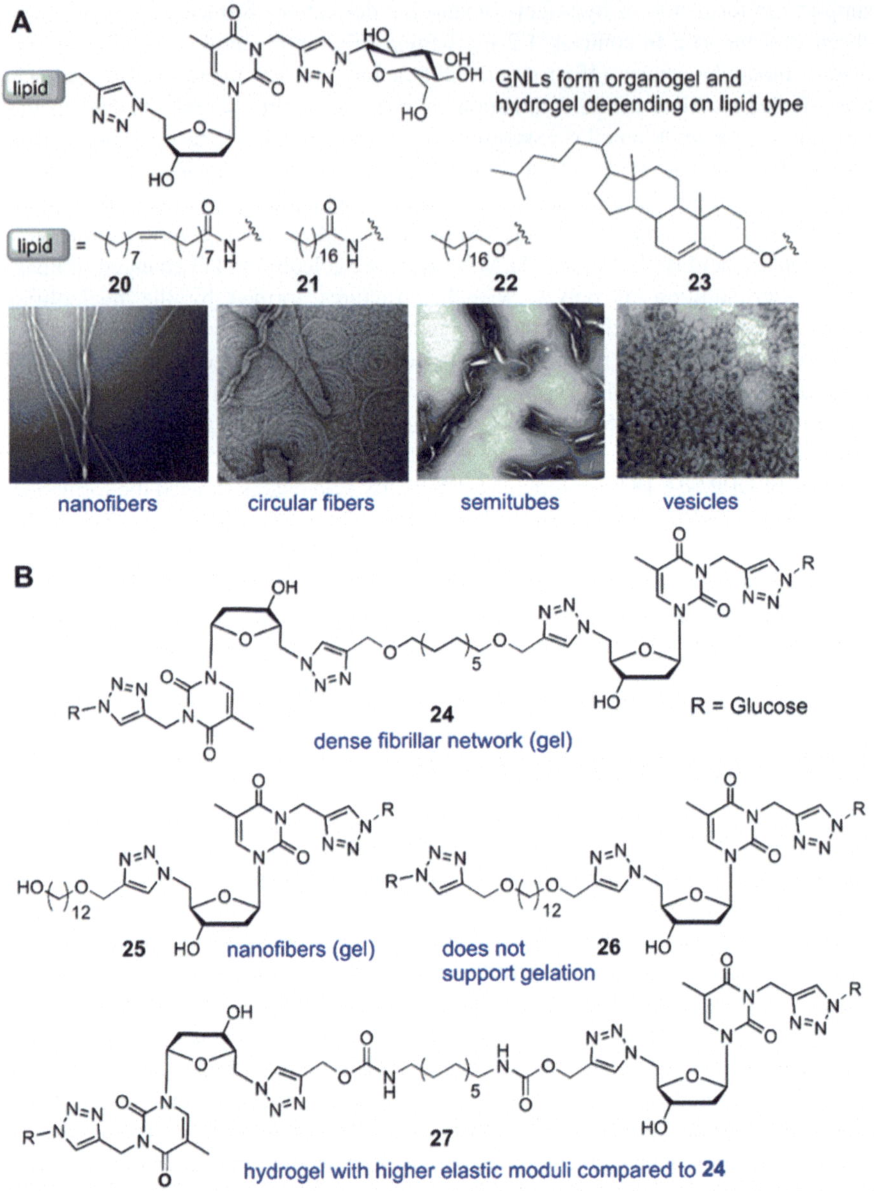

Fig. 13.10 (**a**) Examples of GNLs containing triazole linker and different lipid groups. TEM images have been produced with permission of the journal [56]. (**b**) Structures of GNBAs

fine fibres were observed in the case of saturated octadecylamide derivative **21**. Saturated ether derivative **22** formed fibres composed of semitubular objects and the cholesterol derivative **23** self-assembled into a dense network made of compact vesicles.

Barthélémy and coworkers reported GNLs containing double lipid chains connected via glycerol and hydroxybutanamide linkages [57, 58]. Unlike single-chain versions, double-chain GNLs did not support gel formation, but instead formed liposomes. While GNL-liposomes containing a glycerol linkage efficiently internalized into adipose-derived stromal cell (ADSC) stem cells, liposomes formed by non-nucleoside glycolipids and naked liposomes formed by lecithin could not invade the stem cells [57]. Alternatively, a fluorinated GNL with sugar attached to the 3′ hydroxyl group of the ribose ring instead of the nucleobase formed very dense colloidal suspensions in water as opposed to gels [58].

Combining the useful features of bolaamphiphiles and GNLs, Barthélémy and coworkers came up with the design of glycosyl-nucleoside bolaamphiphiles (GNBAs) as shown in Fig. 13.10b [59]. Here, the hydrophobic chain connected to the head groups and the sugar attached to the nucleoside, all had the triazole linkage derived by click chemistry. The symmetric bolaamphiphile (**24**) and asymmetric bolaamphiphile (**25**) formed stable hydrogels, whereas a bolaamphiphile containing two glucose moieties (**26**) did not support the gel formation, indicating that nucleoside is required for the self-assembling process. TEM images of hydrogels obtained from symmetric GNBA **24** revealed the formation of an anisotropic fibrillar network, while under similar conditions, asymmetric bolaamphiphile **25** exhibited bundles of nanofibres. Further, to improve the rheological properties of hydrogels formed by GNBA **24**, a carbamate moiety at the 5′ position of thymidine (**27**), which could provide an additional hydrogen bonding interaction, was introduced. Rheology analysis showed that carbamate-containing bolaamphiphile stabilized hydrogels with higher elastic modulus ($G' \approx 57$ kPa) than the ether analogs ($G' \approx 30$ kPa) [60].

13.3 Applications of Nucleolipid Assemblies

It is clear that engineered nucleoside-lipid hybrids exhibit dynamic self-assembly processes and supramolecular architectures, which can be tailored by chemical design. Nucleolipid assemblies discussed here are formed by noncovalent reversible interactions involving nucleobase, sugar and lipid moieties. Hence, by virtue of their inherent reversibility and dynamism, they respond well to physical and chemical stimuli. Further, many of these assembles are biologically benign and therefore have found huge application in the biomedical field as liposome delivery vehicles, injectable hydrogels and tissue engineering scaffolds. More recently, functionalized nucleolipid assemblies with interesting recognition and optical properties have been developed with promising potentials in material and sensor applications. In this book chapter, we discuss only the recent examples.

13.3.1 Nucleolipid Delivery Vehicles, Injectable Gels and Tissue Engineering Scaffolds

A drug delivery platform was devised by combining the self-assembling properties of anionic (**12**) and cationic (**7**) nucleolipids to deliver cisplatin [50]. In the initial experiments, a 1:1 colloidal suspension of **12** and DOPC containing cisplatin showed reduced stability in the presence of serum, which was overcome by stabilizing the core shell nanoparticle using cationic nucleolipid **7** that formed an outer shell. This formulation not only increased the serum stability but also enhanced the cytotoxicity of cisplatin and reduced drug resistance. Recently, this system was used to increase the plasmatic bioavailability of an anti-malarial drug, methylene blue. The nanoparticle-drug formulation provided protection against light and reduction, thereby increasing its serum half-life and anti-malarial activity [61]. A series of supramolecular assemblies including entangled fibre and lamellar structures were formed by pairing nucleolipid **12** with different cations. In particular, **12**-Na^+ combination formed injectable as well as biocompatible hydrogels. A gradual release of FITC-BSA from the gelator matrix indicated a controlled release of protein in vivo [51]. Lattices when 3D-printed with a biomaterial ink of **12**-DMEM@+ (Dulbecco's modified Eagle's medium (DMEM) enriched with NaCl) showed structures to be exactly replicable. The formulation upon injection into mice showed steady biodegradation pattern, and upon cell culture, cell survival and proliferation were observed in the gel matrix, which suggests that this self-assembled scaffold could be useful in tissue engineering-based regeneration of connective tissue [62].

Assemblies of GNBAs exhibit mechanical and biological properties that are essential for constructing injectable gels and tissue engineering scaffolds. Rheology study of the solid-like gel GNBA **24** containing the ether linkage showed thixotropic behaviour, and hence, its injectability in a mice model was evaluated. Skin biopsy post-injection indicated that the bulk of the gel in the tissue was stable. The cytotoxicity of the gel was assessed by MTT assay using human mesenchymal stem cells and was found to be non-toxic till 50μM concentration. Further, the gel network served as an efficient matrix for cell adhesion and cell proliferation, which opens a way towards tissue engineering [59]. Rheological properties (e.g. elastic modulus) of the GNBA hydrogel were significantly improved when ether linkage (**24**) was replaced with a carbamate group (**27**) (Fig. 13.10b) [60]. However, the gel-sol transition temperature (T_{gel}) was observed to be close to the human body temperature hampering its in vivo applications. Subsequently, the same group developed urea- and amide-based GNBAs **28** and **29**, respectively, which displayed fast gelation kinetics and better in vivo stability [63] (Fig. 13.11). In particular, bis-urea containing nucleolipid exhibited T_{gel} around 78 °C and fast elastic modulus recovery without changes in gel stability over three cycles. The gel degradation kinetics was evaluated in mouse model by injecting the gel mixed with cyanine 5.5 probe. Fluorescence imaging after administering lucigenin to **28** gel-implanted mice showed low inflammation response. Histological examination after 3 weeks of implantation showed the gel to be the most stable. Taken together, **28** features a

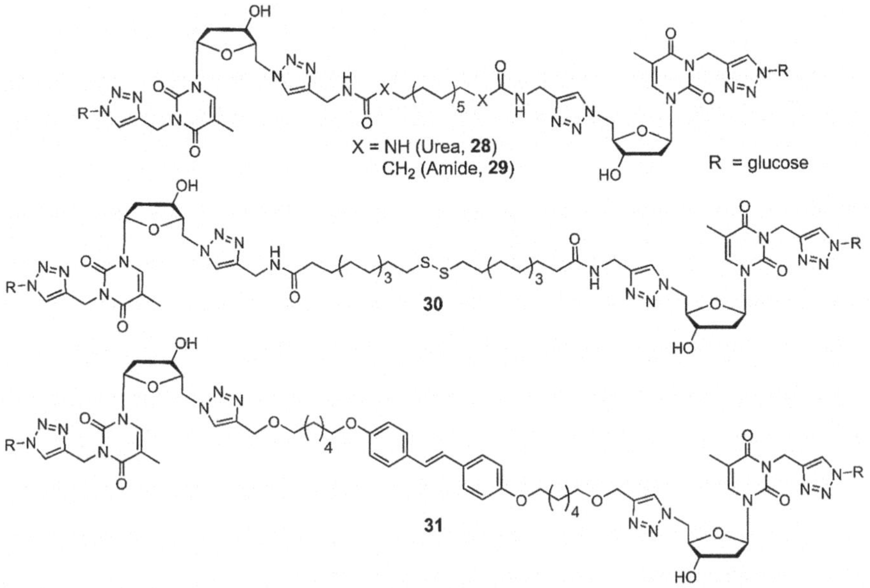

Fig. 13.11 GNBAs containing urea (**28**), amide (**29**), disulphide (**30**) and stilbene (**31**) moieties

combination of optimum supramolecular and rheological properties and also is non-immunogenic after injection in vivo.

More recently, Barthélémy and coworkers incorporated a redox-cleavable disulfide (–SS–) linker into the nucleolipid gelator (**30**) to achieve a controlled release of drug molecules [64]. The CGC for **30** was very low (0.5% w/v), and importantly, sol-gel temperature and rheological properties were amenable for in vivo injection. In vitro experiments in the presence of dithiothreitol (DTT) showed slower rate of degradation for the gel as compared to the non-disulphide control gel. The drug release kinetics was evaluated by using bovine serum albumin (BSA) tagged with Fluorescein isothiocyanate (FITC) and a control of similar molecular weight (dextran-FITC) loaded into the hydrogel. BSA being a thio-containing protein was released much slower than dextran (without any thio group) as a result of disulfide reshuffling between the protein and gel matrix. Further, the hydrogel passed the cytotoxicity test and was found to be injectable and biodegradable. A light-sensitive GNBA (**31**) was constructed by incorporating a stilbene moiety, which formed a gel in a hydroalcoholic mixture (1:1 methanol and water). TEM images of *E*-isomer of stilbene exhibited entangled network composed of nanofibres, which broke into smaller nanofibres as a result of photoisomerization of *E*-stilbene to *Z*-stilbene [65]. It is expected that this light-responsive system could serve as an alternative to the existing polymer-based drug delivery vehicles triggered by light (Fig. 13.11).

Grinstaff and coworkers reported a mechanoresponsive oleoylamide GNL **20**, which was shown to release entrapped cargos upon mechanical shear. The release rates determined using different FITC-dextrans entrapped in the gel network were found to depend on the molecular weights [66]. Further, the shear-mediated macromolecule release was studied using a rabbit immunoglobulin G (IgG) antibody (~150 kDa) directed against human TNFα. The increased release of the anti-TNFα from GNL hydrogel showed efficacy in neutralizing the cytotoxicity of human TNFα in the L929 murine fibroblast cell line. Since this gel has thixotropic properties, it could be extended to in vivo systems also. Marlow and coworkers reported another neutral nucleolipid *N*4-octanoyl-2′-deoxycytidine forming self-healing hydrogels [67]. To check the drug delivery potential of the gel, in vitro release of fluorescently labelled dextrans was evaluated, and based on the diffusion coefficients, it was inferred that smaller molecules diffused significantly faster as compared to macromolecules. Later, erosion-based release profiles of insulin and lysozyme suggested that the gel matrix does not affect the secondary structure and function of the proteins. Notably, in vivo injection of the hydrogel showed that it is biodegradable and non-inflammatory, suggesting its potential use in protein delivery [68].

13.3.2 Fluorescent Nucleolipids and Sensors

Nucleolipids that combine fluorescence and interesting self-assembling properties are very useful in devising responsive materials and sensors. Though not well explored, nucleolipids with such multiple features have gathered significant interest recently. Srivatsan group described the construction of a new family of supramolecular nucleolipid synthons by using environment-sensitive fluorescent nucleoside analogs (Fig. 13.12) [69]. In this design, heterobicycles, namely, benzofuran and benzothiophene, were conjugated to the 5′ position of uridine to impart fluorescence and additional stacking interaction. 2′ and 3′ OH groups were coupled to fatty acids of different chain lengths. The heterocycle-modified nucleoside served as the head group and fatty acid acyl chains served as the tail group. In dilute conditions, the fluorescence properties like emission maximum, quantum yield and anisotropy were found to be significantly influenced by solvent polarity changes. In terms of self-assembly, both the modifications supported organogel formation, but the hierarchical assembling process was distinctly different. Though myristoyl chain containing derivatives formed nanotubes with similar outer diameters, the wall thickness of benzothiophene-modified nucleolipid nanotubes was larger than the tubes formed by benzofuran-modified nucleolipid. Notably, when the chain length was increased, both the analogs self-assembled into long-range entangled fibrous network with lower CGC values (Fig. 13.12a–d).

In this nucleolipid design, the heterocycles were purposely attached to the uracil ring via a rotatable aryl-aryl bond to integrate a molecular rotor element. This aspect was evident from higher fluorescence quantum yield and anisotropy exhibited by the nucleolipids in solvent of higher viscosity. In the gel state also the nucleolipids exhibited intense fluorescence, and in particular, the benzothiophene-modified nucleolipids displayed aggregation-induced enhanced emission (AIEE) due to

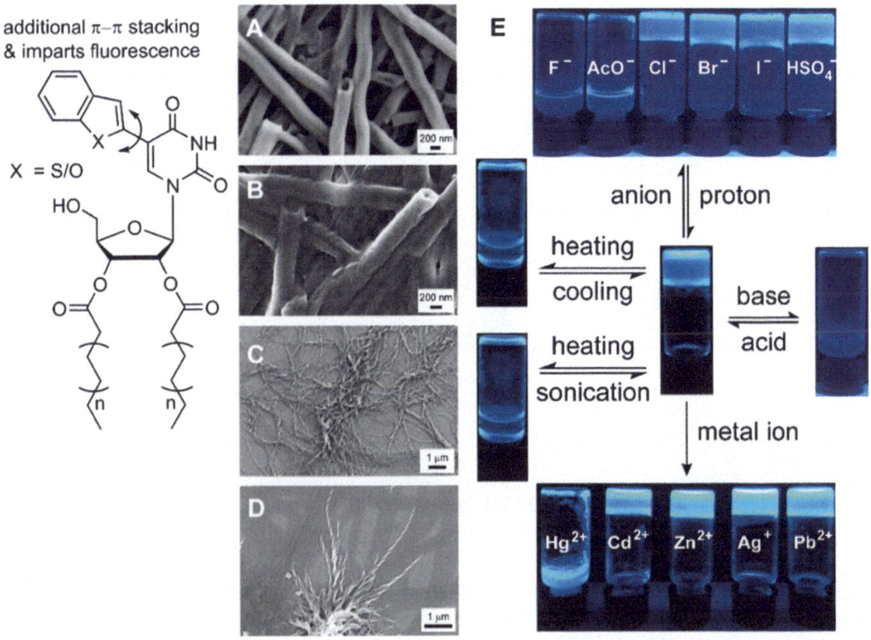

Fig. 13.12 (**a–d**) Field emission scanning electron microscopy (FESEM) images of xerogels of nucleolipids. (**a**, **b**) Benzofuran- and benzothiophene-modified nucleolipids, respectively, containing myristoyl chains. (**c**, **d**) Benzofuran- and benzothiophene-modified nucleolipids, respectively, containing palmitoyl chains. (**e**) Multi-stimuli responsiveness of a benzothiophene-modified nucleolipid gel containing palmitoyl chains. This figure is reproduced by permission of The Royal Society of Chemistry (RSC) [69]

rigidification of the fluorophore in the gel state. Systematic analysis of the self-assembling process by using microscopy, X-ray and NMR indicated that base pairing and stacking interactions along with a coordinated interplay of H-bonding network between adjacent nucleosides set the path for the hierarchical assemblies. Importantly, these interactions and ensuing fluorescence outcome could be reversibly switched by the application of multiple external stimuli such as physical and chemicals (Fig. 13.12e). For example, the gel-sol transition was sensitive to protonation-deprotonation process. F^- and Hg^{2+} ions specifically deprotonated H atoms important for network, which allowed the selective sensing of these ions by fluorescence and morphological analysis. This design approach could be used on other nucleoside analogs as well to expand the structural and functional repertoire of nucleoside-lipid hybrid materials.

Taking forward this approach, Sun and coworkers developed carbazole-modified nucleolipids (Fig. 13.13) [70]. Alkyl chains of different lengths were attached to the carbazole moiety. C12 and C16 containing nucleolipids formed stable fluorescent organogels in nonpolar solvents as a result of fibrous network. The gel network was sensitive to chemical stimuli, particularly F^- ions, which disassembled the gel resulting in quenching in fluorescence intensity. More recently, Gao and coworkers

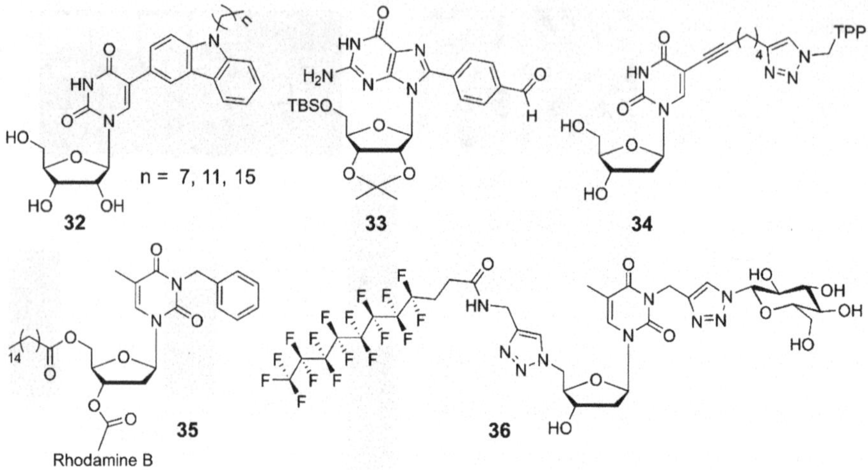

Fig. 13.13 Fluorescent nucleolipids (**32–35**) and fluorocarbon-containing GNL (**36**)

reported fluorescent organogels made of an amphiphilic C8-aryl guanosine derivative (**33**) [71]. In the presence of KPF_6, **33** formed a G_8-octamer complex, which further aggregated to form an organogel in the presence of diamine linker in different solvents such as dimethyl formamide (DMF), tetrahydrofurane (THF) and dichloroethane (DCE). ^{1}H NMR and circular dichroism studies indicated that the driving force for the gelation process was the formation of an imine bond between the diamine linker and aldehyde group. Interestingly, depending on the solvent and the type of amine linker used for gelation, the organogel showed an intense green fluorescence emission that is expected to be useful in material-related research. Xiong and coworkers tethered a tetraphenylethene (TPP) moiety, a well-known system that exhibits AIE, at the 5′ position of 2′-deoxyuridine via a triazole linker [72]. H-bonding interaction between the nucleoside and triazole group and π stacking interaction between triazole and TPP facilitated the aggregation process (sponge-like structures) to support the gelation process. Notably, the gel system made of **34** displayed AIE, which was found to be thermo-reversible.

Desvergnes and coworkers designed hybrid fluorescent nucleolipids by introducing a benzyl group at the N3 position of thymidine, and palmitoyl and rhodamine B groups were coupled to the ribose sugar [73]. These nucleolipids (e.g. **35**) were formulated in nanoemulsions using Miglyol 812, lecithin E80 and Tween 80 and used as nanovectors to target neuronal cells and intracellular organelles. All formulations were monodispersed with an average diameter under 200 nm and were stable for a nearly 2 weeks. Cytotoxicity as studied in a human neuroblastoma cell line indicated very low toxicity over 48 h. The nanovectors were efficiently internalized into cells, and colocalization experiments using rhodamine B fluorescence indicated that these vectors could serve as tools to target lysosomes.

The ability of nucleosides to interact specifically with certain metal ions has been elegantly used in developing metal ion sensors and oligonucleotide-based molecular machines [74]. Taking advantage of these interactions, metal-mediated self-assemblies of nucleolipids have also been designed for specific detection of metal ions. 3′,5′-*O*-di-fatty acid-substituted thymidines developed by Srivatsan and coworkers exhibited hierarchical self-assembling process, wherein a strong H-bonding interaction involving N3H atom was important for the supramolecular gel formation. The gelation process was specifically responsive to the presence of Hg^{2+} ions. Addition of Hg^{2+} (0.02 equivalents) ions disintegrated the gel, which was confirmed by inverted vial method and SEM analysis (Fig. 13.3) [21]. NMR analysis proved that the Hg^{2+} ion deprotonated the N3H, and subsequent formation of T-Hg-T base pair was confirmed by mass analysis. Notably, even five equivalents of other metal ions did not affect the gelation ability of the nucleolipids. Similarly, gelation of 2′,3′-*O*-distearoyl-modified ribothymidine was specifically sensitive to Hg^{2+} ion among the tested metal ions. However, in this case the Hg^{2+} ion formed a T-T metallo-base pair by deprotonating the 5′-OH group, which resulted in the degelation of the nucleolipid [25].

In a similar approach, 1,2 dipalmitoyl-sn-glycerol containing cytidine nucleolipid **14** was used to sense Ag^{+} ion [75]. In the absence of Ag^{+} ions, the nucleolipid in an aqueous buffer formed liposomes, which rigidified upon addition for the metal ion. Since thioflavin T (ThT) is known to show enhanced fluorescence in a rigid network, a combination of the nucleolipid assembly in the presence of the fluorescent probe was utilized in detecting Ag^{+} ion concentration (limit of detection ~3 ppb). However, the mode of interaction of the metal ion with the nucleolipid assembly was not ascertained.

13.3.3 Nucleolipid Assemblies for Environmental Remediation

Gels are highly useful in removing toxic metals and organic pollutants from the aqueous wastes and soil via specific interactions with the assemblies. Barthélémy and coworkers demonstrated the use of fluorinated GNF **36** assembly to remove nanoparticle wastes (e.g. quantum dots, Au and TiO_2 nanoparticles) from aqueous solution [76]. GNF **36** added to an aqueous solution of the nanoparticles slowly formed hydrogels and during this process entrapped the particles in the gel network. UV and fluorescence measurements showed no detectable amounts of the particles in the supernatant, indicating efficient removal of the wastes. In another study, anionic (**12**) and cationic DOTAU (**7**) nucleolipids were used in the removal of pharmaceutical and pesticide pollutants by exploiting the complementary electrostatic interaction [77]. While a gel of **12** showed 93% efficacy in decontaminating an aqueous solution of propranolol, a cationic active pharmaceutical ingredient (API), a gel of **7** displayed 93% efficacy for diclofenac. The decontamination efficiency was found to be significantly higher compared to a commonly used adsorbent, Trivorex. Further, nucleolipid **7** was also used in removing a cocktail of 13 organic contaminants from aqueous medium.

Srivatsan and coworker used heterotypic hydrogels formed by the partial hydrolysis of 5′-*O*-fatty acid-conjugated ribothymidine and uridine nucleolipids to

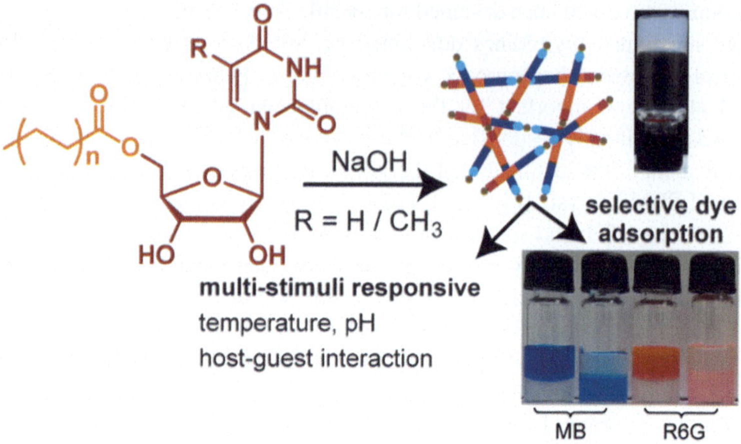

Fig. 13.14 Heterotypic hydrogels 5′-*O*-fatty acid-substituted ribothymidine and uridine nucleolipids show multi-stimuli responsiveness and adsorb cationic dye pollutants. This figure is reproduced by permission of The Royal Society of Chemistry (RSC) [23]

selectively adsorb cationic dyes (Fig. 13.14) [23]. The negatively charged heterotypic hydrogels formed specifically in the presence of NaCl trapped methylene blue (MB) and rhodamine 6G (R6G) in the gel network. Importantly, the gel surface could be recycled to adsorb the dyes for 8 cycles with almost 75% efficiency. Further, the dye from the gel network could be removed by simply adding HCl, which precipitated the nucleolipid. Filtration followed by the addition of NaOH regenerated the gel, which was again ready for decontamination of the dyes.

13.4 Conclusions and Outlook

The examples illustrated in this chapter demonstrate how chemical design could provide easy access to various types of hybrid nucleolipid supramolecular assemblies. Many such assemblies have been fabricated into smart functional materials as they can be tuned to adapt and respond to biological and artificial (chemical and physical) stimuli. Notably, recent use of nucleolipid materials in drug delivery, tissue engineering and 3D bio-printing applications could be seen as a significant step forward in advancing modern medicine. However, there are several challenges and shortcomings in the currently available tools. Immediate challenges that need to be addressed are biocompatibility, immune response behaviour, selectivity in targeting, spatiotemporal control of assembling and disassembling process, controlled release of cargos and clearance of materials once the job is accomplished. Therefore, the next-generation materials, which could be called as "super smart" materials, should be driven to advance gene and immune therapies, 3D bio-printing and precision medicines customized to the individual patient need. While the majority of nucleolipid designs are directed towards biomedical applications, it would be also interesting to see how the potential of nucleolipid assemblies could be harnessed to construct smart probes and materials for technological applications.

Acknowledgements S.G.S is grateful to the past and present lab members and collaborators who have worked in the area of nucleolipids. M.B.W is grateful to CSIR, India, and Wellcome Trust-DBT India Alliance for a graduate research fellowship. Prime Minister's Research Fellowship (PMRF) to S.D. is greatly appreciated. S.G.S thanks Wellcome Trust-DBT India Alliance (IA/S/16/1/502360) and CSIR, India (02-0086/12/EMR-II), for research grants.

References

1. Rothemund PWK (2006) Folding DNA to create nanoscale shapes and patterns. Nature 440:297–302
2. Saccà B, Niemeyer CM (2012) DNA origami: the art of folding DNA. Angew Chem Int Ed 51 (1):58–66
3. Chakraborty K, Veetil AT, Jaffrey SR, Krishnan Y (2016) Nucleic acid-based nanodevices in biological imaging. Annu Rev Biochem 85:349–373
4. Tapio K, Bald I (2020) The potential of DNA origami to build multifunctional materials. Multifunct Mater 3(3):032001
5. Li X, Feng K, Li L, Yang L, Pan X, Yazd HS, Cui C, Li J, Moroz L, Sun Y, Wang B, Li X, Huang T, Tan W (2020) Lipid–oligonucleotide conjugates for bioapplications. Natl Sci Rev 7 (12):1933–1953
6. Halley PD, Patton RA, Chowdhury A, Byrd JC, Castro CE (2019) Low-cost, simple, and scalable self-assembly of DNA origami nanostructures. Nano Res 12:1207–1215
7. Sivakova S, Rowan SJ (2005) Nucleobases as supramolecular motifs. Chem Soc Rev 34:9–21
8. Peters GM, Davis JT (2016) Supramolecular gels made from nucleobase, nucleoside and nucleotide analogs. Chem Soc Rev 45(11):3188–3206
9. Rosemeyer H (2005) Nucleolipids: natural occurrence, synthesis, molecular recognition, and supramolecular assemblies as potential precursors of life and bioorganic materials. Chem Biodivers 2(8):977–1063
10. Allian V, Bourgauz C, Couvreur P (2012) Self-assembled nucleolipids: from supra-molecular structure to soft nucleic acid and drug delivery devices. Nucleic Acids Res 40(5):1891–1903
11. Gissot A, Camplo M, Grinstaff MW, Barthélémy P (2008) Nucleoside, nucleotide and oligonucleotide based amphiphiles: a successful marriage of nucleic acids with lipids. Org Biomol Chem 6(8):1324–1333
12. Latxague L, Dalila M-J, Patwa A, Ziane S, Chassande O, Godeau G, Barthélémy P (2012) Glycoside nucleoside lipids (GNLs): an intrusion into the glycolipid's world. C R Chimie 15 (1):29–36
13. Baillet J, Desvergnes V, Hamoud A, Latxague L, Barthélémy P (2018) Lipid and nucleic acid chemistries: combining the best of both worlds to construct advanced biomaterials. Adv Mater 30(11):1705078
14. Pu F, Ren J, Qu X (2018) Nucleobases, nucleosides, and nucleotides: versatile biomolecules for generating functional nanomaterials. Chem Soc Rev 47(4):1285–1306
15. Chen M, Lin W, Hong L, Ji N, Zhao H (2019) The development and lifetime stability improvement of guanosine-based supramolecular hydrogels through optimized structure. Biomed Res Int 2019:1–18
16. Barthélémy P (2009) Nucleoside-based lipids at work: from supramolecular assemblies to biological applications. C R Chime 12(1–2):171–179

17. Dias E, Battiste JL, Williamson JR (1994) Chemical probe for glycosidic conformation in telomeric DNAs. J Am Chem Soc 116(10):4479–4480
18. Berti D, Montis C, Baglioni P (2011) Self-assembly of designer biosurfactants. Soft Matter 7 (16):7150–7158
19. Barthélémy P, Prata CA, Filocamo SF, Immoos CE, Maynor BW, Hashmi SA, Lee SJ, Grinstaff MW (2005) Supramolecular assemblies of DNA with neutral nucleoside amphiphiles. Chem Commun 10:1261–1263
20. Mulet X, Kaasgaard T, Conn CE, Waddington LJ, Kennedy DF, Weerawardena A, Drummond CJ (2010) Nanostructured nonionic thymidine nucleolipid self-assembly materials. Langmuir 26(23):18415–18423
21. Nuthanakanti A, Srivatsan SG (2017) Surface-tuned and metal-ion-responsive supramolecular gels based on nucleolipids. ACS Appl Mater Interfaces 9(27):22864–22874
22. Nuthanakanti A, Walunj MB, Arun T, Badiger MV, Srivatsan SG (2019) Self-assemblies of nucleolipid supramolecular synthons show unique self-sorting and cooperative assembling process. Nanoscale 11(24):11956–11966
23. Nuthanakanti A, Srivatsan SG (2020) Multi-stimuli responsive heterotypic hydrogels based on nucleolipids show selective dye adsorption. Nanoscale Adv 2(9):4161–4171
24. Yun YJ, Park SM, Kim BH (2003) Novel thymidine-based organogelators and their gelation behaviours. Chem Commun 2:254–255
25. Nuthanakanti A (2019) Cytidine and ribothymidine nucleolipids synthesis, organogelation, and selective anion and metal ion responsiveness. New J Chem 43(34):13447–13456
26. Köstler K, Werz E, Malecki E, Montilla-Martinez M, Rosemeyer H (2013) Nucleoterpenes of thymidine and 2′-deoxyinosine: synthons for a biomimetic lipophilization of oligonucleotides. Chem Biodivers 10(1):39–61
27. Werz E, Viere R, Gassmann G, Korneev S, Malecki E, Rosemeyer H (2013) Synthesis of thymidine, uridine, and 5-methyluridine nucleolipids: tools for a tuned lipophilization of oligonucleotides. Helv Chim Acta 96(5):872–888
28. Lepeltier E, Bourgaux C, Rosilio V, Poupaert JH, Meneau F, Zouhiri F, Lepêtre-Mouelhi S, Desmaële D, Couvreur P (2013) Self-assembly of squalene-based nucleolipids: relating the chemical structure of the bioconjugates to the architecture of the nanoparticles. Langmuir 29 (48):14795–14803
29. Losensky L, Goldenbogen B, Holland G, Laue M, Petran A, Liebscher J, Scheidt HA, Vogel A, Huster D, Klipp E, Arbuzova A (2016) Micro- and nano-tubules built from loosely and tightly rolled up thin sheets. Phys Chem Chem Phys 18(2):1292–1301
30. Pescador P, Brodersen N, Scheidt HA, Loew M, Holland G, Bannert N, Liebscher J, Herrmann A, Huster D, Arbuzova A (2010) Microtubes self-assembled from a cholesterol-modified nucleoside. Chem Commun 46(29):5358–5360
31. Snip E, Koumoto K, Shinkai S (2002) Gel formation properties of a uracil-appended cholesterol gelator and cooperative effects of the complementary nucleobases. Tetrahedron 58 (43):8863–8873
32. Davis JT, Spada GP (2007) Supramolecular architectures generated by self-assembly of guanosine derivatives. Chem Soc Rev 36(2):296–313
33. Bhattacharyya T, Saha P, Dash J (2018) Guanosine-derived supramolecular hydrogels: recent developments and future opportunities. ACS Omega 3(2):2230–2241
34. Stefan L, Monchaud D (2019) Applications of guanine quartets in nanotechnology and chemical biology. Nat Rev Chem 3:650–668
35. Wang X, Zhou L, Wang H, Luo Q, Xu J, Liu J (2011) Reversible organogels triggered by dynamic K^+ binding and release. J Colloid Interface Sci 353(2):412–419
36. Meng L, Liu K, Mo S, Mao Y, Yi T (2013) From G-quartets to G-ribbon gel by concentration and sonication control. Org Biomol Chem 11(9):1525–1532
37. Kaucher MS, Harrell WA, Davis JT (2006) A unimolecular G-quadruplex that functions as a synthetic transmembrane Na^+ transporter. J Am Chem Soc 128(1):38–39
38. Debnath M, Chakraborty S, Kumar YP, Chaudhuri R, Jana B, Dash J (2020) Ionophore constructed from non-covalent assembly of a G-quadruplex and liponucleoside transports K^+-ion across biological membranes. Nat Commun 11(1):469

39. Simeone L, Milano D, De Napoli L, Irace C, Di Pascale A, Boccalon M, Tecilla P, Montesarchio D (2011) Design, synthesis and characterisation of guanosine-based amphiphiles. Chem Eur J 17(49):13854–13865
40. Das RN, Kumar YP, Kumar SA, Schütte OM, Steinem C, Dash J (2018) Self-assembly of a guanosine derivative to form nanostructures and transmembrane channels. Chem Eur J 24 (16):4002–4005
41. Berger O, Adler-Abramovich L, Levy-Sakin M, Grunwald A, Liebes-Peer Y, Bachar M, Buzhansky L, Mossou E, Forsyth VT, Schwartz T, Ebenstein Y, Frolow F, Shimon LJW, Patolsky F, Gazit E (2015) Light-emitting self-assembled peptide nucleic acids exhibit both stacking interactions and Watson-Crick base pairing. Nat Nanotechnol 10(4):353–360
42. Basavalingappa V, Bera S, Xue B, Azuri I, Tang Y, Tao K, Shimon LJW, Sawaya MR, Kolusheva S, Eisenberg DS, Kronik L, Cao Y, Wei G, Gazit E (2019) Mechanically rigid supramolecular assemblies formed from an Fmoc-guanine conjugated peptide nucleic acid. Nat Commun 10:5256
43. Du X, Zhou J, Li X, Xu B (2017) Self-assembly of nucleopeptides to interact with DNAs. Interface Focus 7:20160116
44. Wang H, Feng Z, Qin Y, Wang J, Xu B (2018) Nucleopeptide assemblies selectively sequester ATP in cancer cells to increase the efficacy of doxorubicin. Angew Chem Int Ed 57 (18):4931–4935
45. Chabaud P, Camplo M, Payet D, Serin G, Moreau L, Barthélémy P, Grinstaff MW (2006) Cationic nucleoside lipids for gene delivery. Bioconjug Chem 17(2):466–472
46. Luvino D, Khiati S, Oumzil K, Rocchi P, Camplo M, Barthélémy P (2013) Efficient delivery of therapeutic small nucleic acids to prostate cancer cells using ketal nucleoside lipid nanoparticles. J Control Release 172(3):954–961
47. Yang HW, Yi JW, Bang E-K, Jeon EM, Kim BH (2011) Cationic nucleolipids as efficient siRNA carriers. Org Biomol Chem 9(1):291–296
48. Patil SP, Yi JW, Bang E-K, Jeon EM, Ki BH (2011) Synthesis and efficient siRNA delivery of polyamine-conjugated cationic nucleoside lipids. Med Chem Commun 2(6):505–508
49. Khiati S, Pierre N, Andriamanarivo S, Grinstaff MW, Arazam N, Nallet F, Navailles L, Barthélémy P (2009) Anionic nucleotide-lipids for in vitro DNA transfection. Bioconjug Chem 20(9):1765–1772
50. Khiati S, Luvino D, Oumzil K, Chauffert B, Camplo M, Barthélémy P (2011) Nucleoside–lipid-based nanoparticles for cisplatin delivery. ACS Nano 5(11):8649–8655
51. Ramin MA, Sindhu KR, Appavoo A, Oumzil K, Grinstaff MW, Chassande O, Barthélémy P (2017) Cation tuning of supramolecular gel properties: a new paradigm for sustained drug delivery. Adv Mater 29(13):1605227
52. Alies B, Ouelhazi MA, Patwa A, Verget J, Navailles L, Desvergnes V, Barthélémy P (2018) Cytidine- and guanosine-based nucleotide–lipids. Org Biomol Chem 16(26):4888–4894
53. Tonelli G, Oumzil K, Nallet F, Gaillard C, Navailles L, Barthélémy P (2013) Amino acid-nucleotide-lipids: effect of amino acid on the self-assembly properties. Langmuir 29 (18):5547–5555
54. Pan D, Tang C, Fan X, Li Y, Yang X, Jin H, Guan Z, Yang Z, Zhang L (2013) Thymidine-based amphiphiles and their bonding to DNA. New J Chem 37(4):1122–1127
55. Pan D, Sun J, Jin H, Li Y, Li L, Wu Y, Zhang L, Yang Z (2015) Supramolecular assemblies of novel aminonucleoside phospholipids and their bonding to nucleic acids. Chem Commun 51 (3):469–472
56. Godeau G, Barthélémy P (2009) Glycosyl-nucleoside lipids as low-molecular-weight gelators. Langmuir 25(15):8447–8450
57. Latxague L, Ziane S, Chassande O, Patwa A, Dalila M-J, Barthélémy P (2011) Glycosylated nucleoside lipid promotes the liposome internalization in stem cells. Chem Commun 47:12598–12600
58. Latxague L, Patwa A, Amigues E, Barthélémy P (2013) Glycosyl-nucleolipids as new bioinspired amphiphiles. Molecules 18(10):12241–12263

59. Latxague L, Ramin MA, Appavoo A, Berto P, Maisani M, Ehret C, Chassande O, Barthélémy P (2015) Control of stem-cell behavior by fine tuning the supramolecular assemblies of low-molecular-weight gelators. Angew Chem Int Ed 54(15):4517–4521
60. Latxague L, Gaubert A, Maleville D, Baillet J, Ramin MA, Barthélémy P (2016) Carbamate-based bolaamphiphile as low-molecular-weight hydrogelators. Gels 2(4):25
61. Kowouvi K, Alies B, Gendrot M, Gaubert A, Vacher G, Gaudin K, Mosnier J, Pradines B, Barthélémy P, Grislaine L, Millet P (2019) Nucleoside-lipid-based nanocarriers for methylene blue delivery: potential application as anti-malarial drug. RSC Adv 9(33):18844–18852
62. Dessane B, Smirani R, Bouguéon G, Kauss T, Ribot E, Devillard R, Barthélémy P, Naveau A, Crauste-Manciet S (2020) Nucleotide lipid-based hydrogel as a new biomaterial ink for biofabrication. Sci Rep 10:2850
63. Ramin MA, Latxague L, Sindhu KR, Chassande O, Barthélémy P (2017) Low molecular weight hydrogels derived from urea based-bolaamphiphiles as new injectable biomaterials. Biomaterials 145:72–80
64. Bansode ND, Sindhu KR, Morel C, Rémy M, Verget J, Boiziau C, Barthélémy P (2020) A disulfide based low molecular weight gel for the selective sustained release of biomolecules. Biomater Sci 8(11):3186–3192
65. Baillet J, Gaubert A, Bassani DM, Verget J, Latxague L, Barthélémy P (2020) Supramolecular gels derived from nucleoside based bolaamphiphiles as a light-sensitive soft material. Chem Commun 56(23):3397–3400
66. Kaplan JA, Barthélémy P, Grinstaff MW (2016) Self-assembled nanofiber hydrogels for mechanoresponsive therapeutic anti-TNFα antibody delivery. Chem Commun 52 (34):5860–5863
67. Skilling KJ, Kellam B, Ashford M, Bradshaw TD, Marlow M (2016) Developing a self-healing supramolecular nucleoside hydrogel. Soft Matter 12(43):8950–8957
68. Angelerou MGF, Markus R, Paraskevopoulou V, Foralosso R, Clarke P, Alvarez CV, Chenlo M, Johnson L, Rutland C, Allen S, Brasnett C, Seddon A, Zelzer M, Marlow M (2020) Mechanistic investigations into the encapsulation and release of small molecules and proteins from a supramolecular nucleoside gel in vitro and in vivo. J Control Release 317:118–129
69. Nuthanakanti A, Srivatsan SG (2016) Hierarchical self-assembly of switchable nucleolipid supramolecular gels based on environmentally-sensitive fluorescent nucleoside analogs. Nanoscale 8(6):3607–3619
70. Jia X, Zhao J, Xu S, Zhang F, Sun J, Lu R (2018) Luminescent organogels generated from nucleosides functionalized with carbazole: synthesis and probing for F^-. Eur J Org Chem 2018 (16):1910–1915
71. Zhang Y, He Y, Wojtas L, Shi X, Guo H (2020) Construction of supramolecular organogel with circularly polarized luminescence by self-assembled guanosine octamer. Cell Rep Phys Sci 1 (10):100211
72. Zhao X, Zhao L, Xiao Q, Xiong H (2020) Intermolecular hydrogen-bond interaction to promote thermoreversible 2′-deoxyuridine-based AIE-organogels. Chin Chem Lett 32:1363–1367
73. Cunha A, Prévot G, Mousli Y, Barthélémy P, Crauste-Manciet S, Dehay B, Desvergnes V (2020) Synthesis and intracellular uptake of rhodamine–nucleolipid conjugates into a nanoemulsion vehicle. ACS Omega 5(11):5815–5823
74. Scharf P, Müller J (2013) Nucleic acids with metal-mediated base pairs and their applications. ChemPlusChem 78(1):20–34
75. Alies B, Ouelhazi MA, Noireau A, Gaudin K, Barthélémy P (2019) Silver ions detection via nucleolipids self-assembly. Anal Chem 91(3):1692–1695
76. Patwa A, Labille J, Bottero J-Y, Thiery A, Barthélémy P (2015) Decontamination of nanoparticles from aqueous samples using supramolecular gels. Chem Commun 51 (13):2547–2550
77. Sicard M, Crauste-Manciet S, Dole F, Verget J, Thiery A, Barthélémy P (2020) Decontamination of organic pollutants from aqueous media using polymer-free bioinspired materials. ACS Sustainable Chem Eng 8(30):11052–11057

Chapter 14
Nucleobase- and DNA-Functionalized Hydrogels and Their Applications

Apurba K. Das, Ankan Biswas, Sourav Bhowmik, and Tapas Ghosh

14.1 Introduction

Molecular self-assembly is a unique process where molecules acquire certain arrangement under suitable conditions (pH, concentration, temperature, pressure, ionic strength, solvent medium, etc.) through several noncovalent interactions such as H-bonding, hydrophobic interaction, metal coordination, electrostatic forces, π–π stacking interactions, electrostatic interactions, etc. [1–10]. Self-assembly plays vital roles in biomolecules such as proteins, peptides, DNA, RNA and enzymes to create, protect and maintain balance in biological systems [11–16]. One of the outcomes of molecular self-assembly is the gelation or the formation of gels. Gels are formed by 3D cross-linking of molecules and the formation of fibrils followed by entrapment of water or organic solvents [17–19]. The gels are called hydrogels when the medium is water, and in the case of organic solvents, it is called organogels.

Two nucleotide-functionalized biopolymeric chains interact with each other through H-bonding to build the double-helical supramolecular structure in DNA. The double-helix structure of DNA was first discovered by James Watson and Francis Crick in 1953 [20]. In DNA, the double-helix structures are held together by interactions of complimentary purine/pyrimidine base pairs [20]. The complementary base pairs in DNA are adenine/thymine and guanine/cytosine. Among all the above-discussed base pairs, guanine can engage in Hoogsteen H-bonding through its Hoogsteen face (Fig. 14.1). Four guanines can interact with each other by Hoogsteen H-bonding to form G-quartet or G4 self-assembled structures [21–23]. Low molecular weight G4 and oligonucleotide-based DNA hydrogels are most common and highly significant for different applications.

A. K. Das (✉) · A. Biswas · S. Bhowmik · T. Ghosh
Department of Chemistry, Indian Institute of Technology Indore, Indore, India
e-mail: apurba.das@iiti.ac.in

T. Govindaraju, K. Ariga (eds.), *Molecular Architectonics and Nanoarchitectonics*, Nanostructure Science and Technology,
https://doi.org/10.1007/978-981-16-4189-3_14

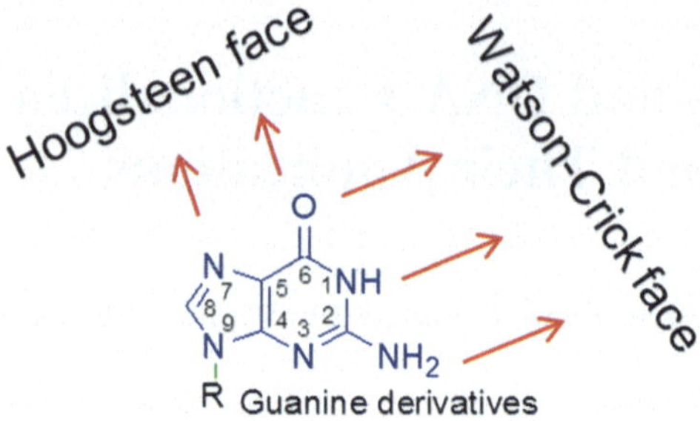

Fig. 14.1 Hoogsteen and Watson–Crick faces of guanine derivatives

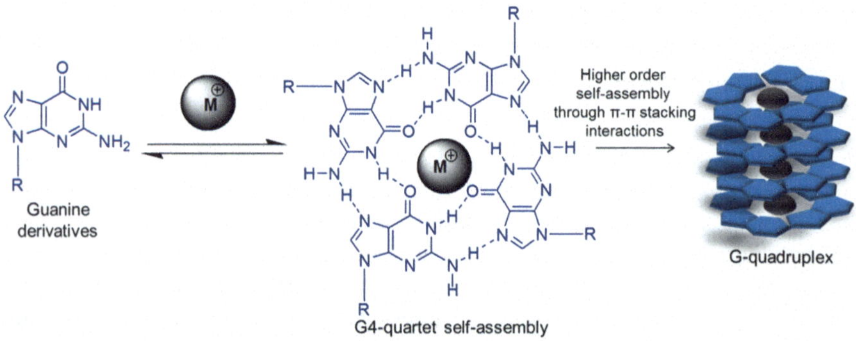

Fig. 14.2 Mechanism of the formation of G-quartet and G-quadruplex

Four guanines interact through Hoogsteen-type H-bonding to form G-quartet (Fig. 14.1). Several G-quartets stabilize by binding with central metal ions, mostly with K^+ and Na^+ ions. Further, G-quadruplex structures are formed with repeated G-quartet units (Fig. 14.2) [21–23]. Most of the DNA is located in the chromosome, inside the nucleus of cells. There are specific areas in the chromosome, genes and genomic region where G4s are found. In human cells, telomeres and immunoglobulin switch regions are most common [24]. G4 plays several vital roles including replication, maintaining the length of telomere, and in transcription, and so it becomes a highly interesting subject to researchers for drug targeting motifs and in nanotechnology [24].

Besides G-quadruplex, oligonucleotide-based molecules are also highly significant. Oligonucleotides are short DNA or RNA molecules, which are usually made up of 13 to 25 nucleotides. Oligomers have found applications in various fields like genetic testing, forensics and research [25–27]. Oligonucleotides (ONs) are prepared in laboratory by common solid-phase chemical synthesis [28]. Particularly, these ONs play a pivotal role in the polymerase chain reactions (PCR), molecular cloning, artificial gene manufacturing and DNA sequencing as it is quite easy to synthesize small chain of nucleic as single-stranded molecules with desired sequence [29–32]. Naturally occurring oligonucleotides are generally observed as small RNA molecules which play a vital role in the gene regulation. Oligonucleotides are degradation intermediates, which are obtained from the cleavage of long nucleic acid molecules.

14.2 G-Quadruplex Hydrogel

G-quartet units of Hoogsteen H-bonded guanine derivatives are stabilized by Na^+ and K^+ as the ionic radii of these ions are suitable to fit the G-quartet cavity [21–23]. Further, the G-quartet units interact through π–π stacking interactions to form columnar G-quadruplex structure. In the presence of suitable conditions, several G-quadruplex structures further self-assemble to form nanofibrillar structures [21–23]. These nanofibrillar structures could encapsulate a large number of water molecules to form G-quadruplex hydrogels [21–23].

14.2.1 Brief History of G-Quadruplex Hydrogel

The formation of gels with guanine derivatives was first observed by Bang in 1910 [33]. After that, in 1962, Gellert et al. explored the G-quadruplex structure by the formation of hydrogel with 5′-guanosine monophosphate (5′-GMP) [34]. Guschlbauer et al. further studied the G-quadruplex with guanosine derivatives in the 1970s [35, 36]. Further, Davis et al. have extensively studied G-quadruplex mechanism in the solid phase as well as in the gel phase [37–42]. In the past decades, several G-quadruplex hydrogels have been reported and potentially have been applied for various applications including cell growth, drug delivery, enzyme reaction and sensing [43–46].

14.2.2 G-Quadruplex Hydrogels from Binary Systems

In the early stage of development, the easy crystallization nature of the G-quadruplex gels was the major drawback. It was a challenging task to stabilize the gels. One of the methods to stabilize the system is to use binary guanosine derivatives to form hydrogels.

McGown et al. reported a binary G-quadruplex gel system with G and 5′-GMP [47]. In this work, several ratios of mixing were tried to improve the stability of the hydrogels (Fig. 14.3). Also, the temperature dependency of the system was studied. Mariani et al. also used G and 5′-GMP in a different ratio to stabilize the gels. Targeted molecules were encapsulated in this gel medium [48]. Rowan et al. found that binary mixtures of G and 2′,3′,5′-tri-O-acetylguanosine (TAcG) form different types (physical appearance, stability, mechanical strength) of gels within the ratio of 60:40–40:60 (Fig. 14.4) [49]. The higher concentration of G than the optimized ratio

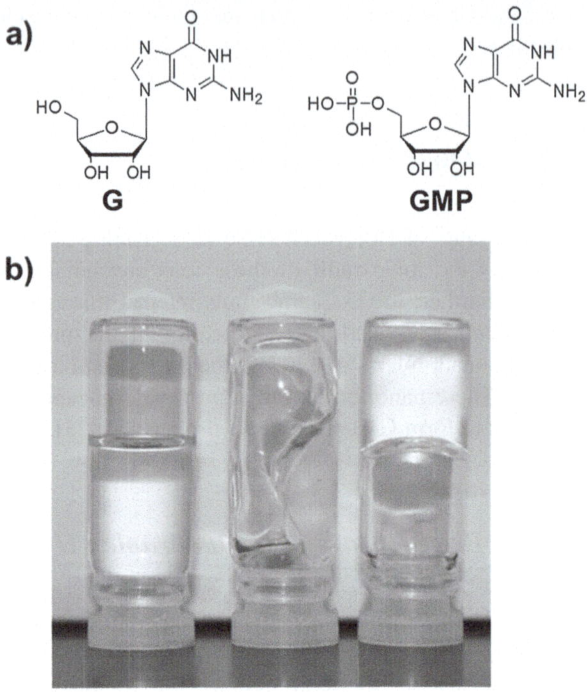

Fig. 14.3 (**a**) Structure of G and GMP. (**b**) Formation of binary G4 hydrogel by mixing with G and GMP. Reprinted with permission from Ref. [47]

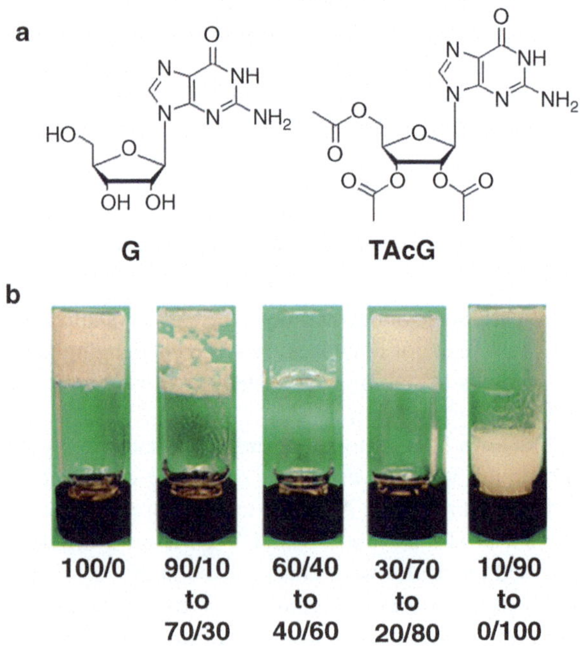

Fig. 14.4 (**a**) The structure of G and TcAG. (**b**) The formation of binary G4 hydrogel by mixing with G and TcAG at a different ratio. Reprinted with permission from Ref. [49]

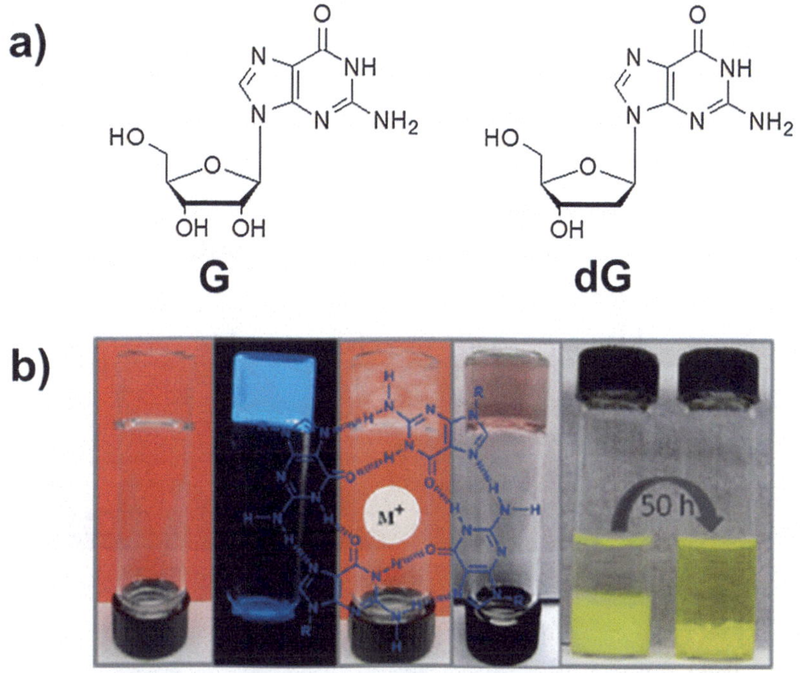

Fig. 14.5 (**a**) The structure of G and dG. (**b**) The use of suitable binary G4 hydrogel for the delivery of vitamins. Reprinted with permission from Ref. [50]

caused crystallization and a higher concentration of TAcG could not result in the gelation.

Kraatz et al. utilized G and 2′-deoxyguanosine (dG) binary mixture for the gelation (Fig. 14.5) [50]. The mechanical strength, viscoelastic nature and thermal stability of this binary gel could be tuned by using various metal ions that were used for G-quartet template. Suitable hydrogel was used for the delivery of vitamins. Chen et al. developed guanosine (G)- and isoguanosine (IsoG)-based binary G4 hydrogel (Fig. 14.6) [51]. The hydrogel condition is screened by using several alkali metal ions and the ratio of G and IsoG. The hydrogel shows shear thinning, injectable and self-healing properties.

14.2.3 Boronate Ester Functionalized Dynamic G-Quadruplex Hydrogels and Their Applications

In guanine derivative, guanosine contains *cis*-1,2-diol that binds with several boronic acids to form reversible boronate ester and produces several varieties of dynamic hydrogels. These hydrogels have been used for several applications

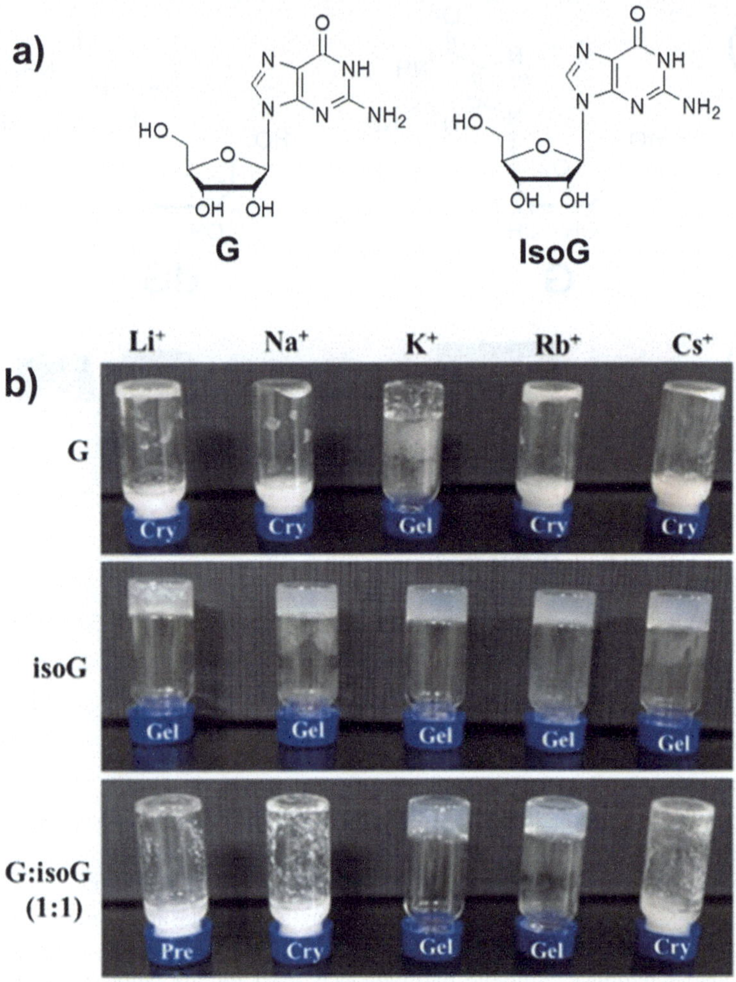

Fig. 14.6 (**a**) The structure of G and IsoG. (**b**) The formation of binary G4 hydrogel by mixing with G and IsoG in the presence of different alkali metal ions. Reprinted with permission from Ref. [51]

including drug delivery, cellular growth, 3D bioprinting, enzyme mimetic reactions and wastewater remediations [42–46, 52–55].

Davis et al. reported G-quadruplex hydrogel formed with G and boric acid in the presence of KOH (Fig. 14.7) [40]. During the gelation, the dynamic boronate ester forms which is stabilized by the borate anion. The G-quartet structure is stabilized by K^+ ions. Other different metal ions (Li^+, Na^+, Rb^+, Cs^+) have also been tested to study their effect on gelation. However, K^+ is the most suitable ion as the size of K^+ is well fitted with G-quartet cavity. The gel is a selectively encapsulated cationic dye like methylene blue over anionic dye (Rose bengal), due to its anionic nature. In another study, Davis et al. reported that a cationic dye thioflavin T (ThT) strengthens

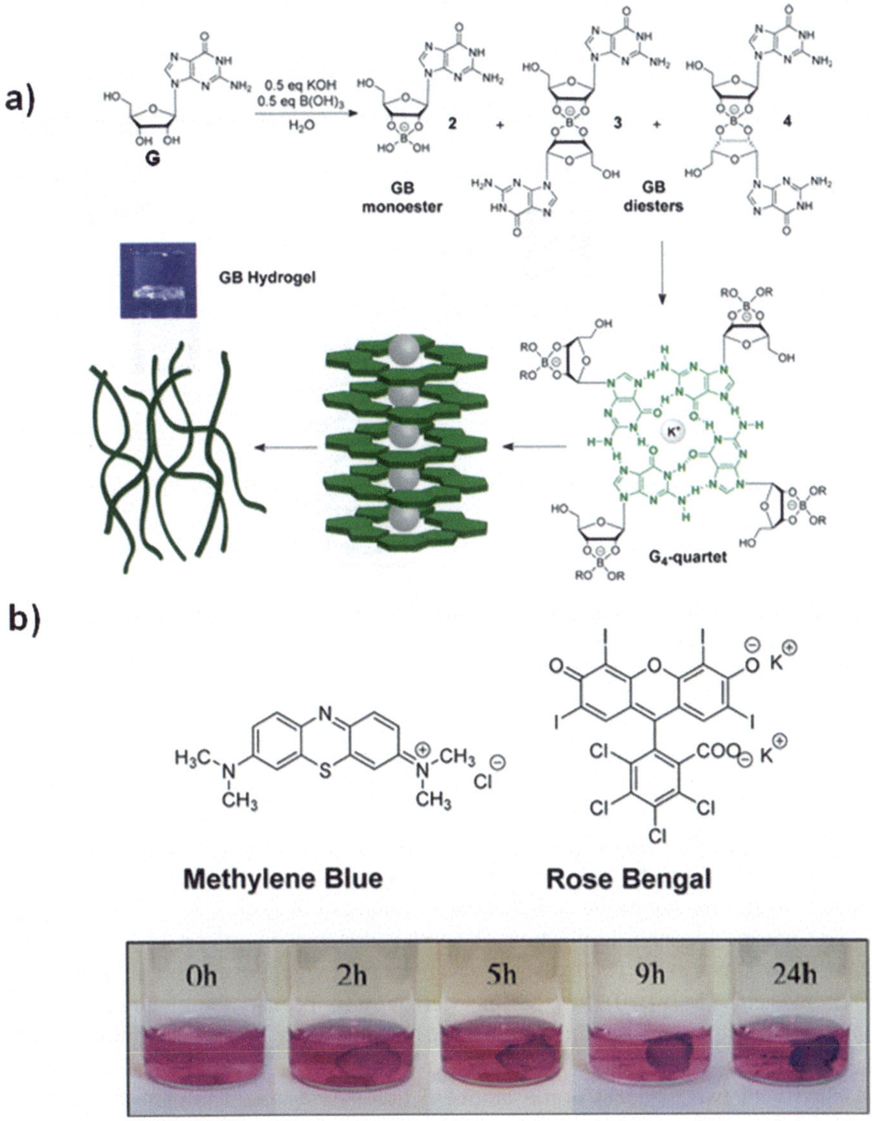

Fig. 14.7 (**a**) The formation of G-quadruplex hydrogel with G and boric acid in the presence of KOH. (**b**) The selective dye absorption of the G-quadruplex hydrogel. Reproduced with permission from Ref. [40]

the weaker Li^+ template G-quartet gel (Fig. 14.8) [41]. The cationic dye ThT interacts with anionic G-quartet by π–π stacking interaction, which was explained in detail in this work. Also, due to charge transfer between two oppositely charged species, the fluorescence intensity of ThT was enhanced intensively. These experiments are highly useful to detect G4-quartet self-assembled structures.

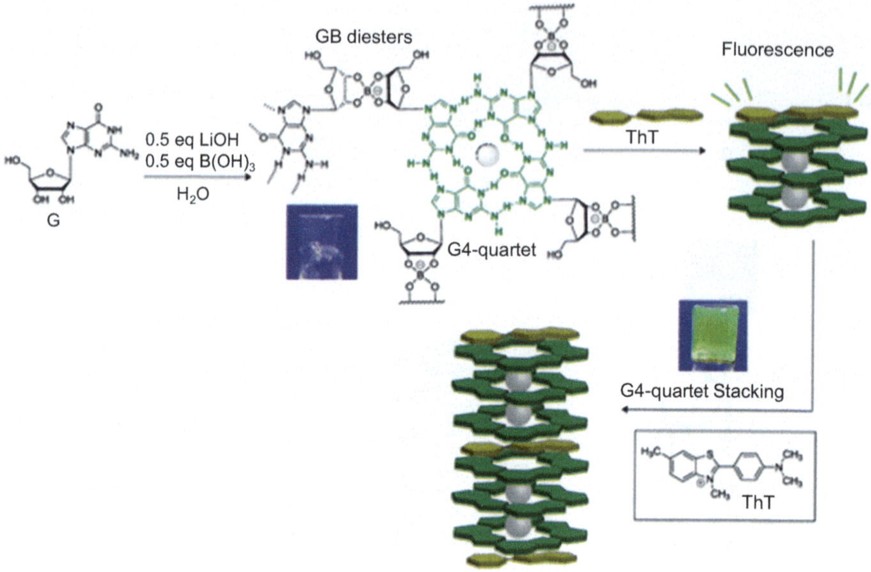

Fig. 14.8 The formation of G-quadruplex hydrogel with G and boric acid in the presence of LiOH and its fluorescent property and strength enhancement mechanism after binding with ThT. Reproduced with permission from Ref. [41]

Davis et al. found the self-destructive nature of 5′-iodoguanosine borate hydrogel (Fig. 14.9) [42]. The gel is destroyed slowly due to nucleophilic substitution reaction between C′5 carbon of guanosine and N5 in aromatic guanine ring where –I is a leaving group. The hydrogel was mixed with anti-viral drug acyclovir. Acyclovir contains guanine moiety and takes part in the formation of G-quartet. The drug releases with time due to the self-destructive nature of the gel.

The preparation and application of another different G-quartet hydrogel were reported by Davis et al. The hydrogel was prepared by guanosine and benzene-1,4-diboronic acid (Fig. 14.10) [45]. K^+ ions were used for primary templating cations to stabilize the G-quartet structure. Further, the physical cross-linking was enhanced by the addition of Mg^{2+}. The Mg^{2+} interacts with two adjacent hydroxyl groups attached with two adjacent boron atoms of boronate esters. The enhanced cross-linking increased water content, robustness and mechanical strength in the hydrogel. Due to higher water content, the hydrogel was used as a cell growth matrix. The cell viability (73%) of the hydrogel was observed after 24 h of incubation.

Sadler et al. developed photoactivatable anticancer hydrogel (Pt-DA-B-G) by incorporating photoactivatable dopamine-functionalized Pt(IV) drug (Pt-DA) inside the G4 borate hydrogel (Fig. 14.11) [44]. The hydrogel shows more potent photocytotoxicity against A2780Cis human ovarian cancer cells that is resistant towards popular cisplatin drug. The hydrogel shows selective photocytotoxicity to cancer cells over normal fibroblast cells. Stimuli-sensitive dynamic G-quartet hydrogel was applied for zero-order drug delivery application by Shi et al. Hydrogel was

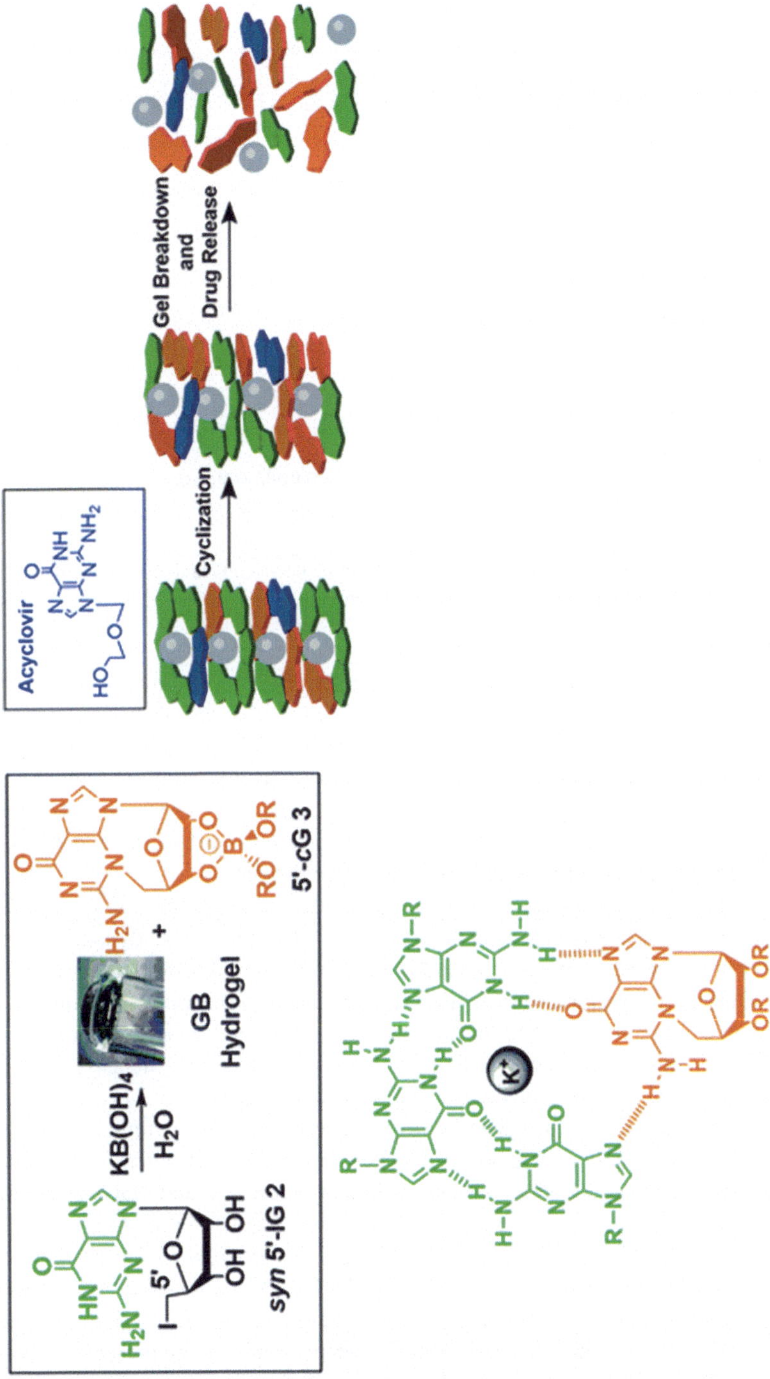

Fig. 14.9 The formation of self-destructive G4 hydrogels and its acyclovir drug release mechanism. Reproduced with permission from Ref. [45]

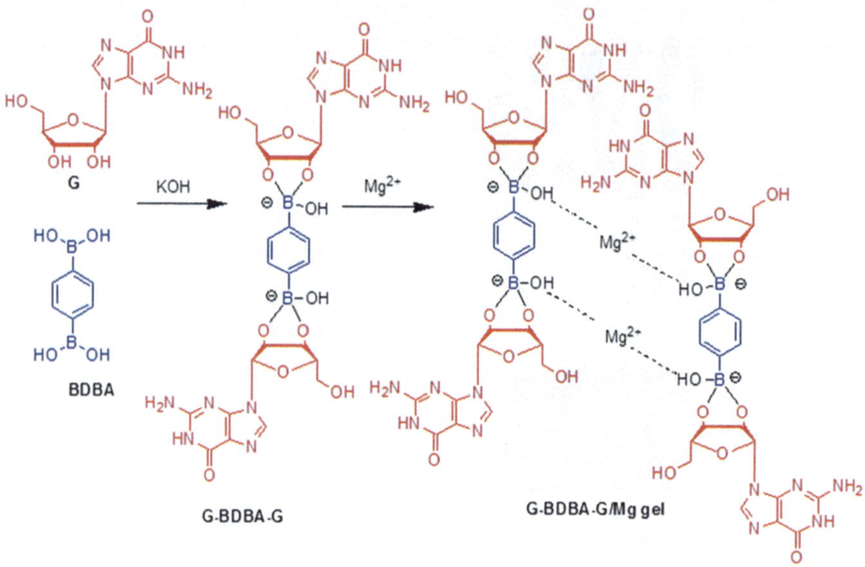

Fig. 14.10 The formation of G4 hydrogel using the reaction between G and BDBA and its further cross-linking mechanism with Mg^{2+}. Reprinted with permission from Ref. [45]

prepared by using guanosine (G), 2-formylphenylboronic acid (2-FPBA), tris (2-aminoethyl)amine (TAEA) and aqueous KCl [43]. The imino-boronate bonds were formed during the gelation (Fig. 14.12). The imino-boronate bonds are sensitive towards acidic pH and in the presence of glucose. The drug delivery from the gel was monitored in the presence of glucose and acidic pH. The zero-order drug delivery profile was achieved due to layer-by-layer peeling off of the hydrogels in the presence of external stimuli like glucose and acidic pH.

Ferroceneboronic acid-functionalized dynamic redox-active G-quadruplex hydrogels were developed by Das et al. (Fig. 14.13) [52]. The dynamic cyclic boronate ester of the gelators generates the injectable and thixotropic behaviour of the gel. During the preparation of the hydrogel, ferrocene moiety was turned to ferrocenium by aerobic oxidation to form gel (green). Initial oxidation of ferrocene to ionic ferrocenium helps to dissolve the gelators. After gelation, the less stable gel (green) was spontaneously converted to gel (brown) to achieve a more stable 18e configuration in the Fe^{II} centre. The hydrogels due to their thixotropic and robustness were applied in 3D printing study. The gel (brown) was stronger, 3D-printable and water stable. This new approach may help to synthesize water-stable structures by using the 3D-printing process for various applications. Further, guanosine–arylboronate ester-mediated dynamic G-quadruplex hydrogels in the presence of K^+ ions were demonstrated by Das et al. (Fig. 14.14) [53]. Guanosine was reacted with phenylboronic acid, 4-nitrophenyl boronic acid and 4-methoxyphenylboronic acid to form GPBA, 4-GNPBA and 4-GMPBA, respectively, in the presence of K^+ at physiological pH. The G-quadruplex assembly of the guanosine purine moiety

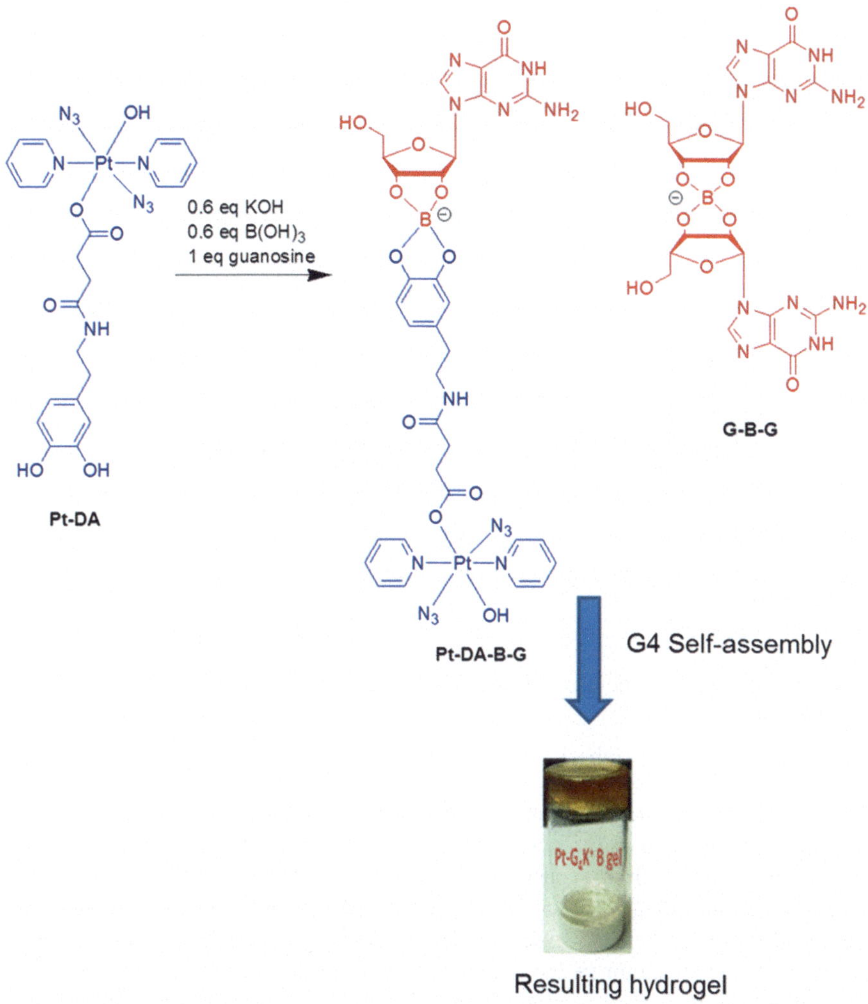

Fig. 14.11 The Pt-DA incorporated photoactivatable G4 hydrogel. Reprinted with permission from Ref. [44]

generates essential nanofibrillar networks for the encapsulation of water molecules to produce hydrogels. The dynamic nature of the boronic acid–boronate ester equilibrium produces essential injectable, self-healing and thixotropic properties for 3D printing (Fig. 14.15). The gel is non-toxic and the printing process does not damage the cell viability. This gel is ready to use and does not require additional cross-linking steps during post-printing process. Guanosine and PBA are relatively low cost and readily available materials. Research on novel bio-inks for 3D printing is vital for the progress of this technology for biomedical applications. This work

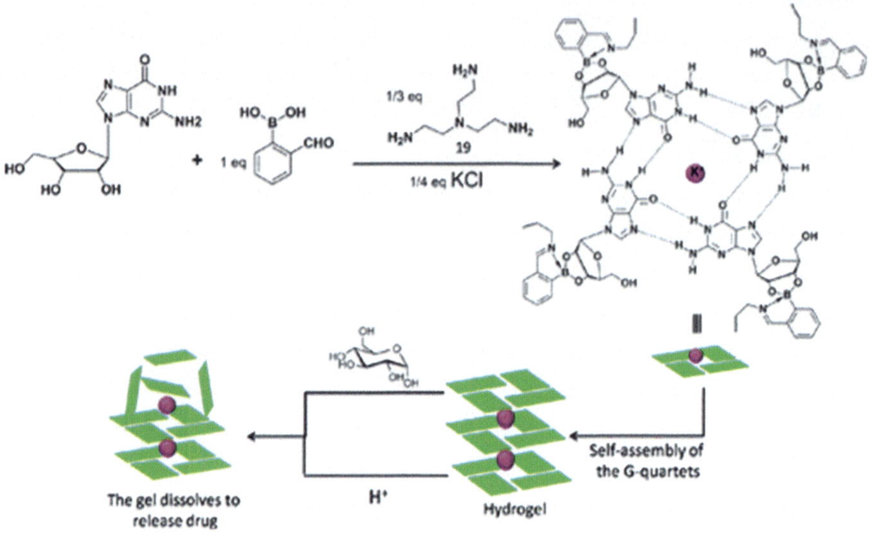

Fig. 14.12 Multi-stimuli-responsive G4 hydrogel for zero-order drug release. Reprinted with permission from Ref. [43]

thus paves the way for the use of low molecular weight biocompatible hydrogels for 3D bioprinting.

The imino-boronate ester-mediated G-quadruplex-based G4PEG hydrogel was synthesized and applied as anticancer drug delivery system by Das et al. (Fig. 14.16) [54]. G-quadruplex structure was confirmed by the optimization of gelation with different alkali metal ions (Li^+, Na^+ and K^+), CD spectrum of the hydrogel and ThT binding study using fluorescence spectroscopy. The self-healing property of the hydrogel was investigated by connecting pieces of hydrogel treated with ThT and a piece of untreated G4PEG hydrogel. The thixotropic nature of the hydrogel was evaluated by cyclic strain sweep experiment. During this experiment the hydrogel retained 90% of its storage modulus after 4 cycles. The hydrogel showed tissue-like self-healable properties that can help to adhere with tissues and cells for wound-healing applications. The pH-dependent delivery of the anticancer drug doxorubicin was achieved successfully. The IC_{50} values of the hydrogel and DOX-hydrogel were calculated as 2.27 and 1.3 mM, respectively. The drug release profiles showed slow and steady release of drug which is beneficial for the steady supply of the drug to the affected tissue over a long time. Overall, the hydrogel is promising as injectable drug delivery system for biomedical applications.

Recently, Das et al. have reported dynamic injectable G-quartet hydrogel for the delivery of vitamin and drug [55]. The gel was formed by G, 1-naphthaleneboronic acid (NapBA) in the presence of K^+. The gel was injectable, thixotropic and self-healing in nature due to the presence of dynamic boronate ester bonds between G and NapBA. The injectable nature of the gel makes it useful for practical applications (Fig. 14.17). The hydrophobic naphthalene group stabilizes the gel for a long

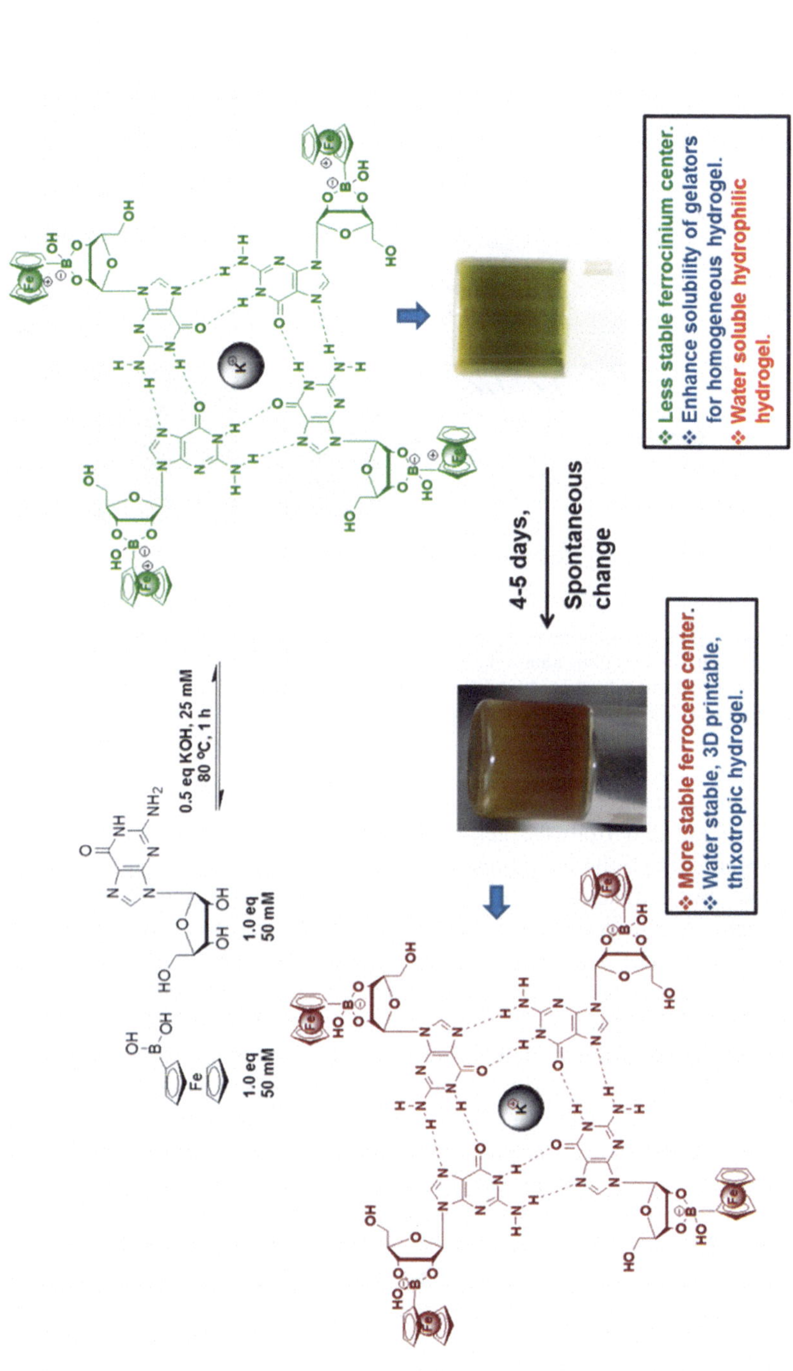

Fig. 14.13 The schematic representation for the preparation of ferrocene-functionalized G-quadruplex hydrogel and spontaneous change from gel (green) to gel (brown) with their properties. Reprinted with permission from Ref. [52]

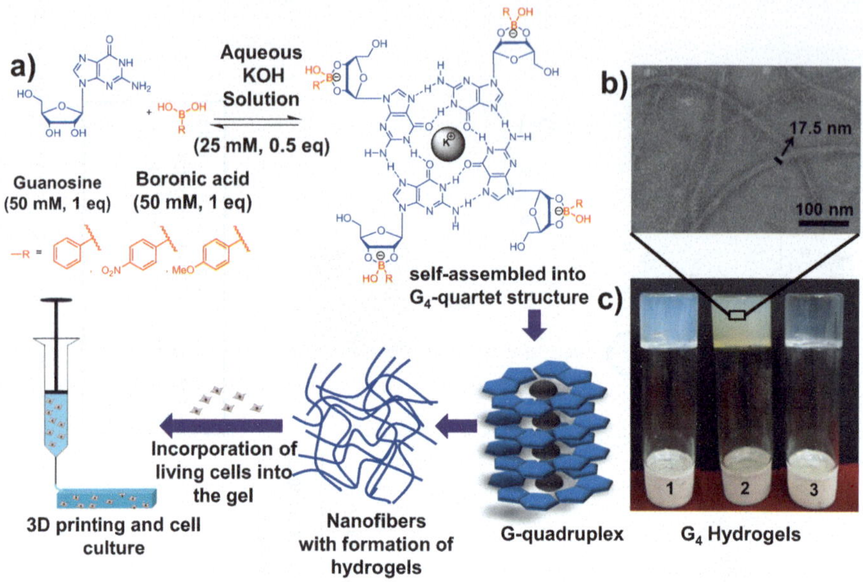

Fig. 14.14 (**a**) Schematic representation of the G4-quadruplex self-assembly process and 3D printing and its cell culture applications. (**b**) TEM image of the **4-GNPBA** hydrogel. (**c**) Optical images of hydrogels: **GPBA** (vial 1), **4-GNPBA** (vial 2) and **4-GMPBA** (vial 3). Reproduced with permission from Ref. [53]

duration in the aqueous medium during in vitro drug release study. Different vitamins were released from the gel. The pH-responsive release of doxorubicin was also observed with the gel. G-quartet hydrogel was prepared using guanosine and different boronic acids for enzyme catalytic reactions (Fig. 14.18) [46]. The hydrogels were prepared by K^+ and Pb^{2+}. The hydrogel containing K^+ was able to form G-quartet. This G-quartet hydrogel binds with Fe(III)-hemin. The system shows peroxidase activity by catalyzing the oxidation of 3,3′,5,5′-tetramethylbenzidine (TMB) in the presence of hydrogen peroxide (H_2O_2).

Li et al. reported an artificial enzyme borate G4 hydrogel that was prepared using guanosine and $KB(OH)_4$ [56]. The hemin was incorporated within G4 hydrogel to prepare artificial enzyme hydrogel (Fig. 14.18). The hydrogel system was used for highly sensitive and selective detection of Pb^{2+} ions. The strategy was established on the Pb^{2+}-induced decrease in the enzyme activity of the hydrogel system. Here, the Pb^{2+} ions showed higher efficiency over K^+ to stabilize the G4 structure. So, the addition of Pb^{2+} replaced the K^+ ions and hemin from G4 which further distorted the artificial enzyme structures and decreased its efficiency. The high response range of the artificial enzyme for Pb^{2+} was found from 1 pM to 50 nM with a detection limit of 0.32 pM.

Pei et al. developed borate G4 hydrogel-based flexible electrochemical sensors which were prepared by guanosine and $KB(OH)_4$ (Fig. 14.19) [46]. Here, hemin was

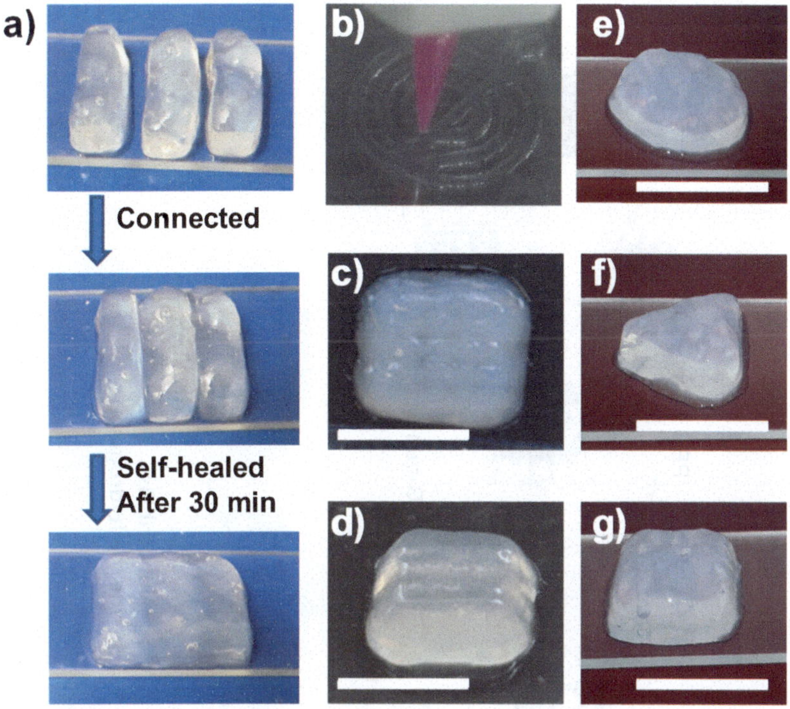

Fig. 14.15 (**a**) Visual observation of self-healing property of **GPBA** hydrogel. Three pieces of gels are self-healed to an integral piece after 30 min. (**b**) Injection of **GPBA** through 3D printing inkjet. (**c**, **d**) 3D printed block with **GPBA** (scale = 10 mm). (**e–g**) Different 3D geometrical patterns drawn by **GPBA** (scale = 10 mm). Reprinted with permission from Ref. [53]

incorporated in the G4 scaffold to produce enzyme-like activity. The dynamic boronate bonds create rapid sol–gel transition and shear-thinning behaviour of the enzyme-like hydrogel that helps to print it as "ink". The enzyme-like hydrogel ink was used to fabricate different patterns. Further polyaniline was incorporated into the G4 nanowires and printed as a flexible electrochemical electrode. Also, a flexible glucose biosensor was developed by loading glucose oxidase into the G4 hydrogel. A multifunctional G4 hydrogel was reported by Pang et al. [57]. The hydrogel was prepared by guanosine and pyridine boronic acid in the presence of KCl (Fig. 14.20). Excellent self-healing property, high conductivity and rapid phase-selective gelation were shown by the hydrogel. Due to its high phase-selective gelation, the hydrogel was applied for recovery and purification of polluted oils and organic solvents. The hydrogel was conductive, and further, it was used as an electrochemical sensor for the highly selective detection of the alpha-fetoprotein.

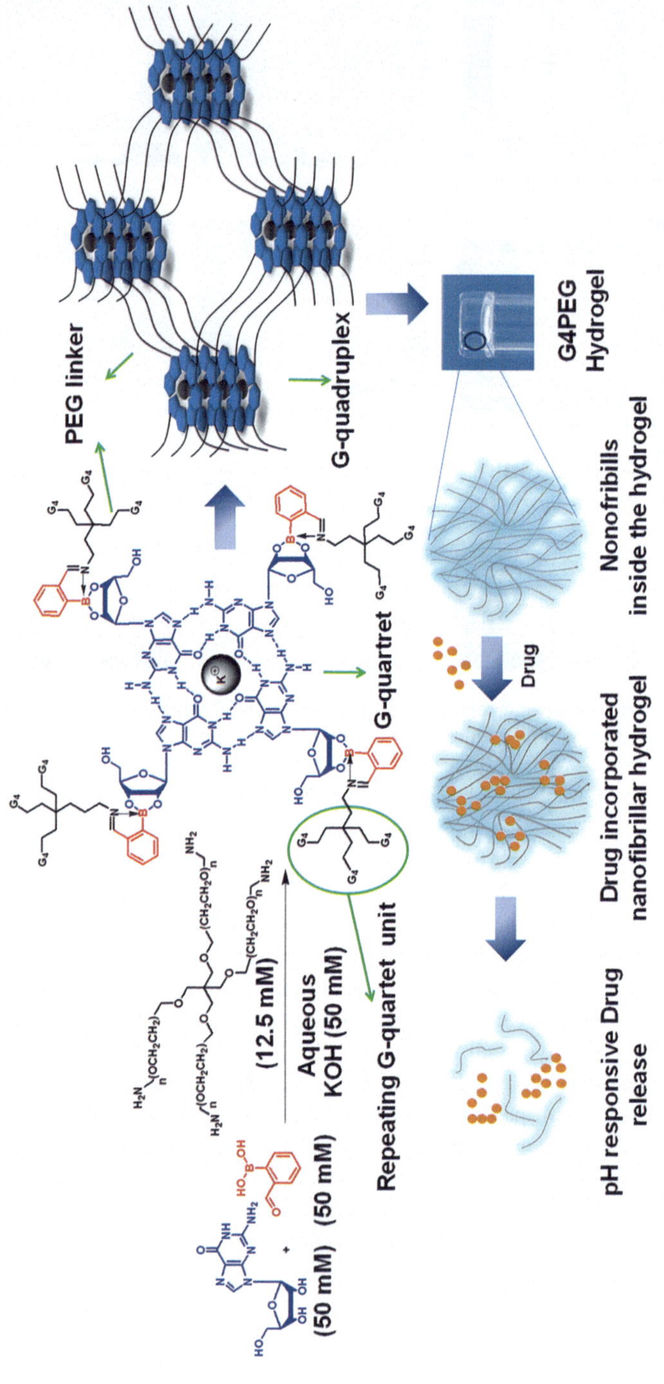

Fig. 14.16 Schematic representation of the formation and drug delivery of G4PEG hydrogel. Reproduced with permission from Ref. [54]

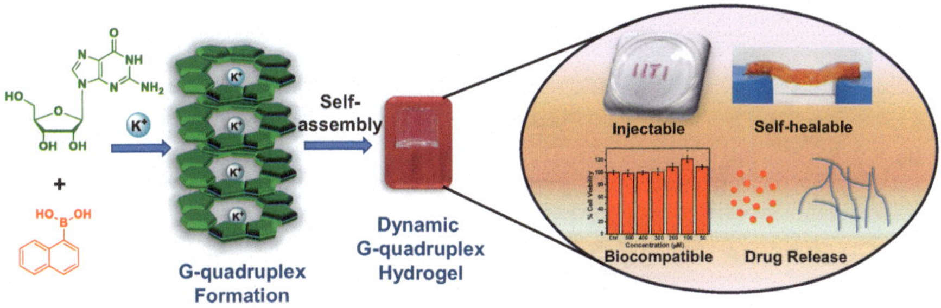

Fig. 14.17 Naphthalene boronic ester mediated G4 hydrogel for slow and sustained release of vitamins. Reproduced with permission from Ref. [55]

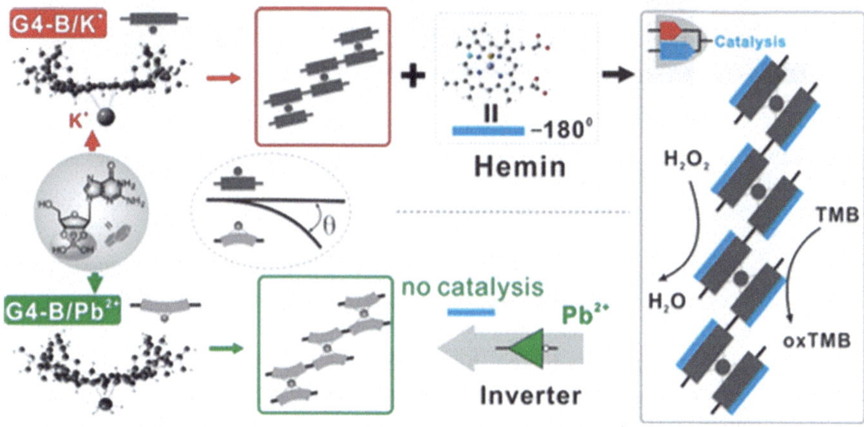

Fig. 14.18 Hemin incorporated G4 borate hydrogel for highly selective Pb^{2+} detection. Reprinted with permission from Ref. [56]

14.3 Oligonucleotide-Based Hydrogel

Oligonucleotide (ON)-based hydrogels belong to the special category of supramolecular hydrogel. The first ON-based hydrogel was reported in 1996 by Nagahara and Matsuda, where they cross-linked polyacrylamide moiety with DNA [58]. On subsequent year, several reports were published with different polymeric backbone like poly(phenylenevinylene), polypeptide and graphene oxide [59–61]. Two different strategies were used to prepare oligonucleotide-based hydrogel. In one of the strategies, hydrogels are prepared by cross-linking nucleic acid subunit, whereas in other strategies, conjugation of hydrophilic polymer with nucleic acid subunit leads to the formation of hydrogel through cross-linking via hybridization of polymeric side chains. Single-stranded DNA (ssDNA)-derived hydrogels have also been proven as responsive hydrogel, where the utilization of nucleic acid Watson–Crick base pairing rules or DNA secondary structures facilitates specific biorecognition [62, 63]. Pure DNA-based oligomeric hydrogels have been synthesized using

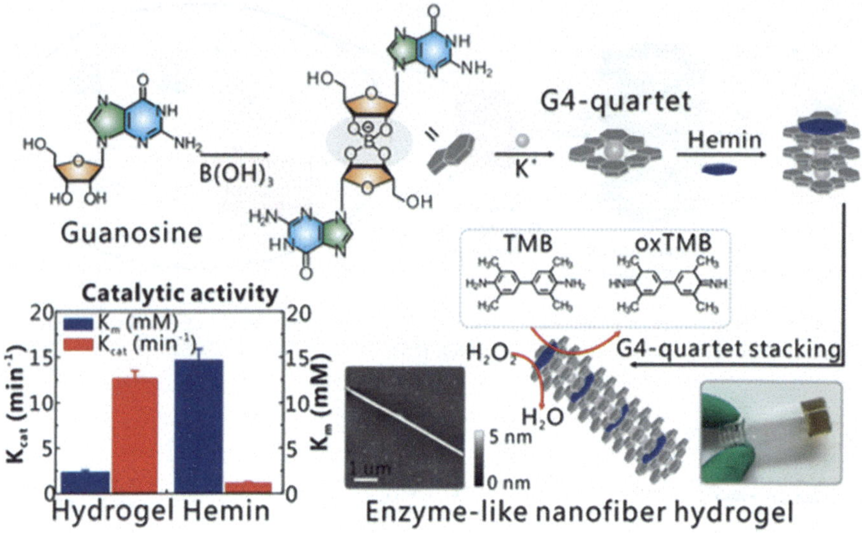

Fig. 14.19 The hemin incorporated G4 borate hydrogel showing enzyme-like activity. Reproduced with permission from Ref. [46]

three- or four-way cross-links, connected to each other by T4 DNA ligase. They formed network-type DNA structure, which degraded by the change of stimuli to release loaded drugs, cells and proteins [64]. Langford et al. reported morpholino oligonucleotide (MO) cross-linked hydrogels for the detection of microRNA sequence. They functionalized MO with an acrylamide moiety and incorporated it into the polymer hydrogel for the detection of mRNA sequence. The MO cross-links are made up of two strands. The "sensor" strand has full complementarity for the chosen "analyte" ssDNA sequence, while the "blocker" strand is partially complementary to the sensor strand, such that the blocker strand will be displaced by the analyte ssDNA sequence (Fig. 14.21) [65].

Liu et al. synthesized a pH-responsive DNA hydrogel. This DNA hydrogel self-assembles through the cross-linking of three-armed DNA Y-unit building block which leads to the generation of inter-Y unit i-motif nanostructures.

The DNA hydrogel is very sensitive to pH. It transforms to non-gel state quickly with the change of environmental pH (Fig. 14.22) [66]. Willner et al. reported a reversible Ag^+ cross-linked DNA base oligomeric hydrogel. This DNA hydrogel consists of acrylamide-functionalized nucleic acid where Ag^+ and cysteamine act as triggering stimuli. The hydrogel is formed by bridging between Y-shaped DNA subunits and the duplex DNA units by the C–Ag^+–C bridges. This hydrogel dissociates reversibly in the presence of cysteamine ligand due to the formation of stable Ag^+ cysteamine complex [67]. Tan et al. reported a photo-responsive DNA-cross-linked hydrogels for controllable release of doxorubicin. Photoresponsive azobenzene cross-linker has been incorporated into DNA molecule. This controllable release of loads is achieved as the hybridization of their hybridization to

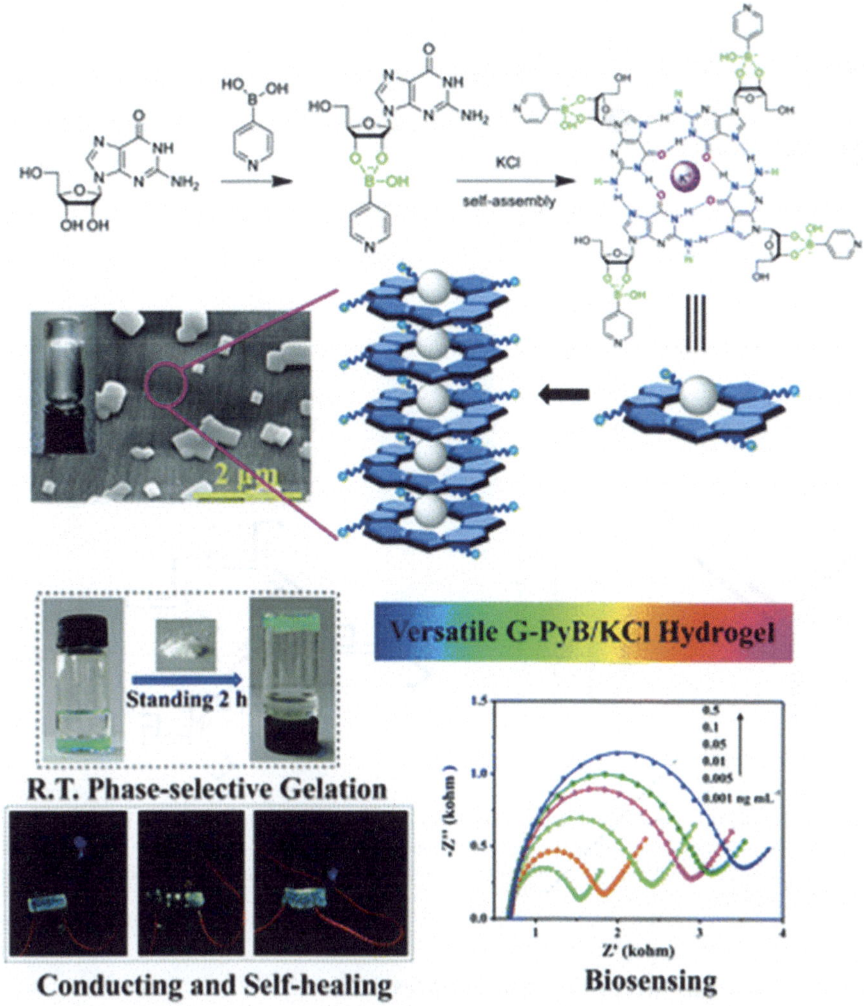

Fig. 14.20 Pyridine boronate ester functionalized conductive, phase-selective G4 hydrogel for biosensing application. Reproduced with permission from Ref. [57]

complementary DNAs (cDNAs) responds differently to different wavelengths of light [65]. Liu et al. reported DNA magnetic nanoparticle hydrogel, where magnetic nanoparticle (MNP) was incorporated into the hydrogel matrix. The DNA–MNP hydrogel was able to move into a particular direction, which easily deformed into various shapes. The DNA–MNP hydrogel could easily jump between two phases. The DNA–MNP hydrogel easily undergoes sol–gel transition upon addition of external stimuli such as enzyme, temperature and magnetic field [68].

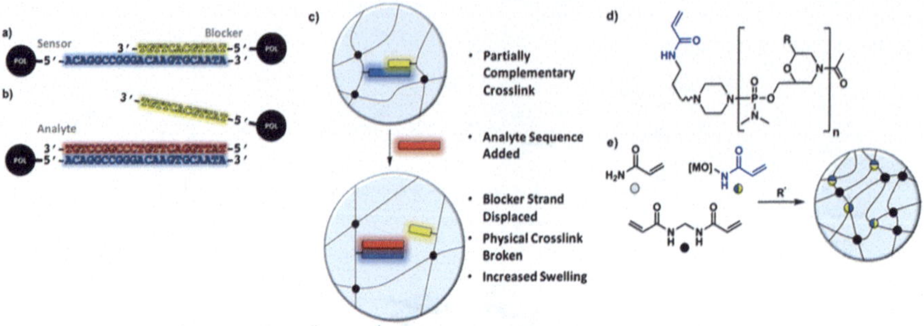

Fig. 14.21 (**a**) Schematic representation of morpholino oligonucleotide sequences where polymer is attached with 5′-end. (**b**) Analyte strand displaced the blocker strand to break the cross-link. (**c**) Morpholino oligonucleotide cross-linked hydrogel. (**d**) Structure representation of morpholino oligonucleotide conjugates. (**e**) Free radical polymerization by UV light. Reproduced with permission from Ref. [65]

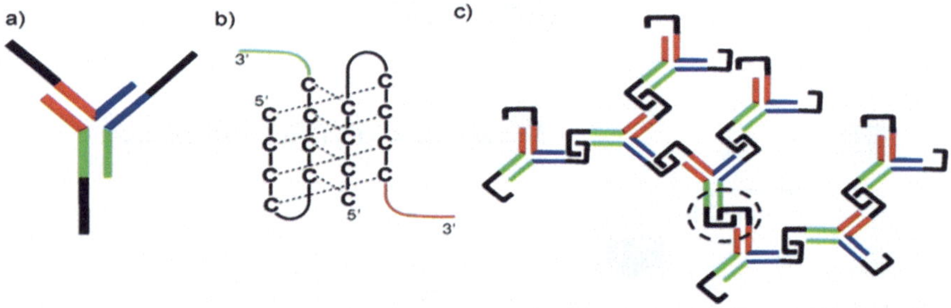

Fig. 14.22 The pH-responsive DNA hydrogel. (**a**) Pictorial description of DNA Y-motif. (**b**) The enlargement of circled region in (**c**) to show the formation of inter-Y-unit i-motif. (**c**) The formation of hydrogel by cross-linking of free interlocking domain of DNA Y units. Reprinted with permission from Ref. [66]

14.3.1 *Conjugated Oligonucleotides*

The chemical structures of biological molecules and their functions have attracted researchers since their discovery. The solid-phase synthesis of oligonucleotides and their analogs from phosphoramidite monomers followed by their chemical conjugation have revolutionized structural and functional studies and enabled the design of novel materials. More precisely, these methods incorporate chemical moieties that are not found in nature. The conjugation with oligonucleotide is carried out to enhance the therapeutic efficiency and intracellular delivery of these moieties. The conjugation of oligonucleotides with ligands that bind with the particular receptors affords certain interaction to the cells. In another way, antisense oligonucleotide or siRNA can be coupled with cell-penetrating peptides to release conjugate from intracellular vesicular compartments.

14.3.2 Peptide–Oligonucleotide Conjugates (POCs)

The conjugated peptide–oligonucleotides (POCs) are covalent molecules which are formed by the conjugation between DNA with a targeted synthetic peptide sequence. Peptide–oligonucleotide conjugated (POCs) materials merge self-assembled properties of oligonucleotide with chemical, biological and structural diversity of synthetic peptides. These materials find application in various fields such as functional biomaterials for interacting and controlling the nature of cells, templated-enhanced coupling of POCs, displaying the peptides on self-assembled oligonucleotide scaffolds and hybrid self-assembling systems that balance peptide and oligonucleotide intermolecular forces. In most of the cases, POCs are synthesized by solution phase, but in some cases, it is synthesized by solid-phase method. Different types of linkages are used for conjugation such as amide, disulfide, ester, thioether and thiol–maleimide. The reversed phase HPLC is used to purify noncharged oligonucleotide conjugates of peptides where charged oligonucleotide conjugates are usually purified by ion exchange. Hamilton et al. synthesized POC where peptides conjugate with oligo-guanosine strands via click chemistry. The peptide–oligo-guanosine self-assembled on a G-quadruplex in the presence of potassium ion [69]. Here two peptides loop assembled on one face of a scaffold through G-quadruplex formation, which is similar to the antibody binding domain. This multicomponent noncovalent synthesis allowed the facile formation of parallel and antiparallel combination (Fig. 14.23) [69].

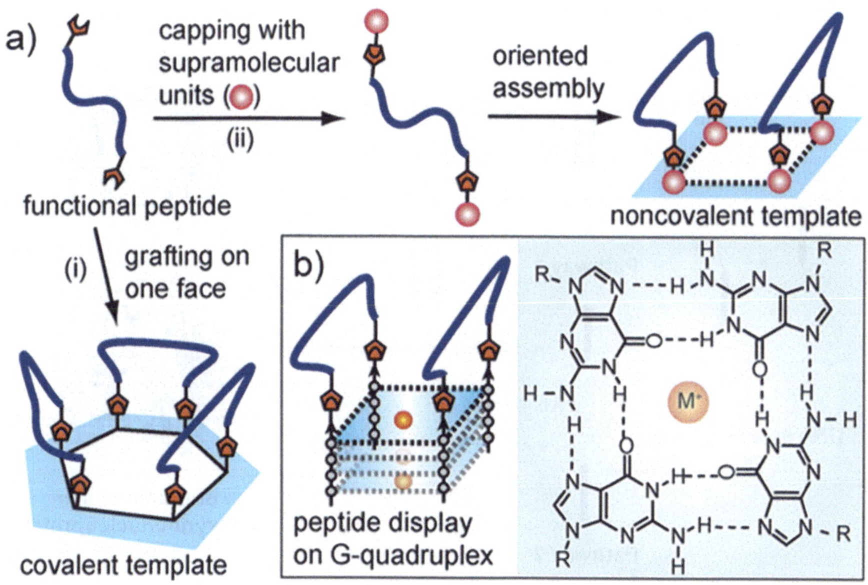

Fig. 14.23 (**a**) Covalent (i) and noncovalent (ii) strategies for mimicking multiloop surfaces. (**b**) Display of two constrained peptides on one face (5′) of G-quadruplex. Reprinted with permission from Ref. [69]

Wengel et al. reported peptide–oligonucleotide conjugates (POCs) where peptides self-assemble into a coiled coil structure and DNA self-assembles into a triple-helical structure. Here the click chemistry was utilized to bind the DNA and peptide molecules [70]. Vebert-Nardin et al. reported a POC that self-assembled into a vesicular structure. The vesicular structure may serve as delivery scaffold as it responds quickly with change of environmental pH. They elaborated the conjugation of diphenylalanine (FF) dipeptide ligand with a C6 amino-modified CTCTCTCTCTTT nucleotide sequence. The fibrillar morphology of diphenylalanine changed to vesicular form upon binding with 12 nt CT-rich nucleotide sequences. The DNA–peptide hybrid self-assembled to give rise to oligonucleotide decorated self-assembled structures in aqueous solution [71]. Lim et al. reported reciprocal self-assembly of a peptide–oligonucleotide conjugate into supramolecular deoxyribonucleoprotein (suDNP). The synergistic effect of peptide and oligonucleotide constrained it to a specific supramolecular nanoarchitecture. A β-sheet peptide arginine–glycine–aspartic acid (β-suRGD) was combined to 20-mer antisense (AS) DNA to generate β-suRGD–AS. The β-suRGD was also combined with 18-mer sense (S) DNA to generate β-suRGD–S. The self-assembly was observed at equimolar mixing of β-suRGD–AS and β-suRGD–S, respectively (Fig. 14.24) [72]. In 2015, Liu et al. reported oligonucleotide-based bio-printable hydrogel. The hydrogel was constituted by polypeptide tagging DNA handles and its complementary cross-linker. The peptide segment helps to form hydrogel. The DNA binds with the peptide segment by Watson–Crick pairing (Fig. 14.25) [73].

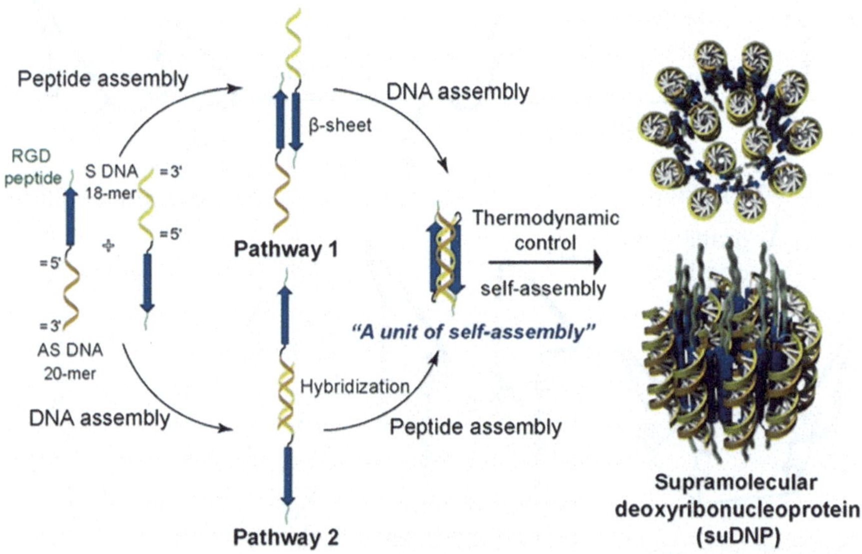

Fig. 14.24 The formation of supramolecular deoxyribonucleoprotein (suDNP) following either pathway 1 or pathway 2. Reprinted with permission from Ref. [72]

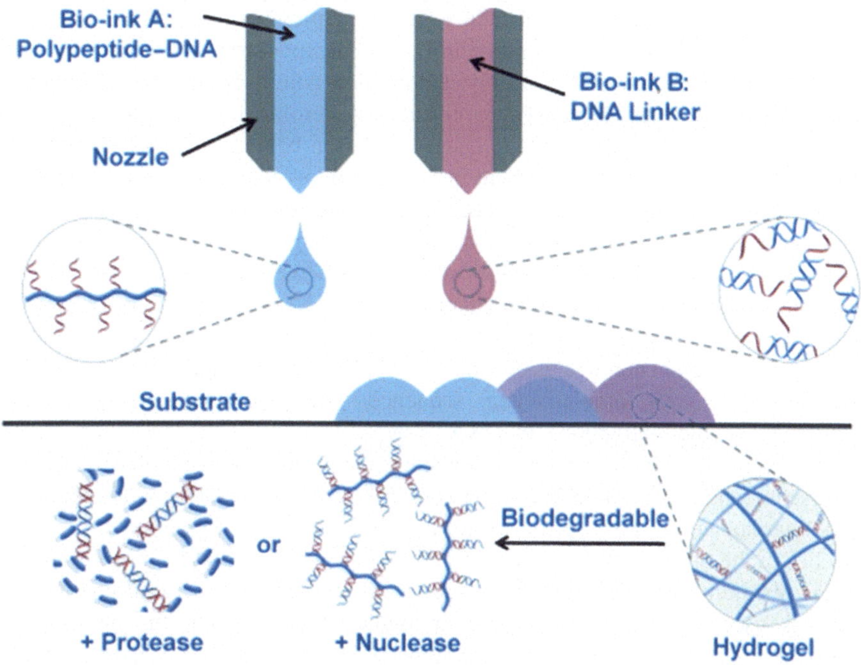

Fig. 14.25 The polypeptide–DNA hydrogel for 3D bioprinting. Bio-ink A (blue), polypeptide–DNA; bio-ink B (red), DNA linker. Cross-linking caused by the hybridization of complementary strand of bio-ink A and bio-ink B leads to the formation of hydrogel. Reprinted with permission from Ref. [73]

14.3.3 Lipid–Oligonucleotide Conjugates

Lipids are biomacromolecules, which are constituted from self-aggregation of monomeric fatty acid. They act as cell signalling and store energy in the cell. Nuclei acids are essential biomolecules, which carry genetic information of living organism. The chemical conjugation of a lipid to an oligonucleotide is carried out with the aim of improvement and optimizing of the existing properties of both biomolecules. Pharmacodynamic and pharmacokinetic properties of oligonucleotides can be improved by incorporating lipophilic moieties to antisense and siRNA. More precisely, the incorporation of cholesterol aids the interaction of the oligonucleotide to the lipoprotein albumin and serum, hence increasing the potentiality of tissue uptake with the help of the hepatic lipoprotein receptors [74]. The conjugation of the tocopherol (vitamin E) with oligonucleotide also serves similar properties [75]. These amphiphiles have the molecular recognition capabilities and the ability to store and transfer encoded information. They self-assemble to give aggregated molecular assemblies like micelles [76], nanoparticles [77] and liposomes [78]. LONs are extensively used in the fields like medicine and cell biology. In the case of siRNA, 3′ extremity position of the double-stranded siRNA was linked covalently to the cholesterol molecule with the help of hydroxyprolinol support. Here in this case, to study apolipoprotein B (apoB) expression, siRNA–cholesterol conjugate, which targets apoB, was injected in mice. After a systemic injection, this results in a decrease of plasma apoB level and cholesterol in the blood. It also results in the suppression of apoB mRNA. The main reason behind this was interactions between the cholesterol group present in LONs and low-density lipoprotein (LDL) along with high-density lipoprotein (HDL). More precisely, a pre-assembly between cholesterol–siRNA and HDL was potentially 8 to 15 times more effective apoB gene silencing in vivo than similar dosing paradigm of free cholesterol–siRNA. The principle of lipoprotein receptor-mediated endocytosis for cholesterol-LON cell penetration is emphasized from this result [79]. Goto et al. demonstrated that lipid–oligonucleotide conjugates can be used as phase transfer agents for the extraction of nucleic acid complementary sequence. They had used reverse micelle technique for the extraction of DNA oligonucleotides with DNA–surfactant. Lipid–oligonucleotide conjugates distribute itself at the interface of biphasic solvent. Reverse micelle construction between hydrophilic complementary nucleic acid and lipid–oligonucleotide takes place at the interface. After the formation of reverse micelles in the organic medium, the desired ON can be extracted from water to the organic medium through a sequence-specific manner. In general, solid-phase synthesis is used to incorporate simple lipophilic moieties using the appropriate phosphoramidites, some of which are commercially available [80]. Herrmann et al. reported an amphiphilic DNA block copolymer (DBCs), which is used as an effective and selective dispersion agent for single-wall nanotubes (SWNT). Here the mentioned DBC comprises of 22-mer single-stranded (ss) oligodeoxynucleotides (ODNs) which was covalently linked with poly(9,9-di-n-octylfluorenyl-2,7-diyl) (PFO). The dispersion of particular semiconducting SWNTs from the raw SWNTs was carried out by the interaction of hydrophobic counterpart of PFO–ODN conjugate [81].

14.3.4 Carbohydrate–Oligonucleotide Conjugates

The conjugation of carbohydrate moieties in oligonucleotides (ONs) not only enhance the existing properties but also develop new properties. Carbohydrates in recognition of sugar-binding proteins (i.e. lectins), which are present on the cell surface, help in cell adhesion and signal transduction process. So, carbohydrate–oligonucleotide conjugates (COCs) have been synthesized which effectively improve the cellular uptake capability as well as tissue target delivery of ONs by the process of receptor-mediated endocytosis. In addition to the ONs delivery, COCs are also used for the preparation of carbohydrate biochips where the DNA-directed immobilization technique (DDI) has been utilized to gain the surface coating with carbohydrate. For this reason, the binding of ONs with carbohydrate groups is a powerful concept to target sugar-binding proteins (e.g. lectin) that are present on many cell types. The conjugation of a monosaccharide with an oligonucleotide is

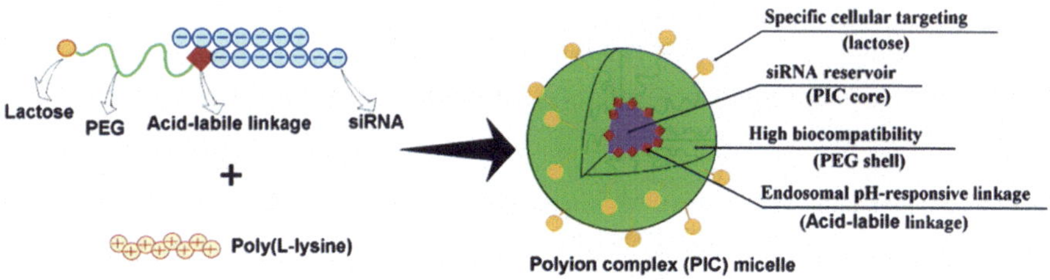

Fig. 14.26 Schematic representation showing the formation of polyion complex (PIC) micelle by the mixing of lactosylated-PEG–siRNA conjugates with poly(l-lysine). Reproduced with permission from Ref. [84]

comparatively easier and can be processed through the synthesis of carbohydrate containing phosphoramidites. The inclusion of oligosaccharide moieties to oligonucleotides is a difficult process. Oligosaccharide-conjugated ONs are required for optimal lectin detection [82]. In recent times, the concept of "click" chemistry has been utilized for the synthesis of different types of COCs including some complex oligosaccharides [83]. Nagasaki et al. reported the synthesis of carbohydrate–oligonucleotide conjugates which were used to regulate the gene expression. Sugar molecule was attached at the 3′-terminus of the oligonucleotide (Fig. 14.26) [84]. Montesarchio et al. reported the formation of glycocomplex by using the concept of DNA–DNA interactions via the generation of G-quadruplexes. For this purpose, a TGGGAG sequence of oligonucleotide was connected through its 3′- or 5′-end with a carbohydrate moiety (glucose, mannose or sucrose) with the help of carbohydrate solid support. The resulting conjugates were capable of forming a parallel quadruplex conformation exhibiting the four carbohydrates at the 3′- or 5′-end [85].

14.4 Conclusion

A nucleobase guanine-functionalized G-quartet hydrogel has been studied extensively. Most of the G-quartet hydrogels are low molecular weight hydrogels (LMWHs). Although G-quartet hydrogels were first discovered early of the twentieth century (1910), real-life applications of the hydrogels took almost a century after extensive research and development. In early stage, single guanine analogues or binary guanine derivatives were used to prepare the G-quartet hydrogels which were less stable and required large amount of alkali metal ions (Na^+ or K^+). Further, functionalization of the guanines with various boronic acids made them highly stable and thixotropic. Due to long-term stability, biocompatibility and thixotropic nature, boronate ester-functionalized G4 hydrogels are most useful in broad field of applications. Oligonucleotide-based compounds and hydrogels are highly capable for functionalization with the different biomolecules including peptides, carbohydrates and lipids. The biomolecule-functionalized oligonucleotides have played an important role in various supramolecular and self-assembly studies. As the oligonucleotides are short sequences of DNA, they are mostly used in biomedical applications including gene expression and receptor binding agents.

Acknowledgement AKD sincerely acknowledges CSIR, New Delhi, India (Project No. 01(2936)/18/EMR-II), for the financial support. AB acknowledges IIT Indore and SERB OVDF for his PhD fellowships. SB and TG thank DST-Inspire for their PhD fellowship.

References

1. Du X, Zhou J, Shi J, Xu B (2015) Supramolecular hydrogelators and hydrogels: from soft matter to molecular biomaterials. Chem Rev 115:13165
2. Lehn JM (1990) Perspectives in supramolecular chemistry – from molecular recognition towards molecular information-processing and self-organization. Angew Chem Int Ed 29:1304
3. Whitesides GM, Mathias JP, Seto CT (1991) Molecular self-assembly and nanochemistry - a chemical strategy for the synthesis of nanostructures. Science 254:1312
4. Luisi PL (2015) Chemistry constraints on the origin of life. Isr J Chem 55:906
5. Terech P, Weiss RG (1997) Low molecular mass gelators of organic liquids and the properties of their gels. Chem Rev 97:3133
6. Du X, Zhou J, Xu B (2004) Supramolecular hydrogels made of basic biological building blocks. Chem Asian J 9:1446
7. Bronstein LM, Platonova OA, Yakunin AN, Yanovskaya IM (1998) Complexes of polyelectrolyte gels with oppositely charged surfactants: interaction with metal ions and metal nanoparticle formation. Langmuir 14:252
8. Baek S, Kim J, Kim H, Park S (2019) Controlled grafting of colloidal nanoparticles on graphene through tailored electrostatic interaction. ACS Appl Mater Interfaces 11:11824
9. Chen T, Li M, Liu J (2018) π–π stacking interaction: a nondestructive and facile means in material engineering for bioapplications. Cryst Growth Des 18:2765
10. Yu X, Yuan X, Zhao Y, Ren L (2019) From paramagnetic to superparamagnetic ionic liquid/poly(ionic liquid): the effect of π–π stacking interaction. ACS Macro Lett 8:1504
11. Tan LL, Zhang Y, Li B, Wang K (2014) Selective recognition of "solvent" molecules in solution and the solid state by 1,4-dimethoxypillar[5]arene driven by attractive forces. New J Chem 38:845
12. Hasanbašić S, Taler-Verčič A, Puizdar V, toka V (2009) Prolines affect the nucleation phase of amyloid fibrillation reaction; mutational analysis of human Stefin B. ACS Chem Neurosci 10:2730
13. Ebbinghaus S, Gruebele M (2011) Protein folding landscapes in the living cell. J Phys Chem Lett 2:314
14. Fujii N (2017) Potential strategies to target protein–protein interactions in the DNA damage response and repair pathways. J Med Chem 60:9932
15. Alam KK, Tawiah KD, Lichte MF, Porciani D (2017) A fluorescent split aptamer for visualizing RNA–RNA assembly in vivo. ACS Synth Biol 6:1710
16. Mahasenan KV, Bastian M, Gao M, Frost E (2017) Exploitation of conformational dynamics in imparting selective inhibition for related matrix metalloproteinases. ACS Med Chem Lett 8:654
17. Deng G, Ma Q, Yu H, Zhang Y (2015) Macroscopic organohydrogel hybrid from rapid adhesion between dynamic covalent hydrogel and organogel. ACS Macro Lett 4:467
18. Yang L, Zhang C, Ren C, Liu J (2019) Supramolecular hydrogel based on chlorambucil and peptide drug for cancer combination therapy. ACS Appl Mater Interfaces 11:331
19. Ma D, Tu K, Zhang L (2010) Bioactive supramolecular hydrogel with controlled dual drug release characteristics. Biomacromolecules 11:2204

20. Watson J, Crick F (1953) A structure for deoxyribose nucleic acid. Nature 171:737
21. Miyoshi D, Nakao A, Sugimoto N (2002) Molecular crowding regulates the structural switch of the DNA G-quadruplex. Biochemistry (50):15017
22. Dominick PK, Jarstfer MB (2004) A conformationally constrained nucleotide analogue controls the folding topology of a DNA G-quadruplex. J Am Chem Soc 126:5050
23. Fernando H, Rodriguez R, Balasubramanian S (2008) Selective recognition of a DNA G-quadruplex by an engineered antibody. Biochemistry 47:9365
24. Maizels N, Gray LT (2013) The G4 genome. PLoS Genet 9:e1003468
25. Fusco DD, Dinallo V, Marafini I et al (2019) Antisense oligonucleotide: basic concepts and therapeutic application in inflammatory bowel disease. Front Pharmacol 10
26. Sharma VK, Watts JK (2010) Oligonucleotide therapeutics: chemistry, delivery and clinical progress. Med Chem 121:10030
27. Chan JHP, Lim S, Wong WSF (2006) Antisense oligonucleotides: from design to therapeutic application. Clin Exp Pharmacol Physiol 33:533–540
28. Glazier DA, Liao J, Roberts BL, Li X, Yang K, Stevens CM, Tang W (2020) Chemical synthesis and biological application of modified oligonucleotides. Bioconjug Chem 31:1213–1233
29. Lu K, Duan QP, Ma L, Zhao DX (2010) Chemical strategies for the synthesis of peptide-oligonucleotide conjugates. Bioconjug Chem 21:187–202
30. Marks JD, Tristem M, Karpas A, Winter G (1991) Oligonucleotide primers for polymerase chain reaction amplification of human immunoglobulin variable genes and design of family-specific oligonucleotide probes. Eur J Immunol 21:985–991
31. Broude NE (2002) Stem-loop oligonucleotides: a robust tool for molecular biology and biotechnology. Trends Biotechnol 20:249–256
32. Saiki RK, Walsh PS, Levenson CH, Erlich HA (1989) Genetic analysis of amplified DNA with immobilized sequence-specific oligonucleotide probes. Proc Natl Acad Sci U S A 86:6230–6234
33. Bang I (1910) Untersuchungen über die Guanylsäure. Biochemische Zeitschrift Biochem Z 26:293
34. Gellert M, Lipsett MN, Davies DR (1962) Helix formation by guanylic acid. Proc Natl Acad Sci U S A 48:2013
35. Chantot JF, Sarocchi MT, Guschlbauer W (1971) Physico-chemical properties of nucleosides: 4.- Gel formation by guanosine and its analogues. Biochimie 53:347
36. Chantot JF, Haertle T, Guschlbauer W (1974) Nucleoside conformations: XIII. Circular dichroïsm of guanosine gels and the conformation of GpG and poly (G). Biochimie 56:501
37. Davis JT (2004) G-quartets 40 years later: from 5′-GMP to molecular biology and supramolecular chemistry. Angew Chem Int Ed 43:668
38. Shi X, Mullaugh KM, Fettinger JC, Jiang Y, Hofstadler SA, Davis JT (2003) Lipophilic G-quadruplexes are self-assembled ion pair receptors, and the bound anion modulates the kinetic stability of these complexes. J Am Chem Soc 125:10830
39. Davis JT, Spada GP (2007) Supramolecular architectures generated by self-assembly of guanosine derivatives. Chem Soc Rev 36:296
40. Peters GM, Skala LP, Plank TN, Hyman BJ, Reddy GNM, Marsh A, Brown SP, Davis JT (2014) A G4·K^+ hydrogel stabilized by an anion. J Am Chem Soc 136:12596
41. Peters GM, Skala LP, Davis JT (2016) A molecular chaperone for G4-quartet hydrogels. J Am Chem Soc 138:134
42. Plank TN, Davis JT (2016) A G4·K^+ hydrogel that self-destructs. Chem Commun 52:5037
43. Li Y, Liu Y, Ma R, Xu Y, Zhang Y, Li B, An Y, Shi L (2017) A G-quadruplex hydrogel via multicomponent self-assembly: formation and zero-order controlled release. ACS Appl Mater Interfaces 9:13056
44. Venkatesh V, Mishra NK, Canelon IR, Vernooij RR, Shi H, Coverdale JPC, Habtemariam A, Verma S, Sadler PJ (2017) Supramolecular photoactivatable anticancer hydrogels. J Am Chem Soc 139:5656

45. Rotaru A, Pricope G, Plank TN, Clima L, Ursu EL, Pinteala M, Davis JT, Barboiu M (2017) G-quartet hydrogels for effective cell growth applications. Chem Commun 53:12668
46. Zhong R, Tang Q, Wang S, Zhang H, Zhang F, Xiao M, Man T, Qu X, Li L, Zhang W, Pei H (2018) Self-assembly of enzyme-like nanofibrous G-molecular hydrogel for printed flexible electrochemical sensors. Adv Mater 30:1706887
47. Yu E, Nakamura D, DeBoyace K, Neisius AW, McGown LB (2008) Tunable thermoassociation of binary guanosine gels. J Phys Chem B 112:1130
48. Carducci F, Yoneda JS, Itri R, Mariani P (2018) On the structural stability of guanosine-based supramolecular hydrogels. Soft Matter 14:2938
49. Buerkle LE, Li Z, Jamieson AM, Rowan SJ (2009) Tailoring the properties of guanosine-based supramolecular hydrogels. Langmuir 25:8833
50. Adhikari B, Shah A, Kraatz HB (2014) Self-assembly of guanosine and deoxyguanosine into hydrogels: monovalent cation guided modulation of gelation, morphology and self-healing properties. J Mater Chem B 2:4802
51. Tang F, Feng H, Du Y, Xiao Y, Dan H, Zhao H, Chen Q (2018) Developing a self-healing supramolecular nucleoside hydrogel based on guanosine and isoguanosine. Chem Asian J 13:1962
52. Biswas A, Maiti S, Kalaskar DM, Das AK (2018) Redox-active dynamic self-supporting thixotropic 3D-printable G-quadruplex hydrogels. Chem Asian J 13:3928
53. Biswas A, Malferrari S, Kalaskar DM, Das AK (2018) Arylboronate esters mediated self-healable and biocompatible dynamic G-quadruplex hydrogels as promising 3D-bioinks. Chem Commun 54:1778
54. Biswas A, Ghosh T, Gavel PK, Das AK (2020) PEG functionalized stimuli responsive self-healable injectable dynamic imino-boronate G-quadruplex hydrogel for the delivery of doxorubicin. ACS Appl Bio Mater
55. Ghosh T, Biswas A, Gavel PK, Das AK (2020) Engineered dynamic Boronate Ester-mediated self-healable biocompatible G-quadruplex hydrogels for sustained release of vitamins. Langmuir
56. Zhong R, Xiao M, Zhu C, Shen X, Tang Q, Zhang W, Wang L, Song S, Qu X, Pei H, Wang C, Li L (2018) Logic catalytic interconversion of G-molecular hydrogel. ACS Appl Mater Interfaces 10:4512
57. Li J, Wei H, Peng Y (2019) A multifunctional self-healing G-PyB/KCl hydrogel: smart conductive, rapid room-temperature phase-selective gelation, and ultrasensitive detection of alpha-fetoprotein. Chem Commun 55:7922
58. Nagahara S, Matsuda T (1996) Hydrogel formation via hybridization of oligonucleotides derivatized in water-soluble vinyl polymers. Polym Gels Networks 4:111–127
59. Xu Y, Wu Q, Sun Y (2010) Three-dimensional self-assembly of graphene oxide and DNA into multifunctional hydrogels. ACS Nano 4:7358–7362
60. Tang H, Duan X, Feng X (2009) Fluorescent DNA-poly(phenylenevinylene) hybrid hydrogels for monitoring drug release. Chem Commun 6:641–643
61. Chen P, Li C, Liu D, Li Z (2012) DNA-grafted polypeptide molecular bottlebrush prepared via ring-opening polymerization and click chemistry. Macromolecules 45:9579–9584
62. Li S, Chen N, Zhang Z, Wang Y (2013) Endonuclease-responsive aptamer-functionalized hydrogel coating for sequential catch and release of cancer cells. Biomaterials 34:460–469
63. Liu J (2011) Oligonucleotide-functionalized hydrogels as stimuli responsive materials and biosensors. Soft Matter 7:6757–6767
64. Um SH, Lee JB, Park N (2006) Enzyme-catalysed assembly of DNA hydrogel. Nat Mater 5:797–801
65. Langford GJ, Raeburn J, Ferrier DC et al (2019) Morpholino oligonucleotide cross-linked hydrogels as portable optical oligonucleotide biosensors. ACS Sensors 4:185–191
66. Cheng E, Xing Y, Chen P (2009) A pH-triggered, fast-responding DNA hydrogel. Angew Chem Int Ed 48:7660–7663

67. Guo W, Lu CH, Orbach R (2014) pH-stimulated DNA hydrogels exhibiting shape-memory properties. Adv Mater 27:73–78
68. Ma X, Yang Z, Wang Y (2017) Remote controlling DNA hydrogel by magnetic field. ACS Appl Mater Interfaces 9:1995–2000
69. Ghosh PS, Hamilton AD (2012) Noncovalent template-assisted mimicry of multiloop protein surfaces: assembling discontinuous and functional domains. J Am Chem Soc 134:13208–13211
70. Lou C, Martos-Maldonado MC, Madsen CS et al (2016) Peptide-oligonucleotide conjugates as nanoscale building blocks for assembly of an artificial three-helix protein mimic. Nat Commun 7:1–9
71. Gour N, Kedracki D, Safir I (2012) Self-assembling DNA-peptide hybrids: morphological consequences of oligonucleotide grafting to a pathogenic amyloid fibrils forming dipeptide. Chem Commun 48:5440–5442
72. Kye M, Lim Y (2016) Reciprocal self-assembly of peptide–DNA conjugates into a programmable Sub-10-nm supramolecular deoxyribonucleoprotein. Angew Chem Int Ed 55:12003–12007
73. Li C, Faulkner-Jones A, Dun AR et al (2015) Rapid formation of a supramolecular polypeptide-DNA hydrogel for in situ three-dimensional multilayer bioprinting. Angew Chem Int Ed 54:3957–3961
74. Wolfrum C, Shi S, Jayaprakash KN (2007) Mechanisms and optimization of in vivo delivery of lipophilic siRNAs. Nat Biotechnol 25:1149–1157
75. Nishina K, Unno T, Uno Y (2008) Efficient in vivo delivery of siRNA to the liver by conjugation of α-tocopherol. Mol Ther 16:734–740
76. Liu H, Zhu Z, Kang H (2010) DNA-based micelles: synthesis, micellar properties and size-dependent cell permeability. Chem Eur J 16:3791–3797
77. Thompson MP, Chien MP, Ku TH (2010) Smart lipids for programmable nanomaterials. Nano Lett 10:2690–2693
78. Banchelli M, Gambinossi F, Durand A (2010) Modulation of density and orientation of amphiphilic DNA on phospholipid membranes. II. Vesicles. J Phys Chem B 114:7348–7358
79. Patwa A, Gissot A, Bestelab I, Barthélémy P (2011) Hybrid lipid oligonucleotide conjugates: synthesis, self-assemblies and biomedical applications. Chem Soc Rev 40:5844–5854
80. Maruyama T, Hosogi T, Goto M (2007) Sequence-selective extraction of single-stranded DNA using DNA-functionalized reverse micelles. Chem Commun 7345:4450–4452
81. Kwak M, Gao J, Prusty DK (2011) DNA block copolymer doing it all: from selection to self-assembly of semiconducting carbon nanotubes. Angew Chem Int Ed 50:3206–3210
82. Lönnberg H (2009) Solid-phase synthesis of oligonucleotide conjugates useful for delivery and targeting of potential nucleic acid therapeutics. Bioconjug Chem 20:1065–1094
83. Karskela M, Helkearo M, Virta P, Lönnberg H (2010) Synthesis of oligonucleotide glycoconjugates using sequential click and oximation ligations. Bioconjug Chem 21:748–755
84. Oishi M, Nagasaki Y, Itaka K (2005) Lactosylated poly(ethylene glycol)-siRNA conjugate through acid-labile β-thiopropionate linkage to construct pH-sensitive polyion complex micelles achieving enhanced gene silencing in hepatoma cells. J Am Chem Soc 127:1624–1625
85. D'Onofrio J, Petraccone L, Erra E (2007) 5′-Modified G-quadruplex forming oligonucleotides endowed with anti-HIV activity: synthesis and biophysical properties. Bioconjug Chem 18:1194–1204

Chapter 15
RNA Nanoarchitectures and Their Applications

Shahaji H. More, Abhishek Bastiray, and Ashwani Sharma

15.1 Introduction

The field of nanoscience and nanotechnology has grown tremendously in the last two decades due to its wide-scale applications including but not limited to drug delivery, detection, diagnostics, agriculture, food industries, and many other diverse fields [1]. Nanoparticles can be divided broadly into two main classes: inorganic and organic. Inorganic nanoparticles are mainly metal-based nanoparticles that are made up of elemental metals (e.g., gold, silver, copper), metal oxides (e.g., iron oxide, zinc oxide, titanium oxide), and metal salts (e.g., zinc sulfate, cadmium sulfide) [2]. These are generally prepared by precipitating inorganic salts. They involve a three-dimensional arrangement with linked atoms through covalent or metallic bonds. On the other hand, most of the organic nanoparticles such as liposomes, micelles, and DNA-/RNA-based nanoparticles are mainly driven together by self-assembly of several organic molecules [3]. The interactions between them are generally governed by electrostatic, hydrophilic, hydrophobic, and van der Waals forces of attraction, and hydrogen bonding [3]. These weak interactions in most self-assembled organic nanoparticles provide them dynamic character to change size and shape with time by fusion among them or by providing external stimuli. For

S. H. More · A. Bastiray
Department of Chemistry, Indian Institute of Science Education and Research (IISER), Tirupati, Tirupati, India

A. Sharma (✉)
Department of Chemistry, Indian Institute of Science Education and Research (IISER), Tirupati, Tirupati, India

Department of Biology, Indian Institute of Science Education and Research (IISER), Tirupati, Tirupati, India
e-mail: a.sharma@iisertirupati.ac.in

T. Govindaraju, K. Ariga (eds.), *Molecular Architectonics and Nanoarchitectonics*, Nanostructure Science and Technology,
https://doi.org/10.1007/978-981-16-4189-3_15

example, liposomes and micelles can be fused together, and DNA-/RNA-based nanostructures having pH-sensitive elements respond to stimuli and change shape or size [4]. These features have been explored to develop various detection platforms and biosensors for disease diagnosis [5]. These nanoparticles can also be made multifunctional scaffolds that can be tuned to different size ranges and could be loaded with different drugs allowing use of nanoparticles for various therapeutic applications [6].

In this chapter, we will restrict our discussion mainly to RNA-based nanoparticles and will provide an overview of RNA-based nanoarchitecture and their diverse applications. Before we move to aspects of RNA nanoarchitecture, it is good to refresh our knowledge about chemical composition of RNA, and how RNA is different from DNA both structurally and in terms of its thermal and chemical stability.

15.2 RNA vs DNA: Structural Differences and Its Implications on Stability

15.2.1 Key Structural Differences Between RNA and DNA

Ribonucleic acid (RNA) and deoxyribonucleic acid (DNA) are structurally comprised of repeating nucleotide units that contain a pentose sugar, a phosphate group, and one of the four nitrogenous bases (Fig. 15.1).

The phosphate group is attached to the hydroxy group of 5′ carbon of one nucleotide and the 3′ carbon of the next nucleotide forming a 5′–3′ phosphodiester linkage. The base is attached to 1′ carbon of pentose sugar. The key differences between RNA and DNA have been shown in Fig. 15.1. Briefly, RNA has a ribose sugar that contains a hydroxyl group (-OH) attached to 2′ carbon, while DNA has a deoxyribose sugar that lacks 2′-hydroxyl group. RNA has adenine (A), guanine (G), cytosine (C), and uracil (U) as four nitrogenous bases, while DNA also has four bases except that there is thymine (T) in DNA in place of uracil (U). Another important difference is that DNA generally in its double-stranded form responsible for storing and transferring genetic information adopts B-helix through Watson-Crick base pairing between two strands, wherein A pairs with T and G pairs with C. On the other hand, RNA is mostly single stranded but shows extensive intramolecular base pairing between complementary sequences and adopts A-helix that is slightly more compact than B-helix [7, 8]. It is to note that in RNA, A pairs with U instead of T as RNA does not contain T. In addition to canonical Watson-Crick base pairing (G-C and A-U), RNA can also have noncanonical base pairing estimated to be approximately 33% in functional RNA that includes G-A and G-U wobble base pairs in addition to a wide range of base pair possibilities [9]. This allows RNA to form a variety of secondary structures including helices, loops, hairpins, motifs,

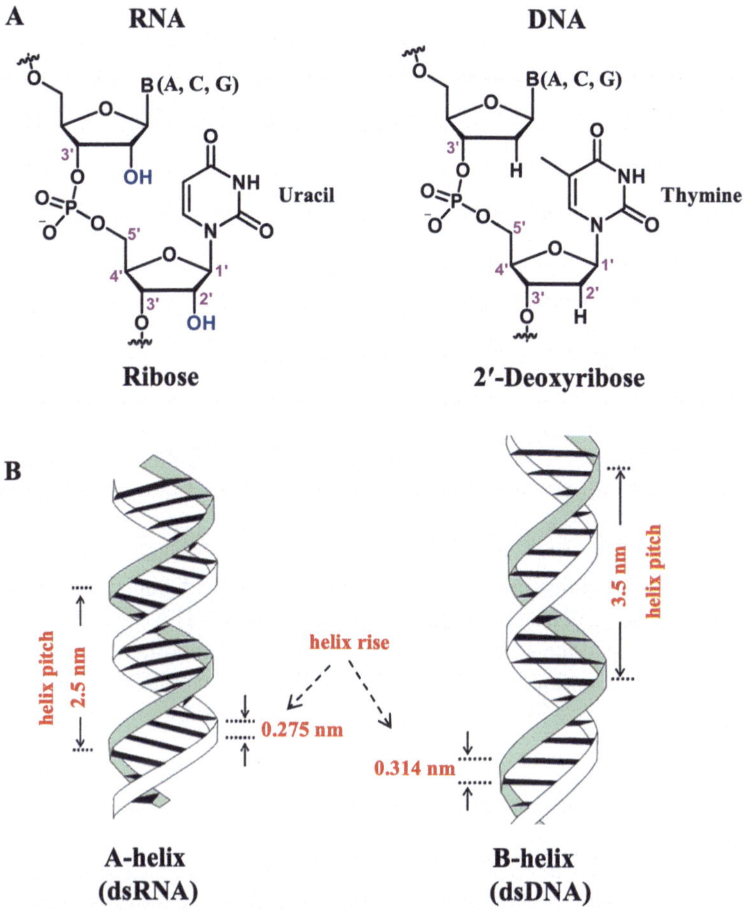

Fig. 15.1 (**a**) Structural differences between RNA and DNA. (**b**) Adapted with permission from [7], Copyright (2014) Oxford University Press

knots, and pseudoknots, which will be discussed in later sections of this chapter, and are important in terms of their utilization in building RNA nanoarchitectures.

15.2.2 Structural Implications on RNA Stability

RNA is often considered less stable than DNA, but that is not completely true. RNA shows considerable difference in its thermodynamic stability and chemical stability because of its structure. In terms of thermal stability, RNA-RNA duplexes are thermally more stable than DNA-DNA duplexes of same sequences [10]. The higher thermal stability of RNA-RNA duplexes compared to DNA-DNA duplexes is attributed to 2′-hydroxyl group that favors C3′-endo puckering of sugar in RNA. This leads to more compact A-helix of RNA rather than B-helix that generally forms in case of DNA with C2′-endo puckering because of lack of 2′-hydroxyl group in DNA. The higher stability of RNA-RNA duplexes is also important for construction of stable RNA nanoarchitectures.

In terms of chemical stability, RNA is chemically or hydrolytically much less stable than DNA at alkaline pH because of additional 2′-hydroxyl group that attacks the adjacent phosphodiester bond of the sugar-phosphate backbone, leading to cleavage of the RNA strand. RNA is extremely susceptible to nuclease digestion by RNase enzymes that exist almost everywhere and specifically recognize 2-′-hydroxyl group in RNA in addition to its structure. This makes RNA unstable in blood plasma that hinders its applications in nanomedicine. However, RNA can be made chemically as well as enzymatically more stable by chemical modifications of RNA [11]. The substitution of the 2′-hydroxyl group with a fluorine (2′-F) [12] or O-methyl (2′-OMe) [13] dramatically increases RNA stability by preventing its degradation by RNases, and is the commonly used method in many of the laboratories working on RNA applications in nanomedicine [12, 13]. In addition, a complete repertoire of RNA-DNA modification (base modification, sugar modifications, backbone modifications) exists that includes locked nucleic acids (LNA), peptide nucleic acid (PNA), and other nucleic acid mimics [14, 15]. Further discussion on this is beyond the scope of this chapter. It is also important to note here that most of these modifications further increase the thermal stabilities of RNA.

15.3 Aspects of RNA Nanoarchitecture

RNA nanoarchitecture generally refers to the construction of various 2D and 3D nanostructures with defined shape that are made entirely of RNA using principles of RNA nanotechnology, and falls in the size range of 1 nm to few hundred nanometers. RNA is able to form complex secondary structures through Watson-Crick base pairing (A-U, G-C) and noncanonical base pairing (e.g., G-U wobble, sheared G-A, Hoogsteen A-U) [9] that allows the formation of large array of structural motifs, which act as building blocks for RNA nanostructures. RNA nanotechnology has seen rapid development in the last decade because of RNA's ability to construct a variety of nanostructures of defined size and shape because of its modular nature and tendency to self-assemble [16]. RNA nanotechnology also holds promise for its diverse applications in therapy as well as in drug delivery [17, 18]. The construction of RNA nanoparticles is usually through bottom-up approach where the small secondary and tertiary RNA structural motifs such as bulges, stems, hairpins, kissing loops, and junctions become the primary structural elements for design of large RNA nanostructures. The defined sizes and shapes of nanostructures are usually achieved by the angles and sequence lengths of the known RNA structural motif used in construction of nanoarchitecture. A unique aspect of RNA nanotechnology is that RNA nanoarchitectures can be decorated with a variety of functional RNA units such as RNA aptamers and ribozymes (catalytic RNAs) or the oligotherapeutic such as siRNA and miRNA. As both the functional units and the nanostructures are made up of RNA, the fusion of both gives a nanoarchitecture made exclusively of RNA that has potential to overcome problems associated with delivery of these functional RNAs. Also, the RNA nanoparticles have recently been reported to be nontoxic and can be rapidly cleared through urine via renal excretion with little accumulation in the body [19]. As mentioned in Sect. 15.2.2, modified RNAs (2′-F, 2′-OMe) can be used in the bottom-up assembly of nanoarchitecture that provides increased thermal and chemical stability to nanoarchitecture.

15.3.1 RNA Nanotechnology in Comparison with DNA Nanotechnology

RNA nanotechnology is still emerging and overcomes some of the deficiencies plaguing DNA nanotechnology, and provides an alternate avenue for construction of nanostructures with ease for various applications [16]. Compared to DNA, RNA can adopt a wider range of structural motifs that further enables construction of more complex nanoarchitecture. A variety of the natural motifs such as junctions, folds, and hairpins are available in reported crystal structure of many of the RNAs that can be utilized as building blocks for nanostructure construction [20]. As DNA is double stranded in biological systems, so there are relatively lesser motifs in DNA compared to RNA that exists mainly as single stranded. Further, lesser noncanonical base pairs in DNA compared to RNA are also responsible for lesser motifs. RNA duplexes are thermally more stable than DNA duplexes; thus, RNA nanostructures can be self-assembled during synthesis by RNA polymerase – a process known as co-transcriptional folding–and have also been reported to be produced in vivo [21]. RNA nanotechnology can also integrate RNA functional elements such as aptamers, siRNA, and miRNA into the nanoarchitecture that display favorable biodistribution and pharmacological profile for drug delivery applications. RNA nanoparticles are nontoxic and can be cleared from the body through renal excretion that is essential for transitioning of RNA-based nanomedicines from bench to bedside. However, the challenge remains for the higher cost of production of RNA compared to DNA.

15.3.2 Building Blocks of RNA Nanoarchitecture: RNA Motifs

RNAs, especially noncoding RNAs, display properties similar to proteins by partaking their various cellular processes. For instance, ribosome is a large structured RNA-protein complex made up from smaller RNA units, which is stabilized by

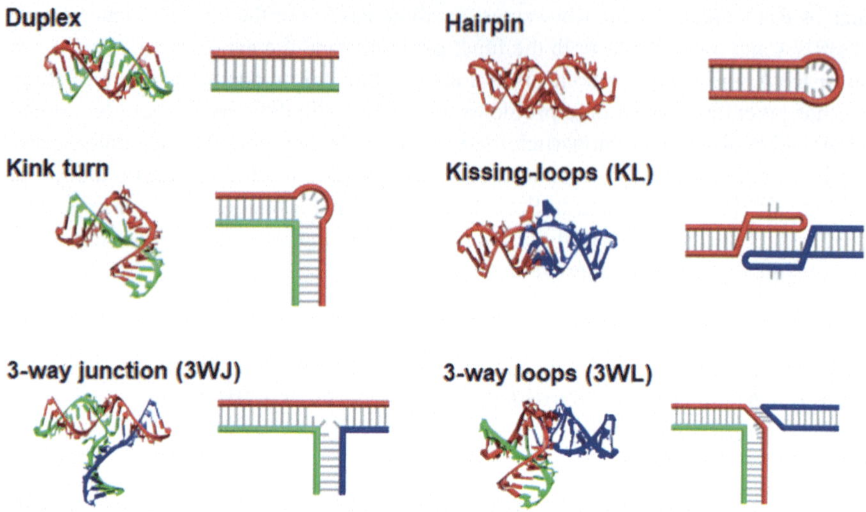

Fig. 15.2 Putative RNA motifs used in RNA nanotechnology. Adapted from [21]

protein units. These complex RNA nanomachines can be viewed as architectures built from simple building blocks of RNA through self-assembly. These existing natural RNA structures inspire the design and architecture of synthetic RNA nanoparticles through bottom-up approach [16].

There are varieties of motifs which have been employed as scaffolds for construction of RNA nanoarchitectures of defined sizes and shapes. These motifs are recurrent structural elements containing multiple RNA-RNA interactions observed in the atomic resolution of RNA structures in their crystal form. Few of the common motifs that have been used for the construction of RNA nanoarchitecture have been shown in Fig. 15.2 [21].

Duplex RNA duplexes that form A-helix are the fundamental motifs for generation of RNA nanostructures and are the essential building blocks as they provide support and combine different modules of RNA in three dimensions.

Hairpin An RNA hairpin is an essential structural motif that guides RNA folding and helps folding a single-stranded RNA to give rise to junctions and other motifs. It also serves as motif to interact with various RNA binding proteins [22].

Kink Turn The RNA kink-turn motif (usually abbreviated as k-turn) is a widespread RNA structural motif that has a short (3–4 bases) bulge giving it a tight kink of certain angle. Many kink-turn motifs interact with proteins. Figure 15.2 shows a 90° kink found in domain IIa bulge in the IRES (internal ribosome entry site) of the hepatitis C virus RNA genome [23].

Kissing Loops (KLs) The kissing loops are formed through Watson-Crick base pairs usually between unpaired nucleotides of two hairpin loops. The KL interactions provide the necessary cohesion for oligomerization without the dependence on free ends. The KLs have also been recently used for generation of higher-ordered structures known as tiles [24].

Three-Way Junction (3WJ) Three-way junction (3WJ) motifs are commonly found in a lot of different folded RNA structures, and are one of the primary motifs for construction of RNA nanoparticles of defined shape and size [25]. 3WJ motif derived from packaging RNA (pRNA) of phi29 bacteriophage has been extensively used for construction of a variety of highly stable 2D and 3D RNA nanostructures [26]. The pRNA-3WJ has been further extended to 4WJ with the addition of an extra duplex stem. Many 4WJ, 5WJ, or higher junctions exist in nature that can be utilized as building blocks for construction of RNA nanoarchitectures.

Three-Way Loops (3WL) Three-way loops are similar to kissing loops wherein instead of two hairpins, a bulge or a kink interacts with a hairpin through base pairing of unpaired nucleotides. RNA nanostructures made from 3WL have also been shown to be produced in vivo [21].

15.3.3 Strategies for Building RNA Nanoarchitecture

Nature precisely designs RNA molecules that folds and assembles into RNA machineries, a complex three-dimensional architecture that performs complex functions in cells. Structural investigations of these natural RNA machineries such as ribosomes or spliceosome suggest that they are comprised of various structural motifs including but not limited to dsRNA stems, helices, loops, and turns as discussed in the previous section. This inspired scientists to create bottom-up assemblies of RNA nanoarchitecture by using these natural or engineered motifs that can perform a wide variety of functions akin to natural system. The platform set by DNA nanotechnology was extremely helpful, and RNA followed the trails of DNA [27]. As inspired from nature, there are many self-assembling strategies that have been applied to construct artificial RNA nanoarchitectures. Figure 15.3 reviews few of the different strategies that have been used to build a wide variety of RNA nanoarchitectures in literature.

(A) RNA Tectonics The word "tectonics" can be translated to "the art of designing buildings". RNA tectonics deals with utilizing well-characterized RNA structural motifs (through X-ray or NMR) that serve as building blocks to construct complex RNA architecture following the rules governing the assembly of biological macromolecules. These RNA building blocks also called RNA tectons can be compared to Lego pieces that can be plugged together to build a large architecture [32–34]. Figure 15.3a shows the strategy for construction of RNA tectonics also called architectonics. First, RNA tectons are carefully designed from known motifs. As shown in Fig. 15.3, RNA tectons have three components a right-angle motif (RA motif) identified in ribosomes at the center and two kissing loop motifs (KL motif) carefully attached to both sides of RA motif having a tail at one of the

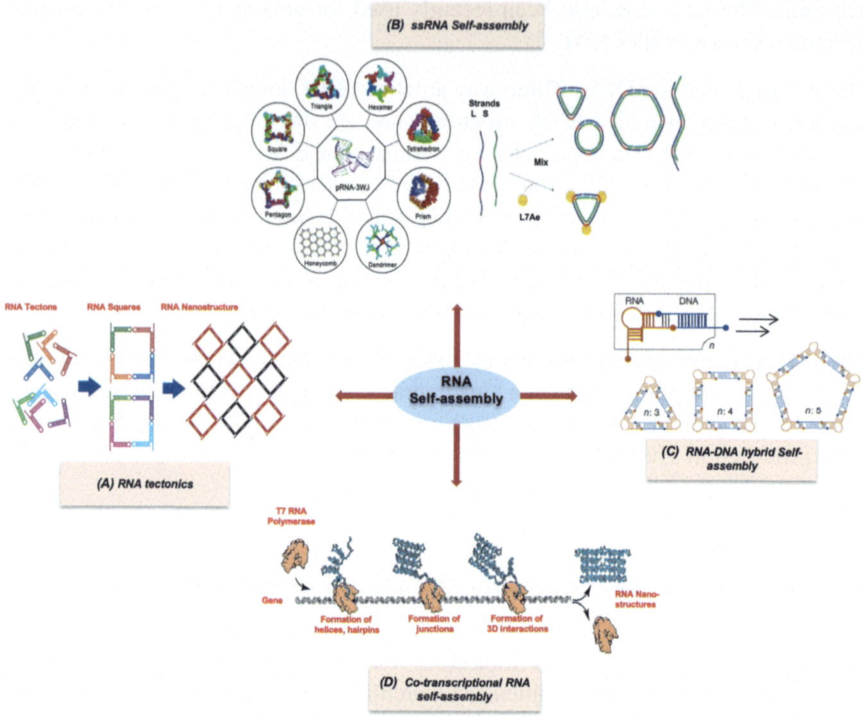

Fig. 15.3 Different strategies for building RNA nanoarchitecture: (**a**) RNA tectonics. (**b**) ssRNA self-assembly. (**c**) RNA-DNA hybrid self-assembly. (**d**) Co-transcriptional RNA assembly. (**a**) Reprinted with permission from [28], Copyright (2014) American Chemical Society. (**b**) Reprinted with permission from [20], Copyright (2017) John Wiley & Sons. Reprinted with permission from [29], Copyright (2011) Springer Nature. (**c**) Reprinted from [30]. (**d**) Reprinted from [31]

ends (3′-end). The assembly of these tectons through KL interactions forms a tectosquare that can further be used to construct larger RNA nanostructures [28]. The broader applicability and versatility of RNA tectons used as a building blocks yielded a vast range of nanostructures such as filaments [35], triangle [29], square [32, 36], hexagon [37, 38], polyhedrons [39], 2D arrays [40], and so on. Each tecton has its own intrinsic folding property; therefore, it is necessary to pre-assemble it in an isolated form, and then it can be mixed with other tectons to generate complex RNA nanoarchitecture.

(B) ssRNA Self-Assembly Single-stranded RNA (*ss*RNA) molecules hold the ability to fold into a variety of secondary/tertiary structures due to facilitating canonical and noncanonical base pairs as well as *intra*- and *inter*molecular base stacking. In *ss*RNA self-assembly, nanostructures are constructed by mixing a number of *ss*RNAs together in one pot. The *ss*RNA sequences are designed so that these different RNAs in unstructured form on mixing assemble through base pairing into a well-defined nanostructure. The defined shape of nanostructures is guided by highly stable natural or engineered RNA motifs that are formed only through intermolecular base pairing of different designed strands. One of the extensively used RNA motifs in construction of nanostructures is a highly stable pRNA-3WJ that consists of three individual strands of RNA that anneal together to form a highly stable 3WJ. The pRNA-3WJ has been taken from packaging RNA (pRNA) of bacteriophage phi29, and engineered to construct a wide variety of 2D and 3D nanostructures (e.g., triangle, square, honeycomb, tetrahedron, prism, dendrimer) utilizing 3WJ as corners of these nanostructures (Fig. 15.3b) [20]. Some of the nanoconstructs remain intact up to boiling temperature [41].

With the help of multistrand association through hybridization principle, diverse RNA nanostructures have been generated. Kink turn of bulge IIa in the IRES (internal ribosome entry site) of the hepatitis C with 90° angle has been used to form corners of a self-assembling RNA square [42]. In silico designing also helped to construct various nanostructures, e.g., RNA cube with a diameter of 10 nm has been assembled in silico [43]. Additionally, ssRNA can be used to construct RNA origami similar to the DNA origami [44]. RNA-protein self-assembly has also been evolving. High affinity of L7Ae protein toward kink-turn RNA motif has been shown to form triangular-shaped RNA-protein complex. K-turn motif is flexible, but L7Ae protein binding produces 60° bent in it to give it a shape of equilateral triangle that forms exclusively in the presence of protein (Fig. 15.3b) [29].

(C) RNA-DNA Hybrid Self-Assembly RNA-DNA hybrids can also be self-assembled to construct functional nanoarchitecture similar to those of DNA only or RNA only [27]. In terms of structure, RNA on hybridization with complementary DNA readily adopts A-helix similar to RNA duplex rather than B-helix adopted by DNA duplex. RNA-DNA hybrids in general are assumed to be enzymatically more stable than RNA and so are of great interests as they provide higher stabilities in blood serum as well as provide more circulation time [45]. Thus, RNA-DNA hybrid nanoarchitecture might have the potential to improve pharmacological profiles of RNA-based nanoparticles. Moreover, RNA molecules in the nanoarchitecture can be fused to diverse functional RNA motifs such as RNA aptamers, ribozymes, and riboswitches, and thus its complementary DNA part can be chemically modified to enhance the stability of the nanoarchitecture [45]. Monferrer et al. designed a strategy to combine both RNA and DNA for building polygonal RNA-DNA hybrid nanoarchitectures such as triangle, square, and pentagon. A bent RNA motif taken from an internal loop RNA of hepatitis C virus genome was used. Straight DNA modules were used as connectors to glue RNA joints. The polygonal shapes tolerate wide ranges of DNA modules, which can be chemically modified for stability, and protein binding rendering the RNA-DNA nanoarchitecture for various applications (Fig. 15.3c) [30].

Large nucleic acid structures have also been created using DNA as staple strands to fold long ssRNA [46–48]. With the guidance of the staple strands, various simple shapes like ribbon and triangle, as well as more complex square tiles, seven-helix bundled tiles have been generated successfully [46]. Double crossovers, three-point stars, and four-point stars have been adopted using RNA-DNA hybrid motifs

[49]. RNA-DNA strategy has been utilized to create functional RNA moieties formed from RNA-DNA hybrid through selective DNA strand displacement [47]. RNA-DNA hybrids have also been engineered and programed for initially inactive form by splitting functionality into two parts that on undergoing triggered release yielded active functional system [50]. It is possible to prepare RNA-DNA hybrid or RNA origami structures by applying similar construction principle as like that of DNA origami; moreover, nanostructure generated from the newly emerging field such as RNA-DNA hybrid systems carries great potential for biomedical application with resistance to nuclease, and precise controllability.

(D) Co-transcriptional RNA Self-Assembly One of the advantages of RNA nanostructures over DNA nanostructures is that RNA can be encoded genetically and thus can be transcribed inside cells using endogenous transcription machinery. As natural RNA can fold during its synthesis in cells, this inspired researchers to assemble large nanostructures during transcription. Co-transcriptional RNA self-assembly in a simple way is an enzymatic synthesis of RNA nanoparticles and is a relatively new strategy to assemble RNA nanostructures (Fig. 15.3d) [31, 51]. Compared to other RNA engineering techniques, this technique gained greater attention because of its production of high packing density of RNA molecules, low cost, and one-pot synthesis [52, 53]. Geary et al. demonstrated first-time formation of RNA tiles that assembles into hexagonal lattices. Formation of tiles can be scaled up to the size comparable to that of natural ribozymes [51]. There are other examples of RNA nanostructures co-transcriptionally assembled into nanocubes from 12 ssRNAs [43], filaments and tubes [52], and hexagonal rings [54, 55]. RNA nanostructures have also recently been shown to be produced in vivo via programmed folding of ssRNA [44].

15.4 Applications of RNA Nanoarchitecture

RNA field is rapidly growing in the last few years. Many new RNA motifs have been characterized. In vitro selected functional RNAs and their applications in various fields are being reported. RNA because of its ability to fold into a variety of different secondary structures can be in vitro evolved to have specific functions. These functional RNAs either natural or evolved in a lab such as aptamers, ribozymes, and riboswitches alone or in combination with the field of RNA nanotechnology provide useful application in the area of nanomedicine, drug delivery, imaging, and diagnostics. In this section, we will try to review the application of various RNA nanoarchitectures [56].

15.4.1 RNA Nanoarchitectures in Drug Delivery

RNA nanoarchitectures have been explored enormously for drug delivery applications that include small molecule chemotherapeutic drugs to oligotherapeutic such as siRNA or miRNA. This has been discussed briefly as follows:

(a) *Small molecule drug delivery*: RNA nanoarchitectures have been utilized mainly for delivery of small molecule chemotherapeutic drugs for the treatment of different cancers. These chemotherapeutic drugs often used for cancer treatment generally include taxane derivatives such as paclitaxel and docetaxel and anthracycline derivatives such as doxorubicin and daunorubicin. These chemotherapeutic agents also have several drawbacks. The first drawback is that these drugs after systematic injection get distributed in the whole body with poor accumulation at the target site. This leads to many side effects as it targets both fast-growing normal cells and the cancer cells. The solubility of the hydrophobic small molecule drugs is also of concern [57].

RNA nanoarchitectures have been able to solve some of the drawbacks associated with use of small molecule drugs alone. Small molecule chemotherapeutic drugs can be chemically conjugated to RNA architecture using different bioconjugation chemistries that allows the controlled release of drug at specific sites. The conjugation chemistry can be optimized to release the drug under certain stimulus, e.g., high esterase activity in some of the cancer cells, low pH of cancer cells, or even using light as stimuli.

Most successes achieved using RNA nanoarchitectures for the drug delivery have been from nanostructures constructed using pRNA-3WJ as the structural motif that has been reviewed in detail elsewhere [26]. Recently, photo-controlled release of cancer drug paclitaxel has been shown that was attached to an RNA pyramid constructed from pRNA-3WJ using a linker that can be cleaved by triggering light. In this study, irradiation of light showed rapid release of paclitaxel from the nanocages and considerable toxicity to the MDA-MB-231 breast cancer cells (Fig. 15.4a) [58]. RNA nanotechnology has also been applied to enhance the solubility of a hydrophobic drug camptothecin (CPT) for cancer treatment. Multiple units of hydrophobic CPT drug were attached to two RNA strands of pRNA-3WJ ($3WJ_a$ strand with 3 CPT units and $3WJ_b$ strand with 4 CPT units) using azide-alkyne chemistry (click reaction) through cleavable ester bonds. The drug-loaded individual strands self-assembled in the presence of third strand to form a stable 3WJ that remained soluble in aqueous solutions without any detectable aggregation at therapeutic dose (Fig. 15.4b) [57].

Apart from covalent attachment of drug molecules to RNA nanoarchitecture using different conjugation chemistries, intercalation of drug molecules within base pairs of RNA has also been shown. This is considerably an easy method to load drugs to nanoarchitecture and usually gives higher ratio of drug to RNA nanoparticle. Doxorubicin is an intercalating anticancer drug that is known to bind to 5′-GC-3′ region specifically. Doxorubicin has been used widely as a model drug for targeted delivery to different cancers by intercalating in GC-rich

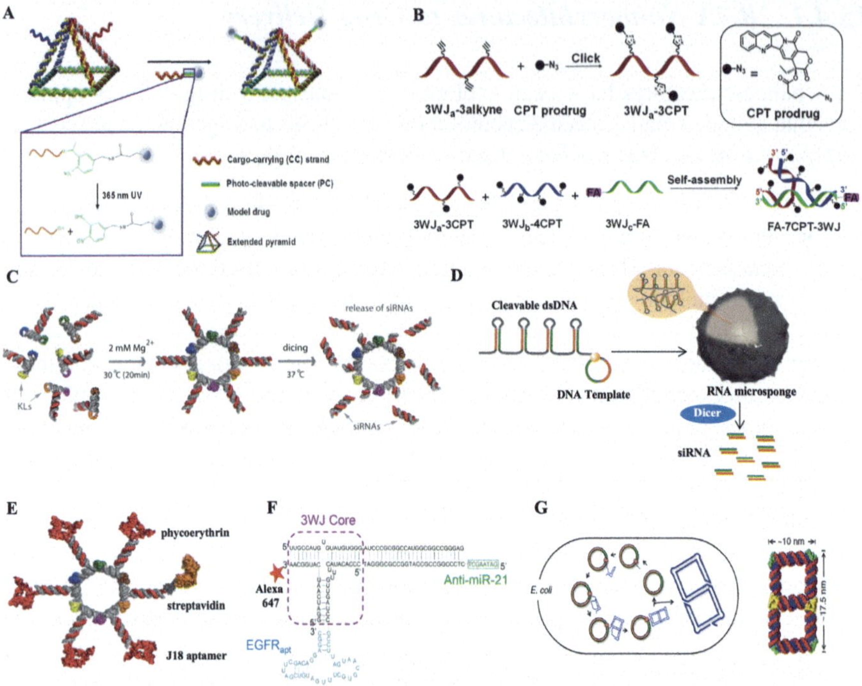

Fig. 15.4 RNA nanoarchitectures for delivery of small molecule drugs (**a**, **b**), oligotherapeutic drugs (**c**, **d**), targeted drug delivery (**e**, **f**), and in vivo assembly of nanoarchitectures (**g**). (**a**) Reprinted with permission from [58], Copyright (2014) Springer Nature. (**b**) Reprinted from [57], (**c**) Reprinted from [59]. (**d**) Reprinted with permission from [60], Copyright (2012) Springer Nature. (**e**) Reprinted from [59]. (**f**) Reprinted from [61]. (**g**) Reprinted from [21]

sequences of RNA aptamers or RNA nanostructures or combination of both [62, 63].

(b) *Oligotherapeutic (siRNA and miRNA) drug delivery*: The discovery of RNA interference (RNAi) in the year 1998 opened a new avenue of treating many diseases that were previously thought to be undruggable. RNAi is the process by which gene expression can be controlled in a cell by targeting its mRNA with the help of small (typically 21–25 nt) noncoding dsRNA sequences that are often microRNA (miRNA) or small-interfering RNA (siRNA). siRNA and miRNA have been used extensively as drugs and often termed as oligotherapeutic in terms of their biomedical applications. The delivery of synthetic siRNA or miRNA to diseased cells for clinical purpose remained a big challenge in the last two decades because of their small size that leads to renal excretion and poor cell uptake.

RNA nanoarchitecture harbored with siRNA has been used to overcome delivery challenges associated with delivering siRNA alone. In vivo delivery of siRNAs has been successfully achieved using pRNA-3WJ-based RNA nanoparticles [61, 64]. Cui et al. designed pRNA-3WJ-based NPs functionalized with folate as a targeting ligand that targets folate receptors expressed on many cancer cells. BRCAA1 siRNA as a therapeutic for gastric cancer therapy and Alexa 647 for imaging were also attached to nanoparticle. In this study they could able to show siRNA silenced the expression of BRCAA1, thereby observing regression in the growth of the gastric tumors with almost no accumulation of RNA NPs in the healthy organs [65]. By using similar approach, another report demonstrated the efficient silencing of the luciferase gene expression [64].

In order to enhance effectiveness of gene knockdown, one need to overcome the limited efficacy of siRNA, and this can be achieved by increasing the siRNA copy number by constructing multivalent RNA NPs. Haque et al. constructed thermodynamically stable X-shaped RNA NPs to carry four therapeutic siRNA motifs without compromising the shape and size of nanoparticles. It has been observed that the gene silencing effects were progressively enhanced as the number of siRNA in each corner increases [66]. In another approach Afonin et al. demonstrated RNA nanorings functionalized with six different siRNAs each of them targeting a different region of HIV-1. RNA nanorings functionalized with siRNAs are able to show a drastic reduction of virus production in HeLa cells (Fig. 15.4c) [59].

RNA microsponges composed of tandem repeats of luciferase siRNA have also been synthesized by utilizing rolling circle transcription (RCT) and showed that it efficiently gets cleaved by dicer only on cellular internalization. This converts stable hairpin RNA into siRNA that silences luciferase expression in ovarian cancer in mice model (Fig. 15.4d) [60]. Taking forward this approach, the same group reported the possibility of attaching different siRNAs with the nanoparticle for targeting multiple genes at the same time [67].

Ye et al. constructed folate-linked pRNA-based delivery vehicle to which artificial miRNAs were linked, and it has been observed that delivered miRNAs target stem loop of the CVB3 3′UTR and effectively inhibit the replication of CVB3 with no effect on interferon [68]. Guo lab also demonstrated the use of RNA nanotechnology for specific delivery of miRNA to the prostate cancer. Thermodynamically ultra-stable 3WJ of pRNA was used to design and construct RNA NPs containing anti-miR17 and anti-miR21 as therapeutic molecules [69].

(c) *Targeted drug delivery*: One of the major hurdles in drug delivery is to deliver the drug in a cell- or tissue-specific manner. Cells are very distinct from one another by their external appearance. At different stages of disease development, cells express specific proteins that serve as receptors or biomarkers on the cell surface. These receptors can be utilized for targeted drug delivery to only the diseased cells sparing the normal cells. Aptamers have previously shown to be one of the best candidates for the selective delivery of therapeutics by a target-dependent manner [70, 71]. There are both DNA and RNA aptamers available in literature. Aptamers are small RNA or DNA sequences (generally 20–60 nt) that are selected against a target through in vitro selection method known as SELEX (systematic evolution of ligands by exponential enrichment) undergoing several rounds of selection. The final in vitro selected sequences now known as aptamers (either DNA or RNA) have high binding affinity and specificity for the target it is selected for. The targets can be small molecules, peptides, or proteins that could be biomarkers of a specific disease. RNA aptamers have been selected for a wide variety of small molecules as well as proteins, which have been reviewed elsewhere [72].

Aptamers alone could also be used to deliver various oligotherapeutic drugs attached to them. However, because of smaller size, many times RNA aptamers and siRNAs are attached to an RNA nanoarchitecture making the targeting ligand (aptamer), delivery vehicle (RNA nanoarchitecture), and the drug (siRNA) all made up of RNA only. Afonin et al. utilized RNA nanorings for targeted drug delivery to cancer cells by conjugating J18 aptamers specific to human EGFR (epidermal growth factor receptor) that is highly expressed in many different cancer types (Fig. 15.4e). RNA nanorings were fused with multiple copies of aptamers to check the efficacy of cell targeting by flow cytometry. Phycoerythrin was used for visualization by coupling it to the nanorings through biotin-streptavidin linkage. RNA nanoring architecture showed increased fluorescence signal with increase in the number of the J18 aptamer as targeted ligand [59]. In another example, pRNA-3WJ architecture containing EGFR aptamer as targeting ligand was assembled to target miRNA-21 in cancer cells. An 8-mer LNA (locked nucleic acid) sequence was incorporated in pRNA-3WJ core that acted as an anti-miR-21 agent, and Alexa 647 fluorophore was used as an imaging module (Fig. 15.4f) [61]. The assembled RNA nanoarchitecture was used for targeted delivery of anti-miR-21 to cancer cells with very little accumulation of RNA nanoarchitecture in other tissues and organs. The nanoarchitecture was able to penetrate biological barriers and target cancer cells inhibiting their growth in mouse models.

15.4.2 In Vivo Assembly of RNA Nanoarchitecture

Complex nanostructures can be generated in vivo by cloning of nucleic acids in bacterial expression systems such as *E. coli* similar to recombinant proteins [73, 74]. RNA nanostructures in vivo have been demonstrated for catalyzing biosynthetic pathways in *E. coli*. The multidimensional RNA structures act as 1D and 2D scaffold for protein docking sites and consequently control spatial organization of bacterial metabolism [52].

Li et al. developed a strategy for construction of RNA nanostructures in vivo by programmed self-assembly from individual long single-stranded RNAs [21]. The in vivo assembly of thermodynamically stable structures tends to entangle in kinetic traps rendering the process slower. As heating or cooling is not an option during in vivo production of nanostructures, thus ssRNAs were allowed to fold in a sequential and hierarchical pathway. The strategy involved folding of long ssRNA initially into thermodynamically stable hairpin loops. These loop regions further form long-distance interaction subsequently leading to formation of fully folded designed nanostructures. Each nanostructure was self-assembled from an individual ssRNA that was expressed and folded inside bacterial expression system such as *E. coli.* The nanostructures were later isolated and subsequently characterized by gel electrophoresis, atomic force microscopy (AFM), and cryo-electron microscopy (Fig. 15.4g). In vivo assembly of RNA nanoarchitecture is yet a new area that needs more exploration, and many useful applications might come in the near future.

15.4.3 RNA Nanoarchitecture in Detection and Imaging: Light-Up Aptamers

RNA nanoarchitectures have found applications in imaging an array of biomolecules in the cell. RNA functional modules such as aptamers have been used extensively for analyte detection, diagnostics, as well as imaging. For most of these applications, RNA aptamers after selection could be engineered into sensors (known as aptasensors) by attaching fluorophores/quenchers and FRET pairs or by attaching other modules for detection output.

A new class of fluorogenic RNA aptamers also called light-up aptamers has recently attracted the attention of researchers [75, 76]. These fluorogenic RNA aptamers bind to their target that is a small rotor molecule and is usually nonfluorescent as it loses its energy by rotation. Aptamer binding to the target restricts its rotation, and the stored energy is released in terms of fluorescence, making it fluoresce or light up. The ability of target dye to penetrate into cells and ability of RNA aptamer to be expressed in cells make this light-up aptamer system to be used for a variety of applications such as RNA localization and imaging in living cells, small molecule detection, and miRNA detection [76]. Spinach, broccoli, corn, and mango are few of the names given to a few aptamers of this light-up aptamer class depending upon the color they emit as fluorescence emission. For instance, vegetable name spinach was given for aptamer system that gives green emission and the same way names were given to other light-up aptamer system.

Spinach aptamer (Fig. 15.5a) [75, 77] has been extensively used for imaging different RNAs in living cells by tagging the RNA with spinach aptamer. The small molecule target dye DFHBI is known to penetrate into cells and shows fluorescence after binding to its aptamer. Another exciting application of spinach aptamer is that its fluorescence can be turned off by engineering RNA sequences, and the binding of the target can turn on its fluorescence again, thus acting as on-off switch. The strategy has been exploited in the engineering of various biosensors for detection of small molecules and biomacromolecules. Figure 15.5b shows a spinach aptamer-based detection system that can be fused to another aptamer that targets cellular metabolites. Binding of target metabolite (shown in orange; Fig. 15.5b) zips the stem region that now allows binding of small molecule dye DFHBI to bind to spinach aptamer producing fluorescence output for detection [78].

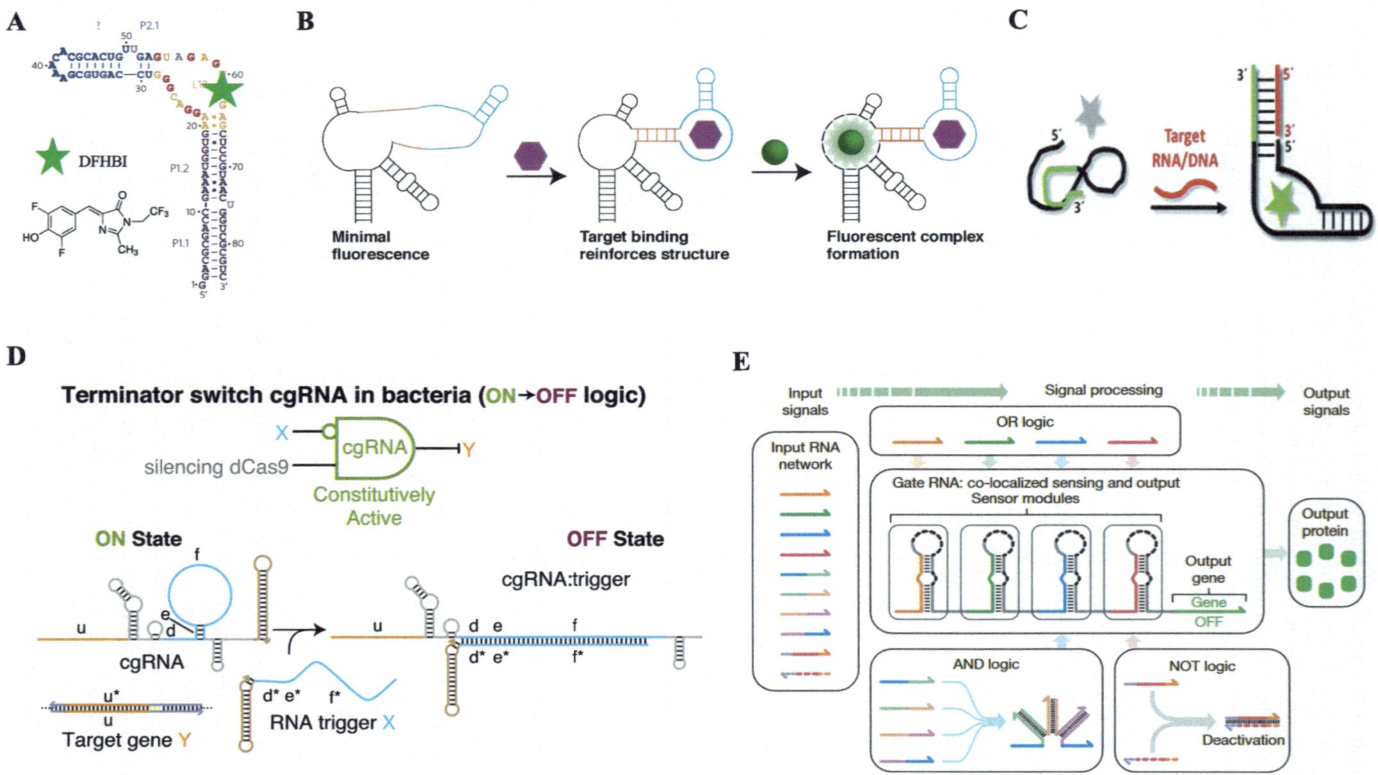

Fig. 15.5 (**a**) Spinach aptamer and its target DFHBI, (**b**) spinach aptamer-based metabolite detection, (**c**) BSMS sensor for miRNA detection, (**d**) conditional regulation in CRISPR-Cas system using RNA nanotechnology. (**e**) Cellular logic computation using RNA computing (**a**). Adapted with permission from [77], Copyright (2014) Springer Nature. (**b**) From [78]. Reprinted with permission from AAAS. (**c**) Reproduced from Ref. [79], with permission from The Royal Society of Chemistry. (**d**) Reprinted from [80]. (**e**) Adapted from [81], Copyright (2017) Springer Nature

As miRNAs serve as biomarkers of different subtypes of cancers [82], Soni et al. utilized structural switching ability of a truncated version of spinach aptamer called baby spinach and designed BSMS (baby spinach-based minimal modified sensor) to detect miRNA. In this design, attaching a sequence of a particular length (between 16 and 36 nucleotides) at 3′-end of baby spinach aptamer (shown in green; Fig. 15.5c) disturbs its folding and turns off its fluorescence. The binding of complimentary analyte DNA/RNA or miRNA sequence (shown in red) induces structural switching of BSMS, thus making it light up again, and thus can detect miRNA in low nM range (Fig. 15.5c) [79]. Tools from RNA nanotechnology have also been used concertedly for miRNA detection. Lee et al. developed a gold nanoparticle-functionalized RNA system by attaching a single copy of pRNA-3WJ for ultrasensitive detection of miRNA-155 using electrochemical surface-enhanced Raman spectroscopy (EC-SERS) [83]. Ying et al. have used spinach aptamer-based biosensors utilizing T7 in vitro transcription process for generation of unmodified spinach sequences for multiplexed and label-free detection of miRNA [84]. Light-up aptamer systems such as malachite green (MG) aptamer and spinach RNA have been shown to fuse to RNA nanoparticles for in vivo real-time monitoring of folding and degradation of RNAs [85].

15.4.4 RNA Nanoarchitecture in Gene Editing: CRISPR-Cas System

CRISPR-Cas (Clustered Regularly Interspaced Short Palindromic Repeat-CRISPR-associated proteins) system has revolutionized genetic manipulation in cells, animals, and plants to delete certain genes or to add newer desirable traits [86]. The system involves a single-guide RNA (sgRNA) that is used to locate genes that need to be edited by Cas endonucleases. As RNAs are an integral part of CRISPR toolbox, there are avenues for application of RNA nanotechnology in many components of CRISPR. sgRNAs can be engineered for the desired function through conditional activation by triggers such as small molecules, nucleases, nucleic acids, and photoactivations using functional modules like aptamers or ribozymes [87].

Hanewich-Hollatz et al. have developed an endogenous RNA-triggered conditional guide RNAs (cgRNA) which in complex with dCas9 controls gene expression. In the absence of trigger, the cgRNA-dCas9 complex remains bound to the dsDNA and inhibits transcription. Addition of an RNA trigger causes it to hybridize with cgRNA which in turn does not allow the formation of cgRNA-dCas9 complex leading the target gene to be efficiently transcribed (Fig. 15.5d) [80].

15.4.5 RNA Computing

Genetic materials such as DNA and RNA store a vast amount of information for processing in cells. The precision and speed at which the information is further used for various biological pathways are astounding. DNA as a medium of storage of information has been explored in recent years due to its high storage density and to overcome the challenges of storing the data explosion in the modern world. For instance, 4 g of DNA could store the annual global digital information [88].

RNA nanotechnology will be instrumental in development and construction of computational devices termed as "ribocomputing" devices in living cells that will carry out its functions using complex logic computation (Fig. 15.5e) [81]. The diversity in the structural and functional motifs makes RNA a better candidate for design of molecular devices. Catalyzing agents such as ribozyme, regulatory element riboswitches, and sensing agents such as aptamers provide multiple option to build logic gates such as AND/OR/NOR/NAND, and the Boolean networks. Specific chemical stimuli act as input in contrast to electronic current in conventional electronic computers.

Smolke and colleagues have used RNA to design various logic gates designed from RNA components where presence or absence of different stimuli will elicit a response that leads to regulation of gene expression using combinatorial approach. The three main components consist of sensors of RNA aptamers, an actuator made up of hammerhead ribozyme, and a transmitter sequence for coupling of sensor and actuator components [89]. In similar fashion, Rinaudo et al. developed a computing core that operates Boolean logic gates in human kidney cells using RNA interference (RNAi). The circuit consisted of two encoded mRNAs that code for the same protein albeit with different noncoding regions. The fluorescent protein was the system output, and was expressed when at least one mRNA was translated representing [90].

15.5 Future Prospective

RNA-based nanoarchitectures are rapidly evolving and are evident from the large number of publications in this field in the last few years from different researchers around the world. From the last few decades, the understanding about RNA and its structural and functional properties has improved quite well, and scientists have recently learnt to design the RNA architectures from different strands of RNA or from different natural or artificial motifs that can self-assemble to large nanostructures. The bonding of RNA nanotechnology with functional RNA modules such as aptamers, ribozymes, or riboswitches has provided a variety of RNA nanoarchitectures that have found applications in diverse fields. New functional RNA modules are continuously being evolved in test tubes. Many folding algorithms (M-Fold, NUPACK) have been developed that can predict RNA oligonucleotides folding on a computer in less than a minute with sufficient accuracy.

However, RNA is not only limited to simple base pairing; the motifs like G-quadruplexes, knots, and pseudoknots make folding of RNA even more complex to predict. It is important to understand these interactions more closely so that complex RNA architecture like ribosomes in cells can be constructed that works as a coordinated machine with many hands that works in sync with each other for useful functions. The effect of chemical modifications on RNA structure also needs to be understood in more detail for designing RNA nanoarchitecture with precise control. Large-scale production of RNAs of longer length and their production cost is also a bottleneck for future applications. The last decade was more about construction of various complex nanoarchitectures with precise control. The future should focus on finding new and useful applications of complex 2D and 3D nanoarchitecture, and efforts for that are underway.

Acknowledgment Authors thank IISER Tirupati for infrastructure and support. The financial support from the Department of Biotechnology, Government of India (DBT grant BT/PR23487/NNT/28/1295/2017), is gratefully acknowledged.

References

1. Nasrollahzadeh M, Sajadi SM, Sajjadi M, Issaabadi Z (2019) Applications of nanotechnology in daily life. In: Interface science and technology. Elsevier, New York, pp 113–143
2. Rao CNR, Ramakrishna Matte HSS, Voggu R, Govindaraj A (2012) Recent progress in the synthesis of inorganic nanoparticles. Dalton Trans 41:5089–5120
3. Fang F, Li M, Zhang J, Lee CS (2020) Different strategies for organic nanoparticle preparation in biomedicine. ACS Mater Lett 2:531–549
4. Blum AP, Kammeyer JK, Rush AM et al (2015) Stimuli-responsive nanomaterials for biomedical applications. J Am Chem Soc 137:2140–2154
5. Pirzada M, Altintas Z (2019) Nanomaterials for healthcare biosensing applications. Sensors 19:5311
6. Srinivasan M, Rajabi M, Mousa SA (2015) Multifunctional nanomaterials and their applications in drug delivery and cancer therapy. Nanomaterials 5:1690–1703
7. Arias-Gonzalez JR (2014) Single-molecule portrait of DNA and RNA double helices. Integr Biol (United Kingdom) 6:904–925
8. Tanaka Y, Fujii S, Hiroaki H et al (1999) A'-form RNA double helix in the single crystal structure of r(UGAGCUUCGGCUC). Nucleic Acids Res 27:949–955
9. Varani G, McClain WH (2000) The G·U wobble base pair: a fundamental building block of RNA structure crucial to RNA function in diverse biological systems. EMBO Rep 1:18–23
10. Sugimoto N, Nakano S, Katoh M et al (1995) Thermodynamic parameters to predict stability of RNA/DNA hybrid duplexes. Biochemistry 34:11211–11216
11. Watts JK, Deleavey GF, Damha MJ (2008) Chemically modified siRNA: tools and applications. Drug Discov Today 13:842–855
12. Liu J, Guo S, Cinier M et al (2011) Fabrication of stable and RNase-resistant RNA nanoparticles active in gearing the nanomotors for viral DNA packaging. ACS Nano 5:237–246
13. Majlessi M, Nelson NC, Becker MM (1998) Advantages of 2'-O-methyl oligoribonucleotide probes for detecting RNA targets. Nucleic Acids Res 26:2224–2229
14. Grünweller A, Hartmann RK (2007) Locked nucleic acid oligonucleotides: the next generation of antisense agents? BioDrugs 21:235–243

15. Kumar VA, Ganesh KN (2005) Conformationally constrained PNA analogues: structural evolution toward DNA/RNA binding selectivity. Acc Chem Res 38:404–412
16. Guo P (2010) The emerging field of RNA nanotechnology. Nat Nanotechnol 5:833–842
17. Shu Y, Pi F, Sharma A et al (2014) Stable RNA nanoparticles as potential new generation drugs for cancer therapy. Adv Drug Deliv Rev 66:74–89
18. Li H, Lee T, Dziubla T et al (2015) RNA as a stable polymer to build controllable and defined nanostructures for material and biomedical applications. Nano Today 10:631–655
19. Guo P, Ghimire C, Wang H et al (2020) RNA nanoparticles as rubber for compelling vessel extravasation to enhance tumor targeting and for fast renal excretion to reduce toxicity. ACS Nano 14:13180–13191
20. Haque F, Pi F, Zhao Z et al (2018) RNA versatility, flexibility, and thermostability for practice in RNA nanotechnology and biomedical applications. Wiley Interdiscip Rev RNA 9
21. Li M, Zheng M, Wu S et al (2018) In vivo production of RNA nanostructures via programmed folding of single-stranded RNAs. Nat Commun 9:1–9
22. Svoboda P, Di Cara A (2006) Hairpin RNA: a secondary structure of primary importance. Cell Mol Life Sci 63:901–918
23. Dibrov SM, Johnston-Cox H, Weng YH, Hermann T (2007) Functional architecture of HCV IRES domain II stabilized by divalent metal ions in the crystal and in solution. Angew Chemie Int Ed 46:226–229
24. Liu D, Geary CW, Chen G et al (2020) Branched kissing loops for the construction of diverse RNA homooligomeric nanostructures. Nat Chem 12:249–259
25. Shu D, Shu Y, Haque F et al (2011) Thermodynamically stable RNA three-way junction for constructing multifunctional nanoparticles for delivery of therapeutics. Nat Nanotechnol 6
26. Shu YI, Haque F, Shu D et al (2013) Fabrication of 14 different RNA nanoparticles for specific tumor targeting without accumulation in normal organs. RNA
27. Kim H, Park Y, Kim J et al (2016) Nucleic acid engineering: RNA following the trail of DNA. ACS Comb Sci 18:87–99
28. Grabow WW, Jaeger L (2014) RNA self-assembly and RNA nanotechnology. Acc Chem Res 47:1871–1880
29. Ohno H, Kobayashi T, Kabata R et al (2011) Synthetic RNA-protein complex shaped like an equilateral triangle. Nat Nanotechnol 6:116–120
30. Monferrer A, Zhang D, Lushnikov AJ, Hermann T (2019) Versatile kit of robust nanoshapes self-assembling from RNA and DNA modules. Nat Commun 10:1–8
31. Geary C, Meunier P-É, Schabanel N, Seki S (2019) Oritatami: a computational model for molecular co-transcriptional folding. Int J Mol Sci 20:2259
32. Jaeger L, Chworos A (2006) The architectonics of programmable RNA and DNA nanostructures. Curr Opin Struct Biol 16:531–543
33. Ishikawa J, Furuta H, Ikawa Y (2013) RNA tectonics (tectoRNA) for RNA nanostructure design and its application in synthetic biology. Wiley Interdiscip Rev RNA 4:651–664
34. Jaeger L, Leontis NB (2000) Tecto-RNA: one-dimensional self-assembly through tertiary interactions. Angew Chemie Int Ed 39:2521–2524
35. Nasalean L, Baudrey S, Leontis NB, Jaeger L (2006) Controlling RNA self-assembly to form filaments. Nucleic Acids Res 34:1381–1392
36. Chworos A, Severcan I, Koyfman AY et al (2004) Building programmable jigsaw puzzles with RNA. Science 306:2068–2072
37. Grabow WW, Zakrevsky P, Afonin KA et al (2011) Self-assembling RNA nanorings based on RNAI/II inverse kissing complexes. Nano Lett 11:878–887
38. Zhang H, Endrizzi JA, Shu Y et al (2013) Crystal structure of 3WJ core revealing divalent ion-promoted thermostability and assembly of the Phi29 hexameric motor pRNA. RNA 19:1226–1237
39. Severcan I, Geary C, Chworos A et al (2010) A polyhedron made of tRNAs. Nat Chem 2:772–779
40. Shu D, Moll WD, Deng Z et al (2004) Bottom-up assembly of RNA arrays and superstructures as potential parts in nanotechnology. Nano Lett 4:1717–1723

41. Khisamutdinov EF, Jasinski DL, Guo P (2014) RNA as a boiling-resistant anionic polymer material to build robust structures with defined shape and stoichiometry. ACS Nano 8:4771–4781
42. Dibrov SM, McLean J, Parsons J, Hermann T (2011) Self-assembling RNA square. Proc Natl Acad Sci U S A 108:6405–6408
43. Afonin KA, Bindewald E, Yaghoubian AJ et al (2010) In vitro assembly of cubic RNA-based scaffolds designed in silico. Nat Nanotechnol 5:676–682
44. Han D, Qi X, Myhrvold C et al (2017) Single-stranded DNA and RNA origami. Science
45. Jasinski DL, Khisamutdinov EF, Lyubchenko YL, Guo P (2014) Physicochemically tunable polyfunctionalized RNA square architecture with fluorogenic and ribozymatic properties. ACS Nano 8:7620–7629
46. Endo M, Yamamoto S, Tatsumi K et al (2013) RNA-templated DNA origami structures. Chem Commun 49:2879–2881
47. Wang P, Ko SH, Tian C et al (2013) RNA–DNA hybrid origami: folding of a long RNA single strand into complex nanostructures using short DNA helper strands. Chem Commun 49:5462–5464
48. Zheng HN, Ma YZ, Xiao SJ (2014) Periodical assembly of repetitive RNA sequences synthesized by rolling circle transcription with short DNA staple strands to RNA–DNA hybrid nanowires. Chem Commun 50:2100–2103
49. Ko SH, Su M, Zhang C et al (2010) Synergistic self-assembly of RNA and DNA molecules. Nat Chem 2:1050–1055
50. Afonin KA, Viard M, Martins AN et al (2013) Activation of different split functionalities on re-association of RNA-DNA hybrids. Nat Nanotechnol 8:296–304
51. Geary C, Rothemund PWK, Andersen ES (2014) A single-stranded architecture for cotranscriptional folding of RNA nanostructures. Science 345:799–804
52. Delebecque CJ, Lindner AB, Silver PA, Aldaye FA (2011) Organization of intracellular reactions with rationally designed RNA assemblies. Science 333:470–474
53. Jasinski DL, Binzel DW, Guo P (2019) One-pot production of RNA nanoparticles via automated processing and self-assembly. ACS Nano 13:4603–4612
54. Afonin KA, Kasprzak WK, Bindewald E et al (2014) In silico design and enzymatic synthesis of functional RNA nanoparticles. Acc Chem Res 47:1731–1741
55. Afonin KA, Kireeva M, Grabow WW et al (2012) Co-transcriptional assembly of chemically modified RNA nanoparticles functionalized with siRNAs. Nano Lett 12:5192–5195
56. Jasinski D, Haque F, Binzel DW, Guo P (2017) Advancement of the emerging field of RNA nanotechnology. ACS Nano 11:1142–1164
57. Piao X, Yin H, Guo S et al (2019) RNA nanotechnology to solubilize hydrophobic antitumor drug for targeted delivery. Adv Sci 6:1900951
58. Xu C, Li H, Zhang K et al (2019) Photo-controlled release of paclitaxel and model drugs from RNA pyramids. Nano Res 12:41–48
59. Afonin KA, Viard M, Koyfman AY et al (2014) Multifunctional RNA nanoparticles. Nano Lett 14:5662–5671
60. Lee JB, Hong J, Bonner DK et al (2012) Self-assembled RNA interference microsponges for efficient siRNA delivery. Nat Mater 11:316–322
61. Shu D, Li H, Shu Y et al (2015) Systemic delivery of anti-miRNA for suppression of triple negative breast cancer utilizing RNA nanotechnology. ACS Nano 9:9731–9740
62. Pi F, Zhang H, Li H et al (2017) RNA nanoparticles harboring annexin A2 aptamer can target ovarian cancer for tumor-specific doxorubicin delivery. Nanomed Nanotechnol Biol Med 13:1183–1193
63. Zhao N, Woodle MC, Mixson AJ (2018) Advances in delivery systems for doxorubicin. J Nanomed Nanotechnol
64. Lee TJ, Haque F, Shu D et al (2015) RNA nanoparticle as a vector for targeted siRNA delivery into glioblastoma mouse model. Oncotarget 6:14766–14776
65. Cui D, Zhang C, Liu B et al (2015) Regression of gastric cancer by systemic injection of RNA nanoparticles carrying both ligand and siRNA. Sci Rep 5:1–14

66. Haque F, Shu D, Shu Y et al (2012) Ultrastable synergistic tetravalent RNA nanoparticles for targeting to cancers. Nano Today 7:245–257
67. Roh YH, Deng JZ, Dreaden EC et al (2016) A multi-RNAi microsponge platform for simultaneous controlled delivery of multiple small interfering RNAs. Angew Chemie Int Ed 55:3347–3351
68. Ye X, Liu Z, Hemida MG, Yang D (2011) Targeted delivery of mutant tolerant anti-Coxsackievirus artificial MicroRNAs using folate conjugated bacteriophage Phi29 pRNA. PLoS One 6:e21215
69. Binzel DW, Shu Y, Li H et al (2016) Specific delivery of MiRNA for high efficient inhibition of prostate cancer by RNA nanotechnology. Mol Ther 24:1267–1277
70. Catuogno S, Esposito CL, de Franciscis V (2016) Aptamer-mediated targeted delivery of therapeutics: an update. Pharmaceuticals 9:69
71. Parlea L, Puri A, Kasprzak W et al (2016) Cellular delivery of RNA nanoparticles. ACS Comb Sci 18:527–547
72. Zhang Y, Lai B, Juhas M (2019) Recent advances in aptamer discovery and applications. Molecules 24:941
73. Ponchon L, Dardel F (2007) Recombinant RNA technology: the tRNA scaffold. Nat Methods 4:571–576
74. Nelissen FHT, Leunissen EHP, Van De Laar L et al Fast production of homogeneous recombinant RNA-towards large-scale production of RNA. Nucleic Acid Res
75. Paige JS, Wu KY, Jaffrey SR (2011) RNA mimics of green fluorescent protein. Science 333:642–646
76. Swetha P, Fan Z, Wang F, Jiang JH (2020) Genetically encoded light-up RNA aptamers and their applications for imaging and biosensing. J Mater Chem B 8:3382–3392
77. Huang H, Suslov NB, Li NS et al (2014) A G-quadruplex-containing RNA activates fluorescence in a GFP-like fluorophore. Nat Chem Biol 10:686–691
78. Paige JS, Nguyen-Duc T, Song W, Jaffrey SR (2012) Fluorescence imaging of cellular metabolites with RNA. Science 335:1194
79. Soni R, Sharma D, Krishna AM et al (2019) A highly efficient baby spinach-based minimal modified sensor (BSMS) for nucleic acid analysis. Org Biomol Chem 17:7222–7227
80. Hanewich-Hollatz MH, Chen Z, Hochrein LM et al (2019) Conditional guide RNAs: programmable conditional regulation of CRISPR/Cas function in bacterial and mammalian cells via dynamic RNA nanotechnology. ACS Cent Sci 5:1241–1249
81. Green AA, Kim J, Ma D et al (2017) Complex cellular logic computation using ribocomputing devices. Nature 548:117–121
82. Tricoli JV, Jacobson JW (2007) MicroRNA: potential for cancer detection, diagnosis, and prognosis. Cancer Res 67:4553–4555
83. Lee T, Mohammadniaei M, Zhang H et al (2020) Single functionalized pRNA/gold nanoparticle for ultrasensitive microRNA detection using electrochemical surface-enhanced Raman spectroscopy. Adv Sci 7:1902477
84. Ying ZM, Tu B, Liu L et al (2018) Spinach-based fluorescent light-up biosensors for multiplexed and label-free detection of microRNAs. Chem Commun 54:3010–3013
85. Shu D, Khisamutdinov EF, Zhang L, Guo P (2014) Programmable folding of fusion RNA in vivo and in vitro driven by pRNA 3WJ motif of phi29 DNA packaging motor. Nucleic Acids Res 42:e10–e10
86. Knott GJ, Doudna JA (2018) CRISPR-Cas guides the future of genetic engineering. Science 361:866–869
87. Adli M (2018) The CRISPR tool kit for genome editing and beyond. Nat Commun 9:1–13
88. Panda D, Molla KA, Baig MJ et al (2018) DNA as a digital information storage device: hope or hype? 3 Biotech 8:239
89. Nyan Win M, Smolke CD (2008) Higher-order cellular information processing with synthetic RNA devices. Science 322:456–460
90. Rinaudo K, Bleris L, Maddamsetti R et al (2007) A universal RNAi-based logic evaluator that operates in mammalian cells. Nat Biotechnol 25:795–801

Part V
Architectonics of Complex Systems and Advanced Objects

Chapter 16
Covalent Organic Frameworks as Tunable Supports for HER, OER, and ORR Catalysts: A New Addition to Heterogeneous Electrocatalysts

Ramanathan Vaidhyanathan

16.1 Introduction to Covalent Organic Framework [COF]

Covalent organic frameworks are crystalline organic frameworks, or they can be seen as organic polymers with highly ordered periodically repeating units [1–17]. Some of the most known forms of crystalline polymers are liquid crystals, but they are typically confined to a substrate or a 2D surface [18–21]. Surface-confined 3D all-organic polymeric architecture is rare [22]. Growing 3D crystalline polymers into single crystals or as polycrystalline materials remained elusive. In 2005, Cote et al. reported the first crystalline polymer, COF, with three-dimensional order [23]. The boronate-linked COF had a polymeric 2D layer with hexagonal rings (Fig. 16.1) resembling graphene sheets and the aromatic rings of the polymer form ordered π-stack columns with graphite-like structure, and this creates nanospaces (27 Å in COF-5) running as 1D channels perpendicular to the plane of the layers. Being creative, the COF layer can be visualized as graphene with well-ordered holes in them.

In 2013, Wuest and co-workers reported the single crystals of an organic polymer [24]. The nitroso-linked polymer had a dense diamondoid structure with a tenfold interpenetration (Fig. 16.2). They exploited the high reversibility of the very weak nitroso bonds (20–30 kcal mol^{-1}). The reversibility is crucial for the formation of a defect-free polymeric structure. More recently, Yaghi and co-workers reported the single-crystal structure of microporous COFs, with permanent porosity [25, 26]. This opened a new avenue toward the possibility of weaving organic

R. Vaidhyanathan (✉)
Department of Chemistry and Centre for Energy Science, Indian Institute of Science Education and Research, Pune, India
e-mail: vaidhya@iiserpune.ac.in

T. Govindaraju, K. Ariga (eds.), *Molecular Architectonics and Nanoarchitectonics*, Nanostructure Science and Technology,
https://doi.org/10.1007/978-981-16-4189-3_16

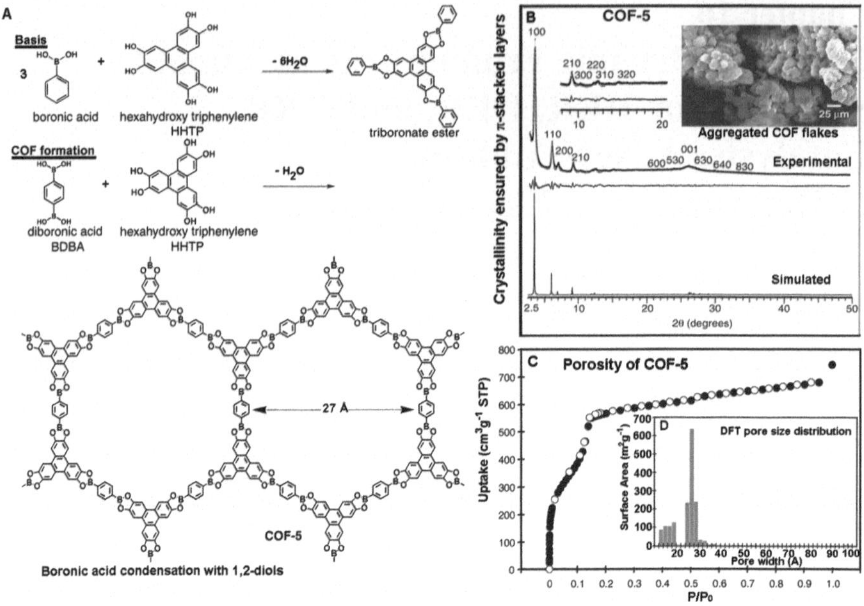

Fig. 16.1 (**a**) Formation of a polymeric crystalline framework via polycondensation. (**b**) Highly ordered structure of the COF-5 evidenced from PXRD and the comparison with a pattern simulated from modeled structure. (**c**) Permanent porosity of COF-5 with uniform 27 Å pores confirmed by 77 K N_2 isotherm. Crystallinity and porosity are signature of any functional COF (Adapted from Ref. 23)

polymers with a high degree of crystallinity and order by choosing building units with the appropriate geometry and chemical functionalities, for example, monomers that promote self-assembly via π-stacking or hydrogen bonding or other cumulatively strong weaker interactions.

16.2 Chemistry of COF Formation

A crucial aspect of achieving high order is the nature of the covalent bonds formed between the monomers. It is well-understood that these all-organic polymers are able to crystallize into COF owing to the reversibility of the covalent bonds formed during polymerization. In COF design principles, this is also called "dynamic covalent chemistry" [27]. For example, during COF formation, as the polymeric conjugated sheet of a COF gets formed simultaneously, they need to π-stack with the adjacent layer, which would mean that the π-rings between the sheets need to orient parallel. To meet this, the covalent bonds that are formed need to be reversible/repairable. Whenever there is a defect (aromatic rings misoriented), the covalent bond breaks and reforms to achieve the energetically favorable orientation. Some of the major defects include slipped oligomers, dangling bonds, missing covalent links,

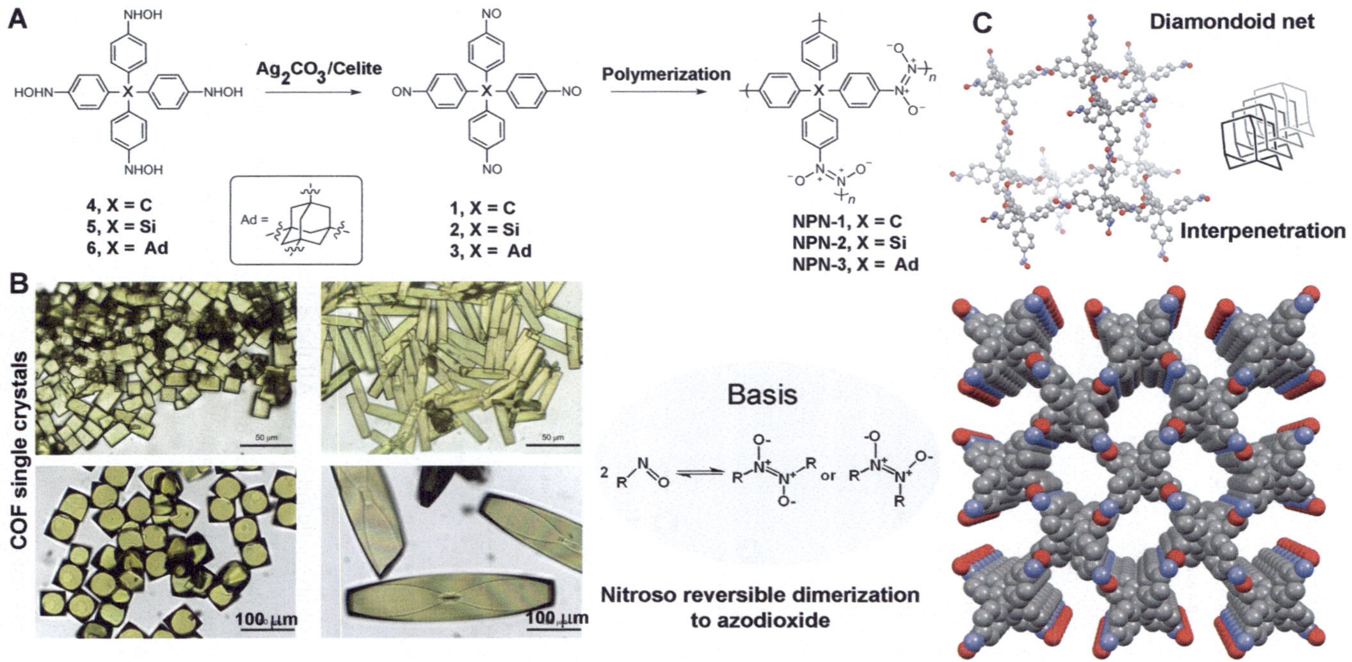

Fig. 16.2 (**a**) Oxidation of hydroxylamines produces tetranitroso monomers, which polymerize to give covalent nitroso polymer networks. (**b**) Photographs of large single crystals of covalent nitroso polymer networks NPN-1, NPN-2, and NPN-3. (**c**) NPN-1, showing part of the diamondoid framework (ball-and-stick image), the degree of interpenetration [grayscale image], and the cross-sections of parallel channels viewed along the c-axis (space-filling image). Atoms of carbon are shown in gray, nitrogen in blue, oxygen in red, and silicon in yellow (Copyright © 2013, Springer Nature)

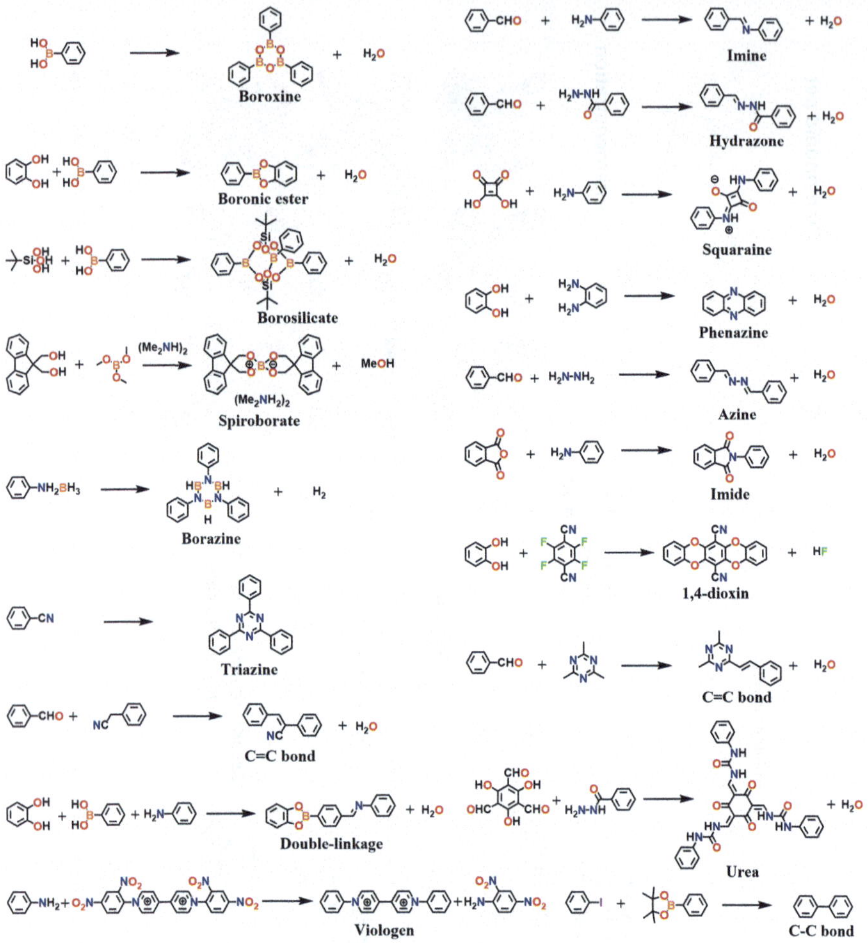

Fig. 16.3 Various linkages explored for the synthesis of COFs [4] (Copyright © 2020, American Chemical Society)

rotated or misaligned aromatic rings, etc. All these lead to off-equilibrium configurations. Such reversibility principles are required to ensure high optimizations of all the intermolecular and intramolecular interactions. However, it is important to bear in mind that the final COF is still a metastable structure formed as the potentially most stable configuration under the reaction condition (temperature, pressure, solvent, additives, and concentration). Unless perfect single crystals are achieved, a defect-free 100% crystalline structure is a distant claim. Such reversible chemistry has been successfully extended to the construction of three-dimensional porous COFs too [28, 29]. With many of these design principles in place still, the growth of COF as a single crystal is not easy.

A library of reversible reactions that can assist COF formation is presented in Figs. 16.3, 16.4, 16.5, and 16.6 [4, 27]. Jiang and co-worker's recent expansive

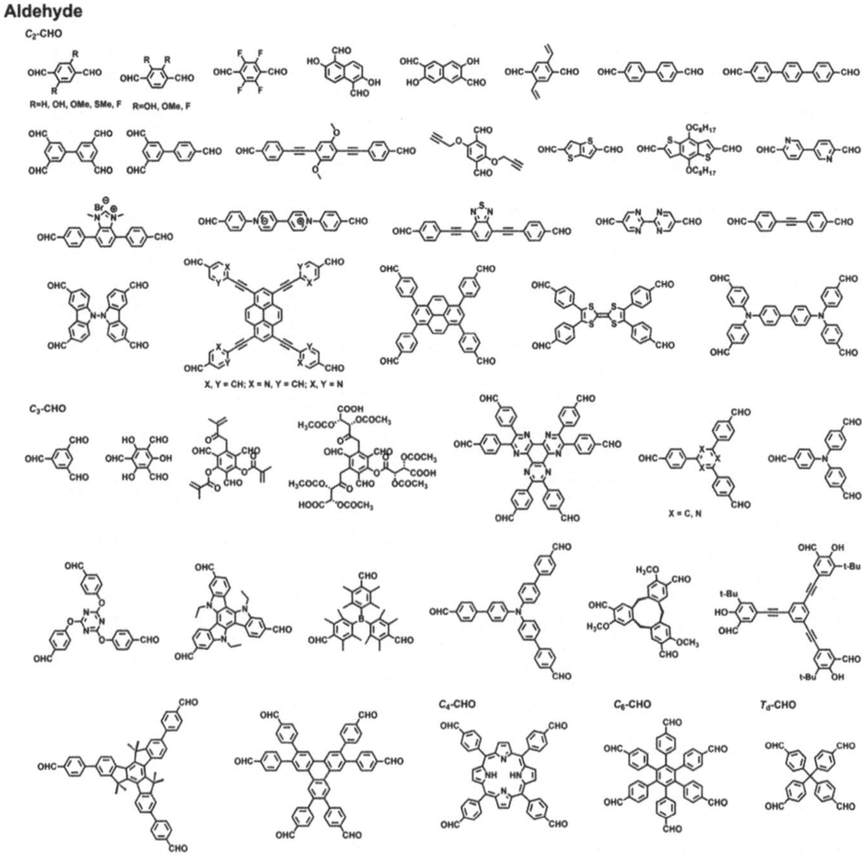

Fig. 16.4 Typical examples of C2, C3, C4, C6, and Td symmetric monomers with aldehyde functional groups for the synthesis of COFs [4] (Copyright © 2020, American Chemical Society)

review covers all different aspects of COF construction, including how the monomer geometries can enforce various COF symmetries, thereby the pore size, shape, dynamic role of the linker orientations, etc. Besides, this review covers the concepts in COF synthesis, characterization, and selected applications of COF-derived materials in energy research [4].

In COF formation, the reversibility of these bonds can challenge the stability of the COF, and hence to achieve sufficient thermal, mechanical, and chemical stability, the extent of polymerization needs to be high. From principles of phase formation, it is expected that the COFs can be made at high temperatures leading to more thermodynamically favorable structures with enhanced stability. Again this leads to another complication. One of the most sought-out properties in a COF is the highly ordered porous structure leading to high surface areas and pore volumes. COFs synthesized at high temperatures are most unlikely to satisfy this demand. Realizing this decisive reversibility in the covalent bond formation, researchers are constantly

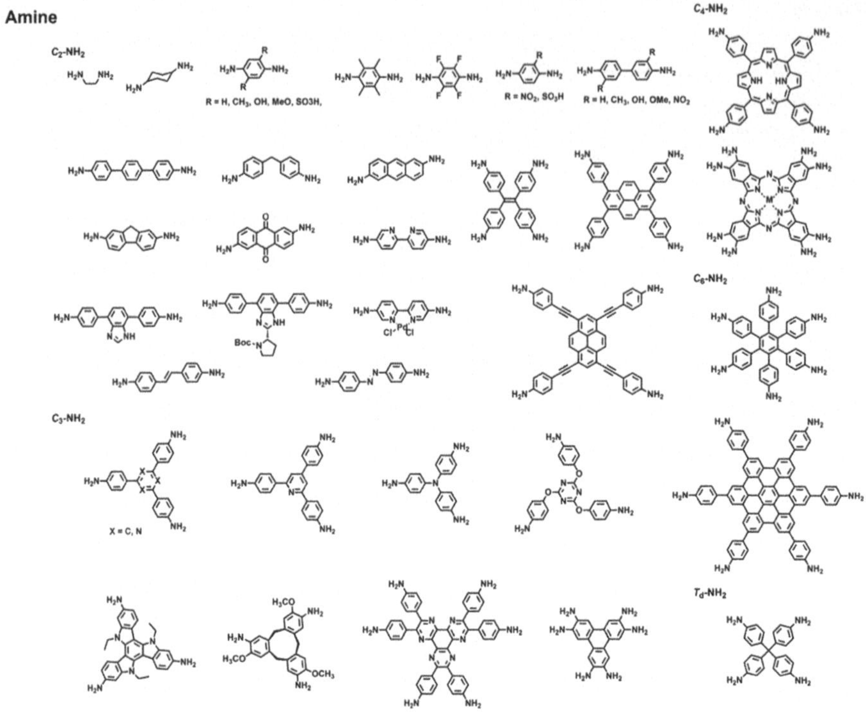

Fig. 16.5 Typical examples of C2, C3, C4, C6, and Td symmetric monomers with amine functional groups for the synthesis of COFs [4] (Copyright © 2020, American Chemical Society)

coming up with chemistries and strategies that favor a high degree of reversibility (Figs. 16.3 and 16.7a). The approaches to obtain crystalline COFs can be classified into three strategies: [i] reversible reaction, [ii] pre-orientation of building blocks, and [iii] the single reaction pathway [27]. A schematic explanation of these three different strategies is presented in Fig. 16.7b. Lotsch et al. provide a detailed explanation of these in their recent review [27]. Several factors such as the presence of water and catalytic additives such as acetic acid, trifluoroacetic acid, and toluenesulfonic acids impact the COF formation and crystallinity (Fig. 16.7d) [30–33]. Slow supply of building block in COFs based on reversible COF linkages, direct synthesis of COFs based on irreversible reactions, having substitutions on the building blocks or the monomers which assist their orientation during the COF formation via weak interactions (H-bonding, van der Waal and agostic CH–π interactions, etc.) can also influence the crystallinity of the COF [34–36].

Asserting the fact that COF can have a robust framework, even strongly acidic phosphonate groups and sulfonic acid units have been incorporated into the azine-linked COF, and it is found to work as a sponge to remove U(VI) and Pu(IV) ions from acidic solutions with pH = 1 (Fig. 16.8). However, when demonstrating the stability of COF in acidic or basic solutions, more than the pH, the absolute molarity

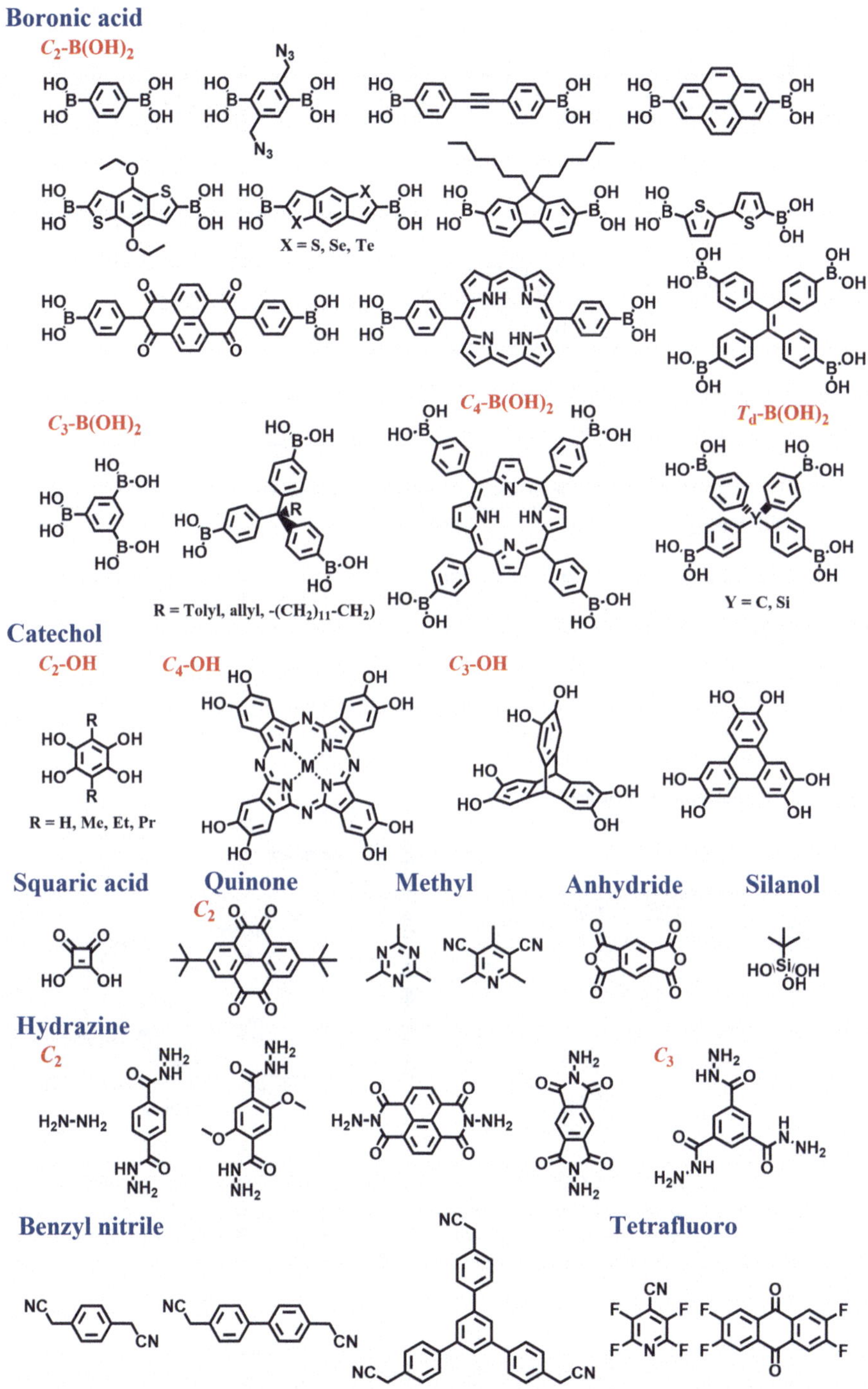

Fig. 16.6 Typical examples of C_2, C_3, C_4, C_6, and T_d symmetric monomers with other functional groups for the synthesis of COFs [4] (Copyright © 2020, American Chemical Society)

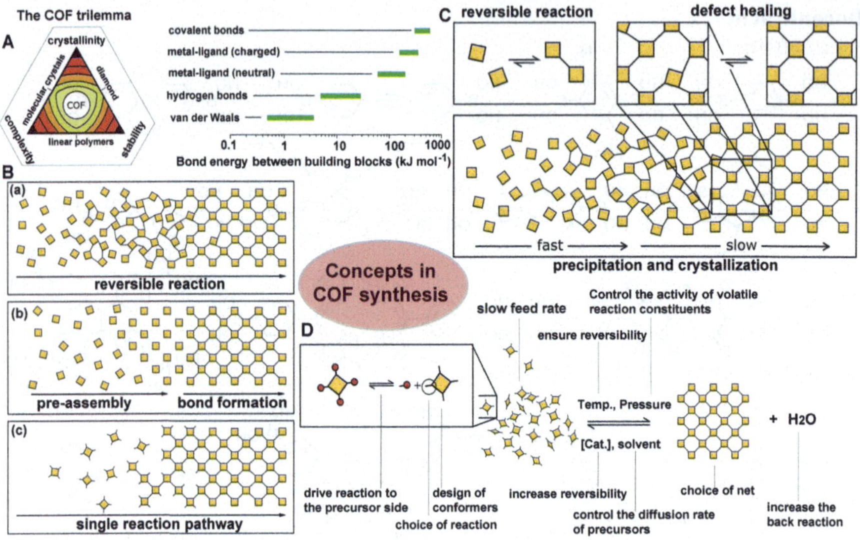

Fig. 16.7 (**a**) The trilemma of materials design for COF synthesis based on reversible reactions (left). Strengths of bonds that are used for the synthesis of molecularly defined materials [right] [27]. (**b**) Three approaches to forming a crystalline covalently bound polymer: (a) reversible reactions, (b) pre-orientation, and (c) single reaction pathway. (**c**) Reversible COF formation by slow crystallization. First, an amorphous gel is formed that then slowly crystallizes. This crystallization mechanism corresponds to the formation of imine COFs. (**d**) Overview of the strategies used to enable crystallization of COFs (Copyright © 2007, Royal Society of Chemistry)

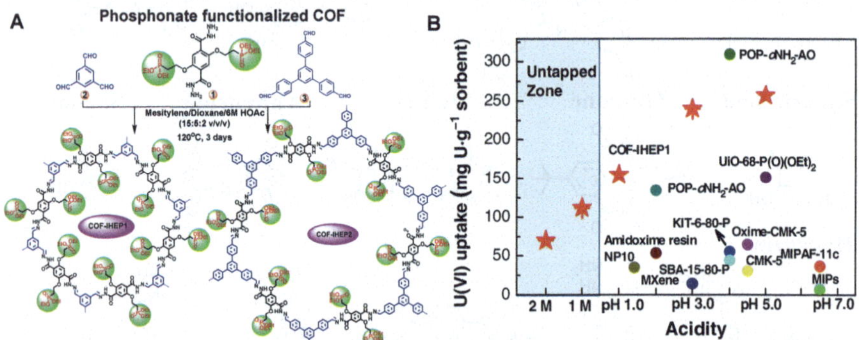

Fig. 16.8 (**a**) Construction of COF-IHEP1 via condensation of 2,5-bis[2-[diethoxyphosphoryl] ethoxy]terephthalohydrazide (**1**) and 1,3,5-triformylbenzene (**2**). Formation of COF-IHEP2 was accomplished by treating (**1**) with 1,3,5-tris[4-formylphenyl]benzene (**3**) under solvothermal conditions. (**b**) Comparison of the U(VI) sorption of COF-IHEP2 for a wide range of acidity with benchmark materials. ★ denotes the value of COF-IHEP (Copyright © 2019 Chinese Chemical Society)

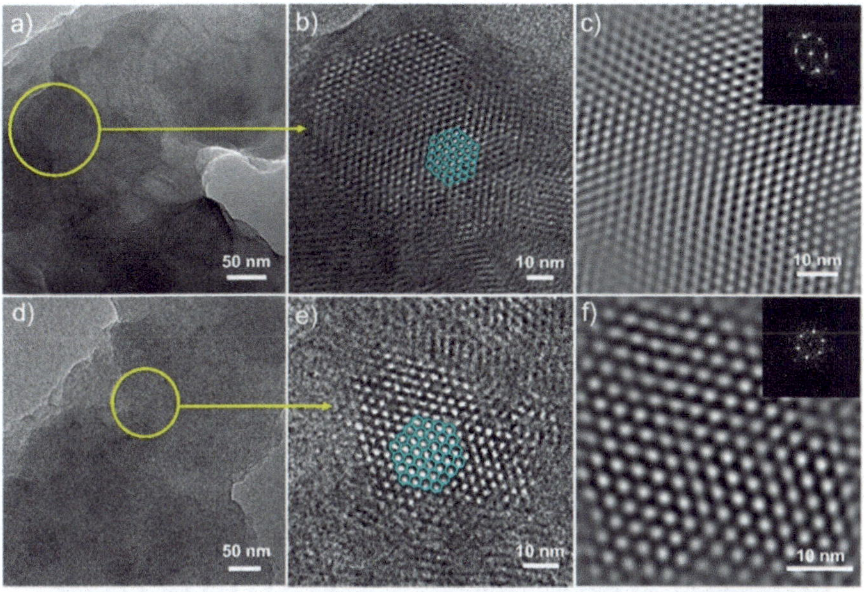

Fig. 16.9 (**a**) TEM image of B-COF-2. (**b**) TEM image of B-COF-2, enlarged view of a selected area in panel (**a**). (**c**) Fourier-filtered image of selected areas of B-COF-2, inset: fast Fourier transform (FFT) from the selected areas; for T-COF-2 see (**d–f**), respectively. The imine-COF has been converted to d π-conjugated fused-aromatic thieno(3,2)pyridine-linked COFs: this gives a high degree of crystallinity inferred from HRTEM (Copyright © 2020, American Chemical Society)

of the solutions is more relevant. The presence of hydrogen bonding substituents on the monomers or the coexistence of guests such as ionic liquids or heterocycles in the COF along with acidic functionalities [phosphonate, sulfonate] brings sufficient electrostatic and dispersive forces which source stability to the otherwise weaker COF linkages [35, 37–40]. However, in all these cases, incorporating the highly acidic or reactive groups in the framework often leads to a visible broadening of their PXRD reflections [37–41]. Retaining a high degree of crystallinity in highly functionalized COFs is a serious challenge. However, with the state-of-the-art HRTEM, when high-quality COF samples are synthesized, brilliant atomically resolvable in-plane images of COFs with indexable reflections in the selected area electron diffraction (SAED) images have been possible (Figs. 16.9 and 16.10) [42, 43]. The SEM and TEM microscopy images suggest that in many cases, the COFs grow with various morphologies (flakes, fibers, thick or thin-wrinkly sheets, or even rods). Typically, many regions of the COF exhibit high crystallinity displayed as well-imaged lattice fringes or, in some cases, moiré fringes. However, poorly grown COFs in most cases appear as amorphous spheres or random sheets [no specific symmetry and of nonuniform sizes] with no appreciable lattice fringes or no reflections in the diffracted images. The methoxy-substituted monomers tend to yield very crystalline and robust COFs. The methoxy groups form strong interlayer – C–H•••π interactions, in addition to the common π π ones.

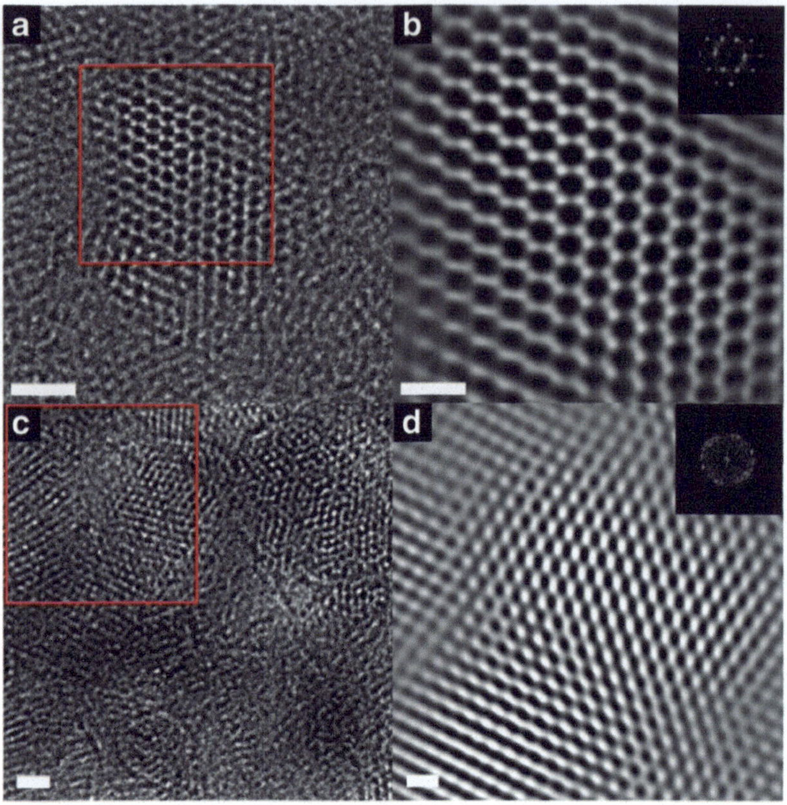

Fig. 16.10 HRTEM characterization of COF-1 and MF-1a. (**a**) Low-dose, high-resolution TEM image of COF-1 (scale bar, 10 nm). (**b**) The Fourier-filtered image of selected red square areas (scale bar, 5 nm). Inset: fast Fourier transform (FFT) from the red square on the COF-1. (**c**) Low-dose, high-resolution TEM image of MF-1a (scale bar, 10 nm). (**d**) The Fourier-filtered image of the selected red square area (scale bar, 5 nm). Inset: FFT from the red square on the MF-1a (Copyright © 2018, Li et al., Nature Comm)

16.3 Selected Notable Chemistries for COF–MOF Construction

Many chemistries have been employed over the last two decades in making COFs (Fig. 16.3) [4]. For example, boronic acid and boronate ester-derived COF with high crystallinity are known. Nitroso chemistry, borazine, and ionic spiroborate have also been used (Fig. 16.3). As an alternate approach, even inorganic tetrahedrally linking heteroatoms such as silicon [7] or the zeolitic secondary building units have been incorporated into COF construction to craft highly crystalline COF–MOF hybrids (Fig. 16.11) [44]. The zeolitic COF–MOF structures comprising inorganic secondary building units have a high chance of rendering the final material as a well-

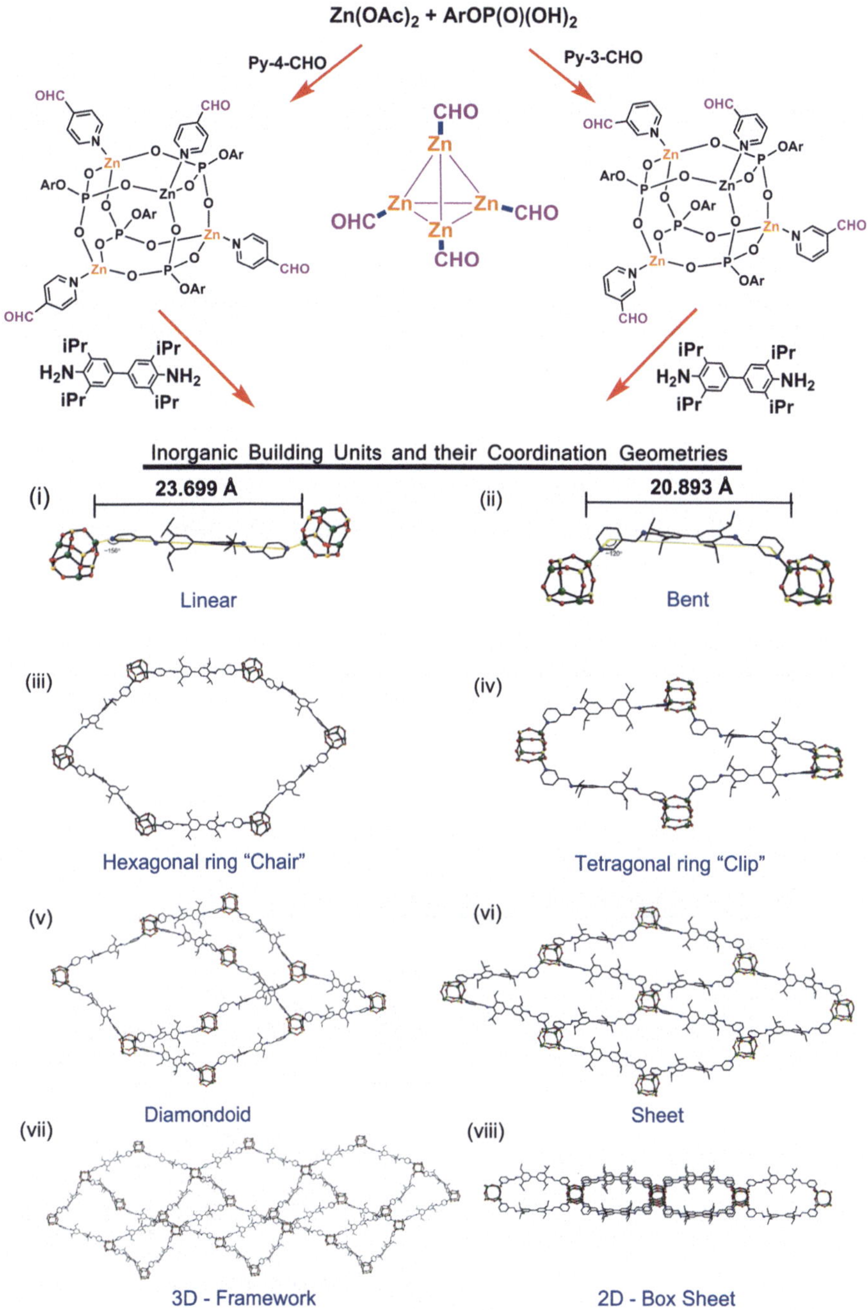

Fig. 16.11 A combined MOF–COF approach to zinc phosphate frameworks. Linking of D4R cages by a ditopic ligand to form a diamondoid basic unit (v). Linking of D4R cages by a ditopic ligand with an angle of 120° to form a 2D sheet via a clip-like basic unit (vi) (Adapted from Ref. 44)

diffracting single crystal from which rich structural insights can be gathered to secure future design principles. However, the imine [–C=N–] bond derived from Schiff-base chemistry has been the most implemented reversible chemistry for COF formation [45–49]. This is due to the vast library of cross-linking aldehydes and amines that can be synthesized in two to three steps in reasonably good yields, and also the convenient reversibility and tautomerism-assisted stability of the imine bonds [45–50], leading to the formation of many targeted COF as polycrystalline powders in good yields. But they do not exhibit the high degree of crystallinity possessed by the other boronate or silicate COFs. This is attributed to the easy hydrolyzability of the imine bonds [47, 49, 50], which causes more defects to develop during synthesis and even after synthesis. In our experience, many imine-COFs do lose their crystallinity upon standing on a shelf without any special protection.

16.4 Self-Exfoliation and Functionalizing Exfoliation Agent [FEA] (Fig. 16.12)

In distinct cases, when poorly π-stacking motifs or monomers are employed in the COF formation, they tend to form self-exfoliated COFs [51–53], wherein the COF does not exhibit long-range stacking in the *c-direction* leading to few-layer-thick nanosheets. Alternatively, exfoliation of COF has been achieved by solvent-assisted sonication [54, 55], mechanical ball milling [56, 57], in-situ charge assistance [58], and acid impregnation [59]. Structural changes during exfoliation of a multi-layer stacked COF into few-layer-thick nanosheets can be tracked by analyzing their layer

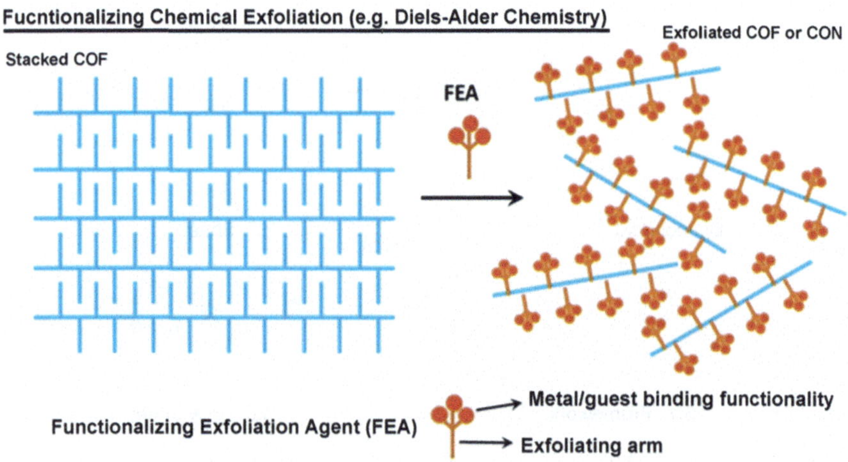

Fig. 16.12 Concept of functionalizing exfoliation agent (FEA) (Copyright @ 2019, John Wiley & Sons, Inc.)

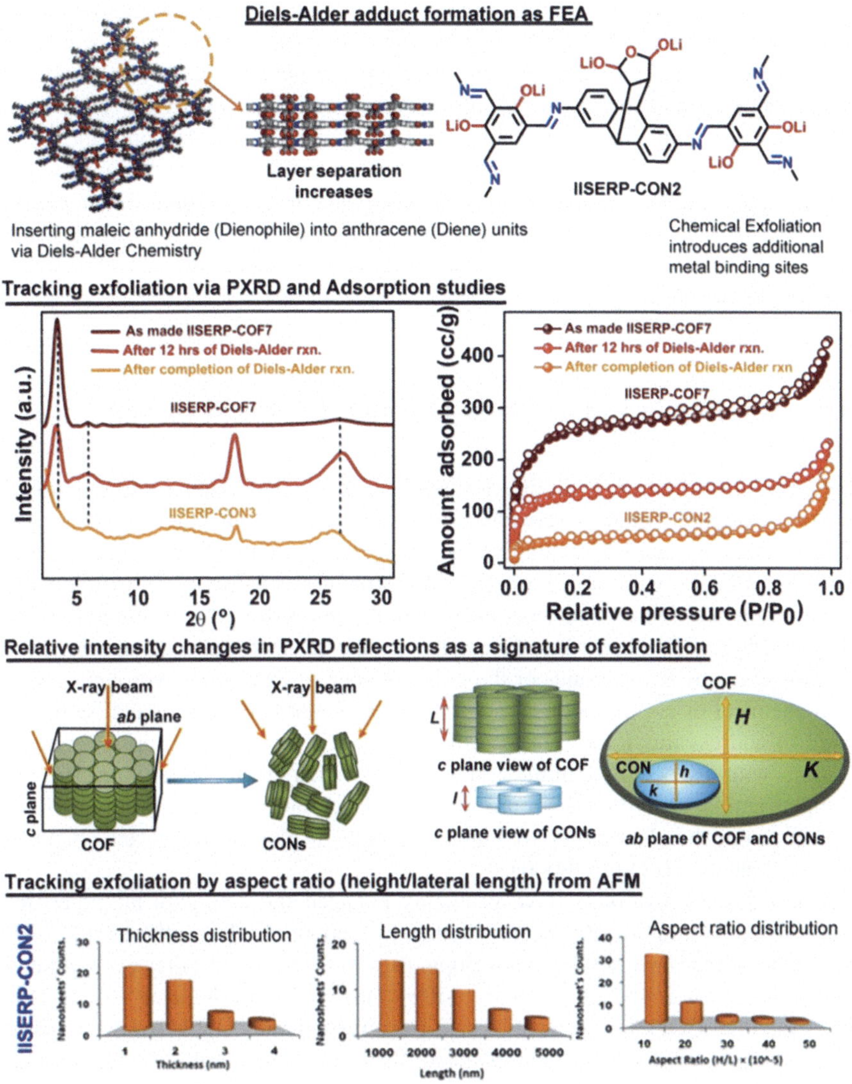

Fig. 16.13 The use of Diels–Alder cyclo-adduct formation as a chemical exfoliation tool in COF. Tracking the structural changes during the transformation of a covalent organic framework (COF) to a covalent organic nanosheet (CON) (Copyright @ 2019, John Wiley & Sons, Inc.)

thickness [height profile] and lateral dimensions from AFM [54]. Also, breaking down the heavily π-stacked COF structure to few-layer-thick covalent organic nanosheets (CON) can significantly alter the atoms contributing to low-angle reflections vs. those contributing to the high-angle *001* reflections, creating characteristic changes in their relative intensity in the PXRD (Fig. 16.13). Suppose the exfoliation happens by covalent linking of a new chemical group into the polymeric layer of the COF, it can give rise to new characteristic peaks in the PXRD [60]. This was observed in a COF to CON transformation achieved from a simple 4 + 2 cycloaddition of maleic anhydride onto the anthracene ring of a COF via Diels–Alder chemistry (Fig. 16.13). Notably, the maleic anhydride added onto the CON brings additional oxygen atoms into the structure, which can serve as binding groups. Thus, the dienophile serves as a *functionalizing exfoliation agent [FEA]* [60] (Fig. 16.12). This has resulted in the chemically exfoliated COF showing enhanced Li^+ binding compared to the as-made COF when employed as the anode in Li-ion battery [60].

16.5 Stability in Imine-COFs Through Chemical Design

Though achieving a high level of crystallinity with minimal defects and maximum surface areas/pore volumes remains critical for many applications such as gas storage, catalysis, sportive separations, sensing, etc., there are other energy applications where COFs with amply polymerized structure constructed from relatively weaker imine and azine bonds become potentially impactful. This is due to the rich chemistry they add to bring functional sites to interact with the guest species [56–59]. However, the Schiff bonds or the imine linkages are prone to hydrolysis under harsh chemical or electrochemical environments. Several strategies have been promoted to address this issue. The use of pre-oriented functional group at strategic positions with respect to the imine bonds improves their chemical stability [58]. For example, amine or hydroxyl groups positioned 1,2 on a benzene ring can form cyclized imine bonds leading to higher strength displaying improved chemical stability (Scheme 16.1) [45, 60–64].

Banerjee and co-workers have profitably engaged the keto-enol tautomerism-assisted stabilization of the Schiff bonds in various COF syntheses [41, 48, 65]. Their findings have served in developing several imine COFs; they even pointed out that the keto-form is not easily reversible to the enol form (Scheme 16.1d). This makes the final COFs formed with a keto-form substantially stable [65]. Likewise, Jiang and co-workers [66] demonstrated that the substitution of electron-rich methoxy groups in the imine-forming aromatic rings could lead to exceptional chemical stability. Such COFs withstand exposure to acids, bases, boiling solvents, water, and electrolytes [67, 68]. This makes them suitable for several energy applications. The methoxy groups strengthen the framework through increasing the electro-density [push–pull] in the plane and by forming favorable out-of-plane –C–H•••π interactions between the COF layers. Recently, a single monomer carrying both aldehyde and amine groups has been shown to undergo self-polycondensation leading to exceptionally stable imine-bonded COFs as inferred from PXRD and porosity measurements [69]. Most importantly, as mentioned earlier, the accessibility of a vast library of building units or monomers and their relatively easy scalability [at least in polycrystalline forms] makes them available for wide-

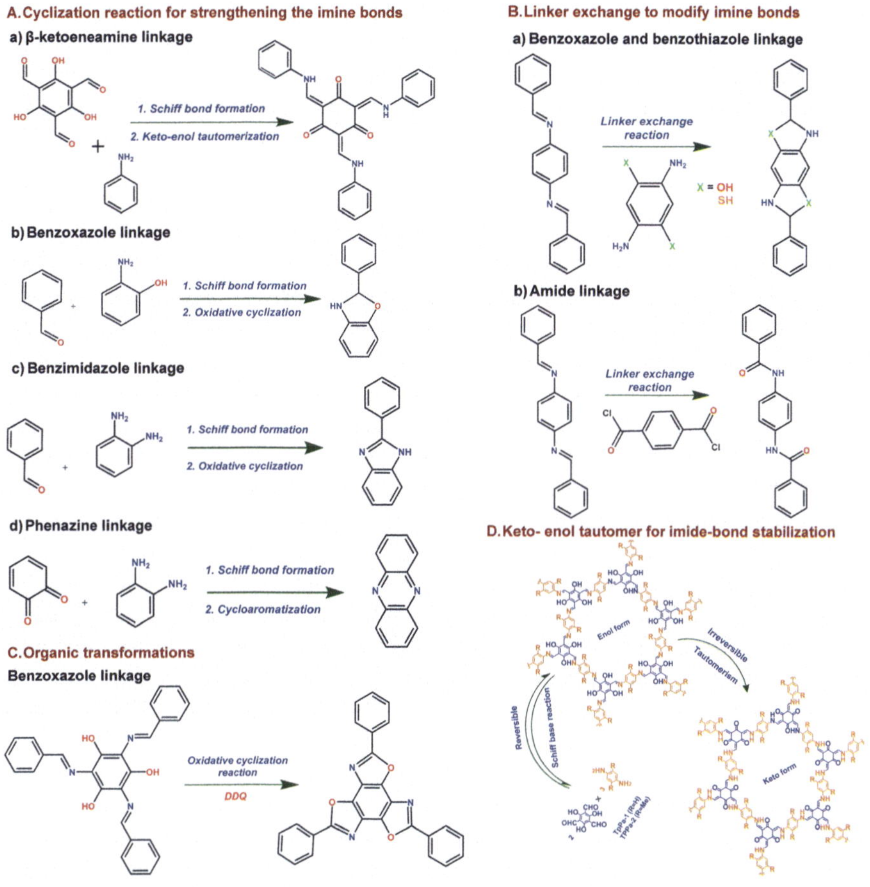

Scheme 16.1 Representative strategies used for the stabilization of otherwise hydrolyzable imine bonds in COF construction. More can be found in Ref. [45] (Adapted from Ref. 45)

range electrochemical employment [49, 56–58, 60, 62]. More recently, Samori and co-workers have compiled a choice of organic reactions and linker exchange that can be employed in converting imine bonds into more stable cyclized forms (Scheme 16.1) [45].

16.6 Imparting Nanoparticle Binding Units and Conductivity into COF

A major advantage of COF is its ability to incorporate lone-pair-rich heteroatoms such as nitrogen and sulfur with a stoichiometric control and exact periodicity. Such heteroatoms drastically improve metal binding [62, 67, 70–80] embedding heteroatoms with stoichiometric control and in an ordered fashion into carbonaceous support is not easy to achieve via classical high-temperature pyrolysis. However, this lower degree of conjugation in a COF makes them much less electronically conducting than carbon supports such as graphite, graphene, or CNT. To overcome this as a practical solution, (i) highly conducting carbon is being added to the COF as an additive to realize composites with high synergistic conductivity and having favorable guest-binding/redox-active heteroatoms supplied by the COF [81–87]. (ii) Improving the conjugation-assisted conductivity of the COF, predefined or pre-oriented polymerizing groups such as 1,2-substituted amines or aldehydes are incorporated to make cyclized imine rings, which improve conjugation and minimize defects. (iii) Inclusion of thiophene and phthalocyanine type of units which are known to have intrinsic molecular conductivity [56, 88–92]. Alternatively, linear conducting polymers have been incorporated into COF pores to enhance its overall conductivity [93–95]. In a fascinating article Jiang and co-workers showed how pyrene tetraldehyde can be reacted with linear dialkylcyanide linkers to form a 2D COF with periodically dispositioned pyrene knots connected exclusively by sp^2-carbon containing C=C bonds, yielding a semi-conducting COF with a 1.9 eV bandgap. This construct shows higher conjugation than imine and propagates as a superior electronic conductor when oxidized with iodine (COF with a conductivity of 6.1×10^{-14} S/cm becomes highly conducting when oxidized by iodine, 7.1×10^{-2} S/cm). Notably, the iodine-enhanced conductivity of this COF was superior over the discrete monomeric model compounds and sp^2-C containing 1D polymers. Interestingly, the iodine-doped sp^2-C COF behaves as a bulk magnet. ESR studies prove that iodine doping generates charge carriers that maintain the spin degree of freedom. It exhibits a spin concentration of 0.7 per pyrene and a Weiss temperature of 8 K. The COF exhibits a ferromagnetic transition below 10 K [6]. Following this, the idea of combining acetonitrile comprising monomers with aldehydes to form C=C bonds that enable the polymerization is becoming the most promising method for generating highly conjugated semi-conducting COFs with interesting optical, photocatalytic, and even chiral properties [96–98].

Another advantage of COF comes from the ability to introduce *salt-mediated* hierarchical porosity. Arne Thomas and co-workers have a simple approach to introducing macroporosity into COF without altering its intrinsic porosity [99]. The protonatable amine monomer is made into a salt suspension with the para-toluene sulfonic acid (p-TSA). Now, this salt solution is reacted with the cross-linking trialdehyde. The resulting COF has the para-sulfonic acid trapped in the inter-polymer spaces acting as a macropore template (Fig. 16.14). The ready solubility of the p-TSA in polar organic solvents enables their removal via a Soxhlet wash leading to uniform macropores on the COF. These macropores are visible in SEM, but, as expected, do not contribute to the microporosity measured from gas sorption. In some cases, this can lower the atomic-level crystallinity of the COF, but can drastically enhance the mass-transfer capabilities of the COF. This approach has visibly shown enhancement in OER kinetics (discussed later) and is bound to have impact in other guest mobility-controlled processes. In another embodiment, the COF has been composited with rGO and made into lightweight aerogels

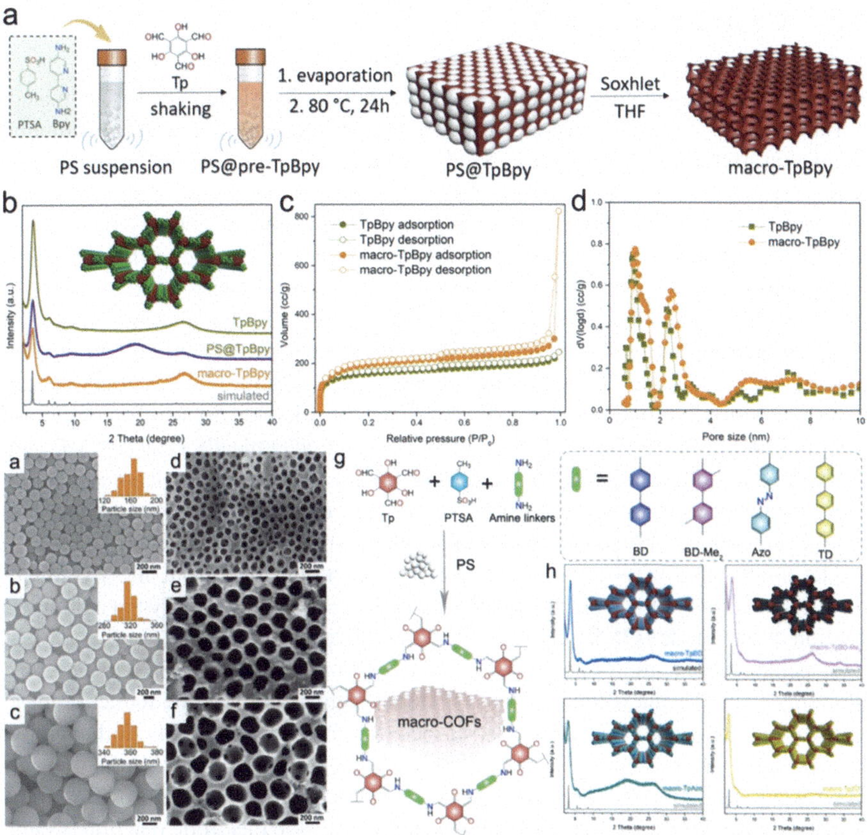

Fig. 16.14 Synthesis and characterization of macro–microporous COF. (**a**) Schematic representation of macro-TpBpy fabrication in the presence of PSs. (**b**) PXRD pattern of TpBpy, PS@TpBpy, and macro-TpBpy showing the preservation of crystallinity in macro-TpBpy with and after removing the PSs. (**c**) N_2 adsorption–desorption isotherms of the TpBpy and macro-TpBpy. (**d**) Pore size distribution of the TpBpy and macro-TpBpy, calculated from the adsorption branch of isotherms by the QSDFT model. SEM images of PSs and macro-TpBpy with macropore sizes of (**a**, **d**) ~160 nm, (**b**, **e**) ~320 nm, and (**c**, **f**) ~360 nm. (**g**) General synthesis of macro-COFs from different amines (BD, BD-Me_2, azo, and TD) in the presence of PSs. (**h**) The corresponding PXRD patterns of macro-TpBD, macro-TpBD-Me_2, macro-TpAzo, and macro-TpTD, and their respective simulated structures. Inset: Space-filling packing models of respective COFs (Copyright © 2020, American Chemical Society)

[83]. Visualizing COF as lightweight tunable support can have many advantages even in futuristic defense and energy applications [76, 100–105]; however, numerous advances are required in the appropriate designing of the COF-derived composites [104, 106].

16.7 Concepts in HER

Oxygen and hydrogen gas manufacturing in the US market size in 2021 is estimated at 8.6 billion USD, while the UK produces 100,000 tonnes/year [107]. Ammonia production in the Al-Jubail plant in Saudi Arabia alone consumes 1300 tonnes/day of H_2. Hydrogen is produced in multiple ways: natural gas reforming/gasification, electrolysis, renewable liquid reforming, and fermentation. Energy and cost-effective alternates are being developed. High-temperature water splitting: High temperatures generated by solar concentrators or nuclear reactors drive chemical reactions that split water to produce hydrogen. Photobiological water splitting: Microbes, such as green algae, consume water in the presence of sunlight, producing hydrogen as a byproduct. Photoelectrochemical water splitting: Photoelectrochemical systems produce hydrogen from water using special semiconductors and energy from sunlight [108, 109]. In the heart of all these are electrochemically catalyzed oxygen and hydrogen production [110]. In the pursuit of carbon-free energy, electrical energy is promisingly used to electrolyze water to hydrogen, accounting for 4% of the world's hydrogen production compared to more than 90% driven by thermal energy [111].

The splitting process consists of two half-reactions: one from the hydrogen evolution reaction at the cathode (**HER**: $2H^+ + 2e^- = H_2$) and the oxygen evolution reaction at the anode (**OER**: $2H_2O = O_2 + 4H^+ + 4e^-$) [112, 113].

In acidic media, the water reduces from hydronium ions (H_3O^+) to gaseous dihydrogen (H_2) at the cathode. The first step of this reaction is the reduction of a proton on an active site of the catalyst surface (Volmer step, Eq. 16.1), followed by the evolution of molecular H_2, either through a second proton/electron transfer (Heyrovsky step, Eq. 16.2) or through the recombination of two adsorbed protons (Tafel step, Eq. 16.3). By thermodynamics, this multi-step electrode reaction occurs at the potential of the reference hydrogen electrode [RHE, 0 V]. But there is always a slight overpotential associated with the reactions.

$$[\text{Volmer}]\ H_3O^+ + e^- + * = H^* + H_2O$$

$$[\text{Heyrovsky}]\ H^* + H_3O^+ + e^- = H_2 + H_2O$$

$$[\text{Tafel}]\ 2H^* = H_2$$

* denotes an active site on the catalyst surface, and H^* a hydrogen atom adsorbed on an active site. Any one of these three steps can limit the process, and it becomes the rate-determining one.

The HER kinetics depends on the electrode material; for instance, a mercury (Hg) electrode exhibits slow kinetics, while the HER on platinum is one of the fastest electrocatalytic processes known [114]. Other parameters such as the electrolyte or the structure and morphology of the electrode also impact the HER kinetics

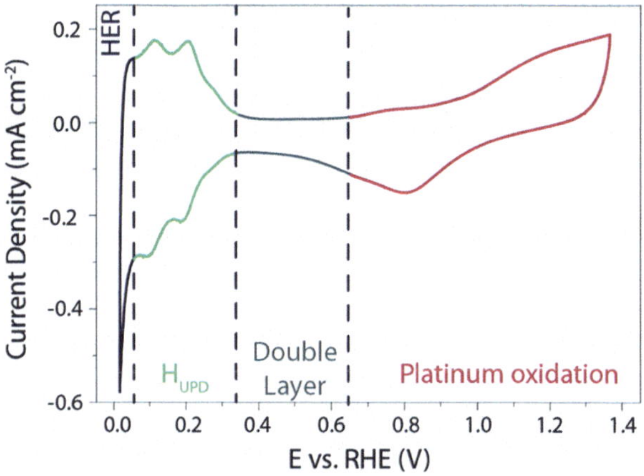

Fig. 16.15 Cyclic voltammogram recorded on a $Pt_{(pc)}$ electrode at 50 mV s^{-1} in a 0.5 M H_2SO_4 solution degassed with argon. H_{UPD} = hydrogen underpotential deposition. The large envelope is related to the H_{UPD} on Pt[111] facets, while the peaks at ~0.10 V vs. RHE and ~0.20 V vs. RHE are related to H_{UPD} on the Pt[110] and the Pt[100] sites, respectively (Copyright © 2019, Royal Society of Chemistry)

[112]. Notably, the kinetics is subject to variations of parameters such as the nature of the electrolyte or the crystalline nature and orientation of the electrode (single-crystal, polycrystalline, amorphous, etc.). Typically, crystalline materials could manifest higher charge and mass transfer, and this is where COFs could have a role.

Figure 16.15 displays a classical cyclic voltammogram (CV) of a polycrystalline platinum ($Pt_{(pc)}$) electrode recorded in acidic media. There are four distinct regions in this CV. At potentials >0.65 V vs. RHE, the surface of Pt is oxidized/reduced forming either Pt–OH or Pt-oxide with the specific adsorption of anions on the surface of the platinum. The second region (0.35 and 0.65 V vs. RHE) is referred to as the "double layer region" with predominant non-chemisorptive surface charge storage. The third region (0.05 and 0.35 V vs. RHE) is referred to as the hydrogen underpotential deposition (H_{UPD}) region. Here the discharge of protons on the Pt surface occurs following the reaction: $H^+ + e^- + Pt^* = Pt - H_{UPD}$. This phenomenon may involve cation adsorption as well as OH desorption [112]. Finally, at more negative potentials, H_2 is evolved. Notably, the H_{UPD} is not unique to platinum and is known to occur on other metallic surfaces [112]. The kinetics of the HER on platinum electrodes is ultrafast, but the mass transport of the generated H_2 is relatively slow. Thus, the faradaic current generated from the redox reactions is often restricted by the latter, and this issue remains even when a rotating-disk electrode (RDE) apparatus is employed. This limitation precludes a fine understanding of the HER mechanisms under acidic conditions [112].

16.8 Concepts in Oxygen Evolution Reaction

The electrocatalytic production of molecular fuels like hydrogen from water or hydrocarbons from CO_2 is based on reduction reactions which, in turn, require an electron-donating counter-reaction [115, 116]. In this context, the electrocatalytic oxidation of water to molecular oxygen, the oxygen evolution reaction [OER], is the most promising candidate concerning availability and sustainability [117, 118]. While the mechanism of HER is well understood, the mechanism of OER is still not completely clear. But even the simplest thermodynamics-based mechanisms [not considering the kinetics] involve multi-step proton–electron transfer, which imposes significantly higher overall *overpotential* for evolving oxygen from water [119]. Overpotential is the excess potential needed over the thermodynamic potential of 1.23 V. Typically, catalysts are required to bring down this overpotential, and the most effective catalysts are quite expensive precious metals. Additionally, the inherent high electrode potentials during the OER, compared to HER, are demanding for the stability of the catalyst.

16.9 Acidic and Alkaline Polymer Catalyst for OER

The OER mechanism is quite different depending on the pH of the medium and the applied electrical potential will alter the path. Low-temperature electrolyzers can be subdivided into [acidic] polymer electrolyte water electrolyzers (PEWEs) and alkaline water electrolyzers (AWEs) [115, 120]. PEWE employs a solid polymer electrolyte, which conducts only the positive ions such as protons and creates an acidic local environment. In PEWEs, the cathodic hydrogen evolution reaction occurs with fast kinetics and achieves high-voltage efficiencies at high current densities. Furthermore, they can produce pure hydrogen at relatively high pressures [150 bar] [121], offering the possibilities of storing hydrogen directly without or only with small further mechanical compression [122]. Only a few electrodes are stable under the typical operating conditions of a PEWE (high acidity and potential). This limits the anodic and cathodic reactions to be catalyzed by only a few noble metal-based compounds such as Pt, Ru, and Ir. In contrast, the main advantage of AWEs is the possibility of using alternative catalysts to noble metals and cheap flow fields since several materials present adequate stability when in contact with an alkaline electrolyte. Furthermore, the alkaline medium allows more favorable oxygen electrocatalysis at the anode side compared to the acidic one. However, while OER catalysis profits from alkaline conditions, the cathode reaction, the hydrogen evolution reaction (HER), is usually impeded in an alkaline environment [115]. This circumstance lowers the possible gain on the overall system level. Thus, to profit from the numerous advantages of PEM electrolyzers and facilitate their large-scale application, the noble metal amount required in the anode catalyst needs to be drastically reduced. For this purpose, an in-depth fundamental understanding of the OER mechanism and the applied catalyst materials in an acidic environment is required [115–118].

16.10 Analyzing OER Mechanism: A Thermodynamic Perspective

Several different but related mechanisms have been proposed. Rossmeisl et al. first proposed one of the most recognized routes based on density functional theory (DFT), which adopts a concerted proton–electron transfer pathway.

$$H_2O[l] + * \rightleftharpoons H^+ + HO * + e^- \qquad \Delta G1. \tag{16.1}$$

$$HO* \rightleftharpoons H^+ + O * + e^- \qquad \Delta G2. \tag{16.2}$$

$$O * + H_2O[l] \rightleftharpoons H^+ + HOO * + e^- \qquad \Delta G3. \tag{16.3}$$

$$HOO* \rightleftharpoons H^+ + O_2[g] + e^- + * \qquad \Delta G4. \tag{16.4}$$

$$2H_2O[l] \rightleftharpoons 4H^+ + O_2[g] + 4e^- \qquad \Delta G_{\mathrm{overall}} = 4.92\ \mathrm{eV}. \tag{16.5}$$

* indicates activated at catalyst sites.

As shown in the steps above (Eqs. 16.1–16.5 and in Fig. 16.16), firstly, a water molecule is adsorbed on the active site and it dissociates a proton to form HO*, followed by the removal of the second proton to yield O*. After that, another water molecule performs a nucleophilic attack on the oxygen species (O*), leading to the formation of HOO*, desorbed O_2, and a released active site in sequence. The overall oxygen evolution half-reaction in acid is shown in Eq. (16.5) [124, 125]. The theoretical potential of oxygen evolution is 1.23 V under standard conditions (Fig. 16.16b), with the Gibbs free energy (ΔG) of the overall reaction calculated to be 4.92 eV under equilibrium conditions.

16.11 OER Mechanism Based on Kinetic Measurements: Challenges

Various reaction mechanisms have been proposed for the heterogeneous catalyst-assisted OER. They are derived from either kinetic studies or DFT calculations (Fig. 16.16). However, none of the OER mechanisms proposed for heterogeneous catalysts have been fully validated based on experimental results. The kinetic models advised by Bockris for a variety of different conceivable OER mechanisms are shown in Fig. 16.16a I–III [126]. Importantly, he pointed out that the Tafel slope obtained from an electrocatalytic measurement is determined by the actual rate-determining step (rds) within a specific reaction mechanism [126]. This analysis was

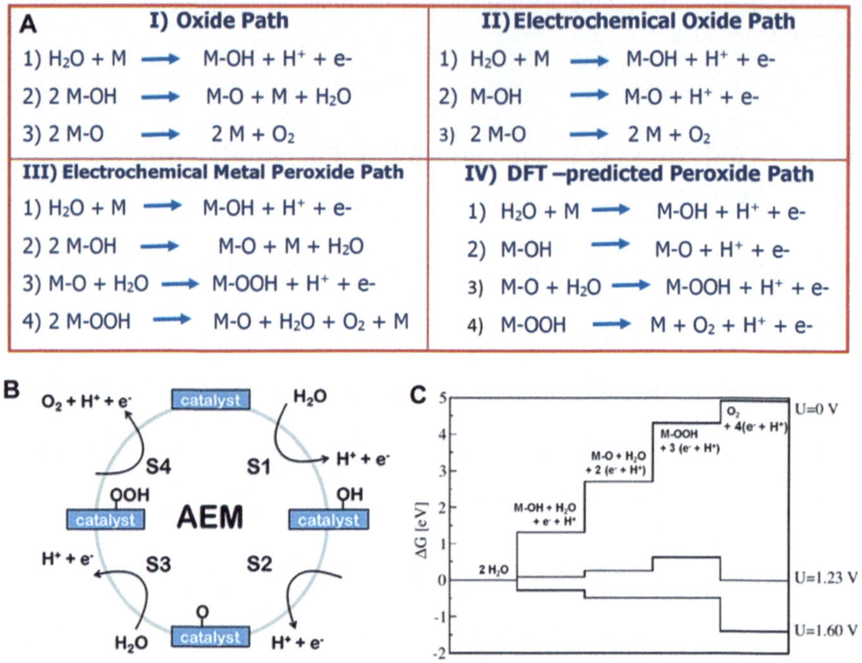

Fig. 16.16 (**a**) The oxygen evolution reaction mechanism in acidic environments—current state of knowledge (Adapted from Ref. 123). (**b**) Primary pathway of the adsorbate evolution mechanism proposed by Rossmeisl et al. applicable to any heterogeneously catalyzed OER (Copyright © 2020, Royal Society of Chemistry). (**c**) The free energies of the intermediates on O^-*-covered RuO_2 [112] at three different potentials ($U = 0$, $U = 1.23$, and $U = 1.60$ V vs. SHE) are depicted. At the equilibrium potential ($U = 1.23$ V), the reaction steps are uphill in free energy. At 1.60 V the OER becomes thermodynamically feasible ($\Delta_R G \leq 0$ eV under the considered conditions) (Copyright @ 2017, John Wiley & Sons, Inc.)

based on the assumption that one step in each reaction mechanism is the rds and that only the reactant of the rds can build up a high surface coverage [concentration] [126]. However, it is reasoned that the rds cannot unambiguously be identified based on the Tafel slope alone. Different rds in different mechanisms can result in similar Tafel slopes. Moreover, the Tafel slope itself is a somewhat unspecific measure that can be altered by factors besides the electrocatalytic reaction. For instance, a semiconducting oxide layer, located between catalyst and substrate [current collector], can increase the Tafel slope [127]. Thus, precise knowledge about the electrode's material properties, especially with respect to conductivity and possible buried interfaces, is required to obtain valid mechanistic insights from a Tafel slope analysis.

Rossmeisl et al., using DFT calculations, demonstrated a mechanism to explain OER (see Fig. 16.16a IV) [128, 129]. The Gibbs free energy of every reaction (ΔRGx) for every step in the mechanism is measured as a function of the electrode potential (Fig. 16.16b). Only the reactions involving an electron exchange with the electrode are considered as they depend directly on the electrode potential. As a thermodynamic requirement, (i) ΔRGx of every reaction step must be compulsorily ≤ 0 J mol^{-1} (compare Fig. 16.16b). (ii) Although the sum of ΔRGx of the individual reaction steps has to equal ΔRG of the overall reaction (water's oxidation to O_2), each reaction step can have a different ΔRGx within this limit. (iii) For every step involving an electron transfer to the catalyst, the ΔRGx will change equally with electrode potential. (iv) The reaction step, with the largest ΔRGx (at the reference potential of 0 V SHE), demands the highest electrode potential to become downhill in ΔRG (step 3 in Fig. 16.16b requires 1.60 V). This becomes the potential determining step (PDS) [123].

To meet the condition of ΔRGx ≤ 0 J mol^{-1}, the reaction steps might require a higher electrode potential than the standard potential of the overall reaction ($E^0 = 1.23$ V). The ΔRGx of the individual reaction steps is decided by the adsorption energy of the intermediates, making it a function of the catalyst. Thus, the overpotential required to facilitate the overall reaction is dependent on the nature of the catalyst. The reactivity trends of different catalysts are described well by this approach. Hence, the proposed mechanism appears reasonable, although the model is based on thermodynamics alone and does not account for any kinetic barrier. Also, it points out that the binding strength of the oxygen-containing species with the catalyst surface largely determines the overpotential. Thus, following the Sabatier principle, the best catalyst displaying the minimum overpotential binds oxygen on its surface neither too strongly nor too weakly [116, 130]. For example, in the case of RuO_2, particularly known for high OER activity [129], the overpotential 0.37 V can be lowered by doping cobalt or nickel [128].

From Fig. 16.16b, the largest free energy difference (1.60 eV) is found for the third step, which is splitting the second water molecule. We will use the minimum potential at which all reaction steps become downhill in free energy as the measure of the electro-catalytic activity of the different surfaces. Notice that the lower limit is defined by the equilibrium potential for the overall reaction, which in this case is 1.23 V. Figure 16.16c illustrates that the first three reaction steps are endothermic for the equilibrium potential, and that an overpotential of 0.37 V is necessary to make all steps downhill in free energy. In order to minimize the potential which is required to trigger the reaction thermodynamically on a given catalyst, ΔRGx of the reactions in mechanism [IV] has to be optimized by tuning the binding energy of the M–OH, M–O, and M–OOH intermediates. Mixed-metal catalyst might be effective in this [123].

16.12 Concepts in ORR: Different Pathways with Varying Thermodynamics

Another important reaction inspired by biological systems includes the oxygen reduction reaction (**ORR**) [131–142], which involves the reduction of oxygen to water in alkaline or acidic medium. ΔE value of ORR is different at different pH values.

The ORR is alkaline media versus reversible hydrogen electrode (RHE) at 25° (pH = 14, $(OH^-_{(aq)}) = 1$ mol dm^{-3}), and the thermodynamic potentials at standard conditions for this four-electron process are presented as follows [133–138]:

$$O_2 + 2H_2O + 4e^- \rightarrow 4OH^- \qquad E^0 = 0.401\ \text{V}. \tag{16.6}$$

The pathway involving the hydrogen peroxide is expressed as follows [141, 142]:

$$O_2 + H_2O + 2e^- \rightarrow HO_2^- + OH^- \qquad E_0 = 0.080\ \text{V}. \tag{16.7}$$

$$HO_2^- + H_2O + 2e^- \rightarrow 3OH^- \qquad E_0 = 0.880\ \text{V}. \tag{16.8}$$

The ORR is acidic media versus RHE at 25° (at pH = 0, $(H^+_{(aq)}) = 1$ mol dm^{-3}), and its thermodynamic potentials at standard conditions are presented as follows [139]:

Direct four-electron reduction:

$$O_2 + 4H^+ + 4e^- \rightarrow 2H_2O \qquad E_0 = 1.230\ \text{V}. \tag{16.9}$$

Indirect reduction:

$$O_2 + 2H^+ + 2e^- \rightarrow H_2O_2 \qquad E_0 = 0.680\ \text{V}. \tag{16.10}$$

$$H_2O_2 + 2H^+ + 2e^- \rightarrow 2H_2O \qquad E_0 = 1.776\ \text{V}. \tag{16.11}$$

Or chemical decomposition:

$$2H_2O_2 \rightarrow 2H_2O + O_2 \qquad E_0 = 1.776\ \text{V}. \tag{16.12}$$

Thus, hydrogen reduction is a two-electron process, while the oxygen generation requires four electrons; for these reactions to proceed under ambient conditions, they require catalysts. Semiconductors with a suitable bandgap that match the energy needed for the redox reactions catalyze light-driven water splitting. Typically, the catalyst absorbs a photon and splits exciton into electron and hole; thus, the generated electron aids to reduce water into hydrogen, while the hole sets off oxidation of water to oxygen. The former reduction step requires the catalyst to have a high enough lowest unoccupied molecular orbital (LUMO) level (> −4.02 eV) and the highest occupied molecular orbital (HOMO) level (< −5.25 eV) needs to lie low enough for water oxidation [30]. In terms of the experimentally controlled potential, the conduction band needs to lie at a potential less than 0 V vs. NHE (H^+/H_2) for the water-reduction reaction. The valence band needs to have a potential of more than 1.23 V that corresponds to light of 1008 nm. Most of the reported photocatalysts work under ultraviolet (UV) irradiation. But, UV (<400 nm) accounts for only 4% of the total solar energy, while visible light (400–800 nm) and infrared light (>800 nm) envelop 53% and 43% of the total solar spectrum, respectively [140]. Henceforth, even a less efficient photocatalyst

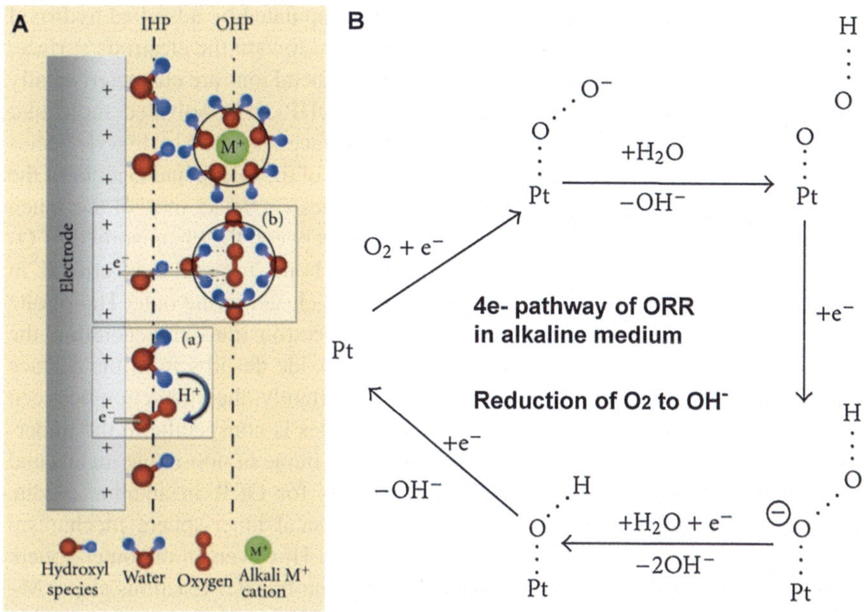

Fig. 16.17 (**a**) Schematic illustration of the double-layer structure during ORR in alkaline media. Insets (**a**, **b**) illustrate the inner- and outer-sphere electron transfer processes. (**b**) Electrocatalytic inner-sphere electron transfer mechanism for ORR on Pt in alkaline media assuming end-on adsorption of O_2 on a single Pt active site [141] (Copyright @ 2012, Hindawi Publishing Corporation)

that absorbs visible light is more beneficial than a more efficient one absorbing solely UV light. For efficient photocatalysis, the semiconductor needs to be small in size and should have minimal heterogeneity and defects to minimize electron-hole recombinations. Besides, an active site is required in conjugation with the semiconductor to provide a place for water to reside and react with the electron/hole. Practical support with assembled donor–acceptor backbone can play an essential role in absorbing the photogenerated holes, a hole scavenger.

Electrocatalytic ORR is performed in an acidic medium using Pt electrodes [141]. This is because it involves an inner-sphere electron transfer process that exhibits a strong dependence on the nature of the electrode surface and involves strong chemisorption of molecular O_2 [with or without bond breaking] on oxide-free Pt active sites followed by $4e^-/4H^+$ transfer. This inner-sphere mechanism depicted in Fig. 16.17a (inset (a)) is common to both acidic and alkaline media, wherein all the reaction intermediates remain adsorbed on the active site until the final product is desorbed into the bulk electrolyte. The non-noble metals do not favor the adsorption of molecular oxygen and several other intermediates involved in the inner-sphere mechanism; this limits their application in acidic media.

Shown in Fig. 16.17 (inset b) is the outer-sphere electron transfer mechanism that appears to be unique to the alkaline conditions. During water splitting in alkaline media, water molecules serve both as the solvent and the proton source for ORR. In alkaline media, the inner Helmholtz plane (IHP) is populated by adsorbed hydroxyl species, solvent water with their oxygen atom oriented toward the electrode surface, and chemisorbed O_2. Meanwhile, the solvated alkali metal ions are characteristically expected to populate the outer Helmholtz plane (OHP). The solvated molecular oxygen, a cluster $O_2{\cdot}(H_2O)_n$, can interact with the surface-adsorbed hydroxyl species (OH_{ads}) through hydrogen bonds between the H atom of the OH_{ads} and O atom of the solvent water. The associated hydrogen bond energies (< 35 kJ mol^{-1}) are much lower than the energy of the covalent bond formed between the chemisorbed of O_2 and Pt (> 300 kJ mol^{-1}) [142]. This hydrogen bond formation is crucial in stabilizing the solvated molecular oxygen $O_2{\cdot}(H_2O)_n$ cluster in the outer Helmholtz plane (OHP) and promotes an outer-sphere $2e^-$ electron transfer generating the superoxide species. This product peroxide or superoxide desorbs from the surface and is eventually detected at the ring electrode. Importantly, the interaction between the $O_2{\cdot}(H_2O)_n$ cluster and the surface hydroxyl species is nonspecific to the underlying metal electrode. This opens the door for a wide range of non-noble metals and their oxides to be employed as electrode materials for ORR in alkaline media. Further, the alkaline medium can support the classical inner-sphere mechanism when Pt type of electrocatalyst is used (Fig. 16.17b). However, in the outer-sphere mechanism (Fig. 16.17a) [a], one of the multi-electron transfer reactions steps, M–OH + $(O_2{\cdot}(H_2O)_n)_{aq}$ + e $- \rightarrow$ M–OH + $(HOO\bullet)_{ads}$ + OH^- + $(H_2O)_{n-1}$, involved in the peroxide formation, requires transfer [or tunneling] from the electrode surface across a thin oxide film and at least one layer of solvation shell to solvated O_2. Rendering of the non-noble metal catalysts in nanostructured forms can improve this e- transfer/tunneling. The non-noble metals operating predominantly on the outer-sphere mechanism lead to hydrogen peroxide formation. Instead, to enable water to undergo efficient $4e^-$ conversion of O_2, it is essential to facilitate direct adsorption of desolvated molecular O_2 on the OH^- free active site and prevent the precipitous reaction of solvated molecular O_2 with the OH^- covered active site. This makes ORR with non-noble metals highly challenging.

First-row transition-metal ions stabilized on the surface of nitrogen-functionalized graphite and phthalocyanine complexes as a self-standing catalyst have been studied intensely [143–147]. The environment of the active site on the carbon support (edge vs. basal plane) [148], coordination number (Fe–N_4 vs. non-Fe–N_4 environment) [149], and chemical identity of the nitrogen functional groups (pyridinic, pyrrolic, and quaternary) [135, 150] largely control the activity. ORR has been conducted on sites with surface nitrogen groups devoid of metal ion centers [151, 152]. Still, it is widely accepted that the transition-metal ion coordinated to four nitrogen groups (Me–N_4) on graphitic surfaces constitutes the active site [148, 149, 153], wherein advantageously, the chelation primarily serves to prevent the metal center from passivation/corrosion under electrochemical conditions [154]. This is where the designability of COFs with tailor-made chelating binding sites for transition metals becomes advantageous. One can exploit their pore shapes and sizes to define the geometry of the nanoclusters. As a bonus, the nano-confinement could also alter the facets that get exposed, impacting the ratio of hydroxide adsorbed vs. hydroxide-free sites, which is key to the electron transfer process as well as in preventing unwanted abrupt reactions between the adsorbed reactive species. And rendering these electrocatalysts as reactive nanoparticles in COF pores will enhance the kinetics of the reactions preventing the undesired precipitation of solvated oxygen and hydroxyl species.

16.13 COF as Active Porous Support to Improve Catalyst Activity

As can be seen, all these three important processes HER, OER, and ORR require a suitable catalyst that satisfies Sabatier's principle [155, 156]—here you can find detailed tables on performances of COF-derived electrocatalysts and electrodes for supercaps/batteries. Though the non-noble metal-based catalysts are desirable for realizing the large-scale deployment of electrosplitting of water, their sluggish kinetics is a bottleneck. Porous supports become ideal molecular platforms to contain these nanocatalysts and provide rapid diffusion for active species [157]. Several porous supports ranging from silica, silicates, carbons, mofs, and metal oxides are being employed for this purpose. Or in some cases, the catalysts such as oxides, sulfides, and nitrides themselves have been rendered porous [158]; importantly, structuring the catalyst to expose different active facets is critical to assimilating high activity. In this class, COF, with its tailorable flexible framework, qualifies as a designer molecular sponge for water-splitting catalyst hosting. Different facets of the catalytic nanoparticles or nanostructures can be exposed when confined in the COF pores [67, 70–80, 159, 160].

For photoelectrochemical water splitting, one can utilize the COF pores to host and orient semiconducting nanoparticles (nPs) to confine the water and other reactive intermediates proximal to the active sites of the nPs. When constructed with suitable donor–acceptor modules [161, 162], it can do the scavenger (hole/electron) role. COF nanopores (10 to 70 Å) are suited to grow the nPs to sizes limited by its pore volume—capping agent-free method to grow nanoparticles. Their in-plane covalently linked donor–acceptor appendages offer super heterojunction for splitting excitons into electron and hole, while the donor and acceptor π-columns enable the transport of holes and electrons, respectively. For example, semiconducting metal oxide, metal nitride, sulfide, and carbide trapped in these COF pores can interact synergistically with the COF through the lone-pair-rich atoms and π-electron clouds leading to lowered bandgaps which can lead to a favorable flow of electrons and holes between the active site and the water-generated intermediates. In 2014, Lotsch and co-workers reported a triazine-hydrazone mesoporous COF with a bandgap of 2.8 eV, which was employed as efficient support for hydrogen evolution. They could achieve an HER rate of 230 μmol h^{-1} g^{-1} in the presence of sodium ascorbate as the sacrificial donor and a rate of 1.97 mmol h^{-1} g^{-1} for a system with triethanolamine (TEOA) as the sacrificial donor [163]. Jiang and co-workers reported a series of low-bandgap semiconducting COFs containing pyrene, vinylene, and azine modules with bandgaps in the range of 1.69–2.7 eV, all capable of showing impressive chemosensing and HER [30, 164–166]. Photo-to-chemical energy conversion requires a systematic and complex molecular design to manage consecutive photochemical processes, including energy harvesting, exciton migration, electron transfer, charge separation, and charge transport. Their sp^2-carbon-conjugated pyrene-based COF manages all this through efficient push–pull electronics. For the HER [164], they employ Pt as the active metal and it synergistically with the COF generates 2.12 mmol h^{-1} g^{-1} upon irradiation with light ≥420 nm and 1.24 mmol h^{-1} g^{-1} upon irradiation with light ≥495 nm (Fig. 16.18). They use $AgNO_3$ or triethanolamine as the hole scavenger. A valuable comparative analysis of the performance of the COF-based HER photo-/electrocatalysts is presented by Jiang and co-workers (Table 16.1) [164]. Typically, having chemically distinguishable π-rich nodes connected by e-deficient linkers in the polymeric structure sets up an efficient push–pull mechanism enabling light absorption–emission in the UV–Vis region [168]. However, an sp^2-carbon or a conjugated C–C bond is far superior in rendering a less defective planar system and in photogeneration of free radicals at the nodes [164, 167].

16.14 Visible Light HER by Sulfone-Functionalized COF [169]

The above example represents the role of COF as support for metal-based electrocatalyst; Cooper and co-workers [169] have reported a metal-free covalent organic framework (COF) based on a benzobis(benzothiophene sulfone) moiety exhibiting photocatalytic HER with quantum efficiency rivaling benchmark amorphous or semicrystalline counterparts (Fig. 16.19). The COF is stable under long-term visible irradiation and shows steady photochemical hydrogen evolution with a sacrificial electron donor for at least 50 h. The inclusion of these predefined sulfone modules helps them tune the all-organic framework to behave hydrophilic with 3.2 nm mesopores. These pores allow the framework to be dye-sensitized, leading to a further 61% enhancement in the hydrogen evolution rate up to 16.3 mmol g^{-1} h^{-1} [169], which is remarkable. The hydrophobicity could be playing a role in making the COF support non-interfering with the water and the reactive intermediates formed during the HER process at the dye. Meanwhile, the high activity suggests that the hydrophobicity of the COF does not hamper the electrode–electrolyte wettability or the reaction kinetics. The inherent porosity of the support favors mass and charge transfer.

Molecular-level tuning of building units and translating them as properties through the COF is well pictured by the approaches wherein increasing the number of pyridinic nitrogen in the building unit and thereby in the backbone of the COF, a systematic increase in the hydrogen evolution capability was observed.

The COF systems, the N0, N1, N2, and N3–COF (Fig. 16.20) [170], in about 8 h in a photocatalytic reaction, produced 23, 90, 438, and 1703 μmol h^{-1} g^{-1}, respectively. The hydrogen evolution experiments were performed by taking a suspension

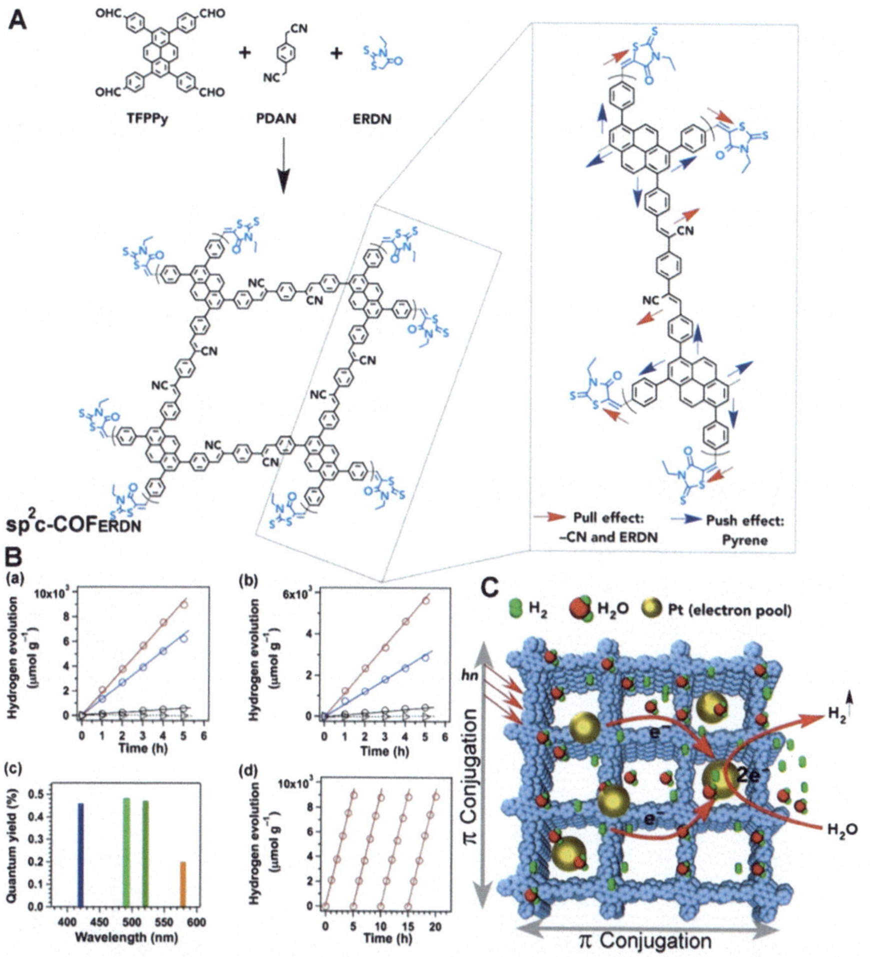

Fig. 16.18 (**a**) Schematic representation of the synthesis of sp^{2}c-COFERDN with electron-deficient ERDN end groups. Inset: electron donor–acceptor pull–push effects on the 2D skeletons. (**b**) Photocatalytic hydrogen production monitored over 5 h with sp^{2}c-COF (blue circles), sp^{2}c-COF$_{ERDN}$ (red circles), sp^{2}c-CMP (black circles), and imine-linked pyrene COF (triangles) as photocatalysts under irradiation with (a) wavelengths $\geq$ 420 nm and (b) $\geq$ 495 nm. (c) Apparent quantum yields (AQYs) of sp^{2}c-COF$_{ERDN}$ under irradiation with monochromatic light at 420 nm (blue bar), 490 nm (green bar), 520 nm (deep green bar), and 578 nm (orange bar). (d) Stability of sp^{2}c-COF$_{ERDN}$ over four prolonged repeated photocatalytic operations under irradiation $\geq$ 420 nm. (**c**) Assembled light-driven interlocked system for hydrogen production from water based on a crystalline sp^2 carbon-conjugated framework and Pt water-reduction centers (Copyright @ 2019, Elsevier)

Table 16.1 Photocatalysis using COFs. Summary of photocatalytic performances and bandgap structures

Photocatalyst	Abs (nm)	Bandgap (eV)	H_2 evolution[a] (mmol h^{-1} g^{-1})	Irradiance	Sacrificial donor	Performance stability	Co-catalyst
TFPT-COF	300–600	2.8	1.9	≥420 nm	TEOA	pH sensitive	Pt
N_3-COF	200–475	2.6	1.7	≥420 nm	TEOA	Stable	Pt
TP-BDDA COF	300–525	2.31	0.3	≥395 nm	TEOA	Stable in water and HCl (3 M)	Pt
TpPa-2 COF	300–800	2.52	0.028	≥420 nm	Lactic acid	Stable	Pt
10 wt% TpPa-2 COF + 90 wt% CdS	300–800	–	3.7	≥420 nm	Lactic acid	Stable	Pt
N_2-COF	200–475	2.6–2.7	0.78	≥420 nm	TEOA	Stable	Co
PTP-COF	300–500	2.1	0.083	≥420 nm	TEOA	Stable	Pt
A-TEBPY-COF	300–600	1.94	0.098	≥420 nm	TEOA	Stable	Pt
BDT-ETTA COF[b]	350–550	2.47	–	Am 1.5	–	Stable	–
BDT-ETTA COF[c]	350–550	2.47	–	Am 1.5	–	Stable	Pt
FS-COF	300–670	1.85	16.3	≥420 nm	Ascorbic acid	Stable	Pt

[a]We do not recommend a direct use of the H_2 evolution rate for comparing the photocatalytic activity; this comparison may lead to a misunderstanding because of the great difference in light source, irradiance, composition, and content of the systems [167]
[b]Photoelectrochemical device using AM 1.5 without filter; this is different from the other photochemical systems. The BDT-ETTA COF thin-film (500 nm) photoelectrode reaches currents up to 4.3 μA cm^{-2} at 0.3 V vs. RHE
[c]The BDT-ETTA COF thin-film (100 nm) photoelectrode with Pt co-catalyst reaches currents up to 3.6 μA cm^{-2} at 0.4 V vs. RHE. No H_2 evolution rate was reported for this system

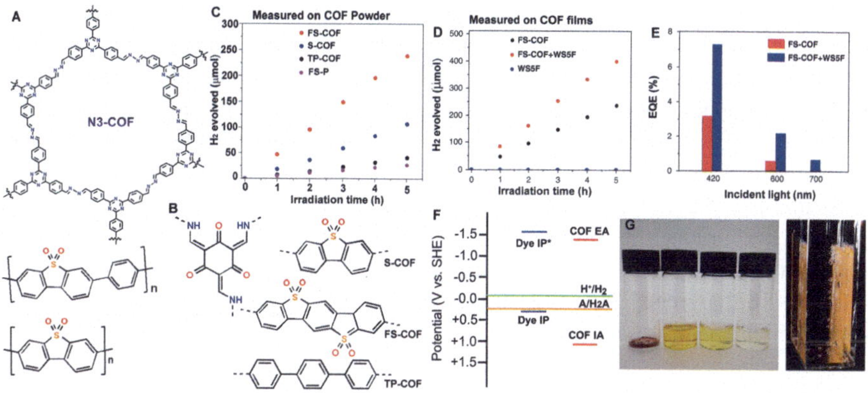

Fig. 16.19 (**a**) Chemical structures of COF photocatalysts **N$_3$-COF**35, **P7**16, and **P10**. (**b**) Chemical structures of the sulfone-COF photocatalysts reported by Cooper et al.: **S-COF**, **FS-COF**, and **TP-COF**. (**c**) Hydrogen evolution from water catalyzed by FS-COF. (**d**) Time course for photocatalytic H_2 production using visible light for **FS-COF**, a neat, near-infrared dye (WS5F) and a dye-sensitized COF (**FS-COF + WS5F**). (**e**) External quantum efficiencies at three different incident light wavelengths for **FS-COF** and **FS-COF + WS5F**. (**f**) Relative energy levels as calculated for ascorbic acid, **FS-COF**, and a near-infrared dye, WS5F; dashed green and orange lines indicate potentials for proton reduction and the two-hole oxidation of ascorbic acid in solution, respectively. (**g**) Photographs showing (left to right) solid **FS-COF** and colloidal dispersions in DMF, water, and acetone, respectively, and **FS-COF** film on glass-producing hydrogen (Adapted from Ref. 169)

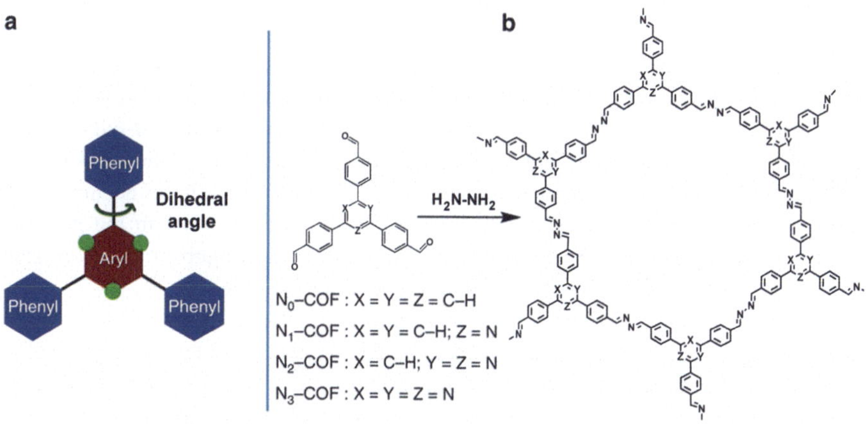

Fig. 16.20 (**a**) A tunable triphenylarene platform for photocatalytic hydrogen evolution. Replacement of "C–H" by "nitrogen atoms" at the green dots changes the angle between central aryl and peripheral phenyl rings, which leads to varied planarity in the platform. (**b**) Synthesis of N$_x$–COFs from N$_x$–aldehydes and hydrazine (Copyright © 2015, Vyas et al.)

of the COFs in phosphate-buffered saline (PBS) at pH 7 and irradiating with visible light (≥420 nm) at 25 °C. As a convenient method, the hexachloroplatinic acid was added for the in-situ formation of the platinum [Pt] co-catalyst, which reduces the HER overpotential. Also, triethanolamine (TEOA) was used as the sacrificial electron donor. The amounts are comparable to carbon nitride photocatalysts and outperform benchmark systems such as Pt-modified amorphous melon (720 μmol h^{-1} g^{-1}) [171], "g–C_3N_4" (840 μmol h^{-1} g^{-1}) [172], or crystalline poly(triazine imide) (864 μmol h^{-1} g^{-1}) [171]. One can decorate the COFs with transition-metal-binding pockets, such as porphyrin [173–178], phthalocyanine [83, 179–182], pyrene [183], and tetraazannulene [184], to ensure strong interactions and to prevent catalyst leaching.

16.15 Electrocatalysis Using COF with Transition Metals

Photocatalytic processes using COFs in the presence of co-catalysts (Pt) and sacrificial scavengers (TEA, $AgNO_3$) utilize the photogenerated electrons for the reductive half-reaction [185]. Few of them can realize the much more challenging reaction: photogenerated hole-involved water oxidation reaction for O_2 evolution. As the water oxidation reaction requires a four-electron-transfer process (see Eq. 16.5 in the OER concept section), which involves a O–O bond formation as one of its steps, this has a high overpotential and sluggish kinetics. It has been widely recognized as the key and bottleneck in artificial photosynthesis [186]. Yang and co-workers [187] developed a COF by incorporating 2,2′-bpy units within its framework, comparable to the COF reported for OER by Kurungot et al. [188]. The bpy units are well-known to bind to M^{2+} cations, and seizing this understanding, they loaded Cobalt[II] salts into the framework. They authenticated its structural closeness to Co–bpy complexes via XPS, and from near-edge X-ray absorption fine structure (NEXAFS), they could determine the oxidation of Co^{2+} to Co^{3+} under irradiation which happens when the Co^{2+} collects the holes accumulated during the photoirradiation ($\lambda \geq 420$ nm), while the $AgNO_3$ used in the process scavenges the electrons. This evolves 152 μmol/g/h of hydrogen at an optimized loading of 1 wt% of Co^{2+} but drops when the loading increases.

Numerous COF-derived catalysts have been investigated in recent times for OER [189]. An effective approach has been to integrate nanoparticles of catalytically active non-noble metals or their compounds within the COF [184, 190–197]. For example, Vaidhyanathan and co-workers anchored relatively cheaper cobalt and nickel hydroxides as electrocatalysts into a flexible COF backbone to enable very high activity for the relatively more challenging oxygen evolution reaction [198]. Co$(OH)_2$/Ni$(OH)_2$ nanoparticles with sizes <2 nm were grown on the sp^3 N-rich COF support without any additional capping agent, and the Co-Ni(4:12)@COF catalyst oxidized water with an overpotential of 258 mV at a current density of 10 mA cm^{-2} [198], while an even superior catalytic system contains metallic Ni_3N nanosheets cased in a conjugated COF [62] constructed from building units with a low number of structural degrees of freedom (Fig. 16.21a, b) [27]. The direct growth of the Ni_3N nanoparticles on the surface of the COF was visualized from an HRTEM image (Fig. 16.21c). The pristine COF exhibited significant conjugation-assisted conductivity (Fig. 16.21d). This could be crucial for the smooth transfer of electrons and reactive species between these entities. This was reflected in high OER activity with an overpotential of 230 mV @10 mA cm^{-2} (Tafel slope, 79 mV/dec) (Fig. 16.21e)

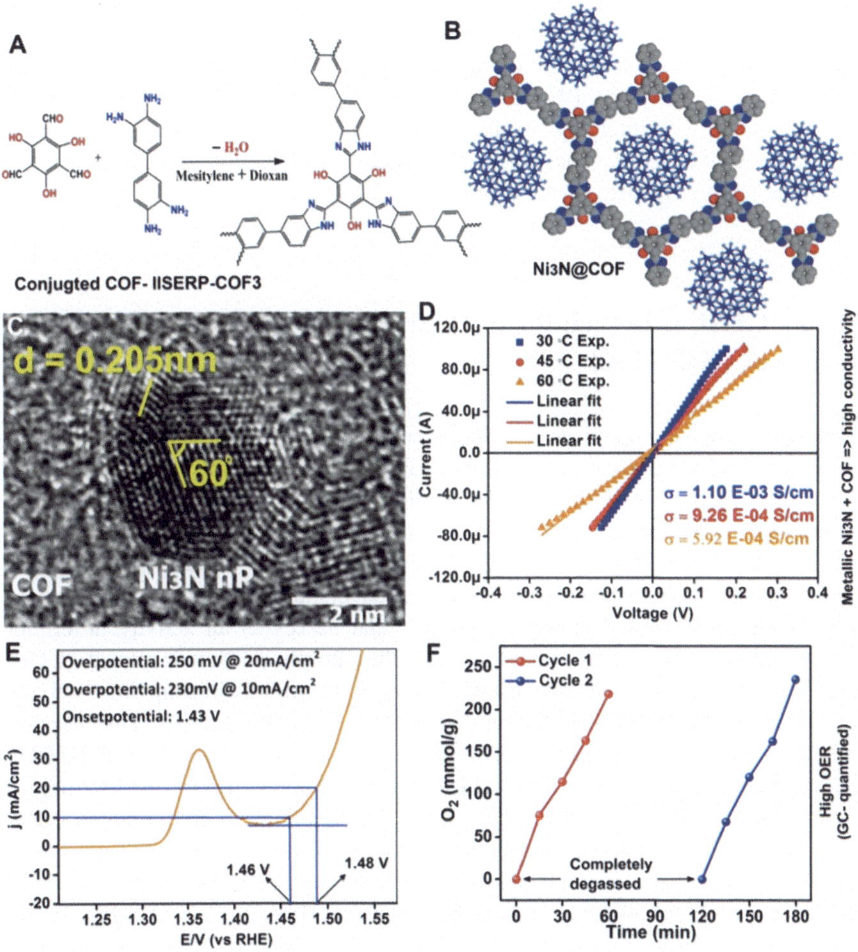

Fig. 16.21 (**a**) A conjugated COF construction from predefined monomers. (**b**) Modeled structure of the COF with Ni_3N nanoparticles/nanosheets loaded into it. (**c**) HRTEM image showing the Ni_3N nanoparticle directly grown on the surface of the COF. (**d**) Four-probe conductivity measurements revealing the enhancement of conductivity of the COF when loaded with Ni_3N (μA to mA and optical bandgap lowers: neat COF = 1.84 eV; Ni_3N@COF = 1.59 eV). (**e**) Exceptionally low overpotential for OER from this composite COF-derived catalyst. (**f**) Oxygen evolved from alkaline water catalyzed by Ni_3N@COF— among the highest O_2 evolution reported for a COF-derived catalyst (Copyright @ 2016, John Wiley & Sons, Inc.)

attributed to the synergistically lowered bandgap of the Ni_3N@COF (neat COF = 1.84 eV; Ni_3N@COF = 1.59 eV). Nitrides are known to perform extremely well as electrocatalyst for water splitting [199, 200]. This composite catalyst evolved ~225 mmol g^{-1} h^{-1} of O_2 (Fig. 16.21f), which is among the highest reported for COF-based OER catalysts [62].

Similarly, introducing macroporosity into the COF was shown to improve OER kinetics, as evidenced from Tafel slopes (54 mV/dec) [99]. Borrowing notes from biological systems, cobalt-, nickel-, iron-, and manganese-based electrocatalysts stabilized on typically porphyrin- or phthalocyanine-knitted COF supports are being investigated continuously owing to their versatile redox activity and ready accessibility [83, 184, 189–192, 194, 195, 197, 198, 201, 202]. There could be room for more synergistic interplay of mixed-transition metals if they can be stabilized within the COFs [203]. The below theoretical investigation unveils that other non-noble metals can also be a good choice.

16.16 CFSE a Descriptor to Predict the Catalytic Activity of COF [204]

The energy level of a metal atom's d-band center could be an activity descriptor for metal surfaces [205], while the e_g-filling could serve as the descriptor for transition-metal-oxide perovskites [206]. For the family of metal-free carbonaceous catalysts, the product of electronegativity and electric affinity of dopants was employed as the activity descriptor for predicting their bifunctional ORR/OER capability [207]. Recently, Xia and co-workers identified crystal field stabilization energy (CFSE)—intrinsic orbital energy of atoms—that serves as an activity descriptor for predicting ORR/OER activities of the porphyrin-functionalized COFs (Fig. 16.22a). They have also established a volcano relationship between the descriptor and the ORR/OER catalytic activity of the COF catalysts. They support their predictions with experimental reports from the literature. The TM@COF (porphyrin unit containing COF) revealed their relative energies for the four important processes involved in ORR/OER. In this study, using the first-principle calculation, they have simulated the elementary steps of ORR on a transition-metal site by introducing O_2 molecules and then protons near the metal site. They have calculated the free energies for both two-electron ($2e^-$) transfer ($H_2 + O_2 \rightarrow H_2O_2$) (Fig. 16.4b) and four-electron ($4e^-$) transfer ($2H_2 + O_2 \rightarrow 2H_2O$) (Fig. 16.4c) during the ORR under acidic medium. Which pathway is followed seems to depend on the types of transition metals. Free energy diagrams for ORR (Fig. 16.1c) suggest the transformation of OH* to water to be the rate-limiting step for ORR. This appears to be the case for all 3d transition metals except for Zn, Cu, and Ni. The rate-limiting step for Zn, Cu, and Ni is the first electron transfer ($O_2 \rightarrow OOH^*$). Among the 3d transition metals, similar to two-electron transfer, Fe and Co could spontaneously lead to four-electron transfer. In contrast, all other 3d transition metals have energy barriers in the

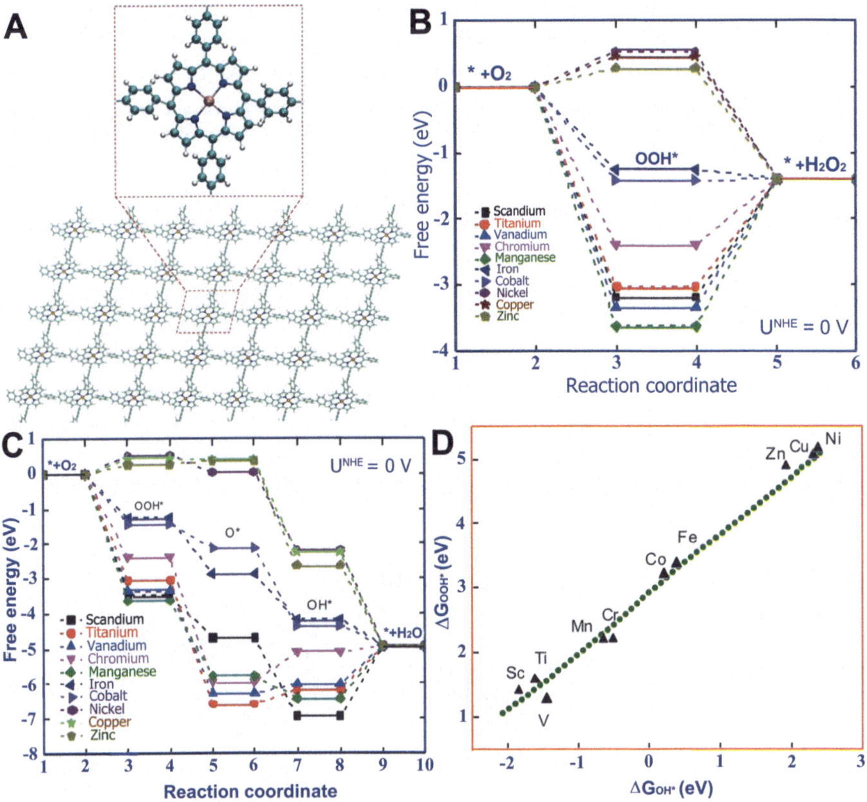

Fig. 16.22 Transition-metal-COFs and their free energy diagrams. (**a**) Schematic of a TM-COFs (TM-Sc, Ti, V, Cr, Mn, Fe, Co, Ni, Cu, and Zn). The green, blue, deep gold, and silver colors represent C, N, TM, and H, respectively. Free energy diagrams of TM–porphyrin COFs for (**b**) two-electron transfer ORR and (C) four-electron transfer ORR in acid medium at $U^{NHE} = 0$ V. D: Absorption energy of OOH^* versus that of OH^* (Adapted from Ref. 204)

elementary reactions. They also identify the Fe and Co as the best metals to embed in COF for obtaining H_2O to H_2O_2 conversion, a two-electron process [204]. The identified rate-determining steps in the ORR processes in these TM@COF are different from those reported for the neat TM electrocatalysts. Thus, the inclusion of COF seems to alter the scenario.

16.17 OER by Semi-crystalline Highly Conjugated Phenazine COFs

Graphenic substrates with C_3N_4 or C_4N type units are known to be very efficient in water splitting [208]. Thus, designing COF with such units could be an effective approach to forming metal-free electrocatalysts. Zhang and co-workers, through a computational design approach, designed a phenazine-linked 2DCOF (COF–C_4N) synthesized by combining triphenylenehexamine (TPHA) and hexaketocyclohexane (HKH) (Fig. 16.23a, b) [209]. It has a COF–C_4N graphene-like crystalline structure with 11.8 Å micropores. Notably, the phenazine modules are pre-disposed with amines in 1,2 positions that tend to form rigid cyclized five-membered rings when combined with the aldehyde enriching the structural stability and the conjugation in the polymer. This design strategy works very efficiently for both COF and MOF [62, 91, 94, 210–213]. Interestingly, this highly conjugated N-rich COF records the lowest overpotential of 349 mV at 10 mA cm^{-2} and a Tafel slope of 64 mV dec^{-1} among all metal-free organic materials. The high performance is attributed to its low interfacial charge-transfer resistance, which decreases as the voltage increases. DFT band structure calculations establish the COF as a direct bandgap semiconductor with the valence band centered on carbon (hole) and the conduction band on the nitrogens (electron) of the five-membered rings (Fig. 16.23c). Also, specific carbon sites of the framework are found to be more active in binding to OER intermediates owing to their hole-rich character consistent with N-doped graphene systems

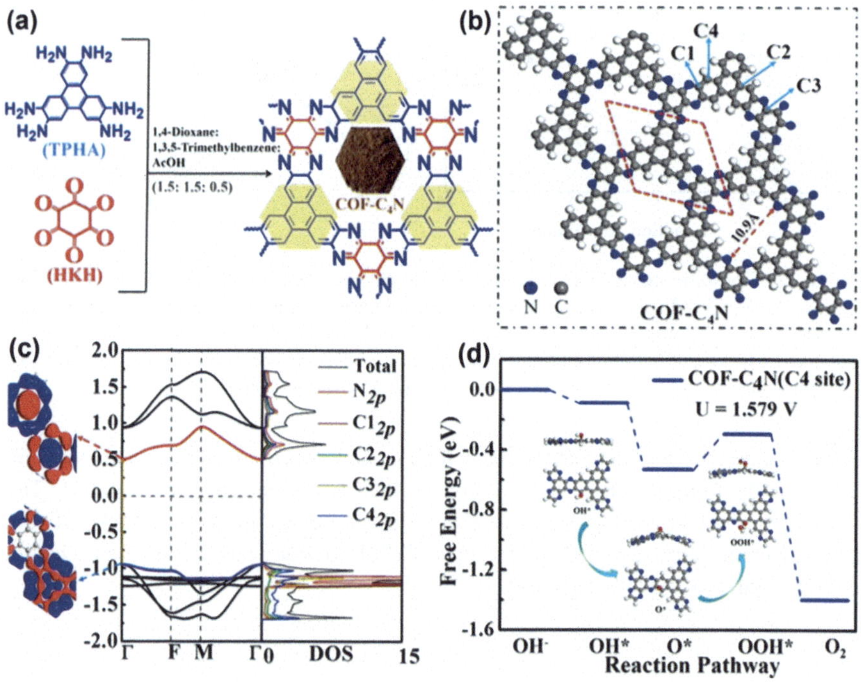

Fig. 16.23 (**a**) Schematic representation of the COF–C_4N synthesis. (**b**) A 1 × 3 × 3 super cell of a theoretically optimized COF–C_4N monolayer. The four C sites are marked. (**c**) Band structures, DOS, and wave functions of the CBM and VBM for monolayer COF–C_4N. (**d**) Free energy profile for the OER pathway at an applied bias of 1.579 V for COF–C_4N; the model structures of adsorption intermediates OH*, O*, and OOH* on the C_4 site are in the inset (Copyright © 2020, American Chemical Society)

[208, 214–216], leading to a favorable downhill pathway with a low overpotential of 1.579 V. The interactions of the OER intermediates with various C_4N sites of the COF at this potential help them identify the lowest energy active sites (Fig. 16.23d). The bulk characterizations of the COF reveal that it is not crystallized to the highest quality, and even in this form, it outdoes the completely amorphous counterparts (CMP–C_4N).

16.18 Modeling the Potential of a COF as a Bifunctional Catalyst

In another computation-driven study, Wu and co-workers [217] have carried out detailed DFT-MD studies to investigate the ability of imine-, aza-, and azo-linked COFs to serve as a bifunctional catalyst for HER and OER (Fig. 16.24). Recent computational analyses, including theirs, reveal that the benzene rings, triazine rings, and other aromatic rings in the system can serve as sites for OER. At the same time, the linking units mentioned here can act as catalytic sites for HER. They conclude that many COFs are semiconductors and have bandgaps in the range of 1.92–3.23 eV. Notably, the positions of both conduction and valence band edges of the investigated COFs match well with the chemical reaction potential of H_2/H^+ and O_2/H_2O. As a proof of concept, they have shown that the imine-linked 2,4,6-tris [4-methylphenyl]-1,3,5-triazine (TST) COF does show noticeable HER and OER, and their activity gets enhanced when coupled with Pt and Co for HER and OER, respectively. However, it becomes evident that there is a lot of improvement required for applying COF as a self-standing metal-free electrocatalyst. One of the reasons for this lowered catalytic activity is the reduced crystallinity of the COFs, particularly when subjected to potential under the aqueous/alkaline/acidic electrolyte system. Independently, the intrinsic structural defects such as slippage of layers, aperiodically missing building units, presence of insufficiently cross-polymerized substructures, and loss of π-stacking through exfoliation [sometimes even upon standing in the air] need to be addressed to be able to make the COF a truly stable electrocatalyst capable of reaching its theoretical capacities.

Oxygen reduction reaction using COF-derived catalyst is the most challenging and is being explored using experimental and computational methods [192–197, 218–224].

Exfoliated COF for ORR/zinc–air cell: Exfoliation of COF seems to be another exciting strategy that opens up the formation of covalent organic nanosheets with unique properties. Exfoliation can be achieved during synthesis itself via the use of (i) weakly pi-stacking building units (self-exfoliation), (ii) a well-designed controllable "functionalizing exfoliation agent (FEA)" concept—a chemical exfoliation, and (iii) a charge assistance (making the COF layers charged). Most relevant to the present discussion is a recent example wherein the concept of charge-assisted exfoliation is effectively exploited. They have prepared an iron-embedded porphyrin

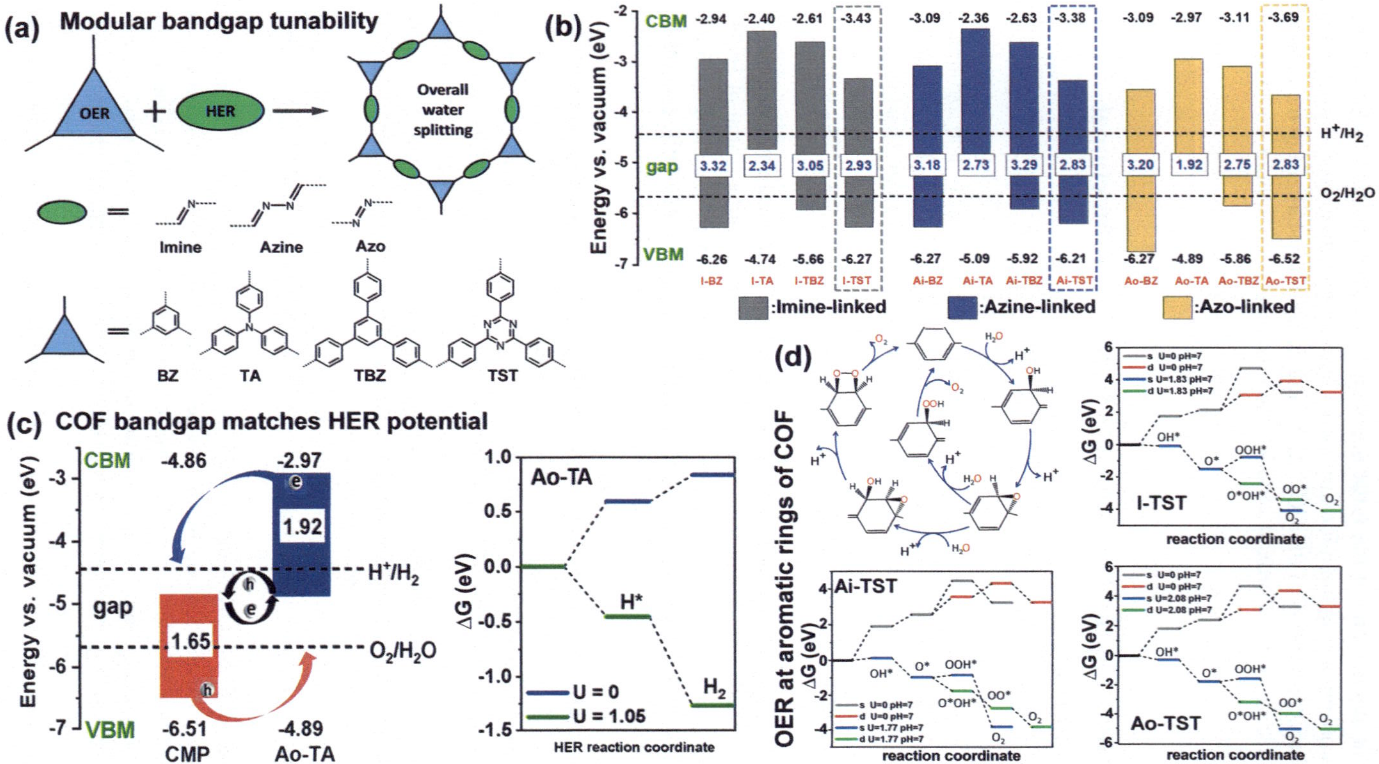

Fig. 16.24 (**a**) Schematic illustration of structure and design principles of 2D NL COFs for photocatalytic overall water splitting. Nitrogen-containing linkages (imine, azine, and azo groups) are used as HER active sites, and building units with C_{3v} molecular symmetry, denoted as BZ, TA, TBZ, and TST, respectively, are used as OER active sites. (**b**) The calculated energy positions of VBM and CBM of 2D COFs relative to vacuum level. Horizontal dashed lines are redox

potentials of water at pH = 0. (**c**) Gibbs free energy change with and without light-induced external bias potential for HER on Ao-TAT. (**d**) Illustrates two possible processes of OER via single-site or dual-site reactions on TST segment in 2D COFs, labeled as "s" or "d," respectively. The calculated Gibbs free energy changes of intermediate states involved in OER processes for I-TST, Ai-TST, and Ao-TST at pH = 7 are displayed in (**b–d**), respectively (Copyright © 2020, American Chemical Society)

COF [58] (check out the series of videos in the SI of this *ACS nano*). Now the iron centers carry a positive charge and are desirable for interacting with OH^- ions in an alkaline solution (Fig. 16.25a). Due to the electrostatic or ionic nature of the COF, they are exfoliated when exposed to a high concentration of OH^- in an aqueous solution. The COF dissolved when suspended in a highly basic solution (pH = 13) (Fig. 16.25b). The most visual evidence for the suspension having the COF's polymeric structure intact comes from the highly ordered iron–porphyrin units seen in the HRTEM image. Also, the PXRD reveals that the COF is a mixture of orthorhombic and hexagonal phases and the HAADF-STEM beautifully images it (Fig. 16.25c).

Further, the electrophoretic mobility and zeta potential (ζ) of the COF.OH solution have a negative value ($\zeta = -1.26$) under neutral conditions; on the contrary, charge reversal was observed in an acidic environment ($\zeta = +6.27$ at pH 1). The increase in the absolute ζ value along with the rise in pH ($\zeta = -21.6$ at pH 13) resulted in the enhancement of surface charge, particle repulsion, and stability of the true solution [26]. The dispersion is stable even after a year (Fig. 16.25b). The UPS spectrum reveals the low energetic barrier with a work function as low as 4.84 eV, which denotes the ease with which the catalyst can transfer the electron to the adsorbed molecular oxygen to generate OOH-type intermediates involved in the ORR process. The XANES study reveals that the Fe–N_4 centers in the COF transform to an (H)O–Fe–N_4 intermediate during the catalysis process, suggesting the Fe as the ORR active site. The increased ORR activity of the dissolved COF in alkaline medium shows a favorable onset potential, half-wave potential (965 mV vs. RHE). The high solubility of the COF.OH permits assembling a unique Zn–air-flow battery in an alkaline solution (Fig. 16.25d), which exhibited better cycling stability than its counterpart with a Pt/C-coated air electrode (Fig. 16.25e). When combined with graphene, this material delivers exceptional kinetic current density in oxygen reduction catalysis (four times higher than the benchmark Pt/C) and superior power density and cycling stability in Zn–air batteries compared with Pt/C as air electrodes [225]. The study secures the exceptional chemical and electrochemical stability that the highly conjugated COF frameworks can possess, and their solution processability via such exfoliation is hugely advantageous. In general, exfoliation of the COFs seems to help ORR activity [226].

Yao and co-workers demonstrated the influence of sulfur-rich thiophene units in the assembling metal-free ORR active COFs (Fig. 16.26) [227]. An aqueous Zn–air battery (ZAB) was assembled using JUC-528 (most S-rich) as the cathode. The JUC-528 ZAB exhibits a current density (J) of ~38 mA cm^{-2} at a voltage of 1.0 V and a power density of ~0.164 W cm^{-2} at 0.45 V (see Fig. S34 of Ref. 227), a voltage of 1.12 V at 10 mA cm^{-1} after 60 h, and excellent rate performance (see Figs. 3g and S35 of Ref. 227). The half-wave potential of JUC-528 is 0.70 V vs. RHE, which is more positive than those of PDA-TAPB-COF and JUC-527 (Fig. 16.26b). The electron transfer numbers (*n*) of PDA-TAPB-COF,

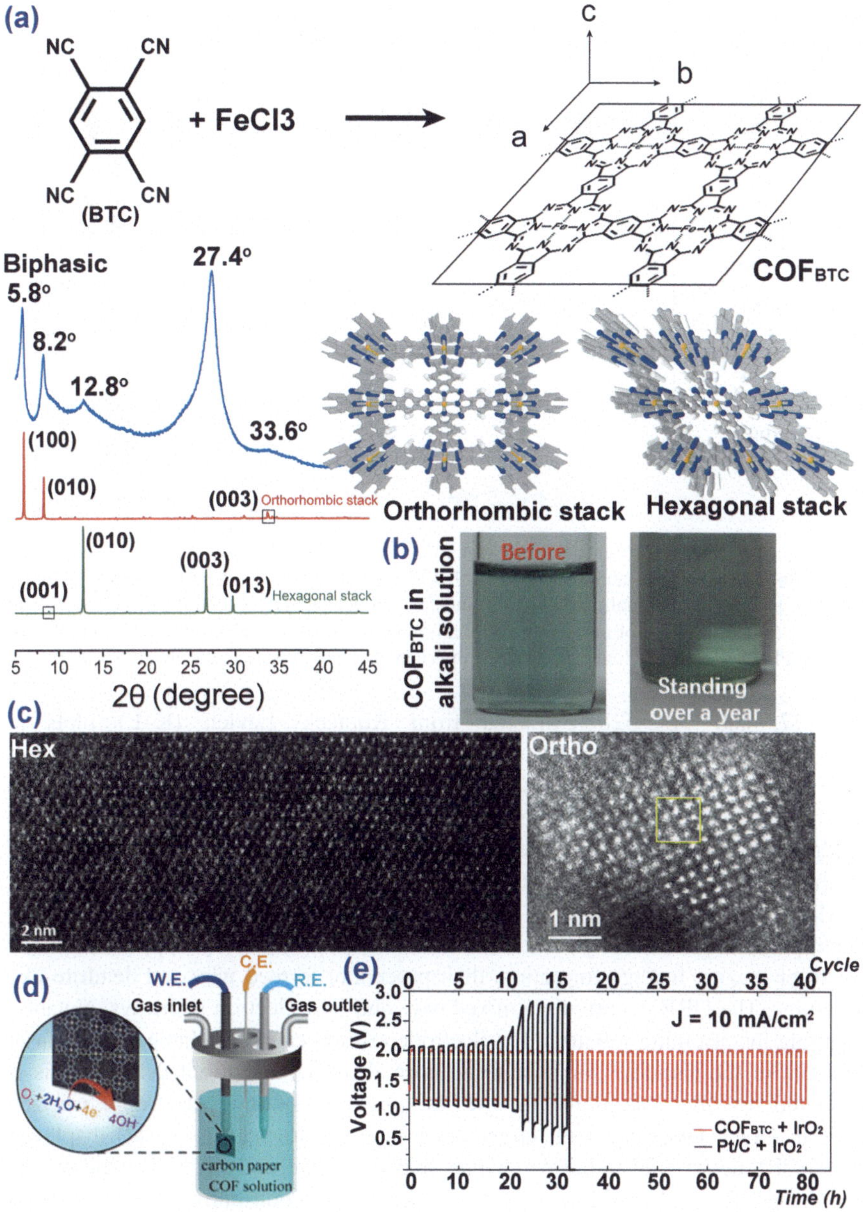

Fig. 16.25 (**a**) Synthesis scheme of COF_{BTC}. PXRD comparisons showing the biphasic (ortho + hex) nature of the sample. (**b**) The Fe–N_4 center due to their cationic nature enables the COF to dissolve in highly alkaline solutions and stay stable for over a year. (**c**) HADDT-STEM images showing the hexagonal and orthorhombic phases coexisting. (**d**) A zinc–air flow cell using COF_{BTC}. (**e**) COFBTC solution showing more durability under charge–discharge than Pt/C (Copyright © 2019, American Chemical Society)

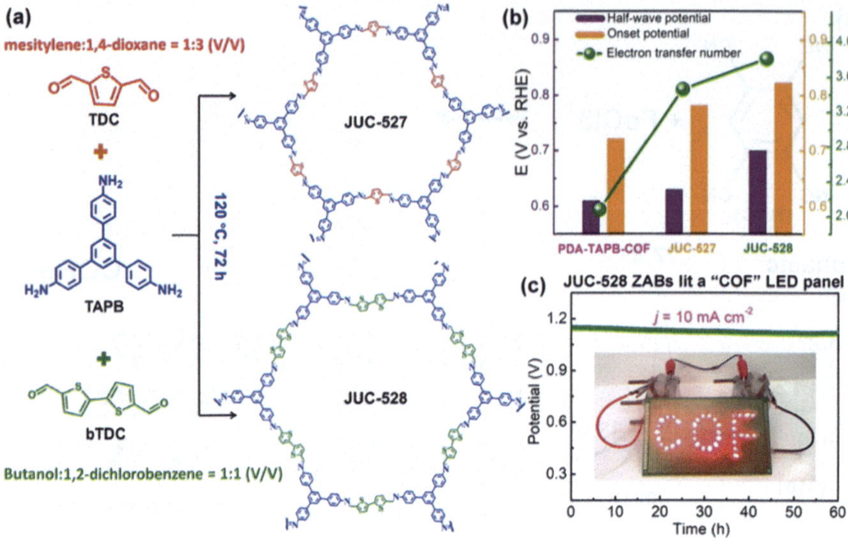

Fig. 16.26 (**a**) Syntheses and structures of JUC-527 and JUC-528. (**b**) Comparison of onset potential, half-wave potential, and electron transfer number of PDA-TAPB-COF, JUC-527, and JUC-528. (**c**) Comparison of onset potential, half-wave potential, and electron transfer number of PDA-TAPB-COF, JUC-527, and JUC-528. Copyright © 2020, American Chemical Society)

JUC-527, and JUC-528 extracted from Koutecky–Levich (K–L) plots at 0.2 V vs. RHE (Fig. S28 of Ref. 227) are 2.08, 3.46, and 3.81, respectively (Fig. S27 of Ref. 227). The density of states (DOS) and Gibbs free energy (ΔG) diagrams calculated for the thiophene-free and thiophene-rich COF revealed the lowered bandgap for the latter, while the C-atoms closer to the pentacyclic thiophene-S exhibit low overpotentials for ORR.

Meanwhile, Nejati and co-workers grew porphyrin-based COFs via in-situ electropolymerization on glass carbon surfaces (Fig. 16.27) [228]. Through the in-situ phenazine linkage formation, their procedure formed wire and dendrite-like structures. The ORR occurred via mixed two- and four-electron pathways. Notably, changing the scanning rate used for electrodeposition caused a change in the onset potential of the cathodic waves and the normalized ORR current density. This is attributed to the variations in the morphology of the electropolymerized Py-POR-COF. The films with larger crystalline dendrites (Fig. 16.27) displayed higher ORR activity. In addition to temperature, scan rate, and electrode material, the stabilizing effect of pyridine is yet another factor contributing to the ordered networks of (pyridine-mediated) $\pi-\pi$ stacked COFs. There are many scopes to improve these COF-derived catalysts' charge and mass transfer by developing them as structured 2D films [229].

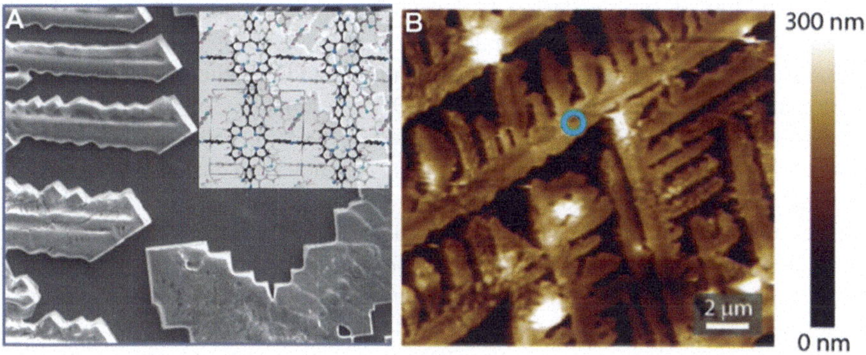

Fig. 16.27 (**a**) SEM images of dendritic pTAPP films, electrodeposited with scan rates of 2 mV/s at 25 °C on GC (scale bar is 50 µm). (**b**) AFM height image of pTAPP, electrodeposited on a GC electrode (Copyright © 2019, American Chemical Society)

16.19 Conclusion

This chapter has tried to give you a flavor of the latest happenings in covalent organic framework chemistry, particularly in photo-/electrocatalysis for water splitting. Given the rapidly exploding literature on the use of COF in various applications, it is impossible to cover every report, apologies for the missed ones. The above representative works display the ability of COF to serve as size–shape material with added electronic advantages coming from the choice of nodes and linkers. The polymeric structure enables the expression of the molecular-level HOMO–LUMO constructs into bandgaps that get tuned efficiently to deliver COFs as semiconductors with variable bandgaps. Wide-to-narrow bandgaps favor the harvesting of light in the entire UV–Vis-near-IR spectrum. Meanwhile, the p-rich monomers facilitate the generation of organic radicals, and the conjugated backbone transports it efficiently. The presence of well-defined electron-rich and deficient modules in the framework ensures the splitting of the excitons. The ordered columns of such modules enable the facile migration of charges. Recent studies reveal that the conjugated C–C or C=C bonds as linkers are more efficient in this.

One major advantage of COF is its crystalline structure with a well-defined formula. This enables the normalization or absolute quantification of their catalyst efficiency expressed as TOF, TON, quantum yield, gas evolution, active radical/COF ratio (from EPR), gas uptakes, and so on. This is not easy with amorphous polymers and this feature obliges future designability. When COFs with these advantageous traits are employed as "active" supports for electrocatalytic noble and non-noble metal catalysts, they result in composites with exceptional activity for water splitting. However, COF as a metal-free catalyst has not been as effective for OER and ORR due to their higher thermodynamic and kinetic demands. One way to look at this is that the active radicals or charges generated by COF as a self-standing catalyst are not in sufficient quantum, and this is partly due to the intrinsically low electronic/electrical conductivity of these all-organic frameworks. And the defects in the polymeric framework can create nodes where the charges/ radicals collapse [230, 231]. However, some recent studies indicate that this can be improved by design and by engineering COFs as thin films or membranes with reduced defects. It is well-understood that the effectiveness of these catalysts is in how many excitons they can generate, how long can they keep the charges separated, how far can they be transported within the material, and how well can they be transferred across the interface between the catalyst and the active substrates. Given the current developments that have been discussed in this article, it would be fascinating to see the electron-charge transfer capabilities of a highly conjugated COF engineered into thin films in HER, OER, and ORR.

As a more simplistic approach, the COFs have been combined with conducting materials such as CNT, graphene, super P carbon, etc. And they do work effectively. The recently developed highly conjugated sp^2-c COFs to such materials are going to prove big. A fascinating prospect includes the bandgap tuning of COFs via the inclusion of phosphorene [232–234] or BCN-type components [235–238]. These are recognized to have semiconductor or even semimetal-type behavior [237] and enable more topological crystal engineering. The biggest challenge would be that COFs are synthesized via a chemie douze method, while the Xenes or BCN are all typically prepared under relatively harsher conditions. Integrating them at the atomic level poses an invigorating challenge. Also, given all this tremendous advantage of COF as a yarn-by-yarn weavable organic polymer [101], its organic backbone facilitates its integration with bulk materials, making it much closer to real-life applications. Yet, the most abundant and cost-effective imine-derived COFs are faced with challenges of stability and durability in open air for long periods. In contrast, the more robust and high-activity COFs involve monomers complicated to synthesize and tend to be expensive. Future research would address these issues and find niche applications for COF where the performance becomes so advantageous that many other disadvantages become irrelevant. COF's future is very bright.

References

1. Diercks CS, Yaghi OM (2017) The atom, the molecule, and the covalent organic framework. Science 355(6328):923
2. Lyle SJ, Waller PJ, Yaghi OM (2019) Covalent organic frameworks: organic chemistry extended into two and three dimensions. Trends Chem 1(2):172–184
3. Waller PJ, Gandara F, Yaghi OM (2015) Chemistry of covalent organic frameworks. Acc Chem Res 48(12):3053–3063
4. Geng K, He T, Liu R, Dalapati S, Tan KT, Li Z, Tao S, Gong Y, Jiang Q, Jiang D (2020) Covalent organic frameworks: design, synthesis, and functions. Chem Rev 120 (16):8814–8933
5. Lohse MS, Bein T (2018) Covalent organic frameworks: structures, synthesis, and applications. Adv Funct Mater 28(33):1705553

6. Ji E, Asada M, Xu Q, Dalapati S, Addicoat MA, Brady MA, Xu H, Nakamura T, Heine T, Chen Q, Jiang D (2017) Two-dimensional sp^2 carbon–conjugated covalent organic frameworks. Science 357(6352):673–676
7. Roeser J, Prill D, Bojdys MJ, Fayon P, Trewin A, Fitch AN, Schmidt MU, Thomas A (2017) Anionic silicate organic frameworks constructed from hexacoordinate silicon centres. Nat Chem 9(10):977–982
8. Ascherl L, Sick T, Margraf JT, Lapidus SH, Calik M, Hettstedt C, Karaghiosoff K, Döblinger M, Clark T, Chapman KW (2016) Molecular docking sites designed for the generation of highly crystalline covalent organic frameworks. Nat Chem 8(4):310–316
9. Li Y, Chen W, Xing G, Jiang D, Chen L (2020) New synthetic strategies toward covalent organic frameworks. Chem Soc Rev 49(10):2852–2868
10. Lukose B, Kuc A, Frenzel J, Heine T (2010) On the reticular construction concept of covalent organic frameworks. Beilstein J Nanotechnol 1(1):60–70
11. Ding SY, Wang W (2013) Covalent organic frameworks (COFs): from design to applications. Chem Soc Rev 42(2):548–568
12. Tilford RW, Mugavero SJ III, Pellechia PJ, Lavigne JJ (2008) Tailoring microporosity in covalent organic frameworks. Adv Mater 20(14):2741–2746
13. Zhang G, Tsujimoto M, Packwood D, Duong NT, Nishiyama Y, Kadota K, Kitagawa S, Horike S (2018) Construction of a hierarchical architecture of covalent organic frameworks via a postsynthetic approach. J Am Chem Soc 140(7):2602–2609
14. Waller PJ, Lyle SJ, Osborn Popp TM, Diercks CS, Reimer JA, Yaghi OM (2016) Chemical conversion of linkages in covalent organic frameworks. J Am Chem Soc 138 (48):15519–15522
15. Zhao Y, Guo L, Gándara F, Ma Y, Liu Z, Zhu C, Lyu H, Trickett CA, Kapustin EA, Terasaki O, Yaghi OM (2017) A synthetic route for crystals of woven structures, uniform nanocrystals, and thin films of imine covalent organic frameworks. J Am Chem Soc 139 (37):13166–13172
16. Li X, Yang C, Sun B, Cai S, Chen Z, Lv Y, Zhang J, Liu Y (2020) Expeditious synthesis of covalent organic frameworks: a review. J Mater Chem A 8(32):16045–16060
17. Segura JL, Royuela S, Ramos MM (2019) Post-synthetic modification of covalent organic frameworks. Chem Soc Rev 48(14):3903–3945
18. Díaz U, Corma A (2016) Ordered covalent organic frameworks, COFs and PAFs. From preparation to application. Coord Chem Rev 311:85–124
19. Thakur VK, Kessler MR (2016) Liquid crystalline polymers. Cham, Switzerland, Springer
20. Liu D, Broe DJ (2014) Liquid crystal polymer networks: preparation, properties, and applications of films with patterned molecular alignment. Langmuir 30(45):13499–13509
21. Yang SH, Hsu CS (2009) Liquid crystalline conjugated polymers and their applications in organic electronics. J Polym Sci Part A Polym Chem 47(11):2713–2733
22. Chen C, Poppe M, Poppe S, Tschierske C, Liu F (2020) Liquid organic frameworks: a liquid crystalline 8-connected network with body-centered cubic symmetry. Angew Chem Int Ed 59 (47):20820–20825
23. Côté AP, Benin AI, Ockwig NW, O'Keeffe M, Matzger AJ, Yaghi OM (2005) Porous, crystalline, covalent organic frameworks. Science 310(5751):1166–1170
24. Beaudoin D, Maris T, Wuest JD (2013) Constructing monocrystalline covalent organic networks by polymerization. Nat Chem 5(10):830–834
25. Ma T, Kapustin EA, Shawn X, Liang L, Zhou Z, Niu J, Li LH, Wang Y, Su J, Li J, Wang X, Wang WD, Wang W, Sun J, Yaghi OM (2018) Single-crystal x-ray diffraction structures of covalent organic frameworks. Science 361(6397):48–52
26. Liu Y (2020) Yi Liu answers questions about 15 years of research on covalent organic frameworks. Nat Commun 11(1):5333
27. Haase F, Lotsch BV (2020) Solving the COF trilemma: towards crystalline, stable and functional covalent organic frameworks. Chem Soc Rev 49(23):8469–8500

28. El-Kaderi HM, Hunt JR, Mendoza-Cortés JL, Côté AP, Taylor RE, O'Keeffe M, Yaghi OM (2007) Designed synthesis of 3D covalent organic frameworks. Science 316(5822):268–272
29. Fang Q, Gu S, Zheng J, Zhuang Z, Qiu S, Yan Y (2014) 3D Microporous Base-functionalized covalent organic frameworks for size-selective catalysis. Angew Chem Int Ed 53 (11):2878–2882
30. Tao S, Jiang D (2020) Covalent organic frameworks for energy conversions: current status, challenges, and perspectives. CCS Chem 3(1):2003–2024
31. Feriante CH, Jhulki S, Evans AM, Dasari RR, Slicker K, Dichtel WR, Marder SR (2020) Rapid synthesis of high surface area imine-linked 2D covalent organic frameworks by avoiding pore collapse during isolation. Adv Mater 32(2):1905776
32. Matsumoto M, Dasari RR, Ji W, Feriante CH, Parker TC, Marder SR, Dichtel WR (2017) Rapid, low temperature formation of imine-linked covalent organic frameworks catalyzed by metal triflates. J Am Chem Soc 139(14):4999–5002
33. Evans AM, Parent LR, Flanders NC, Bisbey RP, Vitaku E, Kirschner MS, Schaller RD, Chen LX, Gianneschi NC, Dichtel WR (2018) Seeded growth of single-crystal two-dimensional covalent organic frameworks. Science 361(6397):52–57
34. Yan D, Wang Z, Cheng P, Chen Y, Zhang Z (2021) Rational fabrication of crystalline smart materials for rapid detection and efficient removal of ozone. Angew Chem 60(11):6055–6060
35. Li S, Liu Y, Li L, Liu C, Li J, Ashraf S, Li P, Wang B (2020) Enhanced proton conductivity of imidazole-doped thiophene-based covalent organic frameworks via subtle hydrogen bonding modulation. ACS Appl Mater Interfaces 12(20):22910–22916
36. Wang S, Zhang Z, Zhang H, Rajan AG, Xu N, Yang Y, Zeng Y, Liu P, Zhang X, Mao Q, He Y, Zhao J, Li BG, Strano MS, Wang WJ (2019) Reversible polycondensation-termination growth of covalent-organic-framework spheres, fibers and films. Matter 1(6):1592–1605
37. Yu J, Yuan L, Wang S, Lan J, Zheng L, Xu C, Chen J, Wang L, Huang Z, Tao W, Liu Z, Chai Z, Gibson JK, Shi W (2019) Phosphonate-decorated covalent organic frameworks for actinide extraction: a breakthrough under highly acidic conditions. CCS Chem 1(3):286–295
38. Peng Y, Xu G, Hu Z, Cheng Y, Chi C, Yuan D, Cheng H, Zhao D (2016) Mechanoassisted synthesis of sulfonated covalent organic frameworks with high intrinsic proton conductivity. ACS Appl Mater Interfaces 8(28):18505–18512
39. Yao BJ, Wu WX, Ding LG, Dong YB (2021) Sulfonic acid and ionic liquid functionalized covalent organic framework for efficient catalysis of the Biginelli reaction. J Org Chem 86 (3):3024–3032
40. Yang Z, Chen P, Hao W, Xie Z, Feng Y, Xing G, Chen L (2021) Sulfonated 2D covalent organic frameworks for efficient proton conduction. Chem Eur J 27(11):3817–3822
41. Halder A, Karak S, Addicoat M, Bera S, Chakraborty A, Kunjattu SH, Pachfule P, Heine T, Banerjee R (2018) Ultrastable imine-based covalent organic frameworks for sulfuric acid recovery: an effect of interlayer hydrogen bonding. Angew Chem Int Ed 57(20):5797–5802
42. Wang Y, Liu H, Pan Q, Wu C, Hao XJ, Chen R, Liu J, Li Z, Zhao Y (2020) Construction of fully conjugated covalent organic frameworks via facile linkage conversion for efficient photoenzymatic catalysis. J Am Chem Soc 142(13):5958–5963
43. Li X, Zhang C, Cai S, Lei X, Altoe V, Hong F, Urban JJ, Ciston J, Chan EM, Liu Y (2018) Facile transformation of imine covalent organic frameworks into ultrastable crystalline porous aromatic frameworks. Nat Commun 9(1):2998
44. Jangir R, Kalita AC, Kaleeswaran D, Gupta SK, Murugavel R (2018) A [4+2] condensation strategy to imine-linked single-crystalline zeolite-like zinc phosphate frameworks. Chem Eur J 24(23):6178–6190
45. Cusin L, Peng H, Ciesielski A, Samorì P (2021) Chemical conversion and locking of the imine linkage: enhancing the functionality of covalent organic frameworks. Angew Chem Int Ed 60:2–17
46. Vitaku E, Dichtel R (2017) Synthesis of 2D imine-linked covalent organic frameworks through formal transimination reactions. J Am Chem Soc 139(37):12911–12914

47. Belowicha ME, Stoddart JF (2012) Dynamic imine chemistry. Chem Soc Rev 41 (6):2003–2024
48. Gallant AJ, Yun M, Sauer M, Yeung CS, MacLachlan MJ (2005) Tautomerization in naphthalenediimines: a keto-enamine Schiff Base macrocycle. Org Lett 7(22):4827–4830
49. Segura JL, Mancheñoa MJ, Zamora F (2016) Covalent organic frameworks based on Schiff-base chemistry: synthesis, properties and potential applications. Chem Soc Rev 45 (20):5635–5671
50. Liu J, Han G, Zhao D, Lu K, Gao J, Chung T (2020) Self-standing and flexible covalent organic framework (COF) membranes for molecular separation. Sci Adv 6(41):eabb1110
51. Mitra S, Kandambeth S, Biswal BP, Khayum MA, Choudhury CK, Mehta M, Kaur G, Banerjee S, Prabhune A, Verma A, Roy S, Kharul UK, Banerjee R (2016) Self-exfoliated guanidinium-based ionic covalent organic nanosheets (iCONs). J Am Chem Soc 138 (8):2823–2828
52. Haldar S, Roy K, Nandi S, Chakraborty D, Puthusseri D, Gawli Y, Ogale S, Vaidhyanathan R (2018) High and reversible lithium-ion storage in self-exfoliated triazole-triformyl phloroglucinol-based covalent organic nanosheets. Adv Energy Mater 8(8):1702170
53. Singh H, Devi M, Jena N, Iqbal MM, Nailwal Y, Sarkar AD, Pal SK (2020) Proton-triggered fluorescence switching in self-exfoliated ionic covalent organic nanosheets for applications in selective detection of anions. ACS Appl Mater Interfaces 12(11):13248–13255
54. Berlanga I, Ruiz-González ML, González-Calbet JM, Fierro JLG, Mas-Ballesté R, Zamora F (2011) Delamination of layered covalent organic frameworks. Small 7(9):1207–1211
55. Berlanga I, Mas-Ballesté R, Zamora F (2012) Tuning delamination of layered covalent organic frameworks through structural design. Chem Commun 48(64):7976–7978
56. Wang Z, Li Y, Liu P, Qi Q, Zhang F, Lu G, Zhao X, Huang X (2019) Few layer covalent organic frameworks with graphene sheets as cathode materials for lithium-ion batteries. Nanoscale 11(12):5330–5335
57. Wang S, Wang Q, Shao P, Han Y, Gao X, Ma L, Yuan S, Ma X, Zhou J, Feng X, Wang B (2017) Exfoliation of covalent organic frameworks into few-layer redox-active nanosheets as cathode materials for lithium-ion batteries. J Am Chem Soc 139(12):4258–4261
58. Peng P, Shi L, Huo F, Zhang S, Mi C, Cheng Y, Xiang Z (2019) In situ charge exfoliated soluble covalent organic framework directly used for Zn-air flow battery. ACS Nano 13 (1):878–884
59. Burke DW, Sun C, Castano I, Flanders NC, Evans AM, Vitaku E, McLeod DC, Lambeth RH, Chen LX, Gianneschi NC, Dichtel WR (2020) Acid exfoliation of imine-linked covalent organic frameworks enables solution processing into crystalline thin films. Angew Chem Int Ed 59(13):5165–5171
60. Haldar S, Roy K, Kushwaha R, Ogale S, Vaidhyanathan R (2019) Chemical exfoliation as a controlled route to enhance the anodic performance of COF in LIB. Adv Energy Mater 9 (48):1902428
61. Baldwin LA, Crowe JW, Pyles DA, McGrier PL (2016) Metalation of a mesoporous three-dimensional covalent organic framework. J Am Chem Soc 138(46):15134–15137
62. Nandi S, Singh SK, Mullangi D, Illathvalappil R, George L, Vinod CP, Kurungot S, Vaidhyanathan R (2016) Low band gap benzimidazole COF supported Ni_3N as highly active OER catalyst. Adv Energy Mater 6(24):1601189
63. Lanni LM, Tilford RW, Bharathy M, Lavigne JJ (2011) Enhanced hydrolytic stability of self-assembling alkylated two-dimensional covalent organic frameworks. J Am Chem Soc 133 (35):13975–13983
64. Zhang B, Wei M, Mao H, Pei X, Alshmimri SA, Reimer JA, Yaghi OM (2018) Crystalline dioxin-linked covalent organic frameworks from irreversible reactions. J Am Chem Soc 140 (40):12715–12719
65. Kandambeth S, Mallick A, Lukose B, Mane MV, Heine T, Banerjee R (2012) Construction of crystalline 2D covalent organic frameworks with remarkable chemical (acid/base) stability via a combined reversible and irreversible route. J Am Chem Soc 134(48):19524–19527

66. Xu H, Gao J, Jiang D (2015) Stable, crystalline, porous, covalent organic frameworks as a platform for chiral organocatalysts. Nat Chem 7(11):905
67. Mullangi D, Chakraborty D, Pradeep A, Koshti V, Vinod CP, Panja S, Nair S, Vaidhyanathan R (2018) Highly stable COF-supported Co/Co(OH)$_2$ nanoparticles heterogeneous catalyst for reduction of nitrile/nitro compounds under mild conditions. Small 14(37):1801233
68. Gottschling K, Savasci GK, Vignolo-González H, Schmidt S, Mauker P, Banerjee T, Rovó P, Ochsenfeld C, Lotsch BV (2020) Rational design of covalent cobaloxime–covalent organic framework hybrids for enhanced photocatalytic hydrogen evolution. J Am Chem Soc 142 (28):12146–12156
69. Yan X, Liu H, Li Y, Chen W, Zhang T, Zhao Z, Xing G, Chen L (2019) Ultrastable covalent organic frameworks via self-polycondensation of an A_2B_2 monomer for heterogeneous photocatalysis. Macromolecules 52(21):7977–7983
70. Zhang F, Yang Z, Hao J, Zhao K, Hua M, Yang Y, Wei J (2019) Dynamic covalent chemistry steers synchronizing nanoparticle self-assembly with interfacial polymerization. Commun Chem 2(1):1–10
71. Chakraborty D, Nandi S, Mullangi D, Haldar S, Vinod CP, Vaidhyanathan R (2019) Cu/Cu$_2$O nanoparticles supported on a phenol–pyridyl COF as a heterogeneous catalyst for the synthesis of unsymmetrical Diynes via Glaser–Hay coupling. ACS Appl Mater Interfaces 11 (17):15670–15679
72. Ghosh S, Khan TS, Ghosh A, Chowdhury AH, Haider MA, Khan A, Islam SM (2020) Utility of silver nanoparticles embedded covalent organic frameworks as recyclable catalysts for the sustainable synthesis of cyclic carbamates and 2-oxazolidinones via atmospheric cyclizative CO_2 capture. ACS Sustain Chem Eng 8(14):5495–5513
73. Benyettou F, Das G, Nair AR, Prakasam T, Shinde DB, Sharma SK, Whelan J, Lalatonne Y, Traboulsi H, Pasricha R, Abdullah O, Jagannathan R, Lai Z, Motte L, Gándara F, Sadler KC, Trabolsi A (2020) Covalent organic framework embedded with magnetic nanoparticles for MRI and chemo-thermotherapy. J Am Chem Soc 142(44):18782–18794
74. Park E, Jack J, Hu Y, Wan S, Huang S, Jin Y, Maness PC, Yazdi S, Ren Z, Zhang W (2020) Covalent organic framework-supported platinum nanoparticles as efficient electrocatalysts for water reduction. Nanoscale 12(4):2596–2602
75. Ramanathan V, Mullangi D, Nandi S (2019) Covalent organic frameworks as porous supports for non-noble metal-based water splitting electrocatalysts. US Patent 10,301,727, 28 May 2019
76. Kushwaha R, Kaleeswaran D, Haldar S, Chakraborty D, Mullangi D, Borah A, Vinod CP, Murugavel R, Vaidhyanathan R (2020) Nanoporous covalent organic framework embedded with Fe/Fe$_3$O$_4$ nanoparticles as air-stable low-density nanomagnets. ACS Appl Nano Mater 3 (9):9088–9096
77. Sun P, Hai J, Sun S, Lu S, Liu S, Liu H, Chen F, Wang B (2020) Aqueous stable Pd nanoparticles/covalent organic framework nanocomposite: an efficient nanoenzyme for colorimetric detection and multicolor imaging of cancer cells. Nanoscale 12(2):825–831
78. Zhu J, Hu L, Zhao P, Lee LYS, Wong KY (2019) Recent advances in electrocatalytic hydrogen evolution using nanoparticles. Chem Rev 120(2):851–918
79. Mullangi D, Nandi S, Shalini S, Sreedhala S, Vinod CP, Vaidhyanathan R (2015) Pd loaded amphiphilic COF as catalyst for multi-fold heck reactions, C-C couplings and CO oxidation. Sci Rep 5(1):10876
80. Sharma RK, Yadav P, Yadav M, Gupta R, Rana P, Srivastava A, Zbořil R, Varma RS, Antonietti M, Gawande MB (2020) Recent development of covalent organic frameworks (COFs): synthesis and catalytic (organic-electro-photo) applications. Mater Horiz 7 (2):411–454
81. Lei Z, Yang Q, Xu Y, Guo S, Sun W, Liu H, Lv LP, Zhang Y, Wang Y (2018) Boosting lithium storage in covalent organic framework via activation of 14-electron redox chemistry. Nat Commun 9(1):1–13

82. Xiong Y, Liao Q, Huang Z, Huang X, Ke C, Zhu H, Dong C, Wang H, Xi K, Zhan P, Xu F, Lu Y (2020) Ultrahigh responsivity photodetectors of 2D covalent organic frameworks integrated on graphene. Adv Mater 32(9):1907242
83. Li C, Yang J, Pachfule P, Li S, Ye M-Y, Schmidt J, Thomas A (2020) Ultralight covalent organic framework/graphene aerogels with hierarchical porosity. Nat Commun 11:4712
84. Wang W, Kale VS, Cao Z, Kandambeth S, Zhang W, Ming J, Parvatkar PT, Abou-Hamad E, Shekhah O, Cavallo L, Eddaoudi M, Alshareef HN (2020) Phenanthroline covalent organic framework electrodes for high-performance zinc-ion supercapattery. ACS Energy Lett 5 (7):2256–2264
85. Kushwaha R, Haldar S, Shekhar P, Krishnan A, Saha J, Hui P, Vinod CP, Subramaniam C, Vaidhyanathan R (2021) Exceptional capacitance enhancement of a non-conducting COF through potential-driven chemical modulation by redox electrolyte. Adv Energy Mater 11 (13):2003626
86. Sun B, Zhu C-H, Liu Y, Wang C, Wan LJ, Wang D (2017) Oriented covalent organic framework film on graphene for robust ambipolar vertical organic field-effect transistor. Chem Mater 29(10):4367–4374
87. Gomes R, Bhattacharyya AJ (2020) Carbon nanotube-templated covalent organic framework nanosheets as an efficient sulfur host for room-temperature metal–sulfur batteries. ACS Sustain Chem Eng 8(15):5946–5953
88. Guillaume HV, Bertranda VKM, Ta-Chung O, Robert GG, Mircea D (2013) Thiophene-based covalent organic frameworks. PNAS 110(13):4923–4928
89. Zhang MD, Si DH, Yi JD, Zhao SS, Huang YB, Cao R (2020) Conductive phthalocyanine-based covalent organic framework for highly efficient electroreduction of carbon dioxide. Small 16(52):2005254
90. Keller N, Bessinger D, Reuter S, Calik M, Ascherl L, Hanusch FC, Auras F, Bein T (2017) Oligothiophene-bridged conjugated covalent organic frameworks. J Am Chem Soc 139 (24):8194–8199
91. Meng Z, Stolz RM, Mirica KA (2019) Two-dimensional chemiresistive covalent organic framework with high intrinsic conductivity. J Am Chem Soc 141(30):11929–11937
92. Rotter JM, Guntermann R, Auth M, Mahringer A, Sperlich A, Dyakonov V, Medina DD, Bein T (2020) Highly conducting wurster-type twisted covalent organic frameworks. Chem Sci 11 (47):12843–12853
93. Mulzer CR, Shen L, Bisbey RP, McKone JR, Zhang N, Abruña HD, Dichtel WR (2016) Superior charge storage and power density of a conducting polymer-modified covalent organic framework. ACS Cent Sci 2(9):667–673
94. Liua S, Yaoa L, Lub Y, Huac X, Liuc J, Yanga Z, Wei H, Mai Y (2019) All-organic covalent organic framework/polyaniline composites as stable electrode for high-performance supercapacitors. Mater Lett 236:354–357
95. Wu Y, Yan D, Zhang Z, Matsushita MM, Awaga K (2019) Electron highways into nanochannels of covalent organic frameworks for high electrical conductivity and energy storage. ACS Appl Mater Interfaces 11(8):7661–7665
96. Han X, Yuan C, Hou B, Liu L, Li H, Liu Y, Cui Y (2020) Chiral covalent organic frameworks: design, synthesis and property. Chem Soc Rev 49(17):6248–6272
97. Bi S, Yang C, Zhang W, Xu J, Liu L, Wu D, Wang X, Han Y, Liang Q, Zhang F (2019) Two-dimensional semiconducting covalent organic frameworks via condensation at arylmethyl carbon atoms. Nat Commun 10(1):1–10
98. Yuan C, Fu S, Yang K, Hou B, Liu Y, Jiang J, Cui Y (2021) Crystalline C—C and C=C bond-linked chiral covalent organic frameworks. J Am Chem Soc 143(1):369–381
99. Zhao X, Pachfule P, Li S, Langenhahn T, Ye M, Schlesiger C, Praetz S, Schmidt J, Thomas A (2019) Macro/microporous covalent organic frameworks for efficient electrocatalysis. J Am Chem Soc 141(16):6623–6630

100. Kandambeth S, Jia J, Wu H, Kale VS, Parvatkar PT, Czaban-Jóźwiak J, Zhou S, Xu X, Ameur ZO, Abou-Hamad E, Emwas AH (2020) Covalent organic frameworks as negative electrodes for high-performance asymmetric supercapacitors. Adv Energy Mater 10(38):2001673
101. Liu Y, Ma Y, Zhao Y, Sun X, Gándara F, Furukawa H, Liu Z, Zhu H, Zhu H, Suenaga K, Oleynikov P, Alshammari AS, Zhang X, Terasaki O, Yagh OM (2016) Weaving of organic threads into a crystalline covalent organic framework. Science 351(6271):365
102. Wang Z, Zhang S, Chen Y, Zhang Z, Ma S (2020) Covalent organic frameworks for separation applications. Chem Soc Rev 49(3):708–735
103. Wu FF, Chen QY, Ma XJ, Li T, Wang LF, Hong J, Sheng YH, Ye ML, Zhu Y (2019) N-doped magnetic covalent organic frameworks for preconcentration of allergenic disperse dyes in textiles of fall protection equipment. Anal Methods 11(27):3381–3387
104. Jarju JJ, Lavender AM, Espiña B, Romero V, Salonen LM (2020) Covalent organic framework composites: synthesis and analytical applications. Molecules 25(22):5404
105. Mullangi D, Shalini S, Nandi S, Choksi B, Vaidhyanathan R (2017) Super-hydrophobic covalent organic frameworks for chemical resistant coatings and hydrophobic paper and textile composites. J Mater Chem A 5(18):8376–8384
106. Song Y, Sun Q, Aguila B, Ma S (2019) Opportunities of covalent organic frameworks for advanced applications. Adv Sci 6(2):1801410
107. Refinery hydrogen production capacities by country. H_2 tools. https: //h2tools.org/node/820
108. https: //afdc.energy.gov/fuels/hydrogen.production.html
109. IRENA (2018) Hydrogen from renewable power: technology outlook for the energy transition. International Renewable Energy agency, Abu Dhabi
110. McCrory CL, Jung S, Ferrer IM, Chatman SM, Peters JC, Jaramillo TF (2015) Benchmarking hydrogen evolving reaction and oxygen evolving reaction electrocatalysts for solar water splitting devices. J Am Chem Soc 137(13):4347–4357
111. Tee SY, Win KY, Teo WS, Koh LD, Liu S, Teng CP, Han MY (2017) Recent progress in energy-driven water splitting. Adv Sci 4(5):1600337
112. Dubouis N, Grimaud A (2019) The hydrogen evolution reaction: from material to interfacial descriptors. Chem Sci 10(40):9165
113. Zeradjanin AR, Grote JP, Polymeros G, Mayrhofer KJ (2016) A critical review on hydrogen evolution electrocatalysis: re-exploring the volcano-relationship. Electroanalysis 28 (10):2256–2269
114. Gileadi E (2011) Physical electrochemistry. Wiley-VCH, Weinheim
115. Fabbri E, Habereder A, Waltar K, Kötz R, Schmidt TJ (2014) Developments and perspectives of oxide-based catalysts for the oxygen evolution reaction. Catal Sci Technol 4 (11):3800–3821
116. Shi Z, Wang X, Ge J, Liu C, Xing W (2020) Fundamental understanding of the acidic oxygen evolution reaction: mechanism study and state-of-the-art catalysts. Nanoscale 12(25):13249
117. Wang H, Wu J, Dolocana A, Lia Y, Lüg X, Wu N, Park K, Xin S, Lei M, Yang W, Goodenough JB (2019) Short O–O separation in layered oxide Na0.67CoO2 enables an ultrafast oxygen evolution reaction. PNAS 116(47):23473–23479
118. Anantharaj S, Ede SR, Karthick K, Sankar SS, Sangeetha K, Karthikc PE, Kundu S (2018) Precision and correctness in the evaluation of electrocatalytic water splitting: revisiting activity parameters with a critical assessment. Energy Environ Sci 11(4):744–771
119. Rossmeisl J, Qu ZW, Zhu H, Kroes GJ, Nørskov JK (2007) Electrolysis of water on oxide surfaces. J Electroanal Chem 607(1–2):83–89
120. Schalenbach M, Zeradjanin AR, Kasian O, Cherevko S, Mayrhofer KJ (2018) A perspective on low-temperature water electrolysis-challenges in alkaline and acidic technology. Int J Electrochem Sci 13(2):1173–1226
121. Ayers KE, Anderson EB, Capuano C, Carter B, Dalton L, Hanlon G, Manco J, Niedzwiecki M (2010) Research advances towards low cost, high efficiency PEM electrolysis. ECS Trans 33 (1):3

122. Millet P, Ngameni R, Grigoriev S, Mbemba N, Brisset F, Ranjbari A, Etiévant C (2010) PEM water electrolyzers: from electrocatalysis to stack development. Int J Hydrog Energy 35 (10):5043–5052
123. Reier T, Nong HN, Teschner D, Schlögl R, Strasser P (2017) Electrocatalytic oxygen evolution reaction in acidic environments–reaction mechanisms and catalysts. Adv Energy Mater 7(1):1601275
124. Koper MT (2013) Theory of multiple proton–electron transfer reactions and its implications for electrocatalysis. Chem Sci 4(7):2710–2723
125. Koper MT (2011) Thermodynamic theory of multi-electron transfer reactions: implications for electrocatalysis. J Electroanal Chem 660(2):254–260
126. Bockris JOM (1956) Kinetics of activation controlled consecutive electrochemical reactions: anodic evolution of oxygen. J Chem Phys 24(4):817–827
127. Scheuermann AG, Prange JD, Gunji M, Chidsey CE, McIntyre PC (2013) Effects of catalyst material and atomic layer deposited TiO_2 oxide thickness on the water oxidation performance of metal–insulator–silicon anodes. Energy Environ Sci 6(8):2487–2496
128. Man IC, Su HY, Calle-Vallejo F, Hansen HA, Martínez JI, Inoglu NG, Kitchin J, Jaramillo TF, Nørskov JK, Rossmeisl J (2011) Universality in oxygen evolution electrocatalysis on oxide surfaces. ChemCatChem 3(7):1159–1165
129. Halck NB, Petrykin V, Krtil P, Rossmeisl J (2014) Beyond the volcano limitations in electrocatalysis–oxygen evolution reaction. Phys Chem Chem Phys 16(27):13682–13688
130. Fabbri E, Schmidt TJ (2018) Oxygen evolution reaction—the enigma in water electrolysis. ACS Catal 8(10):9765–9774
131. Bock C, Halvorsen H, MacDougall B (2008) Catalyst synthesis techniques. In: PEM fuel cell electrocatalysts and catalyst layers. Springer, London, pp 447–485
132. Damjanovic A, Sepa D, Vojnovic M (1979) New evidence supports the proposed mechanism for O2 reduction at oxide free platinum electrodes. Electrochim Acta 24(8):887–889
133. Raghuveer Y, Viswanathan B (2002) Nanocrystalline lead ruthenium pyrochlore as oxygen reduction electrode. Indian J Eng Mater Sci 9:137–140
134. Li Y, Li Q, Wang H, Zhang L, Wilkinson DP, Zhang J (2019) Recent progresses in oxygen reduction reaction electrocatalysts for electrochemical energy applications. Electrochem Energy Rev 2(4):518–538
135. Shao M (2013) Electrocatalysis in fuel cells: a non-and low-platinum approach, vol 9. Springer, New York. isbn:978-1-4471-4911-8
136. Sui S, Wang X, Zhou X, Su Y, Riffat S, Liu C (2017) A comprehensive review of Pt electrocatalysts for the oxygen reduction reaction: nanostructure, activity, mechanism and carbon support in PEM fuel cells. J Mater Chem A 5(5):1808–1825
137. Khotseng L (2018) Oxygen reduction reaction. In: Electrocatalysts for fuel cells and hydrogen evolution-theory to design. IntechOpen, New York
138. Damjanovic A, Genshaw MA, Bockris JOM (1967) The mechanism of oxygen reduction at platinum in alkaline solutions with special reference to H_2O_2. J Electrochem Soc 114(11):1107
139. Singh SK, Takeyasu K, Nakamura J (2019) Active sites and mechanism of oxygen reduction reaction electrocatalysis on nitrogen-doped carbon materials. Adv Mater 31(13):1804297
140. Maeda K, Domen K (2010) Photocatalytic water splitting: recent progress and future challenges. J Phys Chem Lett 1(18):2655–2661
141. Ramaswamy N, Mukerjee S (2012) Fundamental mechanistic understanding of electrocatalysis of oxygen reduction on Pt and non-Pt surfaces: acid versus alkaline media. Adv Phys Chem 2012:491604
142. Janin E, Von Schenck H, Göthelid M, Karlsson UO, Svensson M (2000) Bridge-bonded atomic oxygen on Pt (110). Phys Rev B 16(19):13144–13149
143. Yeager E (1984) Electrocatalysts for O_2 reduction. Electrochim Acta 29(11):1527–1537
144. Jasinski R (1964) A new fuel cell cathode catalyst. Nature 201(4925):1212–1213
145. Zagal JH (1992) Metallophthalocyanines as catalysts in electrochemical reactions. Coord Chem Rev 119:89–136

146. Zagal JH (2003) Macrocycles. In: Handbook of fuel cells-fundamentals. Wiley, Chichester, UK, pp 544–554
147. Wang B (2005) Recent development of non-platinum catalysts for oxygen reduction reaction. J Power Sources 152(1):1–15
148. Zagal JH, Bedioui F, Dodelet JP (2006) N4-macrocyclic metal complexes. oxygen reduction in PEM fuel cell conditions: heat-treated non-precious metal-N4 macrocycles and beyond, vol 83. Springer, Berlin, Germany
149. Ziegelbauer JM, Olson TS, Pylypenko S, Alamgir F, Jaye C, Atanassov P, Mukerjee S (2008) Direct spectroscopic observation of the structural origin of peroxide generation from co-based pyrolyzed porphyrins for ORR applications. J Phys Chem C 112(24):8839–8849
150. Artyushkova K, Levendosky S, Atanassov P, Fulghum J (2007) XPS structural studies of nano-composite non-platinum electrocatalysts for polymer electrolyte fuel cells. Top Catal 46 (3–4):263–275
151. Amalnerkar DP, Radhakrishnan S, Minoura H, Sugiura T, Ueno Y (1989) Origin of the photocathodic effect at CdS+CuCl electrodes. J Electroanal Chem 260(2):433–440
152. Gruenig G, Wiesener K, Gamburzev S, Iliev I, Kaisheva A (1983) Investigations of catalysts from the pyrolyzates of cobalt-containing and metal-free dibenzotetraazaannulenes on active carbon for oxygen electrodes in an acid medium. J Electroanal Chem 159(1):155–162
153. Van Veen JAR, Van Baar JF, Kroese KJ (1981) Effect of heat treatment on the performance of carbon-supported transition-metal chelates in the electrochemical reduction of oxygen. J Chem Soc Faraday Trans 77(11):2827–2843
154. Vasudevan P, Santosh MN, Tyagi S (1990) Transition metal complexes of porphyrins and phthalocyanines as electrocatalysts for dioxygen reduction. Transit Met Chem 15(2):81–90
155. Che M (2013) Nobel prize in chemistry 1912 to Sabatier: organic chemistry or catalysis. Catalogue 218:162–171
156. Wang D-G, Qiu T, Guo W, Liang Z, Tabassum H, Xia D, Zou R (2021) Covalent organic framework-based materials for energy applications. Energy Environ Sci 14(2):688–728
157. Sun T, Xie J, Guo W, Li DS, Zhang Q (2020) Covalent–organic frameworks: advanced organic electrode materials for rechargeable batteries. Adv Energy Mater 10(19):1904199
158. Zhao G, Rui K, Dou SX, Sun W (2018) Heterostructures for electrochemical hydrogen evolution reaction: a review. Adv Funct Mater 28(43):1803291
159. Guo F, Guo JH, Wang P, Kang YS, Liu Y, Zhao J, Sun WY (2019) Facet-dependent photocatalytic hydrogen production of metal–organic framework NH2-MIL-125(Ti). Chem Sci 10(18):4834–4838
160. Zhang Y, Hu Y, Zhao J, Park E, Jin Y, Liu Q, Zhang W (2019) Covalent organic framework-supported Fe–TiO_2 nanoparticles as ambient-light-active photocatalysts. J Mater Chem A 7 (27):16364–16371
161. Feng X, Chen L, Honsho Y, Saengsawang O, Liu L, Wang L, Saeki A, Irle S, Seki S, Dong Y, Jiang D (2012) An ambipolar conducting covalent organic framework with self-sorted and periodic electron donor-acceptor ordering. Adv Mater 24(22):3026–3031
162. Haldar S, Kaleeswaran D, Rase D, Roy K, Ogale S, Vaidhyanathan R (2020) Tuning the electronic energy level of covalent organic frameworks for crafting high-rate Na-ion battery anode. Nanoscale Horizons 5(8):1264–1273
163. Stegbauer L, Schwinghammer K, Lotsch BV (2014) A hydrazone-based covalent organic framework for photocatalytic hydrogen production. Chem Sci 5(7):2789
164. Jin E, Lan Z, Jiang Q, Geng K, Li G, Wang X, Jiang D (2019) 2D sp2 carbon-conjugated covalent organic frameworks for photocatalytic hydrogen production from water. Chem 5 (6):1632–1647
165. Chen X, Geng K, Liu R, Tan KT, Gong Y, Li Z, Tao S, Jiang Q, Jiang D (2020) Covalent organic frameworks: chemical approaches to designer structures and built-in functions. Angew Chem Int Ed 59(13):5050–5091
166. Dalapati S, Jin S, Gao J, Xu Y, Nagai A, Jiang D (2013) An Azine-linked covalent organic framework. J Am Chem Soc 135(46):17310–17313

167. Jin E, Li J, Geng K, Jiang Q, Xu H, Xu Q, Jiang D (2018) Designed synthesis of stable light-emitting two-dimensional sp2 carbon-conjugated covalent organic frameworks. Nat Commun 9(1):4143
168. Haldar S, Chakraborty D, Roy B, Banappanavar G, Rinku K, Mullangi D, Hazra P, Kabra D, Vaidhyanathan R (2018) Anthracene-resorcinol derived covalent organic framework as flexible white light emitter. J Am Chem Soc 140(410):13367–13374
169. Wang X, Chen L, Chong SY, Little MA, Wu Y, Zhu WH, Clowes R, Yan Y, Zwijnenburg MA, Sprick RA, Cooper AI (2018) Sulfone-containing covalent organic frameworks for photocatalytic hydrogen evolution from water. Nat Chem 10:1180–1189
170. Vyas VS, Haase F, Stegbauer L, Savasci G, Podjaski F, Ochsenfeld C, Lotsch BV (2015) A tunable azine covalent organic framework platform for visible light-induced hydrogen generation. Nat Commun 6(1):8508
171. Schwinghammer K (2013) Triazine-based carbon nitrides for visible-light-driven hydrogen evolution. Angew Chem Int Ed 52(9):2435–2439
172. Zhang J (2010) Synthesis of a carbon nitride structure for visible-light catalysis by copolymerization. Angew Chem Int Ed 49(2):441–444
173. Sheng JL, Dong H, Meng XB, Tang HL, Yao YH, Liu DQ, Bai LL, Zhang FM, Wei JZ, Sun XJ (2019) Effect of different functional groups on photocatalytic hydrogen evolution in covalent-organic frameworks. ChemCatChem 11(9):2313–2319
174. Patra BC, Khilari S, Manna RN, Mondal S, Pradhan D, Pradhan A, Bhaumik A (2017) A metal-free covalent organic polymer for electrocatalytic hydrogen evolution. ACS Catal 7 (9):6120–6127
175. Chen R, Shi JL, Ma Y, Lin G, Lang X, Wang C (2019) Designed synthesis of a 2D porphyrin-based sp2 carbon-conjugated covalent organic framework for heterogeneous photocatalysis. Angew Chem Int Ed 58(19):6430–6434
176. Fan Z, Nomura K, Zhu M, Li X, Xue J, Majima T, Osakada Y (2019) Synthesis and photocatalytic activity of ultrathin two-dimensional porphyrin nanodisks via covalent organic framework exfoliation. Commun Chem 2(55):1–8
177. Lin G, Ding H, Chen R, Peng Z, Wang B, Wang C (2017) 3D porphyrin-based covalent organic frameworks. J Am Chem Soc 139(25):8705–8709
178. Feng X, Liu L, Honsho Y, Saeki A, Seki S, Irle S, Dong Y, Nagai A, Jiang D (2012) High-rate charge-carrier transport in porphyrin covalent organic frameworks: switching from hole to electron to ambipolar conduction. Angew Chem Int Ed 124(11):2618–2622
179. Spitler EL, Dichtel WR (2010) Lewis acid-catalysed formation of two-dimensional phthalocyanine covalent organic frameworks. Nat Chem 2(8):672–677
180. Wang M, Ballabio M, Wang M, Lin HH, Biswal BP, Han X, Paasch S, Brunner E, Liu P, Chen M, Bonn M, Heine T, Zhou S, Cánovas E, Dong R, Feng X (2019) Unveiling electronic properties in metal–phthalocyanine-based pyrazine-linked conjugated two-dimensional covalent organic frameworks. J Am Chem Soc 141(42):16810–16816
181. Wang M, Wang M, Lin HH, Ballabio M, Zhong H, Bonn M, Zhou S, Heine T, Cánovas E, Dong R, Feng X (2020) High-mobility semiconducting two-dimensional conjugated covalent organic frameworks with p-type doping. J Am Chem Soc 142(52):21622–21627
182. Wang H, Ding H, Meng X, Wang C (2016) Two-dimensional porphyrin and phthalocyanine based covalent organic frameworks. Chin Chem Lett 27(8):1376–1382
183. Bhunia S, Das SK, Jana R, Peter SC, Bhattacharya S, Addicoat M, Bhaumik A, Pradhan A (2017) Electrochemical stimuli-driven facile metal-free hydrogen evolution from pyrene-porphyrin-based crystalline covalent organic framework. ACS Appl Mater Interfaces 9 (28):23843–23851
184. Yan Y, He T, Zhao B, Qi K, Liu H, Xia BY (2018) Metal/covalent–organic frameworks-based electrocatalysts for water splitting. J Mater Chem A 6(33):15905–15926
185. Yang S, Hu W, Zhang X, He P, Pattengale B, Liu C, Cendejas M, Hermans I, Zhang X, Zhang J, Huang J (2018) 2D covalent organic frameworks as intrinsic photocatalysts for visible light-driven CO_2 reduction. J Am Chem Soc 140(44):14614–14618

186. Inoue H, Shimada T, Kou Y, Nabetani Y, Masui D, Takagi S, Tachibana H (2011) The water oxidation bottleneck in artificial photosynthesis: how can we get through it? An alternative route involving a two-electron process. ChemSusChem 4(2):173–179
187. Chena J, Taoa X, Lia C, Mab Y, Taoa L, Zhenga D, Zhud J, Lia H, Lia R, Yang Q (2020) Synthesis of bipyridine-based covalent organic frameworks for visible-light driven photocatalytic water oxidation. Appl Catal B 262:118271
188. Aiyappa HB, Thote J, Shinde DB, Banerjee R, Kurungot S (2016) Cobalt-modified covalent organic framework as a robust water oxidation electrocatalyst. Chem Mater 28(12):4375–4379
189. Li J, Jing X, Li Q, Li S, Gao X, Feng X, Wang B (2020) Bulk COFs and COF nanosheets for electrochemical energy storage and conversion. Chem Soc Rev 49(11):3565–3604. Many references and comparative tables on electrochemical energy conversion using COF can be found in this review
190. Gao Z, Gong LL, He XQ, Su XM, Xiao LH, Luo F (2020) General strategy to fabricate metal-incorporated pyrolysis-free covalent organic framework for efficient oxygen evolution reaction. Inorg Chem 59(7):4995–5003
191. Lu S, Hu Y, Wan S, McCaffrey R, Jin Y, Gu H, Zhang W (2017) Synthesis of ultrafine and highly dispersed metal nanoparticles confined in a thioether-containing covalent organic framework and their catalytic applications. J Am Chem Soc 139(47):17082–17088
192. Jia H, Sun Z, Jiang D, Du P (2015) Covalent cobalt porphyrin framework on multiwalled carbon nanotubes for efficient water oxidation at low overpotential. Chem Mater 27 (13):4586–4593
193. Ren S, Wang J, Xia X (2016) Highly efficient oxygen reduction electrocatalyst derived from a new three-dimensional polyporphyrin. ACS Appl Mater Interfaces 8(39):25875–25880
194. Ma W, Yua P, Ohsaka T, Mao L (2015) An efficient electrocatalyst for oxygen reduction reaction derived from a co-porphyrin-based covalent organic framework. Electrochem Commun 52:53–57
195. Zhao X, Pachfule P, Li S, Langenhahn T, Ye M, Tian G, Schmidt J, Thomas A (2019) Silica-templated covalent organic framework-derived Fe-N-doped mesoporous carbon as oxygen reduction electrocatalyst. Chem Mater 31(9):3274–3280
196. Liu W, Wang C, Zhang L, Pan H, Liu W, Chen J, Yang D, Xiang Y, Wang K, Jiang J, Yao X (2019) Exfoliation of amorphous phthalocyanine conjugated polymers into ultrathin nanosheets for highly efficient oxygen reduction. J Mater Chem A 7:3112–3119
197. Zuo Q, Chenga G, Luo W (2017) Reduced graphene oxide/covalent cobalt porphyrin framework for efficient oxygen reduction reaction. Dalton Trans 46(29):9344–9348
198. Mullangi D, Dhavale V, Shalini S, Nandi S, Collins S, Woo T, Kurungot S, Vaidhyanathan R (2016) Low-overpotential electrocatalytic water splitting with noble-metal-free nanoparticles supported in a sp^3 N-rich flexible COF. Adv Energy Mater 6(13):1600110
199. Theerthagiri J, Lee SJ, Murthy AP, Madhavan J, Choi MY (2020) Fundamental aspects and recent advances in transition metal nitrides as electrocatalysts for hydrogen evolution reaction: a review. Curr Opin Solid State Mater Sci 24(1):100805
200. Shalom M, Ressnig D, Yang X, Clavel G, Fellinger TP, Antonietti M (2015) Nickel nitride as an efficient electrocatalyst for water splitting. J Mater Chem A 3(15):8171–8177
201. Gao Z, Yu Z, Huang Y, He X, Su X, Xiao L, Yu Y, Huang X, Luo F (2020) Flexible and robust bimetallic covalent organic frameworks for the reversible switching of electrocatalytic oxygen evolution activity. J Mater Chem A 8(12):5907–5912
202. Feng X, Gao Z, Xiao L, Laia Z, Luo F (2020) A Ni/Fe complex incorporated into a covalent organic framework as a single-site heterogeneous catalyst for efficient oxygen evolution reaction. Inorg Chem Front 7(20):3925–3931
203. Zhang B, Zheng X, Voznyy O, Comin R, Bajdich M, García-Melchor M, Han L, Xu J, Liu M, Zheng L, de Arquer FP (2016) Homogeneously dispersed multimetal oxygen-evolving catalysts. Science 352(6283):333–337

204. Lin Chun Y, Zhang L, Zhao Z, Xia Z (2017) Design principles for covalent organic frameworks as efficient electrocatalysts in clean energy conversion and green oxidizer production. Adv Mater 29(17):1606635
205. Hammer B, Norskov JK (1995) Why gold is the noblest of all the metals. Nature 376 (6537):238–240
206. Suntivich J, Gasteiger HA, Yabuuchi N, Nakanishi H, Goodenough JB, Shao-Horn Y (2011) Design principles for oxygen-reduction activity on perovskite oxide catalysts for fuel cells and metal-air batteries. Nat Chem 3(7):546–550
207. Zhao Z, Li M, Zhang L, Dai L, Xia Z (2015) Design principles for heteroatom-doped carbon nanomaterials as highly efficient catalysts for fuel cells and metal-air batteries. Adv Mater 27 (43):6834–6840
208. Wang X, Maeda K, Thomas A, Takanabe K, Xin G, Carlsson Johan M, Domen K, Antonietti M (2009) A metal-free polymeric photocatalyst for hydrogen production from water under visible light. Nat Mat 8(1):76–80
209. Yang C, Yang ZD, Dong H, Sun N, Lu Y, Zhang FM, Zhang G (2019) Theory-driven design and targeting synthesis of a highly-conjugated basal-plane 2D covalent organic framework for metal-free electrocatalytic OER. ACS Energy Lett 4(9):2251–2258
210. Guan X, Li H, Ma Y, Xue M, Fang Q, Yan Y, Valtchev V, Qiu S (2019) Chemically stable polyarylether-based covalent organic frameworks. Nat Chem 11(6):587–594
211. Wang S, Xu X, Yue Y, Yu K, Shui Q, Huang N, Chen H (2020) Semiconductive covalent organic frameworks: structural design, synthesis and application. Small Struct 1(2):2000021
212. Vitaku E, Gannett CN, Carpenter KL, Shen L, Abruña HD, Dichtel WR (2019) Phenazine-based covalent organic framework cathode materials with high energy and power densities. J Am Chem Soc 142(1):16–20
213. Zhong H, Ly KH, Wang M, Krupskaya Y, Han X, Zhang J, Zhang J, Kataev V, Büchner B, Weidinger IM, Kaskel S (2019) A phthalocyanine-based layered two-dimensional conjugated metal–organic framework as a highly efficient electrocatalyst for the oxygen reduction reaction. Angew Chem 131(31):10787–10792
214. Wang Q, Ji Y, Lei Y, Wang Y, Wang Y, Li Y, Wang S (2018) Pyridinic-N-dominated doped defective graphene as a superior oxygen electrocatalyst for ultrahigh-energy-density Zn–air batteries. ACS Energy Lett 3(5):1183–1191
215. Yang HB, Miao J, Hung SF, Chen J, Tao HB, Wang X, Zhang L, Chen R, Gao J, Chen HM, Dai L, Liu B (2016) Identification of catalytic sites for oxygen reduction and oxygen evolution in N-doped graphene materials: development of highly efficient metal-free bifunctional electrocatalyst. Sci Adv 2(4):e1501122–e1501132
216. Jiao Y, Zheng Y, Jaroniec M, Qiao SZ (2014) Origin of the electrocatalytic oxygen reduction activity of graphene-based catalysts: a roadmap to achieve the best performance. J Am Chem Soc 136(11):4394–4403
217. Wan Y, Wang L, Xu H, Wu X, Yang J (2020) A simple molecular design strategy for two-dimensional covalent organic framework capable of visible-light-driven water splitting. J Am Chem Soc 142(9):4508–4516
218. Zhai L, Yang S, Yang X, Ye W, Wang J, Chen W, Guo Y, Mi L, Wu Z, Soutis C, Xu Q (2020) Conjugated covalent organic frameworks as platinum nanoparticle supports for catalyzing the oxygen reduction reaction. Chem Mater 32(22):9747–9752
219. Iwase K, Nakanishi S, Miyayama M, Kamiya K (2020) Rational molecular design of electrocatalysts based on single-atom modified covalent organic frameworks for efficient oxygen reduction reaction. ACS Appl Energy Mater 3(2):1644–1652
220. Roy S, Mari S, Sai MK, Sarma SC, Sarkar S, Peter SC (2020) Highly efficient bifunctional oxygen reduction/evolution activity of a non-precious nanocomposite derived from a tetrazine-COF. Nanoscale 12(44):22718–22734
221. Chen S, Zheng Y, Zhang B, Feng Y, Zhu J, Xu J, Zhang C, Feng W, Liu T (2018) Cobalt, nitrogen-doped porous carbon nanosheet-assembled flowers from metal-coordinated covalent organic polymers for efficient oxygen reduction. ACS Appl Mater Interfaces 11(1):1384–1393

222. Yang C, Tao S, Huang N, Zhang X, Duan J, Makiura R, Maenosono S (2020) Heteroatom-doped carbon electrocatalysts derived from nanoporous two-dimensional covalent organic frameworks for oxygen reduction and hydrogen evolution. ACS Appl Nano Mate 3 (6):5481–5488
223. Li Y, Peng P, Huo F, Shao X, Xiang Z (2019) Two-dimensional closed conjugated covalent organic polymers for oxygen reduction reaction. Front Mater Sci 6:244
224. Bhat SA, Das C, Maji TK (2018) Metallated azo-naphthalene diimide based redox-active porous organic polymer as an efficient water oxidation electrocatalyst. J Mater Chem A 6 (40):19834–19842
225. Peng P, Shi L, Huo F, Mi C, Wu X, Zhang S, Xiang Z (2019) A pyrolysis-free path toward superiorly catalytic nitrogen-coordinated single atom. Sci Adv 5(8):aaw2322
226. Royuela S, Martínez-Periñán E, Arrieta MP, Martínez JI, Ramos MM, Zamora F, Lorenzo E, Segura JL (2020) Oxygen reduction using a metal-free naphthalene diimide-based covalent organic framework electrocatalyst. Chem Commun 56(8):1267–1270
227. Li D, Li C, Zhang L, Li H, Zhu L, Yang D, Fang Q, Qiu S, Yao X (2020) Metal-free thiophene-sulfur covalent organic frameworks: precise and controllable synthesis of catalytic active sites for oxygen reduction. J Am Chem Soc 142(18):8104–8108
228. Tavakoli E, Kakekhani A, Kaviani S, Tan P, Ghaleni MM, Zaeem MA, Rappe AM, Nejati S (2019) In situ bottom-up synthesis of porphyrin-based covalent organic frameworks. J Am Chem Soc 141(50):19560–19564
229. Zhou D, Tan X, Wu H, Tian L, Li M (2019) Synthesis of C–C bonded two-dimensional conjugated covalent organic framework films by Suzuki polymerization on a liquid–liquid interface. Angew Chem Int Ed 58(5):1376–1381
230. Wu H, Zeng M, Zhu X, Tian C, Mei B, Song Y, Du XL, Jiang Z, He L, Xia C, Dai S (2018) Defect engineering in polymeric cobalt phthalocyanine networks for enhanced electrochemical CO_2 reduction. ChemElectroChem 19:2717–2721
231. Haase F, Troschke E, Savasci G, Banerjee T, Duppel V, Dörfler S, Grundei MM, Burow AM, Ochsenfeld C, Kaskel S, Lotsch BV (2018) Topochemical conversion of an imine-into a thiazole-linked covalent organic framework enabling real structure analysis. Nat Commun 9 (1):1–10
232. Gogotsi Y, Anasori B (2019) The rise of MXenes. ACS Nano 13(8):8491–8494
233. Anasori B, Lukatskaya MR, Gogotsi Y (2017) 2D metal carbides and nitrides (MXenes) for energy storage. Nat Rev Mater 2(2):1–17
234. Hart JL, Hantanasirisakul K, Lang AC, Anasori B, Pinto D, Pivak Y, van Omme JT, May SJ, Gogotsi Y, Taheri ML (2019) Control of MXenes' electronic properties through termination and intercalation. Nat Commun 10(1):1–10
235. Rao CNR, Chhetri M (2019) Borocarbonitrides as metal-free catalysts for the hydrogen evolution reaction. Adv Mater 31(13):1803668
236. Rao CNR, Lingampalli SR (2016) Generation of hydrogen by visible light-induced water splitting with the use of semiconductors and dyes. Small 12(1):16–23
237. Meinzer N (2017) Catalysis: topology does the water splits. Nat Rev Mater 2(4):17021
238. Chhetri M, Maitra S, Chakraborty H, Waghmare UV, Rao CN (2016) Superior performance of borocarbonitrides, $B_xC_yN_z$, as stable, low-cost metal-free electrocatalysts for the hydrogen evolution reaction. Energy Environ Sci 9(1):95–101

Chapter 17
Ligand-Functionalized Nanostructures and Their Biomedical Applications

Koushik Debnath and Nikhil R. Jana

17.1 Introduction

Nano-biotechnology is one of the rapidly developing subjects of the twenty-first century which has gained enormous attention due to the fascinating properties of nanoparticles such as small size (nanometer ~10^{-9} m) with large surface area, size-dependent property, shape-dependent self-assembly, surface chemistry-dependent function, etc. If we closely observe the dimensions of nanoparticles and compare with atomic and macroscopic worlds, it can be observed that colloidal nanocrystals closely resemble various units of biological world (Fig. 17.1a). This is particularly because the nanometer length scale is similar to many biological entities and biological languages are often written at nanoscale. For example, there are many biological entities such as cellular organelles, antibodies, DNA, and viruses in the nanometer length scale. Similarly, many biological events such as molecular self-assembly of bio-membranes, molecular recognition events, cellular endocytosis, and DNA replication occur in nanometer dimension. As the dimensions of different types of organic and inorganic nanostructures are very similar to biological entities and biological events, they are very useful to study, monitor, and control of many biological processes. Different types of nanostructure have distinct properties. For example, the quantum dot has fluorescence property, iron oxide nanoparticle has magnetic property, silver and gold nanoparticle has plasmonic property, TiO_2 nanoparticle has photocatalytic property, barium titanate nanoparticle has piezoelectric property, and polymeric vesicles have cargo loading capacity (Fig. 17.1b). However, these nanoparticles need to be transformed to functional nanoparticle prior to biomedical application. This is particularly important as nanoparticle either

K. Debnath · N. R. Jana (✉)
School of Materials Science, Indian Association for the Cultivation of Science, Kolkata, India
e-mail: camnrj@iacs.res.in

T. Govindaraju, K. Ariga (eds.), *Molecular Architectonics and Nanoarchitectonics*, Nanostructure Science and Technology,
https://doi.org/10.1007/978-981-16-4189-3_17

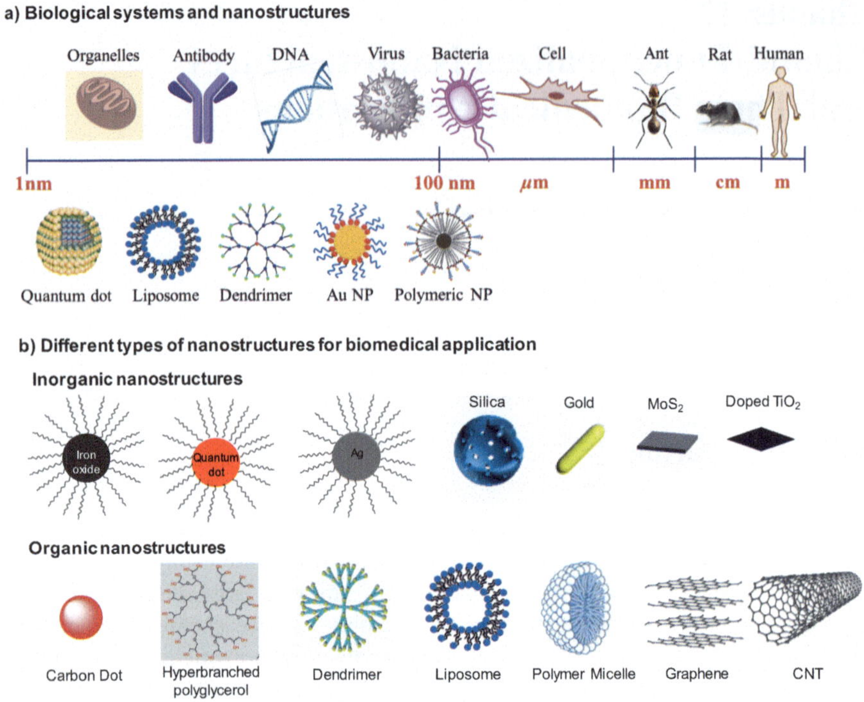

Fig. 17.1 (**a**) Synthetic nanostructures vs. nanostructures in biological world. (**b**) Examples of different synthetic nanostructures that are potentially useful in biomedical application

does not interact with biological interface or interacts non-specifically, but we want to control this interaction as per our requirement. Considering this aspect, most of the as-synthesized nanostructures without appropriate ligand functionalization are useless for any biomedical application.

17.2 Why Ligand Functionalization Is Important for Biomedical Application?

Generally, bare nanoparticles do not exist. Most nanoparticles exist in their colloidal form, or in some limited cases, they exist in a powder form or composite form. Most of the nanoparticle surfaces are covered with ions/molecules or polymer which are responsible for their colloidal stability. Since nanoparticle interacts with the biological interfaces through their surfaces and in particular through the chemical functional groups present on their surfaces, it can be said that the interaction of a nanoparticle with the biological interface is dictated by the surface molecule. Thus, in order to control the interaction, appropriate surface modification is necessary, and this can be achieved only by appropriate ligand functionalization.

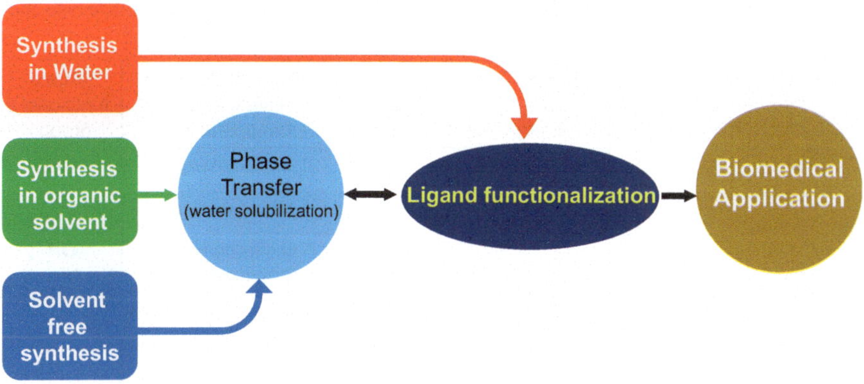

Fig. 17.2 General strategies for the preparation of various nanomaterials

General strategies for the preparation of nanostructure which can be used for biomedical applications are shown in Fig. 17.2. Nanoparticles can be synthesized broadly in three different ways. It can be synthesized in water or organic solvent or in a solvent-free process. The nanoparticle synthesized in water as a medium can be directly functionalized with a proper ligand prior to biomedical applications. In case of synthesis in organic solvent or solvent-free approach, nanoparticles have to undergo additional step of water solubilization prior to ligand functionalization. This is particularly important since the biomedical application requires nanoparticles in their colloidal form in water for the interaction with the biological interface. In the next section of the chapter, we will discuss these two challenges.

17.3 Coating Chemistry for Nanoparticle

Mainly eight types of coating strategies are most powerful to transfer hydrophobic nanostructure into hydrophilic one (Table 17.1). These coating chemistries produce core-shell-type nanoparticle where the shell protects nanoparticle core and offers option for ligand functionalization.

Ligand Exchange This technique is suitable for the nanomaterial having an affinity for thiols. In thiol-based ligand exchange, ligands that are already adsorbed on the surface of the nanomaterial are replaced by thiol as ligands of interest. If the nanomaterial does not have affinity for thiol, the thiolated group cannot replace the ligands from the surface that is already present. This method is applied to CdSe/ZnS quantum dot (QD), Ag/Au nanoparticle, and Au nanorod as all these nanomaterials have affinity for thiol. Typically, hydrophobic ligands are exchanged by hydrophilic thiol-based ligands of interest. Thiol-functionalized polyethylene glycol (PEG), polyaspartic acid, and surfactants are commonly used as a ligand of

Table 17.1 List of coating strategies developed for nanoparticle along with nanoparticle type and chemicals used

Method	Core	Ligand/Polymer used	Interaction between ligand and nanoparticle	Surface functional group
Ligand exchange	CdSe/ZnS, Au, MoS_2	PEG-SH, polymer-SH	Cd-S bonding Au-S bonding Mo-S bonding	-OH $-NH_2$
Silica coating	Fe_3O_4, Au, Ag, CdSe-ZnS, TiO_2, $ZnSnO_3$	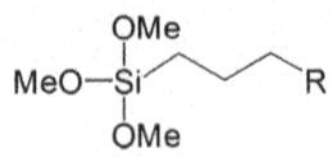 R = $NHCH_2CH_2NH_2$, PO_3Me, PEG	chemisorption, physisorption	-OH $-NH_2$ -SH
Polyacrylate coating	QD, Fe_2O_3, Fe_3O_4, Ag, Au, carbon dot, TiO_2, $BaTiO_3$, Mn doped ZnS	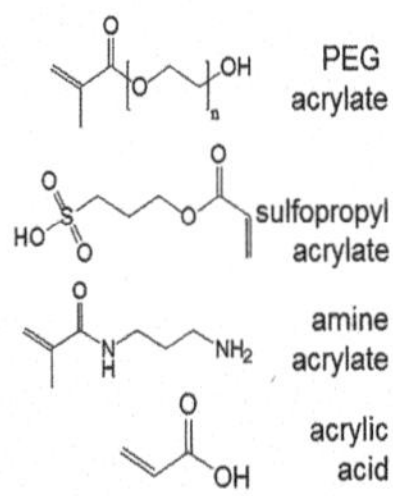	chemisorption, physisorption	-OH $-NH_2$ $-SO_3^-$ -COOH
Poly(maleic anhydride-alt-1-octadecene) coating	QD, Fe_2O_3/Fe_3O_4, Ag, Carbon dot.	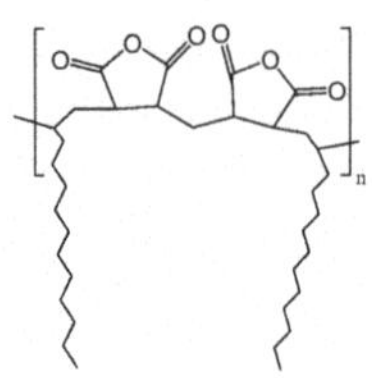	hydrophobic interaction	-COOH $-NH_2$
Poly (aspartic acid) coating	QD, Fe_2O_3/Fe_3O_4, Ag.	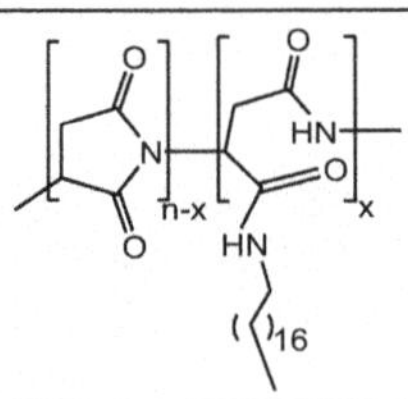	hydrophobic interaction	-COOH $-NH_2$
Poly-dopamine coating	Fe_2O_3, Fe_3O_4, organic nanoparticle.	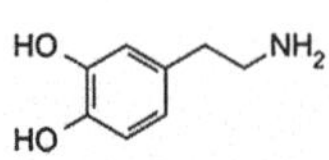	weak interaction	$-NH_2$
Hyperbranched polyglecarol coating	CdSe-ZnS, Fe_2O_3, Mn-ZnSeS, Au	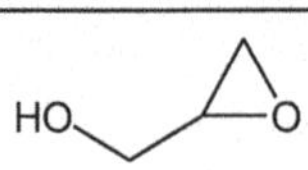	chemical interaction	-OH -COOH

interest in this process. This strategy gives amine or hydroxyl-terminated nanoparticles which can be further functionalized with other biomolecules.

Silica Coating This method is applied to iron oxide, Au, Ag, CdSe/ZnS, TiO_2, $ZnSnO_3$ nanoparticle, etc. This approach has been developed using commercially available different silanes, which contain trimethoxysilyl or silanol groups at one end and amino or phosphonate group at the other end. Trimethoxysilyl groups is hydrolyzed in basic/acidic medium at the nanoparticle surface to produce silica shell around nanoparticle. This method introduces a cross-linked silica shell to protect the core of the nanoparticles from the external environment. For oxide nanoparticles, hydroxyl groups on the nanomaterial undergo condensation with functional trimethoxysilane and form the functional silica-coated nanomaterials. For the silica coating of Au/Ag/QD nanoparticle, mercaptopropyl silane is used that adsorbs at nanoparticle surface. After that, the silica shell formation is initiated in the presence of another functional trimethoxy silanes. This strategy also gives amine- or hydroxyl-terminated nanoparticles which can be further functionalized with other biomolecules.

Poly-acrylate Coating This is a reverse micelle-based coating approach where the hydrophobic nanoparticle can be transformed into core-shell-type hydrophilic nanoparticle. This approach can introduce amine, PEG, sulfate, and carboxylate groups on the surface of the nanoparticles, and their ratio can be tuned as per the requirement and so is the surface charge. The hydrodynamic diameter of the particles can also vary depending upon the polymerization time. In this method the hydrophobic nanoparticles and hydrophilic monomers are taken in reverse micelle of Igepal and cyclohexane mixture. The polymerization is started with the addition of persulfate. The shell thickness is controlled by the duration of the polymerization. Typically, 10, 20, and 40 nm spherical poly-acrylate-coated nanoparticle can be synthesized by varying the polymerization time 15 min, 30 min, and 1 h, respectively. At the end of the polymerization, ethanol is added to break the reverse micelle, and after that, the hydrophilic nanoparticles are separated by centrifugation and dissolved in water. By this approach, almost every hydrophobic nanoparticle such as QD, Fe_2O_3/Fe_3O_4, Ag, Au, TiO_2, graphene, $BaTiO_2$, or doped nanomaterial can be converted into a hydrophilic nanoparticle. This strategy gives amine- or carboxylic acid-terminated nanoparticles which can be further functionalized with other biomolecule.

Polydopamine Coating This coating technique gains special attention because of its unique adhesion property that mimics the protein found in mussels. Monomer dopamine has a catechol moiety and a functional amine group. It can easily undergo self-polymerization through oxidation in a mild alkaline condition to form a polydopamine structure and spontaneously deposit on many surfaces including an organic polymers like cellulose and inorganic materials like Fe_2O_3/Fe_3O_4, Ag, Au, carbon dot, etc. Furthermore, due to the presence of a large number of catechol and imines group in the polydopamine skeleton, amine- and thiol-terminated molecule can be easily grafted on polydopamine-coated materials through Schiff base or Michael addition reaction. One important feature in this method is that the polydopamine coating thickness can be tuned by varying the amount of dopamine and reaction time.

Poly(aspartic Acid) Coating and Poly(maleic Anhydride) Coating Both these polymer coatings are quite similar. Most of the hydrophobic nanoparticles having an oleyl or octadecyl surface group can easily undergo this coating process by hydrophobic-hydrophobic interaction. Both these strategies give –COOH-terminated nanoparticles which can be further functionalized with other biomolecules. One interesting feature in these two coating methods is that in situ biomolecule functionalization can be done at the time of coating.

Hyperbranched Polyglycerol Coating This technique has been developed to overcome the colloidal stability issues in silica coating. In this method, amine-terminated silica-coated nanoparticles are incubated with glycidol that modify surface with hyperbranched polyglycerol. It gives hydroxyl group functionalized nanoparticle with good colloidal stability in physiological conditions with minimal cytotoxicity.

So far we have discussed the methods by which we can overcome our first challenge that is the preparation of dispersed nanoparticle in water with good colloidal stability. Now let us address the second challenge, which is ligand functionalization of nanostructure.

17.4 Bioconjugation Chemistry for Ligand Functionalization of Nanoparticle

There are three critical aspects that need to be kept in mind for the functionalization chemistry. First, mild reaction condition is required because most of the biochemicals and nanoparticles are unstable under harsh reaction condition. In particular, reactions should be performed at <25 °C and preferably in aqueous phase. Second, selection of an appropriate organic reagent is important for efficient chemical conjugation. In particular chemicals should be commercially available. Third is the efficient purification after conjugation. Common methods employed for purification include high-speed centrifuge-based separation of larger nanoparticle, dialysis-based separation of unbound small molecules, and gel permeation chromatography-based purification.

There are five organic chemical reactions which are commonly used for ligand functionalization (Fig. 17.3). EDC coupling involves reaction between primary amine and carboxylic acid. Maleimide reagent-based coupling involves reaction with thiol-based ligand. NHS coupling involves reaction with primary amine-based ligand. Glutaraldehyde-based coupling followed by cyanoborohydride reduction reaction occurs between primary amine and aldehyde. Click chemistry involves reaction between cyanide and azide group. There are other approaches that utilize specific interaction between biotin and streptavidin.

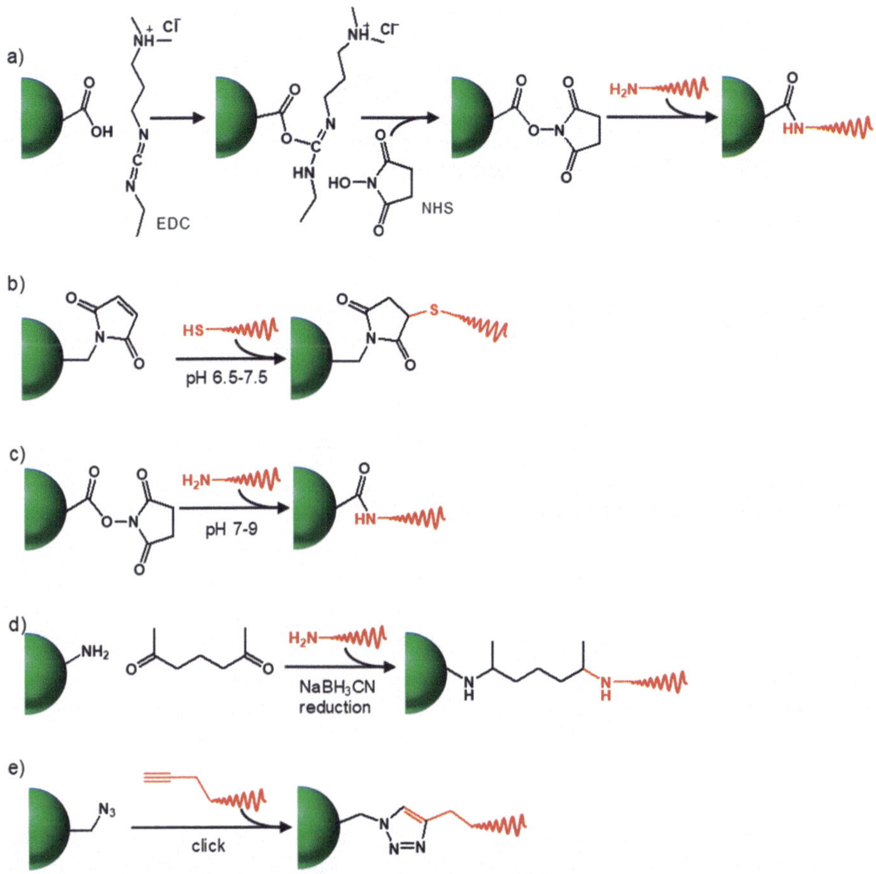

Fig. 17.3 Five common organic chemical reactions that are used for ligand functionalization. In these cases primary amine-/carboxylate-terminated nanoparticle is used for functionalization. (**a**) EDC coupling that involves reaction between primary amine and carboxylic acid using 1-ethyl-3-(3-dimethylaminopropyl)carbodiimide. (**b**) Maleimide-based coupling that involves reaction with thiol-based ligand. (**c**) NHS coupling that involves reaction of N-hydroxysuccinimide with amine-based ligand. (**d**) Glutaraldehyde-based coupling for linking between two primary amines. (**e**) Click chemistry that involves between cyanide/alkynes and azide groups

17.5 Biomedical Applications of Ligand-Functionalized Nanostructures [1–25]

The biomedical applications of ligand-functionalized nanostructure can be broadly divided into six categories (Fig. 17.4). They include *bio-imaging* by using cell/subcellular organelle-specific ligand-functionalized optical nanoprobe, *bio-sensing* by using an antibody-/peptide-functionalized fluorescent nanoprobe, *bio-therapeutics* by using cell/tumor-specific ligand, d*rug protection/delivery* by using cell/tissue/tumor-targeting ligand, *DNA/RNA delivery* by using cell receptor-specific ligand or

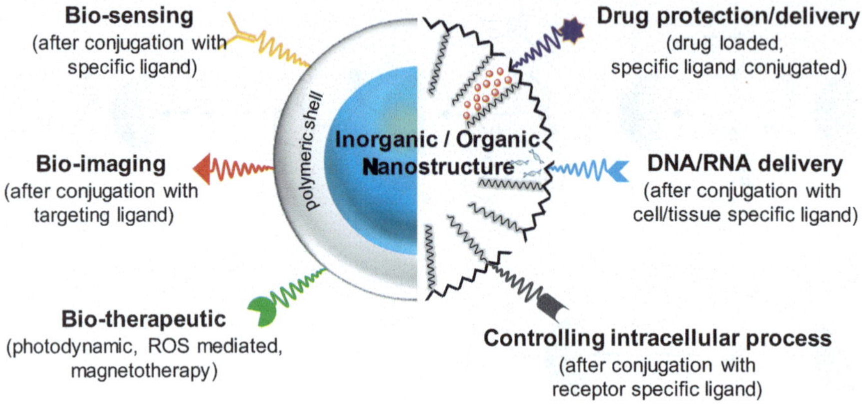

Fig. 17.4 Six most common biomedical applications of ligand-functionalized nanostructure

Table 17.2 Examples of functional nanoparticles developed for bio-imaging application

Nanoparticle	Functional ligand	Molecule/cell/tissue/organ targeting
Cdse/ZnS quantum dot	Oleyl amine	Lipid droplet imaging
	Folate	Targeting/imaging perinuclear region
	Triphenylphosphine (TPP)	Mitochondria imaging
Carbon dot	Riboflavin	Cancer cell imaging
Gold nanoparticle	Glutathione	Malignant tumor imaging
AIE nanoparticle	Lactose	Cancer cell imaging

DNA/RNA protecting ligand-functionalized polymeric vesicle, and finally, *controlling the intracellular process* by using ligand-functionalized nanoparticle.

Bio-imaging Applications This is the most important technique for visualization of selected molecule/cell/tissue/organ and is routinely used for a mechanistic understanding of drug performance, protein functioning, and cellular activity. For this purpose various nanoparticle-based imaging probes are developed with specific advantages over conventionally used molecular probes. In particular QD-based fluorescent nanoprobes are developed with superior brightness and photostability and are expected to be used as new generation medical diagnostic tools (Table 17.2). For example, subcellular imaging of lipid droplet is performed by using oleyl-functionalized QD nanoprobe, QD-folate nanoprobe is used for imaging of the perinuclear region, and QD-TPP nanoprobe is used for mitochondria imaging. For specific cancer cell imaging, riboflavin-functionalized fluorescent carbon dot is used. Glutathione-decorated gold nanoprobe is used for tumor imaging via in vivo mice model. AIE-based nanoprobe with lactose functionality is used for HepG2-based cancerous cell imaging.

Bio-sensing Applications Nanotechnology-based bio-sensing promisingly upgrades the biomedical science. Various biomolecules such as toxic metal ion,

Table 17.3 Examples of nanoparticles that are used for bio-sensing application

Nanoparticle	Ligand used	Method of sensing	Biomarker of disease
Fe_2O_3-Au	4-Mercapto pyridine (SERS probe)	SERS based	Glutathione sensing as cancer biomarker
AIE-peptide	RGKLVFFGR peptide	Fluorescent turn on based	Alzheimer's
QD, Au	DNA receptor and sequence unit	FRET based	Bacteria
Graphene	β-Cyclodextrin	Fluorescent turn on based	Cholesterol
Au nanoparticle	Antisense oligonucleotides	Naked-eye colorimetric detection	Severe acute respiratory syndrome-coronavirus-2 (SARS-CoV-2)

protein, enzyme, cholesterol, glutathione, and viruses are detected. Different detection approaches are explored that include surface-enhanced Raman spectroscopy (SERS), fluorescence, and electrochemical (Table 17.3). For example, iron oxide-gold composite functionalized with 4-mercapto pyridine is used as SERS probe for glutathione sensing. AIE-peptide probe is used for amyloid-beta fibril sensing which is a biomarker for Alzheimer's disease. DNA receptor-conjugated QD and short-DNA sequence-labeled gold nanoparticles are used for specific detection of bacteria and virus via fluorescence resonance energy transfer (FRET).

Bio-therapeutic Applications More than 75 nanomedicines or nanocarriers are already approved by US-FDA and more than 100 are in the pipeline with clinical phase 3 trials. Figure 17.4 highlights some of the examples. Notable examples are folic acid and CD44-functionalized polymeric nanoparticles which are used for targeted delivery of doxorubicin (DOX) in specific tumor cells. Dextran-functionalized reduced graphene oxide is used for efficient delivery of antigens to dendritic cells, potent antigen-presenting cells, and subsequent antigen presentation to initiate the production of activated cytotoxic T cells for cancer immunotherapy. Trehalose-functionalized iron oxide nanoparticles are designed in such a way that not only it can cross the blood-brain barrier but also minimize the required trehalose concentration for the treatment of Huntington disease. Photosensitizer molecule-loaded, folic acid-functionalized PLGA polymeric nanoparticles are used for photodynamic therapy for specific cancer cells (Table 17.4).

Targeted Cargo Delivery and DNA/RNA Delivery Applications Nanoparticles are unique carriers for in vitro/in vivo delivery of drugs, nucleotides, and protein (Fig. 17.5 and Table 17.5). In particular, they can be used to carry water-insoluble/soluble drugs and deliver them at a particular organ with subcellular precision. Generally, liposome, micelle, dendrimer, and porous nanomaterials are used as drug delivery nanocarriers. These carriers are promising, owing to their nanometer size, biocompatibility, colloidal stability, high drug loading capacity, and versatile surface chemistry. For example, in drug delivery application, dopamine-functionalized polymeric micelle is used for the cancerous cell-targeted delivery of hydrophobic curcumin drug. Similar polymeric micelle is used for the protection of

Table 17.4 Examples of functional nanoparticles for therapeutic application

Nanoparticle	Ligand used	Method of therapy	Disease
Hyaluronic acid-based polymer	Folic acid and CD44	Drug therapy	Cancer
Graphene oxide	Dextran	Immunotherapy	Cancer
Poly-acrylate-coated iron oxide nanoparticle	Trehalose	Drug therapy	Huntington's disease
PLGA	Folic acid	Photodynamic therapy	Cancer
Superparamagnetic iron oxide nanoparticles	PEG silane	Magnetic hyperthermia therapy	Cancer
Gold nanorod	RRLA peptide	Photodynamic therapy	Malignant tumor

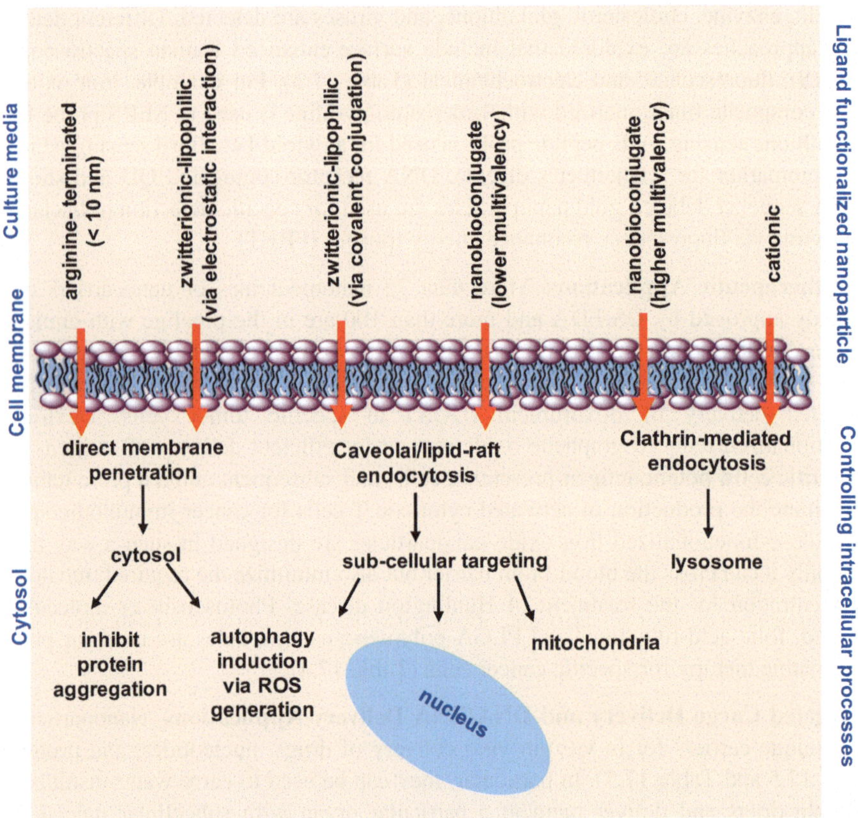

Fig. 17.5 Chemically designed ligand-functionalized nanoparticle to control the cellular uptake mechanism followed by controlling the intracellular process

Table 17.5 Examples of functional nanoparticles for drug delivery application

Nanoparticle	Targeting ligand	Cargo used	Application
Poly(amino acid)-based micelle	Dopamine	Curcumin, doxorubicin	Targeted drug delivery in cancerous cell
Polyaspartic acid-based polymeric micelle	Dopamine	Green tea polyphenol	Increased drug efficiency and targeted drug delivery in neuronal cell
Gold nanoparticle (<10 nm)	Arginine	BSA, lysozyme	Protein delivery to cytosol and nucleus
Chitosan nanoparticle	Tripolyphosphate (TPP)	siRNA	Gene silencing
Poly(amine-co-ester) polymer nanoparticles	Lexiscan	Gene delivery	Gene therapy in brain tumor

green tea polyphenol (EGCG). This system not only protects the drug from degradation but also increases its efficiency as an anti-amyloidogenic agent. Here dopamine molecules are functionalized to target the neuronal cell. In another example different types of lipid-based vesicles are used for targeted si-RNA delivery. These types of nanovesicles not only deliver si-RNA to a targeted organ but also protect si-RNA from enzymatic degradation. Arginine-functionalized small-sized gold nanoparticles are used for protein delivery by direct membrane penetration within a very short period of time.

Controlling Intracellular Process Nanostructures can be chemically designed in such a way that it not only controls the cellular uptake mechanism but also controls the intracellular process (Fig. 17.5). For example, nanoparticles are conjugated non-covalently or covalently with specific ligand in such a way that they can directly penetrate the cell membrane (first two nanostructures from left to right in Fig. 17.5). This can be achieved by electrostatic binding of hydrophobic long chain on nanoparticle surface or by arginine functionalization. Alternatively, zwitterionic lipophilic nanoparticle or nanobioconjugate having lower multivalency is designed for caveolae or lipid raft endocytosis. Moreover, nanobioconjugate having higher multivalency or cationic nanoparticle is designed for clathrin-mediated endocytosis. Direct cell penetration helps for direct cytosolic delivery of cargo like drug, protein, and gene. Caveolae or lipid raft endocytic pathway is important for subcellular targeting, intracellular site-specific reactive oxygen species (ROS) generation, and autophagy induction. In contrast, clathrin-mediated endocytic pathway offers lysosomal trafficking as well as lysosome degradation.

17.6 Challenges and Future Aspect of Ligand-Functionalized Nanostructure for Biomedical Applications

Despite the promising biomedical applications of nanostructures, toxicity of nanomaterial limits its usage. The toxicity of nanostructures depends on a wide range of parameters, like composition, surface charge, renal clearance, and long-term stability. Although various nanoprobes for bio-imaging as well as sensing are already on the market, undoubtedly these technologies are expensive. There are other two challenges in this field. One is large-scale production and another, most difficult challenge is the reproducibility. Various methods have been standardized to overcome such challenges but still it needs more improvement. Apart from these, there are other challenges which need to be addressed like control of multivalency that is the number of biomolecules conjugated per nanoparticle. The number of biomolecules per nanoparticle at surface dictates the interaction with biological interface. For example, it has been observed that higher multivalency (>100 per nanoparticle) leads to nonspecific interaction and substandard targeting option, whereas lower multivalency (<10 per nanoparticle) shows better performance. However, synthesis of nanoprobe with controlled multivalency in the range of 1–100 is a difficult task. Another challenge in nanocarrier-based drug delivery system is inadequate delivery option in brain. Since most of the nanocarriers do not cross the blood-brain barrier, drug delivery in brain is still limited. Another challenging aspect of nanoparticle based therapy is renal clearance. Since 50–100 nm sized nanocarriers are best for tumor-specific drug/gene delivery-based therapy, they do not degrade and hence do not clear from the body after treatment. The accumulation of such nanocarriers can be hazardous and their use is hence limited.

References

1. Saha A, Jana NR (2013) Detection of cellular glutathione and oxidized glutathione using magnetic–plasmonic nanocomposite-based "turn-off" surface enhanced Raman scattering. Anal Chem 85:9221–9228
2. Chakraborty A, Jana NR (2015) Design and synthesis of triphenylphosphonium functionalized nanoparticle probe for mitochondria targeting and imaging. J Phys Chem C 119:2888–2895
3. Pradhan N et al (2015) Detection and monitoring of amyloid fibrillation using a fluorescence "switch-on" probe. ACS Appl Mater Interfaces 7:25813–25820
4. Das P, Jana NR (2015) Dopamine functionalized polymeric nanoparticle for targeted drug delivery. RSC Adv 5:33586–33359
5. Debnath K et al (2016) Efficient inhibition of protein aggregation, disintegration of aggregates, and lowering of cytotoxicity by green tea polyphenol-based self-assembled polymer nanoparticles. ACS Appl Mater Interfaces 8:20309–20318
6. Debnath K, Mandal K, Jana NR (2016) Phase transfer and surface functionalization of hydrophobic nanoparticle using amphiphilic poly (amino acid). Langmuir 11:2798–2807

7. Sinha A et al (2017) Carbohydrate-functionalized rGO as an effective cancer vaccine for stimulating antigen-specific cytotoxic T cells and inhibiting tumor growth. Chem Mater 29:6883–6892
8. Debnath K et al (2017) Poly(trehalose) nanoparticles prevent amyloid aggregation and suppress polyglutamine aggregation in a Huntington's disease model mouse. ACS Appl Mater Interfaces 9:24126–24139
9. Mandal S, Jana NR (2017) Quantum dot-based designed nanoprobe for imaging lipid droplet. J Phys Chem C 121:23727–23735
10. Pal S, Dalal C, Jana NR (2017) Supramolecular host–guest chemistry-based folate/riboflavin functionalization and cancer cell labeling of nanoparticles. ACS Omega 2:8948–8958
11. Mandal K, Jana NR (2018) Galactose-functionalized, colloidal-fluorescent nanoparticle from aggregation-induced emission active molecule via polydopamine coating for cancer cell targeting. ACS Appl Nano Mater 1:3531–3540
12. Ali H, Ghosh S, Jana NR (2018) Biomolecule-derived fluorescent carbon nanoparticle as bioimaging probe. MRS Adv 15:779–788
13. Chakraborty A, Dalal C, Jana NR (2018) Colloidal nanobioconjugate with complementary surface chemistry for cellular and subcellular targeting. Langmuir 45:13461–13471
14. Panja P et al (2020) Surface chemistry-and intracellular trafficking-dependent autophagy induction by iron oxide nanoparticles. ACS Appl Bio Mater 3:5974–5983
15. Debnath K, Jana NR, Jana NR (2019) Designed polymer micelle for clearing amyloid protein aggregates via up-regulated autophagy. ACS Biomater Sci Eng 1:390–401
16. Haliza K, Oya Alpar H (2006) Development and characterisation of chitosan nanoparticles for siRNA delivery. J Control Release 115:216–225
17. Jang B et al (2011) Gold nanorod-photosensitizer complex for near-infrared fluorescence imaging and photodynamic/photothermal therapy in vivo. ACS Nano 2:1086–1094
18. Liu J et al (2013) Passive tumor targeting of renal-clearable luminescent gold nanoparticles: long tumor retention and fast normal tissue clearance. J Am Chem Soc 135:4978–4981
19. Draz MS et al (2014) Nanoparticle-mediated systemic delivery of siRNA for treatment of cancers and viral infections. Theranostics 4:872–892
20. Han L et al (2016) Increased nanoparticle delivery to brain tumors by autocatalytic priming for improved treatment and imaging. ACS Nano 10:4209–4218
21. Bobo D et al (2016) Nanoparticle-based medicines: a review of FDA-approved materials and clinical trials to date. Pharm Res 33:2373–2387
22. Lee JY et al (2016) Dual CD44 and folate receptor-targeted nanoparticles for cancer diagnosis and anticancer drug delivery. J Control Release 236:38–46
23. Tay ZW et al (2018) Magnetic particle imaging-guided heating in vivo using gradient fields for arbitrary localization of magnetic hyperthermia therapy. ACS Nano 12:3699–3713
24. Clement S et al (2018) X-ray radiation-induced and targeted photodynamic therapy with folic acid-conjugated biodegradable nanoconstructs. Int J Nanomed 13:3553–3570
25. Moitra P et al (2020) Selective naked-eye detection of sars-cov-2 mediated by n gene targeted antisense oligonucleotide capped plasmonic nanoparticles. ACS Nano 14:7617–7627

Chapter 18
Biomimetic Composite Materials and Their Biological Applications

Svetlana Batasheva, Ekaterina Naumenko, and Rawil Fakhrullin

18.1 Overview of Drug Delivery with Particulate Vehicles

In the organism, the concentration of a drug compound has to be maintained in a narrow concentration range (therapeutic range). Importantly, the lower concentration is ineffective, while the higher concentration is toxic. Most currently available drug delivery systems aim at sustained supply of a therapeutic agent to maintain its plasma concentration within the therapeutic range. Ideally, the drug should only be delivered to the diseased site in order not to affect the healthy tissues and avoid potential side effects, necessitating the development of the targeted drug delivery systems.

Nanoparticles have emerged as a new and powerful tool for drug transfer to organs and tissues and bioimaging, because of their ability to improve the delivery of poorly soluble, unstable or otherwise insufficiently bioavailable compounds [1]. Versatile nano- and micro-sized vehicles have been proposed, e.g. micelles, liposomes, hydrogels, carbon nanomaterials, polymeric nanoparticles and inorganic nanoparticles either engineered (such as gold, silver, iron, etc.) or available from natural deposits or biological objects (such as clays and bacterial magnetosomes). Increased drug delivery efficacy can be achieved by using composite materials rather than their pure counterparts as drug carriers, because, arguably, only a combination of various functionalities can allow the targeted drug delivery and prolonged or stimulus-responsive drug release.

Nanoparticulate drug delivery vehicles can be built exclusively from synthetic materials or composed of both natural and synthetic materials. The nanoparticles offer protection to the carried drugs, which can be further increased by sealing the

S. Batasheva · E. Naumenko · R. Fakhrullin (✉)
Institute of Fundamental Medicine and Biology, Kazan Federal University, Kazan, Republic of Tatarstan, Russian Federation

T. Govindaraju, K. Ariga (eds.), *Molecular Architectonics and Nanoarchitectonics*, Nanostructure Science and Technology,
https://doi.org/10.1007/978-981-16-4189-3_18

porous or tubular particles orifices with polymeric coatings. The nanoparticle coating provides a physical barrier to the drug outflow; moreover, the surface modification can additionally impart the carrier with stimulus-responsive properties to achieve the site-specific or external stimuli-responsive drug release when the appropriate conditions are encountered by the drug carriers [2]. This can be illustrated by the enzyme-activated intracellular delivery of brilliant green into cancer cells using clay nanotubes (Fig. 18.1) [3]. The antiseptic and malignant cell mitochondria-

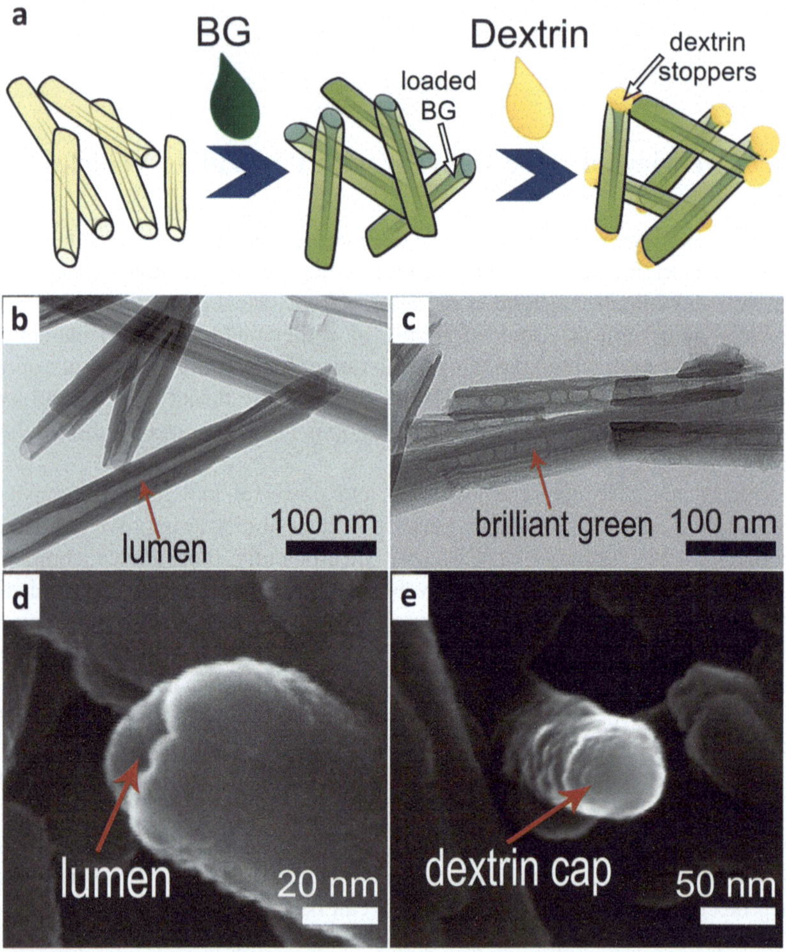

Fig. 18.1 A scheme demonstrating the halloysite nanotube loading with brilliant green and formation of dextrin stoppers (**a**); TEM images of pristine (**b**) and drug-loaded (**b**) halloysite nanotubes; SEM images of a pristine halloysite nanotube (**c**) and nanotube sealed with a dextrin cap (**d**) [3]. The images are reproduced under Creative Commons Attribution 4.0 International License

targeting organic dye brilliant green was loaded into the lumen of the biocompatible halloysite clay nanotubes (HNTs) using the vacuum-facilitated loading procedure. The nanotube ends were closed by physically adsorbed polysaccharide (dextrin) stoppers to secure the intercellular release of the drug. After the addition of the dextrin stoppers, a twofold reduction in the brilliant green release was observed, as compared with the uncoated tubes. The sealed drug-loaded nanotubes penetrated through the cellular membranes into the cell interior, where the intercellular glycosyl hydrolases mediated degradation of the dextrin tube-end stoppers triggering the release of brilliant green from the nanotube lumen. The uptake of the drug carrier nanotubes was related to the cell proliferation rate, and since the drug release depended on the activity of intercellular enzymes, the faster proliferating cells suffered more from the drug than the slower proliferating ones. As the high proliferation rate is a characteristic feature of malignant cells, the formulation based on the drug encapsulation into a nanoparticulate carrier with intercellular stimulus-triggered drug release can be a promising approach for selective elimination of cancer cells, while the non-cancerous cells are predominantly not targeted.

Once a smart and effective drug release system is obtained, the next challenge is to transport it to the desired destination point in the organism. Apparently, the most important determinant of the therapeutic potential of drug-bearing particulate carriers is their ability to remain in blood circulation, which determines both the ability of nanoparticles to accumulate at the target site and the duration of drug delivery by nanoparticles [1]. Certain particles are quickly cleared from blood circulation by the reticuloendothelial system (RES), consisting of circulating monocytes and macrophages of the liver, spleen, lung and bone marrow. These immune cells recognise circulating nanoparticles covered with adsorbed plasma proteins (opsonins) and eliminate them by phagocytosis. Thus, much effort in designing nanoparticulate drug carriers is put into producing nanoparticles with prolonged blood circulation time. Currently, the most extensively applied approaches to increase the nanoparticle blood circulation include modifications of nanoparticles with polyethylene glycol (PEG), PEG-derived polymers and biomimetic polysaccharides, which can alter the binding of plasma proteins to nanoparticles (the 'opsonisation' process) and thus the phagocytosis of nanoparticles by macrophages [1]. The inspiration from biological entities with their naturally evolved mechanisms of immune cell evasion could be of great help in designing long circulating drug carriers.

This chapter is limited exclusively to non-living composite drug carriers, although they can comprise some components of natural origin. The living mammalian and bacterial cells bearing surface-conjugated or internalised drug-loaded nanoparticles (and therefore called 'Trojan horses' and 'microbots', respectively) have been reviewed elsewhere as well as artificial drug carriers reconstituted exclusively from components of natural cells or viruses [4].

18.2 Particles Mimicking Mammalian Cell Architecture and Morphology

One of the ways for efficient drug delivery is by using a kind of stealth technology, where composite vehicles employed for targeted drug delivery mimic the constituent cells of the organism. Such biomimetic composites are fabricated by coating the drug-loaded nanoparticulate core with a natural cell membrane. An artificial vehicle camouflaged by the cell membrane can escape clearance by reticuloendothelial system, which results in prolonged persistence in blood circulation. The coating of nanoparticles with cell membranes derived from erythrocytes, leukocytes, stem cells and platelets [5, 6] allowed obtaining high-performance biocompatible core-shell delivery systems with an extended circulation time. During the study of the interaction between erythrocyte-derived membranes and polymeric nanoparticles, it was found that the nanoparticle cores of varying sizes, ranging from tens of nanometres to several hundred nanometres, could be successfully coated with cell membranes, and the negatively charged nanoparticle cores are more suitable for membrane coating, than the positively charged ones [7].

Cell membrane-coated core-shell drug delivery vehicles can be produced using several different methods [5]:

1. Mechanical extrusion in which the drug-loaded nanoparticulate cores and membrane vesicles obtained from purified membranes of freshly harvested cells are co-extruded through a porous membrane
2. Sonication based technique, in which a core–shell nanostructure is spontaneously formed as a result of ultrasonic treatment of components.
3. A microfluidic technique combining rapid mixing and electroporation
4. In situ packaging of single or multiple nanomaterials into live cells, which then secrete vesicles containing the exogenous nanoparticles.

Polymeric, gelatine, gold, iron oxide, silica nanoparticles and quantum dots have been used as core particles [5]. Additional functionalities can be rendered into the cell-mimicking composites, such as magnetic components for magnetic field-guided delivery, or responsive moieties endowing the vehicle with ability to release the payload in response to natural environmental changes (pH, light, temperature, presence of specific compounds).

Moreover, the natural membrane coating can provide targeting to certain cell populations, for example, tumour cells. High tumour affinity can be achieved by covering the nanoparticulate core with the membranes of bone marrow-derived mesenchymal stem cells, thanks to specific molecular recognition moieties contained in the outer stem cell membrane. Remarkably, mesenchymal stem cells isolated from different species are suitable for tumour targeting, because the mechanism of mesenchymal stem cell targeting is tumour-specific but not species-specific Thus, the membranes of bone marrow-derived mesenchymal stem cells were used to deliver gelatine nanogels, loaded with doxorubicin (DOX) (a widely known anti-cancer drug) to tumour cells [8]. Compared with uncoated nanogels, the coated particles exhibited remarkable stability, tumour-targeting ability and antitumour efficiency both in vitro and in vivo.

The membranes of malignant cells can also be exploited to target tumours because of the intrinsic ability of cancer cells to interact with each other via surface antigens such as carcinoembryonic antigen (CEA), epithelial cell adhesion molecule (EpCAM), N-cadherin, galectin-3 and others, with this homotypic adhesion playing a critical role in metastasis and tumour formation. Cancer cell membrane-coated nanoparticles were used for tumour delivery of anticancer drugs, imaging probes, light-responsive agents for photodynamic therapy and oxygen for overcoming intratumoural hypoxic conditions, and it was noticed that the most efficient vehicles are obtained when the nanoparticles are coated with the membranes derived from the same source cells as the targeted tumour [5]. These homotypic binding molecules can also be applied separately as ligands decorating the surface of synthetic nanoparticles if targeted delivery to tumours is required.

Membrane-cloaked drug-loaded nanoparticles can target not only peer mammalian cells but also pathogenic microorganisms, for instance those which bind to host cells to hide themselves from the immune system. Thus, antibiotic-containing polymeric nanoparticles coated with the platelet membrane allowed for increased binding, better targeting and improved bactericidal efficacy of vancomycin against methicillin-resistant *Staphylococcus aureus* (MRSA) in vitro, because MRSA tends to bind to platelets by their serine-rich adhesin molecules [9]. Mimicry of natural cells also promotes the biding of pathogen-derived toxins to membrane-cloaked nanosponges made of polymeric nanoparticles, allowing the scavenging of bacterial toxins [10].

Engineered nanoparticles that mimic cell external and internal morphology have also been synthesised. The appealing properties of red blood cells (RBCs) such as long blood persistence (~120 days) and flexibility have already inspired the creation of RBC-like particles from organic materials (e.g. PEG or hydrogels), which imitated the size, shape and elasticity of RBCs as well as their oxygen-carrying ability and increased blood circulation time [4]. Biocompatible particles resembling RBC were built by layer-by-layer deposition of alternating layers of positively and negatively charged polymers such as haemoglobin (Hb)/bovine serum albumin (BSA) or poly(allylamine hydrochloride) (PAH)/BSA onto hollow polystyrene spheres or onto a poly(lactic acid-co-glycolide) (PLGA) template [11]. Additionally, the particles made of only Hb layers were fabricated. Then, the shells were cross-linked by glutaraldehyde, and the cores were dissolved. The resulting hollow cell-mimicking particles were similar to the natural RBCs in shape, size, mechanical properties and oxygen-carrying ability and were capable of encapsulating drugs (heparin) and imaging agents (iron oxide nanocrystals used as contrast agents for magnetic resonance imaging).

Another approach to prolong blood circulation of drug-loaded nanoparticles is to attach an RBC membrane protein, CD47, to the nanoparticle surface. This protein serves as a 'marker of self' and prevents the particle phagocytosis by macrophages, which was proved by significantly elongated blood circulation of polystyrene particles after attachment of a recombinant CD47 domain to their surface [4].

Oppositely, a 'marker of non-self' can be added to nanoparticles if induction of effective immune response is needed for the purpose of vaccination. A key role in the immune response to pathogens is played by antigen-presenting cells, which engulf and digest the pathogens and expose the parts of crushed pathogens on their surface together with molecules of major histocompatibility complex. Recognition of these pathogen-associated antigens on the surface of APCs by lymphocytes leads to their activation. Artificial antigen-presenting cells (aAPCs) have been fabricated from nondegradable paramagnetic beads and polystyrene latex as well as biodegradable PLGA microparticles, modified with T-cell receptor-binding recognition ligands and various costimulatory ligands. Occasionally, aAPCs are functionalised with additional adhesion ligands or loaded with soluble cytokines to enhance signalling and T-cell proliferation. The ability of these synthetic APCs to activate and cause expansion and differentiation of naïve T lymphocytes has broad applications for vaccination, cancer immunotherapy and immunosuppression [12].

Other blood cells can also be successfully imitated during fabrication of drug delivery vehicles. Synthetic leukocytes have been obtained by modifying lipid nanoparticles, polymeric vesicles and porous silicone with leukocyte surface proteins or coating them with the leukocyte membrane to target and deliver drugs to inflamed tissues [10]. Artificial platelets built from various materials conjugated with peptide ligands or platelet surface glycoproteins adhered to activated native platelets at the bleeding site and promoted haemostasis [6]. As wounding is often associated with a high risk of infection, the platelet-mimetic particles can be used also to deliver antimicrobial agents. Antimicrobial gold was covalently or non-covalently included into plateletlike deformable hydrogels to make composite particles, having simultaneously blood clotting and antimicrobial activities [13]. The composite hydrogels were synthesised from N-isopropylacrylamide (NIPAm)/co-acrylic acid and decorated with a fibrin-specific antibody resulting in particles morphologically and mechanically similar to native platelets (Fig. 18.2).

Other drug delivery systems imitate the compartmental organisation of eukaryotic cells, which allows concurrent delivery of several spatially separated drugs within a single particulate carrier. A compartmented delivery system may be composed of a number of similar subunits, as in a 'vesosome' (a liposome comprising multiple internal bilayer-enclosed drug-loaded compartments), or may be assembled from blocks with distinct physical and chemical properties, as in micellar structures and multicompartmental solid particles [4]. The complex internal structure of such particles is designed to combine drugs with differing properties and/or to influence the drug release profile.

The sequence of drug release can be regulated by loading one drug in the outer layer and the other in the internal space of the particle. A sophisticated system was developed to allow sequential release of two antitumour compounds, a cytotoxic agent, doxorubicin, and anti-angiogenesis agent, combretastatin [14]. Combretastatin was confined within the lipid bilayer surrounding the inner core of doxorubicin conjugated to biodegradable PLGA polymer. A rapid release of the anti-angiogenesis agent resulted in the collapse of tumour blood vessels, thus trapping the nanocells within the tumours. The subsequent steady release of the

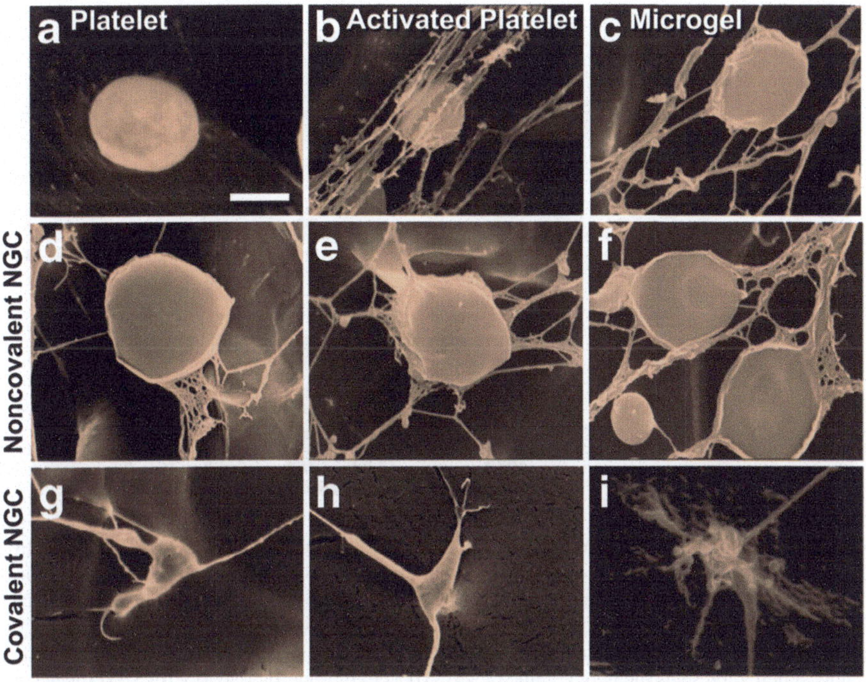

Fig. 18.2 Cryogenic scanning electron microscopy images of natural platelets and platelet-like nanogold composites. The morphology of native circulating platelets (**a**), thrombin-activated platelets forming spindle-like projections (**b**), unloaded platelet-like microgel particles (**c**); microgels with non-covalently incorporated gold nanospheres of varying diameters: 5 nm (**d**), 50 nm (**e**) and 100 nm (**f**); microgels with covalently bound gold, synthesized at three different concentrations of gold (III) chloride (**g–i**), displaying a spindle-like morphology. Reprinted by permission from Springer Nature: Springer Nature, Regenerative engineering and translational medicine (Development of biomimetic antimicrobial platelet-like particles comprised of microgel nanogold composites, Sproul et al.), Copyright (2019)

cytotoxic doxorubicin from the nanocell efficiently killed the tumour cells, thus suppressing the tumour growth.

18.3 Composites Mimicking Bacterial Cells

Similarly to mammalian cell-mimetic composites, fake bacterial cells can be prepared by coating the nanoparticles with membranes derived from bacterial membrane vesicles, the nano-sized lipid spherical bilayer structures secreted by Gram-positive bacteria (extracellular vesicles, EV) and Gram-negative bacteria (outer membrane vehicles, OMV). These false bacteria bear various molecules of parent bacterial cell and thus are capable of targeted delivery of therapeutic cargo and vaccine into immune cells [15]. Some bacterial pathogens like *Mycobacterium*

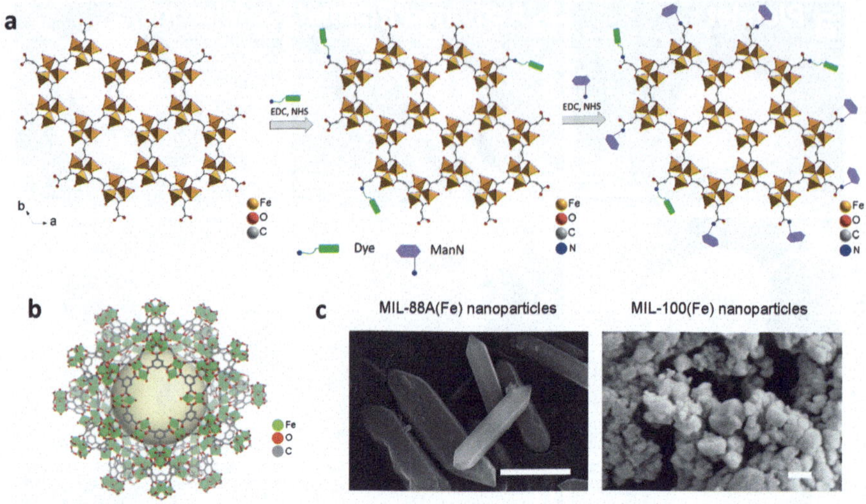

Fig. 18.3 Structure and images of bacteria-mimetic metal-organic framework (MOF) particles. (**a**) Scheme of the structure and post-synthetic modifications of rod-shaped MIL-88A(Fe) particles (orange, Fe(III) octahedra; red, oxygen; gray, carbon; blue, nitrogen; purple, mannosamine; green, Cy5 dye; H atoms have been omitted). (**b**) Scheme of the structure of spherical MIL-100(Fe) particles (green, Fe(III) octahedra; red, oxygen; gray, carbon; H atoms have been omitted. The yellow sphere represents the pore volume of MIL-100(Fe) mesoporous cavity). (**c**) SEM images of MIL-88A(Fe) (scale bar is 3μm) and MIL-100(Fe) (scale bar is 100 nm) nanoparticles. Reprinted by permission from Springer Nature: Springer Nature, Pharmaceutical Research (Metal organic framework (mof) particles as potential bacteria-mimicking delivery systems for infectious diseases: characterization and cellular internalization in alveolar macrophages, Guo et al.), Copyright (2019)

tuberculosis, *Mycobacterium leprae*, *Legionella pneumophila*, *Toxoplasma gondii* and *S. aureus* are capable of surviving inside host macrophages, the phagocytic cells of the immune system, resulting in persistent infections that are very difficult to treat. Different strategies were proposed to kill these bacteria hiding inside the macrophages, using the propensity of macrophages to engulf bacteria-mimicking particles [15]. One approach has used the antibiotic-loaded PLGA nanoparticles coated with the membrane derived from bacterial extracellular vesicles, secreted by the bacteria of the same species as the hiding pathogen [16]. Another approach is based on the modification of nanoparticles with mannose (a sugar moiety often found on the surface of some bacterial species). Using this approach, rod-shaped and spherical iron-based metal-organic framework particles (crystalline materials comprised of the metal ions or clusters bridged by polydentate organic linkers), modified with mannose, were synthesised (Fig. 18.3) [17]. Both approaches resulted in successful internalisation of nanoparticles by macrophages and their delivery into the intracellular location of pathogens.

Bacterial ghosts, which are empty bacterial cell envelopes, bearing characteristic surface components (fimbriae, flagella, polysaccharides) as well as membrane components such as lipopolysaccharides and peptidoglycans, have been also used for targeted delivery of a range of cargos to mammalian cells [18]. The bacterial cytoplasmic contents are removed by formation of a transmembrane tunnel induced by a plasmid-encoded bacteriophage-derived protein E, resulting in formation of bacterial ghosts which can be efficiently absorbed by cancer cells, dendritic cells and macrophages through endocytosis. However, due to the intrinsic immunogenicity of bacterial ghosts, they are preferably used as vaccines rather than conventional drug delivery carriers [4].

Synthetic carriers presenting pathogen-associated traits to immune cells are safer than traditional vaccines based on inactivated microbes. Instead of coating the delivery system with the whole bacterial envelope, individual components of the bacterial pathogen ('pathogen-associated molecular patterns') can be attached to the surfaces of liposomes and polymeric nanoparticles for vaccination purposes [19]. The vaccine delivery platforms can be synthesised from a variety of materials, for example, polyacrylates, polyanhydrides and polyesters, including highly biocompatible poly(lactic acid) (PLA), poly(glycolic acid) (PGA) and their copolymer, PLGA as well as biodegradable natural polymers like chitosan, alginate and gelatine [19]. The particle-attached antigens are much more effective in inducing the immune response than soluble antigens, and different combinations of pathogen-associated traits can be engineered in a single vaccine delivery platform to co-stimulate and guide the response of immune cells [4].

The proteins of the outermost layer of bacterial cells (the surface layer or S-layer) are used for production of biomimetic drug delivery vehicles [18]. For release of S-layer proteins, the bacterial cell envelopes are treated with detergents and hydrogen bond-disrupting agents at a high salt concentration. S-layer proteins can be then reassembled on a drug vehicle and bound to its surface by covalent and non-covalent interactions. Attachment of S-layer proteins has been used to stabilise the solid lipid nanoparticles, encapsulating poorly water-soluble drugs, the lipid membranes of liposomes and 'emulsomes', consisting of a solid lipid core covered with phospholipid bilayers and capable of carrying both hydrophilic and hydrophobic drugs.

Some prokaryotic and eukaryotic cells are motile, using flagella or cilia for swimming; their various swimming mechanisms can be copied in synthetic medical microbots. The behaviour of bacterial and eukaryotic flagella has been already reproduced in synthetic systems, with magnetically driven propulsion, which can potentially be powered by a clinical magnetic resonance imaging (MRI) system. The further development of this technology can lead to fabrication of devices that perform complex tasks, such as targeted drug delivery and biopsy sampling [20].

18.4 Virus-Mimicking Synthetic Delivery Systems

Viruses have evolved as crafty cell invaders, effectively penetrating the cell membrane, delivering their nucleic acids and replicating inside the cell. Thus, viral particles are a good inspiration to scientists involved in designing drug carriers.

Reproducing merely some morphological traits of viral particles in synthetic drug carriers has proven to be efficient to facilitate drug delivery. Filamentous viruses inspired the creation of 'filomicelles'—long filamentous micelles self-assembled from amphiphilic block di-copolymers of PEG-polyethylethylene or PEG-polycaprolactone [21]. The hydrophobic core of filomicelles allowed loading of hydrophobic drugs, like paclitaxel, or fluorescent dyes. These filomicelles were not eliminated by reticuloendothelial system and had an exceptionally long blood circulation time (about 1 week), effectively delivering the anticancer drug paclitaxel to tumour cells in a mouse model, inducing cancer cell apoptosis and tumour shrinkage.

Particulate drug carriers internalised by the cell usually get into the endosomes, while their intracellular targets are located outside the endosomes, so the first task for the carrier is to find the way out. Synthetic virus-like systems have been engineered to enable endosomal escape of vehicles carrying therapeutic nucleic acids. Such systems exploit polymer-attached fusogenic viral or synthetic peptides, as well as polymers with a large number of secondary and tertiary amines ('proton-sponge' polymers) to destabilise and disrupt endosomal membranes [22]. Both of these approaches take advantage of the lower endosomal pH compared with the cytosolic one. Amphipathic fusogenic peptides help viruses to escape the endosomes, because at low endosomal pH, they undergo structural changes, allowing them to penetrate the endosomal membrane and facilitate the fusion of the viral membrane with the endosomal membrane. 'Proton-sponge' polymers prevent acidification of endosomes, causing the endosome to uptake more protons together with counter-ions, which eventually causes osmotic swelling and rupture of endosomal membrane.

Silica nanoparticles mimicking the surface topography of viruses were synthesised. The core silica particles (ca. 200 nm) were studded with smaller shell silica particles (ca.10 nm) to obtain a rough surface, similar to the envelope-spike structure in some viruses, such as herpes simplex virus [23]. The nanoparticles were further modified with polyethylenimine (PEI) to induce a proton-sponge effect. The comparison between the efficiency of delivery of siRNA to KHOS cells (osteosarcoma cell line) by smooth vs. rough, virus-like, particles in order to silence the polo-like kinase 1 (PLK1) gene and decrease cell viability revealed that the nanoscale surface roughness enhanced both binding of the cargo and cellular uptake efficacy.

Moreover, the mechanisms of viral infection can also be reproduced for efficient drug delivery. A pH-sensitive virus-mimetic nanogel [24] consisting of doxorubicin-loaded hydrophobic core and two outer hydrophilic layers with tumour-targeting moieties was synthesised to deliver doxorubicin to cancer cells. The nanogel reversibly swelled at low endosomal pH, causing the rupture of the endosome and the drug-loaded carrier escape into the cytosol. After killing the cell with doxorubicin, the nanogel moved to the adjacent cells and the whole cycle repeated.

18.5 Drug Delivery Vehicles Imitating Antibody-Antigen Interactions

Antibodies are protein molecules that play a key role in the immune response by recognising and selective binding to specific epitopes, the characteristic parts of larger molecules manifesting the presence of pathogens or malignant cells. Due to exceptionally high specificity of antibody-antigen interactions, the attachment of antibodies to synthetic drug carriers is a frequently used way to achieve a targeted drug delivery [25]. Obtaining antibodies to a certain antigen usually includes immunisation of an animal with the antigen, followed by selection and production of a specific antibody in cultured immortal cells.

Alternatively, artificial antibodies can be attached to drug carriers. Artificial antibodies are synthetic materials created by molecular imprinting process, involving the molecular template-guided formation of selective sites in materials. The imprinted materials 'remember' their template and thus are capable of selective recognition of corresponding epitopes, much alike the natural antibodies. There are two main epitope imprinting techniques, differing in the way the template is presented: epitope bulk imprinting and epitope surface imprinting [26].

In epitope bulk imprinting, the imprinted materials are prepared by polymer self-assembly or radical polymerisation occurring in the mixture of the solvated epitope, functional monomers, initiators and cross-linkers. After polymerisation, the templates are extracted, and the obtained imprinted materials are crushed and sieved to achieve the desired particle size and expose the recognition sites. In epitope surface imprinting, the epitope template is immobilised on a supporting matrix, which is either retained or deleted from the resulting imprinted material after template removal.

Artificial antibodies can replace their more costly natural counterparts currently used in procedures aimed at the detection and examination of biological targets, such as antibody-based affinity chromatography or immunosorbent assays. Moreover, they significantly extend the diversity of recognised molecular targets, including even those biological targets that are not easily addressed by natural antibodies. For example, artificial antibodies against post-translationally modified proteins such as glycoproteins, phosphoproteins, acetylated and hydroxylated proteins can be obtained, while it is very difficult currently to obtain natural antibodies against these proteins [26]. The comprehensive mapping of post-translationally modified proteins is a task of uttermost importance in modern biology, because such protein modifications play a crucial regulatory role in the organism, being involved in cell signalling, gene expression, immune responses and cell adhesion, and thus also implicated in several diseases, including cancer. In epitope imprinting, glycans, monosaccharides and polyhistidine-tags (His-tags) can be applied as templates, while they are usually not considered epitopes in general immunology [26]. Thus, protein recognition with artificial antibodies can find numerous biological applications, such as disease biomarker detection, selective depletion of high-abundance proteins as a part of proteomic sample pretreatment, purification of recombinant proteins (e.g. those marked with a His-tag), designing new sensors and others.

Selective recognition of biomarkers is crucially important in targeted drug delivery. Similarly to natural antibodies, the imprinted materials can be applied for targeting drug-loaded nanoparticles to receptors associated with specific cell populations. Thus, molecularly imprinted polydopamine was synthesised on the surface of silica nanoparticles using DOX and an epitope of HER2 protein (human epidermal growth factor receptor overexpressed in 30% of ovarian cancers) as templates [27]. The imprinted nanoparticles adsorbed DOX and delivered it in vivo to cancer cells in an ovarian cancer mouse model, resulting in suppressed tumour growth and increased lifespan of mice. Being modified with fluorescent dyes or carbon nanodots, the epitope-imprinted materials are used as imaging tools for probing cancer cell [26].

Even the whole cells, either prokaryotic or eukaryotic, can be used as a template [28] to construct the so-called colloid antibodies. In this case, the methods of nanoarchitectonics are coupled with cell surface engineering, a chemical methodology to deliberately modify cell walls and membranes. First, an artificial shell is created around the cell template, and the characteristic features of the cell surface become imprinted in the shell. After the destruction of the cell, an empty shell is obtained, which is capable of selective recognition of template-resembling cells in a mixed cell population. Other components can be also included to impart the shells with desirable properties, such as magnetic field guidance or delivery of therapeutic and imaging agents. When human cervical cancer (HeLa) cells were used as a template, the obtained halloysite-modified silica imprints could successfully detect human cells in a mixture with yeast cells [29]. HNTs were included into these imprints to improve the imprint mechanical properties and model an attached drug container. Colloid antibodies against microbial cells can even selectively recognise their template microorganism in a mixture of two microorganism species. Selective photothermal killing of yeast cells mixed with *Bacillus subtilis* cells was achieved by binding of colloid antibodies formed from silica shells with deposited gold nanoparticles, followed by irradiation with laser light [30].

18.6 Biomimetic Materials for Tissue Engineering

Tissue engineering is an emerging interdisciplinary research area in biomedicine aimed to developing the biological substitutes for replacing damaged or lost tissues. Donor organ transplantation has been utilised as a conventional method for replacement of damaged organs or tissues. However, certain limitations, such as donor organ shortage, risk of rejection due to immunological incompatibility, high costs and postsurgery complications, severely limit donor organ transplantation [31]. In addition, ethical issues and doubtful biological safety of transplants together with lifelong immunosuppressive therapy drugs limit the wide-scale use of donor organ transplantation in clinical practice [32, 33].

The emerging concept of using engineered tissue-like structures for regenerating or replacing the lost or damaged biological tissues led to a rapidly growing interdisciplinary field of science potentially capable of revolutionising surgery and transplantology. Tissue engineering combines cells and growth factors with engineering strategies that use synthetic and natural materials for maintaining and guiding the cellular growth and proliferation. Artificial tissues and tissue-like constructs are commonly produced using cell support scaffolds, which may be fabricated from biocompatible and/or biodegradable polymers seeded with certain types of cells depending on the desired tissue-engineered construction. These scaffolds allow for creating a specific niche for cells, which, due to the combination of physicochemical properties, affects the manifestation of the functional properties of cells and contributes to their correct configuration. Furthermore, a properly selected scaffold material serves as an artificial extracellular matrix, which is subsequently replaced by natural matrix during cell activity. The following materials' properties, e.g. porosity, pore size, surface area, interconnectivity and diameter of fibres, strongly affect cellular behaviour [34]. Polymeric composites made using renewable natural materials are excellent platforms for the fabrication of tissue engineering scaffolds. Two widespread groups of scaffold materials are proteins and polysaccharides due to their biosafety, bioresorbability, good immune compatibility with biological tissues, thermal stability, the ability to form fibrous and porous structures as well as thin films [35]. The porosity of scaffolds plays an important role in the transport of nutrients within the matrix, diffusion of gases and cell proliferation [34, 36]. Pore size has an important effect on cell proliferation; moreover, it can promote differentiation of stem cells in a certain direction [37].

The key stage in artificial tissue formation is the selection of a suitable culture medium containing growth factors, followed by incubation of the scaffold with seeded cells in it. A good medium should support the growth and proliferation of cells and, in the case of stem cells, promote their desirable differentiation. The distribution of cells according to a given pattern, which is characteristic of the replaced tissue, is achieved by using scaffolds with certain mechanical characteristics. After spreading of cells within the scaffold, the artificial tissue is ready to be implanted into the human body. The ultimate fate of the scaffold depends on the tissue type and localisation; it can eventually be either absorbed or dissolved in the surrounding native tissues. The materials used for the fabrication of scaffolds need to be biocompatible and non-toxic, lacking any pronounced immune response after implantation into the body. Biodegradable materials should be selected taking into account their degradation kinetics to maintain an appropriate balance between scaffold resorption and natural tissue regeneration. Adequate porosity is an important parameter, essential for the delivery of nutrients to the growing cells, and can form the appropriate nanotopography to promote cell adhesion and proliferation [32].

Overall, tissue engineering demonstrates great potential, which is reflected by a growing number of publications and case reports of successful clinical applications. The main directions of the formation of tissue-engineered structures are the fabrication of artificial structures similar to the natural ones, patching up the damaged area of tissue before its natural regeneration with the replacement of its functions and guiding of recovery processes due to the growth of certain types of cells and the selection of the optimal composition for the scaffolds. Alternatively, scaffold-free approaches are developed. Herein we introduce the reader to (1) magnetic cell surface engineering for fabrication of scaffold-free multicellular tissue-mimicking constructs and (2) application of HNTs for fabrication of biocompatible and robust polymer scaffolds for tissue engineering.

Cells can be directly modified with nanomaterials to modulate their mechanical properties and obtain structures capable of spatial control due to various impulses. MNPs have attracted the attention of researchers in the field of tissue engineering due to their ability to spatially organise the cells [38–40]. In addition, the mechanical forces induced by a magnetic field contribute to the differentiation of stem cells in various directions (chondrogenic [41, 42], adipogenic [43] and mesodermal cardiac pathways [44]). MNPs are a promising tool to assemble the magnetically labelled cells into tissue prototypes [38, 40] and multicellular clusters [3, 45]. Tissue engineering based on magnetically responsive cells offers a scaffold-free arrangement of the cells, allowing to avoid using any scaffolds, both natural and synthetic, and fabricating of artificial tissues with elaborate morphologies guided by the application of an external magnetic field [40]. Magnetically labelled cells and tissue-like constructs can be easily delivered or immobilised within the body controlled by an external magnetic field [46, 47] or visualised using MRI [46–48]. In particular progenitor neuronal cells functionalised using polycation-stabilised MNPs (Fig. 18.4a) can be fully controlled in 2D culture before attachment by an external magnetic field with subsequent formation of clusters with a predetermined shape [39].

Magnetic nanoparticles can be used as an advanced technique in human tissue engineering methodology to drive tissue regeneration. Viable MNP-coated lung carcinoma cells (A549) and human skin fibroblast (HSF) cells were employed to fabricate two-layered scaffold-free lung tissue-mimicking multicellular clusters [38]. Briefly, this technic is based on introduction of MNP-coated HSF into culture plates equipped with 3 mm NdFeB cylindrical magnets under the wells. The cells were incubated for 24 h; then magnetic A549 cells were assembled precisely above the magnetically retained HSF cells under the influence of magnetic field and were incubated for another 24 h (Fig. 18.4b). The resulting two-layered architecture of the clusters is supported by the proliferating viable cells. MNPs can be clearly seen between the cells as brown aggregates which allow for further magnetic manipulation to effectively control the spatial positioning of the clusters. Artificial lung tissue fragments exhibit the typical lung tissue porous alveoli-mimicking morphology with relatively large mature and smaller emerging pores (Fig. 18.4c). This suggests that the magnetic modification of human cells with MNPs does not inhibit the functionality of the cells; therefore, the magnetically functionalised human cells can reconstitute the natural tissue.

Surface modification of human cells with polyelectrolyte-modified MNPs was also applied for the fabrication of three-dimensional multicellular spheroids [3, 49]. 3D-cell culture models or multicellular spheroids recently have found

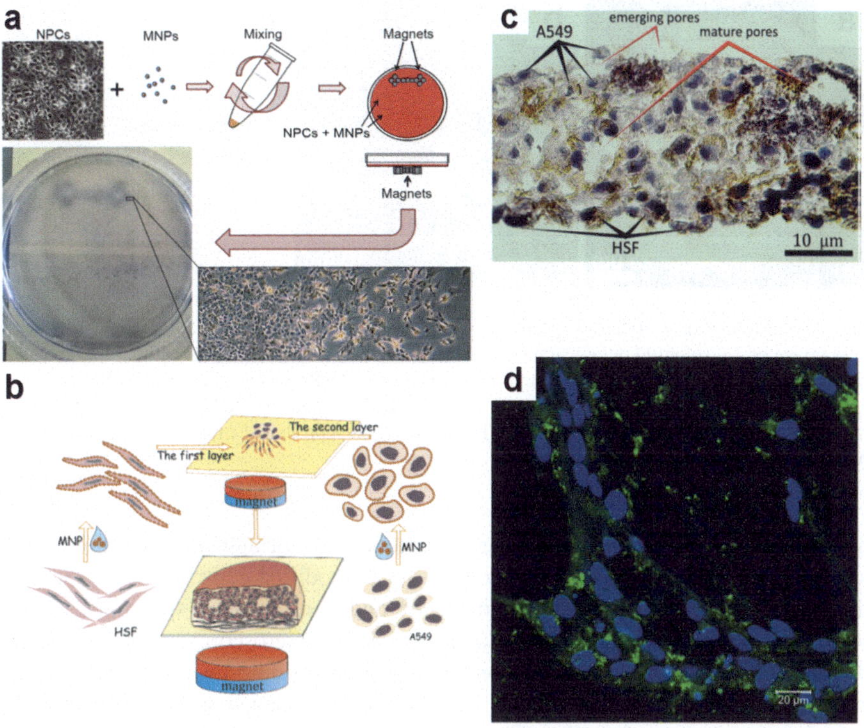

Fig. 18.4 Magnetic modification of cells for the spatial manipulation and tissue engineering. (**a**) A scheme demonstrating the spatial control over the MNP-coated neuronal progenitor cells (NPCs). (**b**) A sketch demonstrating the protocol of two-layered lung tissue-like construct using magnetically modified A549 cells and HSF under the influence of magnetic field. (**c**) Hematoxylin-eosin-stained magnetically facilitated lung-mimicking multicellular cluster; note the brown aggregates of MNPs predominantly distributed in extracellular space. (**d**) FITC-conjugated MNPs distributed inside the cytoplasm and on the cell surface (green); nuclei are stained with DAPI (blue). (**a**, **d**) Reprinted from Nanomedicine: Nanotechnology, Biology and Medicine, V. 20, I. Guryanov, E. Naumenko, S. Konnova, M. Lagarkova, S. Kiselev, R. Fakhrullin, Spatial manipulation of magnetically-responsive nanoparticle engineered human neuronal progenitor cells, 102038, Copyright (2019), with permission from Elsevier. (**b**, **c**) Reproduced from Dzamukova et al. [38] with permission from The Royal Society of Chemistry

application in such biomedical areas as cancer research [50] and pharmaceutical screening [51]. Moreover, multicellular spheroids are used as building blocks for tissue engineering [52]. Artificial tissues can be formed by assembly of multicellular spheroids with magnetic functionalities, wherein the magnetic nanoparticles have been introduced into cells prior to fabrication of spheroids. Previously, it was shown that normal and malignant cells can be modified by magnetic nanoparticles in various ways [53, 54] depending on cellular uptake. Cell membrane surface deposition of MNPs is a universal method of various cell type magnetisations to obtain three-dimensional cellular structures with magnetic behaviour [3]. Magnetic

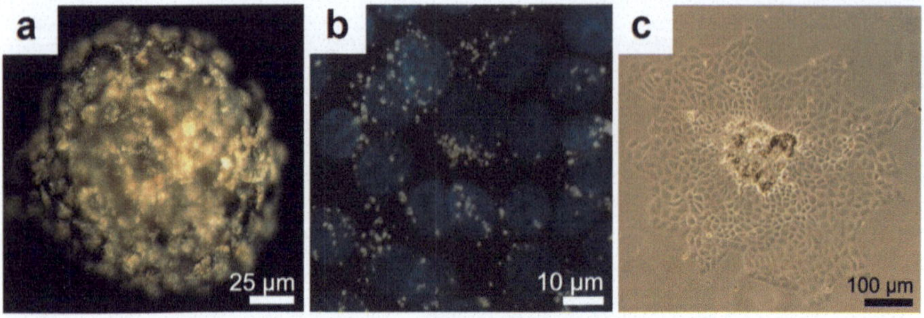

Fig. 18.5 Multicellular spheroids produced using magnetically modified A549 cells. (**a**) A dark-field image of 7-d spheroid demonstrating the diffused MNP distribution at the cellular membranes. (**b**) A merged fluorescence (DAPI) and EDF image of MNPs inside cells. (**c**) Proliferation of spheroid cells after adhesion onto surface of cultural plate (note the MNPs (brown aggregates) remaining at the centre of the spheroid). The images are reproduced under Creative Commons Attribution 4.0 International License

nanoparticles deposited onto cellular membranes do not alter the morphology of spheroids in comparison with spheroids from unmodified cells grown in hanging drops with similar growth rates of magnetic and non-magnetic spheroids. MNPs were imaged in situ using dark-field hyperspectral microscopy, confirming their diffuse distribution on the cell membranes as well as inside cells in the multicellular spheroids, which corresponds well with the results obtained using different cells growing on cultural plates (Fig. 18.5a). Importantly, the MNPs do not penetrate the cellular nuclei, thus reducing the risk of genotoxicity [3, 39] (Figs. 18.4d and 18.5b). The cells within the magnetically modified spheroids remain viable and can be attached onto the surfaces and colonise the substrate with MNP-functionalised cells at the centre of the spheroid, whereas the proliferating cells bear no visible MNPs at the membranes (Fig. 18.5c).

Nanomaterials are helpful to obtain structures with specified mechanical properties that allow eliminating the discrepancy between the artificial inserted construct and the surrounding tissue. Polymer composites supplemented with different types of nanoparticles are becoming a standard platform for fabrication of tissue engineering scaffolds. Nanomaterials can effectively modify the properties of polymer scaffolds, improving considerably a number of practically relevant parameters if compared with the unmodified polymer scaffolds. In general, the goal of tissue engineering is to create universal approaches to the formation of artificial tissues allowing for scalable and reproducible fabrication of biocompatible supports suitable for implantation. Clay nanomaterials are of particular interest for the design of universal scaffolds, with the property modulation by changing the amount of introduced nanoparticles and their modification. Modification of biopolymer composites with platy and tubular nanoclays increases the mechanical strength of scaffolds and stimulates the cell adhesion [55–57].

Among clay nanomaterials HNTs have been studied in recent years as a promising material for various applications, including tissue engineering, as a perspective filler for the fabrication of polymeric tissue engineering scaffolds [58, 59]. Halloysite is a naturally occurring aluminosilicate mineral, having a nanotubular structure with length from 200 to 2000 nm, and the outer and inner diameters are in the range of 50–70 and 10–20 nm, respectively. Biological safety of this nanomaterial has been confirmed for various organisms, from unicellular to multicellular [60–62] within the concentration range of 0.5–2.5 mg mL^{-1}.

When using various nano-additives in the scaffold material, it is necessary to take into account that porosity plays a crucial role in maintaining cell growth and normal functioning prior to implantation into the body. Pure halloysite is a promising filler for scaffolds, because it does not reduce the porosity and pore size of the polymeric scaffolds. Two types of porous hydrogels obtained by lyophilisation based on hyaluronic acid and a mixture of chitosan, agarose and gelatine with the addition of different amounts of halloysite nanotubes demonstrated the retention of the porous structure during nanomodification with pores between 50 and 500μm in diameter. In addition, an increase in the roughness of the pore walls was shown in proportional dependence on the amount of halloysite introduced into the composition of the scaffold material (Fig. 18.6a, b).

Even very high (80 wt%) content of halloysite in scaffold material does not interfere with cell proliferation [57]. At the same time, even low amounts of nanotubes as 3–6 wt% already lead to changes in mechanical stability and moisture capacity. Nanomodified scaffolds combine the properties of a biopolymer matrix and nanodopant, allowing for production of structures with desired parameters and improved physicochemical properties for better cell attachment and migration. Different types of cells, including cancer, immortalised, and primary (healthy), can spread on the nanoclay modified scaffolds, penetrate and migrate within the scaffold material similarly to a native extracellular matrix [56]. Incorporation of halloysite in the scaffold material promoted the increase of thermal and mechanical stability of scaffolds.

Thus, the studies of the last decade allow us to assert that clay nanomaterials are a promising tool for solving the problem facing researchers in the field of tissue engineering to create universal biocompatible scaffolds with the ability to modulate the main physicochemical parameters.

18.7 Conclusion

Biomimetic composite materials have the potential to revolutionise the medical and biological fields by creating new therapeutic approaches and novel research tools to further advance our understanding of living organisms. The limited space of the chapter does not allow a detailed description of the versatility of all biomimetic systems that are under study at the moment. For instance, in addition to the described fields of tissue engineering and drug delivery, biomimetic approach is used to fabricate antimicrobial surfaces that imitate the complex topography and hydrophobicity of insect wings, crab carapace and shark skin, capable of preventing the

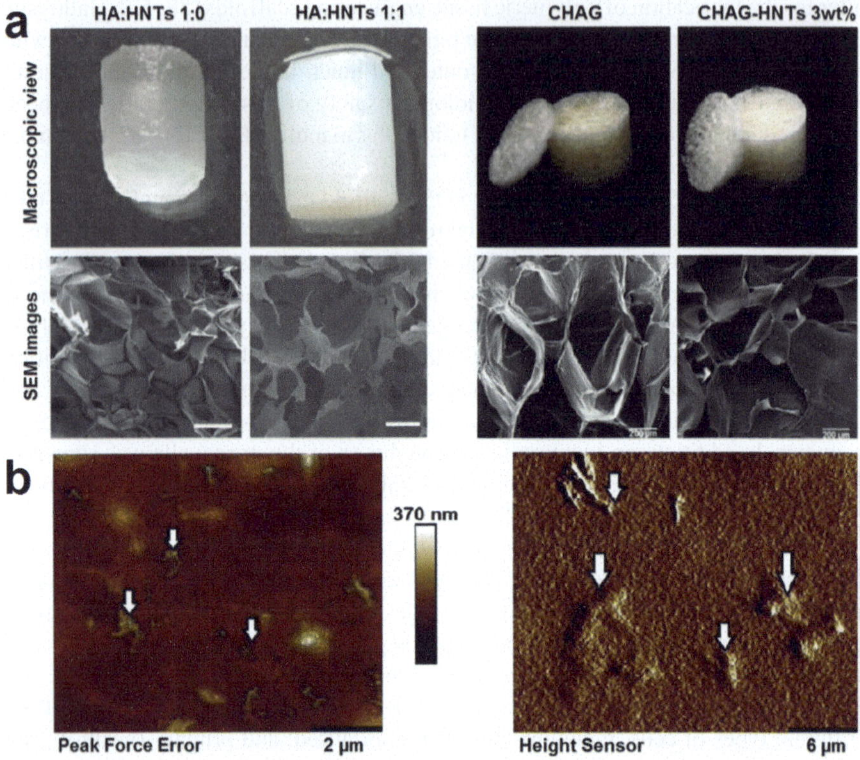

Fig. 18.6 Composite superporous cryogels: (**a**) Macroscopic view and scanning electronic microscopy (SEM) images of superporous hyaluronic acid (HA) (Reprinted from International Journal of Biological Macromolecules, Vol 130, S.S. Suner, S. Demirci, B. Yetiskin, R. Fakhrullin, E. Naumenko, O. Okay, R. S. Ayyala, N. Sahiner, Cryogel composites based on hyaluronic acid and halloysite nanotubes as scaffold for tissue engineering, 627–635, Copyright (2019), with permission from Elsevier) and chitosan/gelatine/agarose (CHAG) cryogels doped with HNTs (reproduced from Naumenko et al., 2013 with permission from The Royal Society of Chemistry). (**b**) HNT distribution within film polymeric scaffolds. Arrows indicate the single halloysite nanotubes and agglomerates incorporated into scaffold material. Reprinted from Biotechnology Journal, Vol 14, E. Naumenko, R. Fakhrullin, Halloysite Nanoclay/Biopolymers Composite Materials in Tissue Engineering, Art. 1900055, Copyright (2019), with permission from John Wiley & Sons

attachment of microbes and biofilm formation [10]. Although some of the characteristics of biological objects have been successfully copied in synthetic carriers and tissue engineering scaffolds, much more work must be done to reach the exceptionally high efficiency of biological systems. Moreover, the progress of biological science will certainly reveal new mechanisms and structure-function relations that could inspire the generation of new biomimetic ideas.

Acknowledgement The reported study was funded by RFBR, project number 20-015-00353 A.

References

1. Yoo JW, Chambers E, Mitragotri S (2010) Factors that control the circulation time of nanoparticles in blood: challenges, solutions and future prospects. Curr Pharm Des 16 (21):2298–2307
2. Lu Y, Aimetti AA, Langer R et al (2017) Bioresponsive materials. Nat Rev Mater 2:16075
3. Dzamukova MR, Naumenko EA, Lvov YM et al (2015) Enzyme-activated intracellular drug delivery with tubule clay nanoformulation. Sci Rep 5:10560
4. Yoo JW, Irvine DJ, Discher DE et al (2011) Bio-inspired, bioengineered and biomimetic drug delivery carriers. Nat Rev Drug Discov 10(7):521–535
5. Fang RH, Kroll AV, Gao W et al (2018) Cell membrane coating nanotechnology. Adv Mater 30 (23):1706759
6. Sen Gupta A (2017) Bio-inspired nanomedicine strategies for artificial blood components. WIREs Nanomed Nanobiotechnol 9(6):e1464
7. Luk BT, Hu CMJ, Fang RH et al (2014) Interfacial interactions between natural RBC membranes and synthetic polymeric nanoparticles. Nanoscale 6(5):2730–2737
8. Gao C, Lin Z, Jurado-Sánchez B et al (2016) Stem cell membrane-coated nanogels for highly efficient in vivo tumor targeted drug delivery. Small 12(30):4056–4062
9. Hu C-MJ, Fang RH, Wang K-C et al (2015) Nanoparticle biointerfacing by platelet membrane cloaking. Nature 526(7571):118–121
10. Chee E, Brown AC (2020) Biomimetic antimicrobial material strategies for combating antibiotic resistant bacteria. Biomater Sci 8(4):1089–1100
11. Doshi N, Zahr AS, Bhaskar S et al (2009) Red blood cell-mimicking synthetic biomaterial particles. Proc Natl Acad Sci U S A 106(51):21495–21499
12. Balmert SC, Little SR (2012) Biomimetic delivery with micro- and nanoparticles. Adv Mater 24:3757–3778
13. Sproul EP, Nandi S, Chee E et al (2020) Development of biomimetic antimicrobial platelet-like particles comprised of microgel nanogold composites. Regen Eng Transl Med 6:299–309
14. Sengupta S, Eavarone D, Capila I et al (2005) Temporal targeting of tumour cells and neovasculature with a nanoscale delivery system. Nature 436(7050):568–572
15. Holay M, Guo Z, Pihl J et al (2021) Bacteria-inspired nanomedicine. ACS Appl Bio Mater 4(5):3830–3848
16. Gao F, Xu L, Yang B et al (2019) Kill the real with the fake: eliminate intracellular *Staphylococcus aureus* using nanoparticle coated with its extracellular vesicle membrane as active-targeting drug carrier. ACS Infect Dis 5(2):218–227
17. Guo A, Durymanov M, Permyakova A et al (2019) Metal organic framework (MOF) particles as potential bacteria-mimicking delivery systems for infectious diseases: characterization and cellular internalization in alveolar macrophages. Pharm Res 36(4):53
18. Farjadian F, Moghoofei M, Mirkiani S et al (2018) Bacterial components as naturally inspired nano-carriers for drug/gene delivery and immunization: set the bugs to work? Biotechnol Adv 36(4):968–985
19. Demento SL, Siefert AL, Bandyopadhyay A et al (2011) Pathogen-associated molecular patterns on biomaterials: a paradigm for engineering new vaccines. Trends Biotechnol 29 (6):294–306
20. Nelson BJ, Kaliakatsos IK, Abbott JJ (2010) Microrobots for minimally invasive medicine. Annu. Rev Biomed Eng 12:55–85
21. Geng YA, Dalhaimer P, Cai S et al (2007) Shape effects of filaments versus spherical particles in flow and drug delivery. Nat Nanotechnol 2(4):249–255
22. Pei D, Buyanova M (2019) Overcoming endosomal entrapment in drug delivery. Bioconjug Chem 30(2):273–283
23. Niu Y, Yu M, Hartono SB et al (2013) Nanoparticles mimicking viral surface topography for enhanced cellular delivery. Adv Mater 25(43):6233–6237

24. Lee E, Kim D, Youn Y et al (2008) A virus-mimetic nanogel vehicle. Angew Chem Int Ed 47:2418–2421
25. Richards DA, Maruani A, Chudasama V (2017) Antibody fragments as nanoparticle targeting ligands: a step in the right direction. Chem Sci 8(1):63–77
26. Yang K, Li S, Liu L et al (2019) Epitope imprinting technology: Progress, applications, and perspectives toward artificial antibodies. Adv Mater 31(50):1902048
27. Hashemi-Moghaddam H, Zavareh S, Karimpour S et al (2017) Evaluation of molecularly imprinted polymer based on HER2 epitope for targeted drug delivery in ovarian cancer mouse model. React Funct Polym 121:82–90
28. Pan J, Chen W, Ma Y et al (2018) Molecularly imprinted polymers as receptor mimics for selective cell recognition. Chem Soc Rev 47(15):5574–5587
29. Rozhina E, Ishmukhametov I, Batasheva S et al (2019) Nanoarchitectonics meets cell surface engineering: shape recognition of human cells by halloysite-doped silica cell imprints. Beilstein J Nanotechnol 2019(10):1818–1825
30. Borovička J, Metheringham WJ, Madden LA et al (2013) Photothermal colloid antibodies for shape-selective recognition and killing of microorganisms. J Am Chem Soc 135 (14):5282–5285
31. Kwansa AL, Freeman JW (2015) Ligament tissue engineering. In: Nukavarapu SP, Freeman JW, Laurencin CN (eds) Regenerative engineering of musculoskeletal tissues and interfaces. Woodhead Publishing, Cambridge, pp 161–193
32. de Mel A, Seifalian AM, Birchall MA (2012) Orchestrating cell/material interactions for tissue engineering of surgical implants. Macromol Biosci 12:1010–1021
33. Katvan E, Doron I, Ashkenazi T et al (2017) Age limitation for organ transplantation: the Israeli example. Age Ageing 46:8–10
34. Bružauskaite I, Bironaite D, Bagdonas E, Bernotiene E (2016) Scaffolds and cells for tissue regeneration: different scaffold pore sizes-different cell effects. Cytotechnology 68:355
35. Bardakova KN, Akopova TA, Kurkov AV et al (2019) From aggregates to porous three-dimensional scaffolds through a mechanochemical approach to design photosensitive chitosan derivatives. Mar Drugs 17(1):48
36. Majima T, Funakosi T, Iwasaki N et al (2005) Alginate and chitosan polyion complex hybrid fibers for scaffolds in ligament and tissue engineering. J Orthop Sci 10(3):302–307
37. Ma T, Li Y, Yang ST et al (2000) Effects of pore size in 3D fibrous matrix on human trophoblast tissue development. Biotechnol Bioeng 70:606–618
38. Dzamukova MR, Naumenko EA, Lannik NI et al (2013) Surface modified magnetic human cells for scaffold-free tissue engineering. Biomater Sci 1:810–813
39. Guryanov I, Naumenko E, Konnova S et al (2019) Spatial manipulation of magnetically-responsive nanoparticle engineered human neuronal progenitor cells. Nanomedicine: NBM 20:102038
40. Ito A, Takahashi T, Kawabe Y et al (2009) Human beta defensin-3 engineered keratinocyte sheets constructed by a magnetic force-based tissue engineering technique. J Biosci Bioeng 108:244–247
41. Fayol D, Frasca G, Le Visage C et al (2013) Use of magnetic forces to stem cell aggregation during differentiation, and cartilage tissue modeling. Adv Mater 25:2611–2616
42. Luciani N, Du V, Gazeau F et al (2016) Successful chondrogenesis within scaffolds, using magnetic stem cell confinement and bioreactor maturation. Acta Biomater 37:101–110
43. Daquinag AC, Souza GR, Kolonin MG (2012) Adipose tissue engineering in three-dimensional levitation tissue culture system based on magnetic nanoparticles. Tissue Eng Part C Methods 19:336–344
44. Du V, Luciani N, Richard S et al (2017) A 3D magnetic tissue stretcher for remote mechanical control of embryonic stem cell differentiation. Nat Commun 8:400
45. Ho VHB, Muller KH, Barcza A et al (2010) Generation and manipulation of magnetic multicellular spheroids. Biomaterials 31:3095–4102

46. Naumenko EA, Dzamukova MR, Fakhrullin RF (2014a) Magnetically functionalized cells: fabrication, characterization, and biomedical applications. In: Katz E (ed) Implantable bioelectronics. Wiley VCH, Weinham, pp 7–26
47. Naumenko EA, Dzamukova MR, Fakhrullina GI et al (2014b) Nano-labelled cells - a functional tool in biomedical applications. Curr Opin Pharmacol 18:84–90
48. Jin R, Lin B, Li D et al (2014) Superparamagnetic iron oxide nanoparticles for MR imaging and therapy: design considerations and clinical applications. Curr Opin Pharmacol 18:18–27
49. Rozhina E, Batasheva S, Gomzikova M et al (2019) Multicellular spheroids formation: the synergistic effects of halloysite nanoclay and cationic magnetic nanoparticles. Colloids Surf A Physicochem Eng Asp 565:16–24
50. Carver K, Ming X, Juliano RL (2014) Multicellular tumor spheroids as a model for assessing delivery of oligonucleotides in three dimensions. Mol Ther Nucleic Acids 3:e153
51. Dufau I, Frongia C, Sicard F et al (2012) Multicellular tumor spheroid model to evaluate spatio-temporal dynamics effect of chemotherapeutics: application to the gemcitabine/CHK1 inhibitor combination in pancreatic cancer. BMC Cancer 12:15–26
52. Mattix B, Olsen TR, Gu Y et al (2014) Biological magnetic cellular spheroids as building blocks for tissue engineering. Acta Biomater 10:623–629
53. Berry CC, Wells S, Charles S et al (2003) Dextran and albumin derivatised iron oxide nanoparticles: influence on fibroblasts in vitro. Biomaterials 24:4551–4557
54. Mahmoudi M, Simchi A, Milani AS et al (2009) Cell toxicity of superparamagnetic iron oxide nanoparticles. J Colloid Interface Sci 336:510–518
55. Liu M, Jia Z, Jia D et al (2014) Recent advance in research on halloysite nanotubes-polymer nanocomposite. Prog Polym Sci 39:1498–1525
56. Naumenko EA, Guryanov ID, Yendluri R et al (2016) Clay nanotube-biopolymer composite scaffolds for tissue engineering. Nanoscale 8(13):7257–7271
57. Suner SS, Sahin D, Yetiskin B et al (2019) Cryogel composites based on hyaluronic acid and halloysite nanotubes as scaffold for tissue engineering. Int J Biol Macromol 130:627–635
58. Mishra S, Sharma S, Javed MN et al (2019) Bioinspired nanocomposites: applications in disease diagnosis and treatment. Pharm Nanotechnol 7(3):206–219
59. Naumenko E, Fakhrullin R (2019) Halloysite nanoclay/biopolymers composite materials in tissue engineering. Biotechnol J 14(12):1900055
60. Fakhrullina GI, Akhatova FS, Lvov YM et al (2015) Toxicity of halloysite clay nanotubes in vivo: a *Caenorhabditis elegans* study. Environ Sci Nano 2:54–59
61. Konnova S, Lvov Y, Fakhrullin R (2016) Magnetic halloysite nanotubes for yeast cell surface engineering. Clay Miner 51:429–433
62. Kryuchkova M, Danilushkina A, Lvov Y et al (2016) Evaluation of toxicity of nanoclays and graphene oxide in vivo: a *Paramecium caudatum* study. Environ Sci Nano 3:442–452

Chapter 19
Combining Polymers, Nanomaterials, and Biomolecules: Nanostructured Films with Functional Properties and Applications

Andressa R. Pereira, Antonio F. A. A. Melo, Frank N. Crespilho, and Osvaldo N. Oliveira Jr

19.1 Introduction

Since the early 2000s when the concept of nanoarchitectonics was proposed [1], the incorporation of inorganic nanoparticles into a polymer matrix has been used to create functional structures with morphologies and functionalities which could never be produced in natural systems. When these polymer architectures are obtained in the form of 2D nanoarchitectures as Langmuir–Blodgett (LB) films and layer-by-layer (LbL) films, they represent a valuable toolbox for high-precision fabrication and tuning of properties through the choice of their constituents [2]. For instance, magnetic nanoparticles (MNPs) functionalized with the peptide melittin (MLT) were used in an electrical impedance sensor for fast and sensitive detection of Gram-negative and Gram-positive bacteria in food samples [3]. MLT comprises 26 amino acids and displays antimicrobial properties against such microorganisms as *Escherichia coli*, *Staphylococcus aureus*, and *Salmonella typhimurium.* Nanostructured interfaces through immobilization of biomolecules (e.g., proteins, enzymes, DNA molecules, and oligonucleotides) onto solid substrates modified with functional polymer films are an important tenet of the nanoarchitectonics concept. More

A. R. Pereira · O. N. Oliveira Jr (✉)
São Carlos Institute of Physics, University of São Paulo, São Carlos, São Paulo, Brazil
e-mail: chu@ifsc.usp.br

A. F. A. A. Melo
São Carlos Institute of Physics, University of São Paulo, São Carlos, São Paulo, Brazil

Materials Engineering Graduate Program, Federal Institute of Education, Science and Technology of Piauí, Teresina, Brazil

F. N. Crespilho
São Carlos Institute of Chemistry, University of São Paulo, São Carlos, São Paulo, Brazil

T. Govindaraju, K. Ariga (eds.), *Molecular Architectonics and Nanoarchitectonics*, Nanostructure Science and Technology,
https://doi.org/10.1007/978-981-16-4189-3_19

specifically, the efficient immobilization of redox proteins and enzymes onto nanoparticle–synthetic polymer films via chemical and physical adsorption methods promotes electron transfer in redox reactions, making it possible to produce biodevices, including high-performance biosensors. Indeed, efficient biomolecule immobilization may enhance charge transport in LbL films, as in films containing gold nanoparticles (AuNP) encapsulated with the polyamidoamine (PAMAM) nanoreactor [4]. AuNP modified with poly(allylamine hydrochloride) (PAH) molecules (AuNP–PAH) can be used to mediate electron transfer of redox protein cytochrome c (Cyt c) on a self-assembled surface [5]. Besides the significant enhancement in electron transfer of Cyt c, AuNP–PAH at the electrode/protein interface impacts the catalytic process.

Synthetic and natural polymers are used in electrochemical sensing [6]. For example, polymer architectures based on microbial nanocellulose sheets adherent to human skin are used on electrochemical wearable sensors to detect metal ions and biomarkers in sweat. Another feature of the nanoarchitectonics concept is the possibility to manipulate, arrange, and investigate the components at the molecular level in a biomimetic situation. Biomimetic structures such as artificial enzymes and membranes can be made from synthetic materials assembled onto conductive substrates. Special attention is paid to amphiphilic molecules in LB films for bioelectronic devices, as in biosensors. Amphiphilic biomolecules have the ability to spontaneously organize into nanostructures that mimic the cell membranes [7]. This is best illustrated in the use of Langmuir monolayers and LB technology as cell membrane models to understand the organization [8] and interaction of biologically relevant molecules with these models.

In this chapter, we focus on polymer nanostructures and biological applications where the nanoarchitectonics principles are employed. Section 19.2 introduces the polymer architectures with 1D, 2D, and 3 D (dimensions), which is followed by the main strategies for fabrication of mono- and multilayer films using the LB and LbL deposition methods in Sect. 19.3. The applications of biointerfaces obtained with nanostructured films are highlighted in Sect. 19.4, which precedes the concluding remarks.

19.2 Polymer Architectures with 1D, 2D, and 3D Dimensions

Polymer architectures have been prepared over a wide range of length scales by arranging polymer chains in soft nanoparticles, in bulk materials, and in ultrathin films. Their morphology and composition may be tuned for specific applications, as can be their dimensions. Figure 19.1a shows examples of polymer 1D nanoarchitectures, including nanowires [9], nanofibers [10], nanorods [11], and nanotubes [12]. Many are the possible applications for these 1D nanostructures, as in miniaturized machines and devices. For instance, polymer nanowires are used in

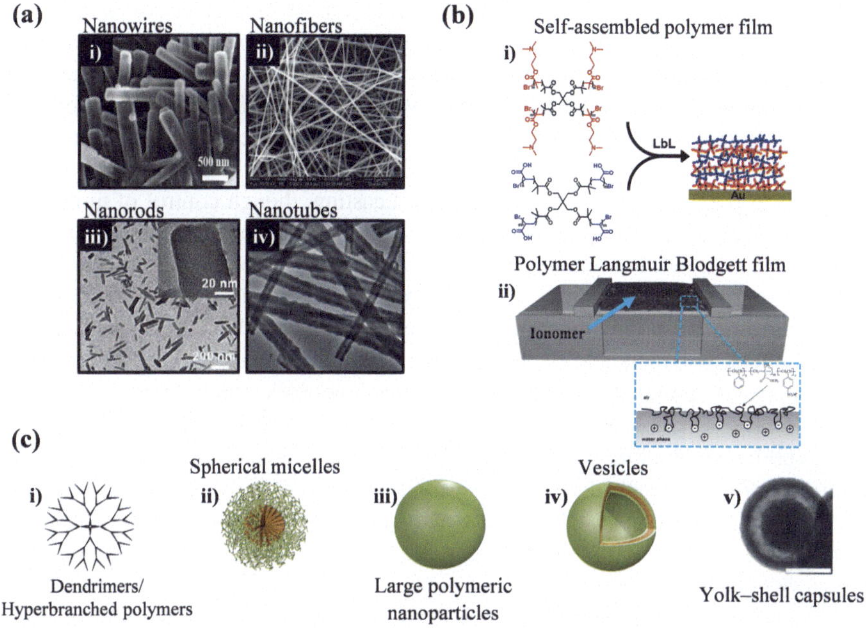

Fig. 19.1 (**a**) Examples of polymer 1D nano-architectures: (i) nanowires [9], (ii) nanofibers [10], (iii) nanorods [11], and (iv) nanotubes [12]. All images are adapted with permission from the publishers: (i) Elsevier 2017, (ii) © Elsevier B.V. 2019, (iii) © Royal Society of Chemistry 2020, and (iv) Royal Society of Chemistry 2018. (**b**) Polymer 2D nano-architectures: (i) self-assembled polymer film [18] and (ii) LB polymer film [19]. Illustrations adapted with permission from the publishers: (i) © Elsevier 2018, and (ii) © Elsevier B.V. 2005. (**c**) Polymeric 3D nano-architectures: (i) dendrimers/hyperbranched polymers [20], (ii) spherical micelles [21], (iii) large polymeric nanoparticles [20], (iv) vesicles [20], and (iv) yolk–shell nanocapsules [22]. Illustrations adapted with permission from the publishers: (i–iv) © Royal Society of Chemistry 2015 and (v) © Royal Society of Chemistry 2016

nanoscale sensors for control of local protein concentration in solution [13]. Polymer nanofibers produced via electrospinning (ES) can be arranged into ultrathin films, thus forming 2D nano-architectures, and then be used in bio-related applications [14]. Electrospun polyamide 6/poly(allylamine hydrochloride) nanofibers functionalized with carbon nanotubes (SWNT) have been used in the electrochemical detection of dopamine [15]. The same strategy was employed with electrospun nanofibers coated with SWNT or gold nanoparticles (AuNPs) serving as matrix for the immobilization of antibodies in immunosensors to detect a pancreatic cancer biomarker [16]. Also, the nanofibers may be produced with solution blow spinning, and this was exploited in fabricating a matrix of poly(lactic acid) for immunosensors to detect the cancer p53 biomarker [17].

Perhaps the most intensive use of polymer nanostructures has been with 2D nanoarchitectures, especially ultrathin films adsorbed onto solid substrates. These films provide an interface with large surface area over micro- and macroscopic length scales, thus facilitating the immobilization of biomolecules and control of physicochemical and biological phenomena. The most used methods to produce these films are self-assembly via chemisorption [23], with which the so-called self-assembled monolayers (SAM) are formed, the LB method [19, 24] and the LbL deposition [18, 25]. Figure 19.1b illustrates the LB and LbL fabrication methods for films whose thickness ranges from 1 to 10 nm. In the next section, further details are provided of these nanostructured films. Other methods worth mentioning are electrochemical polymerization, spin coating, and casting, though control of molecular architecture is not as straightforward [21]. The definition of 0D and 3D polymer architectures may be ambiguous, but a clear distinction can be made by considering as 0D nanostructures only those with a few nanometers at most. The polymeric 3D nanoarchitectures, on the other hand, may be much larger. They are normally made with synthetic polymers and peptides as their building blocks to form dendrimers/hyperbranched polymers, spherical micelles, vesicles, and large polymeric nanoparticles. These four 3D architectures based on self-assembly of polymeric and peptidic amphiphiles differ in shape and size, as indicated in Fig. 19.1c. Dendrimers/hyperbranched polymers are the smallest with diameters in the range from 10 to 20 nm, while vesicles, spherical micelles, and polymeric nanoparticles are significantly larger with diameters ranging from one to several hundred nanometers [20].

The controlled sizes and shapes of these polymeric 3D nanoarchitectures are suitable to fabricate artificial organelles from synthetic polymers and serve as nanocontainers and nanoreactors for enzymatic reactions and as delivery carriers for biomacromolecules. By way of illustration one could mention some possible uses of the dendrimer poly(amidoamine) (PAMAM), which has been employed as nanoreactors to produce platinum nanoparticles (PtNPs) embedded in LbL films for methanol-tolerant cathodes [26]. PAMAM has also been combined with single-walled carbon nanotubes (SWNTs) in LbL films to fabricate field-effect sensors [27]. The hyperbranched polymer network made from poly(vinyl alcohol) (PVA) and poly(acrylic acid) (PAA) was utilized in multifunctional hybrid aerogels that contained mesoporous silica nanoparticles to serve as good host materials for sustained and prolonged drug release [28]. Polymeric yolk–shell capsules of polyaniline (PANI) are used as reactors to form metallic nanoparticles and for the catalysis of CO oxidation [22].

19.3 Polymer 2D Nanoarchitectures from Mono- and Multilayer Films

The preparation of 2D nanoarchitectures from mono- and multilayer films with inorganic nanoparticles incorporated into polymers is a strategy for obtaining controllable supramolecular structures with functional properties. This has been reached with the LB and LbL deposition methods. The LB method has a ca 100-year history, stemming from the seminal works by Irving Langmuir [29] in which he determined the main properties of amphiphilic molecules at the air–water interface. This contribution for surface chemistry led to a Nobel Prize for Langmuir in 1932. These monomolecular films, which is termed as Langmuir monolayers or Langmuir films, can be transferred onto a solid substrate to form an LB film following a methodology pioneered by Katharine Blodgett [30]. Figure 19.2a shows a schematic diagram of a Langmuir trough, the instrument used to spread a monolayer and transfer it as an LB film. The amphiphilic molecules are spread on an aqueous subphase from a solution in a volatile solvent and then compressed to form a closely packed monolayer that can be transferred by immersing and withdrawing a solid substrate. If multiple immersions and withdrawing processes are performed, then a multilayer LB film may be formed.

Though originally conceived for amphiphilic, small molecules, the methodology has also been applied to polymers as indicated in Fig. 19.2b. Examples include monolayer LB films made with conjugated polymer derivatives of poly (o-alkoxyanilines) [33] and pure poly(3,4-dibutoxythiophene) [34]. They are fabricated with no need of adding processing aids or surface-active compounds because these polymer derivatives are soluble in common organic solvents (e.g., chloroform). Several issues must be addressed to fabricate polymeric Langmuir and LB films when the polymers are not readily soluble in the most convenient solvents for spreading a Langmuir monolayer. Strategies employed to obtain such Langmuir films include mixing the polymer with an amphiphilic molecule or pre-dissolving the polymer in a non-volatile solvent and then preparing a solution in a volatile solvent. Yet another method is to produce a Langmuir film from monomers and then perform a 2D polymerization at the air–water interface [32]. The general steps to produce such films on a Langmuir trough are depicted in Fig. 19.2c. A monolayer of the water-insoluble monomer is spread on an aqueous subphase and compressed for a close packing of the monomers into a dense film with highly ordered internal structure (step 1). Then, a water-soluble monomer solution is added into the aqueous subphase in conjunction with metal salts (step 2). With diffusion of these two novel components from the bulk phase to the interface, the 2D polymerization is triggered to yield a polymer film (step 3). The resulting Langmuir film may not be truly a monomolecular film, which is why there is a tendency in the literature to refer to polymeric Langmuir films rather than monolayers. If on one hand this is a limitation in terms of molecular control, on the other hand the polymeric LB films transferred onto substrates films (step 4) may be more mechanically resistant than their counterparts made with small molecules.

The variety of molecules and macromolecules used and the type of application suggested from the 1960s until the early 1990s brought the hope that LB films would be used in commercially available devices [35]. This optimistic view was always accompanied by words of caution owing to the limitations of the methodology, particularly for mass production at low costs. Indeed, the fabrication of LB films requires dedicated equipment in clean environments (preferably clean rooms) and relatively time-consuming procedures. In addition, the poor mechanical properties of the LB films constitute a further hurdle. While attempts were made to mitigate these limitations, alternative techniques emerged to enable the production of organic films

(a)

(b)

(c)

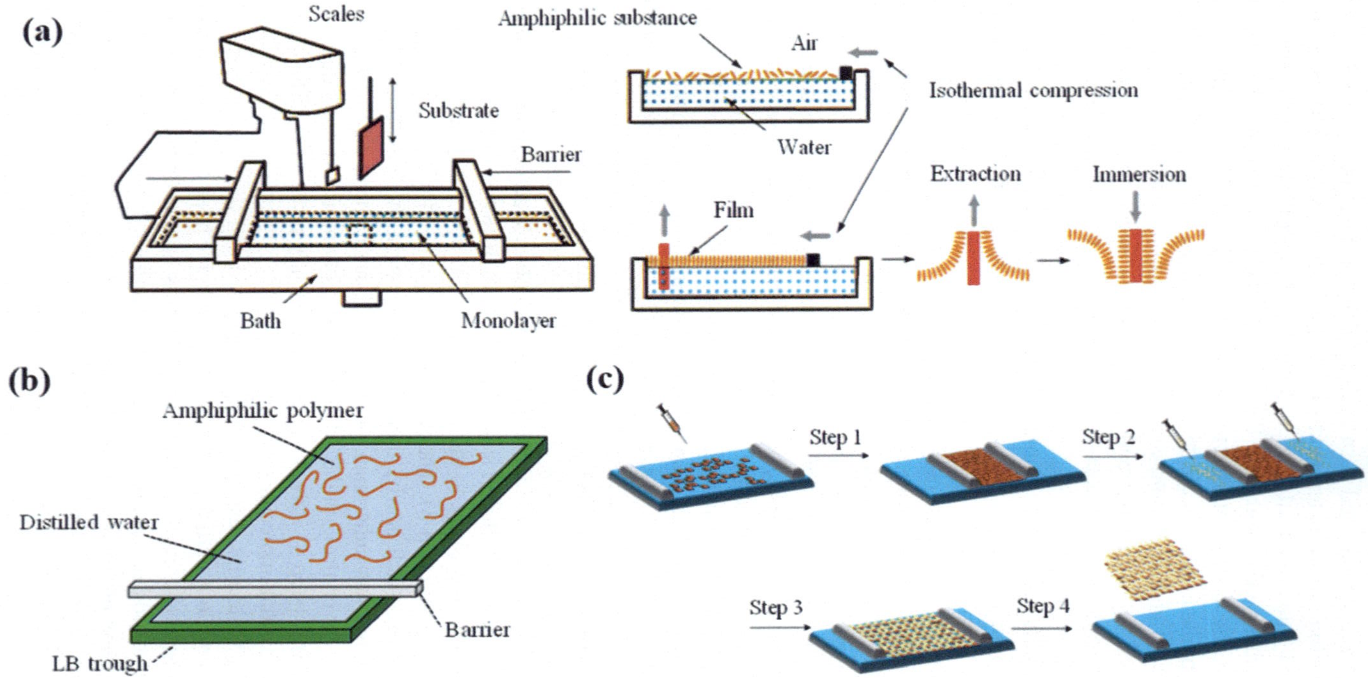

Fig. 19.2 (**a**) Schematic illustration of the preparation of a monomolecular layer from amphiphilic substances at the air–water interface, which is then transferred to a solid substrate in the form of a LB film. (**b**) Spreading of a typical amphiphilic polymer on the water surface for obtaining an LB film. Adapted with permission from [31]. Copyright © 2015 Springer-Verlag Berlin Heidelberg. (**c**) Step by step of LB-assisted 2D polymerization at the air–water interface in a Langmuir trough. Adapted with permission from [32]. Copyright © 2020 The Royal Society of Chemistry

with control at the molecular level, in a layer-by-layer fashion. The most important of such techniques is the LbL method, which we discuss next. As it will become clear in the description, the LbL technique is advantageous in many respects to the LB one with regard to mass production. Nevertheless, the LB technique remains unique for some fundamental studies, mostly related to biointerfaces. In Sect. 19.4.1 we shall describe several examples of the use of Langmuir monolayers and LB films.

The LbL deposition is a bottom-up nanofabrication technique that is especially suitable to produce multilayer polymeric films. It was conceived for polyelectrolytes where electrostatic attraction between oppositely charged polymers would drive adsorption in a LbL fashion by the alternate deposition of polycations and polyanions [25, 36]. One key advantage of LbL deposition is its relatively low cost and simple experimental setup [21]. The method was originally referred to as self-assembly via physical adsorption, but the term LbL technique eventually predominated. A few years after being introduced, LbL films were also fabricated from conducting polymers. Today the method is applied to a variety of charged materials, including amphiphilic molecules, functionalized nanoparticles, colloids, and biomolecules. Furthermore, film deposition can also be based on H-bonding [37] and specific interactions [38], and the experimental procedures have also been diversified. The first and most popular is dip-coating LbL assembly [39], illustrated in Fig. 19.3a. The solid substrate modified with a polycation layer is immersed into a colloidal solution of nanoparticles with opposite charges to those from the first layer. This leads to a bilayer with surface charge reversal. For each new layer adsorbed, surface charge neutralization and reversal occur. Repeated cycles of deposition of positively and negatively charged materials yield the multilayer polymer film. Other approaches include spray-assisted LbL films [40], spin-assisted LbL assembly [41], and low-temperature plasma treatment-assisted LbL self-assembly [42].

A significant change in the experimental setup for the electrostatic LbL technique was achieved with coupling with vacuum-assisted filtration (VA-LbL) [43]. In dip-coating LbL adsorption is based on diffusion and subsequent self-limited electrostatic adsorption of nanometer charged layers. This mechanism usually requires 5–30 min per each adsorption step. In contrast, with the VA-LbL technique, the adsorption of charged nanomaterials and polyelectrolytes is assisted by convection in addition to the electrostatic interactions (see Fig. 19.3b). Depending on the vacuum applied and concentration of the charged materials, one bilayer is adsorbed within only 20–30 s. This has been utilized to fabricate mechanically robust, flexible freestanding supercapacitor electrodes made with polyaniline functionalized with aramid nanofibers (PANI@ANFs). Using these flexible PANI@ANFs electrodes, the mechanical and electrochemical properties were leveraged in a single platform with specific capacitance of 441.0 F g^{-1} at a current density of 1 A g^{-1} and tensile strength of 233.3 MPa [44].

The versatility of the LbL technique will be highlighted in biological applications in Sect. 19.4.2, but here we illustrate the fabrication of hybrid LbL films and polyelectrolyte multilayer (PEM) films with alternate forms of self-assembly or processing. This allows one to obtain more complex polymer architectures, e.g., including block copolymers (BCPs). The formation of PEM films of diblock and

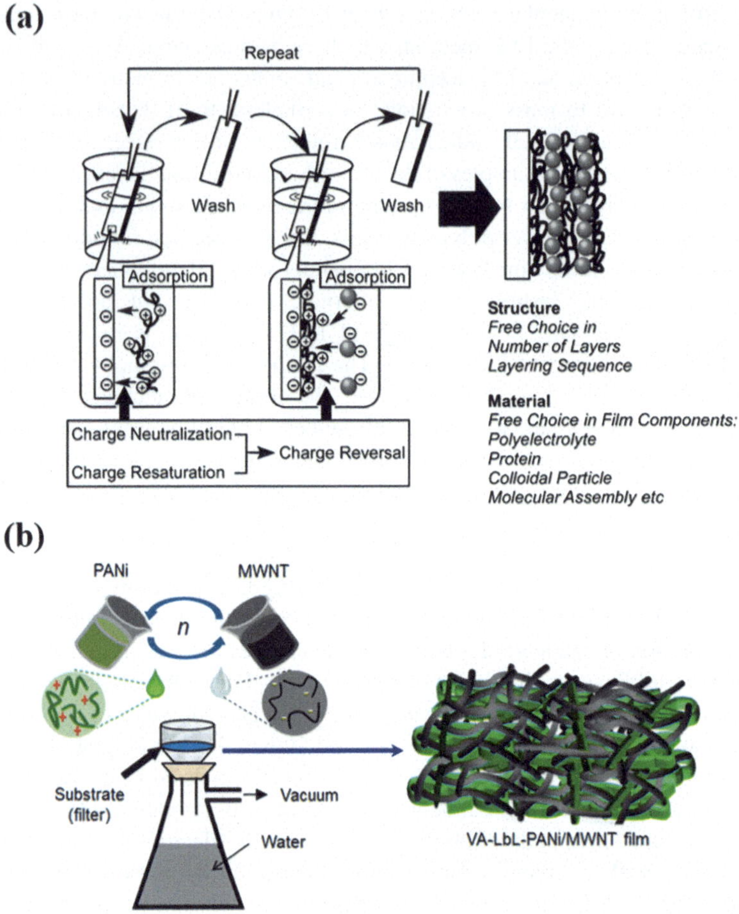

Fig. 19.3 (**a**) Schematic of the dip-coating LbL assembly used for electrostatic deposition of oppositely charged materials as polycations and anionic nanoparticles, which were arranged in the form of a multilayer polymer film adsorbed on a solid substrate. Reproduced with permission from [39]. Copyright © 2007 Royal Society of Chemistry. (**b**) Schematic illustration of the VA-LbL assembly. Based on the electrostatic interaction assisted by convection, multilayer polymeric films of anionic PANI nanofibers and cationic MWNT were assembled on a porous filter substrate used for the deposition of oppositely charged materials, followed by removal of the solvent. Reproduced with permission from [43]. Copyright © 2014 American Chemical Society

triblock copolymers was first explored as a proof-of-concept demonstration by Hustad and co-authors [45]. BCPs containing a block of free H-bond acceptors and a second block with a protected H-donor were used as masked polyampholytes to provide a PEM film obtained via spin-coating LbL assembly. Since these polymers are a class of polyelectrolytes bearing both cationic and anionic functionalities in single polymer chains, they often exhibit a peculiar behavior due to their self-association in solution, thus making them difficult to process. Therefore, efficient protection/deprotection strategies are employed for their synthesis and manipulation. Potential applications of self-assembled BCP films with ordered microdomains of 5–100 nm are now emerging in areas such as microelectronics industry and in high-density data storage [46].

19.4 Biointerfaces: Applications as Mimetic Models and in Biosensing

The methods to produce nanostructured polymer films are also useful to simulate and mimic biointerfaces, which serve not only for fundamental studies but also for biomedical applications. In this chapter we selected two topics related to biointerfaces. In the first, we shall describe the use of Langmuir monolayers and LB films as mimetic models, while in the second topic a few examples will be provided of polymers and nanomaterials at interfaces and in biosensing.

19.4.1 Langmuir Monolayers and Langmuir–Blodgett Films as Mimetic Models

Langmuir monolayers have been used as cell membrane models to understand their organization and the interaction with biologically relevant molecules such as pharmaceutical drugs, toxins, and proteins [8, 47, 48]. Using Langmuir monolayers is advantageous owing to the precise control of packing of the film-forming molecules and to the availability of surface-specific techniques to characterize monolayers at the air/liquid interface. These advantages have placed Langmuir monolayers among the most useful platforms for a detailed study of intermolecular interactions [49]. The LB films are complementary as membrane models, for they can be transferred onto almost any kind of solid substrate and the bilayer structures may have varied compositions [50], including with immobilization of biomolecules in solid films that can be characterized by many experimental techniques.

In this chapter we describe recent works in Langmuir monolayers and LB films for some biologically relevant molecules and nanostructures. The film-forming molecules are mostly phospholipids as they are the main components of the cell membrane framework [51–53]. In fact, in the first studies the Langmuir films used as membrane models had mainly pure or mixtures of phospholipids. Over the last few years, though, other cell components have been introduced, such as cholesterol, sphingomyelin, and membrane proteins. Also, the composition of the Langmuir monolayer is varied to reflect the differences in membranes from different types of cells. More sophisticated models have been used to distinguish mammalian cells from bacteria cells, and healthy cells from cancerous cells. For bacteria, for instance, mimicking the cell membrane in a realistic manner can be challenging since the membrane comprises an inner and an outer layer. In the studies with membrane models, there are generally three main objectives: (i) verify whether the guest molecule (drug, protein, polymer) is adsorbed onto the membrane represented by the Langmuir monolayer, (ii) determine the changes induced by the guest molecules, and (iii) correlate the molecular-level interactions in the Langmuir monolayer with the physiological activity of the guest molecule. In order to achieve these goals, the monolayer is usually characterized by a combination of experimental techniques and in some cases with molecular dynamics simulations [54, 55].

The incorporation of guest molecules into the monolayer can be inferred from surface pressure isotherms, from which changes in the elasticity of the membranes can be obtained. The monolayer structuring can be probed with X-ray diffraction or reflectivity measurements [56]. Spectroscopic techniques are employed to determine the molecular-level interactions, which include polarization-modulated infrared reflection absorption (PM-IRRAS) [57, 58] and sum-frequency generation (SFG) spectroscopy [59]. Film morphology may be investigated with fluorescence [60] or Brewster angle [61, 62] microscopy.

The usefulness of Langmuir monolayers in cell membrane models can be illustrated with studies of interaction involving chitosans where the main aim was to correlate with their physiological action. Chitosans are natural polysaccharides derived from chitin [63] and used in nanoparticle synthesis [64] and in drugs and gene delivery [65] because of their biodegradability, biocompatibility, and low cytotoxicity [66]. Chitosans are also bactericide, being employed in biomedical applications. Since chitosans are positively charged when dissolved in acidic solutions, they showed enhanced interaction with Langmuir monolayers made with negatively charged phospholipids such as dimyristoyl phosphatic acid (DMPA) [67]. In these monolayers chitosan was incorporated even at high surface pressures, which allowed transfer onto solid supports using LB technique. Chitosan induced an expansion in the monolayer and a decrease in elasticity. The possible antilipidemic dietary activity of chitosan has been attributed to its ability to traverse the digestive system practically intact, thus acting like a dietary fiber [68]. Therefore, one should expect chitosan to interact with cholesterol and fatty acids, and this was investigated with Langmuir monolayers of these compounds. Chitosan caused the surface pressure isotherms to shift to higher areas per molecule, and the shift increased with its concentration. This behavior was attributed to the attraction between chitosans and the fatty acids, also depending on whether these acids had saturated chains. For stearic acid, chitosan induced a reduction in the slope of the isotherms and a decrease in the collapse pressure upon increasing the chitosan concentration. In contrast, for the unsaturated fatty acids, both the slopes and collapse pressures increased. That is to say, chitosan caused the stearic acid monolayer to be more fluid, while the reverse effect was observed for the unsaturated fatty acids [68]. Overall, the degree of saturation of the acid hydrocarbon chains is relevant for the architectures adopted by the monolayers. These results showed the formation of lipid–chitosan complexes and allow understanding some interactions of these complexes. Furthermore, they made it possible to validate chitosan as a dietary antilipidemic supplement.

The relevance of membrane composition was highlighted in a work where chitosans were shown to interact more strongly with Langmuir monolayer made of a lipid raft composition than even with negatively charged phospholipids [69]. Lipid rafts are domains in the membranes with lipid and protein compositions that differ from other parts of the membrane [70]. The lipid rafts are simulated using a ternary mixture of a phosphatidylcholine, cholesterol, and sphingomyelin, which is a more realistic model of a mammalian cell membrane than with only pure phospholipids. Three types of chitosans were employed, two of which are soluble in physiological pH. The use of the latter chitosans is relevant because experiments could be done with aqueous subphases with the physiological pH, closer to in vivo systems. The minimum chitosan concentration at which an expansion of the surface pressure isotherms could be observed was the criterion to compare the effects from the different types of chitosan on the Langmuir monolayers of lipid rafts and dipalmitoyl phosphatidyl choline (DPPC). For the three chitosans, the effects appeared at a much lower concentration for the raft composition than for DPPC. For instance, for the high molecular weight chitosan which is soluble at physiological pH [71], the minimum concentration at which effects appeared with the lipid rafts was 10^{-4} mg mL^{-1}, to be compared with 0.5×10^{-2} mg mL^{-1} for DPPC monolayers [69]. It is significant that these chitosan concentrations for the lipid rafts are much lower than in previous reports in the literature. An implication of these results is that caution should be taken in correlating the interaction of molecules with cell membrane models, since the composition may play a significant effect.

Using Langmuir monolayers as cell membrane models may also serve to test possible toxicity of biologically relevant molecules and nanostructures. This is relevant for nanoparticles and other types of nanostructures which have been used in biomedical applications, from diagnosis to therapy and to the so-called theranostic systems. An issue in these applications is whether the nanostructures could be toxic to animals, including humans. It is known that the action of the nanostructures depends on their molecular-level interaction with the cell membranes. Access to this type of interaction is not straightforward in in vivo settings and even in in vitro systems if the whole cell is considered. This is why most researchers resort to cell membrane models such as vesicles, Langmuir monolayer, and LB films. AuNPs are known as inert material, but their biocompatibility remains a key issue, which is a motivation to the investigation of AuNPs with distinct phospholipid monolayers [72]. The AuNPs can be capped with positively and negatively materials, and as expected this capping affected the interaction with the Langmuir monolayers. For instance, Fig. 19.4 shows that the charged AuNPs had a significant effect on dipalmitoyl phosphatidyl glycerol (DPPG) monolayers, which is negatively charged. The compressional modulus (C_S^{-1}) in the inset of Fig. 19.4, calculated with $C_S^{-1} = -A(\partial \pi / \partial A)$ [73], indicates that the in-plane elasticity of DPPG monolayers was affected significantly only for the highest concentration of the nanoparticle, particularly for pressures between 30 and 35 mN/m, which correspond to the lateral pressure of real cell membranes [74]. The effects were more pronounced for the positively charged AuNPs, but the negatively charged ones also affected the monolayers. For zwitterionic 1,2-dipalmitoyl-*sn*-glycero-3-phosphocholine (DPPC), the monolayers were more affected by negatively charged AuNPs. In general, the results confirmed the importance of electrostatic forces in the interaction of nanoparticles

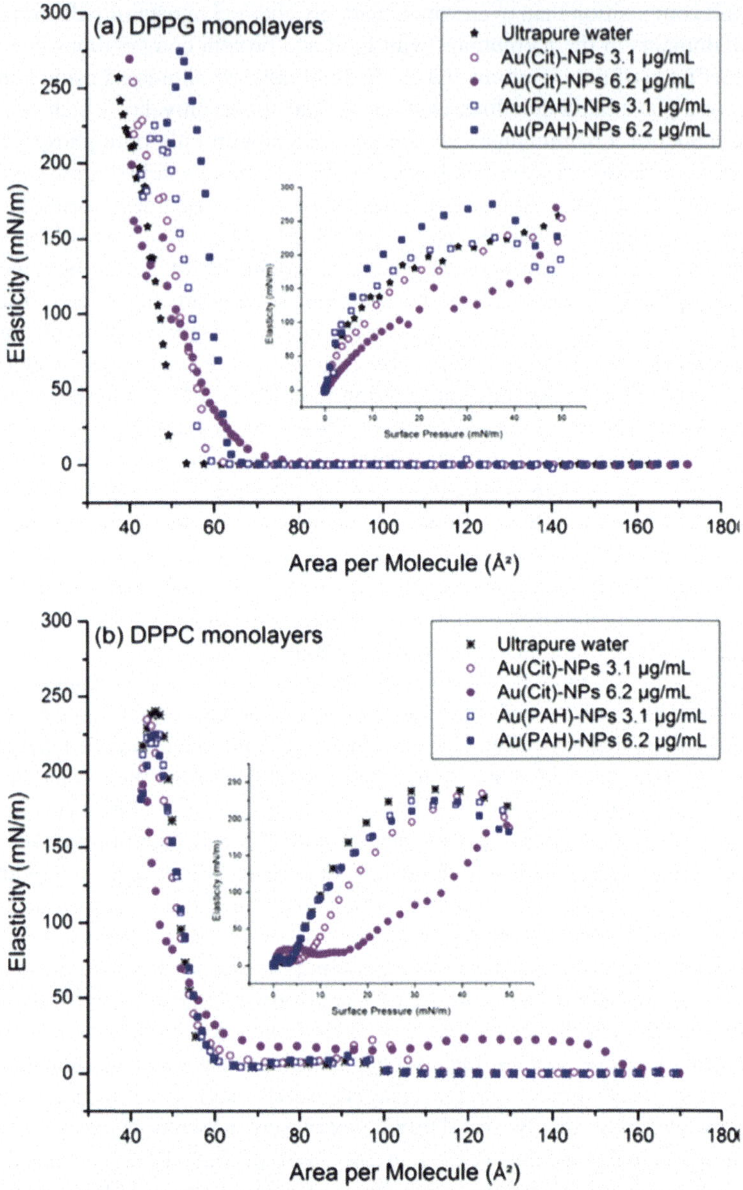

Fig. 19.4 C_S^{-1} of (**a**) DPPG and (**b**) DPPC monolayers. Insets: C_S^{-1} as a function of the surface pressure. Reproduced with permission from Ref. [72]. Copyright 2020 Elsevier

with model membranes, with important implications for the toxicity of nanostructures [72].

Another application of nanoparticles which depends on their interaction with cell membranes is in photothermal therapies. Shell-isolated gold nanoparticles (AuSHINs) can kill cancer cells in vitro via photoinduced heating [75], which requires disrupting the membranes. The interaction of AuSHINs with cell membranes was simulated with Langmuir monolayers of two phospholipids, namely, 1,2-dioleoyl-*sn*-glycero-3-phosphocholine (DOPC) and 1,2-dioleoyl-*sn*-glycero-3-phospho-L-serine (DOPS). The rationale behind this choice of phospholipids was to mimic healthy and cancerous cells, as the latter possess a higher amount of phosphatidyl serines. Surface pressure isotherms and PM-IRRAS showed that the negatively charged AuSHINs adsorbed on the monolayers mainly by attractive electrostatic interactions with the cationic choline group for the DOPC and ammonium head group for DOPS. Therefore, the effects observed in both types of membrane models may indicate that photoinduced heating with AuSHINs may be unspecific, and caution should be taken in the photothermal therapy.

The constraining environment provided by a Langmuir monolayer at the air/water interface can also be exploited as a strategy to synthesize nanoparticles [76]. This idea was implemented to form amorphous iron-containing magnetic nanoparticles [77], where the nanoparticles were obtained via ultraviolet decomposition of a volatile organometallic compound with iron pentacarbonyl in a mixed Langmuir monolayer with stearic acid. Highly anisotropic amorphous metal alloy nanoparticles and one-dimensional nanostructures were obtained under electric and/or magnetic fields.

There are also examples where drugs and nanostructures are studied in conjunction with cell membrane models. The influence of single-walled carbon nanotubes (SWCNTs) modified with an anticancer drug doxorubicin (DOx) was studied with Langmuir monolayers and LB films [78]. SWCNT–DOx was incorporated in a monolayer of the thiolipid 1,2-dipalmitoyl-*sn*-glycero-3-phosphothioethanol (DPPTE), thus indicating that LB films may be a potential drug carrier. The amount of drugs incorporated into the LB film transferred onto gold electrodes was determined using cyclic voltammetry. SWCNT–DOx remained at the electrode with a linear relationship obtained between the scan rate and the peak current.

LB films have been employed in proof-of-principle studies for bioelectronic devices, particularly because amphiphilic biomolecules can self-assemble into lipidic structures that mimic cell membranes and biological phenomena. For instance, an electrochemiluminescent device [79] was developed [80] where a biomimetic proteo-lipid nanostructure was designed to retain acetylcholinesterase monomers (AChE) in an oriented position [81]. These proteo-lipidic nanostructures are produced with the combination of liposome fusion at an air–buffer interface and the LB method. The orientation of AChE in the LB film at the electrode and the contact between two distinct enzyme layers induced metabolite vectorization [80], which made it possible to detect acetylcholine at the micromolar level. Furthermore, the molecular assembly was stable and so was the sensing device which could detect acetylcholine after repeated measurements. LB films made with manganese dioxide

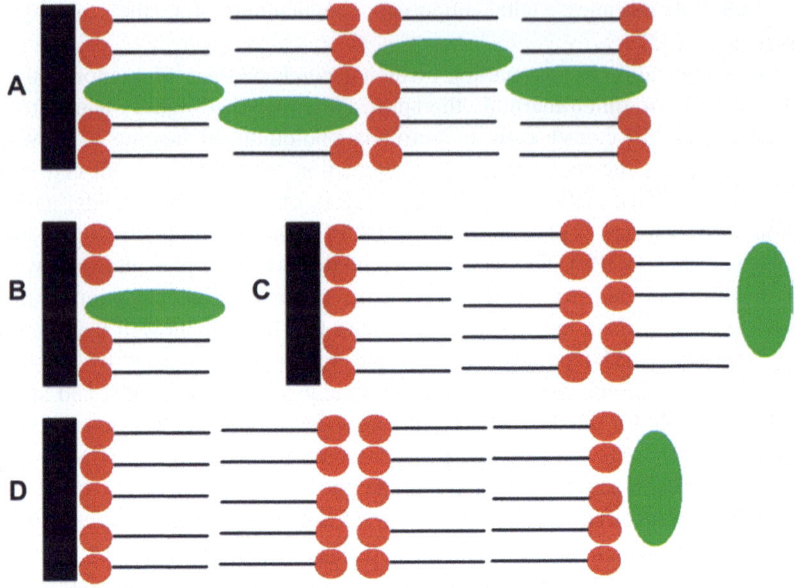

Fig. 19.5 Scheme of the four architectures of uricase co-immobilized in solid supports together with SA: (**a**) SA–uricase LB film (four layers), (**b**) SA–uricase LB film (one layer), (**c**) uricase adsorbed from solution onto a SA LB film (three layers), and (**d**) uricase adsorbed from solution onto a SA LB film (four layers). Reproduced with permission from Ref [84]. Copyright 2020 Elsevier

nanowires transferred onto a membrane of tetrabromobisphenol A (TBBPA) and polydimethylsiloxane (PDMS) were capable of decreasing bacterial invasion by capturing and killing the bacteria [82]. The nanowire monolayers prevented the contact between the antibacterial membrane and the captured bacteria, and the efficiency of this process was enhanced by placing the MnO_2 nanowires horizontally.

The control of molecular architecture inherent in LB films can be exploited to incorporate enzymes and seek synergy upon combining with nanomaterials. In particular, one may control the thickness, molecular density, and composition [7, 83], in addition to obtaining well-ordered enzyme–lipid systems. Various enzymes have been used to form LB films, in most cases with the goal of biosensing-specific analytes [84]. For example, films were obtained with uricase, an enzyme that catalyzes the oxidation of uric acid to alantonin, carbonic acid, and hydrogen peroxide. Uric acid oxidation is important for mammalians because alantoin has enhanced solubility in water, which facilitates excretion. Uricase was proven to adsorb to Langmuir monolayers of stearic acid (SA) by using surface pressure, surface potential, and PM-IRRAS measurements. Such adsorption was sufficient for uricase to be transferred along with SA onto solid supports in LB films. The four film architectures depicted in Fig. 19.5 were produced, with uricase maintaining its catalytic activity in all cases. The optimized performance was found for a film architecture where only one mixed uricase–SA Langmuir monolayer was transferred from the air–water interface. This enhanced performance was attributed to the preservation of uricase structure, which then resulted in an enzyme activity of ~71% of the value in a homogeneous environment. Using this optimized architecture, uric acid could be detected at levels comparable to those in the human blood.

Another example of a biosensor made with LB films incorporating enzymes was presented for detection of lactose. The enzyme was β-galactosidase or lactase that catalyzes the hydrolysis of β-galactosides in monosaccharides by breaking a glycosidic bond [85]. Lactase has an important role in humans as a source of carbon since lactose is broken into galactose and glucose. Because many people are lactose intolerant and cannot digest it, detection in dairy products and other food becomes essential. When immobilized in LB films of stearic acid and the conjugated polymer poly[(9,9-dioctylfluorene)-*co*-thiophene] (PTPF) [86], lactase had its activity preserved in different film architectures. Significantly, the sensitivity of the biosensors depended on the LB film architecture, as in the case of uricase films.

19.4.2 Polymers and Nanostructured Films for Biosensing

Among their many applications, polymers and nanostructures have been used in biosensing for various purposes, especially in attempts to develop affordable technologies for clinical diagnostics. Some of the challenges to achieve these methodologies include the need of cheap and easy to manufacture sensing units, in addition to simple principles of detection that do not require trained personnel to operate the measuring equipment. Biosensors have been developed to solve these challenges [87–89], but in many cases the number of tests that will be commercialized would not justify the large investments, which is why there are not many products in the market. Nevertheless, continued efforts are being made to develop sensing technology which is sufficiently generic for monitoring health conditions and diseases. In some scenarios, a low-cost solution is the only alternative for screening a considerable part of the population for high-incidence diseases. This is the case of prostate cancer which is today detected by combining blood tests and invasive procedures, as the biomarker employed is not very specific for the disease. There is now hope that biosensors can be used for another biomarker, referred to as PCA3 [90], a prostate cancer gene 3 located on the chromosome 9q21-22 [91, 92]. Genosensors for PCA3 can be made to detect prostate cancer using electrochemical methods and optical absorption spectroscopy and through image analysis of the sensing devices. A high sensitivity was achieved by incorporating the DNA probe in a matrix containing gold nanoparticles along with chitosan and chondroitin sulfate. The matrix consists of an LbL film onto which the active layer of PCA3 probe is deposited. The limit of detection of 83 pM should be sufficient to detect PCA3 in body fluids as well as in cells, which was demonstrated by projecting the impedance spectroscopy data with the multidimensional projection technique interactive document mapping (IDMAP)

[93]. The possibility of low-cost diagnosis may be critical to improve the therapy efficacy and prevent relapse by the patient [94]. In addition to electrochemical sensors fabricated with LbL multilayers containing redox-active polycations and DNA [95, 96], there are those based on aptamers [97, 98]. The use of aptamers, which are artificially selected functional DNA or RNA oligonucleotides with high recognition ability to specific targets [99], may be advantageous owing to the increased sensitivity and stability. An example of an aptasensor made with LbL films contained a ferrocene redox probe on the electrode and was capable of detecting proteins [100].

The combination of nanomaterials and polymers in nanostructured films allows for synergy to be achieved upon using distinct types of materials in the sensing units. Such combination has been increasingly exploited for biosensing, as in the detection of a pancreatic cancer biomarker, the antigen CA19-9 [101]. Immunosensors were produced with a matrix made of a self-assembled monolayer (SAM) and an LbL film containing one layer of chitosan and one layer of concanavalin A. The active layer was obtained with anti-C19-9 antibodies. With impedance spectroscopy, these immunosensors could detect CA19-9 with a limit of detection of 0.69 U mL^{-1}, which is sufficient for diagnosis in clinical analysis. No false positives were observed when other antigens were employed as interferents. The high sensitivity and selectivity was attributed to the antigen–antibody-specific interaction, as demonstrated with PM-IRRAS measurements upon examining the amide bands [101].

Another component of sensing devices which can be varied and optimized for specific purposes is the substrate onto which the matrix and active layer are deposited. There is no limitation on the type of material to be employed, for recent technologies allow for electrodes and nanostructured films to be deposited on any material. In order to exploit this versatility, significant attention has been paid to paper-based devices, particularly to lower costs. Issues to be addressed in these types of device include the need to obtain reproducible and robust sensing units and capability to immobilize bioreceptors and enhance the transducing signal to enhance sensitivity. Also relevant for biosensors is to reduce nonspecific adsorption that compromises selectivity, and this is normally made with co-immobilization of bovine serum albumin along with the active layer [102]. A successful biosensor to quantify proteinuria, a condition defined by the excess of protein in the urine, was built which addresses the issues above with an oxidized paper using periodate as substrate [103]. Protein immobilization was quantified in a colorimetric assay with Ponceau dye.

The identification and quantification of different types of biomolecules are important for medical assays and diagnostic tools [104, 105], and this may require the investigation of molecular interactions at interfaces. Various surface-specific techniques are used to study such interactions, including surface-enhanced infrared absorption spectroscopy (SEIRAS) [106]. Since most biomolecules have distinct vibrational fingerprints [107, 108], SEIRA is utilized with top-down nanopatterning techniques of metal substrates to identify the location of the biomolecules. For instance, proteins were identified via SEIRAS with microwires made with Au–Ag nanocages functioning as optical antennas for IR electromagnetic radiation

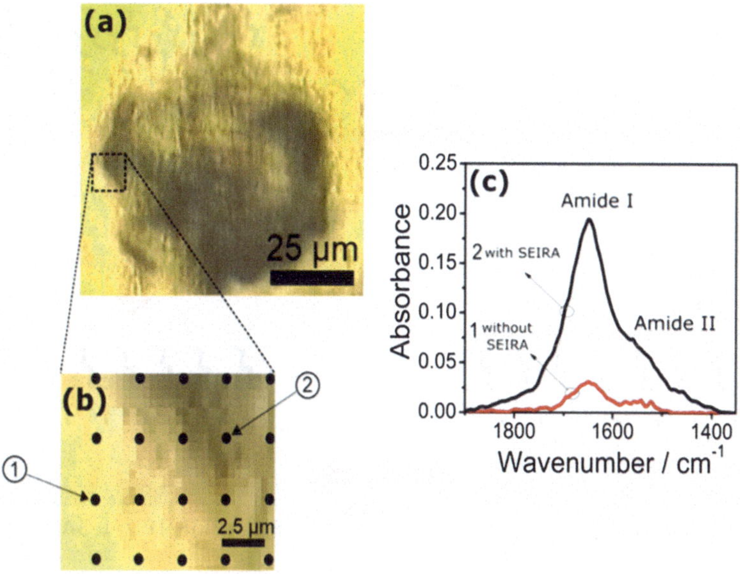

Fig. 19.6 (**a**) Optical microscopy image of a single lithographed square composed of microwires of Au–Ag nanocages onto a gold substrate and adsorbed with Cyt c. (**b**) Closer view for some selected regions for SEIRA analysis. (**c**) Vibrational spectra showing the SEIRA signal enhancement of amide I band when the region with microwires of Au–Ag nanocages was compared with the region without the nanocages. Reproduced with permission from Ref. [106]. Copyright 2020 American Chemical Society

[109]. Detection of proteins adsorbed on the microwires, e.g., cytochrome C, bilirubin oxidase, alcohol dehydrogenase, bovine serum albumin, and glucose oxidase, was made possible with the enhanced infrared absorption from the vibrational modes of these proteins [109]. Figure 19.6 shows the results for cytochrome C with SEIRA related to amide I band, featuring the signal enhancement on the regions containing the microwires of Au–Ag nanocages.

Food quality control can benefit from biosensors that detect pathogenic bacteria which cause infections in animals, including humans. A variety of biosensors have been developed to detect different bacteria, in most cases employing nanostructured films. Some of these biosensors target *Staphylococcus aureus* (*S. aureus*), a frequent cause of diseases owing to tissue invasion by the pathogen or by exotoxin production [110] from ingestion of contaminated food. The dairy industry, in particular, needs to monitor the presence of *S. aureus* to prevent milk contamination, but there is no simple, low-cost procedure for this bacteria detection [111]. Efforts have been thus directed to develop biosensors with the aim of replacing the standard procedures involving counting of bacteria colonies. By way of illustration we mention, immunosensors to detect *S. aureus* with microfluidic electrodes modified with LbL films of the polysaccharide chitosan (CHT) and chondroitin sulfate (CS) and then coated with anti-*S. aureus* antibodies [55]. The choice of CHT/CS films was

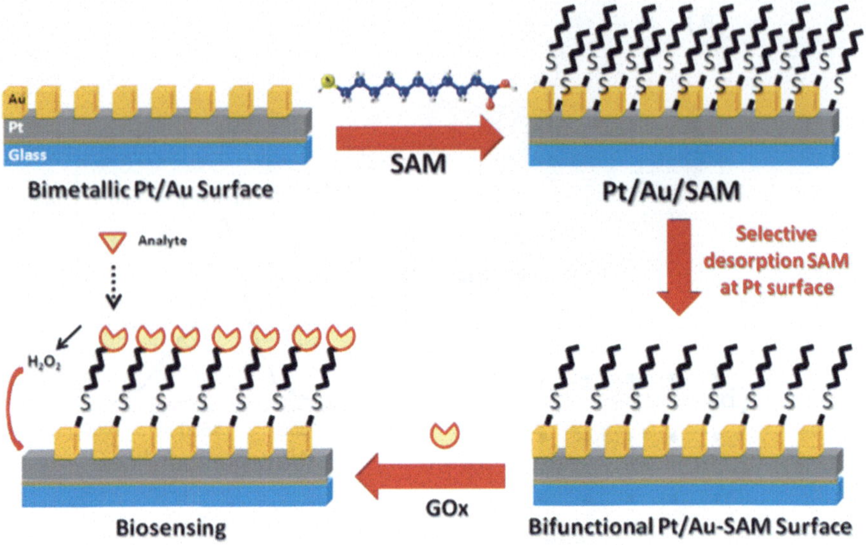

Fig. 19.7 Representation of Pt–Au bimetallic electrode, which was coated with MUA-SAM and contains GOx molecules adsorbed. In the last cartoon (in clockwise direction), the biosensing mechanism was depicted with hydrogen peroxide being generated from the interaction between the glucose analyte and GOx. Reproduced with permission from Ref. [117]. Copyright 2020 Elsevier

made because they are obtained from renewable sources at low cost, in addition to being suitable to preserve the antibody activity. The detection results with impedance spectroscopy indicate that LbL films were efficient as matrices since a high sensitivity and selectivity was achieved to detect the bacteria, including in milk samples. It is also relevant that the sensing performance depends on the number of bilayers in the matrix, which shows an opportunity to optimize molecular architectures.

Bimetallic nanostructures may exhibit enhanced catalysis compared to their monometallic counterparts [112, 113], and this may be explored in electrocatalysts, dehydrogenation catalysts, selective oxidants [114], and label-free biodevices [115]. The bimetallic structure most used for biosensors is produced from gold and platinum because they form a self-assembled monolayer (SAM) which creates a biocompatible environment [116]. A glucose biosensor may be developed with glucose oxidase (GOx) immobilized and a fine control of the Pt catalytic activity by patterning its surface with interdigitated gold fingers [117]. Synergy in biosensing is accomplished with a suitable architecture where GOx was immobilized in a selective way on gold and the hydrogen peroxide was detected on the Pt surface. Figure 19.7 shows a schematic representation of the bimetallic electrode where the GOx molecules are adsorbed. The last cartoon (in the clockwise direction) depicts the biosensing mechanism with hydrogen peroxide generated from the interaction between GOx and the glucose analyte. The Pt/Au–SAM–$GOx_{selective\ desorption}$ structure could be considered bifunctional because it allows the simultaneous catalysis of glucose by GOx immobilized on gold and the detection of hydrogen peroxide on Pt sites. The high catalytic activity led to a detection limit in the nanomolar range for glucose, also permitting determination of glucose in artificial saliva samples.

The finding that enzyme activity can not only be preserved but also enhanced when incorporated in nanostructured films has sparked interest in producing enzymatic biosensors. The LbL technique has been proven suitable for enzyme immobilization with the activity preserved [118, 119], as the adsorption process occurs in aqueous solutions under mild conditions. Biosensors to detect cholesterol, for instance, are fabricated with cholesterol oxidase (COX) alternated with the polycation poly(ethylene imine) in LbL films [120]. Cyclic voltammetric data confirmed the electron transfer involving COX, while amperometric measurements indicated significant immobilization of COX which remained active. A further advantage of LbL films for enzymatic biosensors is in the small quantities of enzymes required, in comparison with other methods such as cross-linking, owing to the nanostructured nature of the films. The precise thickness control of enzyme layers and possibility to vary the film architecture also allow for optimization of biosensing performance [121]. In the amperometric biosensors for glucose with LbL films containing glucose oxidase (GOx) and PAH layers, the ITO (indium tin oxide) electrodes were coated with a layer of Prussian blue and the sensitivity increased with the number of bilayers [121]. The increased sensitivity is attributed to the large amount of enzymes in thicker films, with the ensuing result that these glucose biosensors were more efficient than their counterparts with PB and GOx immobilized by conventional cross-linking.

The control of film architecture in biosensors can also be achieved with the LB technique, including in situations where a well-defined molecular orientation is desirable. DNA biosensors [122, 123] can be made with LB films, as exemplified by the electrochemical detection of trace amounts of methotrexate [124]. In a comparative study, three types of electrodes were used, namely, glassy carbon electrode (GCE), GCE(ox) which was subjected to a surface pre-treatment, and DNA/GCE(ox), containing DNA. A much larger signal was measured for the drug on DNA/GCE(ox) electrode under the same experimental conditions. The larger signal may be ascribed to groups containing oxygen that H-bond to methotrexate molecules. With a DNA/GCE(ox) biosensor, methotrexate is detected by square wave voltammetry with an LOD of 8.0×10^{-9} mol L^{-1}. The drug could also be identified in diluted human urine, thus confirming the suitability of the LB technique to immobilize DNA.

In biosensing, the transduction signal can be increased by modifying the electrodes [125, 126] or changing the structure of biomolecules [127]. For example, a cooperative bioelectrochemical system for ethanol production and its oxidation is obtained by combining alcohol dehydrogenase (ADH) and *Saccharomyces cerevisiae* [128]. The bioelectrodes made with flexible carbon fibers are efficient to evaluate the bioelectrocatalytic activity of ADH. The combination of nanostructures and biomolecules has led to hybrid-modified electrodes with enhanced charge transport and preserved biological activity. For enzymatic biosensors and other devices, enzyme activity depends on the interface with the

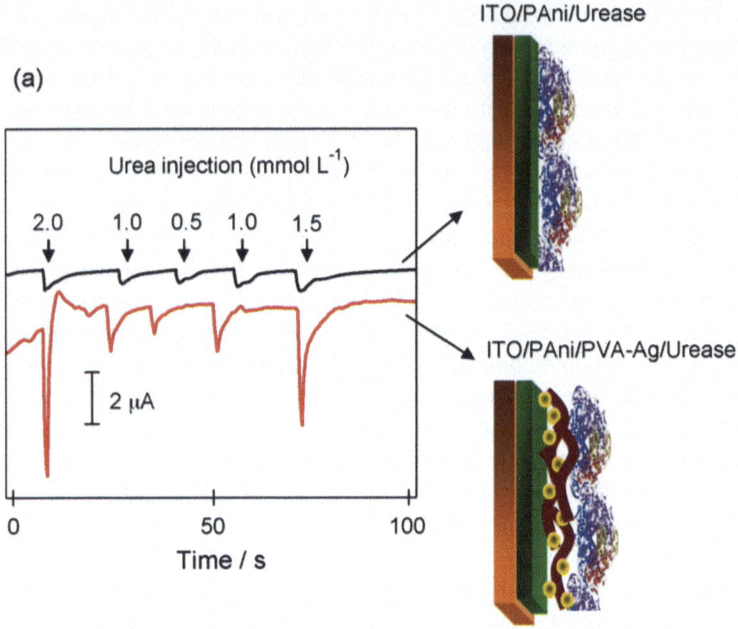

Fig. 19.8 Chronoamperometric curves at different urea concentrations for ITO/PAni/urease electrodes in the absence of PVA–AgNP (black line) and in the presence of PVA–AgNP (red line). Applied potential: 0 V (Ag/AgCl). Electrolyte: phosphate buffer (0.1 mol L^{-1}). Reproduced with permission from Ref [129]. Copyright 2020 Elsevier

nanomaterial, as shown for urease enzyme on ITO electrodes [129]. Various enzyme immobilization strategies were tested for electrodes modified with an electroactive nanostructured membrane containing polyaniline (PAni) and silver nanoparticles (AgNP) stabilized by polyvinyl alcohol (PAni/PVA–AgNP). The best results were obtained using a mixture of PVA–AgNP and urease enzyme deposited by drop-coating on the ITO/PAni electrode. With this configuration there was an efficient catalytic conversion of urea into ammonium and bicarbonate ions. The apparent Michaelis–Menten constant (K_M^{app}) is much lower than for immobilized urease described in the literature, which indicates that the chosen electrode architecture is promising for building enzymatic devices. Figure 19.8 shows that addition of 20μL of urea induces a fast increase in the cathodic current with a well-defined peak, which is larger for electrodes containing PVA–AgNP than for ITO/PAni/urease electrodes.

The synergy in the combination of distinct materials with the LbL technique is demonstrated with electroactive nanostructured membranes (ENMs) [4, 130] obtained by supramolecular manipulation of metallic nanoparticles encapsulated within polyamidoamine dendrimers (PAMAM) [131]. These membranes can be bifunctional (Bi-ENM) and display high electrocatalytic activity for hydrogen peroxide and dopamine [132]. The concepts of molecular control and bifunctionality were demonstrated by fabricating distinct types of LbL films. The films containing PAMAM and Pt nanoparticles alternated with nickel tetrasulfonated phthalocyanine (NiTsPc) layers had enhanced electrochemical oxidation of dopamine, compared to the LbL film without Pt. The membrane made with Pt also exhibited catalytic activity for hydrogen peroxide and could thus be used as a biosensor since hydrogen peroxide is a product of various oxidase-based enzyme reactions. The surface of Pt nanoparticles available for this oxidation is the same as for the dopamine experiments, which confirms the bifunctionality [132]. Another possible use of the LbL technique [5] is in the study of electron transfer from cytochrome c (Cyt c) to gold nanoparticles (AuNP) on a polycrystalline gold surface using poly(allylamine hydrochloride) (PAH) as spacers between Cyt c and the electrode. Using cyclic voltammetry and Marcus theory, a fourfold increase was observed in the electron transfer rate in the presence of AuNP. The distance between the redox center of Cyt c and the electrode was decreased by AuNPs by 20%, which suggests that AuNPs can enhance the electron transfer properties of electrodes containing Cyt c. As it occurs with many other types of LbL films, the bioelectrodes were stable and reproducible.

One of the major challenges in clinical diagnosis is to perform measurements in vivo. Hence, scientists usually utilize in vitro diagnostic devices in tests that often include blood derived from animals [133]. This is made to determine optimized conditions for implantable devices that can monitor substances in vivo. There are, nevertheless, implantable devices whose suitability has been demonstrated. For example, microelectrodes modified with flexible carbon fibers (FCFs) were used for in vivo continuous glucose monitoring in rat blood [133]. This control is important to prevent health complications such as heart diseases and high blood pressure [134]. The FCF electrodes were inserted in the thoracic region of a living rat (*Rattus norvegicus*) using a millimeter catheter and the glucose oxidation in blood was evaluated. The biochip contains two electrodes: the bioelectrode formed by FCF, neutral red and glucose oxidase (GOx), and the FCF electrode. After being implanted as a biosensor, the device had its performance evaluated in a normal concentration of glucose and in a diabetic simulation. It was then possible to compare the biochip's capabilities for in vitro and in vivo glucose monitoring. Since the biochip is inexpensive, this methodology may offer solutions for the long-standing problem of implantable devices.

19.5 Final Remarks

The incorporation of nanoarchitectonics principles in developing materials represented a revolution in the control and tuning of materials properties for a broad range of applications. This applies to any type of material built with bottom-up or top-down approaches. In this chapter we focused on nanostructured films, obtained either with the LB or LbL techniques, for biological applications. With a handful of examples, we hope to have provided the reader with information about strategies to leverage the properties of polymers, nanomaterials, and biomolecules.

Furthermore, we identified a few challenges to move the field forward with nanoarchitectonics, which are mostly related to applying nanostructured films in in vivo conditions, e.g., with implantable devices.

Acknowledgments This work was supported by CAPES, CNPq, and FAPESP (2018/22214-6).

References

1. Weiss PS (2007) A conversation with Dr. Masakazu Aono: leader in atomic-scale control and nanomanipulation. ACS Nano 1:379–383
2. Hensel RC, Moreira M, Riul A, Oliveira ON, Rodrigues V, Hillenkamp M (2020) Monitoring the dispersion and agglomeration of silver nanoparticles in polymer thin films using localized surface plasmons and Ferrell plasmons. Appl Phys Lett 116:103105
3. Wilson D, Materon EM, Ibanez-Redin G, Faria RC, Correa DS, Oliveira ON (2019) Electrical detection of pathogenic bacteria in food samples using information visualization methods with a sensor based on magnetic nanoparticles functionalized with antimicrobial peptides. Talanta 194:611–618
4. Crespilho FN, Zucolotto V, Brett CMA, Oliveira ON, Nart FC (2006) Enhanced charge transport and incorporation of redox mediators in layer-by-layer films containing PAMAM-encapsulated gold nanoparticles. J Phys Chem B 110:17478–17483
5. Luz RAS, Crespilho FN (2016) Gold nanoparticle-mediated electron transfer of cytochrome c on a self-assembled surface. RSC Adv 6:62585–62593
6. Silva RR, Raymundo-Pereira PA, Campos AM, Wilson D, Otoni CG, Barud HS, Costa CAR, Domeneguetti RR, Balogh DT, Ribeiro SJL, Oliveira ON (2020) Microbial nanocellulose adherent to human skin used in electrochemical sensors to detect metal ions and biomarkers in sweat. Talanta 218:121153
7. Girard-Egrot AP, Godoy S, Blum LJ (2005) Enzyme association with lipidic Langmuir-Blodgett films: interests and applications in nanobioscience. Adv Colloid Interf Sci 116:205–225
8. Vollhardt D (2002) Supramolecular organisation in monolayers at the air/water interface. Mater Sci Eng C 22:121–127
9. Hea W, Wanga C, Zhugeb F, Denga X, Xua X, Zhai T (2017) Flexible and high energy density asymmetrical supercapacitors based on core/shell conducting polymer nanowires/manganese dioxide nanoflakes. Nano Energy 35:242–250
10. Abid S, Hussain T, Raza ZA, Nazir A (2019) Current applications of electrospun polymeric nanofibers in cancer therapy. Mater Sci Eng C Mater Biol Appl 97:966–977
11. Feng GX, Mao D, Liu J, Goh CC, Ng LG, Kong DL, Tang BZ, Liu B (2018) Polymeric nanorods with aggregation-induced emission characteristics for enhanced cancer targeting and imaging. Nanoscale 10:5869–5874
12. Mu P, Bai W, Zhang Z, He J, Sun H, Zhu Z, Liang W, Li A (2018) Robust aerogels based on conjugated microporous polymer nanotubes with exceptional mechanical strength for efficient solar steam generation. J Mater Chem A 6:18183–18190
13. Morris JD, Thourson SB, Panta KR, Flanders BN, Payne CK (2017) Conducting polymer nanowires for control of local protein concentration in solution. J Phys D Appl Phys 50:174003
14. Machado-Paula MM, Corat MAF, Lancellotti M, Mi G, Marciano FR, Vega ML, Hidalgo AA, Webster TJ, Lobo AO (2020) A comparison between electrospinning and rotary-jet spinning to produce PCL fibers with low bacteria colonization. Mater Sci Eng C Mater Biol Appl 111:110706

15. Mercante LA, Pavinatto A, Iwaki LEO, Scagion VP, Zucolotto V, Oliveira ON, Mattoso LHC, Correa DS (2015) Electrospun polyamide 6/poly(allylamine hydrochloride) nanofibers functionalized with carbon nanotubes for electrochemical detection of dopamine. ACS Appl Mater Interfaces 7:4784–4790
16. Soares JC, Iwaki LEO, Soares AC, Rodrigues VC, Melendez ME, Fregnani JHTG, Reis RM, Carvalho AL, Correa DS, Oliveira ON (2017) Immunosensor for pancreatic cancer based on electrospun nanofibers coated with carbon nanotubes or gold nanoparticles. ACS Omega 2:6975–6983
17. Soares AC, Soares JC, Paschoalin RT, Rodrigues VC, Melendez ME, Reis RM, Carvalho AL, Mattoso LHC, Oliveira ON (2020) Immunosensors containing solution blow spun fibers of poly(lactic acid) to detect p53 biomarker. Mater Sci Eng C Mater Biol Appl 115:111120
18. Liu WJ, Wijeratne S, Yang L, Bruening M (2018) Porous star-star polyelectrolyte multilayers for protein binding. Polymer 138:267–274
19. Olivati CA, Ferreira M, Carvalho AJF, Balogh DT, Oliveira ON, von Seggern H, Faria RM (2005) Polymer light emitting devices with Langmuir-Blodgett (LB) films: enhanced performance due to an electron-injecting layer of ionomers. Chem Phys Lett 408:31–36
20. Gunkel-Grabole G, Sigg S, Lomora M, Lorcher S, Palivan CG, Meier WP (2015) Polymeric 3D nano-architectures for transport and delivery of therapeutically relevant biomacromolecules. Biomater Sci 3:25–40
21. Correa DS, Medeiros ES, Oliveira JE, Paterno LG, Mattoso LHC (2014) Nanostructured conjugated polymers in chemical sensors: synthesis, properties and applications. J Nanosci Nanotechnol 14:6509–6527
22. Sanchez-Ballester NM, Rydzek G, Pakdel A, Oruganti A, Hasegawa K, Mitome M, Golberg D, Hill JP, Abe H, Ariga K (2016) Nanostructured polymeric yolk-shell capsules: a versatile tool for hierarchical nanocatalyst design. J Mater Chem A 4:9850–9857
23. Sagiv J (1980) Organized monolayers by adsorption. 1. Formation and structure of oleophobic mixed monolayers on solid-surfaces. J Am Chem Soc 102:92–98
24. Ariga K (2020) Don't forget Langmuir-Blodgett films 2020: interfacial nanoarchitectonics with molecules, materials, and living objects. Langmuir 36:7158–7180
25. Decher G (1997) Fuzzy nanoassemblies: toward layered polymeric multicomposites. Science 277:1232–1237
26. Crespilho FN, Huguenin F, Zucolotto V, Olivi P, Nart FC, Oliveira ON (2006) Dendrimers as nanoreactors to produce platinum nanoparticles embedded in layer-by-layer films for methanol-tolerant cathodes. Electrochem Commun 8:348–352
27. Sousa MAM, Siqueira JR, Vercik A, Schoning MJ, Oliveira ON (2017) Determining the optimized layer-by-layer film architecture with dendrimer/carbon nanotubes for field-effect sensors. IEEE Sensors J 17:1735–1740
28. Follmann HDM, Oliveira ON, Lazarin-Bidoia D, Nakamura CV, Huang XX, Asefa T, Silva R (2018) Multifunctional hybrid aerogels: hyperbranched polymer-trapped mesoporous silica nanoparticles for sustained and prolonged drug release. Nanoscale 10:1704–1715
29. Langmuir I (1917) The constitution and fundamental properties of solids and liquids. II. Liquids. J Am Chem Soc 39:1848–1906
30. Blodgett KB, Langmuir I (1937) Built-up films of barium stearate and their optical properties. Phys Rev 51:0964–0982
31. Kobayashi S, Mullen K (2015) Encyclopedia of polymeric nanomaterials. Springer, Berlin, Heidelberg
32. Liu K, Wang L, Dong R (2020) Two-dimensional conjugated polymer films via liquid-interface-assisted synthesis toward organic electronic devices. J Mater Chem C 8:10696–10718
33. Faria RM, Mattoso LHC, Ferreira M, Oliveira ON, Gonçalves D, Bulhoes LOS (1992) Chloroform-soluble poly(o-methoxyaniline) for ultra-thin film fabrication. Thin Solid Films 221:5–8

34. Dhanabalan A, Mattoso LHC, Oliveira ON Jr (2000) Langmuir monolayers and Langmuir-Blodgett films of conducting polymers. Curr Trends Polymer Sci 5:19–39
35. Oliveira ON Jr (1992) Langmuir-Blodgett films: properties and possible applications. Braz J Phys 22:60
36. Decher G, Hong JD (1991) Buildup of ultrathin multilayer films by a self-assembly process. 1. Consecutive adsorption of anionic and cationic bipolar amphiphiles on charged surfaces. Makromol Chem Macromol Symp 46:321–327
37. An Q, Huang T, Shi F (2018) Covalent layer-by-layer films: chemistry, design, and multidisciplinary applications. Chem Soc Rev 47:5061–5098
38. Neto JBMR, Soares AC, Bataglioli RA, Carr O, Costa CAR, Oliveira ON, Beppu MM, Carvalho HF (2020) Polysaccharide multilayer films in sensors for detecting prostate tumor cells based on hyaluronan-CD44 interactions. Cells 9(6):1563
39. Ariga K, Hill JP, Ji QM (2007) Layer-by-layer assembly as a versatile bottom-up nanofabrication technique for exploratory research and realistic application. Phys Chem Chem Phys 9:2319–2340
40. Hsu BB, Hagerman SR, Hammond PT (2016) Rapid and efficient sprayed multilayer films for controlled drug delivery. J Appl Polymer Sci 133:43563
41. Huang J, Qiu X, Yan B, Xie L, Yang JQ, Xu HL, Deng YH, Chen LY, Wang XG, Zeng HB (2018) Robust polymer nanofilms with bioengineering and environmental applications via facile and highly efficient covalent layer-by-layer assembly. J Mater Chem B 6:3742–3750
42. Liu R, Dai JH, Ma LL, Chen JJ, Shi XW, Du YM, Li Z, Deng HB (2019) Low-temperature plasma treatment-assisted layer-by-layer self-assembly for the modification of nanofibrous mats. J Colloid Interface Sci 540:535–543
43. Hyder MN, Kavian R, Sultana Z, Saetia K, Chen PY, Lee SW, Shao-Horn Y, Hammond PT (2014) Vacuum-assisted layer-by-layer nanocomposites for self-standing 3D mesoporous electrodes. Chem Mater 26:5310–5318
44. Yin Q, Jia HB, Mohamed A, Ji QM, Hong L (2020) Highly flexible and mechanically strong polyaniline nanostructure @ aramid nanofiber films for free-standing supercapacitor electrodes. Nanoscale 12:5507–5520
45. Park JK, Zhang JQ, Roy R, Ge SR, Hustad PD (2018) Polyelectrolyte multilayer-like films from layer-by-layer processing of protected polyampholytic block copolymers. Chem Commun 54:9478–9481
46. Qu T, Guan S, Zheng XX, Chen AH (2020) Perpendicularly aligned nanodomains on versatile substrates via rapid thermal annealing assisted by liquid crystalline ordering in block copolymer films. Nanoscale Adv 2:1523–1530
47. Dynarowicz-Latka P, Hac-Wydro K (2014) Edelfosine in membrane environment - the Langmuir monolayer studies. Anti Cancer Agents Med Chem 14:499–508
48. Matyszewska D, Bilewicz R (2015) Interactions of daunorubicin with Langmuir-Blodgett thiolipid monolayers. Electrochim Acta 162:45–52
49. Giner-Casares JJ, Brezesinski G, Mohwald H (2014) Langmuir monolayers as unique physical models. Curr Opin Colloid Interface Sci 19:176–182
50. Girard-Egrot AP, Blum LJ (2007) Langmuir-Blodgett technique for synthesis of biomimetic membranes. In: Martin DK (ed) Nanobiotechnology of biomimetic membranes. Fundamental biomedical technologies. Springer, Boston, MA
51. Leblanc RM (2006) Molecular recognition at Langmuir monolayers. Curr Opin Chem Biol 10:529–536
52. Caseli L, Pavinatto FJ, Nobre TM, Zaniquelli MED, Viitala T, Oliveira ON (2008) Chitosan as a removing agent of beta-lactoglobulin from membrane models. Langmuir 24:4150–4156
53. da Rocha C, Caseli L (2017) Adsorption and enzyme activity of asparaginase at lipid Langmuir and Langmuir-Blodgett films. Mater Sci Eng C Mater Biol Appl 73:579–584
54. Moreira LG, Almeida AM, Camacho SA, Estevao BM, Oliveira ON, Aoki PHB (2020) Chain cleavage of bioinspired bacterial membranes photoinduced by eosin decyl ester. Langmuir 36:9578–9585

55. Soares AC, Soares JC, Rodrigues VC, Oliveira ON, Mattoso LHC (2020) Controlled molecular architectures in microfluidic immunosensors for detecting *Staphylococcus aureus*. Analyst 145:6014–6023
56. Stefaniu C, Brezesinski G (2014) X-ray investigation of monolayers formed at the soft air/water interface. Curr Opin Colloid Interface Sci 19:216–227
57. Blaudez D, Castano S, Desbat B (2011) PM-IRRAS at liquid interfaces. Elsevier, Oxford
58. Blaudez D, Turlet JM, Dufourcq J, Bard D, Buffeteau T, Desbat B (1996) Investigations at the air/water interface using polarization modulation IR spectroscopy. J Chem Soc Faraday Trans 92:525–530
59. Hunt JH, Guyotsionnest P, Shen YR (1987) Observation of C-H stretch vibrations of monolayers of molecules optical sum-frequency generation. Chem Phys Lett 133:189–192
60. Dynarowicz-Latka P, Dhanabalan A, Oliveira ON (2001) Modern physicochemical research on Langmuir monolayers. Adv Colloid Interf Sci 91:221–293
61. Henon S, Meunier J (1991) Microscope at the Brewster angle: direct observation of first-order phase transitions in monolayers. Rev Sci Instrum 62:936–939
62. Honig D, Mobius D (1992) Brewster angle microscopy of LB films on solid substrates. Chem Phys Lett 195:50–52
63. Tharanathan RN, Kittur FS (2003) Chitin - the undisputed biomolecule of great potential. Crit Rev Food Sci Nutr 43:61–87
64. Fernandes ALP, Morais WA, Santos AIB, de Araujo AML, dos Santos DES, dos Santos DS, Pavinatto FJ, Oliveira ON, Dantas TNC, Pereira MR, Fonseca JLC (2005) The influence of oxidative degradation on the preparation of chitosan nanoparticles. Colloid Polym Sci 284:1–9
65. Salmaso S, Bersani S, Semenzato A, Caliceti P (2006) Nanotechnologies in protein delivery. J Nanosci Nanotechnol 6:2736–2753
66. Onishi H, Machida Y (1999) Biodegradation and distribution of water-soluble chitosan in mice. Biomaterials 20:175–182
67. Pavinatto FJ, Caseli L, Pavinatto A, dos Santos DS, Nobre TM, Zaniquelli MED, Silva HS, Miranda PB, de Oliveira ON (2007) Probing chitosan and phospholipid interactions using Langmuir and Langmuir-Blodgett films as cell membrane models. Langmuir 23:7666–7671
68. Wydro P, Krajewska B, Hac-Wydro K (2007) Chitosan as a lipid binder: a Langmuir monolayer study of chitosan-lipid interactions. Biomacromolecules 8:2611–2617
69. Pereira AR, Fiamingo A, Pedro RO, Campana-Filho SP, Miranda PB, Oliveira ON Jr (2020) Enhanced chitosan effects on cell membrane models made with lipid raft monolayers. Colloids Surfaces B Interfaces 193:111017
70. Karnovsky MJ, Kleinfeld AM, Hoover RL, Klausner RD (1982) The concept of lipid domains in membranes. J Cell Biol 94:1–6
71. Fiamingo A, Oliveira ON Jr, Campana-Filho SP (2020) Tuning the properties of high molecular weight chitosans to develop full water solubility within a wide pH range. ChemRxiv
72. Torrano AA, Pereira AS, Oliveira ON Jr, Barros-Timmons A (2013) Probing the interaction of oppositely charged gold nanoparticles with DPPG and DPPC Langmuir monolayers as cell membrane models. Colloids Surfaces B Biointerfaces 108:120–126
73. Davies JT, Rideal EK (1963) Interfacial phenomena. Academic Press, New York
74. Marsh D (1996) Lateral pressure in membranes. Biochim Biophys Acta Rev Biomembr 1286:183–223
75. Camacho SA, Kobal MB, Almeida AM Jr, Toledo KA, Oliveira ON Jr, Aoki PHB (2020) Molecular-level effects on cell membrane models to explain the phototoxicity of gold shell-isolated nanoparticles to cancer cells. Colloids Surf B: Biointerfaces 194:111189
76. Khomutov GB, Obydenov AY, Yakovenko SA, Soldatov ES, Trifonov AS, Khanin VV, Gubin SP (1999) Synthesis of nanoparticles in Langmuir monolayer. Mater Sci Eng C 8–9:309–318
77. Khomutov GB, Gubin SP, Khanin VV, Koksharov AY, Obydenov AY, Shorokhov VV, Soldatov ES, Trifonov AS (2002) Formation of nanoparticles and one-dimensional

nanostructures in floating and deposited Langmuir monolayers under applied electric and magnetic fields. Colloids Surf A Physicochem Eng Asp 198-200:593–604
78. Matyszewska D, Napora E, Zelechowska K, Biernat JF, Bilewicz R (2018) Synthesis, characterization, and interactions of single-walled carbon nanotubes modified with doxorubicin with Langmuir-Blodgett biomimetic membranes. J Nanoparticle Res 20:143
79. Leca B, Blum LJ (2000) Luminol electrochemiluminescence with screen-printed electrodes for low-cost disposable oxidase-based optical sensors. Analyst 125:789–791
80. Godoy S, Leca-Bouvier B, Boullanger P, Blum LJ, Girard-Egrot AP (2005) Electrochemiluminescent detection of acetylcholine using acetylcholinesterase immobilized in a biomimetic Langmuir-Blodgett nanostructure. Sensors Actuators B Chem 107:82–87
81. Godoy S, Chauvet JP, Boullanger P, Blum LJ, Girard-Egrot AP (2003) New functional proteoglycolipidic molecular assembly for biocatalysis analysis of an immobilized enzyme in a biomimetic nanostructure. Langmuir 19:5448–5456
82. Chen BL, Ji F, Wang C, Gao Y, Zhou Z, Li Z, Cao HM, Hao LT, Liu GL, Liu JF, Liang Y (2019) Capture and elimination of *Staphylococcus aureus* based on Langmuir-Blodgett MnO_2 nanowire monolayer promotes infected wound healing. J Mater Chem B 7:4198–4206
83. Siqueira JR, Caseli L, Crespilho FN, Zucolotto V, Oliveira ON (2010) Immobilization of biomolecules on nanostructured films for biosensing. Biosens Bioelectron 25:1254–1263
84. Zanon NCM, Oliveira ON, Caseli L (2012) Immbolization of uricase enzyme in Langmuir and Langmuir-Blodgett films of fatty acids: possible use as a uric acid sensor. J Colloid Interface Sci 373:69–74
85. Juers DH, Hakda S, Matthews BW, Huber RE (2003) Structural basis for the altered activity of Gly794 variants of *Escherichia coli* beta-galactosidase. Biochemistry 42:13505–13511
86. Araujo FT, Peres LO, Caseli L (2019) Conjugated polymers blended with lipids and galactosidase as langmuir-Blodgett films to control the biosensing properties of nanostructured surfaces. Langmuir 35:7294–7303
87. Bhatt G, Bhattacharya S (2019) Biosensors on chip: a critical review from an aspect of micro/nanoscales. J Micromanuf 2:198–219
88. Qian LS, Li QB, Baryeh K, Qiu WW, Li K, Zhang J, Yu QC, Xu DQ, Liu WJ, Brand RE, Zhang XJ, Chen W, Liu GD (2019) Biosensors for early diagnosis of pancreatic cancer: a review. Transl Res 213:67–89
89. Goldsmith BR, Locascio L, Gao YN, Lerner M, Walker A, Lerner J, Kyaw J, Shue A, Afsahi S, Pan D, Nokes J, Barron F (2019) Digital biosensing by foundry-fabricated graphene sensors. Sci Rep 9:434
90. Rodrigues VC, Soares JC, Soares AC, Braz DC, Melendez ME, Ribas LC, Scabini LFS, Bruno OM, Carvalho AL, Reis RM, Sanfelice RC, Oliveira ON Jr (2021) Electrochemical and optical detection and machine learning applied to images of genosensors for diagnosis of prostate cancer with the biomarker PCA3. Talanta 222:121444
91. Rastogi V, Yadav P, Bhattacharya SS, Mishra AK, Verma A, Pandit JK (2014) Carbon nanotubes: an emerging drug carrier for targeting cancer cells. J Drug Deliv 2014:670815
92. Neves AF, Dias-Oliveira JDD, Araujo TG, Marangoni K, Goulart LR (2013) Prostate cancer antigen 3 (PCA3) RNA detection in blood and tissue samples for prostate cancer diagnosis. Clin Chem Lab Med 51:881–887
93. Paulovich F, Moraes M, Maki R, Ferreira M, Oliveira O, de Oliveira M (2011) Information visualization techniques for sensing and biosensing. Analyst 136:1344–1350
94. Kearney AJ, Murray M (2009) Breast cancer screening recommendations: is mammography the only answer? J Midwifery Womens Health 54:393–400
95. Ren KF, Ji J, Shen JC (2006) Construction and enzymatic degradation of multilayered poly-L-lysine/DNA films. Biomaterials 27:1152–1159
96. Rusling JF (2004) Sensors for toxicity of chemicals and oxidative stress based on electrochemical catalytic DNA oxidation. Biosens Bioelectron 20:1022–1028
97. Sano KI, Shiba K (2008) In aqua manufacturing of a three-dimensional nanostructure using a peptide aptamer. MRS Bull 33:524–529

98. Sano KI, Shiba K (2008) Stepwise accumulation of layers of aptamer-ornamented ferritins using biomimetic layer-by-layer. J Mater Res 23:3236–3240
99. Tombelli S, Minunni M, Mascini A (2005) Analytical applications of aptamers. Biosens Bioelectron 20:2424–2434
100. Du Y, Chen CG, Li BL, Zhou M, Wang EK, Dong SJ (2010) Layer-by-layer electrochemical biosensor with aptamer-appended active polyelectrolyte multilayer for sensitive protein determination. Biosens Bioelectron 25:1902–1907
101. Soares AC, Soares JC, Shimizu FM, Melendez ME, Carvalho AL, Oliveira ON (2015) Controlled film architectures to detect a biomarker for pancreatic cancer using impedance spectroscopy. ACS Appl Mater Interfaces 7:25930–25937
102. Imamura AH, Segato TP, de Oliveira LJM, Hassan A, Crespilho FN, Carrilho E (2020) Monitoring cellulose oxidation for protein immobilization in paper-based low-cost biosensors. Microchimica Acta 187:272
103. Barratt J, Topham P (2007) Urine proteomics: the present and future of measuring urinary protein components in disease. Can Med Assoc J 177:361–368
104. Lal S, Grady NK, Kundu J, Levin CS, Lassiter JB, Halas NJ (2008) Tailoring plasmonic substrates for surface enhanced spectroscopies. Chem Soc Rev 37:898–911
105. Ma W, Fu P, Sun MZ, Xu LG, Kuang H, Xu CL (2017) Dual quantification of MicroRNAs and telomerase in living cells. J Am Chem Soc 139:11752–11759
106. Melo AFAA, Hassan A, Macedo LJA, Osica I, Shrestha LK, Ji QM, Oliveira ON, Henzie J, Ariga K, Crespilho FN (2019) Microwires of Au-Ag nanocages patterned via magnetic nanoadhesives for investigating proteins using surface enhanced infrared absorption spectroscopy. ACS Appl Mater Interfaces 11:18053–18061
107. Neubrech F, Pucci A, Cornelius TW, Karim S, Garcia-Etxarri A, Aizpurua J (2008) Resonant plasmonic and vibrational coupling in a tailored nanoantenna for infrared detection. Phys Rev Lett 101:157403
108. Wu CH, Khanikaev AB, Adato R, Arju N, Yanik AA, Altug H, Shvets G (2012) Fano-resonant asymmetric metamaterials for ultrasensitive spectroscopy and identification of molecular monolayers. Nat Mater 11:69–75
109. Ataka K, Stripp ST, Heberle J (2013) Surface-enhanced infrared absorption spectroscopy (SEIRAS) to probe monolayers of membrane proteins. BBA Biomembranes 1828:2283–2293
110. Oliveira D, Borges A, Simoes M (2018) *Staphylococcus aureus* toxins and their molecular activity in infectious diseases. Toxins 10:252
111. Jorgensen PS, Wernli D, Carroll SP, Dunn RR, Harbarth S, Levin SA, So AD, Schluter M, Laxminarayan R (2016) Use antimicrobials wisely. Nature 537:159–161
112. Carrette L, Friedrich KA, Stimming U (2000) Fuel cells: principles, types, fuels, and applications. ChemPhysChem 1:162–193
113. Park S, Wieckowski A, Weaver MJ (2003) Electrochemical infrared characterization of CO domains on ruthenium-decorated platinum nanoparticles. J Am Chem Soc 125:2282–2290
114. Jia HY, Gao PC, Ma HM, Wu D, Du B, Wei Q (2015) Preparation of Au-Pt nanostructures by combining top-down with bottom-up strategies and application in label-free electrochemical immunosensor for detection of NMP22. Bioelectrochemistry 101:22–27
115. Yang L, Zhao H, Fan SM, Deng SS, Lv Q, Lin J, Li CP (2014) Label-free electrochemical immunosensor based on gold-silicon carbide nanocomposites for sensitive detection of human chorionic gonadotrophin. Biosens Bioelectron 57:199–206
116. Manso J, Mena ML, Yanez-Sedeno P, Pingarron JM (2008) Bienzyme amperometric biosensor using gold nanoparticle-modified electrodes for the determination of inulin in foods. Anal Biochem 375:345–353
117. Raymundo-Pereira PA, Shimizu FM, Coelho D, Piazzeta MHO, Gobbi AL, Machado SAS, Oliveira ON (2016) A nanostructured bifunctional platform for sensing of glucose biomarker in artificial saliva: synergy in hybrid Pt/Au surfaces. Biosens Bioelectron 86:369–376
118. Decher G (1996) Comprehensive supramolecular chemistry. In: Sauvage J-P, Housseini MW (eds) Templating, self-assembly and self-organization. Pergamon, Oxford, pp 507–528

119. Mrksich M (1997) Using self-assembled monolayers to understand the biomaterials interface. Curr Opin Colloid Interface Sci 2:83–88
120. Ram MK, Bertoncello P, Ding H, Paddeu S, Nicolini C (2001) Cholesterol biosensors prepared by layer-by-layer technique. Biosens Bioelectron 16:849–856
121. Ferreira M, Fiorito PA, Oliveira ON, de Torresi SIC (2004) Enzyme-mediated amperometric biosensors prepared with the layer-by-layer (LbL) adsorption technique. Biosens Bioelectron 19:1611–1615
122. Wu KB, Fei JJ, Bai W, Hu SS (2003) Direct electrochemistry of DNA, guanine and adenine at a nanostructured film-modified electrode. Anal Bioanal Chem 376:205–209
123. Palecek E, Fojta M, Tomschik M, Wang J (1998) Electrochemical biosensors for DNA hybridization and DNA damage. Biosens Bioelectron 13:621–628
124. Wang F, Wu YJ, Liu JX, Ye BX (2009) DNA Langmuir-Blodgett modified glassy carbon electrode as voltammetric sensor for determinate of methotrexate. Electrochim Acta 54:1408–1413
125. de Souza JCP, Iost RM, Crespilho FN (2015) Nitrated carbon nanoblisters for high-performance glucose dehydrogenase bioanodes. Biosens Bioelectron 77:860–865
126. Martins MVA, Pereira AR, Luz RAS, Iost RM, Crespilho FN (2014) Evidence of short-range electron transfer of a redox enzyme on graphene oxide electrodes. Phys Chem Chem Phys 16:17426–17436
127. Pereira AR, Luz RAS, Dalmati FCDA, Crespilho FN (2017) Protein oligomerization based on Brønsted acid reaction. ACS Catal 7:3082–3088
128. Pagnoncelli K, Pereira A, Sedenho G, Bertaglia T, Crespilho F (2018) Ethanol generation, oxidation and energy production in a cooperative bioelectrochemical system. Bioelectrochemistry 122:11–25
129. Crespilho FN, Iost RM, Travain SA, Oliveira ON, Zucolotto V (2009) Enzyme immobilization on Ag nanoparticles/polyaniline nanocomposites. Biosens Bioelectron 24:3073–3077
130. Crespilho FN, Zucolotto V, Oliveira ON, Nart FC (2006) Electrochemistry of layer-by-layer films: a review. Int J Electrochem Sci 1:194–214
131. Zhao MQ, Crooks RM (1999) Dendrimer-encapsulated Pt nanoparticles: synthesis, characterization, and applications to catalysis. Adv Mater 11:217
132. Siqueira JR, Crespilho FN, Zucolotto V, Oliveira ON (2007) Bifunctional electroactive nanostructured membranes. Electrochem Commun 9:2676–2680
133. Iost RM, Sales FCPF, Martins MVA, Almeida MC, Crespilho FN (2015) Glucose biochip based on flexible carbon fiber electrodes: in vivo diabetes evaluation in rats. ChemElectroChem 2:518–521
134. Koschwanez HE, Reichert WM (2007) In vitro, in vivo and post explantation testing of glucose-detecting biosensors: current methods and recommendations. Biomaterials 28:3687–3703

Chapter 20
Responsive Polymeric Architectures and Their Biomaterial Applications

Lili Chen, Mazaya Najimina, and Mitsuhiro Ebara

20.1 Nano- and Bio-materials

Nanotechnology has already had a notable impact on medicines, health, and safety products, as well as space, entertainment sectors, and IT [1–4]. Nanotechnology covers a wide range of different technologies involved in the manipulation and control of matter on a very small scale, atom by atom and molecule by molecule. Such technology makes it possible to develop materials and products with "nano-scale" structures or to build machines and systems the same size as human cells with highly desirable properties. At the nanoscale scale, the properties of nanomaterials not only depend on composition and structure but also display new phenomena associated with interfaces and the preponderance of surfaces [5, 6]. At the dawn of the twenty-first century, the challenge for nanomaterials is, while their nanoscale dimensions and structures themselves are very unique, to control their property actively on a very small scale. To cite one of the examples that emphasize the importance of controlling interfacial properties of materials on the nanoscale, surfaces of implanted materials (often referred to as "biomaterials") have been designed to control the adsorption of biological molecules onto them [7–9].

The word "biomaterials" was not used just 50 years ago, but they are now widely used throughout dentistry, medicine, and biotechnology. Interestingly, the introduction of tolerated foreign material implants into the human body has been noted far back as early as prehistory, 9000 years ago [10]. Interestingly, as early as 9000 years ago, prehistoric times had noticed the implantation of tolerable foreign bodies into the human body. In 1996, the remains of a man (called "Kennewick") were found

L. Chen · M. Najimina · M. Ebara (✉)
Research Center for Functional Materials, National Institute for Materials Science (NIMS), Tsukuba, Ibaraki, Japan
e-mail: EBARA.Mitsuhiro@nims.go.jp

T. Govindaraju, K. Ariga (eds.), *Molecular Architectonics and Nanoarchitectonics*, Nanostructure Science and Technology,
https://doi.org/10.1007/978-981-16-4189-3_20

near Kennewick, Washington, and a spear point was embedded in his hip. It had already obviously healed in and did not notably hinder his activity. This unexpected implant illustrates the human body's ability to handle the implanted foreign materials. Thus, gaining control of "nano" structures and properties of materials is a matter of great interest in biomedicine and healthcare. In other words, to control the behavior of water near surfaces is a very important challenge in bio-/nanomaterials science because water is the first molecule to contact biomaterials in any clinical applications and is an active participant in biology, which simply could not work the way it does without the special mediating properties of water [11, 12]. Indeed, water suspends living cells as well as the main components of the interstitial fluid that bathes tissues. How can the role of water in biology be controlled, then? The following sections present a perspective on this question.

20.2 "Smart" Polymers

There has been a renewal of interest in controlling "nano" interfaces as mentioned above, and a considerable number of studies have been conducted on designing novel materials to meet these applications [13–15]. Some special types of polymers, for example, have emerged as a very useful class of polymers and have their own special chemical properties and applications in various areas [16–18]. These polymers are coined with different names, based on their physical or chemical properties like "stimuli-responsive" or "environmental-sensitive" or "smart" or "intelligent" polymers. We shall use further on the name "smart" polymers for such polymer systems in this review, because "smart" means "able to adapt to one's environment while "intelligent" comes from the Latin verb intelligence, which means "to understand" [19]. Indeed, "smart polymers" are able to adapt to the change in their environment by changing their interactions with water. Actually making them "smart" characteristic feature is their ability to respond to slight changes in the environment. The singularity of these materials lies not solely in the fast macroscopic changes occurring in their structure but also in these reversible transformations. The responses are displayed as one or more changes, such as solubility, surface characteristics, shape, a sol-to-gel transition formation of an intricate molecular assembly, and others [20]. The environmental trigger that causes these transitions can be either change in pH shift or temperature, presence of certain chemicals, and increase in ionic strength, etc. More recently, change in the magnetic and electric field and light or radiation forces as stimuli for these polymers have also been reported.

There are a multitude of polymers that reveal thermally induced precipitation, and the polymer that has been studied most spaciously is poly(*N*-isopropylacrylamide) (PNIPAAm) [21, 22]. The polymer PNIPAAm is soluble in the water below 32 °C, while it precipitates sharply above 32 °C. The precipitation temperature is called the lower critical solution temperature (LCST). Critical insight into the precipitation mechanism was gained by the Hoffman group's seminal work on PNIPAAm [23–25]. PNIPAAm can be applied in many forms, such as dissolved in aqueous solution, cross-linked in the form of hydrogels, and grafted or adsorbed on aqueous–solid interfaces. Grafting of PNIPAAm onto solid surfaces, as "smart" surfaces, to make the surface energy can be controlled with temperature [26–28]. PNIPAAm can be also functionalized by chemical conjugation to biomolecules such as a targeting ligand to yield a large and diverse family of polymer–biomolecule hybrid systems that can respond to biological as well as to chemical and physical stimuli [29, 30]. Although these grafting or conjugation processes can produce "smart" surfaces or "smart" conjugations, derivatization of PNIPAAm side chains or end group to form reactive groups has resulted in the disappearance of the "smartness" from PNIPAAm (Fig. 20.1). For example, if NIPAAm monomer is copolymerized with a reactive monomer such as acrylic acid (AAc), the LCST increases and may even disappear [25, 31]. Based on the hypothesis that an appropriate proportion of hydrophobicity and hydrophilicity in the molecular structure of PNIPAAm is required for the sharp phase transition to occur, the authors have newly designed a series of monomers which have a parallel structure with NIPAAm monomer to deceive water molecules into behaving like those around PNIPAAm chains even though they have a very hydrophilic reactive group such as carboxyl, amino, and hydroxyl group, i.e., carboxy-IPAAm (CIPAAm) [31–33], amino-IPAAm (AIPAAm) [34], and hydroxyl-IPAAm (HIPAAm) [35], respectively (Fig. 20.2). Use of these monomers for copolymerization with NIPAAm can synergistically combine the individual properties of the two or more components to yield new and desirable properties. One could say that these polymers are "double or multi-smart."

20.3 "Smart" Diagnostic Tools

20.3.1 Early Disease Diagnosis

The first area of intense study in "smart" technology is developing novel diagnostic tools for health delivery services. Early disease diagnosis can advance patient outcomes and lowers overall medical costs, while it is limited by the need for sensitive diagnostic technology that can detect the earliest possible biotransformation process. Protein biomarkers as diagnostic targets (analytes) are diluted in the complex environment of human blood. Separating and enriching biomarkers to easily detectable concentrations will overcome this issue. Microscale diagnostic technologies have been adopted, especially in-home healthcare technologies and distributed diagnostics, successfully meeting this need [36, 37]. Microscale technologies reduce sample volumes and ease fluid handling restrictions while providing modular functionality. Most of the molecular components used in these technologies have remained constant, and there are opportunities to develop new molecular tools to

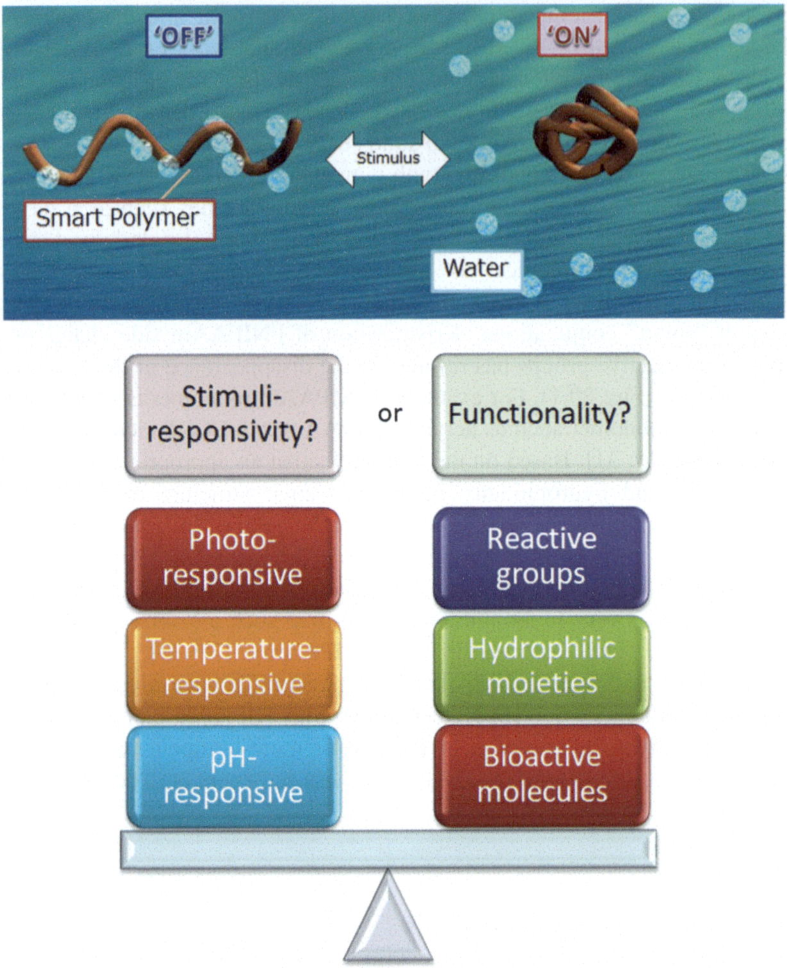

Fig. 20.1 Current problems and challenges in derivatization of smart polymers' side chains to form functional groups. Copolymerization with functional comonomers has resulted in the disappearance of the "smartness" from smart polymers, e.g., if NIPAAm monomer is copolymerized with a hydrophilic monomer, the LCST increases and may even disappear

solve inherent issues. Current technologies rely on traditional chromatographic techniques, which suffer from poor diffusion of large biomarkers and limited activity of surface-bound capture components. Although microfluidic systems mainly operate at the micron level (millionths of a meter), nanotechnology has proposed new concepts and may play an increasing role in the field of diagnosis. A series of nanoscale PNIPAAm–protein conjugates have been developed, which can separate and enrich analytes from solution and enable detection [38, 39]. Since smart conjugates bind the analytes in solution before separation, the binding of conjugate and analytes avoids steric and mass transport limitations associated with

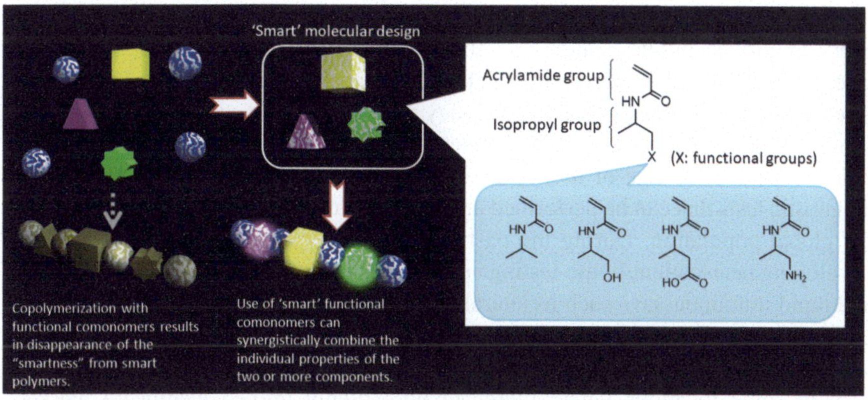

Fig. 20.2 Schematic illustration of a novel strategy for designing "double or multi-smart" polymers that respond to various external stimuli. "Smart" molecular design makes it possible to produce a series of comonomers which have a parallel structure with NIPAAm monomer to deceive water molecules into behaving like those around PNIPAAm chains even though they have a very hydrophilic reactive group such as carboxyl, amino, and hydroxyl group. Use of these NIPAAm-like comonomers for copolymerization with NIPAAm can synergistically combine the individual properties of the two or more components to yield new and desirable properties

surface-based technologies [40, 41]. The author and colleagues have further developed a reversible microchannel surface capture system for stimuli-responsive grafted biological analysis beads [42]. Through UV-mediated graft polymerization, PNIPAAm was grafted onto the surfaces of polydimethylsiloxane (PDMS) surfaces from a photoinitiator that was preadsorbed in the channel wall. The switchable hydrophilic/hydrophobic properties are controlled by changing the initiator concentration and/or the photo-illumination times [43]. The surface trap captured PNIPAAm-grafted nanobeads congruously above the LCST and promoted their rapid release with the temperature reversed to below the LCST. A pH-responsive surface trap has been also constructed in the channel wall using the same methods. The proposed microfluidic systems promise to realize the separation and enrichment of "smart" protein biomarkers by combining with the advanced microfluidic flow and mixing technologies.

20.3.2 Diagnosis in the Developing World

The smart diagnostic systems may be also adopted for civilian healthcare in the developing world since these systems must be inexpensive, besides being accurate, reliable, rugged, and quite suited to the medical and social contexts of the developing world [44, 45]. Many developing countries lack the best medical diagnostic technologies, such as the world's population of 6.1 billion people, 3 billion lack basic sanitation facilities, 2 billion cannot get access to electricity, and more than 1 billion do not have access to basic healthcare services and clean drinking water [46]. For every victory in public health, e.g., the eradication of smallpox, other infectious diseases such as tuberculosis and malaria have reemerged, accompanied by new diseases such as human immunodeficiency virus (HIV) (e.g., in many countries in sub-Saharan Africa, up to 20% of pregnant women are infected with HIV) [47]. In order to take advantage of the treatment options, one crucial factor is the lack of diagnostic tests that can be performed at low-infrastructure sites, which serve most of the global population. During the past decades, many efforts have been made for developing near-patient/home-testing approach in the developing world [48]. Simple and rapid immunoassays such as lateral flow strip tests, for example, have been one of the successfully used diagnostic technologies in the developing world. The use of saliva as a sample is also a suitably challenging issue because blood sampling is not easy in some locations where there are little highly trained healthcare workers [49, 50]. While a salivary test has many advantages over a serum test (i.e., less expensive and less invasive vial), many indicator molecules in saliva are dilute and their concentrations are very low. Taking these problems into account, the "smart" switchable surface trap systems mentioned above will overcome this issue because the proposed systems can separate and enrich indicators with simple "on–off" switch control of the external stimuli such as friction heat, light, and magnet, without using electrical power (Fig. 20.3) [42]. Lai et al. have developed magnetic nanoparticles for diagnostic target isolation because smaller particles display better association and binding properties to the target analytes [51, 52]. The pH-sensitive magnetic particles showed a sharply reversible change in size and magnetospheric mobility, thereby performing particle separation under continuous flow processing conditions. The isolated 5 nm particles do not emerge to have significant magnetophoretic speed, but do show high magnetophoretic speed when aggregated by the action of pH-responsive polymers. The aggregated particles which have captured a diagnostic target analytes magnetophoresis with a simple external magnet from a lower pH laminar flow stream to a higher pH flow stream that induces rapid disaggregation of nanoparticles. This system had been shown to transfer more than 80% of a target protein from the input flow stream to the second flow stream in a continuous flow H-filter device. Therefore, the "smart" polymer-based diagnostic systems could refine dilute salivary samples, overcoming the problem of lack of power supply in the developing world, while retaining the superiority of rapid and reversible capture and concentration capabilities.

20.3.3 "Smart" Microfluidic Flow Control

"Smart" polymers also show considerable promise in microactuators design as an autonomous flow control inside the microfluidic channels due to their ability to undergo abrupt volume changes in response to the surrounding environment without any requirement for external power sources. The flow control system, which mainly consists of a fully functional valve, remains a major challenge in the integrated

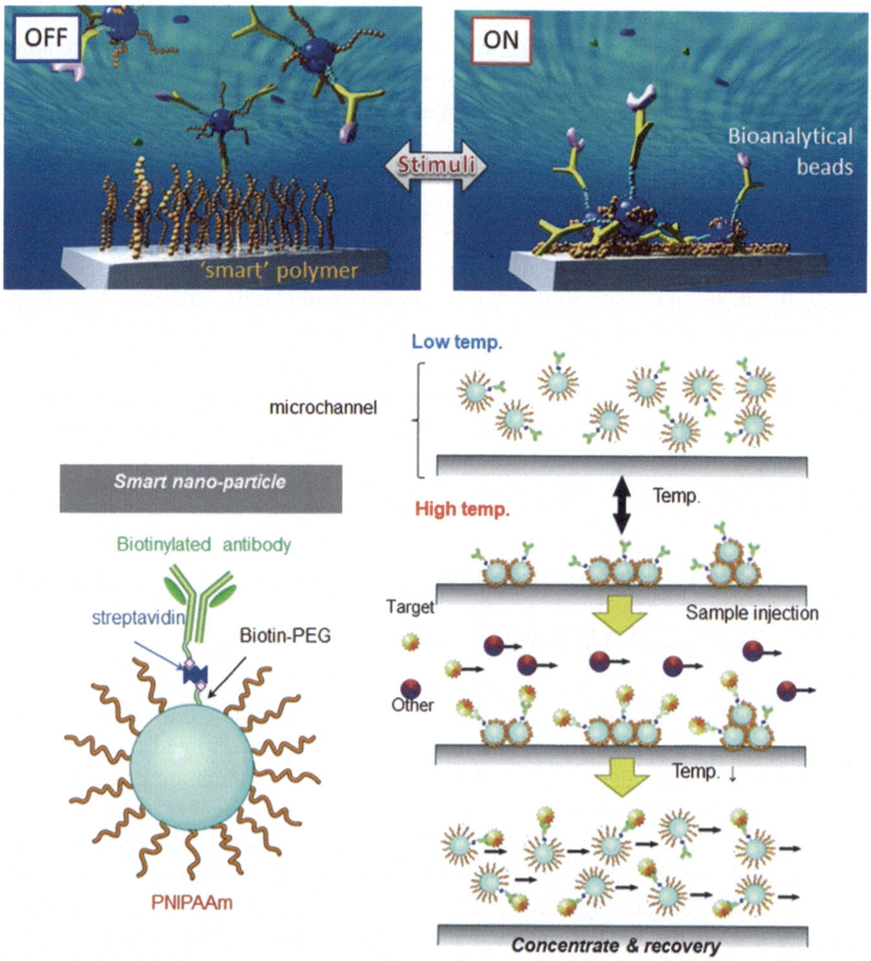

Fig. 20.3 "Smart" separation and enrichment of dilute samples with simple "on–off" switch control of the external stimuli such as friction heat, sun light, magnet, and without using electrical power. The microchannel walls are modified with the smart polymers to minimize nonspecific adsorption of serum components at ambient temperature while allowing the trapping of the beads at selected positions in the channel at higher temperature. This system could be useful for many applications such as affinity-based target isolation and injectable immunoassays, overcoming the problem of low power supply in the developing world

multifunctional microfluidic devices. Beebe et al. utilized a pH-responsive hydrogel-based valve that opened or closed upon the pH changes of the flowing solution [53]. Bi-strip valve and arrowhead-shaped valve have also been reported to allow the fluid to flow in one direction only [54, 55]. Microvalve is, however, not easily fabricated due to its design complexity, and its production reliability for a specific location remained questionable. The triggered control of interfacial properties has also been developed for designing a "smart" microfluidic system by immobilizing the "smart" polymers at the solid–water interface. Several research groups have explored the use of microcapillaries coated with PNIPAAm nano layer for creating an "on–off" valve as a liquid flow controller [56]. Below its LCST, the PNIPAAm-coated capillary allowed the flow of water due to its hydrophilicity; meanwhile, above its LCST, the flow was blocked as the coating switched to hydrophobic. The opposite system has been also developed by Idota et al. where nanometer-sized surface-grafted PNIPAAm layers within microcapillaries stopped the flow below LCST, as the hydration of PNIPAAm-grafted chains at the capillary interfaces increased the microviscosity of the hydration layers at the wall interfaces without physically obstructing the capillary lumen, hence significantly influencing the flow frictional resistance [57]. This system produced both complete and reversible "on–off" flow valving in the microchannels under relevant hydrostatic pressures beneficial for microfluidics approaches. Thus, the nano-/micro-structure design inside the microchannels is useful to regulate the flow, eliminating the need for an external power supply.

20.4 "Smart" Biological Assays

20.4.1 Biological Affinity Measurement

Nanotechnology in the entire area of bioassays will also not only impact early detection, rapid clinical analysis, and home monitoring but also add a new understanding to the molecular biology field. The biological systems are greatly dominated by noncovalent interactions, which are relatively weaker than covalent interactions but more sensitive to the environments [58]. As a result, these interactions provide an excellent mechanism for dynamically regulating the assembly and function in the biological systems [59]. Noncovalent interactions are also ubiquitous in nature and nearly 200,000 interactions between biological molecules have been identified [60]. Many studies have provided various measurement systems and quantitative analysis methods for studying these interactions, such as integrin-mediated adhesive strength of cells to the extracellular matrix (ECM) components [61]. Specific interaction of integrins with ECM components plays an important role in many cellular functions, such as regulation of the cell morphology, growth, differentiation, and motility, because anchorage-dependent cells adhere to a substrate via a series of steps beginning with the binding of cell membrane receptors to the substrate ligands, followed by cytoskeletal and cytoplasmic reorganization [62, 63]. The simplest and most common assay for exploiting integrin–ECM binding strength involves the exposure of cells adhering on the substrates to the force-driven detachment via centrifugation or hydrodynamic fluid flow [64, 65]. Micromanipulation techniques involving the use of microprobes, atomic force microscopy (AFM) cantilevers, or laser tweezers to apply forces to detach cells from substrates also provide sensitive (pico–nano range) measurements of parameters such as low-number receptor–ligand interactions [66, 67]. However, these measurements

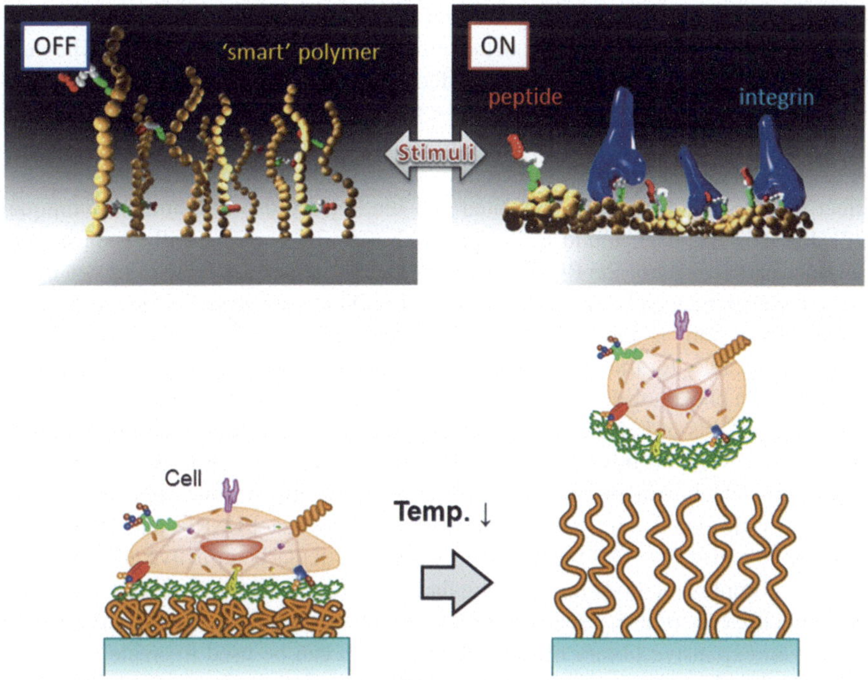

Fig. 20.4 Design concept for an "on–off" affinity switch to measure integrin–peptide binding utilizing smart surfaces. At temperatures above the LCST, the grafted layer dehydrates and shrinks, thus exposing the immobilized peptides to integrins under cell culture conditions. Upon lowering the temperature, the grafted polymer is suddenly hydrated and extends outward to shield the peptides from integrin access, resulting in decreased binding affinity between integrins and peptides, followed by cell detachment from the surface

are time consuming and only able to capture a single point in the continuous "dynamic" spectrum of force-driven detachment.

Given the complexity of the cell detachment process, the authors have proposed a "smart" technique that explores spontaneous and noninvasive cell detachment behavior from substrates as an alternative approach for the analysis of cell–substrate interactions [30, 68]. Over the past two decades, many short peptide sequences in adhesive proteins have been identified as putative adhesion sites to several types of integrins [69]. The tripeptide sequence RGD is the first sequence to have been conclusively identified as the minimal cell-binding domain in fibronectin (FN) [70]. We have developed a novel approach to observe dynamic affinity changes in RGD–integrin binding using RGD-immobilized temperature-responsive nanolayer surfaces (Fig. 20.4). RGD was successfully immobilized on the PNIPAAm copolymer's side chains via reactive groups of comonomer, CIPAAm [30, 71]. The nanodesigned surface can bind cells at 37 °C and spontaneously drives the detachment of cells at 20 °C. The specific design of the grafted polymer is a driving factor of this phenomenon, where peptides were protected from interacting with integrin below the LCST of PNIPAAm. The tight coupling between cells and peptides on the surfaces produces a delay in the detachment of cells from the surfaces. The cell adhesion strength is known to strongly affect the cell detachment time, according to the following tendency: RGD < RGDS < GRGD < GRGDS. This approach can serve as a quick and simple method to evaluate the dynamic affinity changes between cells and peptides. Also, we have successfully investigated the synergistic effect of PHSRN on integrin-mediated cell binding by using a similar technique [72]. PHSRN has been found in FN and is thought to synergistically enhance the cell-adhesive activity of the RGD sequence [73]. The synergistic site is known to be located approximately 3.5 nm away from the RGD loop in native FN, and its precise spatial positioning is thought to play a critical role in the observed synergistic activity, especially for the downstream cell adhesion events involving focal adhesion kinase (FAK) phosphorylation [74]. On the other hand, very little is known about the synergistic effects of PHSRN on the binding strength. Therefore, we have designed PHSRN and RGD co-immobilized PNIPAAm surfaces. The surfaces dramatically retard cell detachment below the LCST only when the peptide sequences were specifically designed with the optimal distance between PHSRN and RGDS, as observed in native FN. This technique provides a particularly attractive alternative to the more complex methods for applying the forces for detachment of cells, owing to its simplicity, ease of use, sensitivity, and noninvasiveness. The ongoing studies will provide an expansive library of interactions, which can be used to create synthetic materials that dynamically alter their assembly, dissociation, shape, volume, porosity, and optical properties in response to the environmental stimuli.

20.4.2 Bio-separations

The separation process of biological products, which is crucial to obtain the high purity product, requires an extra effort, especially to reduce the overall production price and improve the purity of the products. One of the attractive approaches to achieve those purposes is by promoting a selective precipitation of a target molecule from a mixture. Highly selective precipitation is a promising technique for protein purification or enrichment. The early method for precipitation of the target protein was performed upon the addition of large amounts of salts, organic solvents miscible with water, poly(ethylene glycol) (PEG), etc. [75]. Affinity separation of proteins by using "smart" polymers emerged in the early 1980s [23, 24, 76]. Since then, the utilization of "smart" polymers has been contributed to the simple, fast, and cost-effective way to obtain a specific target molecule. "Smart" polymer-based affinity separation consists of the following steps: (1) enhance the strong binding affinity in free solution, (2) induce the precipitation of the affinity reagent–target protein complex from the solution, (3) recover and concentrate the precipitate, (4) dissociate and recover the target molecule from the complex, and (5) recover the affinity-promoting reagent. Polymer–protein conjugates have been extensively investigated over the past 25–30 years [77]. Early publications on "smart" polymer–protein conjugates appeared in the late 1970s and early 1980s. Hoffman et al. reported the synthesis of PNIPAAm–monoclonal antibody conjugates in order to establish a novel thermally induced phase-separation immunoassay [24, 78]. Since then, many researchers have performed the conjugation of smart polymers to proteins for affinity separation in various applications [79]: enzyme bioprocesses [80], drug delivery [81], diagnostics and biosensors [82], cell culture and tissue engineering [83], and also DNA motors [84].

If the amino acid sequence of a protein is known, site-specific mutagenesis ("genetic engineering" of the protein) may be implemented to substitute one amino acid at a specific site with another. For example, a mutant protein with an exposed thiol group can be obtained by introducing the cysteine residue using such techniques. And then, polymers with terminal or pendant maleimide, vinyl, or vinyl sulfone groups can be conjugated to the protein, resulting in a group that reacts preferentially with thiol groups rather than with lysine amino groups. Kochendoerfer et al. have used site-specific mutagenesis plus specific chemistries to conjugate branched PEGs, owing to negatively charged terminal groups, via oxime linkages into two lysine sites glycosylated in the native erythropoietin [85]. In another approach, Tirrell et al. have reported that non-natural amino acids may be incorporated into a protein structure by genetic modification during the protein expression process within cells, by also considering the suitable cell culture conditions [86]. Such non-natural amino acids would play a role as a conjugation site that allows the polymer conjugation to take place by using different chemistries, such as "click chemistry" [87]. The site-specific conjugation enables the conjugation of smart polymers to proteins either near the active site or far away from the active site. When the former case takes place, to avoid the direct effect on the protein activity, the polymer needs to be stimulated to collapse and then to re-hydrate. The collapsed smart polymer coil can firstly "block" (turn "off") the protein active site, and then "unblock" (turn back "on") the active site after re-hydrated. Stayton et al. first reported the versatility of site-specific conjugation by conjugating PNIPAAm far away from the active site of cytochrome-b5 and close to the active site of streptavidin [29, 88]. More recently, "living" free radical polymerization methods have been used to yield polymers with controlled molecular weight (MW), narrow MW distributions, and one reactive end group. Maynard et al. have polymerized PNIPAAm directly from streptavidin by firstly linking a biotinylated initiator group to the streptavidin [89, 90]. One special feature of these controlled, "living" free radical polymerizations is that block copolymers may also be synthesized where one or both of the blocks are smart polymers.

20.5 Conclusions

Although this review focuses solely on the roles of smart polymer-based nanotechnologies within the fields of medicine, they have an extensive range of applications which are expected to assist the effort for addressing various problems facing today's society. These technologies, for example, may make it possible to manufacture programmable materials that require less energy to produce less waste than conventional materials. The "smart" polymer systems have been arising due to the recent advances in polymer synthesis field, as a result of the novel synthesis approach of random, block, or graft copolymers as well as enhanced efficacy control of the molecular weight and its distributions. Furthermore, smart materials focus on many types of novel materials, including not only polymers but also ceramics, hybrid composites, shape memory alloys, magnetorheological fluids, electrorheological materials, nanotubes, etc. These materials will enable us to develop better, faster, cheaper, and more powerful "smart" technology that has the potential for improvements in health, safety, and quality of life, particularly in resource-limited environments such as aerospace, or low-infrastructure sites such as the developing world.

References

1. Kushmerick J (2009) Molecular transistors scrutinized. Nature 462:994–995
2. Robert F Service (2005) Nanotechnology takes aim at cancer. Science 310:1132–1134
3. Robert F Service (2005) Calls rise for more research on toxicology of nanomaterials. Science 310:1609
4. Mirkin CA (1999) Tweezers for the nanotool kit. Science 286:2095–2096
5. Dufrene YF (2008) AFM for nanoscale microbe analysis. Analyst 133:297–301
6. Tambe NS, Bhushan B (2008) Nanoscale friction and wear maps. Philos Transact A Math Phys Eng Sci 366:1405–1424
7. Holly FJ, Refojo MF (1975) Wettability of hydrogels I. Poly(2-hydroxyethyl methacrylate). J Biomed Mater Res 9:315–326
8. Ratner BD, Hoffman AS (1976) Synthetic hydrogels for biomedical applications. In: Andrade JD (ed) Hydrogels for medical and related applications, ACS Symposium Series, vol 31. American Chemical Society, Washington, pp 1–36
9. Hubbell JA, Thomas SN, Swartz MA (2009) Materials engineering for immunomodulation. Nature 462:449–460
10. Ratner BD (2004) A history of biomaterials. In: Ratner BD, Hoffman AS, Schoen FJ, Lemons JE (eds) Biomaterials science: an introduction to materials in medicine, 2nd edn. Elsevier, New York, pp 10–19
11. Baier RE, Dutton RC (1969) Initial events in interactions of blood with a foreign surface. J Biomed Mater Res 3:191–206
12. Vogler EA (2004) Role of water in biomaterials. In: Ratner BD, Hoffman AS, Schoen FJ, Lemons JE (eds) Biomaterials science: an introduction to materials. in medicine, 2nd edn. Elsevier, New York, pp 59–65
13. Pancrazio JJ (2008) Neural interfaces at the nanoscale. Nanomed 3:823–830
14. Lafuma A, Quere D (2003) Superhydrophobic states. Nat Mater 2:457–460
15. Zheng Y, Gao X (2007) Directional adhesion of superhydrophobic butterfly wings. Soft Matter 3:178–182
16. Hoffman AS (2004) Applications of "smart polymers" as biomaterials. In: Ratner BD, Hoffman AS, Schoen FJ, Lemons JE (eds) Biomaterials science: An introduction to materials in medicine, 2nd edn. Elsevier, New York, pp 107–115
17. Ebara M (ed) (2016) Biomaterials nanoarchitectonics. Elsevier, New York
18. Ebara M, Kotsuchibashi Y, Narain R, Idota N, Kim Y-J, Hoffman JM, Uto K, Aoyagi T (2014) Smart biomaterials, NIMS monographs. Springer, Tokyo

19. Perloff R, Sternberg RJ, Urbina S (1996) Intelligence: knowns and unknowns. Am Psychol 51
20. Ebara M, Kikuchi A, Sakai K, Okano T (2004) Fast shrinkable materials. In: Yui N, Mrsny RJ, Park K (eds) Reflexive polymers and hydrogels: understanding and designing fast responsive polymeric systems. CRC Press, Boca Raton, FL, pp 219–244
21. Heskins M, Guillet JE (1968) Solution properties of poly(N-isopropylacrylamide). J Macromol Sci Pure Appl Chem 2:1441–1455
22. Smidsrod O, Guillet JE (1969) Study of polymer-solute interactions by gas chromatography. Macromolecules 2:272–277
23. Hoffman AS (1987) Applications of thermally reversible polymers and hydrogels in therapeutics and diagnostics. J Control Release 6:297–305
24. Monji N, Hoffman AS (1987) A novel immunoassay system and bioseparation process based on thermal phase-separating polymers. Appl Biochem Biotechnol 14:107–120
25. Chen G, Hoffman AS (1995) Graft copolymers that exhibit temperature-induced phase transitions over a wide range of pH. Nature 373:49–52
26. Ebara M, Yamato M, Hirose M, Aoyagi T, Kikuchi A, Sakai K, Okano T (2003) Copolymerization of 2-carboxyisopropylacrylamide with *N*-isopropylacrylamide accelerates cell detachment from grafted surfaces by reducing temperature. Biomacromolecules 4:344–349
27. Uenoyama S, Hoffman AS (1988) Synthesis and characterization of acrylamide-N-isopropylacrylamide copolymer grafts on silicone rubber substrates. Radiat Phys Chem 32:605–608
28. Lahann J, Langer R (2005) Smart materials with dynamically controllable surfaces. MRS Bull 30:185–188
29. Stayton PS, Shimoboji T, Long C, Chilkoti A, Chen G, Harris JM, Hoffman AS (1995) Control of protein-ligand recognition using a stimuli-responsive polymer. Nature 378:472–474
30. Ebara M, Yamato M, Aoyagi T, Kikuchi A, Sakai K, Okano T (2004) Temperature-responsive cell culture surfaces enable "on-off" affinity control between cell integrins and RGDS ligands. Biomacromolecules 5:505–510
31. Aoyagi T, Ebara M, Sakai K, Sakurai Y, Okano T (2000) Novel bifunctional polymer with reactivity and temperature sensitivity. J Biomater Sci Polym Ed 11:101–110
32. Ebara M, Aoyagi T, Sakai K, Okano T (2001) The incorporation of carboxylate groups into temperature-responsive poly(*N*-isopropylacrylamide)-based hydrogels promotes rapid gel shrinking. J Polym Sci Part A Polym Chem 39:335–342
33. Ebara M, Aoyagi T, Sakai K, Okano T (2000) Introducing reactive carboxyl side chains retains phase transition temperature sensitivity in *N*-isopropylacrylamide copolymer gels. Macromolecules 33:8312–8316
34. Yoshida T, Aoyagi T, Kokufuta E, Okano T (2003) Newly designed hydrogel with both sensitive thermo-response and biodegradability. J Polym Sci Part A Polym Chem 41:779–787
35. Maeda T, Yamamoto K, Aoyagi T (2006) Importance of bound water in hydration-dehydration behavior of hydroxylated poly(N-isopropylacrylamide). J Colloid Interface Sci 302:467–474
36. Dupuy A, Lehmann S, Cristol J (2005) Protein biochip systems for the clinical laboratory. Clin Chem Lab Med 43:1291–1302
37. Toner M, Irimia D (2005) Blood-on-a-chip. Annu Rev Biomed Eng 7:77–103
38. Kulkarni S, Schilli C, Grin B, Müller AH, Hoffman AS, Stayton PS (2006) Controlling the aggregation of conjugates of streptavidin with smart block copolymers prepared via the RAFT copolymerization technique. Biomacromolecules 7:2736–2741
39. Kulkarni S, Schilli C, Müller AH, Hoffman AS, Stayton PS (2004) Reversible meso-scale smart polymer–protein particles of controlled sizes. Bioconjug Chem 15:747–753
40. Malmstadt N, Hoffman AS, Stayton PS (2004) "Smart" mobile affinity matrix for microfluidic immunoassays. Lab Chip 4:412–415
41. Malmstadt N, Yager P, Hoffman AS, Stayton PS (2003) A smart microfluidic affinity chromatography matrix composed of poly(N-isopropylacrylamide)-coated beads. Anal Chem 75:2943–2949

42. Ebara M, Hoffman JM, Hoffman AS, Stayton PS (2006) Switchable surface traps for injectable bead-based chromatography in PDMS microfluidic channels. Lab Chip 6:843–848
43. Ebara M, Hoffman JM, Stayton PS, Hoffman AS (2007) Surface modification of microfluidic channels by UV-mediated graft polymerization of non-fouling and 'smart' polymers. Radiat Phys Chem 76:1409–1413
44. Yager P, Edwards T, Fu E, Helton K, Nelson K, Tam MR, Weigl BH (2006) Microfluidic diagnostic technologies for global public health. Nature 442:412–418
45. Urdea M, Penny LA, Olmsted SS, Giovanni MY, Kaspar P, Shepherd A, Wilson P, Dahl CA, Buchsbaum S, Moeller G, Hay Burgess DC (2006) Requirements for high impact diagnostics in the developing world. Nature 1:73–79
46. Black RE, Morris SS, Bryce J (2003) Where and why are 10 million children dying every year? Lancet 361:2226–2234
47. World Health Organization (2005) Making every mother and child count. WHO, Geneva
48. Price CP (2001) Regular review: point of care testing. Br Med J 332:1285–1288
49. Fu E, Chinowsky T, Foley J, Weinstein J, Yager P (2004) Characterization of a wavelength-tunable surface plasmon resonance microscope. Rev Sci Instrum 75:2300–2304
50. Fu E, Foley J, Yager P (2003) Wavelength-tunable surface plasmon resonance microscope. Rev Sci Instrum 74:3182–3184
51. Lai JJ, Hoffman JM, Ebara M, Hoffman AS, Estournes C, Wattiaux A, Stayton PS (2007) Dual magnetic-/temperature-responsive nanoparticles for microfluidic separations and assays. Langmuir 23:7385–7391
52. Lai JJ, Nelson KE, Nash MA, Hoffman AS, Yager P, Stayton PS (2009) Dynamic bioprocessing and microfluidic transport control with smart magnetic nanoparticles in laminar-flow devices. Lab Chip 9:1997–2002
53. Beebe DJ, Moor JS, Bauer JM, Yu Q, Liu RH, Devadoss C, Jo B-H (2000) Functional hydrogel structures for autonomous flow control inside microfluidic channels. Nature 404:588–590
54. Yu Q, Bauer JM, Moore JS, Beebe DJ (2001) Responsive biomimetic hydrogel valve for microfluidics. Appl Phys Lett 78:2589–2591
55. Moorthy J, Beebe DJ (2003) Organic and biomimetic designs for microfluidic systems. Anal Chem 75:292A–301A
56. Saitoh T, Suzuki Y, Hiraide M (2002) Preparation of poly(N-isopropylacrylamide)-modified glass surface for flow control in microfluidics. Anal Sci 18:203–205
57. Idota N, Kikuchi A, Kobayashi J, Sakai K, Okano T (2005) Microfluidic valves comprising nanolayered thermoresponsive polymer-grafted capillaries. Adv Mater 17:2723–2727
58. Kollman PA (1977) Noncovalent interactions. Acc Chem Res 10:365–371
59. Sui ZJ, Murphy WL (2008) Nanoscale mechanisms for assembly of biomaterials. In: Shi D (ed) Nanoscience and its applications to biomedicine. Springer, New York
60. Alfarano C et al (2005) The biomolecular interaction network database and related tools. Nucleic Acids Res 33:D418–D424
61. Garcia AJ, Gallant ND (2000) Stick and grip: measurement systems and quantitative analyses of integrin-mediated cell adhesion strength. Cell Biochem Biophys 39:61–73
62. Ruoslahti E, Pierschbacher MD (1987) New perspectives in cell adhesion: RGD and integrins. Science 238:491–497
63. Hynes RO (1987) Integrins: a family of cell surface receptors. Cell 48:549–554
64. Gallant ND, Capadona JR, Franzier AB, Collard DM, Garcia AJ (2002) Micropatterned surfaces to engineer focal adhesions for analysis of cell adhesion strengthening. Langmuir 18:5579–5584
65. Horbett TA, Waldburger JJ, Ratner BD, Hoffman AS (1988) Cell adhesion to a series of hydrophilic-hydrophobic copolymers studied with a spinning disc apparatus. J Biomed Mater Res 22:383–404
66. Mardilovich A, Kokkoli E (2004) Biomimetic peptide-amphiphiles for functional biomaterials: the role of GRGDSP and PHSRN. Biomacromolecules 5:950–957

67. Litvinov RI, Shuman H, Benett JS, Weisel JW (2002) Binding strength and activation state of single fibrinogen-integrin pairs on living cells. Proc Natl Acad Sci U S A 99:7426–7431
68. Ebara M, Yamato M, Aoyagi T, Kikuchi A, Sakai K, Okano T (2004) Immobilization of cell adhesive peptides to temperature-responsive surfaces facilitates both serum-free cell adhesion and non-invasive cell harvest. Tissue Eng 10:1125–1135
69. Loike JD, Sodeik B, Cao L, Leucona S, Weitz JI, Detmers PA, Wright SD, Silverstein SC (1991) CD11c/CD18 on neutrophils recognizes a domain at the N terminus of the A alpha chain of fibrinogen. Proc Natl Acad Sci U S A 88:1044–1048
70. Pierschbacher MD, Ruoslahti E (1984) Variants of the cell recognition site of fibronectin that retain attachment-promoting activity. Proc Natl Acad Sci U S A 81:5985–5988
71. Ebara M, Yamato M, Aoyagi T, Kikuchi A, Sakai K, Okano T (2008) The effect of extensible PEG tethers on shielding between grafted thermo-responsive polymer chains and integrin–RGD binding. Biomaterials 29:3650–3655
72. Ebara M, Yamato M, Aoyagi T, Kikuchi A, Sakai K, Okano T (2008) A novel approach to observing synergy effects of PHSRN on integrin-RGD binding using intelligent surfaces. Adv Mater 20:3034–3038
73. Garcia AJ, Schwarzbauer JE, Boettiger D (2002) Distinct activation states of $\alpha5\beta1$ integrin show differential binding to RGD and synergy domains of fibronectin. Biochemistry 41:9063–9069
74. Kao WJ, Liu Y, Gundloori R, Li J, Lee D (2002) Engineering endogenous inflammatory cells as delivery vehicles. J Control Release 78:219–233
75. Scopes RK (1994) Protein purification: principles and practice. Springer, New York
76. Shoemaker SG, Hoffman AS, Priest JH (1987) Synthesis and properties of vinyl monomer–enzyme conjugates. Conjugation of L-asparaginase with N-succinimidyl acrylate monomer. Appl Biochem Biotechnol 15:11–23
77. Hoffman AS (1998) A commentary on the advantages and limitations of synthetic polymer–biomolecule conjugates. In: Okano T (ed) Biorelated functional polymers: controlled release and applications in biomedical engineering. Academic Press, New York, pp 231–248
78. Monji N, Cole CA, Hoffman AS (1994) Activated, N-substituted acrylamide polymers for antibody coupling: application to a novel membrane-based immunoassay. J Biomater Sci Polym Ed 5:407–420
79. Luong JHT, Nguyen A (1990) Affinity partitioning of bioproducts. Biotechnology 8:306–307
80. Nguyen A, Luong JHT (1989) Synthesis and applications of water-soluble reactive polymers for purification and immobilization of biomolecules. Biotechnol Bioeng 34:1186–1190
81. Dainiak MB et al (1998) Conjugates of monoclonal antibodies with polyelectrolyte complexes—an attempt to make an artificial chaperone. Biochim Biophys Acta 1381:279–285
82. Gupta MN, Mattiasson B (1992) Unique application of immobilized proteins in bioanalytical systems. In: Suelter CH, Kricka L (eds) Methods of biochemical analysis. Wiley, New York, pp 1–34
83. Morikawa N, Matsuda T (2002) Thermoresponsive artificial extracellular matrix: NIPAAm-graft-copolymerized gelatin. J Biomater Sci Polym Ed 13:167–183
84. Pennadam SS et al (2004) Protein–polymer nano-machines. Towards synthetic control of biological processes. J Nanobiotechnol 2:8–15
85. Kochendoerfer GG et al (2003) Design and chemical synthesis of a homogeneous polymer-modified erythropoiesis protein. Science 299:884–887
86. Dieterich DC, Link AJ, Graumann J, Tirrell DA, Schuman EM (2006) Selective identification of newly synthesized proteins in mammalian cells using bioorthogonal noncanonical amino acid tagging (BONCAT). Proc Natl Acad Sci U S A 103:9482–9487

87. Kolb HC, Finn MG, Sharpless KB (2001) Click chemistry: diverse chemical function from a few good reactions. Angew Chem Int Ed Engl 40:2004–2021
88. Ding ZL, Fong RB, Long CJ, Stayton PS, Hoffman AS (2001) Size-dependent control of the binding of biotinylated proteins to streptavidin using a polymer shield. Nature 411:59–62
89. Bontempo D, Maynard HD (2005) Streptavidin as a macroinitiator for polymerization: in situ protein–polymer conjugate formation. J Am Chem Soc 127:6508–6509
90. Heredia KL, Maynard HD (2007) Synthesis of protein–polymer conjugates. Org Biomol Chem 5:45–53

Index

T. Govindaraju, K. Ariga (eds.), *Molecular Architectonics and Nanoarchitectonics*, Nanostructure Science and Technology,
https://doi.org/10.1007/978-981-16-4189-3

O

S

The manufacturer's authorised representative in the EU is Springer Nature Customer Service Centre GmbH, Europaplatz 3, 69115 Heidelberg, Germany. If you have any concerns regarding our products, please contact ProductSafety@springernature.com

Printed and bound by CPI Group (UK) Ltd, Croydon, CR0 4YY
30/09/2026
02225506-0016